TOXINS AND SIGNAL TRANSDUCTION

Cellular and Molecular Mechanisms of Toxin Action
A series of books on various aspects of toxin research, giving a broader emphasis
on the mechanism of action, structure — function relationship, the use of toxins
as research tools and their therapeutic applications.

Edited by Philip Lazarovici, The Hebrew University of Jerusalem, Israel

Volume 1
Toxins and Signal Transduction
edited by Y. Gutman and P. Lazarovici

Volumes in Preparation

Secretory Systems and Toxins
M. Linial, A. Grasso and P. Lazarovici

Chimeric Toxins
H. Lorberboum-Galski and P. Lazarovici

This book is part of a series. The publisher will accept continuation orders which may be cancelled at
any time and which provide for automatic billing and shipping of each title in the series upon publication. Please write for details.

TOXINS AND SIGNAL TRANSDUCTION

Edited by

Yehuda Gutman

Department of Pharmacology, The Hebrew University — Hadassah School of Medicine Jerusalem, Israel

and

Philip Lazarovici

Department of Pharmacology, The Hebrew University of Jerusalem, Israel, and, Section on Growth Factors National Institute of Child Health and Human Development, NIH, Bethesda, USA

HARWOOD ACADEMIC PUBLISHERS

Australia • Canada • China • France • Germany • India • Japan
Luxembourg • Malaysia • The Netherlands • Russia • Singapore
Switzerland • Thailand • United Kingdom

Amsteldijk 166
1st Floor
1079 LH Amsterdam
The Netherlands

British Library Cataloguing in Publication Data

Toxins and signal transduction. — (Cellular & molecular
    mechanisms of toxin action ; v. 1)
    1. Recombinant toxins   2. Cellular signal transduction
    I. Gutman, Yehuda   II. Lazarovici, Philip
    575.2

ISBN 90-5702-078-5

L'application des toxins est comme des espèces d'instruments physiologiques plus delicats que les moyens mécaniques et destinés à disséquer pour ainsi dire, une à une, les propriétés des éléments anatomiques de l'organisme vivant.

Claude Bernard
Leçons sur les effets des substances toxiques et medicamenteuses 1857

(The use of toxins as physiological tools is much more delicate than mechanical tools and is intended for the dissection, part by part, of the individual features of the living organism.)

To my lovely wife Nina and my precious children Lotan and Limor for their love and support during this project and always!

CONTENTS

PREFACE TO THE SERIES

Pathogenic bacteria, and poisonous animals and plants have been known to mankind for centuries. These organisms produce toxins that act by a variety of mechanisms to immobilize or kill their prey. Recently, toxin research has rapidly expanded as a result of the powerful and productive contributions of recombinant DNA, monoclonal antibodies, microinjection, crystallography, and patch clamp techniques. The number of toxins isolated and identified has increased, and more profound insights into their structure, mode of action, and role in disease has been achieved. The stage is now set to re-examine our previous concepts about toxin action in the light of current findings and to trace new pathways for the future. Accordingly, the purpose of this series is to fill the need for a comprehensive, contemporary work at the cellular and molecular levels of toxin action. Although emphasis will be placed on recent achievements, the new data will be integrated with previous investigations. Stimulating critical evaluations and current views and suggestions for new lines of research have been encouraged. Because of the huge number of toxins now known, a certain degree of selection, of a subjective nature, was necessary, of course.

The aim of this series is to provide a multidisciplinary approach oriented toward an understanding of the basic principles and cellular and molecular mechanisms of the action of toxins and their potential use as research tools. For this reason, each chapter provides a description of a normal physiological cellular structure and function, the interference of toxins with this process, and the use of particular toxins in research. Similarly, the structure of each book in the series was determined partly on scientific, and partly on pedagogic grounds. The first chapter(s) comprise mainly a review of the general principles of the book topic. The chapters that follow present specific reviews of the progress that has been made in different areas of this topic. In each book a glossary is provided, which should appeal to younger students.

We are planning five books in the series: Toxins and Signal Transduction, the first volume, will present selected mechanisms by which toxins affect molecular processes which transduce extracellular signals into intracellular messages regulating cell function. Secretory Systems and Toxins, the second volume, will provide an updated state-of-the-art treatment of vesicle-mediated secretion with special emphasis on the specific action and recognition of the secretory organelle proteins and glycolipids by tetanus, botulinum, and α-latrotoxin neurotoxins. Chimeric Toxins: Mechanisms of Action and Therapeutic Applications, the third book, will focus on toxins affecting protein synthesis, their structure, genetic engineering, mechanism of action, and therapeutic application in medicine. Other books will highlight principles of selectivity in neurotoxicity, toxin action on cytoskeletal proteins, cellular mechanisms of resistence towards toxin action, and other cellular processes.

This book series includes contributions by most of the leading investigators in the field. While each research group has chosen a particular toxin, either at cellular or at molecular level, assembling all efforts into a single series will hopefully provide a unique source of information. Toxin research requires skill, special safety precautions, hard work, and patience. I expect that this field of research will continue to reveal new cellular and molecular processes and provide new, selective research tools and prototypical compounds for drug development. If this series supports this effort in some small way, our work will be rewarded.

This undertaking has been made much easier by the excellent cooperation of the coeditors, Professor Yehuda Gutman, Dr. Michal Linial, Dr. Alfonso Grasso, Dr. Haya Lorberbaum-Galski, and Dr. David Lester. I would like to thank all the authors for their commitment, time, and scholarship. I am indebted and grateful for the support of the National Institute of Child Health and Human Development, NIH, Bethesda, Maryland, and in particular, Dr. Gordon Guroff for his advice, time, and scientific support. We would like to express our gratitude to Harwood Academic Publishers, for the encouragement, advice, and practical assistance during the production of this book series. I would also like to thank Ms. Hana Fibach and Ms. Ellie Hochman for valuable secretarial help, Mr. Salvadore Bru, Mr. Rick Myerchalk and Ms. Linda Brown for the production of the cover pages.

P. Lazarovici
Series Editor

PREFACE

Signal transduction mechanisms are among the most intensively studied topics in modern biology. This is evident from the growing number of meetings dedicated to this topic, as well as from the very fact that this term — signal transduction — has become an established entity in the curriculum of Biochemistry, Pharmacology, and Physiology.

Intellectually, the most stimulating way to study a topic is to analyze the ideas and ways that led to its conception and development. Many aspects and components of signal transduction were the results of experiments with toxins (bacterial and others). Some of these toxins (like botulinum, tetanus, pertussis, and cholera) were long known for their biological effects, but no molecular mechanism was available to explain these effects. Once the cascades of signal transduction began to unfold many other toxins were added as inhibitors of specific steps.

This volume will give the reader a concise view of the major contributions of toxins to the elucidation of signal transduction mechanisms at the molecular level. It emphasizes the use of toxins as pharmacological tools both at the experimental level and, potentially, at the therapeutic level.

Reading the various chapters will be particularly rewarding since they have been written by experts who have made major contributions to the elucidation of the actions of the various toxins discussed.

The book has been organized in four sections, according to the phases of the signal transduction cascade: the first section deals with the actions of toxins on receptors, the second with the actions of toxins on ion channels, the third with toxins and small G-proteins, and the fourth is dedicated to the action of toxins on various effectors, from phospholipase C to the proteins involved in exocytosis.

We hope that this volume will be helpful both to newcomers to the field of toxins and signal transduction, as well as to established scientists in a particular field, who will find interest in widening their scope of experimentation and knowledge.

CONTRIBUTORS

Saleh Abu-Raya
Department of Pharmacology and
Experimental Therapeutics
School of Pharmacy
Faculty of Medicine
The Hebrew University of Jerusalem
Jerusalem 91120
Israel

Klaus Aktories
Pharmakologisches Institut
der Albert-Ludwigs-Universität
Hermann-Herderstrasse 5
79104 Freiburg
Germany

Almunedo Albillos
Departamento de Farmacología
Facultad de Medicina
Universidad Autónoma de Madrid
Arzobispo Morcillo 4
Madrid 28029
Spain

Michal Bejerano
Department of Clinical Microbiology
The Hebrew University — Hadassah
Medical School
Jerusalem 91010
Israel

Torsten Binscheck
Institut für Toxikologie
Medizinische Hochschule Hannover
30625 Hannover
Germany

Eugenia Bloch-Shilderman
Department of Pharmacology and
Experimental Therapeutics
School of Pharmacy
Faculty of Medicine
The Hebrew University of Jerusalem
Jerusalem 91120
Israel

Margarita L. Contreras
Department of Pharmacology and
Toxicology
B403 Life Science Building
Michigan State University
East Lansing MI 48824-1317
USA

Roberto Coronado
Department of Physiology
School of Medicine
University of Wisconsin
Madison WI 53706
USA

Donald W. Fink, Jr.
Laboratory of Cell Biology/
Neurotrophic Factors
Division of Cytokine Biology
Center for Biologics Evaluation and
Research
Food and Drug Administration
Bethesda MD 20892
USA

Sara Fuchs
The Institute of Chemical Immunology
Weizmann Institute of Science
P.O. Box 26
Rehovot 76100
Israel

Luis Gandía
Departamento de Farmacología
Facultad de Medicina
Universidad Autónoma de Madrid
Arzobispo Morcillo 4
Madrid 28029
Spain

Antonio G. García
Departamento de Farmacología
Facultad de Medicina
Universidad Autónoma de Madrid
Arzobispo Morcillo 4
Madrid 28029
Spain

Dalia Gordon
Laboratoire de Biochimie
Faculté de Medicine
Secteur Nord
Institut Federatif de Recherche Jean
Roche
Boulevard Pierre Dramard
13916 Marseille Cedex 20
France

Gordon Guroff
Section on Growth Factors
National Institute of Child Health and
Human Development
National Institutes of Health
Building 49, Room 5A64
Bethesda MD 20892
USA

Yehuda Gutman
Department of Pharmacology
Faculty of Medicine
The Hebrew University of Jerusalem
P.O. Box 12065
Jerusalem 91120
Israel

Emanuel Hanski
Department of Clinical Microbiology
The Hebrew University — Hadassah
Medical School
Jerusalem 91010
Israel

Alan L. Harvey
Strathclyde Institute for Drug
Research
University of Strathclyde
204 George Street
Glasgow G1 1XW
UK

Diana A. Jerusalinsky
Instituto Biología Celular y
Neurociencias
"Professor Dr. Eduardo de Robertis"
Facultad de Medicina
Universidad de Buenos Aires
Paraguay 2155, 3er piso
1121 Buenos Aires
Argentina

Virginia Gray Johnson
Laboratory of Bacterial Toxins
CBER, FDA
Building 29, Room 103, HFM-437
1401 Rockville Pike
Rockville MD 20852-1448
USA

David Kaplan
Montreal Neurological Institute
3801 Rue University
F102 Fieldhouse Labs
Montreal, Quebec
Canada H3A 2B4

Gertrud Koch
Pharmakologisches Institut
der Albert-Ludwigs-Universität
Hermann-Herderstrasse 5
79104 Freiburg
Germany

Edgar Kornisiuk
Instituto Biología Celular y
Neurociencias
"Professor Dr. Eduardo de Robertis"
Facultad de Medicina
Universidad de Buenos Aires
Paraguay 2155, 3er piso
1121 Buenos Aires
Argentina

Philip Lazarovici
Department of Pharmacology and
Experimental Therapeutics
School of Pharmacy, Faculty of
Medicine, The Hebrew University of
Jerusalem, Jerusalem 91120
Israel

David S. Lester
Division of Research and Testing
Center for Drug Evaluation and
Research
Food and Drug Administration
8301 Muirkirk Road, Room 2009
Laurel MD 20708
USA

Manuela G. López
Departamento de Farmacología
Facultad de Medicina
Universidad Autónoma de Madrid
Arzobispo Morcillo 4
Madrid 28029
Spain

Yuzuru Matsuda
Tokyo Research Laboratories
Kyowa Hakko Kogyo Co. Ltd.
3-6-6 Asahimachi, Machida-Shi
Tokyo 194
Japan

Pedro Michelena
Departamento de Farmacología
Facultad de Medicina
Universidad Autónoma de Madrid
Arzobispo Morcillo 4
Madrid 28029
Spain

Carmen Montiel
Departamento de Farmacología
Facultad de Medicina
Universidad Autónoma de
Madrid
Arzobispo Morcillo 4
Madrid 28029
Spain

Jeffery Morrissette
Department of Physiology
School of Medicine
University of Wisconsin
Madison WI 53706
USA

Joel Moss
Pulmonary/Critical Care Medicine
Branch
Building 10, Room 6D03, MSC 1590
National Heart, Lung and Blood
Institute, NIH, 9000 Rockville Pike
Bethesda MD 20892-1590
USA

Israel Nisan
Department of Clinical Microbiology
The Hebrew University — Hadassah
Medical School
Jerusalem 91010
Israel

Walter A. Patton
Pulmonary/Critical Care Medicine
Branch, Building 10, Room 5N307
National Heart, Lung and Blood
Institute, NIH, 9000 Rockville Pike
Bethesda MD 20892
USA

Mordechai Sokolovsky
Laboratory of Neurobiochemistry
The George S. Wise Faculty of Life
Sciences, Tel-Aviv University
Tel-Aviv 69978
Israel

Martha Vaughan
Pulmonary/Critical Care Medicine
Branch
Building 10, Room 5N307
National Heart, Lung and Blood
Institute, NIH, 9000 Rockville Pike
Bethesda MD 20892
USA

Hans H. Wellhöner
Institut für Toxikologie
Medizinische Hochschule Hannover
30625 Hannover
Germany

Michael Whalin
Section on Growth Factors
National Institutes of Child Health and
Human Development
National Institutes of Health
Bethesda MD 20892
USA

Gui-Feng Zhang
Pulmonary/Critical Care Medicine
Branch
Building 10, Room 5N307
National Heart, Lung and Blood
Institute, NIH, 9000 Rockville Pike
Bethesda MD 20892
USA

Eliahu Zlotkin
Department of Cell and Animal
Biology, Institute of Life Sciences
The Hebrew University of Jerusalem
Jerusalem 91904
Israel

1. OVERVIEW OF SIGNAL TRANSDUCTION

YEHUDA GUTMAN

*Department of Pharmacology, The Hebrew University-Hadassah School of Medicine,
Jerusalem 91120, Israel*

The term signal transduction may convey different meanings, according to the system considered. Let us, therefore, define each term, signal and transduction, for the purpose of this book.

Living systems, whether single cells or multicellular organisms, depend for their functioning on constant and repeated adjustments to changes in the environment (light, sound, presence of nutrients, temperature). In multicellular organisms there is an additional condition for coherent function of the organism as a biological unit; this condition is: communication among the various cells, to coordinate the activity of the organism.

The messages arriving from the environment (physical or chemical) and the messages sent from one cell to another (or others) are the signals. In order for a signal to be effective the cells must have the means to *recognize* the signal and to *respond* to it. Thus, the definition of *signal* for our purpose is: *a message or communication between the environment and an organism or among different cells within an organism, a message that can cause a measurable biological change within the signalled cell/organism.*

Since in multicellular organisms the functioning of the organism as a unit is entirely dependent on signalling (communication), it is quite evident that any disruption of this communication can have profound effects on the function of the organism. It is not surprising, therefore, that many toxins (and drugs) act through the communication system, i.e. on the signal transduction mechanism.

We will define signal *transduction* for our purpose as the *series of events between the arrival of the signal at the cell and the biological effect observed at the cellular level.* The effect of each toxin on the specific signal transduction affected will be discussed in the particular chapter concerned. At this stage we will present a concise overview of the major signal transduction mechanisms elucidated to date, alluding here and there to toxins affecting these pathways.

By and large, signals are recognized by proteins and the response to the signal is a change in the protein. Such proteins are: ion channels, enzymes or cytoskeletal (structural) proteins. One could classify the changes in the proteins as of two categories:

A. *Conformational changes.* E.g., the binding of two acetylcholine molecules to the protein (α) of the heteropentamer constituting the nicotinic receptor, results in a conformational change of the protein so that it becomes an open channel for cations (Na, K, Ca). Or, when the $\beta\gamma$ subunit of the heterotrimeric protein G_i binds to the protein, constituting the K channel in the S-A node in the heart, it causes such a conformational change that the channel opens (Clapham and Neer, 1993). A conformational change may result not only from protein-protein interactions, as just described, but also from covalent changes in the protein. E.g., phosphorylation of the Ca channel in cardiac myocytes through the action of cAMP activated protein kinase (PKA) results in such a conformational change that the channel opens to enable increased Ca influx. Another example is phosphorylation of the EGF receptor which activates this protein to function as a protein tyrosine kinase.

B. *Translocation.* The second category of changes induced in proteins by signal transduction is translocation within the cell from one compartment to another. E.g., stimulation of adipocytes by insulin causes translocation of a glucose transporter from cytosolic vesicles to the cell membrane. Stimulation of the kidney collecting duct by antidiuretic hormone (ADH) results in translocation of a water channel protein (aquaporin) from the cytosol to the cell membrane (Nielsen and Agre, 1995). Stimulation of polymorphonuclear leukocytes results in translocation of several proteins (like phox 47, phox 67) to the cell membrane and activation of the phenomenon known as the "respiratory burst" (Leto *et al.*, 1994).

Let us now consider systematically the major transduction mechanisms elucidated to date. Transduction mechanisms can be classified in several ways. Since each transduction pathway is composed of several steps, starting with a receptor and ending with the biological effect at the cellular level, one could classify them by the initial stage (receptor), by the intermediate stage(s) (proteins mediating the conformational change in the receptor to intracellular effectors), or by the effector stage(s) (sometimes a single step, sometimes a series of effector stages, like the MAP kinases cascade). It is customary to classify transduction mechanisms by the enzymatic step involved. E.g., adenylyl cyclase, phospholipase C, tyrosine protein kinase, etc. However, the transduction sequence does not always involve an enzyme. When it does, this does not have to be a single enzyme and the enzyme is not necessarily the final effector involved. Therefore, we will consider these pathways according to the initial step, the receptor, which is obligatory, and introduce the subsequent steps (mediators and effectors) in the appropriate connection. However, this will not be a comprehensive review of all the mechanisms known to date, but it will highlight the major categories.

For any signal to evoke a response there should be a receptor to recognize the signal. There are several categories of cell membrane receptors: The 7 transmembrane (7 TM) receptors, the tyrosine kinase receptors, the guanylyl cyclase receptors, the T-cell receptors (TCR) and the cytokine receptors (growth hormone, prolactin, erythropoietin, etc.).

THE 7 TRANSMEMBRANE RECEPTORS

The 7 transmembrane receptors constitute a widely distributed type of receptor. Evolutionarily, the earliest molecule of this type is bacteriorhodopsin (Zhang and Weinstein, 1994). Many of the receptors for transmittors and hormones belong to this category (e.g., adrenergic receptors, dopaminergic receptors, muscarinic acetylcholine receptors, opioid receptors, cannabinoid receptors and receptors for vasopressin, ACTH, glucagon, angiotensin II, etc.). These receptors are character-ized by an extracellular amino terminus, 7 transmembrane helices, a cytosolic carboxyl terminus, 3 extracellular and 3 intracellular loops connecting the 7 trans-membrane helices and, quite often, a posttranslationally added, membrane embed-ded, palmitate, attached to cysteine at the beginning of the cytosolic carboxyl end.

This type of receptor always acts on a guanine nucleotide binding protein (G protein). The G protein is a heterotrimer, where the α subunit binds the guanine nucleotides (GDP or GTP) and is endowed with a GTPase activity which can be enhanced by binding to the effector (Bourne and Stryer, 1992).

The activated G protein dissociates into an α-GTP and a $\beta\gamma$ heterodimer. Depending on the particular system, either α-GTP or $\beta\gamma$ or both activate the effec-tor(s). This may be a direct effect (e.g., on an ion channel) or an indirect effect — through an enzyme. The enzymes regulated by G proteins are adenylyl cyclase, phopholipase C, phospholipase A_2 and phospholipase D.

G proteins are classified on the basis of the several families of the α subunit (Simon *et al.*, 1991). Functionally, the importance of these families derives from the specificity of the effectors activated by each family. For example, Gs activates adenylyl cyclase (through α_s), Gi inhibits adenylyl cyclase (through α_i), Go closes Ca channels (through α_o), Gq activates phospholipase C (through α_q). However, combinations of activation of two effectors through one G protein have been described: Gi — inhibition of adenylyl cyclase through α_i and activation of phos-pholipase C through $\beta\gamma$ (Simon *et al.*, 1991). Furthermore, the subunit $\beta\gamma$ has been shown to activate the ras and MAP kinase cascade (Crespo *et al.*, 1994).

A major step in the characterization of the G protein family was made possible by the use of bacterial toxins. Thus, cholera toxin, due to its enzymatic activity as an ADP ribosyl transferase, can ADP ribosylate a specific arginine in the α_s protein, thereby inactivating its GTPase activity. This results in a continuous activation of adenylyl cyclase through the α_s-GTP. Pertussis toxin, by a similar enzymatic activity, ADP ribosylates a cysteine (located at position 4 from the carboxyl terminal) in the $\alpha_{i/o}$ family. This results in loss of G-protein-receptor interaction and, thus, inability of agonists to activate the transduction mechanism.

By and large, the most common result of activation of a transduction mecha-nism is the activation of a kinase, either as an early event or at a later stage along the sequence of transduction. E.g., receptors that act through Gs activate adenylyl cyclase. This results in production of cAMP (from ATP). cAMP binds to the regu-latory unit of PKA, relieving the inhibition of the catalytic unit (a serine/threonine protein kinase), which then can phosphorylate a variety of proteins. E.g., cell membrane Ca channels in myocytes, lipase in adipocytes, phosphorylase kinase, etc. Receptors acting through Gq activate phospholipase C. This results in hydrolysis of phosphatidyl-inositol 4,5 bisphosphate (PIP_2) to inositol 1,4,5 trispho-sphate (IP_3) and diacylglycerol (DAG). IP_3 binds to a calcium channel in the endoplasmic reticulum to cause release of calcium into the cytosol. The increase in Ca_i can result in direct activation of various physiologic processes, e.g., exocyto-sis, muscle contraction, etc., but some of the effects of increased Ca_i are caused by Ca-calmodulin (CaM) through the activation of specific kinases (serine/threonine) e.g., CaM activated kinase II. DAG activates another serine/threonine kinase, PKC. A large family of PKC's, differing in their mode of activation, (calcium dependent and independent, activated by DAG or independent of DAG, etc.), as well as in their function (active during cell multiplication or during differentiation, etc.), has been characterized. The activation of some PKC's involves translocation of the enzyme from cytosol to the cell membrane, while others are confined to an intracellular compartment (cytoskeleton) (Kiley *et al.*, 1995). Some substrates of PKC are functionally well characterized, like the Na/H antiporter, whereas others are still awaiting a well defined function (E.g., the protein MARCKS — the myristoylated alanine rich C-kinase substrate) (Nakaoka *et al.*, 1995).

Elucidation of the signal transduction through phospholipase C (PLC) has gradually evolved from a specific single enzyme pathway into an array involving secondarily other transduction pathways acting through hydrolysis of phospholipids. Thus, although phospholipase A_2 (PLA_2) and phospholipase D (PLD) can each be activated directly through a G protein, it turned out that both PLA_2 and PLD can be stimulated secondarily to PLC. This may be through DAG/PKC and Ca, released by IP_3, following activation of PLC. While the substrate of PLC is PIP_2, that of PLA_2 and PLD is mostly phosphatidyl choline (PC). In the case of PLA_2, a further characteristic is arachidonic acid at position sn-2. As a result of hydrolysis by PLA_2, arachidonic acid is released and immediately attacked by a cascade of endoplasmatic enzymes. A variety of products (prostaglandins, leukotrienes, lipoxins, etc.) are released through the cell membrane (apparently through specific membrane transporters) to act on membrane receptors of the 7 TM-G coupled type. In the case of PLD the first product of PC hydrolysis is phosphatidic acid, assumed by some to be biologically active by itself. However, a more plausible outcome is further hydrolysis of phosphatidic acid either by a phosphatase to yield DAG or by a PLA_2 to result in lysophosphatidic acid. The latter has been shown to act on a 7 TM receptor coupled to a G_i protein, which activates the ras pathway (Moolenar, 1994). The rise of Ca_i following PLC activation can also stimulate the activity of constitutive NO synthase (through CaM), thus leading to the production of NO from arginine, and, through stimulation of soluble guanylyl cyclase, to the production of cGMP, which activates a specific serine/threonine kinase (PKG).

These findings underline the futility of the view that each transduction mechanism acts specifically and independently on a single function, or that each particular function is regulated by one transduction mechanism. It seems, rather, that activation of any single transduction pathway evokes a series or array of secondary responses. The functional meaning of this arrangement has yet to be elucidated. A possible result of this arrangement may be the extension of a response over time (e.g., activation of PLD secondary to PLC may provide a larger or more extended production of DAG), or extension of a response to neighboring cells (e.g., activation of PLA2 secondary to PLC can lead to release of arachidonic acid derivatives [prostaglandins or leukotrienes] to the intercellular space and activation of appropriate receptors in neighboring cells), or attenuation of the initial response (e.g., production of NO secondary to activation of PLC, may attenuate muscle contraction or platelet activation), or mobilization of long-term effects, like cell proliferation (e.g., activation of the ras pathway in smooth muscle and cardiac myocytes following PLC activation by angiotensin II) (Butcher *et al.*, 1993; Moolenaar, 1994; Sadoshima *et al.*, 1995).

THE GUANYLYL CYCLASE RECEPTORS

A second type of receptors, widely distributed in nature, from sea-urchin to mammals, are cell membrane receptors endowed with guanylyl cyclase activity. These receptors consist of a single chain of polypeptide: an extracellular portion, which recognizes and binds the specific agonist (resact in sea-urchins, ANP in the nephron, guanylin or Sta, the heat stable bacterial toxin, in the intestinal epithelium), a single transmembrane helix and an intracellular portion. The latter has a protein kinase-like segment (but enzymatically inactive) which binds ATP, and a catalytic segment (guanylyl cyclase). The enzymatic activity of this receptor requires phosphorylation of the intracellular portion of the receptor. An intriguing finding is that some of the receptors for ANP lack the intracellular catalytic segment and, thus, can only bind the agonist. The functional significance of this variant of receptor is unclear, although some investigators ascribe it a "clearance" function (elimination of agonist from the circulation). However, the quantitative aspect (the small number of receptors compared to the number of agonist molecules in the circulation) makes this interpretation rather doubtful.

THE TYROSINE KINASE RECEPTORS

A major class of receptors are those endowed with protein tyrosine kinase activity. These are proteins composed of a single polypeptide chain and serve as receptors for growth factors (like EGF, PDGF). The receptor has an extracellular portion that binds the agonist, a single transmembrane strand and an intracellular portion, part of which is the tyrosine kinase, which is inactive in the absence of agonist binding. The activation process has turned out to be a widely distributed mechanism, i.e. dimerization of two receptor molecules caused by agonist binding. The dimerization induces reciprocal tyrosine phosphorylations of the two receptor molecules on their intracellular portions (Schlessinger and Ullrich, 1992). The phosphorylation of tyrosines on the receptor endows it with two properties: 1) the enzymatically inactive portion becomes active as a protein tyrosine kinase; 2) the various tyrosine phosphorylated domains on the receptor can now bind specific proteins. Proteins eligible for binding are those possessing one or more domains homologous to a sequence in the soluble protein tyrosine kinase src, SH2 (= src homology 2). The SH2 domains have in common a positively charged region (basic amino acids) inducing attraction to the negative charge on the phosphate, bound to tyrosine (Birge and Hanafusa, 1993). Additionally, the binding of each SH2 region depends on the specific amino acids adjoining the phosphorylated tyrosine on the receptor tyrosine kinase (Birge and Hanafusa, 1993). This provides specificity for each of the proteins possessing SH2 sequences to particular receptors and particular sites on the receptor.

The dual result of tyrosine phosphorylation of the receptor is the basis for the functional outcome. One type of functional outcome is the very translocation of the SH2 possessing protein to the membrane. E.g., activation of the MAP kinase cascade by EGF is due to binding of a specific SH2 possessing protein (Grb 2) to the tyrosine phosphorylated receptor (McCormick, 1993). Grb2 in the cytosol binds to another protein, SOS. The binding is through an SH3 domain in Grb2 (SH3 = src homology 3, a sequence which binds to proline-rich regions; SOS has such a proline-rich region) (McCormick, 1993; Feller *et al.*, 1994; Musacchio *et al.*, 1994). The translocation of SOS from the cytosol to the membrane (mediated by Grb2) is all that is necessary to activate ras, since SOS stimulates GDP dissociation from membrane bound ras. GTP immediately replaces GDP at the nucleotide binding site of ras since $[GTP]_i \gg [GDP]_i$. The GTP bound ras activates the cascade of raf, MEK, MAPK sequentially. MAP kinase then produces an array of effects, partly by activating other kinases or proteins, and partly by translocation of the (phosphorylated) activated MAPK to the nucleus. The mechanism by which the diverse, sometimes contadictory (e.g. — cell proliferation vesus differentiation) effects of MAPK are achieved has yet to be elucidated. That translocation of SOS to the membrane is all that is necessary to activate ras, is evident from the finding that isolated SOS (without Grb2) activates isolated ras in solution.

An example of the other result of tyrosine phosphorylation of the growth factor receptor is activation of PLCγ. This phospholipase has an SH2 region and thus translocates from the cytosol to the membrane, binding to a phosphorylated tyrosine on the activated receptor tyrosine kinase. PLCγ is then phosphorylated on tyrosine by the receptor and thus activated (Noh *et al.*, 1995). Being bound to the membrane now exposes PLC to its substrate, the membrane phospholipid PIP_2.

A special case of tyrosine kinase receptors is that of the receptors for nerve growth factor (NGF) and similar neurotrophic factors. In this case the receptor consists of a heterodimer, one, "small" 75 kd protein, devoid of tyrosine kinase activity with relatively low affinity for NGF, the other, "large" 135 kd protein, has tyrosine kinase activity (upon agonist binding) and higher affinity for NGF (Szeberenyi and Erhardt, 1994). The 75 kd receptor can induce apoptosis (when not bound to NGF), apparently through a sequence in its cellular portion similar to the G protein binding sequence in TNFα receptors or in the toxin mastoparan (Chapman, 1995). This is an example of a spontaneously active receptor, i.e. — the receptor causes an effect without being bound to an agonist. In this case agonist (NGF) binding stops the activity of the receptor. Since apoptosis (programmed cell death) is a major process during development of the CNS, its regulation by a receptor is of major significance.

T-CELL AND IMMUNOLOGIC STIMULATED RECEPTORS

Another class of receptors is that characteristic of lymphocytes and T cells. These receptors include a cluster of several proteins, none of which possesses tyrosine kinase or any other enzymatic activity. The most thoroughly studied of this class is the T-cell receptor (TCR) (Chan *et al.*, 1994). It consists of α and β units which can bind to the antigen (when attached to an "antigen presenting cell"), the CD3 complex, consisting of a single γ and single δ units, and two units each of ε and ζ proteins. None of these proteins is endowed with any enzymatic activity. However, when the agonist (antigen) binds to the receptor it causes dimerization, which is very rapidly followed by tyrosine phosphorylation on the ζ and ε proteins. These phosphorylations occur on specific sequences, called ARAMs (antigen recognition activation motives): two sequences of YXXL, separated by 6-8 amino acids. These sequences are recognized by a specific tyrosine kinase, called Lck, which is associated with the CD4 or CD8 T-cell membrane antigens. This antigen is recruited to join the TCR upon stimulation. Lck binds specifically to the ARAMs and phosphorylates them on the tyrosines. This is followed by binding of another tyrosine kinase, ZAP 70. This cytosolic enzyme translocates to the membrane by binding of its two SH2 domains to both phosphorylated tyrosines of the ARAMs (Hatada *et al.*, 1995). The binding of ZAP 70 to the phosphorylated ε causes activation of ZAP as a tyrosine kinase and the sequence of events that follows is reminiscent of that with receptor tyrosine kinases. However, a specific twitch in the initial activation of the signal transduction of TCR involves a tyrosine phosphatase. This phosphatase, another cell membrane antigen, called CD 45, is a phosphatase that recognizes a specific phosphotyrosine on Lck. Lck, similar to the cytosolic tyrosine protein kinases of the src family, has a regulatory tyrosine at its carboxyl terminus. When this tyrosine is phosphorylated, the tyrosine kinase activity is inhibited (in src the inhibition was shown to result from binding of the phosphotyrosine near the carboxyl terminal to an SH2 domain at the amino end of the very same molecule). Dephosphorylation of this tyrosine in Lck by the tyrosine phosphatase activity of CD 45 results in activation of the Lck tyrosine kinase, which, in turn, results in binding of Lck to the ARAMs on the ζ and ε chains and tyrosine phosphorylation (Chan *et al.*, 1994).

This transduction mechanism demonstrates that activation may depend on the concerted action of both phosphatase and kinase and that the assumption of either/or, i.e. either phosphorylation or dephosphorylation, with opposite end results, is too simplistic.

THE CLASS OF GROWTH HORMONE RECEPTORS

This is a special class of receptors that are devoid of tyrosine kinase activity themselves, but act by activation of protein tyrosine kinases. This class includes the receptors for growth hormone, prolactin, erythropoietin, several interleukins and interferons α and γ. These receptors also undergo dimerization upon agonist binding (Kelly *et al.*, 1994). This results in activation of a separate protein, bound to the receptor, which is a tyrosine kinase, leading to tyrosine phosphorylation of the receptor itself, as well as a downstream cascade leading to activation of the raf — MAP kinase cascade. The tyrosine kinase which serves as the intermediate between the receptor and downstream events, in the case of growth hormone receptor is JAK2 (Janus kinase family) (Tourkine *et al.*, 1995). A whole series of such factors, mediating the effect of the receptor in proliferation and differentiation (like the eryhropoietin receptor) inducing receptors have been termed STATs (signal transducing and transforming factors) (Frank *et al.*, 1995).

Some of these receptors, like interferon γ receptors (or bacterial lipopolysacharide) induce, upon activation, the synthesis of the enzyme nitric oxide synthase (NOS). This enzyme produces NO from arginine. It is independent of a rise in Ca_i, unlike the constitutive NOS, which is activated by Ca following PLC stimulation. However, production of NO is dependent on tetrahydrobiopterin (BH_4) supply, which is usually not present at concentrations that saturate NOS. BH_4 binding is essential for the dimerization of NOS, required for its activation, and prevents degradation of the enzyme by NO itself (Cho *et al.*, 1995). Induction of NOS is usually accompanied by induction of GTP cyclohydrolase, the key, rate-limiting step in the production of BH_4. At low BH_4 levels NOS produces H_2O_2 (from H_2O and molecular O_2). The induction of NOS synthesis is typical of macrophages and other cells involved in inflammatory and immunologic responses. NO, a free radical, can act as such when produced in high concentrations, and cause cell death or damage. Therefore, this is a signal transduction pathway that may be suicidal. In addition, NO can activate soluble guanylyl cyclase to produce cGMP, which acts, usually, by activating a specific serine/threonine protein kinase (PKG).

INACTIVATION OF THE TRANSDUCTION PROCESS

Activation of any signal transduction cascade is naturally a temporary process. Adaptability and the ability to respond to future stimluli depends on regaining of the initial state of all the components of the process (schematically: receptor — transducer — effector). This process is also mainly effected through conformational changes in proteins (mostly, but not exclusively, induced by phosphorylation/dephosphorylation) or by translocation of proteins.

A thoroughly studied case is that of the sequential activation of the β-adrenergic receptor — Gs — adenylyl cyclase. The receptor can undergo various changes, all of which result in the reduced activation of adenylyl cyclase by β-adrenergic agonists (like isoproterenol). One process is translocation of the receptor from the cell membrane to an intracellular compartment (endosomes) through endocytosis. The result of this process, which requires the participation of several specific proteins, is usually termed "down regulation" and is practically ubiquitous with a variety of receptor types (G protein coupled receptors, tyrosine kinase receptors, guanylyl cyclase receptors, etc.). A second type of change in receptor characteristics follows activation of the G protein by the receptor, as it can be demonstrated only in the presence of GTP, GTPγS or GppNHp. It results in a rightward shift of the curve for agonist binding to the receptor, i.e. reduced affinity of the receptor for the agonist (antagonist binding is unaffected). This process deserves, appropriately, the name "desensitization". However, this term is frequently applied, inappropriately, to the two other processes which the receptors undergo: down regulation and phosphorylation.

The third type of change is serine/threonine phosphorylation of the receptor in its intracellular carboxyl portion and/or the third intracellular loop. This covalent change results in "receptor-effector uncoupling", i.e. agonist binding to the receptor does not result in activation of the effector. E.g., isoproterenol will cause a smaller activation of adenylyl cyclase or smaller rise of cAMP. A possible mechanism of this "uncoupling" could be charge attraction of the negatively charged phosphate to the positively charged third intracellular loop, assumed to be the essential part of the receptor for activation of G proteins (through its basic amino-acid-rich carboxyl and amino segments). Thus, the negatively charged phosphate could "mask" the essential portion of the receptor necessary for activating G protein. Phosphorylation of the receptor can be effected by PKA (and also PKC). It is then termed "heterologous desensitization" because activation of one type of receptor can result in phosphorylation of other receptors, since PKA (and PKC) are rather promiscuous in the type of receptor recognized as substrate. Phosphorylation can also be effected in a more specific way, i.e. only the activated receptor is phosphorylated. This type of phosphorylation follows the activity of a special class of serine/threonine kinases — GRK's (G protein-coupled receptor kinases). The best known of this class are rhodopsin kinase and βARK (β-adrenergic receptor kinase) (Freedman *et al.*, 1995). This cytosolic kinase has a PH (pleckstrin homology) sequence that binds to the β-subunit of Gs, but only after dissociation of α_s from the $\beta\gamma$ (Pitcher *et al.*, 1995). The $\beta\gamma$ subunit is anchored to the membrane through prenylation (by the hydrophobic hydrocarbon farnesyl or geranyl-geranyl) of cysteine at the carboxyl end of the γ subunit and coiled-coil binding of the β and γ subunits (Simonds *et al.*, 1993). The PH region also contains a sequence rich in basic amino acids, which results in a positive charge. This probably results in attraction to negatively charged phospholipids in the membrane (Frech *et al.*, 1995). This is verified by the significant enhancement of βARK binding to liposomes containing PIP_2 (Pitcher *et al.*, 1995; Onorato *et al.*, 1995).

The form of dual anchoring of proteins that attach to membranes but are not embedded in it (no trans-membrane sequence), seems to become the rule rather than an occasional finding. Thus, a dual mechanism for membrane binding of proteins has also been observed for K-ras (farnesylation, for membrane anchoring, plus a basic amino acids rich segment that binds to acid phospholipids) (Ghomashchi *et al.*, 1995) or MARCKS (the myristoylated alanine-rich C-kinase substrate, which also contains a basic amino acid rich sequence close to the myristic acid binding site) (Nakaoka *et al.*, 1995; Vergeres *et al.*, 1995).

Recently, an additional mechanism of β-adrenergic receptor phosphorylation has been described. The insulin receptor belongs to the tyrosine kinase receptor family. Its activation has been known for a long time to reduce cAMP levels. Recently it was shown that activation of the insulin receptor leads to phosphorylation of the β-adrenergic receptor on tyrosine (Karoor *et al.*, 1995). This is one of a growing body of examples for "cross-talk" between two different transduction pathways. However, as mentioned above, interaction of one transduction pathway with another is the rule rather than the exception.

The various changes which β-adrenergic receptors undergo are not necessarily sequential, i.e. they can occur independently. Thus, mutations in the carboxyl terminus of the receptor can result in loss of down-regulation (endocytosis) without affecting the phosphorylation (Ferguson *et al.*, 1995). Other mutations can cause receptor-effector uncoupling without interfering with GTP induced desensitization (Miller *et al.*, 1988). A similar dissociation of the phenomena of down regulation, uncoupling and desensitization has been reported for other G protein-coupled receptors, like PGE_2 or angiotensin II (Hunyadi *et al.*, 1994; Thomas *et al.*, 1995).

In contrast to the G-protein related receptors, where phosphorylation reduces receptor effectiveness, in tyrosine kinase receptors, phosphorylation is the mechanism of activation. In this type of receptors dephosphorylation results in inactivation. Furthermore, since activation all along the ras-MAPK cascade depends on sequential phosphorylations (serine/threonine and tyrosine) it is appropriate that phosphatases cause inactivation of the cascade (Hunter, 1995; Misra-Press *et al.*, 1995). However, even within this cascade, one step is inactivated by phosphorylation. This step is the initial one — translocation from the cytosol and binding of the complex Grb2-SOS to the membrane-located tyrosine phosphorylated growth factor receptor. MAP kinase (or MEK), when activated, phosphorylates SOS (Waters, Yamauchi and Pessin, 1995; Waters *et al.*, 1995). This leads to dissociation of SOS from Grb2, and consequently, from the membrane, resulting in inactivation of ras.

In view of the major role of phosphorylation and dephosphorylation in cell proliferation and function it is not surprising that many toxins are investigated for their effects on these functions. As a matter of fact, some toxins were instrumental in the elucidation of transduction pathways. These include cyclosporin and FK 506, the immunosuppressants which act, through specific binding proteins, on calcineurin, a serine/threonine phosphatase (Liu, 1993), as well as okadaic acid, which directly inhibits several serine/threonine phosphatases, and genistein which inhibits tyrosine kinase.

Exocytosis, the process of release of vesicle-stored hormones, transmittors, enzymes, etc. by fusion of the vesicular membrane with the cell membrane, is also part of the transduction mechanism. This process, known for a long time to depend on increased Ca_i (e.g., by depolarization of nerve terminals), has recently been explored in detail, partly through the action of several bacterial toxins (botulinum and tetanus toxins). The vesicular membrane protein synaptotagmin, assumed to be the calcium sensor of the exocytotic process, relieves the inhibition of exocytosis upon binding calcium. Having a low affinity for calcium, ($Km \approx 10^{-4}$ M) it responds to the relatively high $[Ca_i]$ in the immediate vicinity of the cell membrane, following membrane depolarization (Littleton and Bellen, 1995). Several of the vesicular and cell membrane proteins involved in the process of exocytosis (e.g., the synaptic vesicle membrane protein synaptobrevin) are substrates of the proteolytic action of some botulinum and tetanus toxins (Zn-dependent proteases) (Montecucco and Schiavo, 1993). The products of proteolysis by the toxins of all the proteins involved are biologically inactive (Ahnert-Hilger and Bigalke, 1995). Some toxins, like brefeldin, inhibit protein translocation between intracellular organelles (Golgi, endoplasmic reticulum).

In recent years it has become more and more evident that some "receptors" are merely structural proteins mediating interactions of the cytoskeleton and ubiquitous intercellular matrix molecules. No soluble agonist in the medium was identified, but the structural elements themselves evoke a cascade of signal transduction events underlying many aspects of cell life. These events may be changes in the cytoskeletal elements as, e.g., dissolution of the actin network essential for exocytotic release (Chang and Grinnell, 1995), or the activation of tyrosine kinase. FAK 125 (focal adhesion kinase), is related to interaction of integrin with intercellular adhesion molecules (ICAMs) (Kanazawa *et al.*, 1995). Also, events such as contact-induced cessation of proliferation of cells in culture, may be the result of such interaction of intracellular and intercellular (matrix) proteins. Shear-induced cell proliferation in the vascular bed may depend on similar signal transduction events, where no soluble chemical agonist can be found (Tseng *et al.*, 1995). An interesting case, recently investigated, is that of dystroglycan. This protein, which is part of a complex that links the extracellular matrix with the cellular cytoskeleton in muscle, plays a crucial role in muscle function. Dystroglycan is the "receptor" for the extracellular protein laminin. It has now been shown that dystroglycan, through an SH3 domain, can bind Grb2 and, thus, possibly initiate an intracellular transduction pathway (Yang *et al.*, 1995). Some of the structural proteins are tyrosine phosphatases, such as LAR or PTPμ. Structural cytoskeletal proteins that interact with the matrix or neighboring cells can modulate, through their phosphatase activity, phenomena such as cell adhesion (Brady-Kalnay and Tonks, 1995), which is of major importance in inflammatory processes and cancer spreading.

As the various processes of signal transduction unfold, it becomes evident that protein-protein interaction is of major importance, either by itself or by causing translocation of proteins from one compartment to another. We have already considered the role of SH2 regions (attachment to phosphorylated tyrosine), SH3 domains (binding to proline-rich regions), PH domains (with dual attraction, to appropriate proteins and to acidic [therefore, membrane] phospholipids). Leucine-rich regions on proteins have similar properties (hydrophobic attachment), as well as the more recently described WD repeat (Neer *et al.*, 1994) and other sequences.

Some proteins, or groups of proteins, seem to have this function as their sole role: binding of specific proteins to provide the "environment" or location for activity. One could regard the "chaperone" molecules of the hsp family as a case of this kind. Recently, a particular example related to signal transduction has been identified. STE 5 is a yeast protein essential for the MAP kinase cascade to be activated (Elion, 1995). However, it is devoid of any enzymatic activity itself. The yeast equivalents of mammalian raf, MEK and MAPK proteins have to be bound to this particular protein to be activated. Also, the $\beta\gamma$ subunit (of the G protein) which activates ras in yeast is bound on STE 5, which apparently serves as a "scaffold" (Whiteway *et al.*, 1995). A similar function is suggested for proteins of the recently described 14–3–3 family (Freed *et al.*, 1994). Such "scaffolding" for active proteins (enzymes) may provide the solution for the many difficulties in explaining "compartmentation" of effects within cells, when the same enzyme, isolated after cell disruption, shows a wide range of activity on many cellular substrates while these substrates are unaffected by the enzyme *in vivo* (Hubbard and Cohen, 1993; Mochly-Rosen, 1995).

REFERENCES

Ahnert-Hilger, G. and Bigalke, H. (1995) Molecular aspects of tetanus and botulinum neurotoxin poisoning. *Prog. Neurobiol.*, **46**, 83–96.

Birge, R.B. and Hanafusa, H. (1993) Closing in on SH2 specificity. *Science*, **262**, 1522–1524.

Bourne, H.R. and Stryer, L. (1992) The target sets the tempo. *Nature*, **358**, 541–543.

Brady-Kalnay, S.M. and Tonks, N.K. (1995) Protein tyrosine phosphatases as adhesion receptors. *Curr. Opin. Cell Biol.*, **7**, 650–657.

Butcher, R.D., Schollmann, C. and Marme, D. (1993) Angiotensin II mediates intracellular signalling in vascular smooth muscle cells by activation of tyrosine- specific protein kinases and c-raf-1. *Biochem. Biophys. Res. Comm.*, **196**, 1280–1287

Chan, A.C., Desai, D.M. and Weiss, A. (1994) The role of protein tyrosine kinases and protein tyrosine phosphatases in T cell antigen receptor signal transduction. *Ann. Rev. Immunol.*, **12**, 555–592.

Chapman, B.S. (1995) A region of the 75 kDa neurotrophin receptor homologous to the death domains of TNFR-1 and Fas. *FEBS Lett.*, **374**, 216–220.

Cheng, B-M. and Grinnell, A.D. (1995) Integrins and modulation of transmitter release from motor nerve terminals by stretch. *Science*, **269**, 1578–1580.

Cho, H.J., Martin, E., Xie, Q-W., Sassa, S. and Nathan, C. (1995) Inducible nitric oxide synthase: Identification of amino acid residues essential for dimerization and binding of tetrahydrobiopterin. *Proc. Natl. Acad. USA*, **92**, 11514–11518.

Clapham, D.E. and Neer, E.J. (1993) New roles for G protein $\beta\gamma$ dimers in transmembrane signalling. *Nature*, **365**, 403–406.

Crespo, P., Xu, N., Simonds, W.F. and Gutkind, J.S. (1994) Ras-dependent activation of MAP kinase pathway mediated by G-protein $\beta\gamma$ subunits. *Nature*, **369**, 418–420.

Elion, E.A. (1995) Ste5: a meeting place for MAP kinases and their associates. *Trends Cell Biol.*, **5**, 322–327.

Feller, S.M., Ren, R., Hanafusa, H. and Baltimore, D. (1994) SH2 and SH3 domains as molecular adhesives: the interaction of Grk and Abl. *Trends Bioch. Sci.*, **19**, 453–458.

Ferguson, S.S.G., Menard, L., Barak, L.S., Koch, W.J., Colapietro, A-M. and.

Caron, M.G. (1995) Role of phosphorylation in agonist-promoted β_2-adrenergic receptor sequestration. *J. Biol. Chem.*, **270**, 24782–24789.

Frank, D.A., Robertson, M.J., Bonni, A., Ritz, J. and Greenberg, M.E. (1995) Interleukin-2 signaling involves the phosphorylation of Stat proteins. *Proc. Natl. Acad. Sci. USA*, **92**, 7779–778.

Frech, M., Ingley, E., Andjelkovic, M.M. and Hemmings, B.A. (1995) Pleckstrin homology domains. *Biochem. Soc. Trans.*, **23**, 616–618.

Freed, E., Symons, M., Macdonald, S.G., McCormick, F. and Ruggieri, R. (1994) Binding of 14–3–3 proteins to the protein kinase raf and effects on its activation. *Science*, **265**, 1713–1716.

Freedman, N.J., Liggett, S.B., Drachman, D.E., Pei, G., Caron, M.G. and

Lefkowitz, R.J. (1995) Phosphorylation and desensitization of the human β_1-adrenergic receptor. *J. Biol. Chem.*, **270**, 17953–17961.

Ghomashchi, F., Zhang, X., Liu, L. and Gelb, M.H. (1995) Binding of prenylated peptides to membranes: affinities and intervesicle exchange. *Biochemistry*, **34**, 11910–11918.

Hatada, M.H., Lu, X., Laird, E.R., Green, J., Morgenstern, J.P., Lou, M., Marr, C.S., Phillips, T.B., Ram, M.K., Theriault, K., Zoller, M.J. and Karas, J.L. (1995) Molecular basis for interaction of the protein tyrosine kinase ZAP-70 with the T-cell receptor. *Nature*, **377**, 32–38.

Hubbard, M.J. and Cohen, P. (1993) On target with a new mechanism for the regulation of protein phosphorylation. *Trends Biochem. Sci.*, **18**, 172–177.

Hunter, T. (1995) Protein kinases and phosphatases: The Yin and Yang of protein phosphorylation and signalling. *Cell*, **80**, 225–236.

Hunyadi, L., Baukal, A.J., Balla, T. and Catt, K.J. (1994) Independence of type I angiotensin II receptor endocytosis from G protein coupling and signal transduction. *J. Biol. Chem.*, **269**, 24798–24804

Kanazawa, S., Ilic, D., Noumura, T., Yamamoto, T. and Aizawa, S. (1995) Integrin stimulation decreases tyrosine phosphorylation and activity of focal adhesion kinase in thymocytes. *Biochem. Biophys. Res. Comm.*, **215**, 438–445.

Karoor, V., Baltensperger, K., Paul, H., Czech, M.P. and Malbon, C.C. (1995) Phosphorylation of tyrosyl residues 350/354 of the β-adrenergic receptor is obligatory for counterregulatory effects of insulin. *J. Biol. Chem.*, **170**, 25305–25308.

Kelly, P.A., Postel-Vinay, M.C., Finidori, J., Edery, M., Sotiropoulos, A., Goujon, L. and Esposito, N. (1994) Growth hormone receptor structure, dimerisation and function. *Acta Paediat. Suppl.*, **399**, 107–111.

Kiley, S.C., Jaken, S., Whelan, R. and. Parker, P.J. (1995) Intracellular targeting of protein kinase C isoenzymes: functional implications. *Biochem. Soc. Trans.*, **23**, 601–606.

Leto, T.L., Adams, A.G. and de Mendez, I. (1994) Assembly of the phagocyte NADPH oxidase: binding of src homology 3 domains to proline-rich targets. *Proc. Natl. Acad. Sci. USA*, **91**, 10650–10654.

Littleton, J.T. and Bellen, H.J. (1995) Synaptotagmin controls and modulates synaptic-vesicle fusion in a Ca^{2+}-dependent manner. *Trends Neurosci.*, **18**, 177–183.

Liu, J. (1993) FK506 and cyclosporin, molecular probes for studyin intracellular signal transduction. Immunol. *Today*, **14**, 290–295.

McCormick, F. (1993) How receptors turn Ras on. *Nature*, **363**, 15–16.

Miller, R.T., Masters, S.B., Sullivan, K.A., Beiderman, B. and Bourne, H.R. (1988) A mutation that prevents GTP-dependent activation of the α chain of Gs. *Nature*, **334**, 712–715.

Misra-Press, A., Rim, C.S., Yao, H., Roberson, M.S. and Stork, P.J.S. (1995) A novel mitogen-activated protein kinase phosphatase. *J. Biol. Chem.*, **270**, 14587–14596.

Mochly-Rosen, D. (1995) Localization of protein kinases by anchoring proteins: a theme in signal transduction. *Science*, **268**, 247–251.

Montecucco, C. and Schiavo, G. (1993) Tetanus and botulism neurotoxins: a new group of zinc proteases. *Trends Biochem. Sci.*, **18**, 324–327.

Moolenaar, W.H. (1994) LPA: a novel lipid mediator with diverse biological actions. *Trends Cell Biol.*, **4**, 213–219.

Musacchio, A., Wilmanns, M. and Saraste, M. (1994) Structure and function of the SH3 domain. *Prog. Biophys. Molec. Biol.*, **61**, 283–297.

Nakaoka, T., Kojima, N., Ogita, T. and Tsuji, S. (1995) Characterisation of the phosphatidylserine-binding region of rat MARCKS (myristoylated alanine-rich protein kinase C substrate). *J. Biol. Chem.*, **270**, 12147–12151.

Neer, E.J., Schmidt, C.J., Nambudripad, R. and Smith, T.F. (1994) The ancient regulatory-protein family of WD-repeat proteins. *Nature*, **371**, 297–300.

Nielsen, S. and Agre, P. (1995) The aquaporin family of water channels in kidney. *Kidney Int.*, **48**, 1057–1068

Noh, D-Y., Shin, S.H. and Rhee, S.G. (1995) Phosphoinositide specific phospholipase C and mitogenic signaling. *Biochim. Biophys. Acta*, **1242**, 99–114.

Onorato, J.J., Gillis, M.E., Liu, Y., Benovic, J.L. and Ruoho, A.E. (1995) The β-adrenergic receptor kinase (GRK2) is regulated by phospholipids. *J. Biol. Chem.*, **270**, 21346–21353.

Pitcher, J.A., Touhara, K., Payne, E.S. and Lefkowitz, R.J. (1995) Pleckstrin homology domain-mediated membrane association and activation of the β-adrenergic receptor kinase requires coordinate interaction with $G_{\beta\gamma}$ subunits and lipids. *J. Biol. Chem.*, **270**, 11707–11710.

Sadoshima, J., Qiu, Z., Morgan, J.P. and Izumo, S. (1995) Angiotensin II and other hypertrophic stimuli mediated by G-protein coupled receptors activate tyrosine kinase mitogen activated protein kinase and 90 kd S6 kinase in cardiac myocytes. *Circ. Res.*, **76**, 1–15.

Schlessinger, J. and Ullrich, A. (1992) Growth factor signalling by receptor tyrosine kinases. *Neuron*, **9**, 383–391.

Simon, M.I., Strathmann, M.P. and Gautam, N. (1991) Diversity of G proteins in signal transduction. *Science*, **252**, 802–808.

Simonds, W.F., Manji, H.K. and Garritsen, A. (1993) G proteins and βARK: a new twist for the coiled coil. *Trends Biochem. Sci.*, **18**, 315–317.

Szeberenyi, J. and Erhardt, P. (1994) Cellular components of nerve growth factor signaling. *Biochim. Biophys. Acta*, **1222**, 187–202.

Thomas, W.G., Thekkumkara, T.J. Motel, T.J. and Baker, K.M. (1995) Stable expression of a truncated AT_{1A} receptor in CHO-K1 cells. *J. Biol. Chem.*, **270**, 207–213.

Tourkine, N., Schindler, C., Larose, M. and. Houdebine, L-M (1995) Activation of STAT factors by prolactin, interferon-γ, growth hormones and a tyrosine phosphatase inhibitor in rabbit primary mammary epithelial cells. *J. Biol. Chem.*, **270**, 20952–20961.

Tseng, H., Peterson, T.E. and Berk, B.C. (1995) Fluid shear stress stimulates mitogen-activated protein kinase in endothelial cells. *Circ. Res.*, **77**, 869–878.

Vergeres, G.,. Manenti, S., Weber, T. and Sturzinger, C. (1995) The myristoyl moiety of myristoylated alanine-rich C kinase substrate (MARCKS) and MARCKS-related protein is embedded in the membrane. *J. Biol. Chem.*, **270**, 19879–19887.

Waters, S.B., Yamauchi, K. and Pessin, J.E. (1995) Insulin-stimulated disassociation of the SOS-Grb2 complex. *Mol. Cell. Biol.*, **15**, 2791–2799.

Waters, S.B., Holt, K.H., Ross, S.E., Syu, L-J., Guan, K-L.,. Saltiel, A.R.,

Koretzky, G.A. and Pessin, J.E. (1995) Desensitization of ras activation by a feedback disassociation of the SOS-Grb2 complex. *J. Biol. Chem.*, **270**, 20883–20886.

Whiteway, M.S., Wu, C., Leeuw, T., Clark, K., Fourest-Lieuvin, A., Thomas, D.Y. and Leberer, E. (1995) Association of the yeast pheromone response G protein $\beta\gamma$ subunits with the MAP kinase scaffold Ste5p. *Science*, **269**, 1572–1575.

Yang, B., Jung, D., Motto, D., Meyer, J., Koretzky, G. and Campbell, K.P. (1995) SH3 domain-mediated interaction of dystroglycan and Grb2. *J. Biol. Chem.*, **270**, 11711–11714.

Zhang, D. and Weinstein, H. (1994) Polarity conserved positions in transmembrane domains of G-protein coupled receptors and bacteriorhodopsin. *FEBS Lett.*, **337**, 207–212.

2. ON THE TOXIN BINDING SITE OF THE NICOTINIC ACETYLCHOLINE RECEPTOR

SARA FUCHS

Department of Immunology, The Weizmann Institute of Science, Rehovot 76100, Israel

INTRODUCTION

The nicotinic acetylcholine receptor (AChR) is a ligand-gated ion channel. It is an integral membrane glycoprotein present in muscle and neuronal systems. AChR forms an oligomeric complex composed of four types of subunits present in a molar stoichiometry of $\alpha_2\beta\gamma\delta$ with an overall molecular mass of 250,000 daltons (reviewed by Karlin, 1980; 1993; Changeux *et al.*, 1984). The complete amino acid sequence of the receptor subunits has been derived from DNA clones of Torpedo, chick, calf, mouse and human receptors (Noda *et al.*, 1982, 1983; Devillers-Thiery *et al.*, 1983; Claudio *et al.*, 1983; Boulter *et al.*, 1985). The receptor subunits are synthesized from independent messenger RNAs (mRNAs).

Among the subunits of AChR, the α-subunit has been shown to contain the cholinergic binding site (Haggerty and Froehner, 1981; Gershoni *et al.*, 1983). Affinity labeling experiments have previously indicated that the binding site is within the α-subunit, in close proximity to a sulfhydryl group (Karlin, 1980), thus focusing attention to the four cysteines (residues 128, 142, 192 and 193) present in the extracellular portion of the α-subunit. The first two cysteines (128 and 142) are common to all AChR subunits, whereas the other two (192 and 193) are present exclusively in the α-subunit.

Snake venom neurotoxin antagonists with curare-like activity, such as α-bungarotoxin (α-BTX), bind to nicotinic AChR with very high affinity (Lee, 1972). The binding site for α-BTX in AChR is at or near the cholinergic binding site. Thus, a great deal of information on the ligand binding site of AChR has been obtained by the application of α-BTX, which can be radiolabeled heavily, and which binds to AChR with a much higher affinity than the natural neurotransmitter, acetylcholine. Nevertheless, there are some cases of nicotinic AChRs in which the structural elements required for acetylcholine binding are not identical to those required for α-BTX binding. For instance, neuronal nicotinic AChRs (Patrick and Stallcup, 1977) as well as snake muscle AChRs (Burden *et al.*, 1975) are not affected by α-BTX although they retain characteristic cholinergic pharmacology.

A number of experimental approaches have been adopted in an attempt to identify the ligand binding site in AChR. In earlier work from our laboratory (Neumann *et al.*, 1985), employing proteolytic fragmentation of the α-subunit along with probing the fragments with antibodies to synthetic peptides and with α-BTX, the ligand binding site has been localized within a fragment of the α-subunit which contains the two tandem cysteines at positions 192 and 193. Later, by employing synthetic peptides, it was shown that residues 185-196 of the

α-subunit contain the essential components of the ligand binding site of AChR (Neumann *et al.*, 1986a, b). This segment contains the two tandem cysteine residues at positions 192 and 193, which are unique to the α-subunit and are present at this position in all α-subunits of muscle and neuronal AChRs thus far sequenced. We have shown that these two cysteines as well as tryptophan at position 187, are required for retaining toxin binding.

Studies by other groups using CNBr derived fragments of affinity labeled α-subunit (Kao *et al.*, 1984), proteolytic fragmentation (Pedersen *et al.*, 1986), synthetic peptides (Wilson *et al.*, 1985; Mulac-Jericevic *et al.*, 1986; Gotti *et al.*, 1987), and genetic constructs (Barkas *et al.*, 1987; Gershoni, 1987), have also localized the ligand binding site in close proximity or contiguous to the two tandem cysteine residues 192 and 193. Recent experiments with other acetylcholine-derived affinity labels (Dennis *et al.*, 1988; Changeux, 1990) confirmed the putative localization of the binding site and demonstrated that other regions in the extracellular domains of the α-subunit and possibly in other subunits as well, may be in close contact with the ligand. It has been demonstrated that aromatic residues from other regions of the α-subunit (Galzi *et al.*, 1991) as well as negatively charged amino acid residues (aspartates and glutamates) in the γ and δ subunits (Blount and Merlie, 1989; Pedersen and Cohen, 1990; Czajkowski and Karlin, 1991) participate also in the acetylcholine binding site. It is proposed from these studies that the two acetylcholine binding sites are formed at the interfaces between the α and γ subunits and between the α and δ subunits (Pedersen and Cohen, 1990).

In an attempt to analyze in detail the structure of the ligand binding site of AChR and to elucidate the structural requirements for agonist vs. α-BTX binding, we have chosen to study nonconventional muscle AChRs of animals that are resistant to α-BTX. It has been interesting to find out how are such animals protected against α-BTX, and what is the structural basis for their resistance.

THE BINDING SITE OF SNAKE AND MONGOOSE AChR

Several elapid snakes were reported to be resistant to the antagonistic effects of α-BTX, whereas their AChR was shown to be sensitive to agonist activation (Burden *et al.*, 1975). As a first step towards understanding the molecular basis of this phenomenon, we examined whether the non-poisonous water snake *Natrix tessellata* is also resistant to α-BTX. Intraperitoneal administration of α-BTX into water snakes in concentrations several orders of magnitude higher than the LD_{50} (LD = lethal dose) for other species did not result in snake death (Table 2.1, and Neumann *et al.*, 1989). The possibility that resistance of the snake to α-BTX was due to neutralizing factors in the serum has been excluded, since preincubation of snake serum with α-BTX did not abolish its lethal effect upon subsequent injection into mice (Neumann *et al.*, 1989).

In search of the snake AChR α-subunit, a genomic clone from *Natrix* genomic DNA has been first isolated and was shown to hybridize to a mouse AChR α-subunit cDNA probe. Sequence analysis of this 1 kb clone revealed significant

Table 2.1   Lethality of α-BTX in various animals species (Barchan *et al.*, 1995)

Species	Dose (μg of α-BTX/g of body weight)	Death (min after injection)
Snake	100	survived (Neumann *et al.*, 1989)
	10	survived
Mongoose	2	survived (Barchan *et al.*, 1992)
Hedgehog	0.2	survived
	0.4	survived
	1.2	30
Shrew	0.1	survived
	0.2	survived
	0.4	120
	0.8	60
Cat	0.2	90
Mouse	0.1	10

homology (75%) to the mouse α-subunit cDNA, on a stretch of 200 nucleotides (Neumann *et al.*, 1989). The deduced amino acid sequence of this fragment revealed considerable homology with residues 96-160 of the mouse AChR α-subunit. This includes the two cysteine residues at positions 128 and 142 of the α-subunit which are present in all the four subunits. The homology was much higher with the α-subunit than with the other subunits.

It was of special interest to identify, in the snake, the region which in other species, such as Torpedo and mouse, was shown to contain essential components of the α-BTX binding site (i.e. the segment containing the tandem cysteines 192 and 193). Polymerase chain reaction (PCR) has been employed for specific amplification of the cDNA encompassing this region. The domain of the α-subunit from *Natrix*, cobra and sand boa AChR, which contains the four extracellular cysteines (128, 142, 192 and 193) of the α-subunit, has been cloned and sequenced. Sequence analysis of the amplified fragments of these snakes revealed 72% homology to the mouse α-subunit cDNA, on the level of nucleotides and 78% homology at the deduced amino acids (Figure 2.1, and Neumann *et al.*, 1989; Fuchs *et al.*, 1992). The presence of two cysteines at positions 192 and 193 verifies that this fragment corresponds to the snake AChR α-subunit. Most differences between the snake and mouse sequences concentrate in the vicinity of these tandem cysteines, and might contribute to the unique pharmacological properties of the snake receptor.

Alignment of the snakes PCR derived sequences and those of the respective segments of muscle AChR α-subunits of other species indicates that the snake sequence is homologous to other muscle AChRs (Figure 2.1, and Neumann *et al.*, 1989). Among the various muscle AChRs, the snake receptor exhibits significant differences, some of which occur in the vicinity of cysteines 192 and 193. Several amino acids which are identical in distant species such as Torpedo, chick, mouse, calf and human are different in snake. Of these, substitution of tryptophan 184 for

```
                 140                   160                   180              200
   *              .     *               .                     .            * *    .
AIFKSYCEIIVTYFPFDEQNCSMKLGTWTYDGTVVAIYPEGPRPDLSNYMQSGEWTLKDYRGFWHSVNYSCCLDTPYLDITYHF
.........................R...........................A.........................
...................................................H............................
........H...Q..T...I.....K.S.S..SD.....TF.E...VM.....WK.W.Y.T..P.................
........Q.................M.V.N..SD.......F.E...VM.....WK.W.Y.A..P................
........H.................S....N..SDQ....NF.E...VI.EA..WK.W.F....PT..............
........H.................S..V.N..SDQ.....F.E...VI.ES..WK.W.F.A..PS..............
........H.................S....N..SDQ.....F.E...VI.ES..WK...T....P...............
```

Figure 2.1 Comparison of amino acid sequence of residues 122–205 of the AChR α-subunit from three snakes [*Naja naja atra* (cobra), *Natrix tessellata* and *Eryx jaculus* (boa)] and from other species. (from Fuchs *et al.*, 1993.)

18

phenylalanine, lysine 185 for tryptophan, and proline 194 for leucine are in close proximity to the two tandem cysteines, and might be relevant for binding properties. Of particular interest is Trp 187 which is conserved in Torpedo, chick, mouse and calf but substituted for Ser in the snake as well as in human. Finally, the PCR-derived fragments of all three snakes contain a putative N-glycosylation site in the binding site domain, at position 189 (Asn 189). This glycosylation site is not present in other known α-subunits. Hence, it is possible that the snake AChR α-subunit might be more glycosylated than that of the other AChRs. These changes or part of them may explain the lack of α-BTX binding to snake AChR.

The cloned fragments corresponding to amino acid residues 122–205 of the snake and, for comparison, the mouse AChR, were expressed employing a pET 8C derived expression vector. Chromatography in SDS-polyacrylamide gel revealed that the expressed protein fragments have the expected molecular weight of 8 kDa. Antibodies against a synthetic peptide corresponding to residues 143–158 of the Torpedo AChR α-subunit stained all three fragments (Figure 2.2, right), indicating that the expressed fragments are indeed from the AChR α-subunit. Overlay of the blotted proteins with ^{125}I-α-BTX showed that the toxin binds only to the mouse fragment and not to snake fragments (Figure 2.2, and Fuchs et al., 1992; Barchan et al., 1992).

As some of the sequence differences observed in the snake AChR could be specific to the group, we have extended our study to a mammal (mongoose) which is

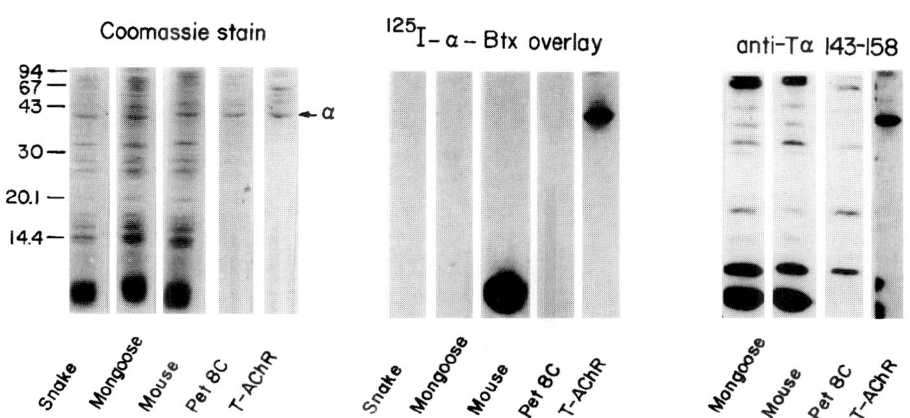

Figure 2.2 Analysis of expressed protein fragments. The expressed protein fragments or purified Torpedo AChR were resolved by polyacrylamide gel electrophoresis. The gels were stained for proteins (left panel) or blotted and overlayed with ^{125}I-α-BTX, or with anti-peptide 143–156, followed by ^{125}I-protein A (from Fuchs et al., 1992).

resistant to neurotoxins (Ovadia and Kochva, 1977) and includes snakes in its diet. Examination of the toxic effect of α-BTX in the mongoose as compared with rabbit and mouse demonstrated that intramuscular administration of α-BTX into mongoose in amounts of 0.3–2 μg/g of body weight, did not kill the mongoose, whereas 0.1 and 0.3 μg/g of body weight were found lethal in mice and rabbits, respectively (Table 2.1, and Barchan *et al.*, 1992). It was then demonstrated that mongoose AChR does not bind α-BTX either in muscle Triton extracts or by histochemical staining (Barchan *et al.*, 1992; Fuchs *et al.*, 1992). Histochemical experiments were carried out utilizing mongoose diaphragm sections. Acetylcholinesterase and AChR were visualized at the endplate regions whereas α-BTX did not bind to these regions (Barchan *et al.*, 1992). AChR in the mongoose endplate was visualized by two different AChR-specific antibodies (Figure 2.3). Both antibodies stained the mongoose endplate whereas no staining was observed with α-BTX. In contrast, both the anti-AChR antibodies and α-BTX stained the endplate regions in sections of the rat diaphragm (Figure 2.3 and Barchan *et al.*, 1992).

Thus, these results have suggested that mongoose AChR differs from other, toxin-binding AChRs in its binding site, as was demonstrated for the snake AChR. The mongoose cDNA fragment which includes the binding site domain (i.e., the

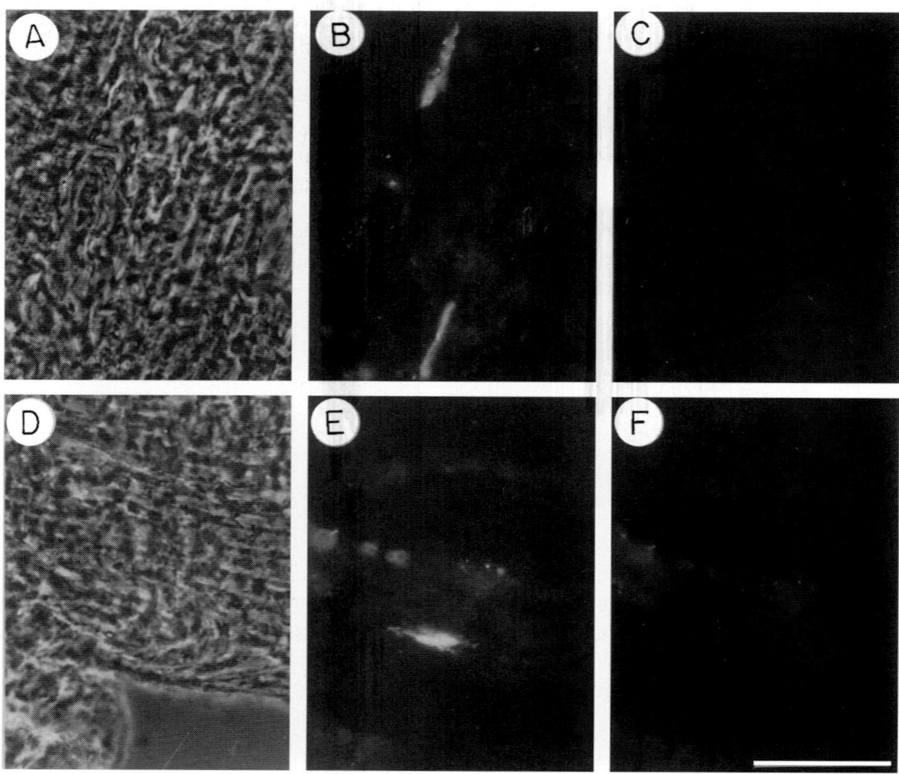

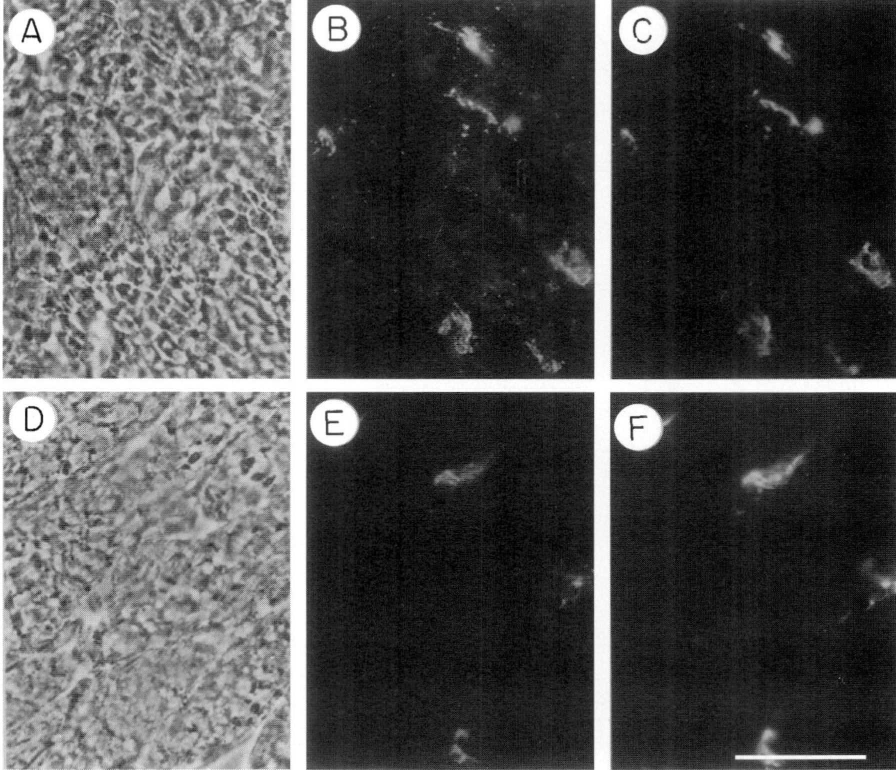

b

Figure 2.3 Staining of AChR in mongoose (*a*) and rat (*b*) diaphragm sections. Diaphragm sections were incubated with two types of purified AChR specific antibodies. **A–C**, rabbit anti-denatured Torpedo AChR antibody, **D–F**, rabbit anti-peptide 351-368 of the human AChR α-subunit. Sections were then incubated with a mixture of TMR-α-BTX and FITC-conjugated goat anti-rabbit IgG. **A** and **D** show phase contrast images. **B** and **E** show FITC fluorescence demonstrating AChR staining with both types of antibodies. **C** and **F**, show TMR fluorescence of the same fields demonstrating that TMR-α-BTX did not stain the mongoose endplate (*a*), and that α-BTX staining colocalized with that of AChR in the rat endplate (*b*). Bar represents 50 μm. (from Barchan *et al.*, 1992).

segment from the α-subunit containing the tandem cysteines 192 and 193) has been cloned and sequenced. The polymerase chain reaction (PCR) was used to amplify the cDNA fragment encompassing the binding site region, from mongoose single stranded cDNA. Sequence analysis of this amplified mongoose fragment (Figure 2.4) revealed high homology with the respective mouse fragment, corresponding to amino acid residues 122–205 of the α-subunit (homology of 89% in nucleotides and 92% in amino acids). The mongoose segment contains the four cysteines at positions 128, 142, 192 and 193, thus verifying that it corresponds to the AChR α-subunit. Interestingly, five of the seven amino acid differences

22

```
          122                           *                                          140           *
          A   I   F   K   S   Y   C   E   I   I   V   T   H   F   P   F   D   E   Q   N   C
Mongoose  GCC ATC TTC AAA AGC TAC TGT GAG ATC ATC GTC ACC CAC TTT CCC TTT GAT GAA CAG AAC TGC
Mouse     ... ... ..T ... ... ... ... ... ... ... ..T ... ..T ... ... ... ..C ... ..G ... ... ...
          .   .   .   .   .   .   .   .   .   .   .   .   .   .   .   .   .   .   .   .   .

                                                                          160
          S   M   K   L   G   T   W   T   Y   D   S   S   V   V   V   I   N   P   E   S   D
Mongoose  AGC ATG AAG CTG GGT ACC TGG ACC TAT GAC AGC TCT GTG GTT GTC ATC AAC CCG GAA AGC GAC
Mouse     ... ... ... ... ..C ... ... ... ... ... G.. ... ... ..G .C. ..T ... ... ... ..T ...
          .   .   .   .   .   .   .   .   .   .   G   .   .   .   A   .   .   .   .   .   .

                                                                  180
          Q   P   D   L   S   N   F   M   E   S   G   E   W   V   I   K   E   A   R   G   W
Mongoose  CAA CCT GAC CTA AGC AAC TTC ATG GAA AGC GGA GAG TGG GTG ATC AAG GAG GCC CGG GGC TGG
Mouse     ..G ..C ... ..G ..T ... ... ... ..G ... ..G ... ... ... ... ... ... ..A ..T ... ... ...
          .   .   .   .   .   .   .   .   .   .   .   .   .   .   .   .   .   .   .   .   .

                                          *   *                           200
          K   H   N   V   T   Y   A   C   C   L   T   T   H   Y   L   D   I   T   Y   H   F
Mongoose  AAG CAC AAT GTG ACC TAC GCC TGC TGC CTC ACC ACC CAC TAC CTG GAC ATC ACC TAC CAC TTC
Mouse     ... ... TGG ... TT. ... T.. ... ... .C. ... ..T .C. ... ... ... ... ... ... ... ...
          .   .   W   .   F   .   S   .   .   P   .   .   P   .   .   .   .   .   .   .   .
```

Figure 2.4 Alignment of nucleotide and deduced amino acid sequences for the mongoose PCR fragment of the AChR α-subunit and the corresponding mouse fragment. Amino acid residues are numbered from 122 to 205 corresponding to their position in the mouse AChR α-subunit. Cysteine residues are marked with an asterisk. Nucleotides or amino acids identical to the mongoose sequence are designated by dots (from Barchan *et al.*, 1992).

between the mouse and the mongoose fragments concentrate in the vicinity of the tandem cysteines in a stretch of eleven amino acid residues (residues 187–197), and most probably contribute to the inability of the mongoose receptor to bind α-BTX. Three of these five differences are at positions (187, 189, 194) where major substitutions take place also in the snake AChR, and one of them (187) creates a potential N-glycosylation site in the mongoose AChR (Barchan et al., 1992). Position 194, adjacent to the tandem cysteines is Pro in the toxin-binders, whereas in the mongoose and snakes it is Leu. This substitution may result in a major conformational change, as proline can form a β-bend or turn. Such a conformational change presumably will not affect the interaction between the receptor and its natural ligand, acetylcholine, but can interfere with the binding of the much larger polypeptide antagonist, α-BTX. It is noteworthy that the mongoose has an additional substitution at position 197, from proline to histidine. This proline is highly conserved in all other muscle AChRs.

The other two major substitutions are from aromatic residues, at positions 187 and 189 in toxin-binders, to non-aromatic residues in the mongoose and snake. Position 189 which is Tyr or Phe in the toxin-binders is changed to Thr in the mongoose, Asn in the snakes and Lys in all neuronal AChRs. At position 187, Trp in the toxin-binding AChRs has been substituted by Asn in the mongoose and Ser in the snake. Asn 187 in the mongoose and Asn 189 in the snake AChR α-subunit are both putative N-glycosylation sites. It has not been shown yet that these asparagines are indeed glycosylated in the intact mongoose and snake receptors, and if they are, whether such a glycosylation contributes to toxin resistance. It should be noted that the non-glycosylated mongoose and snake protein fragments expressed in E. coli as well as the respective non-glycosylated synthetic peptides (residues 185–196), do not bind α-BTX. It is possible, however, that glycosylation which adds a bulky group in the binding site domain may provide additional protection towards the toxin without affecting acetylcholine binding. Recent reports (Kreinkamp et al., 1994) have demonstrated that introducing a putative glycosylation site into the mouse α-subunit at position 187 or 189 resulted in a marked decrease in α-BTX binding to the receptor. Thus, it seems that Pro at position 194, an aromatic residue (Tyr or Phe) at 189, and Trp at 187, are strong requirements for toxin-binding, as mutations at these positions in the mongoose and snake AChR abolish their binding to α-BTX. It was still not clear at this point whether all these changes are required and what is the contribution of each in conferring toxin resistance.

Finally, although mongoose and snake AChR do not bind α-BTX, they still retain their cholinergic properties and therefore, amino acid residues which are essential for binding of acetylcholine should be conserved also in these receptors. Indeed, both the mongoose and snake binding site domains contain the amino acids which were shown to be labeled by dimethyl-aminobenzene diazonium fluoroborate (DDF) (Dennis et al., 1988; Galzi et al., 1991), maleimido benzyltrimetylammonium (MBTA, Kao et al., 1984), and lophotoxin (Abramson et al., 1989) at the same positions as other muscle and neuronal receptors α-subunits. These include Tyr 190, Cys 192 and 193, and Tyr 198. The aromatic residues Trp

149 and Tyr 151, which were reported to be labeled by DDF (Dennis *et al.*, 1988; Galzi *et al.*, 1991), are also conserved in the mongoose and snake. Thus, our study indicates that the requirements for acetylcholine binding are not sufficient for α-BTX binding. Though both bind primarily to the same site in AChR, additional structural elements are necessary for α-BTX binding and those can be manipulated by genetic pressure without the loss of the major physiological function which is acetylcholine binding.

OTHER SPECIES

To further examine the correlation between α-BTX binding and the structure of the binding site domain, the study on the binding site of AChR has been extended to additional animal species: The hedgehog (*Erinaceus europeus*) and shrew (*Crocidura russula*), which belong to the Order Insectivora (a group of primitive mammals) and were suspected to be resistant or partially resistant to α-BTX. The hedgehog was previously reported to exhibit some resistance to snake venoms (Domont *et al.*, 1991). We have also tested the cat (*Felis domestica*) which, like the mongoose, belongs to the Order Carnivora. The domains of the AChR α-subunit (corresponding to amino acid residues 122–205) of these species, as well as of the human AChR were cloned, sequenced and expressed in bacteria (Barchan *et al.*, 1995). Of these, the hedgehog fragment does not bind α-BTX and the human fragment is a partial binder. Both species have substitutions at positions 187 and 189 of their AChR α-subunit, from aromatic to non-aromatic residues, as previously shown by us for the snake and mongoose, confirming the role of these positions in determining resistance to α-BTX.

We have tested the sensitivity of the hedgehog, shrew and cat to α-BTX, by examining the toxic effect of intramuscular administration of α-BTX, in comparison with the effect in mouse, snake and mongoose (Table 2.1). A resistance to α-BTX was observed in the hedgehog, in comparison with the mouse. It survived four times the dose that killed mice within 10 min of α-BTX administration. Some marginal resistance was observed in the shrew which survived at least twice the amount of α-BTX that was lethal in the mouse. Analysis of the binding of ^{125}I-α-BTX to muscle Triton extracts from mouse, cat, shrew and hedgehog, showed just marginal levels of binding of the hedgehog AChR to α-BTX, whereas muscle extracts of the shrew and cat, exhibited specific binding in a concentration dependent manner.

Sequence analysis of the cloned fragments of the various species revealed a high homology with the respective mouse fragment (Figure 2.5). As mentioned above, the hedgehog fragment has two non-conservative amino acid substitutions at positions 187 and 189, from aromatic (tryptophan and tyrosine) to non-aromatic residues (arginine and isoleucine). These are the positions that are also substituted in the snake, mongoose and human AChR. In the shrew fragment, there are non-conservative changes, from the mouse sequence, at positions 155, 160 and 164. The cloned hedgehog, shrew and cat PCR fragments were expressed in

```
          122            140            160            180          200
           .    *          .  *            .              .        **     .
Mouse     AIFKSYCEIIVTHFPFDEQNCSMKLGTWTYDGSVVAINPESDQPDLSNFMESGEWVIKEARGWKHWVFYSCCPTTPYLDITYHF
Cat       ...............................................S.........A.................
Shrew     ..........................K...A..EH............................A................
Hedgehog  ....,.................................N.......................R.I.A...S.........
```

Figure 2.5 Deduced amino acid sequences for the cat, shrew, and hedgehog fragments of the AChR α-subunit, in comparison to the mouse fragment (residues 122–205). Cysteine residues are marked with an asterisk. Amino acid residues identical to the mouse sequence are designated by dots (from Barchan *et al.*, 1995).

E. coli, electrophoresed in SDS/polyacrylamide gel and blotted. Overlay of the blotted proteins with ^{125}I-α-BTX showed that, as with the mongoose fragment, the hedgehog fragment hardly bound α-BTX . On the other hand, the shrew and cat fragments both bound α-BTX to the same extent as the mouse fragment (Barchan *et al.,* 1995).

We and others have previously shown that synthetic peptides from the binding site domain of the human AChR α-subunit, bind α-BTX to a lower extent than the respective peptides from the mouse or Torpedo AChR (Neumann *et al.,* 1986b; Wilson and Lentz, 1988). This may correlate with the fact that human AChR also has two non-aromatic residues at positions 187 and 189, in common with the snake, mongoose and hedgehog AChR. It was therefore interesting to analyze, in addition to the protein fragments of the above mentioned species, the homologous fragment from human AChR. This fragment was amplified by RT-PCR from cDNA from the human TE671 cell line. Overlay of the blotted human AChR fragment with ^{125}I-α-BTX indicated that it is an intermediate binder and binds less than the mouse and more than the mongoose and hedgehog respective fragments (Barchan *et al.,* 1995). Taken together, our results indicate that there seems to be a correlation between the degree of α-BTX binding and amino acid substitutions in the binding site domain and most substitutions in the non-binders are in the putative binding site, close to cysteines 192 and 193.

It appears that positions 187 and 189 represent a hot spot or target for evolutionary modifications, resulting in a functional change, namely sensitivity to α-BTX. All muscle AChRs from species which are sensitive to α-BTX, have a tryptophan at position 187 and an aromatic residue (phenylalanine or tyrosine) at position 189. These two positions are replaced together by non-aromatic amino acid residues in the snake, mongoose, hedgehog and human AChR, although in each of these species the changes are different. We have not found any case of a muscle AChR in which only one position (187 or 189) has been changed. It should be noted that in the neuronal α-BTX binding protein there is an aromatic residue at the position correponding to 189 in the muscle AChR α-subunit, but a glutamic acid at position 187 (Schoepfer *et al.,* 1990; Couturier *et al.,* 1990).

There appears to be several means by which different degrees of resistance to α-BTX have developed during evolution, in taxonomically different groups of animals. Experiments of α-BTX administration have demonstrated that the resistance of the snake and the mongoose to α-BTX is much higher than that of the hedgehog. The nature of the amino acid substituted at positions 187 and 189 is different in the various resistant animals (Figure 2.6). The additional changes at prolines 194 (snake and mongoose) and 197 (mongoose), and the putative glycosylation in both the snake (Asn 189) and mongoose (Asn 187) AChR, probably add to the resistance of these species to α-neurotoxins. On the level of the expressed α-subunit fragments (122–205 residues), the snake, mongoose and hedgehog fragments do not bind α-BTX and the human fragment binds less than the mouse fragment. It should be noted that the binding experiments of ^{125}I-α-BTX to the fragments are performed on protein blots and it is quite likely that under these conditions the fragments do not fully assume their conformation as in the intact

```
                   180        187 189    194 197 200
                    .          |  |   **|   |  .
Mouse α1           EARGWKHWVFYSCCPTTPYLD
Cat α1             .S.........A...........
Torpedo α1         DY.......Y.T...D.....
Calf α1            ...........A...S.....
Chick α1           .........Y.A...D.....
Human α1           .S.....S.T.S...D.....
Hedgehog α1        .......R.I.A...S.....
Mongoose α1        .......N.T.A..L..H...
Snake α1           DY..FW.S.N.S..LD.....

Chick α7           GIP.KRTESF.E..KE-..P.
Chick α2           NAI.RYNSKK.D..TE-I.P.
```

Figure 2.6 Interspecies comparison of amino acid residues 180–200 of muscle AChR α-subunit (from Barchan *et al.*, 1995).

receptor. This may explain some of the differences in binding observed between the fragments, and intact AChR. In addition, changes in other regions of the α-subunit or in other subunits are also known to contribute to the overall structure of the binding site. In conclusion, it appears that aromatic residues at positions 187 and 189 of the muscle AChR α-subunit are necessary for proper α-BTX binding. Changes at these positions to non-aromatic residues are most important in determining resistance to α-BTX, though the nature of these changes, as well as additional structural elements, probably contribute the degree of this resistance.

FINE ARCHITECTURE OF THE α-BTX BINDING DOMAIN OF AChR: ANALYSIS BY SITE-DIRECTED MUTAGENESIS

The strategy of first investigating the binding site domain of AChR from several animal species resistant to α-BTX, made it possible to pinpoint the critical positions of functional importance for α-BTX binding, created by "Nature" during evolutionary development of toxin resistance. With this information in hand, one could now carry out logically plausible point mutations just at the limited number of specific positions, in order to elucidate the relative contribution of each of them to biological function.

As mentioned above, analysis of the AChR binding site from the snake, mongoose, hedgehog and human indicated that positions 187, 189, 194 and 197 in the receptor's α-subunit are most important in determining resistance to α-BTX. To further identify the amino acids participating in α-BTX binding, site-directed mutagenesis was performed on a fragment of the mongoose AChR α-subunit (residues 122-205), and residues 187, 189, 194 and 197 were exchanged, either alone or in combination, with those present in the mouse α-subunit sequence (Table 2.2, and Kachalsky *et al.*, 1995). Only the mongoose fragment, in which all four residues were mutated to the mouse ones, exhibited α-BTX binding similar to that of the mouse fragment. This result demonstrates that changes in amino acid residues at the four positions, 187, 189, 194 and 197 are involved in the formation of the necessary 3-D structure, which allows to restore the binding of α-BTX to the mongoose AChR.

In order to test the possibility that the presence of the two proline residues 194 and 197 together, might be required for assuming the right spacial conformation, to allow toxin binding, a double mutation has been created in the mongoose fragment, converting leucine and histidine at positions 194 and 197, respectively, to prolines, which are highly conserved in these positions in muscle AChR α-subunits. This double mutation resulted in a significant increase of the binding to α-BTX which, when calculated by densitometric analysis, amounted to approximately 60% of the binding of the mouse fragment (Table 2.2, and Kachalsky *et al.*, 1995). It should be noted that there is a striking similarity in the nature of amino acid residues at the four positions 187, 189, 194 and 197 between this mongoose construct and the respective fragment derived from human AChR α-subunit. In both there are prolines at positions 194 and 197 and aliphatic residues at positions 187 and 189, instead of aromatic residues in the toxin binders. Interestingly, the α-BTX binding to the human fragment, is similar to that observed by the mongoose double mutation (mL194P/L197 P) (Kachalsky *et al.*, 1995). To further investigate the role of the two adjacent prolines at positions 194 and 197 in determining

Table 2.2 Mutations introduced in the mongoose AChR α-subunit pragment (Residues 122–205)

	180	187	189	194	197	200	α-BTX binding (%)[a]
Mongoose	EARGWKHNVTYACCLTTHYLD						9 ± 1.4
mN187W	· · · · · · · W · · · · · · · · · · · · ·						13 ± 2.2
mT189F	· · · · · · · · · F · · · · · · · · · · ·						15 ± 2.2
mL194P	· · · · · · · · · · · · · · P · · · · · ·						15 ± 3.1
mN187W/T189F	· · · · · · · W · F · · · · · · · · · · ·						25 ± 2.7
mN187W/L194P	· · · · · · · W · · · · · · P · · · · · ·						32 ± 2.7
mT189F/L194P	· · · · · · · · · F · · · · P · · · · · ·						28 ± 1.3
mL194P/H197P	· · · · · · · · · · · · · · P · · P · · ·						60 ± 1.3
mN187W/T189F/L194P	· · · · · · · W · F · · · · P · · · · · ·						8 ± 0.9
mN187W/T189F/H197P	· · · · · · · W · F · · · · · · · P · · ·						5 ± 1.8
mN187W/T189F/L194/H197P	· · · · · · · W · F · · · · P · · P · · ·						92 ± 2.2
Mouse	· · · · · · · W · F · S · · P · · P · · ·						100

[a] α-BTX binding was calculated from densitometric analysis. Percent binding ± standard error were determined from three different measurements, relative to the binding of the mouse fragment, which was considered as 100% (from Kachalsky *et al.*, 1995).

α-BTX binding, we have analyzed the binding properties of mouse AChR α-subunit fragments (residues 122–205) in which proline 194 or proline 197 have been mutated to serine and histidine, respectively. Mutation of either proline resulted in a marked decrease in α-BTX binding (Kachalsky *et al.*, 1995). This supports the conclusion that the presence of two prolines in the binding site domain are required for creating the appropriate conformation required for α-BTX binding.

The results of this study led us to propose two subsites at the binding domain for α-BTX: The "aromatic subsite" and the "proline subsite". The aromatic subsite includes amino acid residues at positions 187 and 189 and determines the extent of α-BTX binding. The proline subsite includes the two close prolines 194 and 197, which appear to be essential for α-BTX binding in the snake and mongoose AChR. These two prolines are present in this spacing (PXXP) in all muscle AChRs which bind α-BTX (see Figure 2.6). In the absence of one or two of these two prolines, α-BTX binding is impaired, as is the case in the snake and mongoose AChR, as well as in the mouse α-subunit recombinant fragment, in which proline 194 or 197 has been mutated to serine and histidine, respectively (Kachalsky *et al.*, 1995). Similar results were also reported from studies on bacterially expressed fusion proteins containing residues 166–211 (Chatuverdi *et al.*, 1992) or 183–204 (Ohana *et al.*, 1991) of the Torpedo AChR α-subunit. However, recent studies on similar mutations employing AChR expressed in human embryonic kidney cells described only a slight alteration in α-BTX binding, resulting from mutating prolines 194 or 197 in the mouse AChR α-subunit (Krenkamp *et al.*, 1994). Such differences might be explained by the different experimental systems employed.

It appears that whereas the proline subsite is essential for α-BTX binding, the aromatic subsite may provide the fine tuning of the binding and determines the degree of binding, or resistance, to α-BTX. Tryptophan at position 187 and tyrosine at position 189 seem to form the preferred aromatic subsite, as is the case in Torpedo and chicken AChRs. This is followed by tryptophan and tyrosine at positions 187 and 189, respectively, in the mammalian AChRs, which bind α-BTX (Figure 2.6). Lower binding, at least on the level of the recombinant fragment (residues 122–205), is observed with human AChR, as well as in the mongoose double mutation mL194P/H197P, both having the proline subsite and two non-aromatic residues at positions 187 and 189. The hedgehog, snake and mongoose AChRs, all have non-aromatic residues at positions 187 and 189 and they do not bind α-BTX. The nature of the non-aromatic residues at these positions is different and may determine the degree of resistance to α-BTX. It should be noted however that the snake and mongoose are more resistant to α-BTX than the hedgehog (Table 2.1) and in both, one of the non-aromatic residues at positions 187 and 189 is a putative glycosylation site. It is difficult to evaluate the contribution of these latter changes alone to the toxin-resistance of the snake and mongoose, as in both of them there is also an impairment of the proline subsite, which is essential for toxin binding. In conclusion, the study of mongoose and mouse mutations points to the participation of an aromatic subsite and a proline subsite in the α-BTX binding site of muscle AChR. In the absence of an appropriate proline subsite, α-BTX is completely impaired, whereas changes in the aromatic subsite seem to determine the degree and specificity of toxin-resistance.

CONCLUDING REMARKS

The binding site for α-BTX in AChR is at or near the acetylcholine binding site. Thus, analysis of the α-BTX binding site provides a convenient experimental approach for elucidating the fine structure of the ligand binding domain of AChR. The present review describes the analysis of the AChR binding site from animal species that are resistant to α-BTX. It was shown that such toxin-resistant animals differ from toxin-sensitive ones in the ligand binding domain of their AChR. This experimental strategy made it possible to pinpoint the critical positions of functional importance for α-BTX binding. Thus, it has been shown that amino acid substitutions at the binding site domain of the AChR α-subunit, at positioins 187 and 189 (the aromatic subsite) and 194 and 197 (the proline subsite) are important in determining resistance to α-BTX. Furthermore, glycosylation within the binding site may represent an additional evolutionary means for protection against peptide neurotoxins. Finally, the requirements for acetylcholine binding are distinct from those required for α-BTX binding. Though both acetylcholine and α-BTX bind primarily to a similar site in AChR, additional structural elements are required for α-BTX binding and those can be manipulated by genetic pressure without affecting the binding to acetylcholine.

Acknowledgements

This review summarizes studies performed in my laboratory by Drorit Neumann, Dora Barchan, Sylvia Kachalsky and Bo Jensen, and is based on resulting publications referred to in the review. Drs. M. Horowitz, E. Kochva, M. Ovadia and Z. Vogel have collaborated with us in various parts of the work described. The research described in this review was supported by grants from The Basic Research Foundation administered by The Israel Academy of Sciences and Humanities; The Muscular Dystrophy Association of America; The Association Francaise Contre les Myopathies, The Ministry of Science and Arts, Israel and French Ministry of Research and Technology; The Betty and Norman Levy Foundation, The Crown Endowment Fund for Immunological Research; and The Leo and Julia Forshheimer Center for Molecular Genetics at The Weizmann Institute of Science.

REFERENCES

Abramson, S.M., Li, Y. and Culver, P. (1989) An analog of Lophotoxin react covalentely with Tyr-190 in the α-sununit of the nicotine acetylcholine receptor. *Journal of Biological Chemistry*, **264**, 12666–12672.

Balass, M., Heldman, Y., Casbilly, S., Givol, D., Katchalski-Katzir, E. and Fuchs, S. (1993) Identification of a hexapeptide that mimics a conformation-dependent binding site of acetylcholine receptor by use of a phage-epitope library. *Proceedings National Academy of Science USA*, **90**, 10638–10642.

Barchan, D., Kachalsky, S., Neumann, D., Vogel, Z., Ovadia, M., Kochva, E. and Fuchs, S. (1992) How the mongoose can fight the snake: the binding site of the mongoose acetylcholine receptor. *Proceedings National Academy of Science USA*, **89**, 7717–7721.

Barchan, D., Ovadia, M., Kochva, E. and Fuchs, S. (1995) The binding site of the nicotinic acetylcholine receptor in animal species resistant to α-bungarotoxin. *Biochemistry*, **34**, 9172–9176.

Barkas, T., Mauron, A., Roth, B., Alliod, C., Tzartos, S.J. and Ballivet, M.B. (1987) Mapping the main immunogenic region and toxin binding site of the nicotinic acetylcholine receptor. *Science*, **235**, 77–80.

Blount, P. and Merlie, J.P. (1989) Molecular basis of the two nonequivalent ligand binding sites of the muscle nicotinic acetylcholine receptor. *Neuron*, **3**, 349–357.

Boulter, J., Luyten, W., Evans, K., Mason, P., Ballivet, M., Goldman, D., Stengelin, S., Martin, G., Heinemann, S. and Patrick, J. (1985) Isolation of a clone coding for the α-subunit of a mouse acetylcholine receptor. *Journal of Neuroscience*, **5**, 2545–2552.

Burden, S.J., Hartzell, M.C. and Yoshikami, D. (1975) Acetylcholine receptor at neuromuscular synapses: Phylogenetic defferences detected by snake α-neurotoxins. *Proceedings National Academy of Science USA*, **72**, 3245–3249.

Changeux, J.P., Devillers-Thiery, A. and Chemouilli, P. (1984) The acetylcholine receptor: an allosteric protein. *Science*, **225**, 1335–1345.

Changeux, J.P. (1990) The nicotonic acetylcholine receptor: an allosteric protein prototype of ligand-gated ion channels. *Trends in Pharmacological Sciences*, **11**, 485–492.

Chatuverdi, V., Donnelly-Roberts, D.L. and Lentz, T. (1992) Substitution of Torpedo acetylcholine receptor $\alpha 1$ subunit residues with snake $\alpha 1$- and rat nerve $\alpha 3$-subunit residues in recombinant fusion proteins: effect on α-bungarotoxin binding. *Biochemistry*, **31**, 1371–1375.

Claudio, T., Ballivet, M., Patrick, J. and Heinemann, S. (1983) Nucleotide and deduced amino acid sequences of Torpedo Californica acetylcholine receptor γ-subunit. *Proceedings National Academy of Science USA*, **80**, 1111–1115.

Couturier, S., Bertrand, D., Matter, J.M., Hernandez, M., Bertrand, S., Millar, N., Valera, S., Barkas, T. and Ballivet, M. (1990) A neuronal nicotinic acetylcholine receptor subunit ($\alpha 7$) is developmentally regulated and forms a homo-oligomeric channel blocked by α-BTX. *Neuron*, **5**, 847–856.

Czajkowski, C. and Karlin, A. (1991) Agonist binding site of Torpedo electric tissue nicotinic acetylcholine receptor. *Journal of Biological Chemistry*, **33**, 22603–22612.

Dennis, M., Giraudat, J., Kotzyba-Hibert, F., Goeldner, M., Hirth, C., Chang, J.Y., Lazure, C., Chretien, M. and Changeux, J.P. (1988) Amino acids of the Torpedo marmorata acetylcholine receptor alpha subunit labeled by a photoaffinity ligand for the acetylcholine binding site. *Biochemistry*, **27**, 2346–2357.

Devillers-Thiery, A., Giraudat, J., Bentaboulet, M. and Changeux, J.P. (1983) Complete mRNA coding sequence of the acetylcholine binding α-subunit in Torpedo marmorata acetylcholine receptor: A model for the transmembrane organization of the polypeptide chain. *Proceedings National Academy of Science USA*, **80**, 2067–2071.

Domont, G.B., Perales, J. and Moussatché, H. (1991) Natural anti-snake venom proteins. *Toxicon*, **29**, 1183–1194.

Fuchs, S., Barchan, D., Kachalsky, S. and Neumann, D. (1992) Molecular evolution of the binding site of the nicotinic acetylcholine receptor. In *Membrane proteins: structures, interactions and models* (ed. A. Pullman *et al.*), pp. 147–160. Kluger Academic Publishers.

Fuchs, S., Barchan, D., Kachalsky, S., Neumann, D., Aladjem, M., Vogel, Z., Ovadia, M. and Kochva, E. (1993) Molecular evolution of the binding site of the acetylcholine receptor. *Ann. N.Y. Acad. Sci.*, **681**, 126–139.

Galzi, J.L., Revah, F., Bessis, A. and Changeux, J.P. (1991) Functional architecture of the nicotinic acetylcholine receptor: From electric organ to brain. *Annual Review of Pharmacology*, **31**, 37–72.

Gershoni, J.M., Hawrot, E. and Lentz, T.L. (1983) Binding of α-bungarotoxin to isolated α-subunit of the acetylcholine receptor of Torpedo californica: Quantitative analysis with protein blots. *Proceedings National Academy of Science USA*, **80**, 4973–4977.

Gershoni, J.M. (1987) Expression of the α-bungarotoxin binding site of the nicotinic acetylcholine receptor by Escherichia coli transformants. *Proceedings National Academy of Science USA*, **84**, 4318–4321.

Goldberg, G., Mochly-Rosen, D., Fuchs, S. and Lass, Y. (1983). Monoclonal antibodies modify acetylcholine receptor-induced ionic channel properties in cultured cell myoblasts. *Journal of Membrane Biology*, **76**, 123–128.

Gotti, C., Mazzola, G., Longhi, R., Fornasari, D. and Clementi, F. (1987) The binding site for α-bungarotoxin resides in the sequence 188–201 of the α-subunit of the acetylcholine receptor: Structure, conformation and binding characteristics of peptide [Lys] 188–201. *Neuroscience Letters*, **82**, 113–119.

Haggerty, J.G. and Frohner, S.C. (1981) Restoration of ^{125}I-α-bungarotoxin binding activity to the α-subunit of Torpedo acetylcholine receptor isolated by gel electrophoresis in sodium dodecyl sulfate. *Journal of Biological Chemistry*, **256**, 8294–8297.

Kachalsky, S.G., Jensen, B.S., Barchan, D. and Fuchs, S. (1995) Two subsites in the binding domain of the acetylcholine receptor: An aromatic subsite and a proline subsite. *Proceedings National Academy of Science USA*, **92**, 10801–10805.

Kao, P.N., Dwork, A.J., Kaldany, R.R., Silver, M.L., Wideman, J., Stein, S. and Karlin, A. (1984). Identification of the α-subunit half-cystine specifically labeled by an affinity reagent for the acetylcholine receptor binding site. *Journal of Biological Chemistry*, **259**, 11662–11665.

Karlin, A. (1980) Molecular properties of the acetylcholine receptors. In: *In The Cell Surface and Neuronal Function* (ed. G. Poste, C.W. Cotman and G.L. Nicolson), pp. 191–260. Elsevier: North Holland.

Karlin, A. (1993) Structure of the nicotinic acetylcholine receptors. *Curr. Op. Neurobiol.*, **3**, 299–309.

Kreinkamp, H.J., Sine, S.M., Maeda, R.K. and Taylor, P. (1994) Glycosylation sites selectively interfere with α-toxin binding to the nicotinic acetylcholine receptor. *J. Biol. Chem.*, **269**, 1–7.

Lee, C.Y. (1972). Chemistry and pharmacology of polypeptide toxins in snake venoms. *Annual Review of Pharmacology*, **12**, 265–286.

Mochly-Rosen, D. and Fuchs, S. (1981) Monoclonal anti-acetylcholine receptor directed against the cholinergic binding site *Biochemistry*, **20**, 5920–5924.

Mulac-Jericevic, B. and Atassi, M.Z. (1986) Segment alpha-182-198 of the Torpedo californica acetylcholine receptor contains a second toxin-binding region and binds anti-receptor antibodies. *FEBS Letters*, **199**, 68–74.

Neumann, D., Barchan, D., Fridkin, M. and Fuchs, S. (1986b) Analysis of ligand binding to the synthetic dodecapeptide 185–196 of the acetylcholine receptor α-subunit. *Proceedings National Academy of Science USA*, **83**, 9250–9253.

Neumann, D., Barchan, D., Horowitz, M., Kochva, E. and Fuchs, S. (1989) Snake acetylcholine receptor: Cloning of the domain containing the four extracellular cysteines of the α-subunit. *Proceedings National Academy of Science USA.*, **86**, 7255–7259.

Neumann, D., Barchan, D., Safran, A., Gershoni, J.M. and Fuchs, S. (1986a) Mapping of the α-bungarotoxin binding site within the α-subunit of the acetylcholine receptor. *Proceedings National Academy of Science USA*, **83**, 3008–3011.

Neumann, D., Gershoni, J.M., Fridkin, M. and Fuchs, S. (1985) Antibodies to synthetic peptides as probes for the binding site on the α-subunit of the acetylcholine receptor. *Proceedings National Academy of Science USA*, **82**, 3490–3493.

Noda, M., Furutani, Y., Takahashi, H., Toyosato, M., Tanabe, T., Shimizu, S., Kikyotani, S., Kayano, T., Hirose, T., Inayama, S. and Numa, S. (1983) Cloning and sequence analysis of calf cDNA and human genomic DNA encoding α-subunit precursor of muscle acetylcholine receptor. *Nature*, **305**, 818–823.

Noda, M., Takahashi, H., Tanabe, T., Toyosato, M., Furutani, Y., Hirose, T., Takashima, H., Inayama, S., Miyata, T. and Numa, S. (1982) Primary structure of the α-subunit precursor of Torpedo californica acetylcholine receptor deduced from cDNA sequence. *Nature*, **299**, 793–797.

Ohana, B., Fraenkel, Y., Navon, G. and Gershoni, J.M. (1991) Molecular disection of cholinergic binding sites: how do snakes escape the effect of their own toxins. *Biochem. Biophys. Res. Commun.*, **179**, 648–654.

Ovadia, M. and Kochva, E. (1977) Neutralization of viperidae and elapidae snake venom by sera of different animals. *Toxicon*, **15**, 541–548.

Patrick, J. and Stallcup, W. (1977) Alpha-bungarotoxin binding and cholinergic receptor function on a rat sympathetic nerve line. *Journal of Biological Chemistry*, **252**, 8629–8633.

Pedersen, S.E., Dreyer, E.B. and Cohen, J.B. (1986) Location of ligand-binding sites on the nicotinic acetylcholine receptor α-subunit. *J. Biol. Chem.*, **261**, 13735–13743.

Pedersen, S.E., Bridgman, P.C., Sharp, S.D. and Cohen, J.B. (1990) Identification of a cytoplasmic region of the Torpedo nicotinic acetylcholine receptor α-subunit by epitope mapping. *Journal of Biological Chemistry*, **265**, 569–581.

Schoepfer, R., Conroy, W.G., Whiting, P., Gore, M. and Lindstrom, J. (1990) Brain α-bungarotoxin binding protein cDNAs and mAbs reveal subtypes of this branch of the ligand-gated ion channel gene superfamily. *Neuron*, **5**, 35–48.

Scott, J.K. and Smith, G.P. (1990) Searching for peptide ligands with an epitope library. *Science*, **249**, 386–390.

Souroujon, M.C., Mochly-Rosen, D., Gordon, A.S. and Fuchs, S. (1983) Interaction of monoclonal antibodies to Torpedo acetylcholine receptor with the receptor of skeletal muscle. *Muscle and Nerve*, **6**, 303–311.

Souroujon, M.C., Pachner, A.R. and Fuchs, S. (1986) The treatment of passively transferred experimental myasthenia with anti-idiotypic antibodies. *Neurology*, **36**, 622–625.

Wilson, P.T., Lentz, T.L. and Hawrot, E. (1985) Determination of the primary amino acid sequence specifying the α-bungarotoxin binding site on the α-subunit of the acetylcholine receptor from Torpedo californica. *Proceedings National Academy of Science USA*, **82**, 8790–8794.

Wilson, P.T. and Lentz, T.L. (1988) Binding of α-bungarotoxin to synthetic peptides corresponding to residues 173–204 of the α-subunit of Torpedo, calf, and human acetylcholine receptor and restoration of high-affinity binding by sodium-dodecyl sulfate. *Biochemistry*, **27**, 6667–6674.

3. MUSCARINIC TOXINS AND MUSCARINIC ACETYLCHOLINE RECEPTORS

DIANA JERUSALINSKY[1], EDGAR KORNISIUK[1] and ALAN L. HARVEY[2]

[1]*Instituto Biología Celular y Neurociencias, Facultad de Medicina, Universidad de Buenos Aires, Paraguay 2155, 1121 Buenos Aires, Argentina and [2]Department of Physiology and Pharmacology, Strathclyde Institute for Drug Research, University of Strathclyde, 204 George Street, Glasgow G11XW, UK*

INTRODUCTION

Acetylcholine is a major neurotransmitter throughout the body, and it has long been known that there are two distinct classes of acetylcholine receptors, based upon pharmacological evidence: the nicotinic and the muscarinic acetylcholine receptors (Dale, 1914). Muscarinic receptors are widely distributed and are important for the control of many physiological processes including heart rate, regulation of gut and other smooth muscle contractions, motor control, and learning and memory. There are now known to be several molecular subtypes of muscarinic receptors.

This chapter will briefly summarise the characteristics of the different subtypes of muscarinic receptor, and will review the available literature on the snake toxins (the muscarinic toxins) that appear to have the potential to be invaluable tools in the study of the molecular, pharmacological and functional properties of the different subtypes of muscarinic receptor.

THE MUSCARINIC RECEPTORS

Pharmacological Classification

At the beginning of the 1980s it was concluded that the multiple effects of acetylcholine at muscarinic synapses could not be accounted for the existence of only one type of muscarinic receptor. Following the demonstration that pirenzepine exhibited partial selectivity for "neuronal" (high affinity) vs. "heart" (low affinity) muscarinic receptors (Hammer *et al.*, 1980; Hammer and Giachetti, 1982), two distinct pharmacological subtypes, M_1 and M_2, were postulated. However, pirenzepine, as all other available muscarinic ligands, binds to all muscarinic receptors. Subsequently, based on experiments with 4-DAMP, (see Caulfield, 1993) a third subtype, M_3, was proposed for some other receptors on certain smooth muscles. Therefore, the muscarinic receptors have been classified by their differential affinities to these and other antagonists. However, the available pharmacological agents are not selective enough to enable a robust classification. Thus, determining which kind of muscarinic receptor is involved in a particular tissue or response has been difficult and, sometimes, controversial, especially since more than one subtype is often simultaneously present. Consequently, more than one agonist and antagonist

Table 3.1   Antagonist potencies (pK_D = negative logarithm of dissociation constant) from radioligand binding[a] or functional experiments[a']

Pharmacological subtype	M_1	M_2	M_3	M_4
Compound				
Pirenzepine	7.9–8.2	6.3–6.5	6.7–7.1	7.1–7.6
	8.1–8.5	6.7		7.7–8.1
4-DAMP[b]	8.9–9.2	8–8.4	8.9–9.3	8.9
				9.4
AF-DX 116	6.4–6.7	7.1–7.2	5.9	6.6
	6.9		6.6	7
				7.4
Methoctramine	7.1–7.6	7.8–8.3	6.3–6.9	7.8–8.1
				7.6
Himbacine	7–7.2	8–8.3	6.9–7.4	8–8.5
	7.2			8.5–8.8
pFHHSiD[c]	7.2–7.5	6–6.9	7.8–7.9	7.5

[a]Data from radioligand-binding displacement (1st line) and [a'] functional studies and studies on cloned receptors expressed in cell lines. (2nd and 3rd line) For references, see the text.
[b]4-DAMP: 4-diphenylacetoxy-*N*-methyl-piperidine methiodide.
[c]pFHHSiD: *para*-fluoro-hexahydrosiladifenidol.

is usually required to construct a pharmacological profile of the receptor subtype. For example, to define M_2 receptors it should be demonstrated that they have high affinity for methoctramine and AF-DX 116, and low affinity for pirenzepine, 4-DAMP and pFHHSID. M_3 receptors also have low affinity for pirenzepine, but have high affinity for 4-DAMP and pFHHSID (although there is only a two-fold difference between M_1 and M_2. M_4 receptors have high affinity for both pirenzepine and himbacine. Although pirenzepine is one of the most selective ligands for discriminating between M_1 and M_2 receptors, it does bind to all the subtypes. Table 3.1 shows the negative logarithm of apparent dissociation constants (pK_D) for different antagonists in functional or binding assays from data obtained by Mitchelson (1988), Lazareno and Roberts (1989) Lambrecht *et al.* (1989), Dörje *et al.* (1991), Lazareno *et al.* (1990), Hulme *et al.* (1990), Bernheim *et al.* (1992), Caulfield and Brown (1991), Entzeroth and Mayer (1990) and from our own work (for further details, see Caulfield, 1993).

How Many Subtypes of Muscarinic Acetylcholine Receptors?

With the development of techniques for cloning and sequencing of large proteins like membrane receptors, muscarinic receptors were found to belong to the super-family of receptors which have seven transmembrane domains. The response is

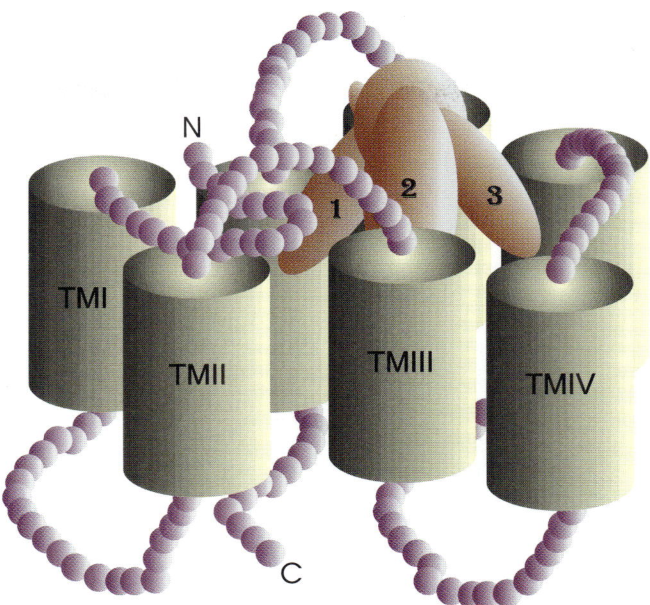

Figure 3.1   Diagram of the three-dimensional representation of a model of the muscarinic acetylcholine receptor with the seven transmembrane domains (TMI-VII), the three intra- and the three extracellular loops, the extracellular C-terminus and the intracellular N-terminus. The three-fingered muscarinic toxin (1, 2, 3) has been represented as binding with the upper regions of TM domains and the exposed loops of the receptor. Loop 2 of the toxin, interacting with the classical ligand binding pocket and loop 1, with the C-terminal region (built taking into account the suggestions made by Max *et al.*, 1993a; Jerusalinsky and Harvey, 1994; Ségalas *et al.*, 1995a; Vandermeers *et al.*, 1995).

transduced by coupling of the receptor to different membrane proteins that bind GTP (G-proteins), either activating or inhibiting different enzymes, and consequently modifying intracellular effector systems.

Muscarinic receptors (Figure 3.1) have a unique polypeptide chain with seven highly hydrophobic segments with an α-helical structure, which is inserted into the lipid bilayer of the plasma membrane. The transmembrane segments are linked by three extra- and three intracelular loops.

Kubo *et al.* (1986) began the cloning of muscarinic receptors by cloning the pig heart receptor (m2), and then that of pig cerebral cortex (m1). A year later, two more muscarinic receptors were cloned, m3 and m4 (Bonner *et al.*, 1987). Subsequently, a fifth gene was identified (m5), and many similar muscarinic receptors have now been cloned from different mammalian species -like rat, pig, mouse and human (Bonner, 1989; Hulme *et al*, 1990; 1991)- and recently from some insect species (Trimmer, 1995).

Through northern blot techniques, the corresponding mRNAs for each muscarinic receptor subtype have been localized mainly as follows (for details see Hulme *et al.*, 1990)

> m1 and m3, in brain and exocrine glands
> m2, heart and smooth muscle
> m4, not widely distributed, but concentrated in some neural tissue
> m5, poorly detected

By *in situ* hybridization in rat brain slices, the subtypes have been localized mainly as follows (for details see Hulme *et al.*, 1990):

> m1, in the forebrain, particularly in cerebral cortex, striatum and hippocampus
> m2, in low concentrations in medial septum, thalamus and pons
> m3, also in the telencephalon, thalamic nuclei and brain stem
> m4, in cerebral cortex, striatum and hippocampus
> m5, in very low concentrations in hippocampus and some nuclei of brain stem

Although something has been learned about muscarinic receptor distribution and structure, much less is known about the receptor-effector coupling and their biological roles. The expression of a single type of receptor in a cell line revealed that m1, m3 or m5 receptors, through coupling to different G-proteins, can activate signalling effectors like phopsholipase A_2, C and D, tyrosine kinase and a voltage-insensitive calcium channel. The inhibitory receptors m2 and m4, mainly attenuate adenylate cyclase activity, but also activate phospholipase A_2 (Felder, 1995).

There is very high homology between receptor subtypes, mostly in the transmembrane domains (90%; Bonner, 1989) but also in the intra- and extracellular loops. In spite of this, there are great differences between subtypes in the third intracellular loop, which is important in the coupling to G-proteins. However, this cannot explain the differential binding of pharmacological agents, which must involve regions exposed to the extracellular environment. The N-terminal region of the receptor is predicted to be extracellular; it is glycosylated at three different points and there are differences in the amino acid sequence between subtypes.

Acetylcholine and other agonists have been proposed to bind to a narrow, flat pocket surrounded mainly by six transmembrane segments and just immediately under the membrane surface. Most of the residues proposed to be involved in agonist binding are located in the upper portion of the transmembrane domains (Hulme *et al.*, 1991; Nordvall and Hacksell, 1993). From directed mutagenesis experiments (Fraser *et al.*, 1989) it has been suggested that the positive charged ammonium head of muscarinic agents might bind to a negatively charged Asp located in the third transmembrane segment. There are some other residues near the surface, shared by all the subtypes of receptor, that have been proposed to contribute to the binding pocket for agonists but not for antagonists (Hulme *et al.*, 1991; Wess *et al.*, 1995).

Our understanding of ligand-receptor interaction is still incomplete because there is little evidence about which residues could account for differential affinities of the receptor subtype for muscarinic agents.

MUSCARINIC TOXINS FROM MAMBA VENOMS

Discovery of the Toxins

Venoms from mamba snakes (*Dendroaspis* species) contain many neurotoxins which are pharmacologically different from toxins found in other related snake venoms. In addition to their classical α-neurotoxins, typical of elapid snake venoms, mamba venoms contain such unusual toxins as the dendrotoxins (blockers of certain neuronal potassium channels), fasciculins (specific inhibitors of acetylcholinesterase), and blockers of calcium ion channels. There was an early report that Eastern green mamba venom, *Dendroaspis angusticeps*, contained acetylcholine (Welsh, 1967) and some pharmacological studies indicated that the venom had cholinomimetic activity (Othman *et al.*, 1973). In studies examining the effects of *Dendroaspis angusticeps* venom and protein-containing subfractions on binding of muscarinic radioligands to brain membranes, Adem *et al.* (1988) found that the specific binding of radiolabelled QNB (quinuclidylnyl benzilate, a non-selective muscarinic antagonist) was partly inhibited by green mamba venom, and they isolated two proteins responsible for most of this activity. Although no toxic actions of these proteins were assessed, they were named muscarinic toxin 1 and muscarinic toxin 2 (MTx1 and MTx2). Both toxins had very similar potency at displacing ^{3}H-QNB, and it was noticeable that the maximum displacement by the toxins of QNB binding to membranes from rat cerebral cortex was only around 50%, leading to the speculation that the toxins might be selective for one subtype of muscarinic receptor.

Similar results were found by Jerusalinsky *et al.* (1992) with MTx1 and MTx2 and ^{3}H-QNB binding to rat and cow cerebral cortex synaptosomal membranes. They found also that the toxins inhibited only about 30% of binding in membranes prepared from rat brain stem. When binding interactions were performed with radiolabelled pirenzepine, at concentrations that mostly bound to m1 and m4 subtypes of muscarinic receptor, MTx1 and MTx2 inhibited more than 80% of the binding.

These results led to the conclusion that the muscarinic toxins MTx1 and MTx2 may bind selectively to central muscarinic receptors of the M_1 subtype (Jerusalinsky *et al.*, 1992).

Subsequently, Jerusalinsky *et al.* (1993) showed that injection of MTx2 into the hippocampus of rats increased their performance in a stepdown inhibitory avoidance test in a manner similar to that of the muscarinic agonist oxotremorine. The effects of the toxin could be blocked by the muscarinic antagonist scopolamine at saturating concentrations. Thus, MTx2 appeared to act like an agonist at some central muscarinic receptors. MTx1 was later shown to have similar effects (Jerusalinsky *et al.*, 1995).

In parallel work, Potter's group isolated a toxin from green mamba venom that they called m1-toxin because of its specificity and potency at m1 muscarinic receptors. In contrast to the findings with MTx2, m1-toxin behaved as an antagonist at m1 receptors (Max *et al.*, 1993 a, b).

The Toxins

There are four species of mamba snakes: *Dendroaspis angusticeps* (Eastern green mamba), *Dendroaspis viridis* (Western green mamba), *Dendroaspis polylepis* (black mamba) and *Dendroaspis jamesonii* (Jameson's mamba). Of these, venom from the Eastern green mamba has been studied most extensively, but there are also muscarinic toxins in *Dendroaspis viridis* and *Dendroaspis polylepis* venoms.

MTx1 and MTx2 have been isolated from *Dendroaspis angusticeps* venom using classical chromatography techniques (Adem *et al.*, 1988; Jerusalinky *et al.*, 1992; Jolkkonen *et al.*, 1995). The venom is dissolved in 0.1M ammonium acetate buffer pH 6.9 and then gel filtered on Sephadex G-50. The protein peak containing components with molecular weights around 7000 contains the muscarinic toxins as well as other components such as dendrotoxins and fasciculins. The next step is ion-exchange chromatography on Bio-Rex 70, again using ammonium acetate buffer. The muscarinic toxins are not bound to this ion-exchange resin and elute in the void fraction. The void fractions are rechromatographed on SP Sephadex, which allows separation of the different muscarinic toxins, which can be rechromatographed to achieve pure fractions. This approach has enabled Karlsson and his colleagues to isolate and characterise seven muscarinic toxins from Eastern green mamba venom, MTx1-MT7, and several homologous toxins from black mamba venom, (MTα, MTβ). Fractions with muscarinic activity have also been isolated from *Dendroaspis viridis* venom (DvMTx) (Jerusalinsky and Harvey, 1994).

In a parallel study, Potter's group used slightly different chromatography techniques to isolate several muscarinic toxins from *Dendroaspis angusticeps* venom. They have characterised m1-toxin, which is apparently highly homologous to MT7 from Karlsson's lab (Max *et al.*, 1993a, b, c; Potter, personal communication), and m4-toxin which is identical to MT3 from Karlsson's lab (Max *et al.*, 1993d; Jolkkonen *et al.*, 1994).

The first sequence of a muscarinic toxin to be determined was that of MTx2. Its sequence was obtained by classical methods (Karlsson *et al.*, 1991) and also by deduction from the isolated cDNA (Ducancel *et al.*, 1991). The toxin has 65 amino acids, giving a formula weight of 7040. It is homologous to a large number of other snake venom toxins including the short chain α-neurotoxins, the fasciculins, the cardiotoxins, and the angusticeps-type toxins of mamba venoms (Table 3.2). Although the disulphide bridges have not been established experimentally, the Cys residues are in corresponding positions to the Cys residues of α-neurotoxins. Therefore, a secondary structure of muscarinic toxins can be predicted. This is consistent with recent structural determinations using NMR spectroscopy (Ségalas *et al.*, 1995a) and X-ray crystallography. The sequence of MTx1 was reported more recently (Jolkkonen *et al.*, 1995); as expected, it is very similar to that of MTx2 (Table 3.2). Two other close homologues, MT3 and MT4, have been described in detail by Karlsson's group (Jolkkonen *et al.*, 1994; Vandermeers *et al.*, 1995). The sequence of m1-toxin was determined to have 64 amino acid residues, but again the positioning of the Cys residues was identical to that in the other toxins (Max *et al.*, 1993a, b, c). The degree of sequence identity between these muscarinic toxins is about 60%. The published sequences are shown in Table 3.2.

Table 3.2 Sequences of muscarinic toxins from *Dendroaspis angusticeps* and *D. polylepis* venoms compared to other three-fingered toxins

				5					10					15					20					25
MTx1	L	T	C	V	T	S	K	S	I	F	G	I	T	T	E	N	C	P	D	G	Q	N	L	C	F
MTx2	L	T	C	V	T	T	K	S	I	G	G	V	T	T	E	D	C	P	A	G	Q	N	V	C	F
m1-tx	L	T	C	V	K	S	N	S	I	W	F	P	T	S	E	D	C	P	D	G	Q	N	L	C	F
MT3	L	T	C	V	T	K	N	T	I	F	G	I	T	T	E	N	C	P	A	G	Q	N	L	C	F
MT4	L	T	C	V	T	S	K	S	I	F	G	I	T	T	E	N	C	P	D	G	Q	N	L	C	F
MTß	L	T	C	V	T	S	K	S	I	F	G	I	T	T	E	D	C	P	D	G	Q	N	L	C	F
DpCM3	L	T	C	V	T	S	K	S	I	F	G	I	T	T	E	D	C	P	D	G	Q	N	L	C	F
cardiotx	L	K	C	N	–	–	K	L	I	P	L	A	Y	K	T	–	C	P	A	G	K	N	L	G	Y

					30					35					40					45					50
MTx1	K	K	W	Y	Y	I	V	P	R	Y	S	D	I	T	W	G	C	A	A	T	C	P	K	P	T
MTx2	K	R	W	H	Y	V	T	P	K	N	Y	D	I	I	K	G	C	A	A	T	C	P	K	V	D
m1-tx	K	R	H	W	Y	I	S	P	R	M	Y	D	F	T	R	G	C	A	A	T	C	P	K	A	E
MT3	K	R	W	H	Y	V	I	P	R	Y	T	E	I	T	R	G	C	A	A	T	C	P	I	P	E
MT4	K	K	W	Y	Y	I	V	P	R	Y	S	D	I	T	W	G	C	A	A	T	C	P	K	P	T
MTß	K	R	R	H	Y	V	V	P	K	I	Y	D	I	T	R	G	C	V	A	T	C	P	I	P	E
DpCM3	K	R	R	H	Y	V	V	P	K	I	Y	D	I	T	R	G	C	V	A	T	C	P	I	P	E
cardiotx	M	F	M	V	S	N	K	T	V	P	V	K	–	–	R	G	C	I	D	A	C	P	K	N	S

41

Table 3.2 *Continued*

						55				60					65	
MTx1	N	V	R	E	T	I	R	C	C	E	T	D	K	C	N	E
MTx2	N	N	–	D	P	I	R	C	C	G	T	D	K	C	N	D
m1-tx	Y	R	–	D	V	I	N	C	C	G	T	D	K	C	N	
MT3	N	Y	–	D	S	I	H	C	C	K	T	D	K	C	N	E
MT4	N	V	R	E	T	I	H	C	C	E	T	D	K	C	N	E
MTß	N	Y	–	D	S	I	H	C	C	K	T	D	K	C	N	E
DpCM3	N	Y	–	D	S	I	H	C	C	K	T	E	K	C	N	N
cardiotx	L	L	–	V	K	Y	V	C	C	N	T	D	R	C	N	

Areas of maximum homology are shaded. MTx1, MTx2, m1-toxin, MT3 and MT4, muscarinic toxins from the venom of *Dendroaspis angusticeps*. MTß and DpCM3, muscarinic and synergistic toxins from the venom of *Dendroaspis polylepis.*.

A Ala	**E** Glu	**H** His	**L** Leu	**P** Pro	**S** Ser	**W** Trp
C Cys	**F** Phe	**I** Ile	**M** Met	**Q** Gln	**T** Thr	**Y** Tyr
D Asp	**G** Gly	**K** Lys	**N** Asn	**R** Arg	**V** Val	

At this point, it may be worth commenting about the rather confusing nomenclature being used for these proteins. In the first place, they are referred to as toxins, but there is no evidence that they have toxic effects. In fact, they are likely to have little gross toxicity on the evidence from early studies on homologous fractions from mamba venoms (see review by Harvey *et al.*, 1984). Secondly, two of the groups that have been isolating and characterising similar toxins have chosen two different naming systems. Karlsson and colleagues have named the components as MT1, 2, 3, etc. as they have been isolated from *Dendroaspis angusticeps* venom and have called equivalent proteins from *Dendroaspis polylepis* venom MTα, β, etc. Potter has chosen to associate the toxin with its supposed primary target, giving for example m1-toxin and m4-toxin. It is not clear how toxins with intermediate specificity (as found in *Dendroaspis polylepis* venom by Jerusalinsky *et al.*, 1994, for example) will be included in such a nomenclature. Recent unpublished work with toxins from *Dendroaspis viridis* has also revealed toxins with similar affinity for m1 and m4 receptors and for m3 and m4 receptors (Cerveñansky, Kornisiuk and Jerusalinsky, unpublished).

Suggestions for a standardised system for naming muscarinic toxins would be welcome. Perhaps it would be possible to agree on a system that would identify the proteins by their origin and by their most favoured molecular target; an arbitrary letter or number could be added to distinguish varients with the same specificity from the same venom. For example, m1-toxin and m4-toxin from *Dendroaspis angusticeps* venom would become DaMT-1 and DaMT-4, respectively; MTx1 and MTx2 from *Dendroaspis angusticeps* venom would be DaMT-1,4a and DaMT-1,4b (because they bind to both m1 and m4 subtypes of receptor -see next section-).

Selectivity of binding

Both MTx1 and MTx2 have been radiolabelled with ^{125}I (Jerusalinsky *et al.*, 1992; Jolkkonen *et al.*, 1995). When tested against cerebral cortex membranes from rat brain, the K_D for both toxins was around 20 nM. Most other binding experiments have used unlabelled toxin in competition with radiolabelled muscarinic antagonists such as pirenzepine, QNB or N-methylscopolamine (NMS). Some caution is needed with MTx2: it is unstable in certain conditions (Karlsson *et al.*, 1994; Jolkkonen *et al.*, 1995) probably because of a labile bond between Asp53 and Pro54 (Ségalas *et al.*, 1995b).

The apparent K_is for MTx1 and MTx2 determined against synaptosomal membranes from cerebral cortex were around 20-100 nM depending on the experimental conditions and the radioligands used (see Jerusalinsky and Harvey, 1994). In contrast, the K_i against binding of radiolabelled AF-DX 384 to membranes from pig cardiac muscle was 14 μM for MTx1, implying very low affinity for m2 muscarinic receptors (Jolkkonen *et al.*, 1995).

MTx1, MTx2 and MT3 have been examined for their interaction with cloned expressed human receptors of the m1, m2, m3, m4 and m5 subtypes. With MTx1 and MTx2, the highest affinity was for m1 receptors, with 2–4 fold less affinity for m4 receptors (Kornisiuk *et al.*, 1995a, b), very little affinity for m5 and no affinity

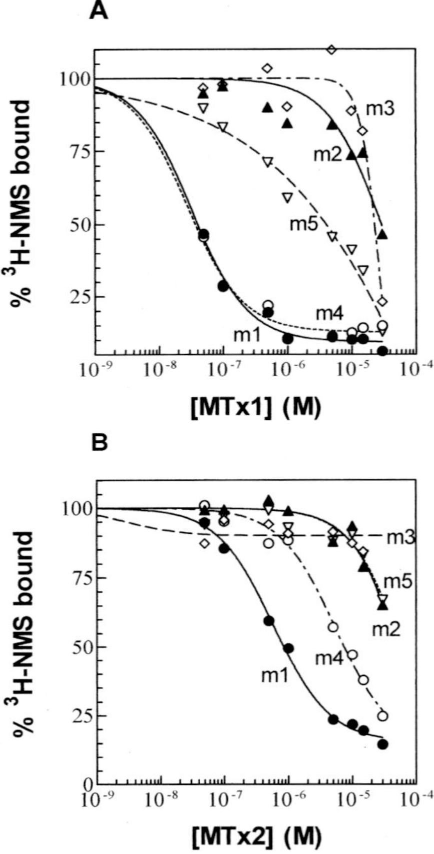

Figure 3.2 Inhibition curves of the binding of ^{3}H-N-methylscopolamine (^{3}H-NMS) to the five subtypes of cloned human muscarinic receptors (m$_1$–m$_5$) by **(A)** muscarinic toxin 1 (MTx1) and **(B)** muscarinic toxin 2 (MTx2). Concentration of ^{3}H-NMS was 275 pM. Curves for m$_1$ and m$_4$ receptors were generated by fitting experimental data to nonlinear regression-competition curve model using GraphPad Prism software. Inhibition constants (K$_i$) were calculated with Cheng-Prusoff equation (Cheng and Prusoff, 1973).

K$_i$ values for MTx1 and MTx2 in m1 and m4 receptors:

	MTx1	MTx2
m1	48 nM	364 nM
m4	72 nM	1200 nM

for the other subtypes (m2, m3) (Figure 3.2). With MT3, the highest affinity was at the m4 subtype (K$_i$ of 2 nM), with 40-fold lower affinity for m1 receptors (K$_i$ of 78 nM) and very little affinity for the other subtypes (Jolkkonen *et al.*, 1994). m1-Toxin and m4-toxin have also been examined on cloned subtypes of muscarinic receptor. m1-Toxin binds irreversibly to m1 receptors and has a reversible interaction at 100 fold higher concentration at m4 receptors but has no affinity for m2, m3 and m5 receptors (Max *et al.*, 1993a). The K$_i$ for m1-toxin at m4 receptors is around 13 nM, and the K$_i$ must, therefore, be extremely low against m1 receptors. However, accurate quantification is difficult because m1-toxin appears to bind noncompetitively and allosterically to m1 receptors (Max *et al.*, 1993c). With m4-toxin, there was little affinity for any of the subtypes of receptors apart from m4, and the K$_i$ at m4 was estimated to be 0.95 nM (Liang *et al.*, 1996).

The inhibition by MTx1 of the ^{3}H-NMS binding to m1 receptors could be fitted to a simple competition curve, with a Hill number close to 1. However, Tucek and

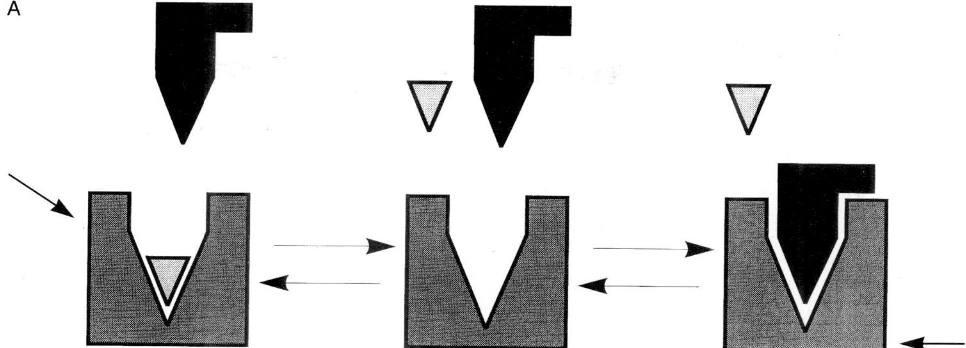

Figure 3.3 Diagram of the possible interaction between the muscarinic receptor, a classical orthosteric ligand and the muscarinic toxins. Muscarinic receptor, dark grey; classical ligand, light grey; muscarinic toxin, black.
Figure 3.3A Putative explanation for competitive inhibition by a muscarinic toxin. The receptor could interact with both as equivalent ligands, but the dissociation rate of the muscarinic toxin-receptor complex would be extremely slow (based on the findings of Kornisiuk and colleagues, 1995a,b).

Proška (1995) have pointed out that a competitive inhibiton is difficult to distinguish from a very strong allosteric effect. A simple scheme with the possible interactions between the receptor, the ligand and the muscarinic toxin is shown in Figure 3.3. Figure 3.3.A shows the possible behavior in the case of competitive inhibition, as could be the case with MTx1 at m1 receptors. Figure 3.3.B shows a more complex behavior assuming a mixed effect, competitive and allosteric, but with a very slow dissociation rate of the muscarinic toxin-receptor complex. This may be the mechanism of action of m1-toxin which slows down the dissociation rate of ^{3}H-NMS at m1 receptor (Max *et al.*, 1993a).

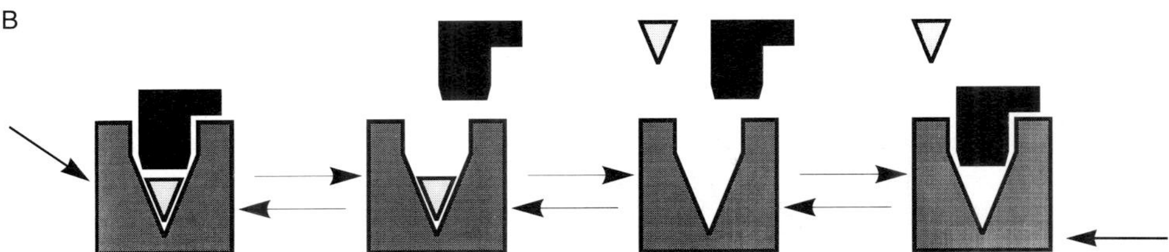

Figure 3.3B Putative mechanism for an allosteric effect of the muscarinic toxin, at a peripheral site, involving interaction with the orthosteric site as well (for details, see Tucek and Proška, 1995). The binding of the toxin would mask the orthosteric site when this is empty (competition-like action), but would slow down dissociation of the classical ligand when it has already bound to the orthosteric site (allosteric action) if the dissociation rate of the toxin-receptor complex is very slow (based on the suggestions of Max *et al.*, 1993c).

The selectivity of muscarinic toxins has been tested by binding experiments with other native membranes. For example, high affinity binding of MTx1 and MTx2 was found in the hippocampus as well as in the cortex. High affinity binding interaction also occurred with membranes from the striatum (with almost complete displacement of ^{3}H-NMS binding), which possesses a high content of the m4 subtype of receptor. There was no interaction of MTx1 and 2 with muscarinic ligands binding to membranes prepared from rat pancreas, which mainly contain the M_3 subtype of receptors (Kornisiuk *et al.*, 1995c).

The muscarinic toxins have not been tested very extensively for the ability to bind to the other types of receptors apart from muscarinic receptors. However, Jerusalinsky *et al.* (1994) reported that neither MTx1 nor MTx2 had any detectable effect on the binding of ^{3}H-flunitrazepam or ^{3}H-AMPA to membranes from rat cerebral cortex. Unexpectedly, MTx1 and MTx2 have been found to displace the specific binding of ^{3}H-prazosin, an α-adrenoceptor antagonist, both to cerebral cortical membranes and to membranes prepared from rat vas deferens. The K_is were 0.3 μM and 0.18 μM for MTx1 in cortex and vas deferens respectively (Harvey *et al.*, 1994). This relatively high affinity for some α-receptors is intriguing and may point to some structural similarity between those receptors and muscarinic receptors.

Functional effects

There are no reports of *in vivo* toxicity following injection of isolated muscarinic toxins. Joubert (1975) isolated the so-called synergistic toxin DpCM3 from *Dendroaspis polylepis*, which has a very similar sequence to the muscarinic toxins (Table 3.2). This author did not find direct lethal effects of DpCM3 in mice, although it enhanced the lethality of other components in the venom through an unknown pharmacological mechanism.

The first reports on functional effects of muscarinic toxins came from Jerusalinsky and colleagues who reported that MTx2, when injected into the striatum of rats, produced behavioural changes similar to those produced by injection of muscarinic agonists. MTx2 also reduced the evoked release of acetylcholine from slices of rat striatum (Jerusalinsky *et al.*, 1991). This preliminary evidence is consistent with MTx2 acting as an agonist at central muscarinic receptors. Subsequently, MTx1 and MTx2 were shown to behave in a manner similar to the muscarinic agonist oxotremorine: when injected into the dorsal hippocampus of rats, the toxins caused memory facilitation in an inhibitory avoidance test. These facilitatory effects were antagonised by high doses of the muscarinic antagonist scopolamine (Jerusalinsky *et al.*, 1993; 1995).

Further studies have looked at the effects of MTx1 and MTx2 on a variety of peripheral tissues that contain different subtypes of muscarinic receptors (Kornisiuk *et al.*, 1995a, b, c; Harvey, Kornisiuk, Cerveñansky and Jerusalinsky, unpublished). For example, neither toxin was found to affect muscarinic receptors on guinea pig atria preparations, which contain m2 receptors and neither toxin affected acetylcholine responses on isolated guinea pig ileum preparations (however, Jolkkonen *et al.*, 1995, reported that MTx1 can cause prolonged contractures of denervated preparations from guinea pig ileum). MTx1 and MTx2 acted in a similar way to the relatively selective M_1 agonists McN-A 343 and CPCP, reducing the twitch responses to nerve stimulation of rabbit vas deferens preparations. There is evidence that presynaptic M_1 receptors on this preparation modulate the evoked release of ATP and noradrenaline (Eltze *et al.*, 1988). The effects of the toxin were not reversed by washing, and they were little affected by preincubation of the preparations with competitive antagonists including pirenzepine, scopolamine, and atropine. However, the effects of MTx1 were altered by compounds known to bind allosterically to muscarinic receptors, gallamine and tacrine (Harvey *et al.*, unpublished results). Also MTx1 still reduced responses to nerve stimulation after pretreatment of the preparations with irreversible blockers of muscarinic receptors. These results imply that the toxins activate an M_1 receptor but not by binding to the classical recognition site for acetylcholine, more probably through an allosteric recognition site.

There is little published information about functional effects of any of the other muscarinic toxins. m1-Toxin has been shown to block the effects of carbachol on phosphatidylinositol turnover in hippocampus slices from rats. m1-Toxin is therefore, an antagonist (Max *et al.*, 1993b).

Structure-activity relationships

Muscarinic toxins are a family of very similar proteins of 64-66 amino acid residues held together by four conserved disulphide bridges. The overall folding pattern is very similar to the super family of so-called three-fingered snake toxins (Figure 3.4). This has been confirmed by NMR spectroscopy of MTx2 (Ségalas *et al.*, 1995a). With the emergence of this information about the 3D structure of the muscarinic toxin, it is likely that comparison of the different sequences of the toxins should provide some insight into the molecular recognition between toxins and subtypes of muscarinic receptors (Table 3.2).

Karlsson and colleagues (Vandermeers *et al.* 1995) have pointed out that the central loop of MTx1 and MTx2 contains 8 or 9 hydrophobic amino acids (Figure 3.4), which is much more than other members of the three-fingered toxin family, such as α-neurotoxins or cardiotoxins (Table 3.2). These authors proposed therefore that the concentration of hydrophobic amino acids between the region of 25–40 probably plays an important functional role in muscarinic toxins. Ségalas *et al.* (1995a) describe the three-dimensional arrangement of two hydrophobic clusters of amino acids in MTx2. However, the hydrophobic residues do not explain why there should be selective binding of, for example, MTx1 and MTx2 for m1 receptors over m2 or m3 receptors. Max *et al.* (1993a) considered this question in relation to the binding properties of m1-toxin. They compared the amino acid sequences of the different subtypes of muscarinic receptors looking for differences between m1 receptors and the other subtypes in the areas of the receptor sequence which are predicted to be exposed on the cytoplasmic membrane. These authors

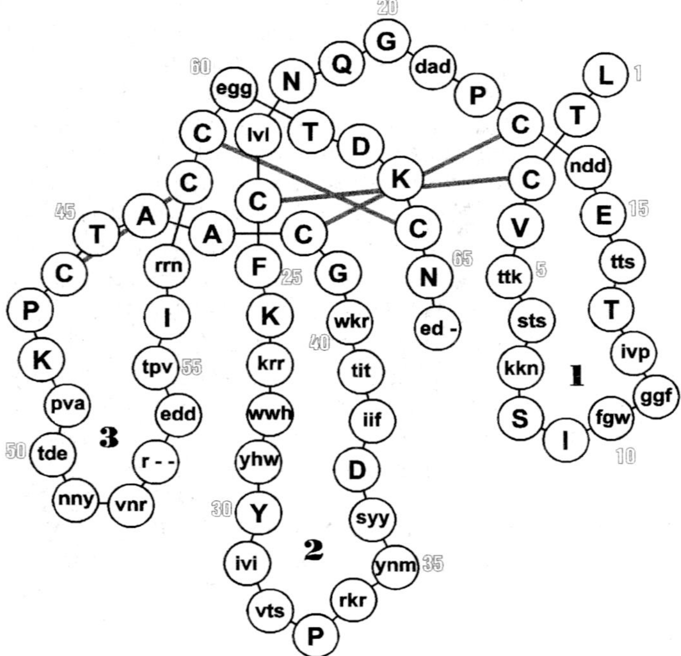

Figure 3.4 Amino acid sequences of MTx1 (Jolkkonen *et al.*, 1994), MTx2 (Ducancel *et al.*, 1991; Karlsson *et al.*, 1991) and m1-toxin (Max *et al.*, 1993a). Each circle represents one residue. They have been arranged according to a representation of the three-dimensional structure of cardiotoxins and α-neurotoxins. Single letters show residues that are identical in all three toxins, while the lower case letters indicate differences between MTx1, MTx2 (behaving as agonists) and m1-toxin (with antagonist action).

pointed out that 13 of the 77 surface amino acids of rat m1 receptors are not found or not conserved in the other subtypes of receptor. They speculate that some of these amino acids will be critical for the selective binding of m1-toxin to m1 receptors. Eight of these thirteen residues are on the amino terminus end of the receptor polypeptide chain. Presumably this region could play an important role in determining the binding of the toxin (Figure 3.1). It would be interesting to examine the interaction of the toxin with receptors mutated at that specific region.

It has also been suggested that Lys 34 of MTx2 interacts with the critical Asp in the receptor's agonist binding pocket (Ségalas *et al.*, 1995a). However, this assumes that the toxin binds to the agonist recognition site, and recent experimental evidence points to an allosteric interaction. Both, the orthosteric and allosteric putative interactions are schematically represented in Figure 3.3 (see previous section).

Other structure-activity relationship information has been obtained from comparisons of the so-called agonist toxins (MTx1 and MTx2) with an antagonist toxin (m1-toxin) (Jerusalinsky and Harvey, 1994). There are 31 positions in which the sequences of the three toxins differ but only eight positions where there is the difference between m1-toxin and the other two. These differences are shown in the representation of the three-dimentional structure of the muscarinic toxins in Figure 3.4. More of the differences are found in the loop 1 (four changes including two involving changes in charge), than in loop 2 (two changes) or loop 3 (two changes and two charge alterations). Hence, loop 1 might be involved in the binding of these toxins to m1 receptors and the differences in amino acids in this local region might determine whether the toxin acts as agonist or antagonist.

Since the cDNA for MTx2 is available (Ducancel *et al.*, 1991), some aspects of these speculations could be tested through the production of appropriate site-directed mutants of muscarinic toxins.

CONCLUSIONS

Muscarinic receptors are found in many different parts of the body and exist in at least five genetically distinct forms. Although the distribution of the subtypes of receptor is becoming known through use of specific antibodies and through *in situ* hybridisation techniques, little is known about the functional significance of individual subtypes, specially in the brain where all five subtypes are expressed to varied degrees in different regions. Existing pharmacological agents whilst plentiful, do not show sufficient discrimination between the different subtypes to be particularly useful in functional studies aimed at elucidating the roles of individual subtypes of muscarinic receptors. Also, the pharmacological agents tend to be readily reversible and are therefore not always useful during *in vivo* experiments. The muscarinic toxins, with their high selectivity, high potency and slow reversibility or irreversibility, seem to offer exceptionally useful tools with which to study muscarinic receptor subtypes. For example, Potter and his colleagues have commented on the potential usefulness of m1-toxin because of its ability to bind to and block the m1 subtype with very high affinity and specificity. The toxin could be used as ligand in very high resolution electromicroscopy, and it also could be useful to help isolate m1 receptors from native membranes (Max *et al.*, 1993c). MTx1 and 2, with their ability to produce long acting activation of m1 receptors, could be used for a variety of studies to probe the effects of cholinergic activation in different brain regions. Such studies could be useful in helping to determine which subtype of muscarinic receptor, if any, may be a potential therapeutic target in a variety of neurodegenerative diseases, including Alzheimer's disease.

Work has continued with other mamba venoms in the expectation that additional muscarinic toxins will be isolated and identified. The hope is that these new toxins will have different specificity from those of the known toxins and therefore represent an addition to the library of muscarinic toxins, available for studying the different subtypes of muscarinic acetylcholine receptors.

Acknowledgements

"The authors' unpublished work was supported by a grant from the British Council and Fundacion Antorchas",

REFERENCES

Adem, A., Åsblom, A., Johansson, G., Mbugua, P. and Karlsson, E. (1988) Toxins from the venom of the green mamba *Dendroaspis angusticeps* that inhibit the binding of quinuclidinyl benzilate to muscarinic acetylcholine receptors. *Biochim. Biophys. Acta,* **968**, 340–345.

Bernheim, L., Mathie, A. and Hille, B. (1992) Characterization of muscarinic receptor subtypes inhibiting Ca^{2+} current and M current in rat sympathetic neurons. *Proc. natn. Acad. Sci. U.S.A.,* **89**, 9544–9548.

Bonner, T. (1989) The molecular basis of muscarinic receptor diversity. *Trends Neurosci.,* **12**, 148–151.

Bonner, T., Buckley, N., Young, A. and Brann, M. (1987) Identification of a family of muscarinic acetylcholine receptor genes. *Science,* **237**, 527–532.

Caulfield, M. (1993) Muscarinic receptors. Characterization, coupling and function. *Pharmac. Ther.,* **58**, 319–379.

Caulfield, M. and Brown, D. (1991) Pharmacology of the putative M_4 muscarinic receptor mediating Ca-current inhibition in neuroblastoma × glioma hybrid (NG 108-15 cells). *Br. J. Pharmac.,* **104**, 39–45.

Cheng, Y. and Prusoff, W. (1973) Relationship between the inhibition constant (K_i) and the concentration of inhibitor which causes 50 per cent inhibition (IC_{50}) of an enzimatic reaction. *Biochem. Pharmac.,* **22**, 3099–3108.

Dale, H. (1914) The action of certain esters and ethers of choline, and their relation to muscarine. *J. Pharmac. exp. Ther.,* **6**, 146–190.

Dörje, F., Wess, J., Lambrecht, G., Tacke, R., Mutschler, E. and Brann, M. (1991) Antagonist binding profiles of five cloned human muscarinic receptor subtypes. *J. Pharmac. exp. Ther.,* **256**, 727–733.

Ducancel, F., Rowan, E., Cassar, E., Harvey, A., Ménez, A. and Boulain, J. (1991) Amino acid sequence of a muscarinic toxin deduced from the cDNA nucleotide sequence. *Toxicon,* **29**, 516–520.

Eltze, M., Gmelin, G., Wess, J., Strohmann, C., Tacke, R., Mutschler, E. and Lambrecht, G. (1988) Presynaptic muscarinic receptors mediating inhibition of neurogenic contractions in rabbit vas deferens are of the ganglionic M_1-type. *Eur. J. Pharmac.,* **158**, 233–242.

Entzeroth, M. and Mayer, N. (1990) Labelling of rat heart muscarinic receptors using the new M2 selective antagonist [^{3}H]AF-DX 384. *Biochem. Pharmac.,* **40**, 1674–1676.

Felder, C. (1995) Muscarinic acetylcholine receptors: signal transduction through multiple effectors. *FASEB J.,* **9**, 619–625.

Fraser, C., Wang, C., Robinson, D., Gocayne, J. and Venter, J. (1989) Site-directed mutagenesis of m1 muscarinic acetylcholine receptors: conserved aspartic acids play important roles in receptor function. *Molec. Pharmac.,* **36**, 840–847.

Hammer, R., Berrie, C., Birdsall, N., Burgen, A. and Hulme, E. (1980) Pirenzepine distinguishes between different subclasses of muscarinic receptors. *Nature,* **283**, 90–92.

Hammer, R. and Giachetti, A. (1982) Muscarinic receptor subtypes: M_1 and M_2 biochemical and functional characterization. *Life Sci.,* **31**, 2991–2998.

Harvey, A., Anderson, A., Mbugua, P. and Karlsson, E. (1984) Toxins from mamba venoms that facilitate neuromuscular transmission. *J. Toxicol. Toxin Rev.,* **3**, 91–137.

Harvey, A., Cerveñansky, C., Kornisiuk, E. and Jerusalinsky, D. (1994) Muscarinic toxin MTx1 from green mamba venom also interacts with α-adrenoceptors. In: *Int. Satellite Symp.: Muscarinic Receptors.* Fort Lauderdale, Florida, 1994. *Life Sci.,* (in press).

Harvey, A., Kornisiuk, E., Cerveñansky, C. and Jerusalinsky, D. (1996) Effects of muscarinic toxins MTx1 and MTx2 on tissues containing different subtypes of muscarinic cholinoceptors. (Submitted)

Hulme, E., Birdsall, N. and Buckley, N. (1990) Muscarinic receptor subtypes. *Ann. Rev. Pharmac. Toxicol.,* **30**, 633–673.

Hulme, E., Kurtenbach, E. and Curtis, C. (1991) Muscarinic acetylcholine receptors: structure and function. *Biochem. Soc. Trans.*, **19**, 133–138.

Jerusalinsky, D., Cerveñansky, C., Kornisiuk, E., Walz, R., Bianchin, M. and Izquierdo, I. (1991) Muscarinic toxin 2 (MTx2), a protein from the green mamba venom with muscarinic agonist-like actions. *J. Neurochem.*, **61**, S19.

Jerusalinsky, D., Cerveñansky, C., Peña, C., Raskovsky, S. and Dajas, F. (1992) Two polypeptides from *Dendroaspis angusticeps* venom selectively inhibit the binding of central muscarinic cholinergic receptor ligands. *Neurochem. Int.*, **20**, 237–246.

Jerusalinsky, D., Cerveñansky, C., Walz, R., Bianchin, M. and Izquierdo, I. (1993) A peptide muscarinic toxin from the Green Mamba venom shows agonist-like actions in an inhibitory avoidance learning task. *Eur. J. Pharmac.*, **240**, 103–105

Jerusalinsky, D. and Harvey, A. (1994) Toxins from mamba venoms: small proteins with selectivities for different subtypes of muscarinic acetylcholine receptors. *Trends Pharmac. Sci.*, **15**, 424–430.

Jerusalinsky, D., Kornisiuk, E., Bernabeu, R., Izquierdo, I. and Cerveñansky, C. (1995) Muscarinic toxins from the venom of *Dendroaspis* snakes with agonist-like actions. *Toxicon*, **33**, 389–397.

Jerusalinsky, D., Raskovsky, S., Kornisiuk, E., Bernabeu, R. and Cerveñansky, C. (1994) Muscarinic toxins from the venom of elapid snakes. In K. Tipton and F. Dajas, (eds.), *Neurotoxins as Tools in Neurobiology*, Ellis Horwood, Chichester, U.K. pp. 89–102.

Jolkkonen, M., Adem, A., Hellman, U., Werndstedt, C. and Karlsson, E. (1995) A snake toxin against muscarinic acetylcholine receptors: amino acid sequence, subtype specificity and effect on guinea-pig ileum. *Toxicon*, **33**, 399–410.

Jolkkonen, M., van Giersbergen, P., Hellman, U., Wernstedt, C. and Karlsson, E. (1994) A toxin from the green mamba *Dendroaspis angusticeps*: amino acid sequence and selectivity for muscarinic m4 receptors. *FEBS Lett.*, **352**, 91–94.

Joubert, F. (1975) The amino acid sequence of protein CM-3 from *Dendroaspis polylepis polylepis* (black mamba) venom. *Int. J. Biochem.*, **17**, 695–699.

Karlsson, E., Jolkkonen, M., Satyapan, N., Adem, A., Kumlin, E., Hellman, U. and Wernstedt, C. (1994) Protein toxins that bind to muscarinic acetylcholine receptors. In D. Suput and R. Zorec, (eds.), *Toxins and Exocytosis*, N.Y.Acad.Sci., USA, pp 153–161.

Karlsson, E., Risinger, C., Jolkkonen, M., Wernstedt, C. and Adem, A. (1991) Amino acid sequence of a snake venom toxin that binds to the muscarinic acetylcholine receptor. *Toxicon*, **29**, 521–526.

Kornisiuk, E., Jerusalinsky, D., Cerveñansky, C. and Harvey, A. (1995a) Binding of muscarinic toxins MTx1 and MTx2 from the venom of the green mamba *Dendroaspis angusticeps* to cloned human muscarinic cholinoceptors. *Toxicon*, **33**, 11–18.

Kornisiuk, E., Jerusalinsky, D., Cerveñansky, C. and Harvey, A. (1995b) Corrigendum: Binding of muscarinic toxins MTx1 and MTx2 from the venom of the green mamba *Dendroaspis angusticeps* to cloned human muscarinic cholinoceptors. *Toxicon*, **33**, 1111.

Kornisiuk, E., Cerveñansky, C., Alfaro, P., Bassarsky, M., Harvey, A. and Jerusalinsky, D. (1995c) Peptide muscarinic toxins with agonist activity that preferentially bind to the m_1 receptor subtype. *Comunicaciones Biológicas*, **13**, 123–133.

Kubo, T., Maeda, A., Sugimoto, K., Akiba, I., Takahashi, H., Haga, T., Haga, K., Ichiyama, A., Kangawa, K., Matsuo, H., Hirose, T. and Numa, S. (1986) Primary structure of porcine cardiac muscarinic acetylcholine receptor deduced from the cDNA sequence. *FEBS Lett.*, **209**, 367–372.

Lambrecht, G., Feifel, R., Wagner-Roder, M., Strohmann, C., Zilch, H., Tacke, R., Waelbroeck, M., Christophe, J., Boddeke, H. and Mutschler, E. (1989) Affinity profiles of hexahydro-sila-difenidol analogues at muscarinic receptor subtypes. *Eur. J. Pharmac.*, **168**, 71–80.

Lazareno, S., Buckley, N. and Roberts, F. (1990) Characterization of muscarinic M_4 binding sites in rabbit lung, chicken heart and NG 108-15 cells. *Molec. Pharmac.*, **38**, 805–815.

Lazareno, S. and Roberts, F. (1989) Functional and binding studies wtih muscarinic M_2-subtype selective antagonists. *Br. J. Pharmac.*, **98**, 309–317.

Liang, J., Carsi-Gabrenas, J., Krajewski, J., McCafferty, J., Purkerson, S., Santiago, M., Strauss, W., Valentine, H. and Potter, L. (1996) Anti-muscarinic toxins from *Dendroaspis angusticeps*. *Toxicon* (in press).

Max, S., Liang, J. and Potter, L. (1993a) Stable allosteric binding of m1-toxin to m1 muscarinic receptors. *Mol. Pharmac.*, **44**, 1171–1175.

Max, S., Liang, J. and Potter, L. (1993b) Purification and properties of m1-toxin, a specific antagonist of m1 muscarinic receptors. *J. Neurosci.*, **13**, 4292–4300.

Max, S., Liang, J., Valentine, H. and Potter, L. (1993c) Use of m1-toxin as a selective antagonist of m1 muscarinic receptors. *J. Pharmac. exp. Ther.*, **267**, 480–485.

Max, S., Liang, J., Purkerson, S. and Potter, L. (1993d) m4-Toxin, a selective, reversible, allosteric antagonist of m4 muscarinic receptors. *Neurosci. Abstr.*, **19**, 462.

Mitchelson, F. (1988) Muscarinic receptor differentiation. *Pharmac. Ther.*, **37**, 357–423.

Nordvall, G. and Hacksell, U. (1993) Binding-site modeling of the muscarinic m1 receptor: a combination of homology-based and indirect approaches. *J. Med. Chem.*, **36**, 967–976.

Othman, O., Ismail, M. and El-Asmar, M. (1973) Pharmacological studies of snake (*Dendroaspis angusticeps*) venom. *Toxicon*, **11**, 185–192.

Ségalas, I., Roumestand, C., Zinn-Justin, S., Gilquin, B., Ménez, R. and Toma, F. (1995a) Solution structure of a green mamba toxin that activates muscarinic acetylcholine receptors, as studied by nuclear magnetic resonance and molecular modeling. *Biochemistry*, **34**, 1248–1260.

Ségalas, I., Thai, R., Ménez, R. and Vita, C. (1995b) A particularly labile Asp-Pro bond in the green mamba muscarinic toxin MTx2. Effect of protein conformation on the rate of cleavage. *FEBS Lett.*, **371**, 171–175.

Trimmer, B. (1995) Current excitement from insect muscarinic receptors. *Trends Neurosci.*, **18**, 104–111.

Tucek, S. and Proška, J. (1995) Allosteric modulation of muscarinic acetylcholine receptors. *Trends Pharmac. Sci.*, **16**, 205–212.

Vandermeers, A., Vandermeers-Piret, M., Rathé, J., Waelbroeck, M., Jolkkonen, M., Oras, A. and Karlsson, E. (1995) Purification and sequence determination of a new muscarinic toxin (MT4) from the venom of the green mamba (*Dendroaspis angusticeps*). *Toxicon*, **33**, 1171–1179.

Welsh, J. (1967) Acetylcholine in snake venoms. In F.E. Russell and P.R. Saunders, (eds.), *Animal Toxins*, Pergamon Press, New York., USA, pp 363–368

Wess, J., Blin, N., Mutschler, E. and Blüml, K. (1995) Muscarinic acetylcholine receptors: structural basis of ligand coupling and G protein coupling. *Life Sci.*, **56**, 915–922.

4. ENDOTHELIN AND SARAFOTOXIN RECEPTORS: LIGAND RECOGNITION AND TRANSMEMBRANE SIGNALING THROUGH MULTIPLE EFFECTORS

MORDECHAI SOKOLOVSKY

Laboratory of Neurobiochemistry, The George S. Wise Faculty of Life Sciences,
Tel Aviv University, Tel Aviv 69978, Israel

INTRODUCTION

The sarafotoxins (SRTXs) are potent vasopressor peptides isolated from the venom of the burrowing asp *Atractaspis enggadensis.* They have powerful cardiac effects in mammals: in mice administration of as little as 5 μg of venom typically causes cardiac arrest and death within minutes (Bdolah *et al.*, 1989). Four SRTXs isopeptides a,b,c and e, each composed of 21 amino acids, have been identified (Figure 4.1). The SRTXs are remarkably similar in structure to the endothelins (ETs) (Figure 4.1) (Kloog *et al.*, 1988). Endothelin-1 (ET-1), a 21-amino-acid peptide originally isolated from cultured porcine aortic endothelial cells, is a potent vasoconstrictor both *in vivo* and *in vitro* (Yanagisawa *et al.*, 1988). Other members of the endothelin (ET) family are the isopeptides ET-2 and ET-3 and the vasoactive intestinal contractor (Figure 4.1). The ETs appear to be produced and to be active in almost all tissues. They are abundant, for example, in brain, lung, intestine, kidney, adrenal gland, thyroid, parathyroid, ovary, uterus, placenta and amnion (for review see Sokolovsky, 1992, 1995; Masaki, 1993). Endothelial cells produce only ET-1. Genes for ET-1, ET-2 and ET-3 are present in human, bovine, porcine, rat and murine tissues (Inoue *et al.*, 1989; Saida *et al.*, 1989), suggesting that ETs are produced in all mammalian species. All members of the ET/SRTX family contain 21 residues with disulfide bridges between cys^1 and cys^{15} and between cys^3 and cys^{11}. These strictly conserved bridges create a rigid "head" with a short tail.

Like the SRTXs, ET-1 causes disturbances in cardiac function at the level of the A-V conduction system and the coronary vessels, leading to cardiac arrest. The LD$_{50}$ for SRTXb and ET-1 in ICR mice is 15 μg/kg body weight (Bdolah *et al.*, 1989). The toxicity of the SRTXs is not unexpected, but that of the ETs (which occur naturally in the plasma of healthy humans) is surprising. Avoidance of the toxic effects is probably achieved by a control mechanism which prevents ET concentrations from increasing above a certain threshold. Some of the unexplained pathophysiological phenomena might be due to a breakdown of this postulated control mechanism.

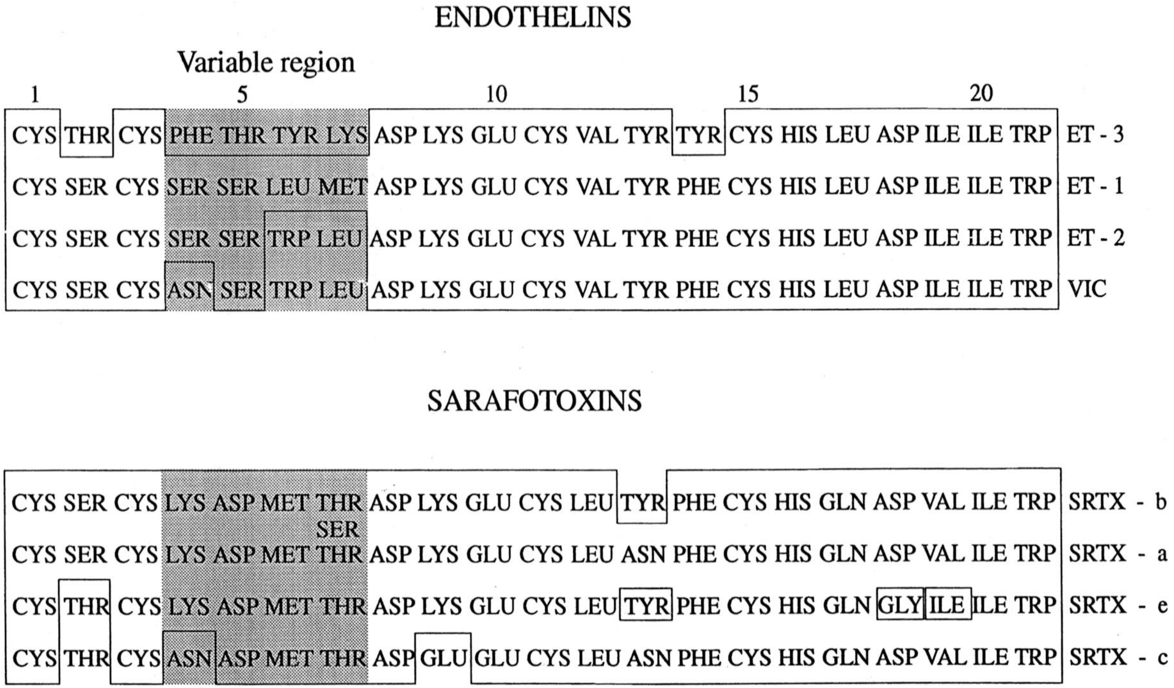

Figure 4.1   Primary structures of endothelins and sarafotoxins.

Considerable research has been devoted to the ET/SRTX family of peptides in recent years, largely because of the postulated causative role of ET in a number of cardiovascular diseases, such as hypertension, myocardial ischemia, congestive heart failure, as well as in renal failure, asthma and atherosclerosis (for review see Sokolovsky, 1992; 1995; Masaki, 1993; Ohlstein and Douglas, 1993). A number of reports have described an increase in plasma ET levels in these pathophysiological states.

Where tissue rather than plasma markers are required, a parameter that might prove a useful indicator of disease states is the density of ET receptors (ET-Rs). In human endometrial carcinoma tissue, for example, ET receptor density was only 356 ± 121 fmol/mg protein as compared with 2029 ± 340 fmol/mg protein in controls (Ben-Baruch et al., 1993). On the other hand, the mean maximal density of ET-R sites was significantly higher in placenta of pre-eclamptic than of normotensive women (905 ± 107 vs. 539 ± 140 fmol/mg protein) (Schiff et al., 1993).

Much of the research has been directed towards elucidating structure-function relationships within the ET/SRTX peptides, with the aim of facilitating the rational design of receptor antagonists that might be useful in the clinical management of ET-related disorders. Early structure-function studies explored the role of conserved residues in ET-1-mediated vasoconstriction, and in particular the contribution of the two disulfide bonds, the free amino group presented by the N-terminus, and the hydrophobic C-terminal region, especially the terminal Trp[21] residue (for review see Sokolovsky, 1992). It was suggested that variations in net charge in the intramolecular loops resulting from the disulfide bonds contribute to differences in receptor affinity and specificity (Sokolovsky, 1992). The emerging picture was that structural elements contributing to peptide function are distributed throughout the length of the molecule and that few residues can be discounted in the search for determinants of receptor binding and biological activity (see also Tam et al., 1994 and refs. therein). Indeed, in recent years ET antagonists have been described, e.g. the peptides BQ-123 (Ihara et al., 1992), FR-139137 (Nirei et al., 1993) and PD-151242 (Davenport et al., 1994). Non-peptide antagonists of ETs include Ro 46-2005 and Ro 47-0203 (Clozel et al., 1993; 1994), SB 209560 (Elliot et al., 1994) and BMS 182874 (Stein et al., 1994).

RECEPTOR HETEROGENEITY

Binding studies have shown that binding of ETs and SRTXs is mutually exclusive, suggesting that they bind to the same receptors. Results of biochemical and pharmacological studies employing SRTX and ET ligands pointed to the possible existence of more than one receptor subtype for ETs and SRTXs (for review, see Sokolovsky, 1992; 1994). This assumption was initially based on studies of ligand binding and later also on cross-linking experiments employing affinity labeling. Studies with cloned receptors revealed the existence of two distinct ET/SRTX receptors (ET-R): ET_A (selective for ET-1, ET-2 and SRTX-b relative to ET-3 and

SRTX-c) and ET_B (nonselective with respect to all five ligands). Karne *et al.* (1993) described the existence of a third subtype, ET_C-R, which was detected on *Xenopus laevis* dermal melanophores and is specific for ET-3. It is not yet clear whether this receptor represents a species variant of ET_A-R or ET_B-R or is a novel subtype with a distinct, as yet undemonstrated mammalian homolog.

The existence of more than two subtypes is supported by recent data from various laboratories. For example Gulati *et al.* (1995) showed that ET-R in the coronary blood vessels is of a different type (neither ET_A-R nor ET_B-R) from those in other vascular beds. Douglas *et al.* (1995) reported the presence of three distinct functional ET-R subtypes in rabbit lateral saphenous vein.

Species differences in ET-R subtypes have been reported by several investigators. Rat and human ET_B-R, for example, despite their high degree of homology, show marked pharmacological differences in binding of both antagonist and agonist (Reynolds *et al.*, 1995 and refs. therein).

We have described a novel ET-R subtype with affinities in the picomolar concentration range (Sokolovsky *et al.*, 1992), in contrast to the 'conventional' receptor which operates at nanomolar concentrations. The novel receptor sites were demonstrated in several rat brain regions, in rat atrium and C6-glioma cells and in Rat-1 and Swiss 3T3 fibroblasts (Ambar and Sokolovsky, 1993). The pharmacological profile of these sites in rat cerebellum, for example, was suggestive of ET_B-R (Sokolovsky *et al.*, 1992) coexisting with a minor population of ET_A-R (Shraga-Levine *et al.*, 1994); in rat atrium, however, it was similar to that of ET_A-R (Sokolovsky, 1993a). These findings support the existence of at least two ET_A-Rs and two ET_B-Rs, one of each characterized by picomolar and the other by nanomolar Kd values.

As to the molecular characteristics of these novel receptors, the following question arises: except for their ability to oscillate between two affinity states (pM and nM sites), are they identical to the conventional receptors in their biological properties, or do they represent distinct receptor subtypes? The answer is suggested by a comparison of their signal transduction activities. A large body of evidence indicates that ETs and SRTXs induce phospholipase C (PLC) activity. Functional studies in cerebellar slices (Sokolovsky, 1993b and refs. therein) have demonstrated, however, that SRTXs and ETs induce phosphoinositide hydrolysis only at concentrations above 500 pM. This suggests the operation of different receptors: those of the picomolar sites are not coupled to the phosphoinositide hydrolysis pathway, whereas those of the nanomolar sites are.

Furthermore, participation of the picomolar sites in signal transduction was shown to occur via adenosine 3′,5′-cyclic monophosphate (cAMP) and guanosine 3′,5′-cyclic monophosphate (cGMP) pathways (Sokolovsky *et al.*, 1994; Shraga-Levine *et al.*, 1994), supporting the assumption that they represent a novel receptor subtype (see below). Several other signals, evoked by low ligand concentrations and therefore probably operating via the pM ET-R subtypes, have recently been reported (Warner *et al.*, 1993; Kasuya *et al.*, 1994; Tabuchi *et al.*, 1994; Etoh *et al.*, 1994).

COUPLING BETWEEN RECEPTORS AND G-PROTEINS

Signal transduction pathways activated by the ET/SRTX-R are depicted in Figure 4.2. Information from the exterior to the interior of a cell is relayed by transmembrane signaling systems via complex protein-protein interactions and second-messenger cascades. One such signaling system involves coupling between receptors and heterotrimeric G-proteins. Activation of the system is triggered by the binding of agonist to receptor. Coupling of receptor to G-protein activates an effector molecule, leading to a second-messenger cascade.

The G-proteins are GTPases. Interaction between receptor and G-protein could therefore be directly assessed by measurement of receptor-stimulated high-affinity GTPase in rat cerebellar membranes (Sokolovsky, 1993b), which are enriched with ET_B-R (Shraga-Levine, 1994 and refs. therein), and in rat myocyte membranes, which are enriched with ET_A-R (Sokolovsky, 1993a). Dose-dependent stimulation of GTPase activity was demonstrated for all four peptides investigated (ET-1, ET-3, SRTXb and SRTXc), with EC_{50} values in the nM range. Pertussis toxin modification was used in these studies to demonstrate that coupling with the receptor involves a Gq-protein (pertussis toxin-insensitive) as well as a Gi-like or Go-like protein (pertussis toxin-sensitive) (Sokolovsky, 1993a, b). Thus, stimulation of GTPase activity by ET-1, ET-3, SRTX-b or SRTX-c directly demonstrates functional coupling between receptors and G-proteins. Such interaction supports the view, now gaining recognition, that receptors which are single polypeptides, and which are likely to traverse the plasma membrane seven times, all achieve regulation of an effector system through activation of heterotrimeric G-proteins. Recent experiments involving cloning of ET-R indicated such an interaction (Sakurai *et al.*, 1990; Arai *et al.*, 1990; Lin *et al.*, 1991; Nakamuta *et al.*, 1991; Ogawa *et al.*, 1991; Hosoda *et al.*, 1991; Sakamoto *et al.*, 1991; Karne *et al.*, 1993).

PHOSPHOINOSITIDE-SPECIFIC PHOSPHOLIPASE C (PLC)

Receptor activated enhancement of inositol lipid turnover is one of the principal molecular mechanisms employed by cells for transmembrane signaling. Phosphodiesteric cleavage of phosphatidylinositol 4,5-bisphosphate, catalyzed by a phosphoinositide-specific phospholipase C (PLC), yields inositol 1,4,5-trisphosphate [Ins(1,4,5)P_3] and diacylclycerol (DAG), which are linked to Ca^{2+} homeostasis and activation of protein kinase C (PKC), respectively (for review see Fisher, 1995). Ins(1,4,5)P_3 binds to a high-affinity receptor on endoplasmic reticulum, triggering the rapid mobilization of intracellular Ca^{2+} (for review see Nishizuka, 1988). An increase in intracellular Ca^{2+} can directly activate Ca^{2+}-dependent protein kinases as well as Ca^{2+}-sensitive PLC and phospholipase A_2 (PLA$_2$), resulting in the release of arachidonic acid. The Ca^{2+}, often in association with calmodulin, triggers or modulates processes such as secretion (including transmitter release) and activates enzymes such as myosin light-chain kinase, which regulates cellular contractile

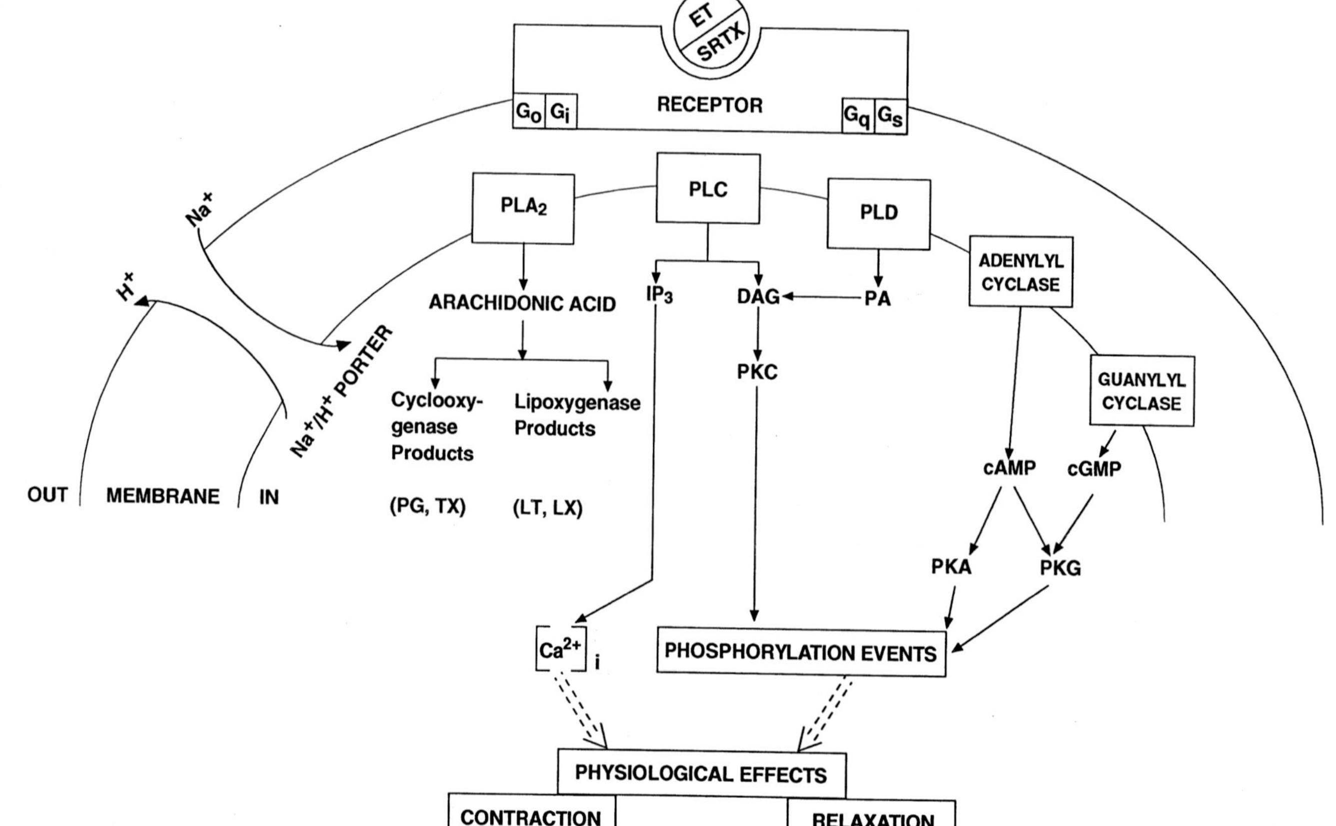

Figure 4.2 Schematic representation of the signal transduction pathways associated with activation of the ET/SRTX receptors. Ins(1,4,5)P_3, inositol-1,4,5-tris-phosphate; LT, leukotriene; LX, lipoxin; PA, phosphatidic acid; PG, prostaglandin; PKA, protein kinase A; PKC, protein kinase C; PKG, protein kinase G; PLA$_2$, phospholipase A$_2$; PLC, phosphoinositide-specific phospholipase C; PLD, phospholipase D, TX, thromboxane.

mechanisms. An increase in $[Ca^{2+}]_i$, as well as enhanced contraction and secretion, are therefore expected responses to ET-R activation. In many studies PLC activation is estimated by the accumulation of total inositol phosphate measured in the presence of Li; such accumulation is independent of the rate at which $Ins(1,4,5)P_3$ is metabolized.

Stimulation of ET_A-R and/or ET_B-R by SRTXs and ETs causes a rapid increase in PLC activity (for review see Sokolovsky, 1992, 1995). In several preparations a dose-dependent increase in phosphoinositide hydrolysis was observed (Sokolovsky, 1993a,b; Ambar and Sokolovsky, 1993 and references therein), yielding EC_{50} values in the nM range; in rat heart myocytes, for example, the values for ET-1, SRTX-b, ET-3 and SRTX-c were 10, 15, 72 and 250 nM, respectively (a typical ET_A-R type response). In a number of preparations, including rat heart myocytes (Sokolovsky, 1993b) and rat cerebellar slices (Sokolovsky, 1993a) phosphoinositide hydrolysis was enhanced by treatment with PT. These data, together with the functional coupling between ET-R and G-proteins mentioned above, support mediation of the process by at least two G-proteins, the one sensitive and the other insensitive to PT. The former might be either Gi-like or Go-like, while the latter is probably a protein of the G_q class (G_q, G_{11}, G_{14}, G_{15} or G_{16}). The fact that phosphoinositide hydrolysis resulting from agonist-induced stimulation of PT-treated preparations is enhanced by PT suggests that a PT-selective inhibitory G-protein, e.g. $Gi_{1/2}$, may normally act to diminish PLC activity. Once this inhibition is removed, stimulation of PLC activity is enhanced.

PHOSPHOLIPASE A_2

In a number of tissues ETs activate phospholipase A_2 (PLA_2) resulting in the release of arachidonic acid. This is metabolized to prostaglandins, thromboxanes and leukotrienes (Resink *et al.*, 1989; Reynolds and Mok, 1990), suggesting that eicosanoid metabolites of arachidonic acid may function as second messengers in mediating some of the biological activities of ETs. Two mechanisms might account for the generation of arachidonic acid induced by ETs in various cells: (i) activation of PLA_2, leading to direct formation of arachidonic acid; and (ii) an indirect pathway via PLC. The first mechanism induces an increase in $[Ca^{2+}]_i$, thereby activating Ca^{2+}-sensitive PLA_2; the second stimulates production of DAG, which might serve as a substrate for lipase-catalyzed production of arachidonic acid.

Activation of phospholipases C and A_2 by ETs might however occur via separate but parallel pathways (Resink *et al.*, 1990). Although vasoconstrictor prostaglandins (PG) such as PG $F_{2\alpha}$ and thromboxane fail to mediate contraction by ET in mesangial cells (Simonson and Dunn, 1990), thromboxane appears to be involved in ET-induced aortal contraction (Reynolds and Mok, 1990) and bronchospasm (Schumacher *et al.*, 1990). Both ET_A-R and ET_B-R can stimulate PLA_2 (Aramori and Nakanishi, 1992) and cytosolic PLA_2 (Schramek *et al.*, 1994a). Using cultured rat mesangial cells, Schramek *et al.* (1994b) showed that ET-1 stimulates cytosolic PLA_2, an intracellular form of the PLA_2 enzyme family. They concluded that ET-1 is capable of stimulating cytosolic PLA_2 in a biphasic manner. The biphasic stimulation probably involves different mechanisms: a rapid activation regulated mainly at the post-translational level, and a slower, chronic stimulation of cytosolic PLA_2 which also occurs transcriptionally.

In line with earlier findings (Fukunaga *et al.*, 1991; Simonson, 1993; Schramek *et al.*, 1993) Schramek *et al.* (1994a) suggest that ET-1-induced short-term and long-term activation of PLA_2 contribute to the ET-1-mediated release of prostaglandins, thus possibly serving as an autocrine and/or paracrine negative regulator for vasoconstrictive, mitogenic and inflammatory stimuli.

Phospholipase D

Phospholipase D (PLD) hydrolyzes phospholipids such as phosphatidylcholine and phosphatidylethanolamine, resulting in production of phosphatidic acid and release of the free polar head-groups, e.g. choline and ethanolamine. Certain biochemical signals, such as the biphasic elevation of intracellular Ca^{2+} triggered by ET-1 and the biphasic elevation of DAG in ET-1-stimulated cells (Griendling *et al.*, 1989; Sunako *et al.*, 1990), might result from the formation of more than one second messenger. Recent evidence suggests that PLD activation by neurotransmitters, hormones and growth factors may represent a novel and ubiquitous signal transduction pathway in mammalian cells, mediated by accumulation of phosphatidic acid and/or DAG (Dennis *et al.*, 1991). MacNulty *et al.* (1990) described ET-1-stimulated generation of choline from phosphatidylcholine in Rat-1 fibroblasts, leading to the suggestion that ET activates PLD via both PKC-dependent and PKC-independent mechanisms.

Activation of PLC and PLD by SRTXs and ETs has been described in C6-glioma cells and in Rat-1 and Swiss 3T3 fibroblasts (Ambar and Sokolovsky, 1993). In all three cell lines, the activation was mediated by sites in the nanomolar but not in the picomolar range. EC_{50} values in all cases were similar and were not significantly affected by either extracellular or intracellular Ca^{2+}. The ET-R-triggered activation of both PLD and PLC has led some authors to suggest that PLC activation, arguably the first event induced by ET-R (MacNulty *et al.*, 1990; Muldoon *et al.*, 1990), can stimulate PLD via its second messengers, DAG and Ca^{2+}. The data reported by Ambar and Sokolovsky (1993) appear to indicate, however, that ET-R induces PLD activation directly and not through PLC activation. This conclusion is supported by the observation that ET-1-induced stimulation of PLD in rabbit iris sphincter smooth muscle (Zhang and Abdel-Latef, 1992) occurs independently of PKC, PLC and PLA_2 activation and intracellular Ca^{2+} mobilization. Similarly, Suzuki *et al.* (1994) demonstrated that in osteoblast-like cells ET-1 induces PLD activation independently of PKC.

PLD-related second messengers, such as choline and phosphatidic acid (which can be metabolized to DAG by phosphatidic acid hydrolase), may be involved in the regulation of various cell functions (Exton, 1990) and participate in ET-mediated signal transduction. Thus, for example, ET-R-induced activation of PLD may account for the sustained generation of DAG and hence the prolonged maintenance of PKC activation. Phosphatidic acid, as a second messenger derived from

PLD activation, may participate in some of the observed ET-1-induced cellular responses, such as the sustained increase in intracellular Ca^{2+} which is not mediated via Ins(1,4,5)P$_3$ (Lin *et al.*, 1992), long-lasting contraction of smooth muscle, and rapid DNA synthesis and cell proliferation (for review see Simonson *et al.*, 1992).

The role of PLD-related second messengers has some intriguing implications, especially in the brain, a rich source of PLD (Chalifour and Kanfer, 1980). For example, choline (a product of PLD activity) can serve as a source of free choline for the synthesis of acetylcholine in the brain (Hattori and Kanfer, 1985). Also, ethanol-intoxicated rats produce large quantities of phosphatidylethanol (an exclusive product of PLD) in various organs, including the brain (Alling *et al.*, 1983). Furthermore, phosphatidylethanol can substitute for phosphatidylserine in activating a brain-specific PKC (Nishizuka, 1988). These observations suggest that PLD-catalyzed transphosphatidylation may be involved in the development of alcohol dependency and alcohol-related pathology. Because of the wide distribution of ET-Rs in the brain, it is worth investigating the effects of ET-R activation in relation to this pathology.

ADENYLATE AND GUANYLYL CYCLASES

In addition to inducing contraction (via phospholipase C and/or D pathways, for example; see Figure 4.2), interaction between the three components (agonist, receptor, G-protein) can lead to relaxation. Two pathways that may be involved in mediating this effect are cAMP and cGMP. The finding that ET-1 at low concentrations possesses vasodilatory properties (Warner *et al.*, 1989; Faraci, 1989) led us to suspect that vasodilation may be a function of the picomolar subsites (Sokolovsky *et al.*, 1992). Early studies on the accumulation of cAMP evoked by relatively high concentrations of ET-1, for example, pointed to the involvement of different and sometimes opposing pathways (as reviewed in Sokolovsky 1992; 1995). ETs inhibit cAMP formation in several tissues, while in a few systems they stimulate it. Inhibition of cAMP formation in some of these pathways was studied by employing a β-adrenergic agonist and/or adenylyl cyclase activator to stimulate cAMP formation in the presence and absence of ETs, rather than via direct ET-induced stimulation.

In rat atrial slices, ET-1 at picomolar concentrations induced a pronounced dose-dependent increase in cAMP (Sokolovsky *et al.*, 1994). Maximal stimulation produced cAMP levels 3- to 4-fold higher than basal levels. Surprisingly, a different pattern of behavior was shown by SRTX-b, SRTX-c and ET-3 (Sokolovsky *et al.*, 1994). With these agonists no stimulation was observed at the picomolar range; however, at the nanomolar range stimulation occurred, though at lower efficiency, with maximal effect at 10^{-7}–10^{-6} M. The EC$_{50}$ values for SRTX-b, SRTX-c and ET-3 are 6, 1 and 5 nM, respectively. Thus, signal transduction of ET$_A$-R appears to be ligand-specific (Sokolovsky *et al.*, 1994), and ET-3, SRTX-b and SRTX-c act as antagonists to ET-1- stimulated cAMP accumulation. However, their ability to stimulate cAMP release at higher concentrations suggests the possible involvement of different pathways. Stimulation of cAMP formation at low ligand concentrations

(10^{-13}–10^{-12} M) has so far been observed only in rat atrial preparation. Measurement of cAMP accumulation induced by very low concentrations of ET-1 (e.g. 1 pM) in the presence of, for example, BQ-123, propranolol, PT treatment, nifedipine or verapamil, prompted the suggestion (Sokolovsky *et al.*, 1994) that ET-1-induced cAMP formation results from catecholamine release via a process mediated by a Ca^{2+} channel coupled to a PT-sensitive G-protein. These results strongly support the notion of cross-talk between signal transduction pathways.

Several hormones and neurotransmitters stimulate cGMP synthesis in a variety of cells and tissues (for review see Garbers, 1989). In many cases the agonist-induced increase in cGMP is mediated by nitric oxide, which binds to the heme moiety of soluble guanylyl cyclase, activating it to synthesize cGMP (Garthwaite, 1991; Moncada *et al.*, 1991). ET-1 and ET-3 appear to mediate cGMP signaling in various cells, such as porcine kidney epithelial cells (Ishii *et al.*, 1991), intact rat glomeruli (Edwards *et al.*, 1992), cultured bovine endothelial cells (Hirata *et al.*, 1993) and rat aorta (Fujitani *et al.*, 1993; Moritoki *et al.*, 1993). (Edwards *et al.*, 1992). Low doses of ET-1 and ET-3 in anesthetized rats elicit transient hindquarter vasodilation, which was shown to be mediated via nitric oxide (Taniguchi *et al.*, 1994).

In an attempt to determine the physiological relevance of the picomolar sites, we examined the possible involvement of a cyclic GMP pathway in the signal transduction occurring in rat cerebellar slices (Shraga-Levine *et al.*, 1994). ET-1 and SRTX-b at picomolar concentrations induced a dose- dependent increase in cGMP, and maximal stimulation induced cGMP levels 1.7- to 2-fold higher than basal levels. The dose response of ET-3 and SRTX-c differed from that of ET-1 and SRTX-b. Thus, cGMP was not produced when ET-3 or SRTX-c was applied at picomolar concentrations; however, it was stimulated by nanomolar ligand concentrations, with maximal effect at 10^{-7}–10^{-6} M. Most of the sites were of the ET_B-R subtype. However, there was also a minority site population of the ET_A-R subtype, and these were the sites that stimulated cGMP production. Thus, these peptides activated the same receptor-binding sites (ET-1 and SRTX-b affecting the pM sites and ET-3 and SRTX-c the nM sites) to produce cyclic GMP, but their signaling pathways differed. Production of cGMP by these ligands was examined in the presence and absence of agents known to be markers for specific responses, namely, BQ-123, N^{ω}-nitro-L-Arg (a specific inhibitor of nitric oxide synthase), zinc protoporphyrin IX (which inhibits the activity of heme oxygenase, the enzyme that releases carbon monoxide) and pertussis toxin. The ETs (ET-1 and ET-3) were found to signal via nitric oxide formation and to involve PT-sensitive G protein(s). The SRTXs (b and c), while also stimulating cyclic GMP production, did so via formation of carbon monoxide, a pathway which is not L-arginine-dependent and does not involve PT-sensitive G-proteins.

Taken together, our findings point to the existence of several signaling pathways induced by ETs and SRTXs via a number of possible mechanisms:

- Ligand-induced coupling between ET-R and specific G-proteins. Binding of ETs to ET-R induces its coupling to a PT-sensitive G-protein, resulting in the production of cGMP via nitric oxide formation. Binding of SRTXs leads to coupling of the ET-R to another G-protein, which is PT- insensitive. This results in cGMP production via the carbon monoxide pathway.

- Pre-existing functional coupling between ET-Rs and PT-sensitive or PT-insensitive G-proteins. This mechanism is similar to the one above, except that the ET-R is coupled to a specific G-protein prior to ligand binding. In this case, ET-R bound to PT-sensitive G-protein interacts specifically with ETs, while ET_A-R coupled with PT-insensitive G- protein(s) binds SRTXs.
- A third, though less likely possibility assumes the existence of two different receptor subtypes: one responds to ETs and operates via nitric oxide formation, while the other interacts with SRTXs and operates via the carbon monoxide pathway.

Whatever the mechanism(s), the novel conclusion is that even ETs and SRTXs produce the same second messenger and share the same binding sites, their signaling pathways may differ.

CONCLUSION

The sarafotoxins (SRTXs) and the endothelins (ETs) are potent vasoactive peptides which participate in diverse biological processes, such as contraction, neuromodulation and neurotransmission. Although the physiological functions of these peptides are not entirely clear, the endothelins are probably involved in certain pathophysiological conditions including cardiac and renal failure, asthma and atherosclerosis. The diversity of action of SRTXs and ETs may be attributed to (i) the existence of a number of receptor subtypes and (ii) the G-protein-mediated activation of different signaling pathways and events, i.e. phospholipases (A2, C and D), Ca^{2+} efflux and adenylyl and guanylyl cyclase. The combined action of these two variables modulates the responses.

Acknowledgment

I thank Ms. Shirley Smith for excellent editorial assistance.

REFERENCES

Alling, C., Gustaavsson, L. and Anggard, E. (1983) An abnormal phospholipid in rat organs after ethanol treatment. *FEBS Lett.*, **152**, 24–28.

Ambar, I. and Sokolovsky, M. (1993) Endothelin receptors stimulate both phospholipase C and phospholipase D activities in different cell lines. *Eur. J. Pharmacol.*, **245**, 31–41.

Arai, H., Hori, S., Aramori, I., Ohkubo, H. and Nakanishi, S. (1990) Cloning and expression of cDNA encoding an endothelin receptor. *Nature*, **348**, 730–732.

Aramori, I. and Nakanishi, S. (1992) Coupling of two endothelin receptor subtypes to differing signal transduction in transfected Chinese hamster ovary cells. *J. Biol. Chem.*, **267**, 12468–12474.

Bdolah, A., Wollberg, Z., Ambar, I., Kloog, Y., Sokolovsky, M. and Kochva, E. (1989) Disturbances in the cardiovascular system caused by endothelin and sarafotoxin. *Biochem. Pharmacol.*, **38**, 3145–3146.

Chalifour, R.J. and Kanfer, J.N. (1980) Microsomal phospholipase D of rat brain and lung tissues. *Biochem. Biophys. Res. Commun.*, **96**, 742–747.

Clozel, M., Breu, V., Burri, K., Cassal, J-M., Fischli, W., Gray, G.A., Hirth, G., Loffler, B-M., Muller, M., Neidhart, W. and Ramuz, H. (1993) Pathophysiological role of endothelin revealed by the first orally active endothelin receptor antagonist. *Nature*, **365**, 759–761.

Clozel, M., Breu, V., Gray, G.A., Kalina, B., Loffler, B-M., Burri, K., Cassal, J-M., Hirth, G., Muller, M., Neidhart, W. and Ramuz, H. (1994) Pharmacological characterization of bosentan, a new potent orally active nonpeptide endothelin receptor antagonist. *J. Pharmacol. Exp. Ther.*, **270**, 228–235.

Davenport, A.P., Kuc, R.E., Fitzgerald, F., Maguire, J.J., Berryman, K. and Doherty, A.M. (1994) [^{125}I]-PD15242, a selective radioligand for human ET_A receptors. *Br. J. Pharmacol.*, **111**, 4–6.

Dennis, E.A., Rhee, S.G., Billah, M.M. and Hannun, Y.A. (1991) Role of phospholipases in generating lipid second messengers in signal transduction. *FASEB J.*, **5**, 2068–2077.

Douglas, S.A., Beck, G.R., Jr., Elliott, J.D. and Ohlstein, E.H. (1995) Pharmacological evidence for the presence of three distinct functional endothelin receptor subtypes in the rabbit lateral saphenous vien. *Br. J. Pharmacol.*, **114**, 1529–1540.

Edwards, R.M., Pullen, M. and Nambi, P. (1992) Activation of endothelin ET_B receptors increases glomerular cGMP via an L-arginine-dependent pathway. *Am. J. Physiol.*, **263**, F1020–F1025.

Elliott, J.D., Lago, M.A., Cousins, R.D., Gao, A., Leber, J.D., Erhard, K.F., Nambi, P., Elshourbagy, N.A., Kumar, C., Lee, J.A., Bean, J.W., DeBrosse, C.W., Eggleston, D.S., Brooks, D.P., Feuerstein, G., Rufolo, R.R. Jr., Weinstock, J., Gleason, J.G., Peishoff, C.E. and Ohlstein, E.H. (1994) 1,3-Diarylindan-2-carboxylic acids, potent and selective non-peptide endothelin receptor antagonists. *J. Med. Chem.*, **37**, 1553–1557.

Etoh, N., Kondoh, M., Ohnishi, J., Murakami, K. and Miyazaki, H. (1994) Characterization of recombinant endothelin receptor type B overexpressed in Chinese hamster ovary cells. *Biomed. Res.*, **15**, 299–309.

Exton, J.H. (1990) Signaling through phosphatidylcholine breakdown. *J. Biol. Chem.*, **265**, 1–4.

Faraci, F.M. (1989) Effects of endothelin and vasopressin on cerebral blood vessels. *Am. J. Physiol.*, **257**, H799–H803.

Fisher, S.K. (1995) Homologous and heterologous regulation of receptor-stimulated phosphoinositide hydrolysis. *Eur. J. Pharmacol.* (Mol. Pharmacol. Section), **288**, 231–250.

Fujitani, Y., Ueda, H., Okada, T., Urade, Y. and Karaki, H. (1993) A selective agonist of endothelin type B receptor, IRL 1620, stimulates cyclic GMP increase via nitric oxide formation in rat aorta. *J. Pharmacol. Exp. Ther.*, **267**, 683–689.

Fukunaga, M., Ochi, S., Takama, T., Yokoyama, K., Fujiwara, Y., Orita, Y. and Kamada, T. (1991) Endothelin-1 stimulates prostaglandin E_2 production in an extracellular calcium-independent manner in cultured rat mesangial cells. *Am. J. Hypertension*, **4**, 137–143.

Garbers, D.L. (1989) Guanylate cyclase, a cell surface receptor. *J. Biol. Chem.*, **264**, 9103–9106.

Garthwaite, J. (1991) Glutamate, nitric oxide and cell-cell signalling in the nervous system. *Trends Neurosci.*, **14**, 60–67.

Griendling, K.K., Tsuda, T. and Alexander, R.W. (1989) Endothelin stimulates diacylglycerol accumulation and activates protein kinase C in cultured vascular smooth muscle cells. *J. Biol. Chem.*, **264**, 8237–8240.

Gulati, A., Sharma, A.C., Robbie, G. and Saxena, P.R. (1995) Endothelin ET_A receptor antagonist, BQ-123, blocks the vasoconstriction induced by sarafotoxin 6b in the heart but not in other bascular beds. *Gen. Pharmac.*, **26**, 183–193.

Hattori, H. and Kanfer, J.N. (1985) Synaptosomal phospholipase D potential role in providing choline for acetylcholine synthesis. *J. Neurochem.*, **45**, 1578–1583.

Hirata, Y., Emori, T., Eguchi, S., Kanno, K., Imai, T., Ohta, K. and Marumo, F. (1993) Endothelin receptor subtype B mediates synthesis of nitric oxide by cultured bovine endothelial cells. *J. Clin. Invest.*, **91**, 1367–1373.

Hosoda, K., Nakao, K., Arai, H., Suga, S-I., Ogawa, Y., Mukoyama, M., Shirakami, G., Saito, Y., Nakanishi, S. and Imura, H. (1991) Cloning and expression of human endothelin-1 receptor cDNA. *FEBS Lett.*, **287**, 23–26.

Ihara, M., Noguchi, K., Saeki, T., Fukuroda, T., Tsuchida, S., Kimura, S., Fukami, T., Ishikawa, K., Nishikibe, M. and Yano, M. (1992) Biological profiles of highly potent novel endothelin antagonists selective for the ETA receptor. *Life Sci.*, **50**, 247–255.

Inoue, A., Yanagisawa, M., Kimura, S., Kasuya, Y., Miyauchi, T., Goto, T. and Masaki, T. (1989) The human endothelin family, Three structurally and pharmacologically distinct isopeptides predicted by three separate genes. *Proc. Natl. Acad. Sci. USA*, **86**, 2863–2867.

Ishii, K., Warner, T.D., Sheng, H. and Murad, F. (1991) Endothelin increases cyclic GMP levels in LLC-PK1 porcine kidney epithelial cells via formation of an endothelium-derived relaxin factor-like substance. *J. Pharmacol. Exp. Ther.*, **259**, 1102–1107.

Karne, S., Jayawickreme, C.K. and Lerner, M.R. (1993) Cloning and characterization of an endothelin-3 specific receptor (ET$_C$ receptor) from Xenopus *laevis* dermal melanophores. *J. Biol. Chem.*, **268**, 19126–18133.

Kasuya, Y., Abe, Y., Hama, H., Sakurai, T., Asada, S., Masaki, T. and Goto, K. (1994) Endothelin-1 activates mitogen-activated protein kinases through two independent signalling pathways in rat astrocytes. *Biochem. Biophys. Res. Commun.*, **204**, 1325–1333.

Kloog, Y., Ambar, I., Sokolovsky, M., Kochva, E., Bdolah, A. and Wollberg, Z. (1988) Sarafotoxin, a novel vasoconstrictor peptide: phosphoinositide hydrolysis in rat heart and brain. *Science*, 242, 268–270.

Lin, H.Y., Kaji, E.H., Winkel, G.K., Ives, H.E. and Lodish, H.E. (1991) Cloning and functional expression of a vascular smooth muscle endothelin 1 receptor. *Proc. Natl. Acad. Sci. USA*, **88**, 3185–3189.

Lin, W.W., Kiang, J.G. and Chuang, D. (1992) Pharmacological characterization of endothelin-stimulated phosphoinositide breakdown and cytosolic Ca^{2+} rise in rat C6 glioma cells. *J. Neurosci.*, **12**, 1077–1085.

MacNulty, E.E., Plevin, R. and Wakelam, M.J.D. (1990) Stimulation of hydrolysis of phosphatidylinositol 4,5-biphosphate and phosphatidylcholine by endothelin, a complete mitogen for Rat-1 fibroblasts. *Biochem. J.*, **272**, 761–766.

Masaki, T. (1993) Endothelins, Homeostatic and compensatory actions in the circulatory and endocrine systems. *Endocr. Rev.*, **14**, 256–268.

Moncada, S., Palmer, R.M.J. and Higgs, E.A. (1991) Nitric oxide, Physiology, pathophysiology and pharmacology. *Pharmacol. Rev.*, **43**, 109–142.

Moritoki, H., Miyano, H., Takeuchi, S., Yamaguchi, M., Hisayama, T. and Kondoh, W. (1993) Endothelin-3-induced relaxation of rat thoracic aorta, A role for nitric oxide formation. *Br. J. Pharmacol.*, **108**, 1125–1130.

Muldoon, L., Pribnow, D., Rodland, K.D. and Magnum, B.E. (1990) Endothelin-1 stimulates DNA synthesis and anchorage-independent growth of Rat-1 fibroblasts through a protein kinase C-dependent mechanism. *Cell Regul.*, **1**, 379–390.

Nirei, H., Hamada, K., Shoubo, M., Sogabe, K., Notsu, Y. and Ono, T. (1993) An endothelin ET(A) receptor antagonist, FR 139317, ameliorates cerebral vasospasm in dogs. *Life Sci.*, **23**, 1869–1874.

Nishizuka, Y. (1988) The molecular heterogeneity of protein kinase C and its implications for cellular regulation. *Nature*, **334**, 661–665.

Ogawa, Y., Nakao, K., Arai, H., Nakagawa, O., Hosoda, K., Suga, S-I., Nakanishi, S. and Imura, H. (1991) Molecular cloning of a non-isopeptide selective human endothelin receptor. *Biochem. Biophys. Res. Commun.*, **178**, 248–255.

Ohlstein, E.H. and Douglas, S.A. (1993) Endothelin-1 modulates vascular smooth muscle structure and vasomotion: Implications in cardiovascular pathology. *Drug Develop. Res.*, **29**, 108–128.

Resink, T.J., Scott-Burden, T. and Buhler, F.R. (1989) Activation of phospholipase A2 by endothelin in cultured vascular smooth muscle cells. *Biochem. Biophys. Res. Commun.*, **158**, 279–286.

Resink, T.J., Scott-Burden, T. and Buhler, F.R. (1990) Activation of multiple signal transduction pathways by endothelin in cultured human vascular smooth muscle cells. *Eur. J. Biochem.*, **189**, 415–421.

Reynolds, E.E. and Mok, L.L. (1990) Role of thromboxane A$_2$/prostaglandin H$_2$ receptor in the vasoconstrictor response of rat aorta to endothelin. *J. Pharmacol. Exp. Ther.*, **252**, 915–921.

Reynolds, E.E., Hwang, O., Flynn, M.A., Welch, K.M., Cody, W.L., Steinbaugh, B., He, J.X., Chung, F-Z. and Doherty, A.M. (1995) Pharmacological differences between rat and human endotheln B receptors. *Biochem. Biophys. Res. Commun.*, **209**, 506–512.

Saida, K., Mitsui, Y. and Ishida, P. (1989) A novel peptide vasocactive intestinal cotnractor of a new endothelin peptide family. *J. Biol. Chem.*, **264**, 14613–14616.

Sakamoto, A., Yanagisawa, M., Sakurai, T., Takuwa, Y., Yanagisawa, H. and Masaki, T. (1991) Cloning and functional expression of human cDNA for the ET_B endothelin receptor. *Biochem. Biophys. Res. Commun.*, **178**, 656–663.

Sakurai, T., Yanagisawa, M., Takuwa, Y., Miyazaki, H., Kimura, S., Goto, K. and Masaki, T. (1990) Cloning of a cDNA encoding a non-isopeptide-selective subtype of the endothelin receptor. *Nature*, **348**, 732–735.

Schramek, H., Marsen, T.A. and Dunn, M.J. (1993) Endothelins, Vasoactive peptides interacting with prostaglandin- and nitric oxide-induced signaling pathways. In, *Thromboxane A_2 and other Vasoconstrictors in Clinical Conditions*, pp. 227–249, Neri Serneri, G.G., Bonomini, V., Gensini, G.F., Pozzi, E. and Prisco, D. (eds.). Scientific Press, Florence.

Schramek, H., Wang, J., Konieczkowski, M., Rose, P.M., Sedor, J.R. and Dunn, M.J. (1994a) Endothelin-1 stimulates cytosolic phospholipase A_2 in Chinese hamster ovary cells stably expressing the human ET_A or ET_B receptor subtype. *Biochem. Biophys. Res. Commun.*, **199**, 992–997.

Schramek, H., Wang, Y., Konieczkowski, M., Simonson, M.S. and Dunn, M.J. (1994b) Endothelin-1 stimulates phospholipase A_2 activity and gene expression in rat glomerular mesangial cells. *Kidney Int.*, **46**, 1644–1652.

Schumacher, W.A., Steinbacher, T.E., Allen, G.T. and Ogletree, M.L. (1990) Role of thromboxane receptor activation in the bronchospastic response to endothelin. *Prostaglandins*, **40**, 71–79.

Shraga-Levine, Z., Galron, R. and Sokolovsky, M. (1994) Cyclic GMP formation in rat cerebellar slices is stimulated by endothelins via nitric oxide formation and by sarafotoxins via formation of carbon monoxide. *Biochemistry*, **33**, 14656–14659.

Simonson, M.S. (1993) Endothelins: Multifunctional renal peptides. *Physiol. Rev.*, **73**, 375–411.

Simonson, M.S. and Dunn, M.J. (1990) Endothelin-1 stimulates contraction of rat glomerular cells and potentiates β-adrenergic-mediated cyclic adenosine monophosphate accumulation. *J. Clin. Invest.*, **85**, 790–797.

Simonson, M.S., Wang, Y. and Dunn, M.J. (1992) Celular signaling by endothelin peptides, Pathways to the nucleus. *J. Am. Soc. Nephrol.*, **2**, S116–S125.

Sokolovsky, M. (1992) Endothelins and sarafotoxins, Physiological regulation, receptor subtypes and transmembrane signaling. *Pharmac. Ther.*, **54**, 129–149.

Sokolovsky, M. (1993a) Functional coupling between endothelin receptors and multiple G-proteins in rat heart myocytes. *Receptors and Channels*, **1**, 295–304.

Sokolovsky, M. (1993b) Endothelin receptors in rat cerebellum, Activation of phosphoinositide hydrolysis is transduced by multiple G-proteins. *Cellular Signaling*, **5**, 473–483.

Sokolovsky, M. (1995) Endothelin receptors subtypes and their role in transmembrane signalling mechanisms. *Pharmac. Ther.* **68**, 435–471.

Sokolovsky, M., Ambar, I. and Galron, R. (1992) A novel subtype of endothelin receptors. *J. Biol. Chem.*, **267**, 20551–20554.

Sokolovsky, M., Shraga-Levine, Z. and Galron, R. (1994) Ligand-specific stimulation/inhibition of cAMP formation by a novel endothelin receptor subtype. *Biochemistry*, **33**, 11417–11419.

Stein, P.D., Hunt, J.T., Floyd, D.M., Moreland, S., Dickinson, K.E.J., Mitchell, C., Liu, E.C-K., Webb, M.L., Murugesan, N., Dickey, J., McMullen, D., Zhang, R., Lee, V.G., Serafino, R., Delaney, C., Schaeffer, T.R. and Kozlowski, M. (1994) The discovery of sulfonamide endothelin antagonists and the development of the orally active ET_A antagonist 5-(dimethylamino)-N- (3,4-dimethyl-5-isoxazolyl)-1-naphthalenesulfonamide. *J. Med. Chem.*, **37**, 329–331.

Sunako, M., Kawahara, Y., Hirata, K., Tsudo, T., Yokoyama, M., Fukuzaki, H. and Takai, Y. (1990) Mass analysis of 1,2-diacylglycerol in cultured rabbit vascular smooth muscle cells. *Hypertension*, **15**, 84–88.

Suzuki, A., Oiso, Y. and Kozawa, O. (1994) Effect of endothelin-1 on phospholipase D activity in osteoblast-like cells. *Mol. Cell Endocrinol.*, **105**, 193–196.

Tabuchi, H., Furuichi, Y. and Miyamoto, C. (1994) Differential regulation of c-fos gene expression by two types of human endothelin receptors in Chinese hamster ovary cells. *J. Mol. Endocrinol.*, **12**, 173–180.

Tam, J.P., Liu, W., Zhang, J-W., Galantino, M., Bertolero, F., Cristiani, C., Vaghi, F. and de Castiglione, R. (1994) Alanine scan of endothelin, Importance of aromatic residues. *Peptides*, **15**, 703–708.

Taniguchi, I., Kageyama, S., Aihara, K. and Isogai, Y. (1994) Effects of N^G-monomethyl-L-arginine, indomethacin, and aspirin on the vasodepressor response to low doses of endothelin-1 and endothelin-3 in rats. *Japanese Circul. J.*, **58**, 69–75.

Warner, T.D., De Nucci, G. and Vane, J.R. (1989) Rat endothelin is a vasodilator in the isolated perfused mesentery of the rat. *Eur. J. Pharmac.*, **159**, 325–326.

Warner, T.D., Allcock, G.H., Mickley, E.J., Corder, R. and Vane, J.R. (1993) Comparative studies with the endothelin receptor antagonists BQ-123 and PD 142893 indicate at least three endothelin receptors. *J. Cardiovasc. Pharmac.*, **22**, S117–S120.

Yanagisawa, M., Kurihara, H., Kimura, S., Tombe, Y., Kobayashi, M., Mitsui, Y., Yazaki, Y., Goto, J. and Masaki, T. (1988) A novel potent vasoconstrictor peptide produced by vascular endothelial cells. *Nature*, **332**, 411–415.

Zhang, Y. and Abdel-Latif, A.A. (1992) Activation of phospholipase D by endothelin-1 and other pharmacological agents in rabbit iris sphincter smooth muscle. *Cell Signal*, **4**, 777–786.

5. THE PROTEIN KINASE INHIBITORS K-252a AND STAUROSPORINE AS MODIFIERS OF NEUROTROPHIN RECEPTOR SIGNAL TRANSDUCTION

PHILIP LAZAROVICI[1], YUZURU MATSUDA[2], DAVID KAPLAN[3]
and GORDON GUROFF[4]

[1]*Department of Pharmacology and Experimental Therapeutics, School of Pharmacy,
Faculty of Medicine, The Hebrew University of Jerusalem, POB 12065,
Jerusalem 91120, Israel*
[2]*Tokyo Research Labs, Kyowa Hakko Kogyo Co. Ltd., 3-6-6 Asahimachi,
Machida-Shi, Tokyo 194, Japan*
[3]*Eukaryotic Signal Transduction Section, ABL-Basic Research Program,
National Cancer Institute, Frederick Cancer Research and Development Facility, P.O. Box B,
Frederick, MD 21702, USA*
[4]*Section on Growth Factors, National Institute of Child Health and
Human Development, National Institutes of Health, Bethesda, MD 20892, USA*

INTRODUCTION

Protein phosphorylation constitutes one of the most important molecular mechanisms by which extracellular signals produce their biological responses in eukaryotic cells. Stimulation of protein kinases is considered to be the most common activation mechanism in signal transduction systems (Walaas and Greengard, 1991). A protein phosphorylation system is composed of three major components: a) protein substrates that undergo changes in their properties during the phosphorylation/dephosphorylation reaction, b) protein kinases that transfer phosphates from ATP to acceptor groups such as alcohol (serine/threonine kinases), phenol (tyrosine kinases), etc. (Hunter, 1991), c) protein phosphatases that dephosphorylate the phosphoproteins thereby restoring the protein phosphorylation system to its basal activity (Cohen, 1989).

The protein serine/threonine kinases can be classified by the nature of their second messenger activators: cAMP-dependent (PKA), cGMP-dependent (PKG), calcium/phospholipid dependent (PKC), calcium/calmodulin dependent (CaMK), etc. (Nairn *et al.*, 1985). The enzymatic activity of these serine/threonine protein kinases is allosterically regulated. The kinases are comprised of several functional regions, such as regulatory and catalytic domains, the latter including the conserved ATP site and the variable substrate-binding site. In the resting state, the regulatory domain keeps the catalytic domain inactive. This inhibition is relieved when a second-messenger activator binds to the regulatory domain. Therefore, the phosphorylation of cellular substrates by these protein kinases serves as an important messenger switch in signal transduction and cross talk of many plasma membrane receptors.

Protein tyrosine kinases mediate the transduction of both extracellular and intracellular signals. The receptor tyrosine kinases (RTK) possess an extracellular ligand binding domain for growth factors and an intracellular catalytic domain with intrinsic tyrosine kinase activity (Hunter, 1991). Binding of a ligand to the receptor leads to a variety of downstream effects including stimulation of other receptor-associated tyrosine kinases (RATK), increase in intracellular calcium levels, activation of cascades of serine/threonine protein kinases, phospholipases A and C, and phosphatidylinositol 3′-kinase, and ultimately changes in gene expression, metabolic effects, proliferation or differentiation, etc. (Fantl *et al.*, 1993).

In order to elucidate receptor signal transduction pathways and to develop therapeutic strategies for the treatment of neurodegenerative disorders, cancer, etc., potent and selective protein kinase inhibitors would be of great value. In the early sixties, Umezawa and his colleagues embarked on the screening of toxic and bioactive substances of microbial origin based on enzyme and receptor assays and evaluation of their antibiotic properties (Umezawa, 1972). This approach resulted in the discovery of many important enzyme inhibitors and receptor antagonists (Hall, 1989). The continuing search for toxic microbial alkaloids led to the discovery of the K-252 family of cytotoxic, indolecarbazole alkaloids that represent a second generation of protein-kinase inhibitors (Matsuda, 1993). A complete survey of the information available on the K-252 compounds is beyond the scope of this review. Therefore, we intend to focus on the kinase inhibitory, neurotrophin-inhibitory, and neurotrophic activities of the K-252 compounds that make them attractive tools in neuroscience studies of protein kinases, neurotrophin receptor signal transduction, and neuroprotection (Koizumi *et al.*, 1990, Rasouly *et al.*, 1995).

K-252 COMPOUNDS AS PROTEIN KINASE INHIBITORS

K-252a, K-252b, and staurosporine (Figure 5.1) were isolated from the culture broth of *Nocardiopsis* sp. and *Streptomyces staurospores* bacteria (Nakanishi *et al.*, 1986; Omura *et al.*, 1995). The culture broth of each of these bacteria was extracted in several steps with different organic solvents and separated by hydrophobic partition chromatography and thin layer chromatography on silica gels (Omura *et al.*, 1995) or by sequential column chromatography on Diaion HP, Sephadex LH, and Lichroprep RP gels (Nakanishi *et al.*, 1986). K-252a and staurosporine were isolated by linear gradient elution with 40–60% methanol and concentrated to dryness under vacuum to yield yellow crystals. The structures presented in Figure 5.1 were determined on the basis of ultraviolet spectral data, infrared spectrum bands, nmr spectral signals measured in DMSO-d$_6$, and the mass spectral analysis of their acetyl derivatives (Nakanishi *et al.*, 1986; Omura *et al.*, 1995). K-252a derivatives and staurosporine display poor water solubility. They are soluble, however, in dimethyl sulfoxide (DMSO), dimethyl formamide (DMF), and slightly soluble in chloroform or methanol. Solutions of these alkaloids in DMSO of DMF are stable for months at

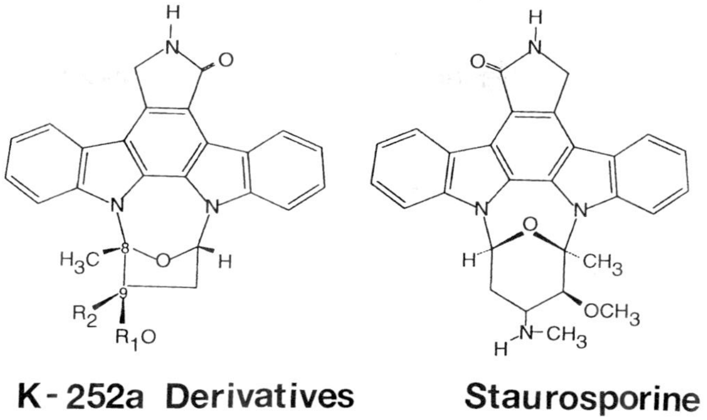

K-252a Derivatives   **Staurosporine**

Figure 5.1 *Structures of staurosporine and K-252 derivatives.* R_1 and R_2 are H and CH_3, and H and H in K-252a and K-252b, respectively. Note the similarity in the indolecarbazole ring structures and the differences in the glycoside linkage and side chain composition. The formulae and molecular weights (daltons) of the derivatives are: K-252a–$C_{27}H_{21}N_3O_5$, M_r-467; K-252b–$C_{26}H_{19}N_3O_5$, M_r-453; staurosporine-$C_{28}H_{26}N_4O_3$, M_r-466.

–20°C in the dark and should be diluted to aqueous solutions just prior to use. These compounds are characterized by the unique presence of the indol [2, 3-α] carbazole chromophore (Figure 5.1) (Omura *et al.*, 1995).

Using the protein histone H1 as a substrate and partially purified protein kinases from different biological sources, the inhibitory potency of these alkaloids was measured *in vitro* (Table 5.1). It is clear from the table that K-252a is a potent (1–20 nM), albeit nonselective protein-kinase inhibitor. K-252b, the 9-carboxylic acid derivative of K-252a, is less potent (10–150 nM) and somewhat more selective for protein kinase C than for other kinases. This characteristic might be related to the accumulation of K-252b in the plasma membrane (Nagashima *et al.*, 1991) resulting in high localized concentrations of the inhibitor, which could prevent PKC translocation and/or activation. Staurosporine is the most potent (2–10 nM) inhibitor of PKC, yet it shows little selectivity toward this particular kinase (Table 5.1). The lack of *in vivo* selectivity of the K-252 alkaloids toward different cellular kinases is further emphasized in Table 5.2. Both serine/threonine and tyrosine kinases are affected by these inhibitors. This lack of selectivity of protein kinase inhibition is also supported by the lack of selective binding of ^{3}H-staurosporine to different cellular protein kinases (Herbert *et al.*, 1990). The limited *in vitro* selectivity of K-252 compounds for the different protein kinases (Table 5.1) might be attributed to their interaction with the ATP binding site on the catalytic domains of the different protein kinases (Ruegg and Burgess, 1989), a site which is likely to be similar from one kinase to another. In general, the K-252 compounds inhibit protein kinases *in vitro* in a competitive manner with ATP (Omura *et al.*, 1995), but it has also been reported that staurosporine inhibits the activity of calcium/calmodulin kinase type II (CaMK-II) in a manner noncompetitive with ATP (Yanagihara *et al.*, 1991).

Table 5.1 *In vitro* inhibitory effects of K-252a, K-252b, and staurosporine on several protein kinases[1]

Protein kinase[2]	K-252a	K-252b	Staurosporine
PKA	10–20	90	20
PKG	15–20	100	8–10
PKC	25	20	0.5–2
MLCK	20	150	2–5
CaMK	2	12	10–100

[1]Inhibitory activity is expressed by inhibition constants, K_i (nM)
[2]PKA-cAMP dependent; PKG-cGMP dependent; PKC-calcium and phospholipid dependent; MLCK-myosin light chain kinase; CaMK-calcium-calmodulin type 2 kinase.

Table 5.2 *In vivo* inhibition of protein kinases by K-252 compounds

Protein kinase[1]	Source	Reference
Serine/threonine kinases		
PKC	brain	Tamaoki, 1991
	PC12	Rasouly *et al.*, 1992
	pancreas	Verme *et al.*, 1989
CaMK	brain	Yanagihara *et al.*, 1991
PKA	platelets	Spacey *et al.*, 1990
MAPK	B lymphoma	Gold *et al.*, 1992
	PC12	Smith *et al.*, 1989
p34[cdc]K	mammary carcinoma	Gadbois *et al.*, 1992
phosphorylase kinase	muscle	Elliot *et al.*, 1990
Tyrosine kinases		
pp60	Rous sarcoma	Nakano *et al.*, 1987
trk	PC12	Ohmichi *et al.*, 1992
PDGFR	keratinocytes	Andrejauskas-Buckdunger and Regenass, 1992
Insulin receptor	muscle	Guma *et al.*, 1992
Unknown tyrosine	hepatoma	Fallon, 1990
kinases	neutrophils	Badwey *et al.*, 1991

[1]PKA-cAMP dependent; PKC-calcium and phospholipid dependent; CaMK-calcium-calmodulin type 2 kinase; MAPK-mitogen activated protein kinase; trk-NGF receptor tyrosine kinase; PDGFR-platelet-derived growth factor receptor; pp60-v-src kinase.

The effect of the K-252 compounds on protein phosphorylation *in vivo* is highly complex. In certain cases, they inhibit serine/threonine kinases (Hashimoto and Hagino, 1989; Robinson *et al.*, 1990, Rasouly *et al.*, 1992) and in other cases they activate serine/threonine kinases such as PKC (Jiang *et al.*, 1992). It was also shown that the K-252 compounds inhibit tyrosine phosphorylations *in vivo* (Berkow, 1992), but staurosporine also stimulates the tyrosine phosphorylation of a 145 kD protein in PC12 cells (Rasouly and Lazarovici, 1994). This apparent discrepancy, that compounds which act as inhibitors of kinases *in vitro* can either activate or inhibit kinases *in vivo*, might be explained by the complex regulation of cellular enzymatic pathways utilized by different protein kinases and phosphatases. Therefore, depending on the cell type and the kinetics and concentrations of the K-252 compounds, different cellular protein kinases may be modulated to different extents.

Another concern is the biological significance of the competitive interaction of K-252 compounds at the ATP binding sites of the protein kinases seen *in vitro*. Since the intracellular concentration of ATP is in the millimolar range and the Km of ATP for protein kinases is in the range of 1–20 μM, concentrations of the K-252a compounds much higher than 20 μM would be required for competitive inhibition with ATP in order to achieve *in vivo* inhibition of protein phosphorylation (Ruegg and Burgess, 1989). Such concentrations are much higher than those at which biological effects occur and are extremely toxic to eukaryotic cells (Rasouly *et al.*, 1995). In conclusion, the identification of the precise protein kinase targets and the elucidation of the *in vivo* mechanism of action of the K-252 compounds remain to be clarified.

NEUROTROPHINS AND SIGNAL TRANSDUCTION BY THE TRK RECEPTORS

The neurotrophins are a family of low molecular weight, basic proteins (Table 5.3) that share about 50% sequence homology and play a central role in the survival, differentiation, plasticity, and repair of the nervous system (Thoenen, 1991). While nerve growth factor (NGF), the first member of this family, was identified and characterized almost a half century ago (Levi-Montalcini and Angeletti, 1968), it is only during the last six years that other neurotrophins have been discovered and characterized (Table 5.3). These include brain-derived neurotrophic factor (BDNF) (Leibrock *et al.*, 1989), neurotrophin-3 (NT-3) (Maisonpierre *et al.*, 1990; Hohn *et al.*, 1990), neurotrophin-4/5 (NT-4/5) (Berkeimeier *et al.*, 1991; Ip *et al.*, 1992), and neurotrophin-6 (NT-6) (Gotz *et al.*, 1994). Other neurotrophic factors include ciliary neurotrophic factor (CNTF) (Sendtner *et al.*, 1994), which is structurally and functionally distinct from the neurotrophins, and glial cell-derived neurotrophic factor (GDNF) (Lin *et al.*, 1993), which is a member of the transforming growth factor β (TGFβ) family.

NGF is required for the survival and development of certain neurons in the sympathetic and sensory nervous systems (Levi-Montalcini and Angeletti, 1968). It also has effects on specific neuronal cell populations in the central nervous system (Gnahn *et al.*, 1983; Hefti *et al.*, 1985), the adrenal medulla (Unsicker *et al.*, 1978), some tumors (Fabricant *et al.*, 1977; Waris *et al.*, 1973), and certain types of lymphoid cells (Matsuda *et al.*, 1988). Although many neuronal cell types are responsive *in vitro* to one or more neurotrophins (Table 5.3), very little is known about the true physiological actions and targets of the neurotrophins other than NGF

Table 5.3 Properties of the neurotrophins

Neurotrophin [1]	Property [2]	Receptor [3]	Targets [4]	Disease [5]
NGF	7S precursor; 2.5S β-subunit; 26 kD; 118 a.a.; pI 9.3; found in many tissues	trkA	trigeminal, dorsal root, sympathetic ganglia, septal cholinergic neurons	Alzheimers, peripheral neuropathies, ischemia
BDNF	119 a.a.; pI 9.9; localized in CNS	trkB	vestibular and retinal ganglia, dopaminergic, GABAergic embryonic, spinal motor, cortical forebrain cholinergic neurons	Parkinsons, Huntingtons, amyotrophic lateral sclerosis (ALS), ischemia
NT-3	119 a.a.; pI 9.5; high concentration in embryos	trkA, trkB, trkC	dorsal root and nodose ganglia, hippocampus and various neuronal cells	Parkinsons, peripheral neuropathies
NT-4/5	Mr = 13,917; pI 10.3; high in placenta and prostate	trkB	motor, septal, cholinergic, dopaminergic, GABAergic neurons	Alzheimers, Huntingtons, Parkinsons, ALS, ischemia
NT-6	143 a.a.; pI 10.8; from Xiphophorus fish	unknown	unknown	unknown
CNTF	200 a.a.; acidic cytoplasmic protein released from injured cells	a complex consisting of three different proteins (CNTFRα, LIFRβ, and GP13)	ciliary, spinal cord, motor neurons	ALS, peripheral neuropathies

Table 5.3 *Continued*

GDNF	homodimeric glycosylated protein of 34–45 kD	unknown	dopaminergic, embryonic spinal cord, motor and sensory neurons	Parkinsons

1. NGF-nerve growth factor; BDNF-brain derived neurotrophic factor; NT-3-neurotrophin-3; NT-4/5-neurotrophin-4/5; NT-6-neurotrophin-6; CNTF-ciliary neurotrophic factor; GDNF-glial cell line derived neurotrophic factor.
2. a. a.-amino acids; pI-isoelectric point; kD-kilodaltons.
3. Reviewed in Barbacid, 1994; Davis and Yancopoulos, 1993.
4. Reviewed in Lindsay *et al.*, 1994; Barbacid, 1995.
5. Reviewed in Lindsay *et al.*, 1994; Lindsay *et al.*, 1993; Vantini, 1992.

(Lindsay *et al.*, 1994). Recently, the advent of gene targeting has made it possible to generate strains of mice lacking specific neurotrophins and/or their receptors (Barbacid, 1995). Analysis of the neurological loss in these mutant mice provides compelling evidence to support the concept that NGF is responsible for the survival of nociceptive and thermoceptive sensory neurons and ganglionic and brain neurons (Table 5.3). BDNF is required to maintain the survival of nodose-petrosal and vestibular ganglia required to control movement coordination and balance as well as autonomic sensory information. Finally, NT-3 is crucial for maintaining sensory proprioception (Barbacid, 1995).

The biological activities of the neurotrophins are mediated by their binding to receptors on the surface of target cells. These receptors belong to the trk family of receptor tyrosine kinases, that are similar to the receptor tyrosine kinases for other growth factors, but are selectively occupied and activated by neurotrophins: trk (also know as trk A) binds NGF; trk B binds BDNF, NT-3, and NT-4/5, and trk C binds NT-3 (Kaplan and Stephens, 1995). In addition to the high-affinity trk receptors, many neurons and nonneuronal cells express low-affinity NGF receptors (named LNGFR or p75 protein) which bind all the neurotrophins (Chao, 1992). Binding of the appropriate neurotrophin induces trk dimerization (Jing *et al.*, 1992) and results in the transautophosphorylation of several of tyrosine residues on trk (Tyr: 490, 670, 674, 675, 785) (Kaplan and Stephens, 1994) (Figure 5.2). Trk tyrosine kinase activity is maximal 5 to 10 min after NGF binding and is rapidly attenuated, most likely due to dephosphorylation and internalization of trk receptors (Hempstead *et al.*, 1992).

The phosphorylation of tyrosines 670, 674, and 675 within the kinase catalytic domain and tyrosines 490 and 785 outside the kinase domain is the first critical event in signal transduction by these receptors. The tyrosine phosphorylation is required for the activation of trk receptor catalytic activity as well as for the recruitment of signalling protein substrates that bind trk. The phosphorylated tyrosines

490 and 785 act as docking sites for effector substrates which contain SH2 (src homology-2) (Figure 5.2). Four intracellular proteins (Figure 5.2) are known to bind to activated trk: phospholipase C-gamma (PLC-γ) (tyrosine 785; Loeb *et al.*, 1994), phosphatidylinositol 3′-kinase (PI-3 kinase) (tyrosine 751; Obermeier *et al.*, 1993), SHC protein (tyrosine 490; Stephens *et al.*, 1994), and extracellular signal-regulated kinase 1 (erk1) (Loeb *et al.*, 1992). The binding of these substrates to the trk receptor enhances their specific enzymatic activity either through receptor-induced tyrosine phosphorylation or alterations in protein conformation, thereby initiating several cascades of signaling events that ultimately lead to neuronal survival, differentiation, metabolic effects, and other biological responses.

The best characterized signaling cascade to date is the ras pathway in which tyrosine phosphorylated SHC-binds to Grb 2, an adaptor protein that bridges SHC to SOS (Kaplan and Stephens, 1994). SOS, one of the guanine nucleotide releasing proteins (GNRP), enhances the rate of GDP-GTP exchange on Ras leading to Ras activation (Figure 5.2). Activated Ras binds to B-Raf which, in turn, phosphorylates and activates MEK (the mitogen-activated protein kinase kinase, MAPKK). MEK is a dual specificity kinase that phosphorylates the mitogen-activated protein kinases (MAPKs) on both tyrosine and threonine residues, leading to their activation. MAP kinase (also known as extracellular relay kinase, or Erk) phosphorylates and activates p90^{rsk} (Figure 5.2) (Johnson and Vaillancourt, 1994) and both enzymes translocate to the nucleus to participate in the activation of transcription factors that may regulate the expression of neurotrophin-inducible genes (Figure 5.2) (Greene and Kaplan, 1995).

Figure 5.2   *Schematic representation of trk receptor signal transduction pathways.* The binding of NGF to the trk receptor stimulates rapid phosphorylation of tyrosine residues and activation of many signal transducing pathways. Activated trk receptors bind to and induce the tyrosine phosphorylation and/or activation of the signaling substrate proteins PLCγ1, PI-3 kinase, and SHC. The latter stimulates Ras and Src activity by binding to Grb and the SOS guanine nucleotide exchange protein. Ras is positively regulated by Grb/SOS and may be negatively regulated by ras GAP. Subsequently, Ras sequentially activates a phosphorylation cascade including B-raf, MAP kinase (MEK), Erks (MAPK) 1 and 2, and p90^{rsk}. Erk and p90^{rsk} may play a role in metabolism, signal transduction, and in activating gene transcription to stop proliferation, activate the differentiation machinery, and/or promote survival. Src activation probably stimulates the focal adhesion kinase (Fak) (Schaller and Parsons, 1994) which will phosphorylate certain cytoskeletal proteins that may, in turn, regulate the assembly/disassembly of the neuritic cytoskeleton. PLCγ1 stimulation will increase intracellular calcium, activate protein kinase C, further activate MAPK, and cross-talk with other signal transduction pathways. PI-3 kinase will interfere with the phosphatidylinositol cycle and will activate S6 kinase signaling pathways. Trk also mediates the ras-independent tyrosine phosphorylation of SNT, which appears to be involved in NGF-induced differentiation. Abbreviations: P-phosphorylated tyrosine residues; DAG-diacylglycerol; PKC-protein kinase C; IP$_3$-inositol trisphosphate; PI-phosphatidylinositol; dashed line indicates action of K-252 derivatives; + activation, −inhibition.

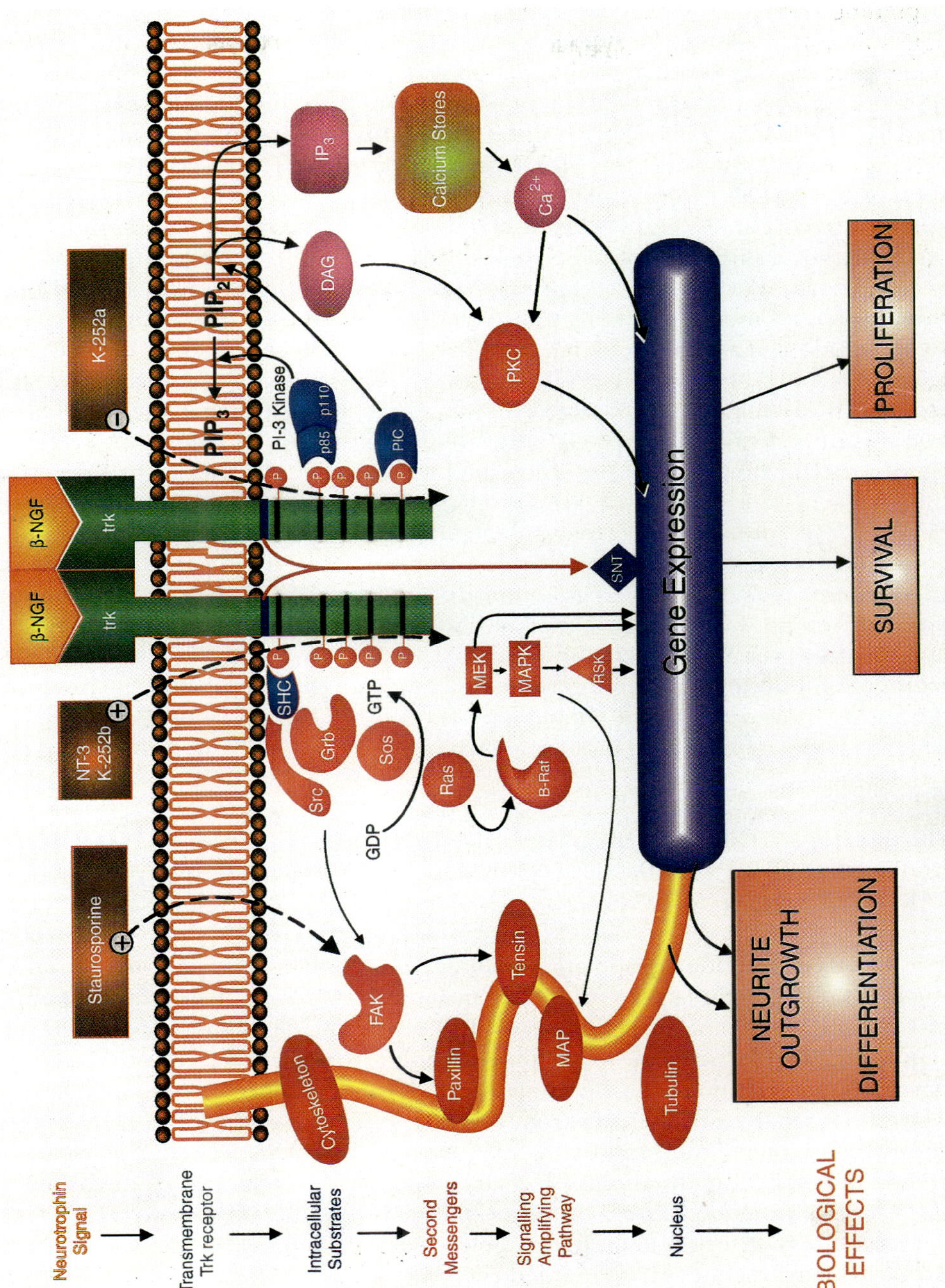
IP₃
Calcium Stores
Ca²⁺
DAG
K-252a
PKC
PIP₃
PIP₂
PI-3 Kinase
p85 p110
PlC
β-NGF
trk
β-NGF
trk
SNT
Gene Expression
PROLIFERATION
SURVIVAL
SHC
Grb
Sos
GTP
Src
Ras
GDP
B-Raf
MEK
MAPK
RSK
NT-3
K-252b
Staurosporine
FAK
Tensin
Paxillin
MAP
Tubulin
Cytoskeleton
NEURITE OUTGROWTH
DIFFERENTIATION
Neurotrophin Signal
Transmembrane Trk receptor
Intracellular Substrates
Second Messengers
Signalling Amplifying Pathway
Nucleus
BIOLOGICAL EFFECTS

Recent studies have revealed another novel target of neurotrophin-mediated signaling events, the suc-associated neurotrophic factor-induced tyrosine-phosphorylated target (SNT; Rabin *et al.*, 1993). In neuronal cells, this protein undergoes rapid tyrosine phosphorylation upon treatment with NGF (Figure 5.2). SNT phosphorylation is independent of any of the known signaling cascades initiated by trk, and its tyrosine phosphorylated form is localized to the nucleus (Rabin *et al.*, 1993). Since this 90 kD protein binds to p13 (Rabin *et al.*, 1993), a subunit of the cell cycle regulatory complex that includes the cyclin-dependent kinase 2 (cdk2) and cyclin D or E it might represent a negative regulator of cell-cycle G1 progression (Sherr and Roberts, 1995).

The tyrosine kinase pp60^{src} is also thought to be involved in NGF-induced neuronal differentiation. PC12 cells transfected with a temperature-sensitive src construct exhibit temperature-dependent neuronal differentiation (Brugge and Halegoua, 1991). Infection of PC12 cells with the v-src oncogene induces neurite outgrowth (Alema *et al.*, 1985; Rausch *et al.*, 1989), while intracellular injection of anti-src antibodies abolished NGF-induced differentiation (Kremer *et al.*, 1991). It is not yet clear how src protein is activated by NGF, but it might occur through a direct association of src with trk using the SH2 domain or indirectly via the SHC/GRB adaptor proteins (Keegan and Halegoua, 1993).

In conclusion, the promotion of neurotrophin-induced biological activities via trk receptors is a cooperative signal transduction process resulting from the stimulation of several different signaling pathways initiated by receptor occupancy and autophosphorylation (Greene and Kaplan, 1995). Most probably, under physiologic conditions, the neurons discriminate between neurotrophins by the intensity, duration, and specificity of trk subtype receptor substrate phosphorylation in order to generate a defined biological response.

SELECTIVE INHIBITION BY K-252 COMPOUNDS OF NEUROTROPHIN ACTIONS RESULTS FROM ATTENUATION OF TRK RECEPTOR TYROSINE KINASE ACTIVITY

The PC12 cells, a clonal population arising from a pheochromocytoma tumor of the rat adrenal medulla, have proven to be a most useful and increasingly popular model for the study of the actions of NGF (Greene and Tischler, 1976). While not requiring NGF for survival in serum-containing media, PC12 cells undergo an observable morphological differentiation under its influence. At nanomolar concentrations, NGF stops the cells from dividing and induces the outgrowth of neurites which elongate and gradually form electrically excitable networks (Guroff, 1984). This is a differentiation process of several weeks which can be described as the *in vitro* conversion of a chromaffin cell into a sympathetic neuron.

Using the PC12 cell system, it was shown that treatment with K-252a at 200 nM prevents neurite generation initiated by NGF (Koizumi *et al.*, 1988; Hashimoto, 1988). Since that time, inhibition of NGF-induced differentiation by K-252a, K-252b, and staurosporine has also been described in different neuronal systems

Table 5.4   K-252 compounds inhibit NGF-induced morphological differentiation

Compound	Concentration (nM)	Neuronal System	Activity	Reference
K-252a	200	PC12 cells	neurite generation	Koizumi *et al.*, 1988
"	10–100	"	"	Tapley *et al.*,1992
"	30	"	neurite outgrowth, regeneration	Ferrari *et al.*, 1991
"	100–1000	Neuroblastoma SY5Y cells	neurite outgrowth and cellular disaggregation	Matsumoto *et al.*, 1995
"	100	NIH 3T3 cells expressing trk A, B, or C	phenotype transforming activity	Tapley *et al.*, 1992
K-252b	100–300	PC12 cells	neurite outgrowth	"
Staurosporine	2–10	"	"	Campbell and Neet, 1995
"	1–10	PC12h subclone	"	Hashimoto and Hagino, 1989

(Table 5.4). Now it is clear that 10–200 nM K-252a, 1 μM K-252b, or 1–10 nM staurosporine block neurotrophin-induced differentiation, expressed as neurite outgrowth in different pheochromocytoma or neuroblastoma clones (Table 5.4). This inhibition is very selective since neurite outgrowth induced by fibroblast growth factor is not affected by K-252a (Koizumi *et al.*, 1988).

NGF actions on PC12 cells can be classified into transcription-dependent and transcription-independent processes. Both types of actions are selectively inhibited by K-252 compounds (Table 5.5). For example, the induction of ornithine decarboxylase and the increase in choline acetyltransferase activity of NGF is blocked (Table 5.5), but comparable enzyme activity induced by fibroblast or epidermal growth factors in parallel cultures is not inhibited. K-252 compounds also selectively block many NGF-induced effects, including the increase in intracellular calcium, the stimulation of the ras signal transduction pathway, the increase in phosphatidylinositol turnover, alterations in phosphorylation events, and the increase in the transcription of the c-fos and erg1/TIS8 genes (Table 5.5). The selective inhibition by K-252 compounds of the effects of other neurotrophins, such as BDNF or NT-3, but not those of basic fibroblast growth factor has also been found on cholinergic neurons of the rat basal forebrain developing in culture (Knusel and Hefti, 1992).

The highly selective inhibition of neurotrophin actions on neuronal cultures (Table 5.4) is in striking contrast to the nonselective inhibitory action of K-252 compounds on different protein kinases in cell-free systems (Table 5.1 and 5.2).

Table 5.5   Selective inhibition of NGF actions by K-252 compounds

Effect	*Neuronal System*	*Compound*	*Concentration (nM)*	*Reference*
Increase in ornithine decarboxylase activity	PC12	K-252a	200	Koizumi *et al.*, 1988
Increased phosphatidyl-inositol breakdown	PC12	K-252a	200	"
Phospho-lipase Cγl tyrosine phosphorylation	PC12	K-252a Staurosporine	1000 1000	Omichi *et al.*, 1992
Increase of choline acetyl-transferase activity	Rat basal forebrain neurons	K-252a K-252b Staurosporine	10–200 1500 100	Knusel and Hefti, 1991
Stimulation of tyrosine hydroxylase phosphorylation	PC12h	K-252a Staurosporine	300 100	Hashimoto, 1988 Hashimoto and Hagino 1989
NGF-induced heterodown regulation of EGF receptors	PC12	K-252a	200	Koizumi *et al.*, 1988
Induction of protein methylation	PC12	K-252a	200	Kujubu *et al.*, 1993
Increase in intracellular calcium	PC12	K-252a	50–200	Lazarovici *et al.*, 1989
ADP-ribosylation of 22 kD rho-like protein	PC12	K-252a	200	Takahashi and Guroff, 1993
p21[ras] activation	PC12	K-252a	1000	Qui and Green, 1991
Activation of protein kinase N	PC12	K-252a	200	Ferrari *et al.*, 1991

Table 5.5 *Continued*

Effect	Neuronal System	Compound	Concentration (nM)	Reference
Decreased phosphorylation of elongation factor 2	PC12 PC12h	K-252a Staurosporine	200 1–10	Koizumi *et al.*, 1988 Hashimoto and Hagino, 1989
Stimulation of MAP kinase	Chick sympathetic and sensory ganglia neurons	K-252a	1000	Klinz and Heumann 1995
	PC12	K-252a	10–30	Smith *et al.*, 1989
Increased c-fos transcription	PC12	K-252a	200	Lazarovici *et al.*, 1989
		K-252a Staurosporine	30 5–100	Ferrari *et al.*, 1991 Campbell and Neet, 1995
Induction of erg-1/TIS8 transcription	PC12	K-252a	200	Kujubu *et al.*, 1993

The finding that K-252 derivatives inhibit all neurotrophin actions tested so far suggests that this inhibition occurs at an early step in the neurotrophin receptor signal transduction mechanism.

One of the first studies presenting K-252a as a specific inhibitor of NGF actions (Koizumi *et al.*, 1988) showed that K-252 derivatives were not acting as classical antagonists of neurotrophin receptors since they didn't affect the binding of ^{125}I-NGF to its receptors. It has been also pointed out in the early studies (Lazarovici *et al.*, 1989) that K-252a acts intracellularly, because when K-252a and NGF are added simultaneously, only partial inhibition of the short-term actions of NGF is seen; complete inhibition requires that K-252a be added somewhat before the addition of NGF.

The identification of the trk-tyrosine kinase family as neurotrophin receptors provided, for the first time, a basic understanding of the early events in the signaling pathways by which neurotrophins promote neuronal differentiation and, in addition, identified the cellular target of the K-252 compounds. K-252a and staurosporine have been shown to be selective *in vivo* inhibitors of neurotrophin-induced tyrosine phosphorylation of trk receptors (Figure 5.3) in PC12 cells (Berg *et al.*, 1992; Ohmichi *et al.*, 1992; Tapley *et al.*, 1992), neuroblastoma SY5Y cells (Matsumoto *et al.*, 1995), and chick sympathetic and sensory ganglionic neurons as well (Kling and Heumann, 1995). This inhibition was achieved at concentrations of 10–1000 nM, comparable to the concentrations required to inhibit neurotrophin actions (Tables 5.4 and 5.5). The tyrosine phosphorylation of EGF receptors, pp60[v-src], insulin receptors (Ohmichi *et al.*, 1992), basic FGF receptors (Berg *et al.*, 1992), or PDGF receptors (Tapley *et al.*, 1992) was not affected by K-252 derivatives. Using NIH 3T3 cells overexpressing trkA, trkB, or trkC a selective *in vivo* inhibition of NGF, BDNF, or NT-3-induced tyrosine phosphorylation of these gp140[trk] receptors, respectively, has also been demonstrated (Tapley *et al.*, 1992). Furthermore, as expected (Table 5.1) K-252a and K-252b also inhibit trk kinase activity *in vitro* (Berg *et al.*, 1992, Knusel *et al.*, 1992). The selective inhibitory effect of K-252 compounds on the catalytic activity of normal and oncogenic trk receptor kinases (Tapley *et al.*, 1992) represents the most dramatic illustration of specificity among the tyrosine protein kinase inhibitors available to date and establish these compounds as important pharmacological tools for the study of neurotrophin action.

TRK-INDEPENDENT NEUROTROPHIC EFFECTS OF K-252 COMPOUNDS

K-252 compounds, in addition to being selective inhibitors of trks and neurotrophin actions in different cells, have significant neurotrophic effects in the absence of neurotrophins (Table 5.6). At concentrations between 1–500 nM, K-252 derivatives induced neurite outgrowth *in vitro* from different pheochromocytoma PC12 and neuroblastoma clones, primary cultures of chromaffin cells, and explants of dorsal root ganglia (Table 5.6). In general, lower concentrations of the K-252 compounds are required to enhance neurite extensions of ganglionic neurons (ED_{50} ~10 nM) than are necessary to elicit neurite outgrowth in pheochromocytoma and neuroblastoma (ED_{50} ~50 nM) cells (Table 5.6). These concentrations are also lower than those necessary to inhibit neurotrophin actions (ED_{50} ~200 nM; Tables 5.4 and 5.5). In addition, within a wide range of concentrations, between 10 fM and 500 nM *in vitro* and about 0.1 mg/kg *in vivo*, K-252 derivatives act as neurotrophins, promoting survival of different neuronal populations in cell cultures of dissociated neurons from the central nervous system (CNS) or in the brain (Table 5.6). K-252a, K-252b, and staurosporine support the survival of spinal cord, sensory and sympathetic ganglionic neurons, and brain mesencephalic and hippocampal neurons in the absence of neurotrophins (Table 5.6). These biological effects are expressed as prolongation of survival and increased number of cultured neurons *in vitro*, increased dopamine uptake, increased choline acetyltransferase (ChAT) activity, stabilization of calcium homeostasis and other survival criteria of primary CNS neuronal cultures (Table 5.6). Furthermore, K-252 derivatives, both *in vitro* and *in vivo*, strongly protect neurons against β-amyloid, iron, ischemia, and other insults (Goodman and Mattson, 1994; Nabeshima *et al.*, 1991; Ohno *et al.*, 1991). Therefore, because K-252 derivatives mimic neurotrophin action at low concentrations, they could be considered as mixed-functional, agonist/antagonist neurotrophic compounds (Rasouly *et al.*, 1992; 1995).

The unique ability of K-252 derivatives to partially mimic neurotrophin action could be due to their binding either to the trk receptor and/or to trk substrates mediating trk receptor signal transduction. Alternatively, K-252 derivatives may bind and activate other unknown cellular systems. Indeed [³H]K-252a and [³H]staurosporine bind to specific protein sites in neuroblastoma (Knight *et al.*, 1995) and PC12 cells (Rasouly *et al.*, 1992). The identity of these binding sites is unknown, but they are not trks since K-252 binding is not competitive with that of the neurotrophins (Rasouly *et al.*, 1992), the number of binding sites is more than

Table 5.6   Neurite-promoting and survival-promoting effects of K-252 compounds

Compound	Concentration (nM)	Neuronal System	Reference
Neurite-promoting Effects			
K-252a	>10	PC12h	Hashimoto and Hagino, 1989
"	60–250	neuroblastoma SH-SY5Y	Maroney *et al.*, 1995
Staurosporine	10–100	PC12, chromaffin cells	Tischler *et al.*, 1990
"	50	PC12	Rasouly *et al.*, 1992
"	15–20	PC12	Campbell and Neet, 1995
	100	PC12D	Sano *et al.*, 1995
"	10–100	chromaffin cells	Demeneix *et al.*, 1990 Maurer and McKay, 1991
"	10–100	neuroblastoma SH-SY5Y	Shea and Beermann, 1991
"	10–100	neuroblastoma SK-N-SH	Slack and Proulx, 1990
"	10–100	neuroblastoma NB-1	Morioka *et al.*, 1985
"	1–20	dorsal root ganglia	Sano *et al.*, 1994
Survival-promoting effects			
K-252a	100–500	dorsal root, ciliary ganglia	Borasio, 1990
"	200–500	embryonic basal forebrain neurons	Knusel and Hefti, 1991
K-252a K-252b	0.01–10	hippocampal, septal, and cortical neurons	Cheng *et al.*, 1994
K-252a Staurosporine	100 0.5	embryonic spinal cord neurons	Glicksman *et al.*, 1993
K-252b	0.001–1	hippocampal neurons	Goodman and Mattson, 1994
Staurosporine	0.1 mg/kg	rat brain	Nabeshima *et al.*, 1991 Ohno *et al.*, 1991

20-fold the number of trk receptors (Knight *et al.*, 1995), and the binding reflected a specific uptake process (Rasouly *et al.*, 1992). Furthermore, direct studies both *in vivo* and *in vitro* failed to demonstrate activation of trk tyrosine phosphorylation by staurosporine (Rasouly and Lazarovici, 1994; Rasouly *et al.*, 1996) or K-252a (Tapley *et al.*, 1992).

Therefore, in order to understand the mechanism of neurotrophin-like actions of the K-252 compounds, we must attempt to explain their additional short-term or long-term actions. K-252a, for example, mimics the actions of NGF by stimulating the release of catecholamines (Nikodijevic *et al.*, 1990), calcium uptake (Nikodijevic *et al.*, 1995), and PI turnover (Isono *et al.*, 1994) in PC12 cells. K-252a induces the tyrosine phosphorylation of the focal adhesion kinase (FAK) in human neuroblastoma SH-SY5Y cells (Maroney *et al.*, 1995) and PCEN pheochromocytoma-endothelial hybrid cells (Rasouly *et al.*, 1996) and staurosporine activates the tyrosine phosphorylation of multiple protein substrates (Cheng *et al.*, 1994), such as the 145 kD protein (Rasouly and Lazarovici, 1994), and activates other kinases such as c-jun amino terminal protein kinase (JNK kinase; Xia *et al.*, 1995). Staurosporine, in contrast to the neurotrophins, inhibits phospholipase C (Bunn and Saunders, 1995), and certain protein kinase C isozymes (Mizuno *et al.*, 1993), and attenuates cytotoxic elevations in intracellular calcium levels (Goodman and Mattson, 1994). Thus, it is reasonable to suggest that, similar to the neurotrophins, K-252 compounds induce changes in specific second messengers such as calcium, IP_3, etc., leading to changes in different protein kinase activities and resulting in changes in the phosphorylation of specific proteins. It is further likely that these changes in the phosphorylation of cytoskeletal proteins mediate the neurite-promoting activity of these compounds (Rasouly *et al.*, 1994; Sadot *et al.*, 1995) and changes in the phosphorylation of nuclear proteins results in alterations in the transcription of specific genes (Tischler *et al.*, 1991) (Figure 5.2). The demonstration of inductions of specific tyrosine phosphorylation by K-252 compounds (Maroney *et al.*, 1995; Rasouly and Lazarovici, 1994; Rasouly *et al.*, 1996) suggest a mechanistic parallel between these compounds and the neurotrophins.

A systematic attempt to understand the mechanism of action of K-252 compounds requires, as a first step, determining whether these compounds, like trks, activate signal-amplifying cascades such as ras-MAPK, SNT, PLCγ, PI-3 kinase, src, etc., presented in Figure 5.2. Recent studies indicate that K-252a and staurosporine do not stimulate the phosphorylation of mitogen-activated kinase, phospholipase Cγ, PI-3 kinase, or SNT suggesting a trk-independent mechanism of action (Sano *et al.*, 1995; Maroney *et al.*, 1995; Campbell and Neet, 1995; Rasouly *et al.*, 1996). In view of the increasingly frequent discovery of new protein kinases, identification of the precise kinase targets and elucidation of the neurotrophic mechanism(s) of action of K-252 compounds remain formidable challenges.

K-252 COMPOUNDS POTENTIATE NEUROTROPHIN ACTION

At concentrations of 50–300 nM, K-252 compounds selectively potentiate neurotrophin effects on different neuronal preparations (Table 5.7). This potentiat-

Table 5.7 K-252 compounds potentiate neurotrophin effects

Compound	Concentration (nM)	Growth Factor	Specific Effect	Mechanism	Reference
K-252a	200	EGF	synergistic potentiation of EGF-induced neurite outgrowth in PC12 clones	ras pathway required	Wu *et al.*, 1993, Wu and Howard, 1995
K-252a	100–300	EGF bFGF	additive potentiation of EGF or bFGF-induced neurite outgrowth in PC12 cells	sustained activation of erks	Isono *et al.*, 1994
K-252a	200	FGF	potentiation of FGF-induced neurite outgrowth in PC12	changes in phosphorylation	Koizumi *et al.*, 1988
K-252a	300	IGF-I bFGF	synergistic increase in ChAT, survival activity in embryonic spinal cord cultures	unknown	Glicksman *et al.*, 1993
K-252b	50	NT-3	synergistic increase in ChAT activity of forebrain cholinergic neurons, synergistic survival of dorsal root ganglion neurons, synergistic outgrowth in PC12	stimulation of trk tyrosine phosphorylation	Knusel *et al.*, 1992
Staurosporine	50	EGF	synergistic upregulation of EGF receptor synthesis, potentiation of EGF induced neurite outgrowth in PC12	sustained EGF receptor and MAP kinase phosphorylation	Raffioni and Bradshaw, 1995
Staurosporine	1–50	forskolin dBcAMP	additive activation of the transcriptional activity of the VIP gene promoter in PC12	unknown	Sasaki *et al.*, 1995

Abbreviations: EGF, epidermal growth factor; bFGF, basic fibroblast growth factor; IGF, insulin-like growth factor; dBcAMP-dibutyryl cyclic AMP; ChAT, choline acetyltransferase; VIP, vasoactive intestinal polypeptide; DRG, dorsal root ganglion; PC12, pheochromocytoma cells; MAP, erk, extracellular signal relay kinases

ing effect of K-252 compounds can be either additive or synergistic. For example, K-252a synergistically or additively potentiates epidermal growth factor or basic fibroblast growth factor-induced neurite outgrowth in different pheochromocytoma PC12 clones (Table 5.7). A most impressive synergistic effect on the survival of rat basal forebrain neurons in primary culture is seen when K-252b is added with NT-3 (Table 5.7). NGF and NT-3 effectively promote the survival of these cultures, as measured by the increase in choline acetyltransferase activity, with ED_{50}s of 0.8 ng/ml and 175 ng/ml, respectively. Dose-response analysis of the effects of NT-3 and NGF, under identical conditions in the presence of 50 nM K-252b revealed a selective synergistic increase in both the potency (ED_{50} = 14.3 ng/ml) and maximal efficacy (two-fold increase) of the trophic actions of NT-3, but not those of NGF on basal forebrain cholinergic neurons (Knusel *et al.*, 1992). This synergistic effect was also observed with primary neuronal cultures from chick dorsal root ganglia from embryonic day 9 in which NGF effectively promotes neuronal survival whereas NT-3 produces only a moderate effect (Knusel, 1992). The number of neurons surviving upon treatment with a combination of NT-3 and 50 nM K-252b was found identical to the number of neurons supported by NGF alone. The same concentration of K-252b failed to influence NGF- or BDNF-mediated survival and K-252b alone did not enhance neuronal survival, indicating again the selectivity of the K-252b/NT-3 potentiation (Knusel and Hefti, 1992). Experiments on dissociated spinal cord neurons (Table 5.7) confirmed the observations made with forebrain cholinergic neurons. K-252 at 300 nM increased the cholinergic expression produced by bFGF or IGF-I on spinal cord cultures resulting in a synergistic increase (5–8-fold) in choline acetyltransferase activity. (Glicksman *et al.*, 1993). Similar potentiation by 50 nM staurosporine of EGF-induced neurite outgrowth and of EGF receptor and MAP kinase phosphorylation in PC12 cells was seen (Raffioni and Bradshaw, 1995). These findings demonstrate that the K-252 compounds selectively potentiate growth factor receptor tyrosine kinase (RTK) effects, and imply that a combination of these protein kinase inhibitors with neurotrophins or other growth factors could produce an enhancement of the signal transduction pathways.

Indeed, the potentiation of neurotrophin effects by K-252 compounds is correlated with changes in phosphorylation, requires a functional ras pathway, involves trk or EGF receptor tyrosine phosphorylation and a sustained phosphorylation activity of Erks (Table 5.7). Most probably, the additive or synergistic effects of K-252 compounds in promoting neurotrophin action indicate that these alkaloids may modulate divergent signal transduction regulatory pathways coupled to trk and/or ras pathways. The ability of staurosporine to enhance cAMP-mediated responses (Table 5.7) also suggests multiple effects of K-252 compounds on different signal transduction cross-talk pathways. We can also consider the possibility that K-252 compounds may activate a new unknown, alternative signal transduction pathway promoting survival and/or differentiation.

The ability of K-252 compounds to potentiate neurotrophin effects is an important addition to the already known neurotrophic functions of these alkaloids. These properties have increased the interest in K-252 compounds, not only in their use as tools to eludicate intracellular targets in neurotrophin signal transduction, but also as potential prototype compounds for the development of drugs for the treatment of neurodegenerative diseases.

SIGNAL TRANSDUCTION PHOSPHORYLATION THERAPY: K-252 COMPOUNDS AS PROTOTYPES OF NEUROTROPHIC DRUGS

In humans, most neuronal populations stop dividing by the second postnatal year. Consequently, any mechanically or chemically injured neuron would not be replaced and the ability of the injured neuronal circuit to regenerate would be totally dependent on the sprouting and development of neurite outgrowth from neighboring neurons. Any compound that could enhance the neurite-regenerating capacity of these adjacent neurons would result in faster restructuring of the injured neuronal circuit and reestablishment of the damaged physiological function.

Neurodegenerative diseases are characterized by the death of specific neuronal populations, many of which are known physiological targets for identified neurotrophins (Table 5.3). Since there are no effective therapies for any neurodegenerative disease, human recombinant neurotrophins are considered as potential biopharmaceutical treatments (Walsh, 1995). CNTF, IGF-I, BDNF, and GDNF are being evaluated in clinical trials for amyotrophic lateral sclerosis (ALS), a disease characterized by the degeneration of spinal and brainstem motor neurons. Chemotherapy-induced or disease-related peripheral neuropathies, characterized by the degeneration of sensory and motor neurons are also potential targets for NGF or IGF-I therapy. Alzheimer's disease is characterized by loss of the cholinergic neurons of the basal forebrain that innervate the hippocampus, the brain's memory center. Degeneration of forebrain cholinergic neurons is also associated with age-dependent decreases in learning and memory. NGF, NT-3, NT-4/5, and BDNF support survival and rescue these neurons, and improve memory and motor function in various animal models and, therefore, have been pursued as potential therapies for Alzheimer's (Kordower *et al.*, 1994; Fischer *et al.*, 1994; Neeper, 1995). Parkinson's disease, characterized by degeneration of the dopaminergic neurons of the substantia nigra also represents a condition for which neurotrophins could prove useful. NT-4/5 and BDNF are known to promote the survival of these cells *in vitro*, and GDNF may have a therapeutic role in stopping or even reversing Parkinsonism (Lindsay, 1995).

In principle, any neurotrophic compound or any strategy which will either increase the levels of neurotrophins in the central nervous system, or stimulate their activity on their cognate trk receptor (Hempstead, 1993) could enhance neuronal survival and maintenance during the process of neuronal degeneration. Unlike diseases affecting the peripheral nervous system, clinical therapy of the neurodegenerative diseases of the central nervous system is rendered complicated by the impermeability of the blood-brain barrier, which precludes the useful administration of neurotrophins to the periphery. While many animal studies involve direct injection of neurotrophins into the brain, this approach is unlikely to be adopted in human therapy. Development of new delivery approaches is a prerequisite to clinically-relevant progress in this area. It is more realistic to consider a strategy of developing small-molecule drugs as effective neurotrophin replacements (Edgington, 1993). Small molecules that act as selective trk agonists are not presently available, although an engineered chimeric neurotrophin containing the active domains of NGF, BDNF, and NT-3 which can bind the neurotrophin receptors trk A, B, and C, and p75 was recently developed (Ilag *et al.*, 1995). K-252 compounds, by virtue of their neurotrophic and neurotrophin-potentiating activities and trk receptor selectivity provide the first prototype of future neurotrophin-like drugs. Low-molecular weight lipophilic alkaloids such as K-252a, K-252b, and staurosporine may reach the brain in sufficient amounts to exert neurotrophic effects, even when administered systemically. A potential clinical danger to the use of these compounds is that they most likely affect both neuronal and nonneuronal cells and this could evoke complicated side effects due to the convergence of phosphorylation substrates in the signal-transduction cascades utilized for cellular growth, maintenance, proliferation, and differentiation.

The signal-transduction phosphorylation therapy concept, using as lead compounds the K-252 alkaloids, is being pursued by several groups. Hundreds of K-252 analogs have been prepared and are being used to explore the structure-activity correlation of these derivatives in a search for drugs that are selectively neuroprotective. Some such compounds have been found and there is a suggestion that one of these drugs works *in vivo* (Saporito *et al.*, 1995).

CONCLUSIONS

K-252 compounds, general and potent kinase inhibitors, are widely used tools in biological research on the role of protein kinases in signal transduction, cellular functions, and the pathophysiology of neoplastic and neurological disorders. The wide neuroscience research interest in these compounds is attributable to their potency and selectivity in inhibiting trk neurotrophin receptor tyrosine phosphorylation activity. They display distinct neurite-promoting and survival effects on a variety of neurons and also potentiate the stimulatory effects of NT-3 and other growth factors on different neurons in culture. Comparing the activities of K-252a, K-252b, and staurosporine it appears that the structural requirements for neurotrophin-like effects are different from those required for the inhibition of neurotrophin actions. Therefore, the synthesis of new K-252 derivatives can be expected to provide a new generation of selective and potent neuroprotective compounds to mimic, potentiate, or inhibit the signal transduction mechanism of the individual neurotrophins. Such compounds would be valuable tools with which to define and characterize the physiological function of the neurotrophins. Further studies of the neuroprotective mode of action of the K-252 compounds and related derivatives are of considerable interest because there is a therapeutic potential for small, polyaromatic, lipophilic neurotrophic molecules for the management of neural injury and neurodegenerative diseases.

REFERENCES

Alema, S., Casalbore, P., Agostini, E., and Tato, F. (1985) Differentiation of PC12 pheochromocytoma cells induced by v-src oncogene. *Nature*, **316**, 557–559.

Andrejauskas-Buchdunger, E., and Regenass, U. (1992) Differential inhibition of the epidermal growth factor, platelet-derived growth factor, and protein kinase C-mediated signal transduction pathways by the staurosporine derivative CGP-41251. *Cancer Res.*, **52**, 5353–5358.

Badwey, J.A., Erickson, R.W., and Curnutte, J.T. (1991) Staurosporine inhibits the soluble and membrane-bound protein tyrosine kinases in human neutrophils. *Biochem. Biophys. Res. Commun.*, **178**, 423–429.

Barbacid, M. (1994) The trk family of neurotrophin receptors. *J. Neurobiol.*, **25**, 1386–1403.

Barbacid, M. (1995) Neurotrophic factors and their receptors. *Curr. Opin. Cell Biol.*, **7**, 148–155.

Berg, M.M., Sternberg, D.W., Parada, L.F., and Chao, M.V. (1992) K-252a inhibits nerve growth factor-induced trk proto-oncogene tyrosine phosphorylation and kinase activity. *J. Biol. Chem.*, **267**, 13–16.

Berkemeier, L.R., Winslow, J.W., Kaplan, D.R., Nikolics, K., Goeddel, D.V., and Rosenthal, A. (1991) Neurotrophin-5: a novel neurotrophic factor that activates trk and trkB. *Neuron*, **7**, 857–866.

Berkow, R.L. (1992) Granulocyte-macrophage colony-stimulating factor induces a staurosporine-inhibitable tyrosine phosphorylation of unique neutrophil proteins. *Blood*, **79**, 2446–2454.

Borasio, G.D. (1990) Differential effects of the protein kinase inhibitor K-252a on the *in vitro* survival of chick embryonic neurons. *Neurosci. Lett.*, **108**, 207–212.

Brugge, J.S., and Halegoua, S. (1991) Induction of neurite outgrowth by v-src mimics critical aspects of nerve growth factor-induced differentiation. *Mol. Cell. Biol.*, **11**, 4739–4750.

Bunn, S.J., and Saunders, H.I. (1995) Staurosporine inhibits inositol phosphate formation in bovine adrenal medullary cells. *Eur. J. Pharmacol.*, **290**, 227–236.

Campbell, X.Z., and Neet, K.E. (1995) Hierarchical analysis of the nerve growth factor-dependent and nerve growth factor-independent differentiation signaling pathways in PC12 cells with protein kinase inhibitors. *J. Neurosci. Res.*, **42**, 207–219.

Chao, M.V. (1992) Neurotrophin receptors: a window into neuronal differentiation. *Neuron*, **9**, 583–593.

Cheng, B., Barger, S.W., and Mattson, M.P. (1994) Staurosporine, K-252a, and K-252b stabilize calcium homeostasis and promote survival of CNS neurons in the absence of glucose. *J. Neurochem.*, **62**, 1319–1329.

Cohen, P. (1989) The structure and regulation of protein phosphatases. *Annu. Rev. Biochem.*, **58**, 453–508.

Davis, S., and Yancopoulos, G.S. (1993) The molecular biology of the CNTF receptor. *Curr. Opin. Cell Biol.*, **5**, 281–285.

Demeneix, B.A., Kley, N., and Loeffler, J.P. (1990) Differentiation to a neuronal phenotype in bovine chromaffin cells is represented by protein kinase C and is not dependent on c-fos oncoproteins. *DNA Cell. Biol.*, **9**, 335–340.

Edgington, S.M. (1993) Neuronal signal transduction: will controlling phosphorylation cure disease? *Biotechnology*, **11**, 1237–1241.

Elliott, L.H., Wilkinson, S.E., Sedwich, A., Hill, H.C., Lawton, G., Davis, D.P., and Nixon, J.S. (1990) K-252a is a potent and selective inhibitor of phosphorylase kinase. *Biochem. Biophys. Res. Commun.*, **171**, 148–154.

Fabricant, R.N., DeLarco, J.E., and Todaro, G.J. (1977) Nerve growth factor receptors on human melanoma cells in culture. *Proc. Natl. Acad. Sci. USA*, **74**, 565–569.

Fallon, R.J. (1990) Staurosporine inhibits a tyrosine kinase in human hepatoma cell membranes. *Biochem. Biophys. Res. Commun.*, **170**, 1191–1196.

Fantl, W.J., Johnson, D.E., and Williams, L.T. (1993) Signalling by receptor tyrosine kinases. *Annu. Rev. Biochem.*, **62**, 453–481.

Ferrari, G., Fabris, M., Fiori, M.G., Gabellini, N., and Volonte, C. (1991) Gangliosides prevent the inhibition by K-252a of NGF responses in PC12 cells. *Brain Res. Devel. Brain Res.*, **65**, 35–42.

Fischer, W., Shirevaag, A., Wiegand, S.J., Lindsay, R.M., and Bjorklund, A. (1994) Reversal of spatial memory impairments in aged rats by nerve growth factor and neurotrophins 3 and 4/5 but not by brain-derived neurotrophic factor. *Proc. Natl. Acad. Sci. USA*, **91**, 8607–8611.

Gadbois, D.M., Hamaguchi, J.R., Swank, R.A., and Bradbury, E.M. (1992) Staurosporine is a potent inhibitor of p34^{cdc2} and p34^{cdc2}-like kinases. *Biochem. Biophys. Res. Commun.*, **184**, 80–85.

Glicksman, M.A., Prantner, J.E., Meyer, S.L., Forbes, M.E., Dasgupta, M., Lewis, M.E., and Neff, N. (1993) K-252a and staurosporine promote choline acetyltransferase activity in rat spinal cord cultures. *J. Neurochem.*, **61**, 210–221.

Gnahn, H., Hefti, F., Heumann, R., Schwab, M.E., and Thoenen, H. (1983) NGF-mediated increase in choline acetyltransferase (ChAT) in neonatal forebrain: evidence for a physiological role of NGF in the brain. *Brain Res.*, **285**, 45–52.

Gold, M.R., Sanghera, J.S., Stewart, J., and Pelech, S.L. (1992) Selective activation of p42 mitogen-activated protein (MAP) kinase in murine B lymphoma cell lines by membrane immunoglobulin cross-linking. Evidence for protein kinase C-independent and dependent mechanisms of activation. *Biochem. J.*, **287**, 269–276.

Goodman, Y., and Mattson, M.P. (1994) Staurosporine and K-252 compounds protect hippocampal neurons against amyloid β-peptide toxicity and oxidative injury. *Brain Res.*, **650**, 170–174.

Gotz, R., Koster, R., Winkler, C., Raulf, F., Lottspeich, F., Schartl, M., and Thoenen, H. (1994) Neurotrophin-6 is a new member of the nerve growth factor family. *Nature*, **372**, 266–269.

Greene, L.A., and Kaplan, D.R. (1995) Early events in neurotrophin signalling via trk and p75 receptors. *Curr. Opin. Neurobiol.*, **5**, 579–587.

Greene, L.A., and Tischler, A.S. (1976) Establishment of a noradrenergic clonal line of rat adrenal pheochromocytoma cells which respond to nerve growth factor. *Proc. Natl. Acad. Sci. USA*, **73**, 2424–2428.

Guma, A., Munoz, P., Camps, M., Testar, X., Palacin, M., and Zorzanon, A. (1992) Inhibitors such as staurosporine, H7 or polymyxin B cannot be used in skeletal muscle to prove the role of protein kinase C in insulin action. *Biosci. Rep.*, **12**, 413–424.

Guroff, G. (1984) PC12 cells as a model of neuronal differentiation. In *"Cell culture in the neurosciences"*. J. Bottenstein and G. Sato, eds. pp. 245–271, Plenum Press, New York.

Hall, M.J. (1989) Microbial product discovery in the biotech age. *Biotechnology*, **7**, 427–430.

Hashimoto, S. (1988) K-252a, a potent protein kinase inhibitor, blocks nerve growth factor-induced neurite outgrowth and changes in the phosphorylation of proteins in PC12h. *J. Cell Biol.*, **107**, 1531–1539.

Hashimoto, S., and Hagino, A. (1989) Blockage of nerve growth factor action in PC12h cells by staurosporine, a potent protein kinase inhibitor. *J. Neurochem.*, **53**, 1675–1685.

Hefti, F., Hartikka, J., Eckenstein, F., Gnahn, H., Heumann, R., and Schwab, M. (1985) Nerve growth factor increases choline acetyltransferase but not survival or fiber outgrowth of cultured septal cholinergic neurons. *Neuroscience*, **14**, 55–68.

Hempstead, B.L. (1993) Strategies for modulating trk receptor activity. *Exp. Neurol.*, **124**, 31–35.

Hempstead, B.L., Rabin S.J., Kaplan, L., Reid, S., Parada, L.P., and Kaplan, D.R. (1992) Overexpression of the trk tyrosine kinase rapidly accelerates nerve growth factor-induced differentiation. *Neuron*, **9**, 883–896.

Herbert, J.M., Soban, E., and Maffrand, J.P. (1990) Characterization of specific binding sites for ^{3}H-staurosporine on various protein kinases. *Biochem. Biophys. Res. Commun.*, **171**, 189–195.

Hohn, A., Leibrock, J., Bailey, K., and Barde, Y.A. (1990) Identification and characterization of a novel member of the nerve growth factor/brain-derived neurotrophic factor family. *Nature*, **344**, 339–341.

Hunter, T. (1991) Protein kinase classification. *Meth. Enzymol.*, **200**, 3–38.

Ilag, L.L., Curtis, R., Glass, D., Funakoshi, H., Tobkes, N.J., Ryan, T.E., Acheson, A., Lindsay, R.M., Persson, H., Yancopoulos, G.S., DiStefano, P.S., and Ibanez. C.F. (1995) Pan-neurotrophin 1: a genetically engineered neurotrophic factor displaying multiple specificities in peripheral neurons *in vitro* and *in vivo*. *Proc. Natl. Acad. Sci. USA*, **92**, 607–611.

Ip, N.Y., Ibanez, C.F., Nye, S.H., McClain, J., Jones, P.F., Gies, D.R., Belluscio, L., Le Beau, M.M., Espinosa III, R., Squinto, S.P., Persson, H., and Yancopoulos, G.D. (1992) Mammalian neurotrophin-4: Structure, chromosomal localization, tissue distribution, and receptor specificity. *Proc. Natl. Acad. Sci. USA*, **89**, 3060–3064.

Isono, F., Widmer, H.R., Hefti, F., and Knusel, B. (1994) Epidermal growth factor induces PC12 cell differentiation in the presence of the protein kinase inhibitor K-252a. *J. Neurochem.*, **63**, 1235–1245.

Jiang, H., Yamamoto, S., and Kato, R. (1992) Staurosporine, a potent protein kinase C inhibitor, augments phorbol ester caused ornithine decarboxylase induction in mouse epidermis. *Carcinogenesis*, **15**, 355–359.

Jing, S., Tapley, P., and Barbacid, M. (1992) Nerve growth factor mediates signal transduction through trk homodimer receptors. *Neuron*, **9**, 1067–1079.

Johnson, G.L., and Vaillancourt, R.R. (1994) Sequential protein kinase reactions controlling cell growth and differentiation. *Curr. Opin. Cell Biol.*, **6**, 230–238.

Kaplan, D.R., and Stephens, R.M. (1994) Neurotrophin signal transduction by the trk receptor. *J. Neurobiol.*, **25**, 1404–1417.

Keegan, K., and Halegoua, S. (1993) Signal transduction pathways in neuronal differentiation. *Curr. Opin. Neurobiol.*, **3**, 14–19.

Klinz, F.J., and Heumann, R. (1995) Time-resolved signaling pathways of nerve growth factor diverge downstream of the p140^{trk} receptor activation between chick sympathetic and dorsal root ganglion sensory neurons. *J. Neurochem.*, **65**, 1046–1053.

Knight, E., Connors, T.J., Hudkins, R., Marony, A.C., and Neff, N. (1995) Membranes of human neuroblastoma SH-SY5Y cells contain specific binding sites for ^{3}H-K-252a. *Biochem. Biophys. Res. Commun.*, **211**, 511–518.

Knusel, B., and Hefti, F. (1991) K-252b is a selective and nontoxic inhibitor of nerve growth factor action on cultured brain neurons. *J. Neurochem.*, **59**, 1987–1996.

Knusel, B., Kaplan, D.R., Winslow, J.W., Rosenthal, A., Burton, L.E., Beck, K.D., Rabin, S., Nikolics, K., and Hefti, F. (1992) K-252b selectively potentiates cellular actions and trk tyrosine phosphorylation mediated by neurotrophin-3. *J. Neurochem.*, **59**, 715–722.

Koizumi, S., Contreras, M.L., Matsuda, Y., Hama, T., Lazarovici, P., and Guroff, G. (1988) K-252a: a specific inhibitor of the action of nerve growth factor on PC12 cells. *J. Neurosci.*, **8**, 715–721.

Koizumi, S., Mutoh, T., Ryazanov, A., Rudkin, B.B., and Guroff, G. (1990) Nerve growth factor sensitive phosphorylations and the action of K-252a on PC12 cells. In "*Trophic factors and the nervous system*". L. Horrocks and N.H. Neff, eds. pp. 195–202, Raven Press, New York.

Kordower, J., Winn, S.R., Liu, Y.T., Mufson, E.J., Sladek, J.R., Hammang, J.P., Baetge, E.E., and Emerich, D.F. (1994) The aged monkey basal forebrain: rescue and sprouting of axotomized basal forebrain neurons after grafts of encapsulated cells secreting human nerve growth factor. *Proc. Natl. Acad. Sci. USA*, **91**, 10898–10902.

Kremer, N.E., DeArcangelo, G., Thomas, S.M., DeMarco, M., Brugge, J.S., and Halegoua, S. (1991) Signal transduction by nerve growth factor and fibroblast growth factor in PC12 cell requires a sequence of src and ras actions. *J. Cell Biol.*, **115**, 809–819.

Kujubu, D.A., Stimmel, J.B., Law, R.E., Herschman, H.R., and Clarke, S. (1993) Early responses of PC-12 cells to NGF and EGF: effect of K-252a and 5′-methylthioadenosine on gene expression and membrane protein methylation. *J. Neurosci. Res.*, **36**, 58–65.

Lazarovici, P., Levi, B.Z., Lelkes, P.I., Koizumi, S., Fujita, K., Matsuda, Y., Ozato, K., and Guroff, G. (1989) K-252a inhibits the increase in c-fos transcription and the increase in intracellular calcium produced by nerve growth factor in PC12 cells. *J. Neurosci. Res.*, **238**, 1–8.

Leibrock, J., Lottspeich, F., Hohn, A., Hofer, M., Hengerer, B., Masiakowski, P., Thoenen, H., and Barde, Y.A. (1989) Molecular cloning and expression of brain-derived neurotrophic factor. *Nature*, **341**, 149–152.

Levi-Montalcini, R., and Angeletti, P.U. (1968) Nerve growth factor. *Physiol. Rev.*, **48**, 534–569.

Lin, L.F., Doherty, D.H., Lile, J.D., Bektesh, S., and Collins, F. (1993) GDNF: a glial cell line-derived neurotrophic factor for midbrain dopaminergic neurons. *Science*, **260**, 1130–1132.

Lindsay, R. (1995) Neuron saving schemes. *Nature*, **373**, 289–290.

Lindsay, R.M., Altar, C.A., Cedarbaum, J.M., Hyman, C., and Wiegand, S.J. (1993) The therapeutic potential of neurotrophic factors in the treatment of Parkinson's disease. *Exp. Neurol.*, **124**, 103–118.

Lindsay, R.M., Wiegand, S.J., Altar, C.A., and DiStefano, P.S. (1994) Neurotrophic factors: from molecule to man. *Trends Neurosci.*, **17**, 182–190.

Loeb, D.M., Stephens, R.M., Copeland, T., Kaplan, D.R., and Greene, L.A. (1994) A trk point mutation affecting interaction with PLCγ abolishes NGF-promoted peripherin induction but not neurite outgrowth. *J. Biol. Chem.*, **269**, 8901–8910.

Loeb, D.M., Tsao, H., Cobb, M.H., and Greene, L.A. (1992) NGF and other growth factors induce an association between ERK1 and NGF receptor gp140^{prototrk}. *Neuron*, **9**, 1053–1065.

Maisonpierre, P.C., Belluscio, L., Squinto, S., Ip, N.Y., Furth, M.E., Lindsay, R.M., and Yancopoulos, G.D. (1990) Neurotrophin-3: a neurotrophic factor related to NGF and BDNF. *Science*, **247**, 1446–1451.

Maroney, A.C., Lipfert, L., Forbes, M.E., Glicksman, M.A., Neff, N.T., Siman, R., and Dionne, C.A. (1995) K-252a induces tyrosine phosphorylation of the focal adhesion kinase and neurite outgrowth in human neuroblastoma SH-SY5Y cells. *J. Neurochem.*, **64**, 540–549.

Matsuda, Y. (1993) Screening of bioactive substances of microbial origin based on enzyme and receptor binding assays. *Actinomycetology*, **7**, 110–118.

Matsuda, H., Coughlin, M.D., Bienenstock, J., and Denburg, J.A. (1988) Nerve growth factor promotes hemopoietic colony growth and differentiation. *Proc. Natl. Acad. Sci. USA*, **85**, 6508–6512.

Matsumoto, K., Wada, R.K., Yamashiro, J.M., Kaplan, D.R., and Thiele, C.J. (1995) Expression of brain-derived neurotrophic factor and p145^{trkB} affects survival, differentiation and invasiveness of human neuroblastoma cells. *Cancer Res.*, **55**, 1798–1806.

Maurer, J.A., and McKay, D.B. (1994) Staurosporine-induced reduction of secretory function in cultured bovine adrenal chromaffin cells. *Eur. J. Pharmacol.*, **253**, 115–121.

Mizuno, K., Saido, T.C., Ohno, S., Tamaoki, T., and Suzuki, K. (1993) Staurosporine-related compounds, K-252a and UCN-01 inhibit both cPKC and nPKC. *FEBS Lett.*, **330**, 114–116.

Morioka, H., Ishihara, M., Shibai, H., and Suzuki, T. (1985) Staurosporine induced differentiation in a human neuroblastoma cell line, NB-1. *Agric. Biol. Chem.*, **49**, 1959–1963.

Nabeshima, T., Ogawa, S., Nishimura, H., Fuji, K., Kameyama, T., and Sasaki, Y. (1991) Staurosporine facilitates recovery from the basal forebrain lesion-induced impairment of learning and deficit of cholinergic neurons in rats. *J. Pharmacol. Exp. Therap.*, **257**, 562–566.

Nagashima, K., Nakanishi, S., and Matsuda, Y. (1991) Inhibition of nerve growth factor-induced neurite outgrowth of PC12 cells by a protein kinase inhibitor which does not permeate the cell membrane. *FEBS Lett.*, **293**, 119–123.

Nairn, A.C., Hemmings, H.S., and Greengard, P. (1985) Protein kinases in the brain. *Annu. Rev. Biochem.*, **54**, 931–976.

Nakanishi, S., Matsuda, Y., Iwahashi, K., and Kase, H. (1986) K-252b, c, and d: potent inhibitors of protein kinase C from microbial origin. *J. Antibiot.*, **39**, 1066–1072.

Nakano, H., Kobayashi, E., Takahashi, I., Tamaoki, T., Kuzuu, Y., and Iba, Y. (1987) Staurosporine inhibits tyrosine specific protein kinase activity of Rous sarcoma virus transforming protein p60. *J. Antibiotics*, **40**, 706–708.

Neeper, S.A., Gomez-Pinilla, F., Choi, J., and Cotman, C. (1995) Exercise and brain neurotrophins. *Nature*, **373**, 109.

Nikodijevic, B., Aschkenasy, M., Dickens, G., Lachance, C., and Guroff, G. (1995) Characteristics of the K-252a-induced increase in calcium uptake in PC12 cells. *J. Neurosci. Res.*, **40**, 494–498.

Nikodijevic, B., Creveling, C.R., Koizumi, S., and Guroff, G. (1990) Nerve growth factor and K-252a increase catecholamine release from PC12 cells. *J. Neurosci. Res.*, **26**, 288–295.

Obermeier, A., Lammers, R., Wiesmuller, K.H., Jung, G., Schlessinger, J., and Ulrich, A. (1993) Identification of trk binding sites for SHC and phosphatidylinositol 3'-kinase and formation of a multimeric signalling complex. *J. Biol. Chem.*, **268**, 22963–22966.

Ohmichi, M., Decker, S.J., Pang, L., and Saltiel, A.R. (1992) Inhibition of the cellular actions of nerve growth factor by staurosporine and K-252a results from the attenuation of the activity of the trk tyrosine kinase. *Biochemistry*, **31**, 4034–4039.

Ohno, M., Yamamoto, T., and Watanabe, S. (1991) Effect of staurosporine, a protein kinase C inhibitor, on impairment of working memory in rats exposed to cerebral ischemia. *Eur. J. Pharmacol.*, **204**, 113–116.

Omura, S., Sasaki, Y., Iwai, Y., and Takeshima, H. (1995) Staurosporine, a potentially important gift from a microorganism. *J. Antibiotics*, **48**, 535–548.

Qui, M.S., and Green, S.H. (1991) NGF and EGF rapidly activate p21^{ras} in PC12 cells by distinct, convergent pathways involving tyrosine phosphorylation. *Neuron*, **7**, 937–946.

Rabin, S.J., Cleghon, V., and Kaplan, D.R. (1993) SNT, a differentiation-specific target of neurotrophic factor-induced tyrosine kinase activity in neurons and PC12 cells. *Mol. Cell. Biol.*, **13**, 2203–2213.

Raffioni, S., and Bradshaw, R.A. (1995) Staurosporine causes epidermal growth factor to induce differentiation in PC12 cells via receptor up-regulation. *J. Biol. Chem.*, **270**, 7568–7572.

Rasouly, D., and Lazarovici, P. (1994) Staurosporine induces tyrosine phosphorylation of a 145 kD protein but does not activate gp140^{trk} in PC12 cells. *Eur. J. Pharmacol.*, **269**, 255–264.

Rasouly, D., Lazarovici, P., and Matsuda, Y. (1995) Biochemical and pharmacological properties of K-252 microbial alkaloids. In *"The toxic action of marine and terrestrial alkaloids"* M.S. Blum, ed. pp. 161–190, Alaken, Inc. Fort Collins, CO.

Rasouly, D., Rahamim, E., Lester, D., Matsuda, Y., and Lazarovici, P. (1992) Staurosporine-induced neurite outgrowth in PC12 cells is independent of protein kinase C inhibition. *Mol. Pharmacol.*, **42**, 35–43.

Rasouly, D., Rahamim, E., Ringel, I., Ginzburg, I., Muarakata, C., Matsuda, Y., and Lazarovici, P. (1994) Neurites induced by staurosporine in PC12 cells are resistant to colchicine and express high levels of tau proteins. *Mol. Pharmacol.*, **45**, 29–35.

Rasouly, D., Shavit, D., Zunigo, R., Elejalde, R.B., Unsworth, B.R., Yayon, A., Lazarovici, P., and Lelkes, P.I. (1996) Staurosporine induces neurite outgrowth in neuronal hybrids (PC12EN) lacking NGF receptors. *J. Cell. Biochem.*, in press.

Rausch, D.M., Dickens, G., Doll. S., Fujita, K., Koizumi, S., Rudkin, B.B., Tocco, M., Eiden, L.E., and Guroff, G. (1989) Differentiation of PC12 cells with v-src: comparison with nerve growth factor. *J. Neurosci. Res.*, **24**, 49–58.

Robinson, J.M., Heyworth, P.G., and Badwey, J.A. (1990) Utility of staurosporine in uncovering differences in the signal transduction pathways for superoxide production in neutrophils. *Biochim. Biophys. Acta*, **1055**, 55–62.

Ruegg, T.V., and Burgess, M.G. (1989) Staurosporine, K-252a and UCN-01: potent but not specific inhibitors of protein kinases. *Trends Pharmacol. Sci.*, **10**, 218–220.

Sadot, E., Barg, J., Rasouly, D., Lazarovici, P., and Ginzburg, I. (1995) Short and long-term mechanisms of tau regulation in PC12 cells. *J. Cell Sci.*, **108**, 2857–2864.

Sano, M., Iwanaga, M., Fujisawa, H., Nagahama, M., and Yamazaki, Y. (1994) Staurosporine induces the outgrowth of neurites from the dorsal root ganglion of the chick embryo and PC12D cells. *Brain Res.*, **639**, 115–124.

Sano, M., Kohno, M., and Iwanaga, M. (1995) The activation and nuclear translocation of extracellular signal-regulated kinases (ERK1 and 2) appear not to be required for elongation of neurites in PC12D cells. *Brain Res.*, **688**, 213–218.

Saporito, M.S., Brown, E.R., Haun, F., Miller, M., Murakata, C., Neff, N., and Carswell, S. (1995) Comparison of the pharmacological actions of CEP1347 to nerve growth factor in rats with lesions of the nucleus basalis magnocellularis. *Soc. Neurosci. Abstr.*, **21**, 551.

Sasaki, K., Tsukada, T., Adachi, I., and Yamaguchi, K. (1995) Staurosporine potentiates cAMP-mediated promoter activity of the vasoactive intestinal polypeptide gene in rat pheochromocytoma PC12 cells. *Biochem. Biophys. Res. Commun.*, **214**, 1114–1120.

Schaller, M.S., and Parsons, J.T. (1994) Focal adhesion kinase and associated proteins. *Curr. Opin. Cell Biol.*, **6**, 705–710.

Sendtner, M., Carroll, P., Holtmann, B., Hughes, R.A., and Thoenen, H. (1994) Ciliary neurotrophic factor. *J. Neurobiol.*, **25**, 1436–1453.

Shea, T.B., and Beerman, M.L. (1991) Staurosporine-induced morphological differentiation of human neuroblastoma cells. *Cell Biol. Int. Rep.*, **15**, 161–168.

Sherr, C.J., and Roberts, J.M. (1995) Inhibitors of mammalian G1 cyclin-dependent kinases. *Genes & Development*, **9**, 1149–1163.

Slack, R.S., and Proulx, P. (1990) Effects of retinoic acid and staurosporine on the protein kinase C activity and the morphology of two related human neuroblastoma cell lines. *Biochim. Biophys. Acta*, **1053**, 89–96.

Smith, D.S., King, C.S., Pearson, E., Gittinger, C.K. and Landreth, G.E. (1989) Selective inhibition of nerve growth factor-stimulated protein kinases by K-252a and 5′-S-methyladenosine in PC12 cells. *J. Neurochem.*, **53**, 800–806.

Spacey, G.D., Bonser, R.W., Randall, R.W., and Gardall, L.G. (1990) Selectivity of protein kinase inhibitors in human intact platelets. *Cell. Signal.*, **2**, 329–338.

Stephens, R.M., Loeb, D.M., Copeland, T.D., Pawson, T., Greene, L.A., and Kaplan, D.R. (1994) Trk receptors use redundant signal transduction pathways involving SHC and PLCγ1 to mediate NGF responses. *Neuron*, **12**, 691–705.

Takahashi, H., and Guroff, G. (1993) Evidence for an indirect effect of nerve growth factor (NGF) on the ADP-ribosylation of a 22 kDa rho-like protein in PC12 cells. *Biochem. Biophys. Res. Commun.*, **190**, 1156–1162.

Tomaoki, T. (1991) Use and specificity of staurosporine, UCN-01 and calphostin C as protein kinase inhibitors. *Meth. Enzymol.*, **210**, 340–347.

Tapley, P., Lamballe, F., and Barbacid, M. (1992) K-252a is a selective inhibitor of the tyrosine protein kinase activity of the trk family of oncogenes and neurotrophin receptors. *Oncogene*, **7**, 377–381.

Thoenen, H. (1991) The changing scene of neurotrophic factors. *Trends Neurosci.*, **14**, 165–170.

Tischler, A.S., Ruzicka, L.A., and Perlman, R.L. (1990) Mimicry and inhibition of nerve growth factor effects: interactions of staurosporine, forskolin and K-252a in PC12 cells and normal rat chromaffin cells *in vitro*. *J. Neurochem.*, **55**, 1159–1165.

Tischler, A.S., Ruzicka, L.A., and Dobner, P.R. (1991) A protein kinase inhibitor staurosporine, mimics nerve growth factor induction of neurotensin/neuromedin N gene expression. *J. Biol. Chem.*, **266**, 1141–1146.

Unsicker, K., Krisch, B., Otten, U., and Thoenen, H. (1978) Nerve growth factor-induced outgrowth from isolated rat adrenal chromaffin cells: impairment by glucocorticoids. *Proc. Natl. Acad. Sci. USA*, **75**, 3498–3502.

Umezawa, H. (1972) "Enzyme inhibitors of microbial origin". University of Tokyo Press, Tokyo, Japan.

Vantini, G. (1992) The pharmacological potential of neurotrophins: a perspective. *Psychoneuroendocrinology*, **17**, 401–410.

Verme, T.B., Velarde, R.T., Cunningham, R.M., and Hootman, S.R. (1989) Effects of staurosporine on protein kinase C and amylase secretion from pancreatic acini. *Am. J. Physiol.*, **257**, G548–G553.

Walaas, S.I., and Greengard, P. (1991) Protein phosphorylation and neuronal function. *Pharmacol. Rev.*, **43**, 299–349.

Walsh, G. (1995) Nervous excitement over neurotrophic factors. *Biotechnology*, **13**, 1167–1171.

Waris, T., Rechardt, L., and Waris, P. (1973) Differentiation of neuroblastoma cells induced by nerve growth factor *in vitro*. *Experientia*, **29**, 1128–1129.

Wu, C.F., and Howard, B.D. (1995) K-252a-potentiation of EGF-induced outgrowth from PC12 cells is not mimicked or blocked by other protein kinase activators or inhibitors. *Dev. Brain Res.*, **86**, 217–226.

Wu, C.F., Zhang, M., and Howard, B.D. (1993) K-252a potentiates epidermal growth factor-induced differentiation of PC12 cells. *J. Neurosci. Res.*, **36**, 539–550.

Xia, L., Dickens, M., Raingeaud, J., Davis, R.J., and Grinberg, M. (1995) Opposing effects of ERK and JNK-p38 MAP kinases on apoptosis. *Science*, **270**, 1236–1331.

Yanagihara, N., Tachikawa, E., Izumi, F., Yasugawa, S., Yamamoto, H., and Miyamoto, E. (1991) Staurosporine: an effective inhibitor for Ca^{2+}/calmodulin dependent protein kinase 2. *J. Neurochem.*, **56**, 294–298.

6. INSECT SELECTIVE NEUROTOXINS FROM SCORPION VENOMS* AFFECTING SODIUM CONDUCTANCE

ELIAHU ZLOTKIN

*Department of Cell and Animal Biology, Life Sciences Institute,
The Hebrew University of Jerusalem, Jerusalem 91904, Israel*

ANIMAL GROUP SPECIFICITY OF VENOM NEUROTOXINS

Venom is defined as a mixture of substances which are produced in specialized glandular tissues in the body of the venomous animal and injected, with the aid of a stinging-piercing apparatus, into the body of its prey in order to paralyze it. The majority of the venomous animals (such as snakes, spiders, scorpions, venomous snails, various coelenterates) are slow, and even static predators which feed on freshly killed prey of mobile and relatively vigorous animals. The locomotory inferiority of the venomous predator is largely compensated by the neurotoxic components of his venom, the neurotoxins. The latter are able to induce a rapid paralysis of the prey at a very low range of concentrations (10^{-9}–10^{-12} M) (Zlotkin, 1988).

In this context a neurotoxin is defined as a substance which interferes with the function of excitable tissues due to a specific 'recognition,' through high affinity binding, to given sites in these tissues (Zlotkin, 1988). A very common characteristic of venom neurotoxins is their polypeptide nature. This chemical characteristic has a double advantage for the survival of the venomous animal. First, polypeptides are the most readily available structures for adaptive modifications by genetic mutations and selection. Second, polypeptides, through their high diversity of covalent structures and the resulting tridimensional conformations, may reveal a highly diverse array of functional specificities.

When classified according to effect and site of action in the nervous system, venom neurotoxins are commonly divided into ion channel toxins, which modify ion conductance, presynaptic toxins, which affect neurotransmitter release, and postsynaptic toxins which block the neurotransmitter receptors (Dolly, 1988). Each of these categories can be further classified according to more specific criteria. For example, the ion channel toxins are subdivided into sodium (such as the below-mentioned scorpion venom toxins), potassium and calcium channel toxins (Dolly, 1988; Zlotkin, 1988). Furthermore, some of the neurotoxins are able to distinguish between subtypes of ion channels such as the ω and μ venomous snail conotoxins which distinguish between the voltage-sensitive axonal and muscle calcium and sodium channels, respectively (Gray *et al.*, 1988).

---

* Abbreviations of scorpion venom toxins are presented in the caption to Figure 6.2.

Where the vital prey-predator relationships serve as a target of preference for evolutionary selective pressure additional specificities occur such as the animal group-selective toxins.

Animal group specificity refers to the phenomenon where an animal venom and/or its derived toxins reveal toxicity to a given group of organisms but are not effective to other groups of organisms. For example, the venoms of marine nemertine worms (Rhynchoceola) and sea anemones (Cnidaria) are capable of affecting a wide range of animals, but their toxicity to crustacea is due substantially to a specific group of polypeptide neurotoxins (Mr = 3–6 kDa) (Kem, 1976; Schweitz *et al.*, 1984) which do not affect other vertebrate and invertebrate animals (Zlotkin, 1988). In cases where a venomous organism feeds on a given and limited group of animals, the animal group specificity is manifested already on the level of the whole venom. For example, the venomous marine Conidae snails are subdivided according to their feeding habits into fish, mollusc and worm eaters. A series of early bioassays (Endean and Rudkin, 1963) showed that the above feeding preference is associated with the specific toxicity of their venoms aimed at the respective groups of prey animals.

An extreme example of insect specificity revealed by the entire whole venom (gland extract) is demonstrated by parasitic wasps of the genus *Bracon* (Braconidae). The braconid wasps do not fit the above mentioned 'slow predator — fast prey' dogma, since they are neither slow nor predators. The venom is employed primarily for securing hosts (moth larvae) for their offspring. The venom is exclusively effective against moth larvae, in which it induces a flaccid paralysis of long duration (weeks) as a consequence of a presynaptic blockage of release of the excitatory transmitter (glutamate) at the neuromuscular junction (Piek and Spanjer, 1986). A distinction between arthropods (affected) and mammals (not affected) has been revealed by various scorpions as well as spider whole venoms (Zlotkin, 1985). However, clear detection of insect specific factors has been essentially obtained by chemical fractionation of arthropod venoms. Thus vertical column electrophoresis of the black widow spider *(Latrodectus)* venom enabled the distinction of several protein fractions (Mr = 60–80 kD) each possessing either fast reversible, or slow progressive irreversible paralysis of flies or toxicity to guinea pigs (Frontali and Grasso, 1964). Similarly, molecular exclusion column chromatography coupled with starch gel electrophoresis enabled the isolation of factors from buthid scorpion venoms which selectively affect insects, vertebrates or crustacea (Zlotkin, 1985; 1988; Zlotkin *et al.*, 1972).

INSECT SELECTIVE NEUROTOXINS FROM SCORPION VENOM

In the past, when the study of Buthinae scorpion venoms was directed mainly by clinical and public health aspects, lethality to laboratory mice served as the exclusive method to monitor their fractionation and purification. This has resulted in the purification and chemical characterization of the so called α neurotoxins (Figure 6.2) responsible for strong toxicity to mammals and human envenomation (Rochat *et al.*, 1967; Miranda *et al.*, 1970; Zlotkin *et al.*, 1978). Zooecological considerations concerning the feeding and hunting behavior of scorpions in the field has motivated the choice of arthropods as test animals in monitoring the fractionation of scorpion venoms. Such an approach coupled with molecular exclusion column chromatography and starch gel electrophoresis enabled the isolation of factors from buthid scorpion venoms which specifically affect insects, vertebrates and crustacea (Zlotkin, 1985; 1988; Zlotkin *et al.*, 1972). The specific symptomatology revealed by blowfly larvae to the injection of various fractions of scorpion venom enabled the distinction among two major categories of neurotoxins which affect insects: The first are the excitatory insect selective neurotoxins which induce in blowfly larvae an immediate fast and reversible contraction paralysis (Figure 6.1A). The second are the depressant insect selective neurotoxins which cause a slow progressively developing flaccid paralysis (Figure 6.1B). It is noteworthy that the selective-exclusive neurotoxicity to insects of the excitatory and depressant toxins was established on three experimental levels: the intact animal, the neuromuscular preparation and binding assays to various neuronal and non-neuronal membranes (Zlotkin, 1988).

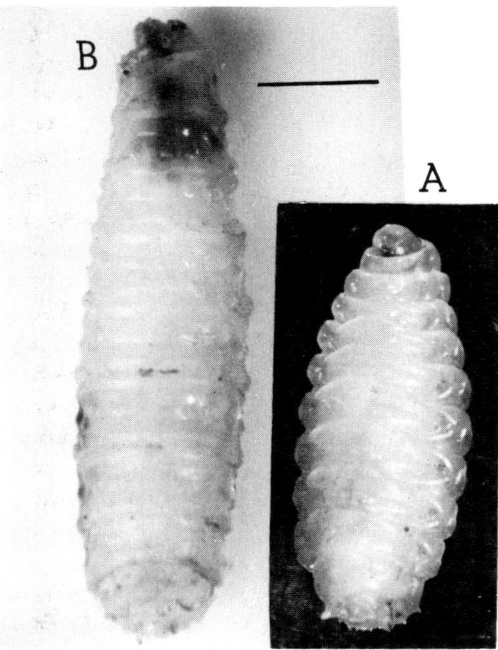

Figure 6.1 Responses of *Sarcophaga falculata* blowfly larva to the injection of various scorpion venoms and their derived insect toxins. **A.** A fast, immediate (seconds) and reversible contraction paralysis induced by excitatory insect selective neurotoxins such as AaIT and LqqIT1. **B.** Typical progressively developing (minutes) flaccid-extended paralysis induced by depressant insect selective toxins such as BjIT2, LqqIT2 and LqhIT2. Bar corresponds to 3.8 mm. Taken from Zlotkin, 1983.

```
                    AMINO ACID SEQUENCES

         1         2         3         4         5         6         7
         0         0         0         0         0         0         0
LqhIT2 ...DGYIKRR DGCKVACLIG NEG.CDKECK AYGG.SYGYC ...WTWGLAC WCEGLPDDET WK..SETNTC G
LqqIT2 ...DGYIRKR DGCKLSCLFG NEG.CNKECK SYGG.SYGYC ...WTWGLAC WCEGLPDEKT WK..SETNTC G
BjIT2  ...DGYIRKK DGCKVSCIIG NEG CRKECV AHGG.SFGYC ...WTWGLAC WCENLPDAVT WK..SSTNTC G
AaIT   .KKNGYAVDS SGKAPECLLS N..YCNNQCT KVHYADKGYC CLL.....SC YCFGLNDDKK VLEISDTRKS YCU'TIIN
LqqIT1 .KKNGYAVDS SGKAPECLLS N..YCYNECT KVHYADKGYC CLL.....SC YCNGLSDDKK VLEISDARKK YCDFVTIN
LqhaIT .VRDAYIAKN YNCVYEC.FR DA.YCNELCT KNGASS.GYC QWAGKYGNAC WCYALPDNVP IR...VPGKC R
Lqq4   GVRDAYIADD KNCVYTC.GS NS.YCNTECT KNGAES.GYC QWLGKYGNAC WCIKLPDKVP IR...IPGKC R
Lqq5   .LKDGYIVDD KNCTFFC.GR NA.YCNDECK KKGGES.GYC QWASPYGNAC WCYKLPDRVS IK...EKGRC N
AaH2   .VKDGYIVDD VNCTYFC.GR NA.YCNEECT KLKGES.GYC QWASPYGNAC YCYKLPDHVR TK...GPGRC H
Ts7    ..KEGYLMDH EGCKLSCFIR PSGYCGRECG .IKKGSSGYC AWP.....AC YCYGLPNMVK VWDRA.TNKC
```

PERCENT IDENTICAL RESIDUES

		LqhIT2	LqqIT2	BjIT2	AaIT	LqqIT1	LqhaIT	Lqq4	Lqq5	AaH2	Ts7
depressant	- LqhIT2	100	87	79	30	31	34	38	43	38	39
depressant	- LqqIT2		100	79	30	30	39	38	44	39	41
depressant	- BjIT2			100	25	28	38	38	43	39	41
excitatory	- AaIT				100	87	27	29	33	34	31
excitatory	- LqqIT1					100	25	29	33	34	33
α-insect	- LqhaIT						100	75	56	59	38
α-mammal	- Lqq4							100	83	83	38
α-mammal	- Lqq5								100	78	43
α-mammal	- AaH2									100	44
β-toxin	- Ts7										100

Figure 6.2 Comparison of scorpion toxin amino acid sequences. The depressant insect selective toxins affecting insects, namely LqhIT2, LqqIT2, BjIT2 are compared to excitatory toxins (AaIT; LqqIT1). As shown the α toxin affecting insects (LqhαIT) closely resembles the α toxins affecting mammals (Lqq4, Laa5 and AaH2). Ts7 is a β-toxin isolated from a South American scorpion. Abbreviation of scorpion venom nomenclature: Aa — the North African scorpion *Androctonus australis Hector*, Bj — the Israeli black scorpion *Buthotus judaicus*, Lqh — the Israeli yellow scorpion *Leiurus quinquestriatus hebraeus*; Lqq — the African scorpion *Leiurus quinquestriatus quinquestriatus*; Ts — the Brazilian scorpion *Tityus serrulatus*. Taken from Zlotkin *et al.*, 1991.

Scorpion toxins were isolated and purified by the aid of various methods of low and high pressure column chromatography and characterized by electrophoresis, amino acid composition and sequence analyses (Eitan *et al.*, 1990; Rochat *et al.*, 1979). Figure 6.2 demonstrates the primary structures of representatives of the three groups of toxins derived from the venom of Buthinae scorpions and, for comparative purposes, one toxin (Ts7) which represents the so-called β toxins derived from scorpions from Central and South America.

As shown (Figure 6.2), scorpion toxins are single chained polypeptides of about 65–70 amino acids, including 8 cysteins of similar allocation involved in disulphide bridges (Eitan *et al.*, 1990; Rochat *et al.*, 1979).

THE INSECT SELECTIVE NEUROTOXINS AFFECT SODIUM CONDUCTANCE

The above (Figure 6.1) and other symptomatological observations suggested that scorpion toxins are neurotoxic and paralyze the insects by affecting their skeletal musculature. Studies on various neuromuscular preparations (Walther *et al.*, 1976; Zlotkin *et al.*, 1991; Tintpulver *et al.*, 1976; Ruhland *et al.*, 1977; Rathmayer *et al.*, 1976) revealed that the insects skeletal muscles are indirectly affected through a presynaptic action (Figure 6.3A) on the peripheral branches of the motor nerves. This aspect is clearly exemplified in Figure 6.3A revealing a synchrony between the axonal (action potentials) and muscular (junction potentials) activities induced by the AaIT. In other words, the train of the junction potentials is caused by the repetitive firing of the motor axon (Walther *et al.*, 1976) induced by the excitatory toxin.

It has been recently shown (Zlotkin *et al.*, 1991) that the depressant insect selective neurotoxins induce in blowfly larvae a short transient phase of contraction preceding the prolonged flaccid paralysis. A study on a prepupal housefly neuromuscular preparation revealed that both phenomena, transient contraction and the prolonged flaccidity, follow from a presynaptic effect of the toxin. The effect of the depressant LqhIT2 toxin on a prepupal housefly neuromuscular preparation mimics the effects of the intact animal; i.e., a brief period of repetitive bursts of junction potentials is followed by suppression of their amplitude and finally by a block of neuromuscular transmission (Figure 6.3). Loose patch clamp recordings indicate that the repetitive activity has a presynaptic origin in the motor nerve and closely resembles the effect of the excitatory toxin AaIT. The final synaptic block is attributed to neuronal membrane depolarization, which results in an increase in spontaneous transmitter release, this effect is not induced by excitatory toxin (Zlotkin *et al.*, 1991).

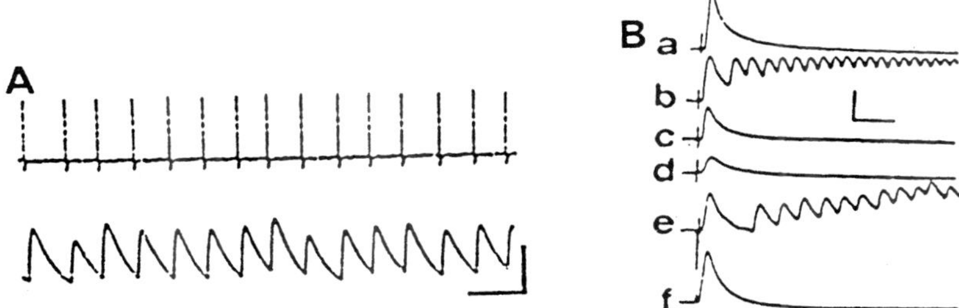

Figure 6.3 Neuromuscular effects of the excitatory (A) and depressant (B) insect selective neurotoxins. **A.** The excitatory toxin AaIT (125 nM) was applied to the locust hindleg *extensor tibiae* preparation. Simultaneous recording form the motor nerve (upper trace, showing a train of action potentials) and the muscle (lower trace, showing a train of junction potentials produced by an activated muscle). Calibration: 50 msec; upper 15 mV, lower 10 mV. Taken from Walther *et al.*, 1976. **B.** The depressant insect toxin LqhIT2 (40 nM) causes a brief phase of repetitive motor neuron activity (b) followed by gradual suppression (b, c, d) of excitatory junction potentials (a) in *prepupal M. domestica.* Upon removal of toxin with a saline wash, repetitive activity reappears again only briefly (trace e) and the amplitude of the EJP partially recovers over a period of 30–60 min (trace f). Vertical calibration is 20 mV; horizontal calibration is 30 ms. Taken from Zlotkin *et al.*, 1991.

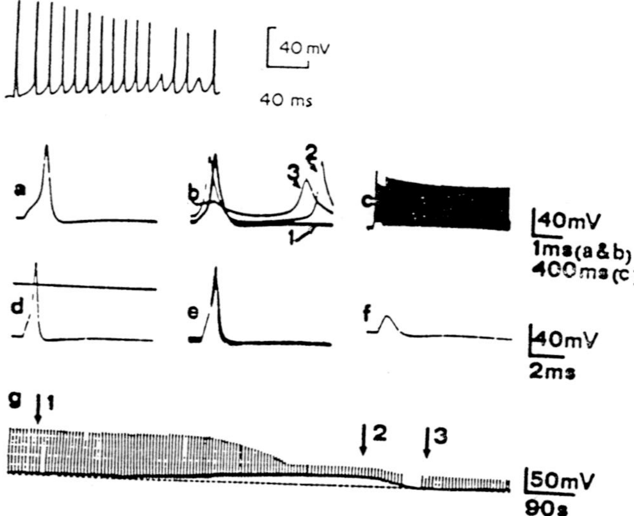

Figure 6.4 The effect of the excitatory (AaIT, BjIT1) and depressant (BjIT2) toxins on the action potentials recorded from the isolated axon of the cockroach. Upper trace: Spontaneous repetitive firing induced by 1.3 μM AaIT. Taken from Pelhate and Zlotkin (1981). (**a–c**) Action of BjIT1 (8.5 μM): (**a**) control action potentials; (b) after 8 min of superfusion with BjIT1 the three superimposed records revealing depolarization and repetitive activity. (**c**) Burst of repetitive activity of action potentials. (**d–g**) Action of BjIT2 (15 μM). (d) control action potential; (**e**) 2 min of superfusion with the toxin resulted in 5 mV depolarization accompanied by a progressive reduction in the amplitude of the action potential; (**f**) 2 min later an additional depolarization of 20 mV and a complete block of the evoked responses were obtained; (**g**) continuous slow sweep recording indicating the gradual depolarization and blockage of the action potential induced by BjIT2 (applied at arrow 1). The depolarization was abolished by saxitoxin (applied at arrow 2) but has slowly reappeared after the removal of the saxitoxin by washing (applied at arrow 3). Taken from Pelhate and Zlotkin 1981 and Lester *et al.*, 1982.

To summarize, the neuromuscular studies have pointed out that the insect neuronal membrane is the target of both the excitatory and the depressant insect selective toxins. This conclusion, which was also supported by binding studies (Teitelbaum *et al.*, 1979; Zlotkin *et al.*, 1979) and microscopical autoradiography (Fishman *et al.*, 1991), has directed our attention to an isolated insect axon as an experimental preparation (Pichon *et al.*, 1967) to study the mode of action of the scorpion toxins.

The isolated giant axon from the central nervous system of the cockroach *Periplaneta americana* in both current and voltage clamp experiments, using a double oil-gap, single fiber technique (Pichon *et al.*, 1967) was employed. Figure 6.4 presents a current clamp experiment which demonstrates the effects of the excitatory and depressant toxin on the action potential of the cockroach giant axon. The excitatory toxin induces a small depolarization (Figure 6.4b) followed by either induced (Figure 6.4c) or spontaneous (Figure 6.4, upper trace) repetitive firing. The depressant toxin (BjIT2, Figure 6.4d–g) induces an obvious depolarization accompanied by a reduction of the action potential amplitude (Figure 6.4d, e) up to a complete blockage (Figure 6.4f). As shown (Figure 6.4g) the depolarization is prevented, in a reversible manner, by saxitoxin, indicating that it follows from an effect on sodium conductance. This depolarization, however, is not the cause or at least not the only cause for the action potential blockage (see below).

Figure 6.5 presents voltage clamp experiments which demonstrate the effects of the excitatory (AaIT) and depressant (LqhIT2) toxins on the sodium currents in

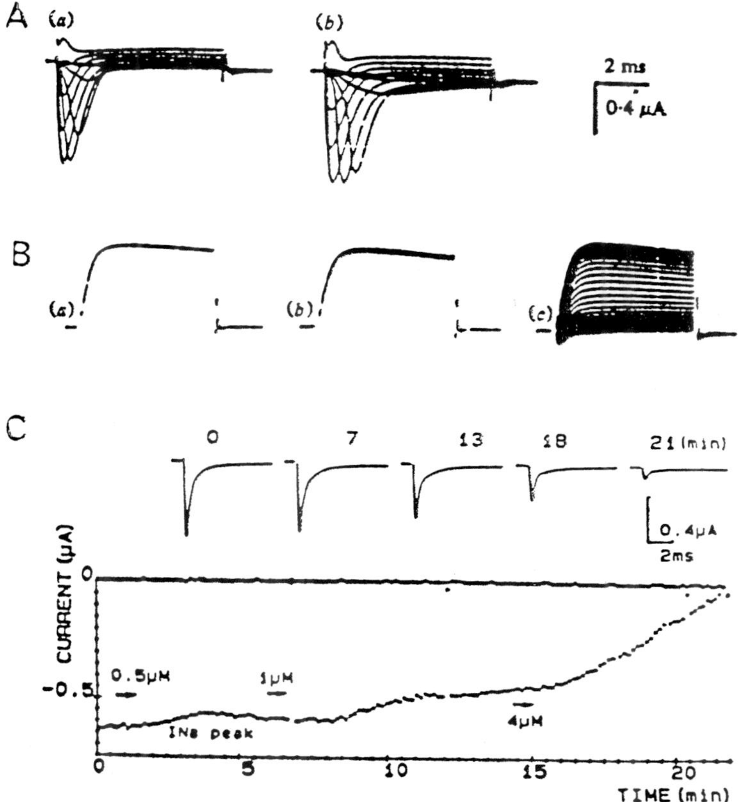

Figure 6.5 The effects of the excitatory (AaIT, A,B) and depressant (LqqIT2, C) insect selective toxins on sodium conductance in the cockroach axonal preparation under voltage-clamp conditions. **A.** AaIT (1.3 μM) increases the sodium peak current and slows its inactivation (b) when compared to control, before toxin (a) recording. **B.** AaIT does not affect potassium currents. (a) before toxin; (b) 3.3 μM AaIT; (c) K current suppressed by 4-AP. **C.** Depressant toxin (0.5–4 μM) progressively suppresses sodium current as revealed by their direct recording (upper trace) and the plotting of peak current against time (bottom). Taken from Pelhate and Zlotkin 1982 and Zlotkin *et al.*, 1985.

the giant cockroach axon. The excitatory toxin does not effect the potassium current (Figure 6.5B). However, as revealed in Figure 6.5A, AaIT increases the sodium peak currents and slows their inactivation process (Figure 6.5Ab) in a voltage dependent manner (not shown). AaIT slowed the Na$^+$ current turnoff most effectively at low negative values of applied voltage and increased by 18% (S.D. ± 3, n = 6) the peak sodium current (Pelhate and Zlotkin, 1981; 1982). It was concluded (Pelhate and Zlotkin, 1981; 1982) that the repetitive activity induced by AaIT results from the voltage dependent modulation of sodium inactivation coupled with an increase in sodium permeability. Figure 6.5C reveals that the depressant toxin blocks the sodium current in a dose and time dependent manner. Thus, the blockage of the action potential (Figure 6.4f) by the depressant toxin is a consequence of two effects on sodium conductance. Firstly, an increase in the resting sodium permeability, responsible for the depolarization (Figure 6.4c, f) and secondly a suppression of the activatable sodium permeability (Figure 6.5).

THE VERTEBRATE AND INSECT VOLTAGE GATED SODIUM CHANNELS

The voltage-sensitive sodium channels are integral membrane proteins responsible for the generation of action potential in excitable cells. Sodium channels isolated in functional form contain a large α subunit with an apparent Mr of 260,000 (Catterall 1986; 1988). In the rat brain, a subunits are associated noncovalently with a β1 subunit (mr = 36,000) and are disulfide linked to a β2 subunit with an Mr of 33,000, which can be removed upon reduction without loss of functional activity. The purified eel electroplax sodium channel is functionally active as a single a subunit (Catterall, 1986; 1988).

Sodium channel polypeptides have been purified from various vertebrate excitable tissues and their distinct α subunit primary structures have been elucidated by cloning and sequence analysis of the cDNA (Catterall, 1988; Gordon, 1990). These sequence determinations reveal homology among the various vertebrate channels. In each case, the cDNA encode a large polypeptide of about 2,000 amino acids containing four conserved repeated domains. The amino acid sequence analysis reveals that each homologous domain contains six or eight (Catterall, 1986; Guy and Conti, 1990) transmembrane segments per domain, connected by internally as well as externally located segments of amino acid sequences (Figure 6.7). A high degree of conservation is present in a short internal segment linking homology domains III and IV, which is suggested to play an important role in sodium channel inactivation (Catterall, 1988; Gordon, 1990). Site-directed antibodies raised against a synthetic peptide (SP 19) which corresponds to the above conserved segment (anti-SP 19 antibodies) were shown to identify sodium channel α subunits in a wide range of vertebrate excitable tissues (Gordon *et al.*, 1988). The externally located amino acid segments are suggested to participate in formation of the binding sites of polypeptide neurotoxins (Thomsen and Catterall, 1989).

As a critical element in excitability, sodium channels serve as specific targets for various neurotoxins included in animal venoms. These toxins were shown to

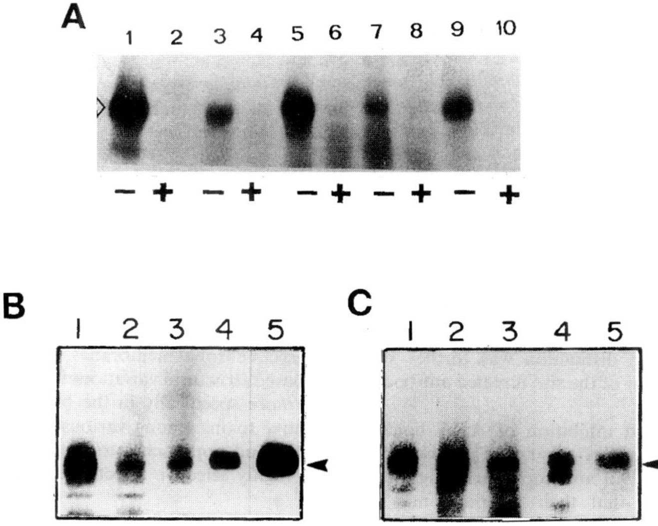

Figure 6.6 Identification of insect channel polypeptides by immunoprecipitation and phosphorylation using site directed antibodies prepared against specific segments of the rat brain sodium channel. Synaptosomal membranes from rat brain and the CNS of various insects were solubilized, immunoprecipitated, radiophosphorylated, resolved by SDS-PAGE and visualized by authoradiography. **A.** The anti-SP19 (1491–1508) antibody is able to recognize the sodium channel proteins from rat brain (1), locust (3), cockroach (5), fly heads (7) and moth larvae (9). (−) or (+) indicate the absence or presence, respectively, of 1 μM of the synthetic polypeptide to block the specific antibody, thus proving its specificity. **B** and **C**: Sodium channel polypeptides of the cockroach (B) and fly head (C) were identified and immunoprecipitated by the various antibodies (Figure 6.7) 382–400, 1429–1449, 1686–1703, 1729–1748 and 1491–1508 (anti SP19) in lanes 1,2,3,4, and 5 respectively. Arrow points to the migration position of the sodium channel polypeptide with a Mr = 260,000 Da. Taken from Gordon *et al.*, 1992 and Moskowitz *et al.*, 1994.

occupy at least four different receptor sites related to the unique functionality of the sodium channel and to serve as valuable tools for its identification and functional-structural characterization (Catterall, 1986). Among them the α and β scorpion venom neurotoxins are well established markers of the voltage dependent sodium channels. The former slow sodium channel inactivation and bind to the receptor site 3 and the latter modify activation and bind to the receptor site 4. It is noteworthy that the binding site of the α toxin on the rat brain sodium channel was recently localized to segments in domain I and IV by a combined use of sodium channel site-directed antibodies and photoaffinity labeling (Thomsen and Catterall, 1989).

The close similarity among the vertebrate and insect sodium channels has been demonstrated by the following data:

1. The electrical activity of insect nerves and its ionic basis are in accordance with the established information of vertebrate neurophysiology (Pichon, 1984).

2. The neuronal sodium conductance in insects is affected by the same blockers (such as tetrodotoxin and saxitoxin) and modifiers (such as the α scorpion and sea anemone toxins, veratridine and non-selective insecticides) as in the vertebrate systems (Pelhate and Sattelle, 1982).

3. In their gross chemical properties sodium channel polypeptides derived from the central nervous systems of insects resemble their vertebrate counterparts by (a) possessing the similar molecular weight (240–280 KDa), (b) being glyco-proteins and (c) serving as a substrate of phosphorylation by cAMP-dependent protein kinase (Gordon *et al.*, 1990; Moskowitz *et al.*, 1991 and Figure 6.6).

4. Sodium channels from locust, cockroach and fly were identified, through immuno-precipitation and radiophosphorylation (Figure 6.6) by certain site directed antibodies which were raised against specific external as well as internal segments of the amino acid sequence of the rat sodium channel (Figure 6.7 Gordon *et al.*, 1992; Moskowitz *et al.*, 1994).

5. In *Drosophila* the para gene has been shown to encode a functional voltage dependent sodium channel (Ganetzky, 1986; O'Dowd *et al.*, 1989) and the amino acid sequence within the four homologous domains (I-IV) of rat brain sodium channels is conserved in the deduced amino acid sequence of the insect sodium channel (39%, 66%, 58% and 62% identity, respectively (Loughney *et al.*, 1989). These transmembrane segments of the conserved homologous domains were shown (by site-directed mutagenesis and expression studies) to be involved in the voltage dependent activation of sodium channels and cation transport (Stuhmer, 1991; Catterall, 1991; Lopez-Barneo, 1993).

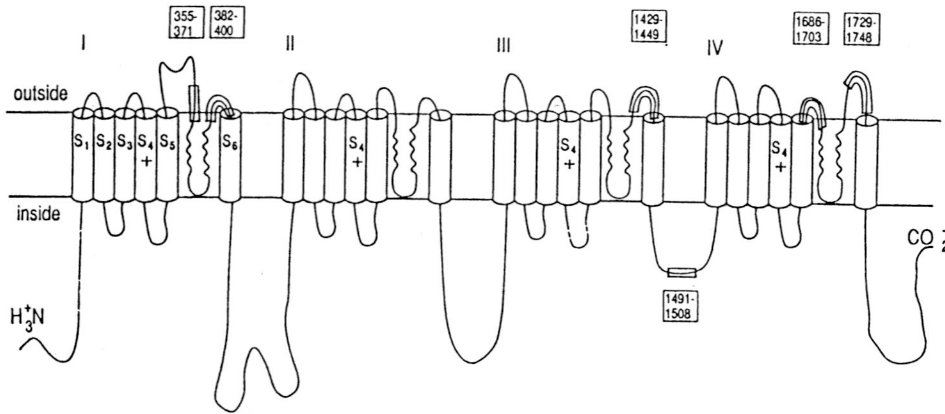

Figure 6.7   Two-dimensional model of the proposed transmembrane folding of a sodium channel α subunit depicting the sites of interaction of the site-directed antibodies (wide boxes). The framed numbers indicate the various site-directed antibodies used. Ab 1491–1508 corresponds to anti SP 19 antibody and represents part of the inactivation site. Taken from Gordon *et al.*, 1992.

With this background of the physiological, pharmacological, immunological and chemical similarities among the insect and vertebrate sodium channels, it appears that the only distinction among them is supplied by the fact that the former, in contrast to the latter, are affected by the scorpion venom derived excitatory and depressant insect selective toxins. Our next question is whether the insect sodium channel is the direct and primary target of the above insect selective neurotoxins.

THE INSECT SODIUM CHANNELS AS THE TARGET OF INSECT SELECTIVE NEUROTOXINS

Several lines of indirect evidence suggest that the insect selective toxins are directed to the insect neuronal sodium channels.

(1) The excitatory (AaIT) as well as the depressant (LqqIT2) insect toxins exclusively affect the sodium conductance of insect neuronal membranes, as demonstrated by voltage clamp studies with an isolated single insect axon (Figure 6.4 and 6.5).
(2) It was shown that the insect neuronal membranes possess high affinity binding sites to STX (K_D = 0.1–0.5 nM) (Figure 6.2B) and TTX, the well-known universal blockers of voltage sensitive sodium channels. The number of receptor sites for AaIT and STX was shown to be practically identical using the same insect synaptosomal membrane preparation (Figure 6.8, Gordon *et al.*, 1985).
(3) There is a similarity in binding properties between the insect selective toxin AaIT (Figure 6.9) and the β scorpion toxins affecting mammals (Figure 6.2), the well-known markers of sodium channels. The binding of both groups of toxins is not dependent on membrane potential and is not affected by veratridine (in contrast to the α scorpion toxin binding [Gordon *et al.*, 1984; Jover *et al.*, 1980]).
(4) The excitatory insect toxin AaIT (De Lima *et al.*, 1986; 1989) as well as the depressant LqhIT2 (unpublished results) were shown to be competitively displaced from their binding sites in insect neuronal membranes by the β scorpion toxin Ts7. The latter was shown to be toxic to mice as well as to insects and to compete with other β scorpion toxins in their mammalian sodium channel binding sites (De Lima *et al.*, 1986; 1989).
(5) Finally, it was shown that the AaIT toxin did not reveal high affinity specific binding to rat brain synaptosomes (Gordon *et al.*, 1984).

The direct evidence, namely the demonstration that the insect sodium channel is the binding receptor for the insect selective neurotoxins, was achieved by a study (Gordon *et al.*, 1992) which was aimed at clarifyiing the curious phenomenon of mutual displaceability among the excitatory and depressant toxins (Figure 6.9).

The data presented in Figure 6.9 indicates that

(1) An excitatory toxin (AaIT) is displaced with high affinity by both the excitatory and the depressant (BjIT2, LqIT2) toxins, with high affinity (Figure 6.9 A, B).

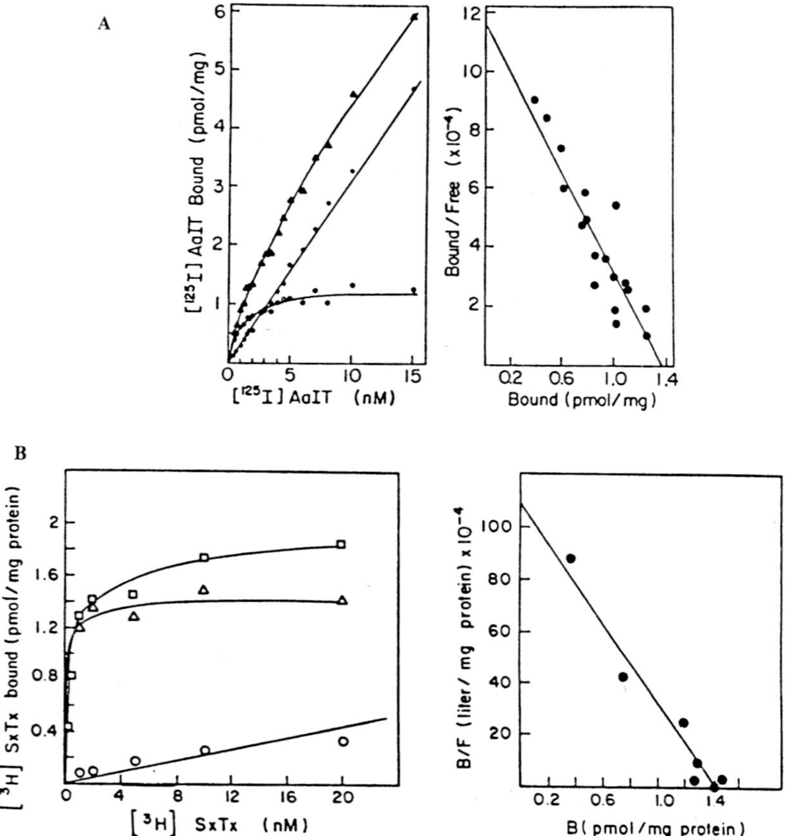

Figure 6.8   The binding of $[^{125}I]$AaIT (A) and $[^3H]$STX (B) to locust neuronal membrane preparation. Saturation equilibrium assays (left) and the respective Scatchard analyses (right) are presented, to determine the K_D and B_{max}. **A:** $[^{125}I]$AaIT binding yielded a K_D = 1.19 nM and B_{max} = 1.37 pmol/mg of membrane protein. **B:** $[^3H]$STX binding yielded a K_D = 0.14 nM and B_{max} = 1.43 pmol/mg protein. Taken from Gordon *et al.*, 1984, 1985.

(2) That the depressant toxin LqhIT2 possesses two binding sites in the locust neuronal membranes: a high affinity and low capacity sites (K_{D1} = 0.9 ± 0.6 nM, B_{max1} = 0.1 ± 0.07 pmol/mg of protein) and a second low affinity and high capacity site (K_{D2} = 185 ± 13 nM, B_{max2} = 10.0 ± 0.6 pmol/mg of protein) (Figure 6.9C). The nature of the low affinity and high capacity LqhIT2 binding sites is presently unknown. The above binding constants, however, can not be related to ion channels.

(3) The excitatory toxin, AaIT, displaces the radioidinated depressant toxin only from its high affinity sites (Figure 6.9D, E).

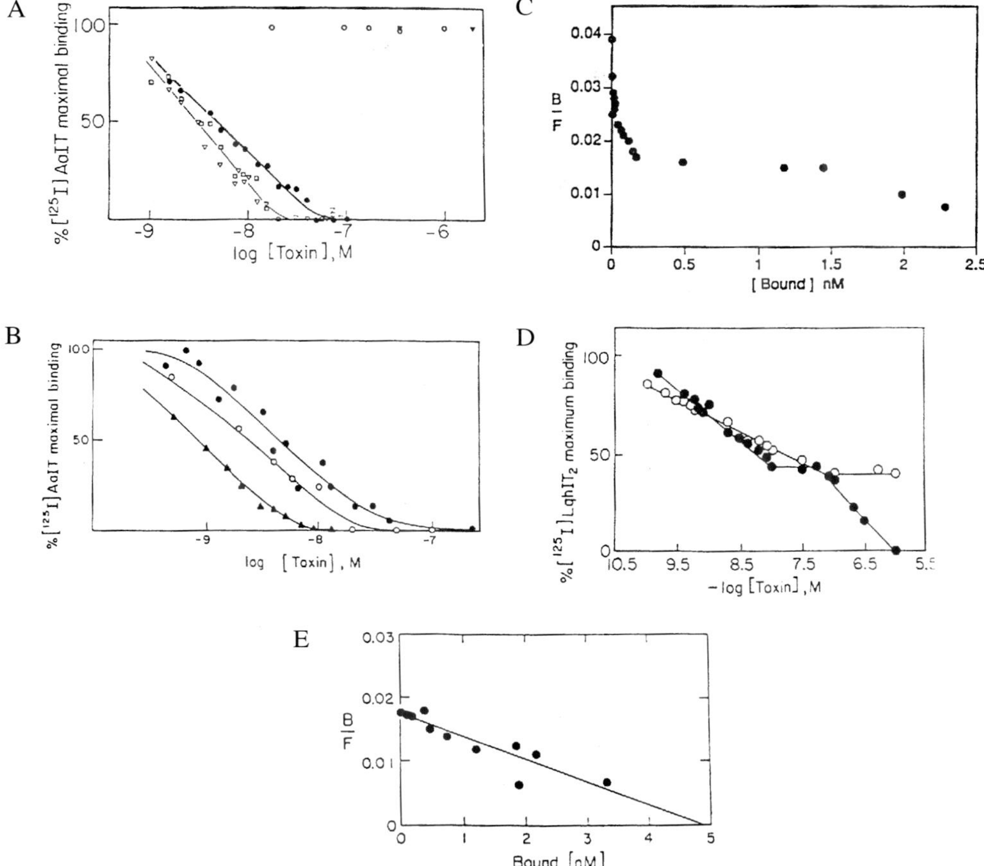

Figure 6.9 Mutual binding displaceability among the excitatory and depressant insect selective neurotoxins. **A, B:** Displaceability of the [^{125}I]AaIT (1.5 nM) by the excitatory and depressant insect toxins: **(A)** AaIT ($\square$); BjIT1 ($\bullet$); BjIT2 ($\triangledown$); AaH2 ($\blacktriangledown$); C. *soffusus* β-mammal toxin 2 ($\bigcirc$); **(B)** AaIT ($\bigcirc$); LqqIT1 ($\blacktriangle$); LqqIT2 ($\bullet$). **(C)** Scatchard analysis of a saturation curve of the depressant toxin LqhIT2 binding to locust neuronal membranes. Scatchard plot analysis yielded the best fit (P < 0.05) using a two binding site model. The K_D values, obtained for three separate experiments, are K_{D1} = 0.9 ± 0.6 nM, B_{max1} = 0.1 ± 0.07 pmol/mg of protein and K_{D2} = 185 ± 13 nM, B_{max2} = 10.0 ± 0.6 pmol/mg of protein. The total concentration of binding sties (R_T) used in the Hill plot was determined from Scatchard analysis nH = 0.996. **D, E:** Displacement of [^{125}I]LqhIT2 binding by LqhIT2 and AaIT from locust neuronal membranes. **(D)** AaIT, the excitatory toxin ($\bigcirc$), displaces the depressant toxin [^{125}I]LqhIT2 (72 pM) only from its high affinity low capacity sites, in contrast to the LqhIT2 ($\bullet$). The figure presents plots of [^{125}I]LqhIT2 bound as a function of log toxin concentration. **(E)** In presence of the unlabeled excitatory AaIT toxin (0.5 μM), the depressant [^{125}I]LqhIT2 toxin reveals only the second low affinity high capacity binding site with binding constants of K_D = 194 ± 36 nM and B_{max} = 13.8 ± 6.9 pmol/mg of protein. Taken from Gordon *et al.*, 1984, 1992.

We have employed sodium channel site-directed antibodies, shown to recognize the insect sodium channel (Figure 6.6), in order to clarify the phenomenon of the mutual displaceability among the insect selective neurotoxins. The specific question was whether the two categories of insect selective toxins, which differ in their induced symptomatology (Figure 6.1), effects on sodium conductance (Figures 6.3 –6.5), and primary structure (Figure 6.2), compete for an *identical* binding site, located on the insect sodium channel.

LOCALIZATION OF RECEPTOR SITES FOR THE INSECT SELECTIVE NEUROTOXINS ON SODIUM CHANNELS BY SITE DIRECTED ANTIBODIES

To probe the binding sites of the depressant and excitatory toxins, we employed five site-directed antibodies that recognize different extracellular regions of the rat brain RII sodium channel α subunit corresponding to amino acid residues 355–371, 382–400, 1429–1449, 1686–1703 and 1729–1748 (Gordon, 1990) (Figure 6.7). The previously described anti-SP-19 antibody, corresponding to the intracellular sequence 1491–1508 (Gordon *et al.*, 1988; Vassilev *et al.*, 1988) was also employed. These sequences resemble the corresponding amino acid segments in the sodium channel polypeptide from *Drosoophila para* locus (55) having 67, 79, 50, 56, 40 and 83% identity, respectively. Preincubation of the locust neuronal membrane with four out of the five externally directed antibodies (Ab) inhibited [^{125}I]LqhIT2 binding in a concentration-dependent manner. Maximum inhibition of about 50% was obtained with Ab 382–400 (55 ± 17%), Ab 1429–1449 (56 ± 18%), and Ab 1729–1748 (42 ± 20%). Ab 355–371 inhibits [^{125}I]LqhIT2 binding by only 23 ± 5% (data not shown, and Figure 6.11).

There was no correlation between the ability of the various site-directed antibodies to recognize the insect sodium channel polypeptide by the immunoprecipitation-radiophosphorylation method (Figure 6.6, and their ability to inhibit toxin binding to insect neuronal membranes. For example, Ab 382–400 significantly inhibited toxin binding (Figure 6.11) and recognized the sodium channel polypeptide by the immunoprecipitation-phosphorylation method (data not shown). Ab 1686–1703, on the other hand, did not reveal any significant inhibition of toxin binding to the locust neuronal membranes (Figure 6.12) but immunoprecipitated the sodium channel polypeptide similarly to Ab 1491–1508 (SP 19) (data not shown). These results are consistent with specific inhibition of toxin binding by a subset of our antibodies that block the toxin receptor site. It is noteworthy that similar results concerning the immunoprecipitation and inhibition of LqhIT2 binding by site-directed antibodies were obtained also using the fly head neuronal preparation (Moskowitz *et al.*, 1992). The SP 19 antibody was previously shown to specifically recognize sodium channel polypeptides in various insects (Figure 6.6).

The inhibitory effect of the site-directed antibodies on the binding of [^{125}I]LqhIT2 may result in a reduction in number of toxin binding sites or a decrease in the toxins' affinity. Figure 6.10 shows Scatchard analysis of saturation curves of [^{125}I]LqhIT2 binding to locust neuronal membranes with and without Ab 382–400. No significant change in K_D values was seen (Figure 6.10). The toxin was used in a low concentration range in order to occupy mainly the high-affinity sites of LqhIT2. Under these conditions, in the presence of a half-maximal concentration of Ab 382–400, about 25% reduction in the number of binding sites was obtained (Figure 6.10).

The competitive interaction observed between the two different insect toxins (Figure 6.9) could be interpreted as either competition at an identical binding site, steric interference between adjacent binding sites, or distant allosteric interactions. In the first possibility, the various site-directed antibodies would be expected to inhibit the binding of the two ^{125}I-labeled toxins to the same extent. Figure 6.11 presents antibody-mediated inhibition of [^{125}I]LqhIT2 and [^{125}I]AaIT binding to insect neuronal membranes. The membranes were preincubated in the presence of maximal amounts of each Ab, and the binding of the two toxins was measured under similar conditions. Antibodies 382–400, 1429–1449 and 1729–1748, which cause over 50% inhibition of [^{125}I]LqhIT2 binding, were shown to inhibit [^{125}I]AaIT binding by less than 20%. Antibody 355–371, which inhibits the depressant toxin by 23%, did not reveal any significant effect on the binding of either toxin (Figure 6.11A).

The data presented in Figure 6.11B demonstrated that the combined application of two Abs, each effective in toxin binding inhibition, was not additive. These results suggest that (a) each antibody was given in a saturable amount, and (b) the partial inhibition of toxin binding (Figure 6.11) indicates that all the antibodies used identify and bind to the same population of sodium channels. Otherwise,

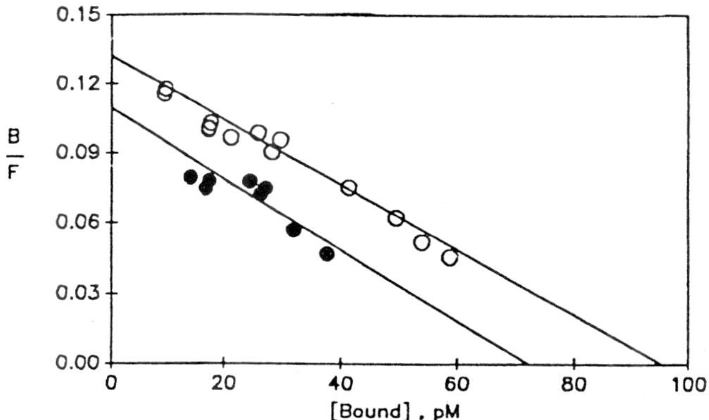

Figure 6.10 Scatchard plots of saturation curves of [^{125}I]LqhIT2 binding to locust neuronal membranes in the presence or absence of site-directed antibody 382–400 indicating mainly a reduction in the toxin's binding capacity. The K_D and B_{max} values obtained in controls (○) were 0.72 nM and 0.21 pmol/mg protein respectively, and in the presence of Ab (○) 0.67 nM and 0.16 pmol/mg protein, respectively. Two separate experiments yielded similar results. Taken from Gordon *et al.*, 1992.

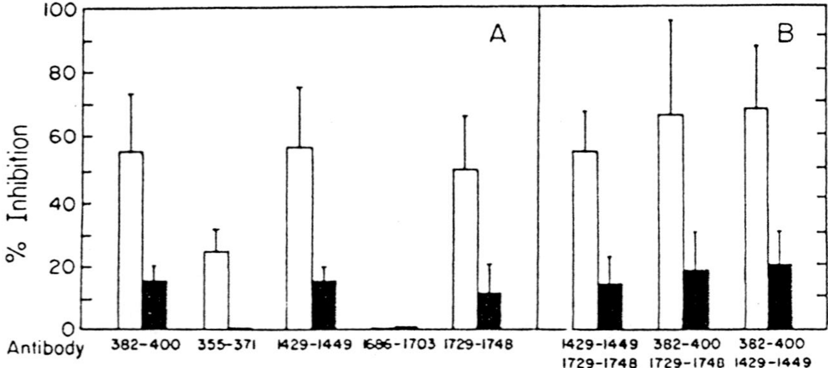

Figure 6.11   Inhibition of [^{125}I]LqhIT2 and [^{125}I]AaIT binding by the site-directed antibodies. Locust neuronal membranes (0.5 mg/ml) were preincubated with 20, 30, and 40 μl of nonimmune antibodies (control, 100% binding) or 20 μl of the nonimmune antibodies together with 10 μl of each of the indicated site-directed antibodies (**A**). In the combined applications of two site-directed antibodies, each antibody was applied in 10 μl in the presence of 20 μl of nonimmune antibodies (**B**). Each bar represents the mean ± SD of three to five independent experiments, where triplicate determinations were made for each antibody. The concentrations of labeled toxins used were 224.7 ± 12.3 pM for [^{125}I]LqhIT2 (open bars) and 249.6 ± 80.4 pM for [^{125}I]AaIT (solid bars). Taken from Gordon et al., 1992.

the combined presence of two effective antibodies would result in an additive effect (Figure 6.11B). It is noteworthy that the possibility of steric hindrance in the binding of the combined antibodies to a single sodium channel is not ruled out by our data. However, it would not alter our suggested interpretation of Figure 6.11B. The main conclusion, however, demonstrated in Figure 6.11 is the ability of the various site-directed Abs to quantitatively differentiate between the binding of the LqhIT2 and AaIT. This indicates that their binding sites are not identical although located on the same insect sodium channel. This conclusion is strongly supported by the below mentioned (see Figure 6.12) information revealing an "asymmetrical" binding interaction in reciprocal assays with the two toxins with different insect neuronal preparations.

To summarize, site-directed antibodies corresponding to conserved putative extracellular segments of sodium channels, coupled with binding studies of radiolabeled insect-selective scorpion neurotoxins, were employed to clarify the relationship between the toxin's receptor sites and the insect sodium channel. The binding of LqhIT2 was significantly inhibited in a dose-dependent manner by each of four site-directed antibodies. As exemplified with Ab 382–400 (Figure 6.10), the binding inhibition resulted from reduction in the number of binding sites. The antibody-mediated inhibition of [^{125}I]AaIT binding differs from that of LqhIT2: three out of the four antibodies which maximally inhibited LqhIT2 binding only partially affected AaIT binding. Two antibodies, one corresponding to extracellular and one to intracellular segments of the channel, did not affect the binding of either toxin.

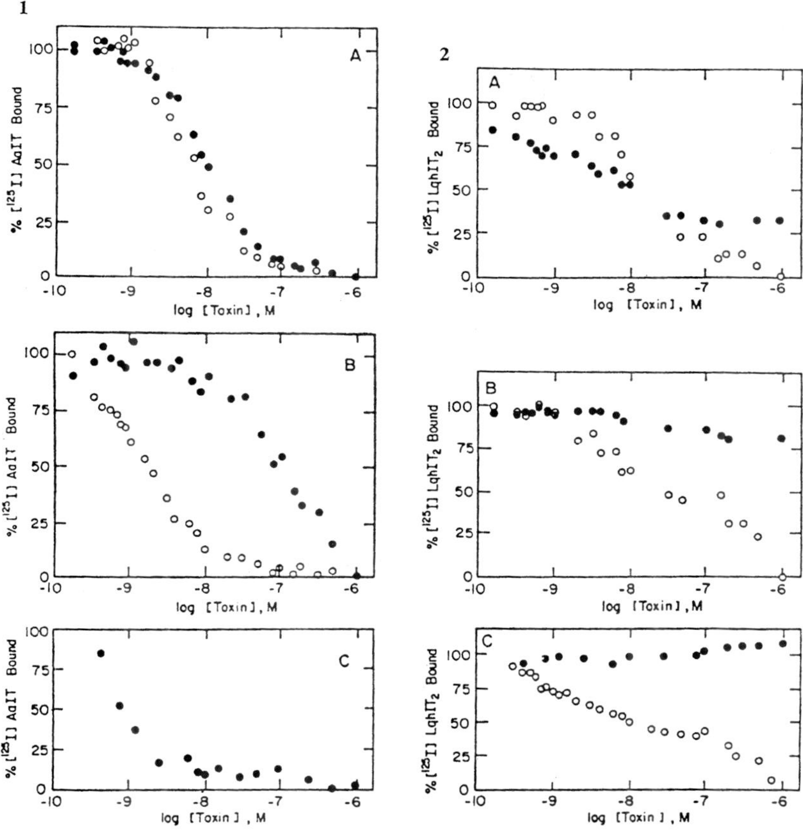

Figure 6.12 Reciprocal competitive displacement assays of the excitatory and depressant toxins. 1. Displacement of [^{125}I]AaIT binding by AaIT and LqhIT2. 2. Displacement of [^{125}I]LqhIT2 binding by LqhIT2 and AaIT. Neuronal membranes of *P. americana* (A) or *Sarcophaga falculata* (B) or *Spodoptera littoralis* (C) were incubated for 60 min in 22°C in the presence of [^{125}I]AaIT (1) or [^{125}I]LqhIT2 (2) and increasing concentrations of AaIT or LqhIT2. Non-specific binding, determined in the presence of 1 μM unlabeled toxin, was subtracted. In (1) O AaIT, ● LqhIT2; in (2) O LqhIT2, ● AaIT. Taken from Moskowitz *et al.*, 1994.

These data suggest that the receptors to the depressant and excitatory insect toxins (a) comprise an integral part of the insect sodium channel, (b) are formed by segments of external loops in domains I, III and IV of the sodium channel, and (c) are localized in close proximity but are not identical in spite of the competitive interaction between these toxins. It is noteworthy that the binding site of an alpha scorpion toxin affecting mammals corresponds to the extracellular loops between the transmembrane segments S5 and S6 in domains I and IV of the rat brain sodium channel (Thomsen and Catterall, 1989; Tejedor and Catterall, 1989).

The present results are in concert with the previously suggested concept of multipoint attachment of scorpion toxins. This concept was based on chemical modification of selected amino acid residues, localized on various regions of the primary structure of the α scorpion toxins (Kharrat *et al.*, 1989) as well as the insect toxins (Loret *et al.*, 1990). Thus, it may be suggested that the different regions of the toxin essential for its attachment are complemented by several points of attachment in the receptor molecule, all simultaneously required to carry out the high-affinity binding reaction. This hypothesis, however, deserves further study.

The requirement of several segments of the sodium channel for toxin binding may reasonably explain the phenomenon of binding competition among chemically and pharmacologically distinct toxins such as the insect and β scorpion toxins. Thus, a partial overlap, even in one of the several attachment points of the various toxins to the receptor, may be sufficient to inhibit toxin binding. This notion is supported by the recent finding of a new scorpion toxin, AaHIT4 (Loret *et al.*, 1991), which has a low homology with other scorpion toxins. AaHIT4, however, displaces both α- and β-toxins, which bind to distinct receptor sites (Catterall, 1980; Cahalan, 1975) from rat brain sodium channels and also displaces the insect toxin AaIT from insect sodium channels.

VARIABILITY AMONG INSECT SODIUM CHANNELS REVEALED BY NEUROTOXINS

With this background of the mutual displaceability of the excitatory and depressant toxins in the locust neuronal membrane and their binding constants (Figure 6.10), neuronal preparations of several other insects were studied (Moskowitz *et al.*, 1994) and it has been shown:

(1) Similarly to locust neuronal membranes the membranes of cockroaches, flies and lepidopterous larvae possessed a single class of binding sites for the excitatory AaIT toxin and two classes of high and low affinities and low and high capacities respectively for the depressant LqhIT2 toxin (Moskowitz *et al.*, 1994).
(2) However, as shown in Figure 6.12, a mutual displaceability between the AaIT and LqhIT2 toxin occurred only in the cockroach preparation but not in the fly head and the *Spodoptera* larvae preparations.

The asymmetry observed in the mutual displacement of the excitatory and depressant toxins (Figure 6.12) can be attributed to the existence of several sodium channel subtypes in neuronal membranes of various insects. This consideration is supported by the notion that the mutual displacement between the depressant the excitatory toxins represents a steric interference between adjacent binding sites, as we suggested previously (Gordon *et al.*, 1992). These putative sodium channel subtypes may differ the spatial location or relative orientation of the binding sites to each of the two toxins. Accordingly, we presume the occurrence of at least three subtypes: the fist, expressed in locust and in cockroach neuronal membranes, in which a mutual steric hindrance between the toxins occurs; the second, expressed in neuronal membranes of lepidopterous larvae, in which no displacement of the depressant toxin by the excitatory toxin takes place; the third subtype of channels, expressed in fly neuronal membranes, in which a decreased ability of the excitatory toxin to interfere with the binding of the depressant toxin is observed (Figure 6.12).

Our data suggest that different insect orders (or even species and developmental stages) express different and possibly variable sodium channels in their nervous system. Such differences and variability may be detected by the selective toxins as well as by the sodium channel site-directed antibodies. Elucidation of the structural basis of the unique pharmacology of insect sodium channels is an important area of future study.

SUMMARY AND CONCLUSIONS

1. The insect voltage gated sodium channels reveal obvious electrophysiological, pharmacological, immunological and chemical-structural similarities to their vertebrate counterparts.
2. The occurrence of the excitatory and depressant insect selective neurotoxins from Buthinae scorpion venoms supplies, so far, the strongest distinction between the vertebrate and insect sodium channels.
3. These insect selective toxins exclusively affect the insect sodium channel through their binding to their receptors which comprise an integral part of the insect sodium channel. The latter are formed by segments of external loops in domains I, III and IV of the sodium channels, and thus form a multi-site attachment to their receptors.
4. We propose that the competitive interaction between the depressant and excitatory insect toxins is a consequence of a partial overlap in their points of attachment in the external segments of the insect sodium channel. Thus, the two groups of toxins may have closely localized, but not identical, binding sites. It may be further suggested that the insect selectivity of the insect toxins may correspond to the recognition of minor modifications in the insect sodium channel when compared to its vertebrate counterpart, in one or more of the several points of attachment required for toxin binding. The difference may be expressed on the amino acid sequence and/or conformational levels. The structural basis of insect selectivity on the level of the toxins as well as their receptors deserves clarification on the molecular level.
5. The attachment sites of the insect selective toxins to the insect sodium channel should represent functionally critical sites which can be targeted by newly designed selective insecticides. The insect selective neurotoxins may serve as valuable tools for the study of the pharmacology of insect neuronal membranes. The understanding of the proposed structural variations in insect

sodium channels, and more specifically in the binding sites of the insect selective toxins among various insects, may suggest new targets and approaches to the design and screening of new highly selective insecticides.

To summarize, the insect selective toxins are able to distinguish between the nervous system of an insect and a non-insect and to identify in the former a functionally critical target. As such, they can serve as research models for the design of insect selective insecticides. This approach may result in one or a combination of the following: their mimicry by synthetic non-peptide substances; the design of modified, metabolically stable neurotoxin polypeptides with oral toxicity, followed by cloning of their genes for the design of transgenic, insect-protected plants; the insertion and association of the insect-toxin genes into microorganisms (such as *Bacillus thuringiensis*) and/or viruses (such as baculoviruses) able to penetrate and inoculate within the insect body and produce the insect toxins there. The feasibility of these approaches was recently exemplified by significantly increasing the rate at which a baculovirus kills an insect pest due to the incorporation of toxin-producing genes from an insectivorous venomous mite (Tomalski and Miller, 1991) or scorpion (Stewart *et al.*, 1991). The latter aspect, of a genetically improved baculovirus bioinsecticide expressing the AaIT toxin, has already reached the, technologically, advanced stage of field trials (Cory *et al.*, 1994).

Acknowledgments

The recent studies presented in this review were supported by grant 90–00186 from the US-Israel Binational Science Foundation, Grant IS-1982–91 from the US-Israel Binational Agricultural Research and Development Fund (BARD), Joint French Israeli Research Grant No. 4403192, and grant 93–294 from the US-Israel Binational Science Foundation.

REFERENCES

Cahalan, M.D. (1975) Modification of sodium gating in frog myelinated nerve fibres by *Centruroides sculpturatus* scorpion venom. *J. Physiol. (Lond.)*, **244**, 511–534.

Catterall, W.A. (1980) Neurotoxins that act on voltage sensitive sodium channels in excitable membranes. *Ann. Rev. Pharmacol. Toxicol*, **20**, 15–43.

Catterall, W.A. (1986) Molecular properties of voltage-sensitive sodium channels. *Ann. Rev. Biochem.*, **55**, 953–985.

Catterall, W.A. (1988) Sstructure and function of voltage sensitive ionic channels. *Science*, **242**, 50–61.

Catterall, W.A. (1991) Structure and function of voltage gated sodium and calcium channels. *Curr. Opin. Neurobiol.*, **1**, 5–13.

Cory, J.S., Hirst, M.L., Williams, T., Halls, R.S., Gouson, D., Green, B.M., Carty, T.M., Possee, R.D., Cayley, P.J. and Bishop, D.H.L. (1994) Field trial of a genetically improved baculovirus insecticide. *Nature (Lond.)*, **370**, 138–140.

De Lima, M.E., Martin, M.F., Diniz, C.R. and Rochat, H. (1986) *Tityus serrulatus* toxin VII bears pharmacological properties of both β-toxin and insect toxin from scorpion venoms. *Biochem. Biophys. Res. Commun.*, **139**, 296–302.

De Lima, M.E., Martin-Eauclaire, M.F., Hue, B., Loret, E., Diniz, C.R. and Rochat, H. (1989) On the binding of two scorpion toxins to the central nervous system of the cockroach *Periplaneta americana*. *Insect Biochem.*, **19**, 413–422.

Dolly, J.E. (Ed.) (1988) *Neurotoxins in Neurochemistry*. Ellis Horwood ltd., Chichester, UK.

Eitan, M., Fowler, E., Herrmann, R., Duval, A., Pelhate, M. and Zlotkin, E. (1990) A scorpion venom neurotoxin paralytic to insects which affects sodium current inactivation: Purification, primary structure and mode of action. *Biochemistry*, **29**, 5941–5947.

Endean, R. and Rudkin, C. (1963) Studies of the venoms of some conidae. *Toxicon*, **1**, 49–64.

Fishman, L., Kagan, M.L. and Zlotkin, E. (1991) Accessibility of the insect nervous system to a neurotoxic polypeptide. *J. Exp. Zool.*, **257**, 10–23.

Frontali, N. and Grasso, A. (1964) Separation of three toxicologically different protein components from the venom of the spider *Latrodectus tredecimgnttatus*. *Arch. Biochem. Biophys.*, **106**, 213–218.

Ganetzky, B. (1986) Neurogenetic analysis of *Drosophila* mutations affecting sodium channels: synergistic effects on viability and nerve conduction in double mutants involving tip E. *J. Neurogenet.*, **3**, 19–31.

Gordon, D. (1990) Ionic channels in nerve and muscle cells. *Curr. Opin. Cell Biol.*, **2**, 695–707.

Gordon, D., Merrick, D., Wollner, D.A. and Catterall, W.A. (1988) Biochemical properties of sodium channels in a wide range of excitable tissues studied with site directed antibodies. *Biochemistry*, **27**, 7032–7038.

Gordon, D., Moskowitz, H. and Zlotkin, E. (1990) Sodium channel polypeptides in central nevous systems of various insects identified with site directed antibodies. *Biochim. Biophys. Acta*, **1026**, 80–86.

Gordon, D., Moskowitz, H., Eitan, M., Warner, C., Catterall, W.A. and Zlotkin, E. (1992) Localization of receptor sites for insect selective toxins on sodium channels by site directed antibodies. *Biochemistry*, **31**, 7622–7628.

Gordon, D., Jover, E., Couraud, F. and Zlotkins, E. (1984) The binding of an insct selective neurotoxin (AaIT) from scorpion venom to locust synaptosomal membranes. *Biochim. Biophys. Acta*, **778**, 349–358.

Gordon, D., Zlotkin, E. and Catterall, W.A. (1985) The binding of and insect selective neurotoxin and saxitoxin to insect neuronal membranes. *Biochim. Biophys. Acta*, **821**, 130–136.

Gray, W.R., Olivera, B.M. and Cruz, L.J. (1988) Peptide toxins from venomous *Conus* snails. *Ann. Rev. Biochem.*, **57**, 665–700.

Guy, H.R. and Conti, F. (1990) Pursuing the structure and function of voltage gated channels. *Trends Neurosci.*, **13**, 201–206.

Jover, E., Couraud, F. and Rochat, H. (1980) Two types of scorpion neurotoxins characterized by their binding to two separate receptor sites on rat brain synaptosomes. *Biochem. Biophys. Res. Commun.*, **95**, 1607–1614.

Kem, W.R. (1976) Purification and characterization of a new family of polypeptide neurotoxins from the Heteronemertine *Cerebratulus lacteus* (Leidy). *J. Biol. Chem.*, **251**, 4184–4192.

Kharrat, R., Darbon, H., Rochat, H. and Granier, C. (1989) Structure-activity relationships of scorpion α-toxins. *Eur. J. Biochem.*, **181**, 381–390.

Lopez-Barneo, J., Hoshi, T., Heinemann, S.H. and Aldrich, R.W. (1993) Effects of external cations and mutations in the pore region on C-type inactivation of shaker potassium channels. *Receptors and Channels*, **1**, 61–71.

Loret, E.P., Mansuelle, P., Rochat, H. and Granier, C. (1990) Neurotoxins active on insects: amino acid sequences, chemical modifications and secondary structure estimation by circular dichroism of toxins from the scorpion *Androctonus australis* Hector. *Biochemistry*, **29**, 1492–1501.

Loret, E.P., Martin-Eauclaire, M.F., Mansuelle, P., Sampieri, F., Granier, C. and Rochat, H. (1991) An anti-insect toxin purified from the scorpion *Androctomus australis* Hector also acts on the α and β sites of the mammalian sodium channel. Sequence and circular dichroism study. *Biochemistry*, **30**, 633–640.

Loughney, K., Kreber, R. and Ganetzky, B. (1989) Molecular analysis of the para locus, a sodium channel gene in *Drosophila*. *Cell*, **58**, 1143–1154.

Miranda, F., Kopeyan, C., Rochat, C., Rochat, H. and Lissitzky, S. (1970) Purification of animal neurotoxins. Isolation and characterization of eleven neurotoxins from the venom of the scorpions *Androctonus australis* Hector, *Buthus occitanus tunetanus* and *Leiurus quinquestriatus*. *Eur. J. Biochem.*, **16**, 514–523.

Moskowitz, H., Herrmann, R., Zlotkin, E. and Gordon, D. (1994) Variability among insect sodium channels revealed by selective neurotoxins. *Insect Biochem. Molec. Biol.*, **24**, 13–19.

Moskowitz, H., Zlotkin, E. and Gordon, D. (1991) Solubilizaiton and characterization of the insct sodium channel. *Neurosci. Lett.*, **124**, 148–152.

Moskowitz, H., Zlotkin, E. and Gordon, D. (1992) The fly neuronal sodium channel as the receptor of the depressant insct selective neurotoxin. In: *Recent Advances in Toxinology Research*, Gopalakrishnakone, P., tan, C.D., Eds., National Univ. Singapore, Vol. 2, 84–99.

O'Dowd, D.K., Germeraad, S.E. and Aldrich, R.W. (1989) Alterations in the expression and gating of *Drosophila* sodium channels by mutations in the paragene. *Neuron*, **2**, 1301–1311.

Pelhate, M. and Sattelle, D.B. (1982) Pharmacological properties of insect axons: a review. *J. Ins. Physiol.*, **28**, 889–903.

Pelhate, M. and Zlotkin, E. (1981) Voltage dependent slowing of the turn off as Na$^+$ current in the cockraoch giant axon induced by the scorpion venom insect toxin. *J. Physiol. (Lond.)*, **319**, 30–31.

Pelhate, M. and Zlotkin, E. (1982) Actions of insect toxin and other toxins derived from the venom of the scorpion *Androctomus australis* in isolated giant axons of the cockraoch *(Periplaneta americana)*. *J. Exp. Biol.*, **47**, 67–77.

Pichon, Y. and Boistel, J. (1967) Current-voltage relations in the isolated giant axon of the cockroach under voltage clamp conditions. *J. Exp. Biol.*, **47**, 343–355.

Pichon, Y. (1984) Ionic basis of electrical activity in insect nerve cells and synapses. In: *Insect Neurochemistry and Neurophysiology*, Borkovec, B., Kelly, T.J., Eds., Plenum Press, New York and London, 23–50.

Piek, T. and Spanjer, W. (1986) Chemistry and pharmacology of solitary wasp venoms. In: *Venoms of the Hymenoptera*, Pick,T., Ed., Academic Press, London, UK, 161–308.

Rathmayer, W., Walther, C. and Zlotkin, E. (1977) The effect of different toxins from scorpion venom on nueromuscular trnasmission and nerve action potential sin the crayfish. *Comp. Biochem. Physiol.*, **56C**, 35–40.

Rochat, H., Bernard, P. and Couraud, F. (1979) Scorpion toxins: chemistry and mode of action. *Adv. in Cytopharmac*, **3**, 325–333.

Rochat, C., Rochat, H., Miranda, F. and Lissitzky, S. (1967) Purification and some poperties of the neurotoxins of *Androctonus australis* Hector. *Biochemistry*, **6**, 578–588.

Ruhland, M., Zlotkin, E. and Rathmayer, W. (1977) The effect of toxins from the venom of the scorpion *Androctonus australis* on a spider nerve-muscle preparation. *Toxicon*, **15**, 157–160.

Schweitz, H., Vincent, J.P., Barhanin, J., Frelin, C., Linden, G., Hugues, M. and Lazdunski, M. (1981) Purification and pharamcological properties of eight sea anemone toxins from *Anemonia sulcata, Anthopleura xanthogrammica Stoichactis giganteus* and *Actinodendron plumosum. Biochemistry*, **20**, 5245–5252.

Stuhmer, W. (1991) Structure-funciton studies of voltage-gated ion channels. *Ann. Rev. Biophys. Chem.*, **20**, 65–78.

Stewart, L.M.D., Hirst, M., Ferber, M.L., Merriweather, A.T., Cayley, P.J. and Possee, R.D. (1991) Construction of an improved baculovirus insecticide containing an insect-specific toxin gene. *Nature (Lond.)*, **352**, 85–88.

Teitelbaum, Z., Lazarovici, P. and Zlotkin, E. (1979) Selective binding of the scorpion venom insect toxin to insect nervous tissue. *Insect Biochem.*, **9**, 343–346.

Tejedor, F.J. and Catterall, W.A. (1989) A site of covalent attachment of alpha-scorpion toxin derivatives in domain I of the sodium channel alpha subunit. *Proc. Natl. Acad. Sci. USA*, **85**, 8742–8746.

Thomsen, W.J. and Catterall, W.A. (1989) Localizaiton of the receptor site for α-scorpion toxins by antibody mapping: implications for sodium chhannel topology. *Proc. Natl. Acad. Sci. USA*, **86**, 10161–10165.

Tintpulver, M., Zerachia, T. and Zlotkin, E. (1976) The action of toxins derived from scorpion venom on the ileal smooth muscle preparation. *Toxicon*, **14**, 370–377.

Tomalski, M.D. and Miller, L.K. (1991) Insect paralysis by baculovirus-mediated expression of a mite neurotoxin gene. *Nature (Lond.)*, **352**, 82–85.

Vassilev, P., Scheuer, T. and Catterall, W.A. (1988) Identification of an intracellular peptide segment involved in sodium channel inactivation. *Science*, **241**, 1658–1661.

Walther, C., Zlotkin, E. and Rathmayer, W. (1976) Action of different toxins from the scorpion *Androctonus australis* on a locust nerve-muscle preparation. *J. Insect Physiol.*, **22**, 1187–1194.

Zlotkin, E., (1985) Toxins derived form arthropod venoms specifically affecting insects. In: *Comoprehensive Insect Physiology, Biochemistry and Pharmacology Vol. 10*, Kerkut, G.A. and Gilbert, L.I. (Eds.), Pergamon, Oxford, UK, 499–546.

Zlotkin, E. (1988) Neurotoxins. In: *Comparative Invertebrate Neurochemistry*, Lunt, G.G. and Olsen, R.W. (Eds.), Croom Helm, London, UK, 256–324.

Zlotkin, E., Eitan, M., Bindokas, V.P., Adams, M.E., Moyer, M., Burkhart, W., and Fowler, E. (1991) Functional duality and structural uniqueness of depressant insect selective neurotoxins. *Biochemistry*, **30**, 4814–4821.

Zlotkin, E., Kadouri, D., Gordon, D., Pelhate, M., Martin, M.F. and Rochat, H. (1985) An excitatory and a depressant insct toxin from scorpion venom-both affect sodium conductance and possess a common binding site. *Arch. Biochem. Biophys.*, **240**, 877–887.

Zlotkin, E., Miranda, F. and Lissitzky, S. (1972) A toxic factor to crustacea in the venom of the scorpion *Androctonus australis* Hector. *Toxicon*, **10**, 211–216.

Zlotkin, E., Miranda, F. and Rochat, H. (1978) Chemistry and Pharmacology of Buthinae scorpion venoms. In: *Arthropod Venoms*, Bettini, S., Ed., Springer, Berlin, Heidelberg, 317–369.

Zlotkin, E., Teitelbaum, Z., Rochat, H. and Miranda, F. (1979) The insect toxin from the venom of the scorpion *Androctonus mauretanicus*. Purification, characterization and specificity. *Insect Biochem.*, **9**, 347–354.

7. SODIUM CHANNELS AS TARGETS FOR NEUROTOXINS
Mode of Action and Interaction of Neurotoxins with Receptor Sites on Sodium Channels

DALIA GORDON

*Faculty of Medicine Nord, Institut Federatif de Recherche Jean Roche,
Laboratory of Biochemistry URA CNRS 1455, Boulevard Pierre Dramard,
13916 Marseille Cedex 20, France*

Several excellent, comprehensive reviews have been published on the structure, physiological effects and binding properties of the various neurotoxins acting on voltage-sensitive sodium channels (e.g. Catterall, 1980; 1986; Strichartz *et al.*, 1987; Baden, 1989; Norton, 1991; Martin-Eauclair and Couraud, 1995). This chapter is focused on the recent developments in the use of neurotoxins as molecular probes of sodium channel structure and function and cites original literature that has appeared subsequently to the previous reviews of this area. The aim of this presentation is to discuss, on the molecular level, some aspects of the interaction between several neurotoxins and their receptor sites, located on the sodium channel protein. The dynamic mutual modulation of the conformational, functional states of the sodium channel and the binding of toxins will be accentuated.

INTRODUCTION

Sodium channels are responsible for the generation and propagation of electrical signals in most excitable cells (such as nerve, muscle and endocrine cells). Being a critical element in excitability, sodium channels serve as specific target for a large variety of chemically distinct neurotoxins produced by many different animals and plants. These neurotoxins target the sodium channels and alter their normal activity, interfering with the electrical signaling, resulting in paralysis and eventual death of the affected animal. The neurotoxins alter each one of the sodium channel properties, including ion conductance, ion selectivity, activation and inactivation (reviewed by Catterall, 1980; 1986; 1992; Barchi, 1988).

This chapter is dedicated to discuss the interactions of neurotoxins with their specific receptor sites on the molecular level. Thus, we will briefly review the molecular structure of the voltage-activated sodium channel protein, which serves as a target for these neurotoxins.

Sodium Channel Structure

The voltage-activated sodium channels belong to the family of voltage-gated ionic channels, which are plasma membrane proteins characterized by the steep voltage dependence of their opening behavior, so called activation (Hille, 1992). Their response to changes in membrane potential is by forming water-filled pores in the cell membrane through which specific ions can diffuse passively down their pre-established gradients. Voltage-gated ionic channels have three fundamental functional properties that enable them to fulfill their role in cellular electrical signaling: selective ion conductance, voltage dependent activation that results in opening of the ion conductive pore, and inactivation, that terminates the ion flux. The voltage activated sodium (Na), calcium (Ca) and potassium (K) channels are members of the same gene superfamily in terms of structural homology at the amino acid level. They are distinguished from other channels by having multiple of six putative trans-membrane segments (S1-S6, Figure 7.1). Most of these channels have been functionally expressed in *Xenopus* oocytes by microinjection of complementary RNAs derived from each channel cDNA, confirming their identity (Catterall, 1988; Hille, 1992; Stuhmer, 1991).

Sodium channels mediate the rapid increase in membrane Na^+ conductance responsible for the rapidly depolarizing phase of the action potential in many excitable cells. This transient increase in Na^+ conductance last only milliseconds and then spontaneously inactivates, and the membrane can be repolarized as K^+ permeability are augmented by potassium channel activities. All sodium channels that have been purified and biochemically characterized contain a large glycoprotein α subunit of 240–280 kDa. In most tissues and species the α subunit is non-covalently associated with a smaller $\beta1$ subunit of 30–40 kDa. In mammalian brain, the sodium channels consist of a heterotrimeric complex of α (260 kDa), $\beta1$ (36 kDa) and a disulfide linked $\beta2$ (33 kDa) subunits. The purified electroplax sodium channel is functional with only a single α subunit, indicating that it contains most, if not all, of the functional domains of the channel. In this chapter I refer to the α subunit as the sodium channel (reviewed by Catterall, 1992; Barchi, 1988; Trimmer and Agnew, 1989; Cohen and Barchi, 1993).

The availability of purified, functional preparations of sodium channels provided the basis for the isolation of cDNA clones encoding the primary structure of the principal α subunit of the sodium channels from various species and tissues (reviewed by Catterall, 1992; Cohen and Barchi, 1993). Oocytes injected with cRNA that encodes for a single subtype of the α subunit express functional voltage-gated sodium channels, confirming that all the functional properties of the channel are contained within this principal subunit.

The primary subunits of voltage-activated sodium channels are single polypeptides of about 2000 amino acid residues. Sodium (and calcium) channels comprise of contiguous repeats of four internally homologous domains, called repeats I–IV. Structural model of voltage-gated ionic channels has been proposed in which the channel is formed by a pseudo-symmetrical assembly of the four repetitive homologous domains (pseudo-subunits) of sodium channel (Figure 7.2) (Guy and Conti, 1990; Noda *et al.*, 1984; 1986). All four repeats are required for the formation of functional sodium channels (Stuhmer *et al.*, 1989). Each sodium channel repeat contains six potential transmembrane segments, S1 to S6, long enough to span the membrane as an α-helix. The largest degree of homology exists in the putative transmembrane regions, with a few charged residues in S2, S3 and S4 being strictly conserved among all voltage dependent channels. The S4 is the most characteristic structure shared by the members of all this superfamily and contains a positively charged residues at every third position and usually hydrophobic residues at the remaining positions. S4 segments are thought to participate in the

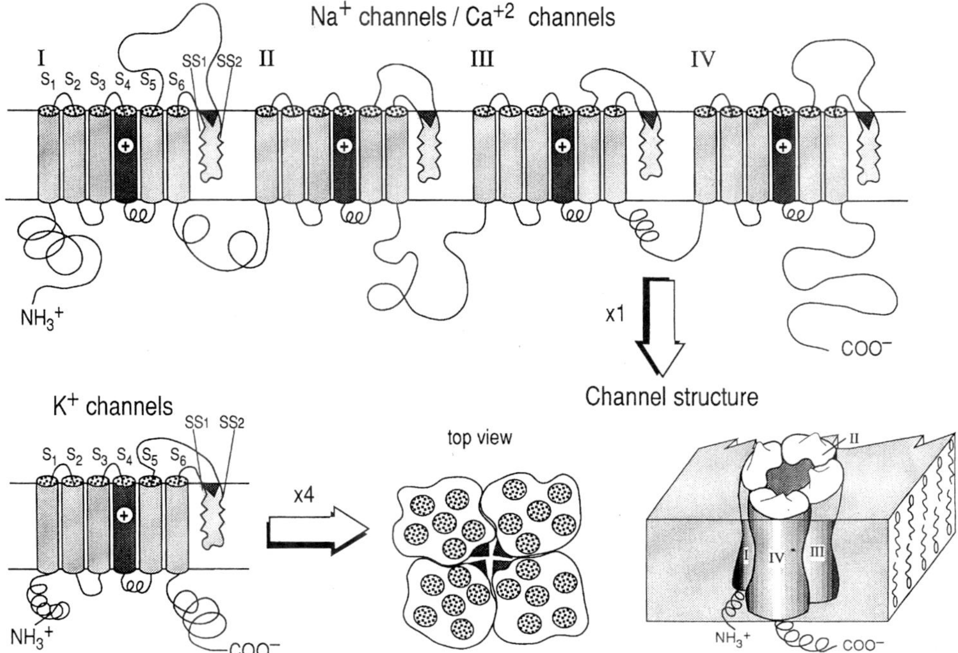

Figure 7.1 Schematic representation of the transmembrane arrangement of the principal submits of sodium (Na+), calcium (Ca+2) and potassium (K+) channels. Sodium and calcium channels share the general structure of four homologous repeated domains (I–IV) that are thought to act as 'pseudo-subunits' coming together in the membrane to form a central ion pore. Such organization brings domain I and IV to close proximity. The potassium channel is composed of four peptides, each homologous to one repeat of the sodium or calcium channel domains (four subunits form a functional potassium channel). Each domain consists of six putative transmembrane α helical segments. The most highly conserved segment is S4, present in each repeat, which contains a unique motif of a positively charged amino acid residues followed by two non polar residues that repeat four to eight times in each S4 helix. These S4 structures, suggested to participate in the voltage sensing mechanism, are indicated in each domain by (+) sign. The short segments (SS1 and SS2), which are part of the extracellular amino acid loop between transmembrane segments S5 and S6, are suggested to form a hairpin structure inside the membrane and to be part of the ion conductive pathway. The intramembrane short segments are referred to as the 'pore region' (indicated by a shaded zigzag line). The black triangle represents the enterance of the pore. The linker connecting domains III and IV in sodium channel represent the region suggested to participate in the fast inactivation process.

transmembrane voltage sensing mechanism, enabling the channel gating (Guy and Conti, 1990; Noda *et al.*, 1986; Stuhmer, 1991).

The polypeptide segment between domains S5 and S6 are thought to form a deep invagination into the membrane to form part of the lining of the channels pore (Figure 7.1) (Hartmann *et al.*, 1991; Stocker *et al.*, 1991). Channels in which amino acids in this region are changed by site-directed mutagenesis, have altered ion permeability and sensitivity to block by non-permeable ions or neurotoxins (Noda *et al.*, 1989; Terlau *et al.*, 1991; Satin *et al.*, 1992; Backx *et al.*, 1992; Heinemann *et al.*, 1992; Kontis and Goldin, 1993; Pusch *et al.*, 1991).

NEUROTOXIN RECEPTOR SITES ON SODIUM CHANNELS

At least six distinct classes of sodium channel neurotoxins have been designated on the basis of physiological activity and competitive binding studies (Catterall 1980; 1986; Barchi, 1988, see Table 7.1 and Figure 7.2). Neurotoxins that exhibit simple competition with one another have been assigned to the same class, although this does not establish that members of a class necessarily bind to the same receptor site; either steric or allosteric (conformational change in the sodium channel protein induced by toxin binding at one receptor site on a physically distinct receptor site) interference can produce competition (for some recent examples see Gordon *et al.*, 1992; 1996a; 1996b; Moskowitz *et al.*, 1994;

Table 7.1   Toxins bound by neurotoxin receptor sites 1–6 and additional unidentified sites on sodium channels

Site	Toxin	Effect
1	Tetrodotoxin (TTX) Saxitoxin (STX) μ-Conotoxin	Inhibition of ion conductance
2	Batrachotoxin (BTX) Veratridine Aconitine Grayanotoxin	Persistent activation
3	α-Scorpion toxins Sea anemone toxin II	Inhibit inactivation; enhance persistent activation
4	β-Scorpion toxins	Shift voltage dependence of activation
5	Brevetoxins Ciguatoxins	Shift voltage dependence of activation
6	δ-Conotoxins (δ TxVIA)	Inhibit inactivation
Unidentified sites	*Goniopora* coral toxin *Conus striatus* toxin	Inhibit inactivation

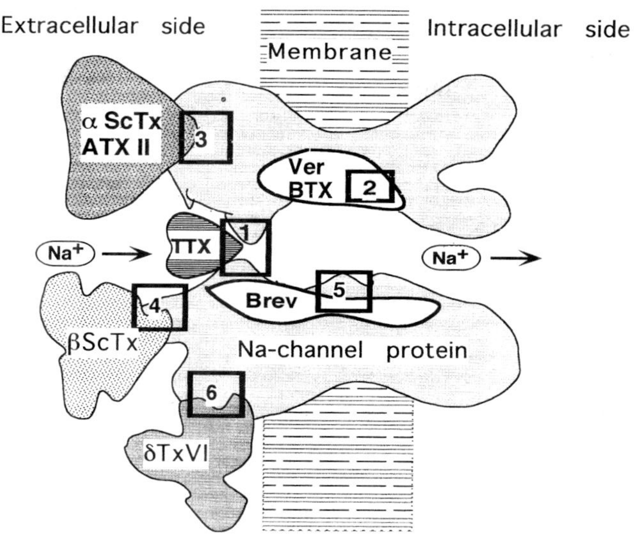

Figure 7.2 Cartoon illustrating neurotoxins bound to their specific, identified receptor sites 1 to 6 on sodium channel α subunit protein. A cross-section through the central part of the protein, traversing the ion-selective pore (and the cell membrane) is illustrated. The numbers in black boxes designate the corresponding neurotoxin receptor site number (see text and Table 7.1). TTX — tetrodotoxin, bound to receptor site 1; Ver — veratridine, BTX — batrachotoxin, alkaloid lipid soluble neurotoxins binding at receptor site 2; αScTx — α-scorpion toxin, ATX II — sea anemone toxin II, polypeptide toxins binding to receptor site 3; βScTx — β-scorpion toxin, binding to receptor site 4; Brev — brevetoxin, ployether lipid soluble toxin binding to receptor site 5; δTxVI — δ-conotoxin TxVIA, binding to receptor site 6. The action of these toxins is described in Table 7.1 and in the text. Na⁺ — illustrate sodium ions passing through the ion-selective pore.

Fainzilber *et al.*, 1994; 1995; Cestele *et al.*, 1995). On the basis of the binding studies, using a known radiolabeled neurotoxin as specific probe, at least six physically distinct neurotoxin receptor sites have been assigned to each of the neurotoxin class and several other, as yet unidentified (by direct radio-toxin binding) receptor sites have been noticed on various sodium channels (see Table 7.1) in different animal phyla.

Definition of Receptor Site

For clarity of discussion, I would like to clarify the use of 'receptor site' in this chapter. Different toxins have been considered to bind to the same or overlapping receptor site if they: 1) induce similar effects on the ion channels in electrophysiological studies; 2) compete in binding assays and 3) reveal similar pharmacological characteristics with their receptor site, such as dependence on membrane potential, competitive or allosteric interactions with other known channel modifiers, etc.

On this background a receptor site refers to the combined points of attachment or recognition sites (namely the different amino acids' functional groups in the surface of the receptor molecule) that are directly involved in the binding interactions with a toxin. Some of these attachment points may be shared by different toxins that bind to the same or similar, partially overlapping receptor sites. In general, simple binding studies are not able to establish identity between receptor sites of chemiaclly different neurotoxins, even if they exhibit identical physiological properties and mutually compete for binding. Detailed structural studies of both the toxin and receptor molecules are required to clarify the identity of receptor sites. The best known example is the receptor site for the sodium channel blockers tetrodotoxin (TTX) and saxitoxin (STX) (Yang and Kao, 1992; Lipkind and Fozzard, 1994, and see below).

Brief Description of Neurotoxin Receptor Sites on Sodium Channels

The neurotoxins that bind with high affinity and specificity to sodium channels at several separate receptor sites greatly facilitated the biochemical studies of sodium channels and reveal new aspects of channel structure and function (reviewed by Catterall, 1980; 1986). The development of specific radioligands for each neurotoxin receptor site enabled and facilitated the detection of conformational changes in the channel protein that occur as a toxin binds, resulting in most cases in allosteric modulation on other toxin binding at distinct receptor sites (see Table 7.2). The remarkable allosteric interactions between various classes of neurotoxin receptors on sodium channels will be discussed.

This section describes some of the main features of neurotoxin-receptor interaction that are relevant for the next molecular description of mode of action and receptor site structure of the most studied toxins. For more comprehensive review of the physiological effects and binding properties of the neurotoxins the reader is referred to the reviews cited in the text.

Neurotoxin receptor site 1 binds the water-soluble heterocyclic guanidines TTX and STX (Figure 7.2). TTX was the first neurotoxin shown to block sodium channels and together with STX, they are the most specific of sodium channel neurotoxins (Fuhrman, 1986). TTX and STX inhibit sodium channel ion transport by binding to a common receptor site that is thought to be located near the extracellular opening of the ion conducting pore of the sodium channel. In addition, peptide μ-conotoxins derived from the marine snail *Conus geographus* were shown to compete for binding with [^{3}H]STX and TTX and to inhibit sodium channels in vertebrate skeletal muscles (Moczydlowski *et al.*, 1986; Ohizumi *et al.*, 1986b).

Receptor site 2 binds several lipid-soluble toxins including grayanotoxin and the alkaloids batrachotoxin, veratridine and aconitine, shown to competitively interact with receptor site 2 (Figure 7.2). These toxins cause persistent activation of sodium channels at resting membrane potential by shifting the voltage-dependence of activation to more negative membrane potentials and blocking the inactivation process (Hille *et al.*, 1987). Their high lipid solubility suggest that their receptor site is located within the transmembrane parts of the channel. Alkaloid toxins binding at receptor site 2 reveals strong allosteric interaction with all other neurotoxin receptor sites (see below and Table 7.2).

Neurotoxin receptor site 3 binds polypeptide toxins derived from scorpion venoms (α-scorpion toxins) and sea anemones. These toxins act from the extracellular side of the membrane (see Figure 7.2) by slowing or blocking sodium channel inactivation process and their binding affinity to mammalian sodium channels was shown to be reduced by depolarization. The voltage-dependence of α-scorpion toxins binding is correlated with the voltage dependence of activation of sodium channels. The latter suggest that receptor site 3 is located on a region that undergoes a conformational change during voltage-dependent channel activation, leading to reduced affinity for toxin binding. The α-scorpion toxins and

Table 7.2 Allosteric interactions among neurotoxins binding at receptor sites 1–6 on vertebrate sodium channels

Site	Toxin	Allosteric modulation *	
		Positive	*Negative*
1	Tetrodotoxin Saxitoxin	3, 5 [a]	2 [b]
2	Veratridine Batrachotoxin Aconitine	3 [c]	6 [d]
3	α-Scorpion toxins Sea anemone toxin	2 [e]	
4	β-Scorpion toxins		
5	Brevetoxins	2, 4 [f]	3 [g]
6	δ-Conotoxins (δ-TxVIA)		

* Allosteric modulations induced by occupancy of a receptor site by the corresponding neurotoxin, on the binding of neurotoxin at the indicated receptor site(s). Positive modulation indicates enhancement of binding of the toxin at its indicated receptor site, and/or stimulation of Na^+ influx. Negative modulation refers to decrease in toxin binding at the indicated receptor site.

References:
[a] Ray *et al.*, 1978; Jover *et al.*, 1980; Tejedor and Catterall, 1990; Poli *et al.*, 1986.
[b] Brown, 1986.
[c] Catterall, 1977; Catterall *et al.*, 1981; Cestele *et al.*, 1995;
[d] Fainzilber *et al.*, 1994.
[e] Ray *et al.*, 1978; Catterall, 1977; Catterall and Beress, 1987; Catterall *et al.*, 1981; Sharkey *et al.*, 1987.
[f] Poli *et al.*, 1986; Sharkey *et al.*, 1987; Catterall and Risk, 1980; Trainer *et al.*, 1993.
[g] Cestele *et al.*, 1995; Gordon *et al.*, 1996a.

sea anemone toxin II also enhance the persistent activation of sodium channels by the lipid-soluble toxins acting at receptor site 2, indicating that receptor site 3 is at least partially allosterically coupled to receptor site 2 (see Table 7.2) (Catterall, 1977; reviewed by Catterall, 1980; 1992; Strichartz *et al.*, 1987; Martin-Eauclaire and Couraud, 1995; Norton, 1991).

Neurotoxin receptor site 4 binds the β-scorpion toxins derived from the venom of American scorpions (review in Strichartz *et al.*, 1987; Martin-Eauclaire and Couraud, 1995). These toxins shift the voltage-dependence of activation to more negative membrane potentials and are active from the extracellular side of the membrane but have no effect on the binding of α-scorpion toxins (Figure 7.2). In contrast to the polypeptide toxin binding at receptor site 3, the binding of β-scorpion toxins to receptor site 4 is not dependent on membrane potential and is not modulated by alkaloid toxins binding at receptor site 2 (reviewed by Catterall, 1992; Martin-Eauclaire and Couraud, 1995).

Receptor site 5 binds two groups of lipid soluble polyether ladder toxins, the brevetoxins and ciguatoxins, produced by special marine dinoflagellates (reviewed by Baden, 1989) (see Figure 7.2). These toxins shift the voltage dependence of activation of sodium channels to more negative membrane potential and inhibit the inactivation, resembling the action of neurotoxins acting at receptor site 2 and 4. Different analogs of brevetoxin and ciguatoxin were shown by direct binding studies to compete for [^{3}H]brevetoxin (PbTx-3) binding (Poli *et al.*, 1986; Lombet *et al.*, 1987; Baden, 1989). Binding of brevetoxin to receptor site 5 was shown to be independent of membrane potential and to allosterically modulate toxins binding at receptor sites 2, 3 and 4 (see Table 7.2) (Catterall and Risk, 1981; Poli *et al.*, 1986; Sharkey *et al.*, 1987; Cestele *et al.*, 1995).

Neurotoxin receptor site 6 has been recently identified by direct binding studies of a new family of peptide conotoxins, designated δ-conotoxins, represented by δ-TxVIA derived from the marine snail-hunting species *Conus textile* (Fainzilber *et al.*, 1994). The δ-conotoxins have been shown to inhibit sodium channel inactivation by binding to a new extracellular receptor site, designated receptor site 6 (see Figure 7.2). The only member of the δ-conotoxins family that was radiolabeled is the mollusc selective δ-TxVIA, shown to bind with high affinity to a single class of receptor sites present on both mollusc and vertebrate sodium channels, despite being completely inactive on vertebrates *in vivo* and *in vitro* (Fainzilber and Zlotkin, 1992; Fainzilber *et al.*, 1994; 1995; Gordon *et al.*, 1996b). δ-TxVIA binding was shown to be independent of membrane polarization and to be negatively modulated by veratridine (Fainzilber *et al.*, 1994). In contrast, other δ-conotoxins that are active on vertebrates, derived from fish-hunting *Conus* species, were shown to induce similar inhibition of sodium current inactivation in electrophysiological studies (Hasson *et al.*, 1993; 1995; Fainzilber *et al.*, 1995; Shon *et al.*, 1994; 1995), but to bind in a voltage-dependent manner and to enhance allosterically the effects of veratridine (similarly to toxins binding at receptor site 3) (Fainzilber *et al.*, 1995). However, all the δ-conotoxins compete for ^{125}I — TxVIA binding, suggesting that they bind to partially overlapping receptor sites (Fainzilber *et al.*, 1994; 1995; Shon *et al.*, 1995; Gordon *et al.*, 1996b).

Other neurotoxin receptor sites have been noticed on sodium channels, by the action of polypeptide neurotoxins derived from the fish-hunting *Conus striatus* (CsTx) and *Goniopora* coral toxin (GPT) (see Table 7.1, Gonoi *et al.*, 1986; 1987). These two different toxins were shown to inhibit sodium channel inactivation in a voltage dependent manner, similarly to α-scorpion toxins. Their binding enhanced the alkaloid toxin binding and action at receptor site 2, but the binding of radiolabeled α-scorpion toxin was only partially inhibited by high concentrations CsTx or GPT (Gonoi *et al.*, 1986; 1987; Fainzilber *et al.*, 1995). In contrast, both polypeptide toxins inhibit completely the binding of ^{125}I — TxVIA to sodium channels. GPT competes for δ-TxVIA binding at high concentration, suggesting that it may sterically interfere with the toxin binding at receptor site 6. CsTx on the other hand, inhibits δ-TxVIA binding at low concentration and its action in rat brain is antagonized by simultaneous injection of δ-TxVIA, suggesting mutual competition between these two different conotoxins at least in rat brain (Fainzilber *et al.*, 1994). However, despite the similarity in action and competition in binding, the opposite allosteric modulation described on δ-TxVIA binding by veratridine (Fainzilber *et al.*, 1994) as compared to CsTx (Gonoi *et al.*, 1987; Fainzilber *et al.*, 1994; 1995) indicate that these two *Conus* toxins bind to similar, but not identical receptor sites that may partially overlap (Fainzilber *et al.*, 1994; 1995; Gordon *et al.*, 1996b).

Other polypeptide toxins had been described to inhibit sodium current inativation, such as different scorpion toxins (α-like toxins, Gordon *et al.*, 1996a) and spider venom toxins (Nicholson *et al.*, 1994). However, competition binding studies revealed that the α-like scorpion toxins do not share receptor site 3 with α-scorpion toxins and their receptor sites are not yet idetified (Gordon *et al.*, 1996a).

NEUROTOXINS THAT INHIBIT ION CONDUCTANCE

The following sections focus on recent developments in understanding neurotoxin mode of actions on the single sodium channel level, using either channels incorporated into planar lipid bilayers or patch clamp techniques in cell or axon membranes. The other approach to study neurotoxin mode of action and receptor site structure has been expressing the cloned sodium channel α-subunit in hetrologous cell expression systems (mainly *Xenopus* oocytes or mammalian cultured cells), using native sodium channel α-subunit clones or after site-directed mutations. This approach follows mainly after the changes in toxin effect on sodium currents measured by electrophysiological methods and not after direct binding of the toxin to its receptor site. The discussion will focus on the best studied toxin representing each neurotoxin receptor class as listed above.

TTX and STX are the most specific sodium channel blockers (Fuhrman, 1986). These toxins have been useful in determining the number and distribution of sodium channels in excitable cells (Richie and Rogart, 1977) and assisting in purification of the small amounts of channel protein from cell membrane (Catterall, 1992; Cohen and Barchi, 1993; Hille, 1992). Some sodium channel subtypes have also been classified according to their toxin affinity levels (Rogart, 1986 and see below).

Mapping the Receptor Site of TTX and STX

Identification of recognition points that comprise the receptor site of TTX within the sodium channel primary structure supported the notion that the region between transmembrane segments S5 and S6 contribute to formation of the ion conductance pore, as suggested by Guy and Conti (1990) (Figure 7.1). The extensive studies by Stuhmer, Numa and colleagues have found that point mutations in this region alter both TTX inhibition and reduce single channel conductance.

Initially, Noda *et al.* (1989) have shown that a single mutation (E387Q) in rat brain sodium channel that neutralizes the negative charge of the conserved glutamic acid residue 387 to glutamine, renders the channel insensitive to TTX and STX and reduces five times its single-channel conductance, but it is still highly selective for Na^+ over K^+ and its macroscopic current properties are only slightly affected. In fact, mutations of charged residues symmetrically located at homologous sites in the S5-S6 loops in each of the four repeated domains were found to dramatically reduces sensitivity to TTX and STX block by at least three orders of magnitude (Terlau *et al.*, 1991). Charge mutations at adjacent positions in that region in each domain or mutations without a change in the net charge at positions 384 (D384E) and 1425 (M1425Q) and at other positions of the SS2 segments had minor or insignificant effects on toxin sensitivity (Terlau *et al.*, 1991). These suggest that portions of all four S5-S6 loops may contribute to the formations of the TTX/STX binding site (see Figure 7.3). Neutralization of aspartic acid at position 1717 of repeat IV (D1717N) abolished STX block, while only partially reducing the block by TTX (Terlau *et al.*, 1991).

Kao and co-workers have identified the active groups in TTX and STX structure, responsible for their high affinity binding and high blocking potency of sodium channels (reviewed by Kao, 1986). In the case of TTX, there are the guanidinium group and the hydroxyls at C9 and C10. In STX, the 7,8,9-guanidinium group and the two C12 hydroxyls were identified as the most important for binding and action (Strichartz, 1984; Kao, 1986). These toxin sites show remarkable stereospecificity, and when the guanidinium groups of TTX and STX are aligned, then the hydroxyls at C9 and C10 for TTX and the two at C12 for STX are also closely aligned (Kao and Walker, 1982). The second positively charged guanidinium group of STX may also be involved in the binding (see below; Lipkind and Fozzard 1994).

On this basis Lipkind and Fozzard (1994) suggested a molecular model of the binding pocket for TTX and STX. For TTX, the guanidium moiety of the toxin formed salt bridges with three carboxyls in repeats I and II (E384, E387 and E942) while two toxin hydroxyls (C9-OH and C10-OH) interact with a forth carboxyl on repeat II (E945). In addition, they also suggest hydrophobic interaction of the toxins with an aromatic ring of phenylalanine or tyrosine residues for brain and skeletal sodium channel subtypes (F385 and Y401, receptively) but not with the cysteine found in the cardiac subtype. In comparison to TTX, there was an additional interaction site for STX through its second guanidinium group with the carboxyl on repeat IV (D1717) (Lipkind and Fozzard, 1994).

All of the mutations involving a decrease in net negative charge that strongly reduced toxin sensitivity also caused a marked decrease in single-channel conductance. The mutants K1422E and A1714E which have an increased negative charge, showed nearly no reduction in single-channel conductance, whereas their toxin

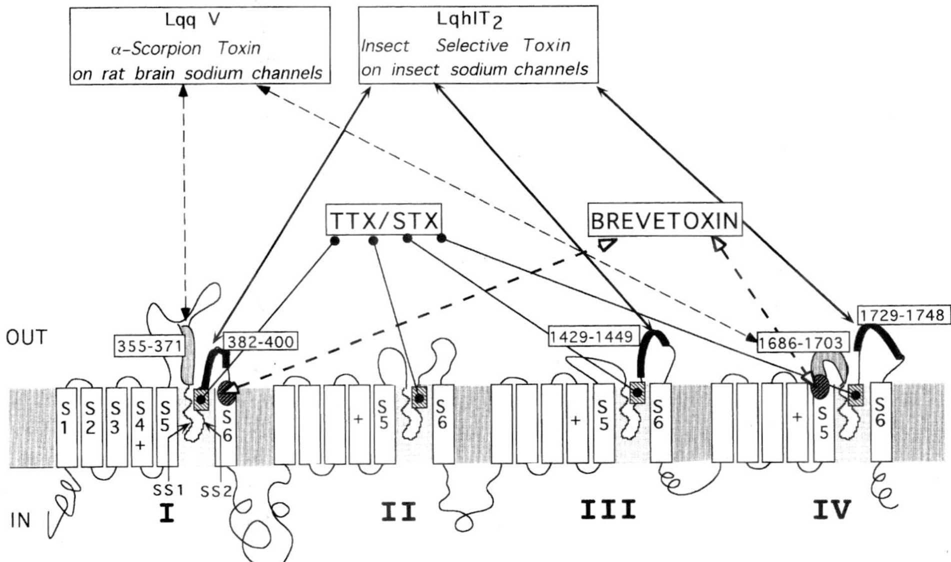

Figure 7.3 Schematic presentation of the mapped and partially localized neurotoxin receptor sites on sodium channel structure. Receptor sites of the two scorpion toxins, Lqq V (an α-scorpion toxin) and LqhIT$_2$ (a depressant insect selective toxin), were partially localized by site-directed antibodies which inhibit the binding of the toxins to rat brain and insect sodium channels, respectively. The synthetic peptides, corresponding to different amino acid segments of the extracellular loops between S5 and S6 in domains I, III and IV are indicated by the wide, shaded areas. The amino acid segments used for each antibody production are indicated by boxed numbers (corresponding to rat brain II amino acid sequence). The arrows indicate the corresponding area recognized by each site-directed antibody that cause maximal binding inhibition of Lqq V (gray area) to rat brain sodium channels (Thomsen and Catterall, 1989) or LqhIT$_2$ (black area) on insect sodium channels (Gordon et al., 1992; Moskowitz et al., 1994). Receptor site 1 that binds TTX and STX was mapped by site-directed mutagenesis of rat brain sodium channel (Noda et al., 1989; Terlau et al., 1991) and was found to consist of mainly negatively charged amino acid residues located in SS2 in equivalent position in the four repeated domains (indicated as rectangular hatched area). Receptor site 5, binding the brevetoxins, was partially localized by photoreactive derivative of brevetoxin and immunoprecipitation of proteolytically cleaved rat brain sodium channels by the site-directed antibodies (Trainer et al., 1994). The extracellular ends of transmembrane segments S6 and S5 in domains I and IV, respectively, were specifically labeled (indicated by hatched black elliptic areas). Other neurotoxin receptor sites have not been presently localized on the sodium channel structure. For details and discussion see text.

sensitivity was strongly reduced. These indicate that the TTX/STX interaction cannot be solely explained by electrostatic attraction between the positively charged toxin guanidinium groups and negatively charged acidic groups in their sodium channel binding site (Satin *et al.*, 1992) and supports the model of Lipkind and Fozzard (1994). These supports the notion that this region participates in forming the extracellular mouth and/or the lining of the ion-conducting pore itself, as modeled by Guy and Conti (1990).

TTX/STX Mode of Inhibition

The binding of TTX and STX is reversible (on seconds or minutes time scale) and, in most cases, of high affinity (equilibrium dissociation constant Kd = 1–10 nM) to a single, extracellular receptor site on the sodium channel molecule (Richie and Rogart, 1977; Schantz, 1986; Kao, 1986).

The inhibition of sodium ion conductance caused by these toxins was originally thought to be the result of toxin binding to the extracellular surface to obstruct the conductance pathway through the selectivity filter of the pore. Because the toxin has a guanidino moiety and guanidine is a permeate ion of the sodium channel, it has been proposed that the toxin guanidinium end enters the extracellular opening of the transmembrane pore of the sodium channel and lodges there, acting somewhat as a plug (Hille, 1975; 1971; Catterall, 1980). However, more recent evidence support models in which TTX occludes the pore by binding to extracellular surface of the channel and not inside the pore, much like a "lid" as opposed to a "plug" (Kao and Walker, 1982; Kao, 1986). Moreover, the binding of TTX and STX may be modulated by the functional state of the sodium channel and the toxin binding has been suggested to induce conformational changes on the channel protein, which may be detected by chemical (Tejedor *et al.*, 1988) or other neurotoxin binding (Brown, 1986; Tejedor and Catterall, 1990 and see below).

The sodium channel has been suggested to undergo several different voltage-dependent conformational changes, yielding different noninactivated closed states shown to have different affinities for TTX or STX (Strichartz *et al.*, 1986; Ruben *et al.*, 1990; Eickhorn *et al.*, 1990; Clarkson *et al.*, 1988; Gonoi *et al.*, 1985; Salgado *et al.*, 1986; Lonnendonker, 1989). Using rat brain RII sodium channels expressed in *Xenopus* oocytes, Patton and Goldin (1991) have shown that the sodium channels are blocked by TTX in a use-dependent manner, namely, the inhibition is accumulated with repetitive stimulation. This use dependence was the result of an increased affinity of the channels for TTX upon depolarization, most likely due to a conformational change in the channel. The transition is probably one occurring during activation of the channel, at potentials more negative than those resulting in channel conductance, suggesting that the conformational change that causes the use-dependent block by TTX is a closed-state voltage dependent gating transition (Patton and Goldin, 1991).

When membrane-bound (Krueger *et al.*, 1983; Green *et al.*, 1987) or purified (Hartshoren *et al.*, 1985) sodium channels are persistently activated by batra-chotoxin and incorporated to planar lipid bilayers, inhibition of the channels by

TTX, STX and congeners is voltage-dependent. This voltage dependence arise from changes of state in the channel itself since it dose not depend on the charge of the toxin (Moczydlowski *et al.*, 1984; Green *et al.*, 1987). The monovalent TTX, divalent STX and other derivatives with charges ranging from 0 to +2, has the same voltage dependence (Moczydlowski *et al.*, 1984), indicating that the toxins do not bind within the transmembrane electric field. These supports the notion that voltage dependent conformational changes in the sodium channel affect TTX affinity.

The binding of a toxin, such as TTX and STX, may induce conformational changes on the sodium channel protein. Direct evidence for a toxin-induced conformational change associated with binding of STX is provided by Tejedor *et al.* (1988). Using purified rat brain sodium channels, the binding of STX enabled the formation of isopeptide bonds within the α subunit by hydrophilic carbodiimides that stabilized the sodium channels in a high affinity state for STX. Under these conditions, when STX protected its receptor site from chemical modification, the overall availability of other carboxyl groups for reaction was increased, providing evidence for a toxin induced conformational change on binding (Tejedor *et al.*, 1988). This reciprocal modulation phenomenon is further exemplified by the interactions of other neurotoxins with their receptor sites on sodium channels (see below).

To summarize, the sensitivity of the sodium channel to TTX and STX is strongly reduced by mutations of specific amino acids residues in the SS2 segments equivalently positioned in the 4 internal repeats. Most mutants reducing net negative charge in these amino acids markedly diminish single-channel conductance. These results provide evidence that the predominantly negatively charged residues line part of the extracellular mouth and/or the pore wall. The toxin binding is outside the electric field of the membrane within the extracellular vestibule of the channel pore. This binding is both influenced by the conformational state of the channel and induce a conformational change upon toxin binding to receptor site 1.

Sodium Channel Subtypes Revealed by Sensitivity to TTX and STX

TTX, STX and congeners can distinguish between sodium channels in cells at different stages of development and from different tissues. Earlier pharmacological studies demonstrated that mammalian skeletal muscle express at least two forms of sodium channel that differ in their sensitivity to TTX (Barchi and Weigele, 1979; Rogart, 1986). Sodium channels that appear at different times during nerve and muscle development have markedly different sensitivity to TTX and STX (Frelin *et al.*, 1983; Sherman and Catterall, 1982; reviewed by Barchi, 1988; Cohen and Barchi, 1993). The sodium channel expressed in adult skeletal muscle is TTX-sensitive, namely is blocked by nanomolar concentrations of these toxins, whereas the channels synthesized in embryonic and denervated muscle are relatively resistant to block by TTX at concentrations several orders of magnitude higher. Similarly, denervation of mature skeletal muscle induces the appearance of TTX-resistant sodium channels (reviewed by Barchi, 1988; Cohen and Barchi, 1993). The TTX-resistant skeletal muscle sodium channel has been shown to be identical to the TTX-resistant sodium channel subtype identified in mammalian heart by molecular cloning (Rogart *et al.*, 1989; Trimmer *et al.*, 1989).

The sodium channels from rat brain and adult skeletal muscle are both sensitive to TTX and STX. They can be distinguished by the small peptide neurotoxins from *Conus geographicus*, so called μ-conotoxins. These μ-conotoxins specifically block the muscle sodium channels, but produce no significant effect on channels from neuronal cells or heart. Binding studies have shown that TTX, STX and μ-conotoxins displace one another, suggesting that they bind to the same or overlapping receptor site on the channel surface (Cruz *et al.*, 1985; Kobayashi *et al.*, 1986; Ohizumi *et al.*, 1986a,b; Moczydlowski *et al.*, 1986). Recent site-directed mutagenesis studies were able to distinguish between receptor sites of TTX, STX and μ-conotoxins. Stephan *et al.* (1994) have demonstrated that mutation of adult muscle sodium channels at glutamate 403 to glutamine (E403Q), that is homologous to glutamate 387 of the rat brain RII channel, shown to be essential for TTX blockade (Noda *et al.*, 1989; Terlau *et al.*, 1991), results in loss of TTX and STX blockade. The mutant channel, however, retains a high level of sensitivity to μ-conotoxins that TTX is no longer able to displace (Stephan *et al.*, 1994). These indicate that the receptor site for TTX, STX and μ-conotoxins may be overlapping, but not identical.

The difference between TTX and μ-conotoxins receptor sites provides an interesting model to understand the specificity of neurotoxin receptor site structure to chemically different but physiologically similar toxin binding. These two groups of toxins revealed close similarity in their properties: 1) Both similarly block single channel conductance of skeletal muscle sodium channels (Cruz *et al.*, 1985; Moczydlowski *et al.*, 1986). 2) They reciprocally compete for binding of radiolabeled STX or μ-conotoxin to sodium channels in rat skeletal muscle and eel electroplax (Cruz *et al.*, 1985; Moczydlowski *et al.*, 1986). 3) Each class of toxin carries a positively charged functional group which is required for sodium channel blockade. Studies of TTX analogs have shown that the positively charged guanidinium group of this toxin is essential for block (Kao, 1986). μ-conotoxins GIIIA and GIIIB are 22 amino acid peptide toxins; site-directed mutagenesis has identified arginine 13 as the critical residue for blockade (Sato *et al.*, 1991). However, despite these similarities Stephan *et al.* (1994) have shown that the two groups of toxins do not have identical receptor binding sites. The E403 of the skeletal muscle sodium channel is not the only candidate for interaction with the arginine 13 of μ-conotoxins (see above; Lipkind and Fozzard, 1994). The interaction sites for μ-conotoxins have not yet been identified, but it is plausible to think that they may involve at least some of the sites responsible for TTX and STX binding.

TOXINS THAT MODIFY SODIUM CHANNEL GATING

Many different neurotoxins modify the kinetics of sodium channel gating, such as various alkaloids and other lipid soluble substances, and peptide toxins from scorpion and spider venoms, sea anemones, *Conus* snails and others. They act on sodium channels by increasing the probability that the channel opens or remain open. These toxins act on sodium channels by binding to at least four distinct receptor sites, designated receptor sites 2–6 (Table 7.1). Strong allosteric interaction have been described among these receptor sites (see Table 7.2), and will be discussed below.

Lipid Soluble Toxins Acting at Receptor Site 2

The lipid soluble compounds isolated from various plants and from tropical frogs have been shown to have dramatic effects on excitable membranes. The best studied of these compounds include the plant alkaloids veratridine, aconitine and grayanotoxin and the steroidal alkaloid isolated from the skin of Colombian frogs, batrachotoxin (reviewed by Catterall, 1980; Narahashi, 1986; Strichartz *et al.*, 1987; Hille *et al.*, 1987). The binding of site 2 lipid soluble toxins result in a net physiological function exhibited by a prolongation of action potential, repetitive firings of nerves and membrane depolarization (reviewed by Catterall, 1980; Narahashi, 1986). The voltage-dependent activation is shifted 20 to 90 mV to more negative membrane potentials, depending on the toxin, and inactivation is slowed or sometimes eliminated. All these toxins cause sodium channels to open more easily and to remain open at rest, but the detailed mechanisms underlying these effects may differ (reviewed by Strichartz *et al.*, 1987; Khodorov, 1985; Hille *et al.*, 1987; Hille, 1992).

Mechanism of action

These toxins reduce both the single channel conductance and the selectivity of the channel to Na^+ ions (Strichartz *et al.*, 1987; Hille, 1992). The channels become less selective to Na^+ over K^+, and in some cases, like in aconitine modification, the channels become more permeate to NH_4^+ than Na^+, and the permeability to other metal ions is changed. The selectivity filter acts as if it had been widened by a few tenths of an Angstrom (Hille, 1992).

All the four toxins have been demonstrated to competitively interact with receptor site 2, as shown by ion flux studies and by direct measurements of specific binding of tritiated derivative of batrachotoxin, [³H]batrachotoxinin A 20α-benzoate ([³H]BTX-B) (Catterall *et al.*, 1981; reviewed by Catterall, 1980; 1992; Hille *et al.*, 1987). These toxins can produce their effects when added to either the extracellular or the cytoplasmic side of the nerve membrane, indicating their membrane-permeating character. Taken together with the structure of these toxins, the latter suggest that their receptor site may be in a strongly hydrophobic region such as the boundary between membrane lipids and the transmembrane segments of the channel, or within a tentative hydrophobic pocket in the membrane crossing segments of the channel.

The lipid-soluble toxins provide another example of state dependent binding. Their binding and action is use-dependent; application of repetitive depolarizing steps on the nerve membrane exposed to one of these toxins results in a use dependent modification of all the sodium channels, and the nerve reveals sodium currents even at the resting potential, at −90 mV (Hille, 1992). These result from preferential binding of the toxins with high affinity to the active state of sodium channels and consequent stabilization of those states. Binding to closed channels is extremely slow whereas binding to open channels is much faster, as evidenced by the rapid appearance of modified channels produced by repetitive depolarization that activate sodium channels (reviewed by Catterall, 1980; Strichartz *et al.*, 1987). Batrachotoxin will serve as an example of these toxins to look at its mode of gating modification on the molecular level.

Batrachotoxin mode of gating modification

In accordance with state-dependent binding, the process of BTX modification of sodium channels is accelerated by holding the membrane at hyperpolarized voltages and then applying depolarizing voltage pulses for short periods of time. The gating kinetics of BTX-modified single channels have been studied using the cut-open squid axon technique (Correa *et al.*, 1992).

During maintained membrane depolarization normal sodium channels inactivate. The gating charge recovered in the "off" of the voltage step, when the channel is in the inactivated state, is less than the gating charge mobilized during the "on" of the pulse, that led to the activated state. In this case the charge is said to be immobilized (Benzanilla and Armstrong, 1974). In the 'ball and chain' model of inactivation, the ball or the inactivation particle, interacts with a receptor contained in the inner mouth of the channel hindering the passage of Na^+ ions completely (reviewed by Armstrong, 1992). In this model charge immobilization is presumably of electrostatic interactions between the inactivation particle bound to the channel mouth and the activation particles making more difficult the return of the latter to their original state. Charge immobilization has been observed in many different sodium channels (reviewed by Armstrong, 1992).

Fast inactivation and charge immobilization are removed by agents like pronase and chloramine-T, known to cause a complete removal of the inactivation particle (Armstrong *et al.*, 1973; Tanguy and Yeh, 1988). In contrast, BTX eliminates fast inactivation, but leaves charge immobilization intact (Tanguy and Yeh, 1988). This suggests that BTX might be interacting with either the receptor for the inactivation particle or causing a conformational change in that area of the channel, still enabling the charge immobilization but without blocking or perhaps partially occluding the channel (Correa *et al.*, 1992). BTX decreases channel conductance by ~50% and modifies ion selectivity (Huang *et al.*, 1982; 1984; 1897; Strichartz *et al.*, 1987; Khodorov, 1985; Hille *et al.*, 1987). A partial occlusion of the sodium channel by BTX, directly or indirectly (by inducing a conformational change) would explain the reduction in channel conductance induced by the toxin (Quandt and Narahashi, 1982; Correa *et al.*, 1991).

BTX shifts the voltage activation curve towards hyperpolarized voltages by ~50 mV, suppresses fast and slow inactivation, and increases the duration of sodium channel open events by more than one order of magnitude (Huang *et al.*, 1982; 1984; 1987; Strichartz *et al.*, 1987). Normally acting sodium channels contain only one open state in the negative potential region (Vandenberg and Benzanilla, 1991). However, the most likely kinetic model for normal sodium channels, which emerges from single channel studies and the information provided by gating and macroscopic current measurements, implies two parallel pathways for channel activation; a pathway between several closed states in parallel to a pathway between closed-inactivated states, based on the ability of the channel to inactivated before opening (Aldrich *et al.* 1983; Horn and Vandenberg, 1984; Scanely *et al.*, 1990). On the basis of this kinetic model, the mode of action of BTX implies that BTX should transform the inactivated states into conducting (open) states so as to keep the process of charge immobilization intact (Correa *et al.*, 1992). Of course, this does not exclude additional modification of the open state by the toxin. Correa and co-workers have suggested that the dramatic increase in the duration of open events after BTX treatment resides in the appearance of additional open state(s), which may have been the inactivated state that would now be conducting and stabilized by the presence of the toxin. Their results reveal the existence of at least three open states, which emerge from the close-inactivated state of the unmodified sodium channel, as discussed above. The molecular identification of the recognition sites for BTX binding within the sodium channel structure are still awaiting to be discovered.

The pharmacology of sodium channels is altered in many ways by BTX and congeners, like veratridine and aconitine. Modified channels have a higher affinity for α-scorpion toxins binding at receptor site 3 (Catterall, 1977; 1980) and lower affinity for δ-conotoxin TxVIA binding at receptor site 6 (Fainzilber *et al.*, 1994). There seem to be an allosteric coupling between the site of action of alkaloid toxins and those of the polypeptide toxins. BTX receptor site seems to be allosterically linked to many other sodium channel receptors such as receptor site 1 and 5 (Table 7.2) as well as to the unidentified receptor sites for polypeptide marine toxins, pyrethroids and local anesthetics (see below, section on allosteric interactions).

PEPTIDE TOXINS THAT INHIBIT SODIUM CURRENT INACTIVATION

Inactivation of sodium current is slowed by numerous neurotoxins, including various polypeptide toxins, such as α-scorpion and sea anemone toxins that act at neurotoxin receptor site 3, δ-conotoxins acting at receptor site 6 and other marine polypeptide toxins (i.e. *Conus striatus* and *Goniopora* toxins) and spider toxins acting at unidentified receptor sites, all on the extracellular side of the sodium channel.

Alpha Scorpion Toxins and Sea Anemone Toxin II, Receptor Site 3

The structure, effects and mechanism of action of scorpion toxins (Martin-Eauclaire and Couraud, 1995; Catterall, 1986; Strichartz *et al.*, 1987) and sea anemone toxins (Norton, 1991) have been recently reviewed.

Scorpion toxins receptor sites

The strong voltage dependence in binding of α-scorpion and sea anemone toxin suggests that the voltage-dependent-conformational changes that lead to channel activation and inactivation involve channel segments that also contribute to the formation of the α-scorpion toxin receptor site (Catterall, 1980; 1992). The use of several anti-peptide antibodies directed against specific sequence regions in the external loop S5–S6 flanking the SS1–SS2 region in rat brain sodium channels enable the identification of sodium channel segments that may associate to form the α-scorpion toxin receptor site. Antibodies that recognized the N-terminal end of SS1 in domain I (antibody 355–371) and IV (antibody 1686–1703) specifically inhibited the binding of an α-scorpion toxin to the channel (Thomsen and Catterall, 1989) (see Figure 7.3). Similarly, the binding of other scorpion toxin, shown to alter Na^+ currents in insect axons, was also effectively inhibited by some of these site-directed antibodies, corresponding to the C-terminal end of SS2 in repeats I, III and IV (Gordon *et al.*, 1992; Moskowitz *et al.*, 1994) (Figure 7.3).

Thus, regions flanking the putative pore domain at the S5–S6 extracellular linker may participate in formation of the receptor binding sites of various neurotoxins that modify sodium channel function. The results support the channel model that the four repeated domains symmetrically assemble around a central pore and the receptor sites for the scorpion neurotoxins are formed by the close apposition of the S5–S6 extracellular loops in domains I and IV (Catterall, 1992; Gordon *et al.*, 1992) (see Figure 7.1). It is not yet understood how binding of α-scorpion toxins to their extracellular site slow the inactivation process. Most recently, other scorpion toxins have been described, that induce inhibition of sodium current decline in mammalian and insect neuronal membranes, similarly to α-scorpion toxins' physiological effect, but do not bind to receptor site 3 on sodium channels (Gordon *et al.*, 1996a). The binding studies strongly suggest that the α-scorpion toxins and sea anemone toxin II do not bind to identical receptor site, but rather to partially overlapping receptor sites (Gordon *et al.*, 1996a).

α-Scorpion toxin mode of action

The prolong action potentials induced by α-scorpion toxins on nerve membranes were attributed to the interference of the toxins with the ability of channels to inactivate upon prolonged depolarization (under voltage clamp condition) (reviewed by Strichartz *et al.*, 1987). Kirsch *et al.* (1989) examined the effects of TsIV-5, an α toxin isolated from the Brazilian scorpion *Tityus serrulatus*, at single channel level in mouse N18 neuroblastoma cells. TsIV-5 prolongs the dwell time of channels in the open state in a voltage dependent manner, by preventing open channels from closing and by facilitating the reopening of closed channels, as indicated by prolong brusting. These effects are consistent with a kinetic model in which TsIV-5 interferes with the ability of the channel to reach the inactivated state (Kirsch *et al.*, 1989). Sea anemone toxin II was shown to have similar effects on the single-channel level, inducing prolongation of the channel open time and repetitive channel opening (bursting) in cardiac muscle and neuroblastoma cells (Schreibmayer *et al.*, 1987; Nagy, 1987; 1988).

The cytoplasmic amino acid segment that links transmembrane domains III and IV (L_{III-IV}) of the sodium channel has an important role in inactivation (reviewed by Catterall, 1992). Application of site-directed anti-peptide antibodies that bind to this highly conserved region in many subtypes of sodium channel proteins (Gordon *et al.*, 1988; 1990; Gordon, 1990; Catterall, 1992) were shown to block inactivation (Vassilev *et al.*, 1988; 1989). Expression of the sodium channel as two polypeptides discontinuous within the L_{III-IV} loop slows inactivation 20-fold (Stuhmer *et al.*, 1989). Fast inactivation is almost completely blocked by mutation of the hydrophobic cluster IFM (isoleucine 1488, phenylalanine 1489, methionine 1490 in rat brain RII sodium channel subtype) within L_{III-IV} to glutamine or by single amino acid substitution of glutamine for phenylalanine (F1489Q), that slows inactivation 5000-fold (West *et al.*, 1992). These results suggest that the IFM residues form the inactivation particle of the sodium channel and enter the intracellular mouth of the pore, thus occluding it during the process of inactivation.

Eaholtz *et al.* (1994) have shown that a synthetic peptide containing the IFM sequence restores fast inactivation to mutant sodium channels having a defective inactivation particle and to wild type sodium channels having inactivation slowed by α-scorpion toxin (Lqq V). When 100 nM α-Lqq V was applied externally to cells expressing the rat brain type IIA sodium channel α subunit, it slowed sodium channel inactivation substantially. Intracellular application of the IFM-containing peptide restored fast inactivation of channels modified by α-Lqq V, causing decay of the sodium current within 1 ms with a time course and voltage dependence that are similar to the intrinsic inactivation (Eaholtz *et al.*, 1994). These results are consistent with the notion that the IFM motif serve as a hydrophobic latch of the inactivation particle in the peptide loop between domain III and IV of sodium channels. The binding of α-scorpion toxin may induce a conformational change in the channel that prevents the interaction of the IFM motif with its receptor, resulting in inhibition of sodium channel inactivation. Freely diffusible IFM-containing peptide, however, may still interact with its receptor and cause inactivation of sodium channels (Eaholtz *et al.*, 1994).

Other peptide toxins (such as *Goniopora* toxin, *Conus striatus* toxin, δ-conotoxins and other scorpion α-like toxins) that cause the apparent effect of inhibition of sodium current inactivation or slowing down the decay of sodium currents, were suggested to have distinct effects on the gating of sodium channels, indicating different mechanism of action (Gonoi *et al.*, 1986; 1987; Hasson *et al.*, 1993; 1995; Gordon *et al.*, 1996a).

TOXINS THAT AFFECT ACTIVATION

β-Scorpion Toxins Acting at Receptor Site 4

Scorpion toxins that bind with high affinity to mammalian sodium channels but do not compete for binding with α-scorpion toxins have been called the

β-scorpion toxins (reviewed by Martin-Eauclaire and Couraud, 1995; Strichartz *et al.*, 1987). Although the different β-scorpion toxins (toxins II and VI from *Centruroides suffusus suffusus*, Css II, VI; Toxins II and IV from *C. sculpturaus*; Cs II and IV; γ-tityustoxin (Ts VII) from *Tityus serrulatus*) were shown to compete in binding studies (Jover *et al.*, 1988; Wheeler *et al.*, 1983; Barhanin *et al.*, 1982), it seems that they do not induce the same effect on sodium currents. Css II and Cs II slow down the activation of sodium currents and reduce the peak current. However, after a depolarizing "conditioning" pulse, the β-toxin-modified currents activated faster than controls, occur at more negative potentials and the pick sodium permeability is increased (Hue *et al.*, 1983; reviewed by Martin-Eauclaire and Couraud, 1995; Strichartz *et al.*, 1987). The effect of Ts VII and Cn II-10 (Toxin II from *C. noxius*) was qualitatively similar to those induced by the cen-truroides toxins after a "conditioning" pulse, namely they shift Na^+ current activa-tion to more negative potentials and slow the onset of Na^+ current inactivation (Zabrodovskaya and Khodorov, 1985; Vijveberg *et al.*, 1984; Barhanin *et al.*, 1985).

In contrast to the voltage-dependent binding of sea anemone and α-scorpion toxin to mammalian neuronal membranes, the binding of β-scorpion toxins, such as Css II and Ts VII, was not modified by membrane depolarization or by alkaloid toxins binding at receptor site 2. The binding of radioiodinated Css II was shown to increase by occupancy of receptor site 5 by brevetoxin (Sharkey *et al.*, 1987 and see below).

β-Scorpion toxins mechanism of action

Yatani *et al.* (1988) studied the effects of the β-scorpion toxins Tss VII and Cn II-10 on single neonatal rat ventricular myocyte, on whole cell and single channel sodium currents. The β-toxins produced two types of effects, depending on the membrane depolarization: both toxins reduced whole cell Na^+ current at test potentials more positive to -40 mV; conversely, the toxins produced an inward current at test potentials lower than -60 mV, where Na^+ currents were negligibly small in controls. Both types of effects were produced at test potentials of -40 mV, where the toxins reduced peak Na^+ current and, at the same time, slowed activa-tion and inactivation kinetics markedly. On a single channel level, the effects of β-scorpion toxins on sodium channel kinetics appear to involve alterations of both the duration of the open state and the latent period between onset of the stimulus pulse and first channel opening, depending on the potential. Ts VII caused a 2.7-fold increase in the mean channel open time at -60 mV. In contrast, at -20 mV the open times of toxin-modified channels were similar to control, suggesting that under condition where channel closure occurs predominantly via inactivation, the toxin is relatively ineffective. The summed single channel records produced the toxin effects obtained in whole cell recordings: thus, at -20 mV the channel activ-ity was reduced as evident by the decrease in peak current, and both the onset and decay phases of the current appeared to be prolonged. The decrease in average current amplitude occurred in the absence of change in single Na^+ current, sug-gesting that the toxin prevents channels from opening. At low depolarization, at a test potential of −80 mV, Ts VII produced significant inward current. The induced prolongation of the decay phase of the macroscopic current at strong depolarization is not the result of longer channel open states, as revealed by the single channel study, but rather is due to delayed openings after the pulse, since the toxins markedly increased the latency to first channel opening (Yatani *et al.*, 1988).

To summarize, the prolong open times at low potentials could be due to a decrease in the transition rate from open to closed states; the prolonged waiting times observed at all potentials suggest that transition rates from the closed to the open state are also decreased by the toxins. The blocking effect, which is most readily observed at higher potentials (owing to the toxin-induced shift in the voltage dependence of activation), can be explained by an increase in number of channels that fail to open at a given potential, consistent with the toxin-induced slowing of transition between close and open states (Yatani *et al.*, 1988).

Joho *et al.* (1990) have shown that α- and β-scorpion toxins induce a typical and additive effects on rat brain sodium channel subtype RIII expressed in *Xenopus* oocytes, consistent with the notion that receptor sites 3 and 4 are located on the α-subunit of the sodium channel (but see also Martin-Eauclaire and Couraud, 1995).

At least two groups of insect-selective scorpion toxins, the excitatory and depressant anti-insect toxins (Zlotkin *et al.*, 1995 and see review by Zlotkin, this volum) have been shown to affect insect axonal sodium conductance in a similar manner to β-scorpion toxins effects on vertebrate sodium channels (reviewed by Stankiewicz *et al.*, 1996). Many β-scorpion toxins have been demonstrated to be active on both insects and mammals and to compete in binding studies (Gordon *et al.*, 1984; 1992; Zlotkin *et al.*, 1985; Lima *et al.*, 1986; Martin-Eauclaire and Couraud, 1995). The high selectivity to insects, revealed by the excitatory and depressant insect selective toxins, indicate the presence of structural peculiarities in insect sodium channels, as compared to their vertebrate counterparts. On the other hand, the similarity in action suggests similarity in function of structurally different sodium channels from various animal phyla. Subtle differences in structure and function may be revealed by the use of such selective neurotoxins and contribute to the understanding of sodium channel structure-function relationship.

Brevetoxins Acting at Receptor Site 5

Brevetoxins (PbTx), produced by the marine dinoflagellate *Ptychodiscus brevis*, are lipid soluble polyether neurotoxins that act at sodium channel receptor site 5 (Poli *et al.*, 1986; Sharkey *et al.*, 1987; Baden, 1989). The structure, toxicity and physiological effects *in vivo* and *in vitro* have been reviewed by Baden (1989). These toxins depolarize nerve membranes in a concentration-dependent manner by causing a shift in sodium channel activation toward negative membrane potentials and by inhibiting normal inactivation (Wu and Narahashi, 1987; Baden, 1989). Brevetoxins have been shown to modify sodium channels in such a way that they open at membrane potentials that normally maintained them in a non-conducting, closed states, namely activation is shifted to a more negative membrane potential (−80 to −160 mV). PbTx-modified channels open much more slowly (about 1000-fold) than do normal channels and do not fast inactivate, thus stay open much longer time than untreated channels, and increase the sodium peak current with no effect on ion selectivity (reviewed by Wu and Narahashi, 1987; Baden, 1989).

Receptor site of brevetoxin

The binding of tritiated PbTx-3 and 9 was shown to be of high affinity (Kd 3–10 nM) and independent of membrane potential in rat brain synaptosomes. All the brevetoxin native analogs as well as ciguatoxin compete for tritiated PbTx-3 binding, suggesting that they share a common receptor site on sodium channels (Poli *et al.*, 1986; Baden, 1989).

Trainer *et al.* (1991; 1994) have demonstrated using a photoactive p-azidobenoyl derivative of [^{3}H]PbTx-3, that the brevetoxin covalently labels the sodium channel α-subunit in rat brain synaptosomes and in purified and reconstituted sodium channels. Regions of the α-subunit specifically photolabeled by this ligand were identified by antibody mapping of proteolytic fragments of the α-subunit. Immunopercipitation of labeled peptides with site-directed anti-peptide antibodies revealed that the antibodies recognized amino acid sequences within or adjacent to sodium channel transmembrane segments IS6 and IVS5, within residues Thr-400 to Lys-443 and Glu-1738 to Lys-1785 or to Lys-1793 on the extracellular side of the transmembrane segments in domain I and IV, respectively (see Figure 7.3). These results provided direct evidence for close association of transmembrane segments S6 and S5 in homologous domains I and IV in the native conformation of the sodium channel α-subunit and implicated their region of interaction as an important component of the brevetoxin receptor site (Trainer *et al.*, 1994). The association between the brevetoxin and α-scorpion toxin receptor sites, both partially localized to the regions near transmembrane segments S5 and S6 in domains I and IV (Thomsen and Catterall, 1989; Trainer *et al.*, 1994) (See Figure 7.3) is further exemplified by the marked allosteric modulations by PbTx-1 of the binding of α-scorpion toxins on both rat brain and insect neuronal sodium channels (Cestele *et al.*, 1995 and see Table 7.2 and below).

ALLOSTERIC INTERACTIONS AMONG NEUROTOXIN RECEPTOR SITES ON SODIUM CHANNELS

Although the various neurotoxin receptor sites on sodium channels are topologically separate, there are strong allosteric interactions among them. Traditionally the positive cooperativity observed between two neurotoxins, such as α-scorpion toxin and alkaloid toxins binding at receptor site 2 (i.e. veratridine, batrachotoxin) was explained in functional terms by adopting models from the activity of allosteric enzymes (Catterall, 1977). Considering the fact that the various neurotoxin receptor sites on sodium channels bind xenobiotics and produce pharmacological effects, but there is no known endogenous ligand for any of them, the conformational changes induced by the binding of one toxin on the channel receptor may be viewed on two levels: 1) The primary functional level, namely some of these conformational changes may be directly responsible for the toxin's action. 2) The secondary mechanistic level, namely some other tentative changes may occur at different regions of the sodium channel protein that are not directly related to the toxin's functional effect. These conformational changes may be detected by the binding of other neurotoxins at different regions of the sodium channel protein. Such alterations may induce positive or negative modulation on other toxins' binding (i.e., the positive and negative modulation by veratridine on α-scorpion toxin and δ-conotoxin TxVIA binding, respectively, on rat brain sodium channels) (Table 7.2). Thus, the use of neurotoxins may elucidate subtle conformational changes in the sodium channels, that could not be easily detected by other means but binding of the various toxins at their specific, high affinity receptor sites.

Allosteric Interactions with Receptor Site 1

The binding of TTX and STX to receptor site 1 has been considered not to induce any significant conformational changes on the sodium channel protein (Catterall, 1980). However, several evidences from various studies point out that the binding of TTX and STX may induce some alterations in the sodium channel structure, as probed by the binding of other neurotoxins that bind to distinct receptor sites (Table 7.2).

Several lines of evidence suggest that receptor site 1 allosterically interacts with receptor site 3. The binding of α-scorpion toxin, Lqq V, has been shown to increase up to 1.5-fold in rat brain synaptosomes by increasing concentrations of STX (Ray *et al.*, 1978). This effect could not be attributed to the blocking effect of STX on sodium channels (and thus preventing possible depolarization), since the experiment was performed in a sodium-free medium. Jover *et al.* (1980) has shown that TTX dose not change the affinity of the α-scorpion toxin AaH II (α-scorpion toxin II from the venom of the scorpion *Androctonus australis* Hector) to rat brain membranes, and the enhancement of AaH II binding is due to the increase in membrane receptor capacity (Bmax). Additional evidences for allosteric modulation of α-scorpion toxin binding by TTX have been provided by Tejedor and Catterall (1990). They have shown that TTX enhanced the binding of ^{125}I-Lqq V to purified and reconstituted sodium channels, similarly to the enhancement of batrachotoxin, and that the effect may be additive when both TTX and BTX were present (Tejedor and Catterall, 1990). Similar allosteric effects were obtained using a photoreactive derivative of Lqq V, and the α-subunit of sodium channels was specifically photolabeled (Tejedor and Catterall, 1990). The change in conformation caused by STX binding, as detected by cross-linking experiments (Tejedor *et al.*, 1988) was discussed above.

The binding of TTX or STX to receptor site 1 has been shown to allosterically alter the binding of neurotoxins to receptor site 2 and 5. Brown (1986) has shown that at 25°C both TTX and STX inhibit [^{3}H]BTX-B binding (in the presence of α-scorpion toxin that increase specific BTX-B binding) to rat brain synaptosomes in a concentration-dependent and noncompetitive manner. The inhibition was due to a 3-fold decrease in the affinity of BTX-B binding with no change in the number of binding sites. The interaction was unidirectional, namely BTX-B did not affect the binding of [^{3}H]STX (Brown, 1986). Poli *et al.* (1986) have shown a small (about 10%) increase in [^{3}H]PbTx-3 binding to receptor site 5 on rat brain synaptosomes by increasing concentrations of STX.

To summarize, the binding of TTX and STX to the extracellular receptor site 1 may induce conformational changes on the sodium channel protein, that are not necessarily directly linked to the blocking effect of these toxins on sodium permeability. These alteration in sodium channel conformation may be detected by enhancement or decrease in the specific binding of other neurotoxins at receptor site 2, 3 and 5 (Table 7.2).

Allosteric Interactions with Receptor Site 2

Binding of alkaloid toxins at receptor site 2 induce profound changes on every aspect of sodium channel function: shift the voltage-dependence of activation, inhibit inactivation and alter the ion selectivity (see above). The dramatic physiological effects must be a result of significant conformational changes induced on the channel protein. Some of these changes were initially revealed by Catterall and co-workers (reviewed by Catterall, 1980; 1986) by studying the action and binding of alkaloid toxins with α-scorpion toxin and sea anemone toxin II.

Interactions with receptor sites of peptide toxins that inhibit sodium current inactivation

Catterall and co-workers have shown that the binding of an α-scorpion toxin to receptor site 3 increased about 2-fold in the presence of veratridine or other alkaloid toxins (Catterall, 1977; Catterall *et al.*, 1981). The positive cooperativity between neurotoxin binding at receptor site 2 and 3 is bidirectional, in a sense that in the presence of α-scorpion toxin or sea anemone toxin the persistent activation of sodium channels by the alkaloids is significantly enhanced, as studied by ion flux assays (Catterall, 1977; Catterall and Beress, 1978), and the specific binding of [^{3}H]BTX-B is increased about 10-fold (Catterall *et al.*, 1981; Sharkey *et al.*, 1987). Veratridine was shown to increase both affinity and receptor concentration (Bmax) of an anti-insect α-scorpion toxin (LqhαIT, α-scorpion toxin highly active on insects, purified from *Leiurus quinquestriatus hebraeus* venom, Eitan *et al.*, 1990) on locust sodium channels (Gordon and Zlotkin, 1993; Cestele *et al.*, 1995), similarly to batrachotoxin effects on an α-scorpion toxin binding on mammalian sodium channels (Catterall, 1977; 1980) (Table 7.2). Interestingly, veratridine has been shown to have no effect on the binding of LqhαIT to cockroach neuronal membranes (Cestele *et al.*, 1995; Gordon *et al.*, 1996a), in contrast to the positive cooperative effect on locust. The differences on the LqhαIT binding modulation by veratridine on sodium channels from the two insect CNS indicate structural differences between them.

O'Leary and Krueger (1989) studied the combined effects of BTX and an α-scorpion toxin from *Leiurus quinquestriatus quinquestriatus* venom, Lqq V (LqqTx) on single rat brain sodium channels in planar lipid bilayers. In the presence of BTX, Lqq V caused the channels to remain open at membrane potentials at where they would normally be closed, at least 50 mV more hyperpolarized than with BTX alone. However, the single channel conductance and block by STX of both BTX-activated and BTX with Lqq V activated channels were similar, indicating that the α-scorpion toxin preferentially affects gating and not ion permeation. In the presence of both toxins, voltage-dependent single channel closing events were eliminated, even at very hyperpolarized potentials.

The synergistic interaction of α-scorpion toxins with BTX was expressed in tracer flux experiments, where α-scorpion toxin increased the apparent potency of BTX to stimulate ^{22}Na influx (Catterall, 1977), and decreased the Kd for [^{3}H]BTX-B binding about 10 fold (Catterall *et al.*, 1981). On the BTX-activated channels in planar lipid bilayer, O'Leary and Krueger (1989) suggested that α-scorpion toxin enhance the BTX-induced hyperpolarizing shift in sodium channel activation, produced in part by BTX binding alone. Since α-Lqq V alone does not significantly alter the voltage dependence of activation and the sodium channels studied are virtually irreversibly activated by BTX, the potentiation of BTX action by Lqq V can be attributed mainly to an increase in the efficacy of BTX once it is bound to the sodium channel and not to an increase in the affinity for BTX, as previously suggested. The results indicated that a cooperative interaction between Lqq V and BTX strongly favor the open state of the sodium channel by causing a large hyperpolarizing shift in the voltage dependence of activation (O'Leary and Krueger, 1989).

Other polypeptide neurotoxins that inhibit inactivation of Na$^+$ currents by binding to distinct, unidentified receptor sites on sodium channels, the *Conus striatus* toxin and *Goniopora* coral toxin, have been shown to enhance the binding of [^{3}H]BTX-B and the persistent activation of sodium channels by veratridine (Gonoi *et al.*, 1986; 1987; Fainzilber *et al.*, 1994; 1995), indicating the presence of positive cooperativity between receptor site 2 and their own.

However, veratridine was shown to inhibit the binding of ^{125}I-δ-conotoxin TxVIA to receptor site 6 on sodium channels of both rat brain and mollusc, due to the increase in dissociation rate constant, resulting in about 2-fold decrease in affinity and consistent with a negative allosteric interaction between δ-TxVIA and veratridine (Fainzilber *et al.*, 1994; Gordon *et al.*, 1996b). Thus, veratridine binding at receptor site 2, in addition to the alteration of sodium channel gating, induces significant conformational changes in different regions of the sodium channel protein, that are probed by the binding of α-scorpion toxin at receptor site 3 (resulting in enhanced binding) as well as by the binding of δ-conotoxin TxVIA at receptor site 6 (resulting in inhibition of binding) (Table 7.2).

Allosteric interactions with other receptor sites

The binding of [^{3}H]BTX-B to receptor site 2 is enhanced 2.9-fold in the presence of PbTx-2 binding to receptor site 5 on sodium channels. The effects of PbTx-2 together with α-scorpion toxin Lqq V on [^{3}H]BTX-B binding are synergistic as demonstrated by a 3-fold increase in binding in the presence of Lqq V over that observed with PbTx-2 alone (Sharkey *et al.*, 1987). Pyrethroids, the synthetic analogs of naturally occurring pyrethrin insecticides act on sodium channels by shifting activation to more negative membrane potentials as well as by inhibiting inactivation (Strichartz *et al.*, 1987). Pyrethroids act at a site distinct from the previously characterized sodium channel receptor sites 1–6. Several pyrethroids (such as deltametrin and cypermethrin) were shown to increase the binding affinity of [^{3}H]BTX-B to rat brain sodium channels (Brown *et al.*, 1988). Similar synergistic interactions have been demonstrated among a synthetic pyrethroid, RU39568, sea anemone toxin II and PbTx-2, which together have been shown to enhance binding of [^{3}H]BTX-B to synaptosomes 100-fold (Lombet *et al.*, 1988). The pyrethroid RU39568 enhance the specific binding of [^{3}H]BTX-B to receptor site 2 on purified and reconstituted sodium channels up to 500-fold, reducing the Kd to 1.5 nM. Brevetoxin and α-scorpion toxins cause further allosteric enhancement of [^{3}H]BTX-B binding (Trainer *et al.*, 1993).

Local anesthetics block sodium channels with complex voltage- and frequency dependent properties (state dependent effect) that are important for the clinical efficacy of the drugs (Butterworth and Strichartz, 1990; Catterall, 1987). The state-dependence of block can be explained by an allosteric model in which the modulated drug receptor has a higher affinity when channels are open or inactivated then when the channels are resting (Catterall, 1987). Ragsdale *et al.* (1994) using site-directed mutations in sodium channel RII from rat brain expressed in *Xenopus* oocytes, defined the location of the local anasthetic receptor site in the pore of the sodium channel. They identify molecular determinants of the state-dependent binding of local anasthetics within the transmembrane segment IVS6 (Ragsdale *et al.*, 1994). Some clinically effective anticonvulsants, believed to share the local anesthetic receptor site, were shown to inhibit the binding of [^{3}H]BTX-B to rat brain synaptosomes, by reducing affinity about 2-fold without altering maximal binding capacity (Willow and Catterall, 1982). The reduction in receptor site 2 affinity for [^{3}H]BTX-B in the presence of these agents appeared to be due to an increased rate of dissociation of [^{3}H]BTX-B from the receptor-ligand complex, suggesting an indirect allosteric mechanism for anticonvulsant inhibition of [^{3}H]BTX-B binding (Willow and Catterall, 1982).

To summarize, the binding of alkaloid neurotoxins to receptor site 2 on sodium channels results, in addition to the profound functional effects on gating and ion selectivity, in alterations of different regions within the sodium channel protein. These allosteric conformational changes are probed by the different neurotoxins, through their high affinity binding interactions with specific recognition sites within the physically distinct receptor sites (Table 7.2). Some of these recognition points might have been altered by the occupancy of an alkaloid toxin at receptor site 2, resulting in increased or decreased affinity of the toxin to its particular receptor site. In parallel, some ligand binding at other regions of the channel are able to induce conformational changes on receptor site 2, expressed in increase or inhibition of [^{3}H]BTX-B binding (Table 7.2).

Allosteric Interactions with Receptor Site 5

The first direct binding studies using tritiated brevetoxin PbTx-3 were performed by Poli *et al.* (1986). They have shown that [^{3}H]PbTx-3 binds to a specific, high affinity receptor site on rat brain sodium channels, and other natural brevetoxin analogs compete with similar affinity for the same receptor site 5. The binding of [^{3}H]PbTx-3 was not dependent on membrane potential and was not affected by batrachotoxin (Poli *et al.*, 1986; Trainer *et al.*, 1993). The binding of [^{3}H]PbTx-9 was increased 2-fold in the presence of both batrachotoxin and RU39568 and by batrachotoxin together with α-scorpion toxin, but not by each of them alone, in purified and reconstituted rat brain sodium channels (Trainer *et al.*, 1993).

Brevetoxin PbTx-2 was shown to increase 2.5-fold the affinity of ^{125}I-Css II, a β-scorpion toxin binding to receptor site 4, on rat brain sodium channels with no effect on Bmax (Sharkey *et al.*, 1987). No effect of brevetoxins was previously reported on the binding of α-scorpion toxins (Sharkey *et al.*, 1987; Catterall and Risk, 1981).

Recently, significant allosteric effects of PbTx-1, the most active brevetoxin analog, have been described on the binding of α-scorpion toxins at receptor site 3 on three different sodium channel preparations: rat brain, locust and cockroach synaptosomes (Cestele *et al.*, 1995). Using ^{125}I-AaH II, the most active α-scorpion toxin on vertebrates, and ^{125}I-LqhαIT, the most active α-scorpion toxin on insects, as specific probes for receptor site 3 on mammal and insect sodium channels, respectively, three types of effects by PbTx-1 were observed: 1) PbTx-1 induced a negative allosteric effect on rat brain sodium channels, by reducing 3.6-fold the affinity of AaH II to receptor site 3, due to the increase in dissociation rate constant, with no change in Bmax (Cestele *et al.*, 1995) (Table 7.2). 2) Positive modulation on LqhαIT binding on locust neuronal sodium channels, similar to the positive effect of veratridine on the same preparation. However, unlike veratridine, that increased both affinity and Bmax (Gordon and Zlotkin, 1993), brevetoxin PbTx-1 increased the affinity of LqhαIT to locust sodium channels about 1.8-fold with no effect on receptor site concentration (Cestele *et al.*, 1995). 3) No effect of PbTx-1 was detected on LqhαIT binding to cockroach neuronal membranes, but brevetoxin has been demonstrated to shift the activation of cockroach axonal sodium current to more negative membrane potentials, similarly to its effects in other preparations (Cestele *et al.*, 1995).

Combined presence of veratridine and brevetoxin PbTx-1 revealed that the resulted modulation on the α-scorpion toxins binding in each preparation could be best described as additive, suggesting that each sodium channel activator, veratridine and brevetoxin, may act independently and affect receptor site 3 on its own way, regardless of the presence of the other neurotoxin. In the presence of saturation concentration of veratridine, the negative modulation of PbTx-1 on AaH II binding to rat brain sodium channels was reduced. Similarly, the presence of saturating concentration of PbTx-1 reduced the increase in binding caused by veratridine alone, resulting in AaH II binding around control level (with no toxin addition) (Cestele *et al.*, 1995). On locust sodium channels, where both veratridine and brevetoxin induced increase in LqhαIT binding, the effect in the presence of both was additive (Cestele *et al.*, 1995; Gordon *et al.*, 1996a).

The dramatic differences in allosteric modulations in sodium channel subtypes in various animal phyla suggest structural differences in receptor sites for brevetoxin, and/or at the coupling regions with α-scorpion toxins receptor site in the different sodium channels. The independent action of each modulator on α-scorpion toxin receptor sites suggest that the conformational changes induced by veratridine and brevetoxin are expressed at different areas of receptor site 3. These results are in concert with the notion that a neurotoxin receptor site is formed by several, discontinuous segments of the channel, closely associated in the tertiary structure of the native channel conformation (Thomsen and Catterall, 1989; Terlau *et al.*, 1991; Gordon *et al.*, 1992; Moskowitz *et al.*, 1994; Tranier *et al.*, 1994). The idea that the allosteric interactions are a result of mechanistic conformational changes induced by the binding of each neurotoxin is also supported by the above results. Elucidation of the molecular structure of the receptor sites on each sodium channel will reveal the dynamic gating activity and the mechanism of allosteric interaction among various neurotoxin receptors, and is an important area of future studies.

SUMMARY

Sodium channels are responsible for the generation and propagation of electrical signals in most excitable cells. Being a critical element in excitability, sodium channels serve as specific target for a large variety of chemically distinct neurotoxins produced by many different animals and plants. The neurotoxins alter each of the sodium channel properties, including ion conductance, ion selectivity, activation and inactivation through binding to at least six distinct classes of receptor sites. Remarkable allosteric interactions occur between various neurotoxins receptor sites, demonstrating the complex dynamic conformational changes within the sodium channel protein, resulting from binding of the neurotoxins. Most neurotoxin binding was shown to be dependent on the functional state of the sodium channel, illustrating the sensitivity of neurotoxin binding to conformational changes within their receptor sites. Only partial mapping of scorpion toxin receptor sites was achieved so far by site-directed antibody mapping. The molecular localization of TTX and STX recognition sites and structure of receptor site 1 has been mainly accomplished by site-directed mutagenesis of sodium channels. Photoaffinity labeling combined with site-directed antibody mapping localized, in part, the receptor site of brevetoxin. In every case, segments from at least two repeated domains participate in formation of a toxin receptor site. Clarification of the mode of action of neurotoxins has been mainly achieved by patch clamp and single channel recordings. Revealing the receptor site and action of different neurotoxins contributed to the understanding of the topology and native tridimentional organization of the sodium channels, the structure of the extracellular pore region and contributed to structure-function relationship of sodium channels.

Several major aspects, however, are still awaiting to be clarified. Among them, the molecular localization of different neurotoxin receptor sites and understanding their allosteric interactions. These will enable the elucidation of the dynamic gating processes and the conformational changes within the channel protein underlying its function. Neurotoxins have been proven to serve as highly sensitive probes to follow after subtle structural and conformational alterations at different regions of the sodium channels. Hence, despite the important progress actualize in our understanding of some neurotoxin mode of action and receptor interactions, much more has to be done. The use of molecular biology methods, coupled with binding and patch clamp as well as new approaches of probing protein structure and molecular interactions will be of great use in this fascinating area of work.

REFERENCES

Aldrich, R.W., Corey, D.P. and Stevens, C.F. (1983) A reinterpretation of mammalian sodium channel gating based on single channel recording. *Nature*, **306**, 436–441.

Armstrong, C.M., Benzanilla, F. and Rojas, E. (1973) Destruction of sodium conductance inactivation in squid axons perfused with pronase. *J. Gen. Physiol.*, **62**, 375–391.

Armstrong, C.M. (1992) Voltage dependent ion channels and their gating. *Physiol. Rev.*, **72(4)** (Suppl), 5–13.

Backx, P.H., Yue, D.T., Lawrence, J.H., Marban, E. and Tomaselli, G.F. (1992) Molecular localization of an ion-binding site within the pore of mammalian sodium channels. *Science*, **257**, 248–251.

Baden, D.G. (1989) Brevetoxins: unique polyether dinoflagellate toxins. *FASEB J.*, **3**, 1807–1817.

Barchi, R.L. and Weigele, J.B. (1979) Characteristics of saxitoxin binding to the sodium channel of the sarcolema isolated from rat skeletal muscle. *J. Physiol.*, **295**, 383–396.

Barchi, R.L. (1988) Probing the molecular structure of the voltage-dependent sodium channel. *Annu. Rev. Neurosci.*, **11**, 455–495.

Barhanin, J., Giglio, J.R., Leopold, P., Schmid, A., Sampaio, S.V. and Lazdunski, M. (1982) *Tityus serrulatus* venom contains two classes of toxins. *J. Biol. Chem.*, **257**, 12553–12558.

Barhanin, J., Meiri, H., Romey, D., Pauron, D. and Lazdunski, M. (1985) A monoclonal immunotoxin acting on the Na⁺ channel, with properties similar to those of scorpion toxin. *Proc. Natl. Acad. Sci. USA*, **82**, 1842–1846.

Benzanilla, F. and Armstrong, C.M. (1974) Gating currents of the sodium channels: three ways to block them. *Science*, **183**, 753–754.

Brown, G.B. (1986) ³H-Batrachotoxinin-A Benzoate binding to voltage-sensitive sodium channels: Inhibition by the channel blockers tetrodotoxin and saxitoxin. *J. Neurosci.*, **6(7)**, 2064–2070.

Brown, G.B., Gaupp, J.E. and Olsen, R.W. (1988) Pyrethroid insecticides: Stereospecific allosteric interaction with the batrachotoxinin-A benzoate binding site on mammalian voltage-sensitive sodium channels. *Mol. Pharmacol.*, **34**, 54–59.

Butterworth, J.F. and Strichartz, G.R. (1990) Molecular mechanism of local anesthesia: a review. *Anasthesiology*, **72**, 711–734.

Catterall, W.A. (1977) Activation of the action potential Na⁺ ionophore by neurotoxins. An allosteric model. *J. Biol. Chem.*, **252**, 8669–8676.

Catterall, W.A. (1980) Neurotoxins that act on voltage-sensitive sodium channels in excitable membranes. *Ann. Rev. Pharmacol. Toxicol.*, **20**, 15–43.

Catterall, W.A. (1986) Molecular properties of voltage-sensitive sodium channels. *Ann. Rev. Biochem.*, **55**, 953–985.

Catterall, W.A. (1987) Common mode of drug action on Na⁺ channels: local anasthetics, antiarrhythmics and anticonvulsants. *Trends Pharmacol. Sci.*, **8**, 57–65.

Catterall, W.A. (1988) Structure and function of voltage-sensitive ionic channels. *Science*, **242**, 50–61.

Catterall, W.A. (1992) Cellular and molecular biology of voltage-gated sodium channels. *Physiol. Rev.*, **72**, 15–48.

Catterall, W.A. and Beress, L. (1978) Sea anemone toxin and scorpion toxin share a common receptor site associated with the action potential sodium ionophore. *J. Biol. Chem.*, **253**, 7393–7396.

Catterall, W.A. and Risk, M. (1980) Toxin T46 from *Ptychodiscus brevis* (formerly *Gymnodinium breve*) enhance activation of sodium channels by veratridine. *Mol. Pharmacol.*, **19**, 345–348.

Catterall, W.A., Morrow, C.S., Daly, J.W. and Brown, G.B. (1981) Binding of Batrachotoxin A 20-α-benzoate to a receptor site associated with sodium channels in synaptic nerve ending particles. *J. Biol. Chem.*, **256**, 9822–9827.

Cestele, S., Ben Khalifa, R., Pelhate, M., Rochat, H. and Gordon, D. (1995) α-Scorpion toxins binding on rat brain and insect sodium channels reveal divergent allosteric modulations by brevetoxin and veratridine. *J. Biol. Chem.*, **270**, 15153–15161.

Clarkson, C.W., Matsubara, T. and Hondeghem, L.M. (1988) Evidence for voltage-dependent block of cardiac sodium channels by tetrodotoxin. *J. Mol. Cell. Cardiol.*, **20**, 1119–1131.

Cohen, S.A. and Barchi, R.L. (1993) Voltage-dependent sodium channels. *Intern. Rev. Cytol.*, **137C**, 55–102.

Correa, A.M., Benzanilla, F. and Latorre, R. (1992) Gating kinetics of batrachotoxin-modified Na+ channels in the squid giant axon. *Biophys. J.*, **61**, 1332–1352.

Correa, A.M., Latorre, R. and Benzanilla, F. (1991) Ion permeation in normal and batrachotoxin-modified Na channels in the squid giant axon. *J. Gen. Physiol.*, **97**, 605–625.

Cruz, L.J., Gray, W.R., Olivera, B.M., Zeikus, R.D., Kert, L., Yoshikami, D. and Moczydlowski, E. (1985) Conus geographicus toxins that discriminate between neuronal and muscle sodium channels. *J. Biol. Chem.*, **260**, 9280–9288.

Eaholtz, G, Scheuer, T. and Catterall., W.A. (1994) Restoration of inactivation and block of open sodium channels by an inactivation gate peptide. *Neuron.*, **12**, 1041–1048.

Eickhorn, R., Weirich, J., Hornung, D. and Antoni, H. (1990) Use dependent of sodium current inhibition by tetrodotoxin in rat heart muscle: influence of channel state. *Pflugers Arch.*, **416**, 398–405.

Eitan, M., Fowler, E., Herrmann, R., Duval, A., Pelhate, M. and Zlotkin, E. (1990) A scorpion venom neurotoxin paralytic to insects that affects sodium current inactivation: purification, primary structure, and mode of action. *Biochemistry*, **29**, 5941–5947.

Fainzilber, M. and Zlotkin, E. (1992) A new bioassay reveals mollusc-specific toxicity in molluscivorous *Conus* venoms. *Toxicon*, **30**, 465–469.

Fainzilber, M., Kofman, O., Zlotkin, E. and Gordon, D. (1994) A new neurotoxin receptor site on sodium channels is identified by a conotoxin that affects sodium channel inactivation in molluscs, and acts as an antagonist in rat brain. *J. Biol. Chem.*, **269**, 2574–2580.

Fainzilber, M., Lodder, H., Kits, K.S., Kofman, O., Vinnitsky, I., Van Reitschoten, J., Zlotkin, E. and Gordon, D. (1995) A new conotoxin affecting sodium current inactivation interacts with the δ-conotoxin receptor site. *J. Biol. Chem.*, **270**, 1123–1129.

Frelin, C., Vigne, P. and Lazdunski, M. (1983) Sodium channels with high and low affinity tetrodotoxin binding sites in the mammalian skeletal muscle cell. *J. Biol. Chem.*, **258**, 7256–7259.

Furhman, F.A. (1986) Tetrodotoxin, tarichotoxin chiriquitoxin: historical perspectives. *Ann. NY Acad. Sci.*, **479**, 1–14.

Gonoi, T., Sherman, S.J. and Catterall, W.A. (1985) Voltage-clamp analysis of tetrodotoxin-sensitive and -insensitive sodium channels in rat muscle cells developing *in vitro*. *J. Neurosci.*, **5**, 2559–2564.

Gonoi, T., Ashida, K., Feller, D., Schmidt, J., Fujiwara, M. and Catterall, W.A. (1986) Mechanism of action of polypeptide neurotoxin from *Goniopora* on sodium channels in mouse neuroblastoma cells. *Mol. Pharmacol.*, **29**, 347–354.

Gonoi, T., Ohizumi, Y., Kobayashi, J., Nakamura, H. and Catterall, W.A. (1987) Actions of a polypeptide toxin from the marine snail *Conus striatus* on voltage-sensitive sodium channels. *Mol. Pharmacol.*, **32**, 691–698.

Gordon, D., Moskowitz, H., Eitan, M., Warner, C., Catterall, W.A. and Zlotkin, E. (1992) Localization of receptor sites for insect-selective toxins on sodium channels by site-directed antibodies. *Biochemistry*, **31**, 7622–7628.

Gordon, D. (1990) Ionic channels in nerve and muscle cells. *Curr. Opin. Cell Biol.*, **2**, 695–707.

Gordon, D. and Zlotkin, E. (1993) Binding of an alpha scorpion toxin to insect sodium channels is not dependent on membrane potential. *FEBS Lett.*, **315**, 125–129.

Gordon, D., Jover, E., Couraud, F. and Zlotkin, E. (1984) The binding of an insect selective neurotoxin (AaIT) from scorpion venom to locust synaptosomal membranes. *Biochim. Biophys. Acta.*, **778**, 349–358.

Gordon, D., Martin-Eauclaire, M.F., Cestele, S., Kopeyan, C., Carlier, E., Ben Khalifa, R., Pelhate, M. and Rochat, H. (1996a) Scorpion toxins affecting sodium current inactivation bind to distinct homologous receptor sites on rat brain and insect sodium channels. *J. Biol. Chem.*, **271**, 8034–8045

Gordon, D., Zlotkin, E., Kofman, O., Kits, K.S. and Fainzilber, M. The δ-conotoxins inhibit sodium current inactivation through binding to distinct but related receptor sites on sodium channels. In: *Biochemical Aspects of Marine Pharmacology*, (Edited by P. Lazarowici, M. Spira and E. Zlotkin), pp. 44–62, Alken, Fort Collins, Colorado, 1996b.

Gordon, D., Merrick, D., Wollner, D.A. and Catterall, W.A. (1988) Biochemical properties of sodium channels in a wide range of excitable tissues studied with site-directed antibodies. *Biochemistry*, **27**, 7032–7038.

Gordon, D., Moskowitz, H. and Zlotkin, E. (1990) Sodium channel polypeptides in central nervous systems of various insects identified with site-directed antibodies. *Biochim. Biophys. Acta*, **1026**, 80–86.

Green, W.N., Weiss, L.B. and Andersen, O.S. (1987) Batrachotoxin-modified sodium channels in planar lipid bilayers. Characterization of saxitoxin- and tetrodotoxin-induced closures. *J. Gen. Physiol.*, **89**, 873–903.

Guy, H.R. and Conti, F. (1990) Pursuing the structure and function of voltage-gated channels. *Trends Neurosci,*. **13**, 201–206.

Guy, H.R. (1988) A model relating the structure of the sodium channel to its function. In: *Current Topics in Membrane and Transport*, Vol. **33**, pp. 289–308.

Hartmann, H.A., Kirsch, G.E, Drewe, J.A., Taglialatela, M, Joho, R.H. and Brown, A.M. (1991) Exchange of conduction pathways between two related K channels. *Science*, **251**, 942–944.

Hartshoren, R.P., Keller, B.U., Talvenheimo, J.A., Catterall, W.A. and Montal, M. (1985) Functional reconstitution of the purified brain sodium channel in planar lipid bilayers. *Proc. Nat. Acad. Sci. USA*, **92**, 240–244.

Hasson, A., Fainzilber, M., Gordon, D., Zlotkin, E. and Spira, M.E. (1993) Alteration of sodium currents by new peptide toxins from the venom of a molluscivorous *Conus* snail. *Eur. J. Neurosci.*, **5**, 56–64.

Hasson, A., Shon, KI-Joon., Olivera, B.M. and Spira, M.E. (1995) Alterations of voltage-activated sodium current by a novel conotoxin from the venom of *Conus gloriamaris*. *J. Neurophysiol.*, **73**, 1295–1302.

Heinemann, S.H., Terlau, H., Stuhmer, W., Imoto, K. and Numa, S. (1992) Calcium channel characteristics conferred on the sodium channel by single mutations. *Nature*, **356**, 441–443.

Hille, B. (1971) The permeability of the sodium channel to organic cations in myelinated nerve. *J. Gen. Physiol.*, **58**, 599–619.

Hille, B. (1975) The receptor for tetrodotoxin and saxitoxin: a structural hypothesis. *Biophy. J.*, **15**, 615–619.

Hille, B. (1992) *Ionic Channels of Excitable Membranes*, Sinauer, Sunderland, MA.

Hille, B., Leibowitz, M.D., Sutro, J.B., Schwarz, J.R. and Holan, G. (1987) State-dependent modification of sodium channels by lipid-soluble agonists. In: *Proteins of Excitable Membranes*, (B. Hille and D.M. Fambrough, Eds.) SGP Series, Vol. 41, New York, pp. 109–124.

Horn, R. and Vandenberg, C. (1984) Statistical properties of single sodium channels. *J. Gen. Physiol.*, **84**, 505–534.

Huang, L-Y.M., Moran, N. and Eherenstein, G. (1982) Batrachotoxin modifies the gating kinetics of sodium channels in internally perfused neuroblastoma cells. *Proc. Natl. Acad. Sci. USA*, **79**, 2082–2085.

Huang, L-Y.M., Moran, N. and Eherenstein, G. (1984) Gating of batrachotoxin-modified sodium channels in neuroblastoma cells determined from single channel measurements. *Biophys. J.*, **45**, 313–322.

Huang, L-Y.M., Yatani, A. and Brown, A.M. (1987) The properties of batrachotoxin-modified cardiac Na$^+$ channels, including state-dependent block by tetrodotoxin. *J. Gen. Physiol.*, **90**, 341–360.

Hue, S.L., Meves, H., Rubly, N. and Watt, D.D. (1983) A quantitative study of the action of *Centruroides sculpturatus* toxins III and IV on the sodium currents of the node of Ranvier. *Pflugers Arch.*, **397**, 90–99.

Joho, R.H., Moorman, J.R., VanDongen, A.M.J., Kirsch, G.E., Silberberg, H., Schuster, G. and Brown, A.M. (1990) Toxin and kinetic profile of rat brain type III sodium channels expressed in Xenopus oocytes. *Mol. Brain Res.*, **7**, 105–113.

Jover, E., Martin-Moutot, N., Couraud, F. and Rochat, H. (1980) Binding of scorpion toxin to rat brain synaptosomal fraction. Effects of membrane potential, ions, and other neurotoxins. *Biochemistry*, **19**, 463–467.

Jover, E., Massacrier, A., Cau, P., Martin, M.F. and Couraud, F. (1988) The correlation between sodium channel subunits and scorpion binding sites. *J. Biol. Chem.*, **263**, 1542–1548.

Kao, C.Y. (1986) Structure-activity relations of tetrodotoxin, saxitoxin, and analogs. *Ann. NY Acad. Sci.*, **479**, 52–67.

Kao, C.Y. and Walker, S.E. (1982) Active groups of saxitoxin and tetrodotoxin as deduced from action of saxitoxin analogs on frog muscle and squid axon. *J. Physiol. (Lond.)*, **323**, 617–637.

Khodorov, B.I. (1985) Batrachotoxin as a tool to study voltage-sensitive sodium channels of excitable membranes. *Prog. Biophys. Mol. Biol.*, **45**, 57–148.

Kirsch, G.E., Skattebol, A., Possani, L.D. and Brown, A.M. (1989) Modification of Na channel gating by an α-scorpion toxin from *Tityus serrulatus*. *J. Gen. Physiol.*, **93**, 67–83.

Kobayashi, M., Wu, C.H., Yoshi, M., Narahashi, T., Nakamura, H., Kobayashi, J. and Ohizumi, Y. (1986) Preferential block of skeletal muscle sodium channel by Geographutoxin II, a new peptide toxin from *Conus geographus*. *Pflugers Arch.*, **407**, 241–243.

Kontis, K.J. and Goldin, A.L. (1993) Site-directed mutagenesis of the putative pore region of the rat IIA sodium channel. *Mol. Pharmacol.*, **43**, 635–644.

Krueger, B.K., Worley, J.F. and French, R.J. (1983) Single sodium channel from rat brain incorporated into planar lipid bilayer membranes. *Nature Lond.*, **303**, 172–175.

Lima, M.E., Martin, M-F., Diniz, C.R. and Rochat, H. (1986). *Tityus serrulatus* toxin VII bears pharmacological properties of both β-toxin and insect toxin from scorpion venoms. *Biochem. Biophys. Res. Commun.*, **139**, 296–302.

Lipkind, G.M. and Fozzard, H.A. (1994) A structural model of the tetrodotoxin and saxitoxin binding site of the Na⁺ channel. *Biophy. J.*, **66**, 1–13.

Lombet, A., Bidard, J.N. and Lazdunski, M. (1987) Ciguatoxin and brevetoxins share a common receptor site on the neuronal voltage-dependent Na⁺ channel. *FEBS Lett.*, **219**, 355–359.

Lombet, A., Mourre, C. and Lazdunski, M. (1988) Interaction of pyrethroid family with specific binding sites on the voltage dependent sodium channel from mammalian brain. *Brain Res.*, **459**, 44–53.

Lonnendonker, U. (1989) Use-dependent block of sodium channel in frog myelinated nerve by tetrodotoxin and saxitoxin at negative holding potential. *Biochim. Biophys. Acta*, **985**, 153–160.

Martin-Eauclaire, M.F. and Couraud, F. (1995) Scorpion neurotoxins: Effects and mechanisms. In *Handk. Neurotoxicology* (L.W. Chang and R.S. Dyer, eds.), pp. 683–716, Marcel Dekker. New York, N.Y.

Moczydlowski, E., Hall, S., Garber, S.S., Strichartz, G.S. and Miller, C. (1984) Voltage-dependent blockade of muscle sodium channels by guanidinium toxins. Effect of toxin charge. *J. Gen. Physiol.*, **84**, 687–704.

Moczydlowski, E., Olivera, B.M., Gray, W.R. and Strichartz, G.S. (1986) Discrimination of neuronal and muscle Na-channel subtypes by binding competition between [³H]saxitoxin and μ-conotoxins. *Proc, Natl. Acad. Sci. USA*, **83**, 5321–5325.

Moskowitz, H., Herrmann, R., Zlotkin, E. and Gordon, D. (1994) Variability among insect sodium channels revealed by binding of selective neurotoxins. *Insect Biochem. Molec. Biol.*, **24**, 13–19.

Nagy, K. (1987) Subconductance states of single sodium channels modified by chloramine-T and sea anemone toxin in neuroblastoma cells. *Eur. Biophys. J.*, **15**, 129–132.

Nagy, K. (1988) Mechanism of inactivation of single sodium channels after modification by chloramine-T, sea anemone toxin and scorpion toxin. *J. Membran Biol.*, **106**, 29–40.

Narahashi, T. (1986) Toxins that modulate the sodium channel gating mechanism. *Ann. NY Acad. Sci.*, **479**, 133–151.

Nicholson, G.M., Willow, M., Howden, M.E.H. and Narahashi, T. (1994) Modification of sodium channel gating and kinetics by versutoxin from the Australian funnel-web spider *Hadronyche versuta*. *Pflugers Arch.*, **428**, 400–409.

Noda, M., Ikeda, T., Kayano, T., Suzuki, H., Takeshima, H., Kurasaki, M., Takahashi, H. and Numa, S. (1986) Existence of distinct sodium channel messenger RNAs in rat brain. *Nature (London)*, **320**, 188–192.

Noda, M., Shimuzu, S., Tanabe, T., Takai, T., Kayano, T., Ikeda, T., Takahashi, H., Nakayama, H., Kanaoka, Y., Minamino, N., Kangawa, K., Matsuo, H., Raftery, M., Hirose, T., Inayama, S., Hayashida, H., Miyata, T. and Numa, S. (1984) Primary structure of *Electrophorus electricus* sodium channel deduced from cDNA sequence. *Nature*, **312**, 121–127.

Noda, M., Suzuki, H., Numa, S. and Stuhmer, W. (1989) A single point mutation confers tetrodotoxin and saxitoxin insensitivity on the sodium channel II. *FEBS Lett.*, **259**, 213–216.

Norton, R.S. (1991) Structure and structure-function relationships of sea anemone proteins that interact with the sodium channel. *Toxicon*, **29**, 1051–1084.

O'Leary, M.E. and Krueger, B.K. (1989) Batrachotoxin and α-scorpion toxin stabilize the open state of single voltage-gated sodium channels. *Mol. Pharmacol.*, **36**, 789–795.

Ohizumi, Y., Minoshima, S., Takahashi, M., Kajiwara, A., Nakamura, H. and Kobayashi, J. (1986a) Geographutoxin II, a novel peptide inhibitors of sodium channels of skeletal muscle and autonomic nerves. *J. Pharmacol. Exp. Ther.*, **239**, 243–248.

Ohizumi, Y., Nakamura, H., Kobayashi, J. and Catterall, W.A. (1986b) Specific inhibition of [^{3}H]Saxitoxin binding to skeletal muscle sodium channels by geographutoxin II, a polypeptide channel blocker. *J. Biol. Chem.*, **261**, 6149–6152.

Patton, D.E. and Goldin, A.L. (1991) A voltage-dependent gating transition induces use-dependent block by tetrodotoxin of rat IIA sodium channels expressed in *Xenopus* oocytes. *Neuron*, **7**, 637–647.

Poli, M.A., Mende, T.J. and Baden, D.G. (1986) Brevetoxins, unique activators of voltage-sensitive sodium channels, bind to specific sites to rat brain synaptosomes. *Mol. Pharmacol.*, **30**, 129–135.

Pusch, M., Noda, M., Stuhmer, W., Numa, S. and Conti, F. (1991) Single point mutations of the sodium channel drastically reduce the pore permeability without preventing its gating. *Eur. Biophys. J.*, **20**, 127–133.

Quandt, F.N. and Narahashi, T. (1982) Modification of single sodium channels by batrachotoxin. *Proc. Natl. Acad. Sci. USA*, **79**, 6732–6736.

Ragsdale, D.S., McPhee, J.C., Scheuer, T. and Catterall, W.A. (1994) Molecular determinants of state-dependent block of Na^{+} channels by local anesthetics. *Science*, **265**, 1724–1728.

Ray, R., Morrow, C.S. and Catterall, W.A. (1978). Binding of scorpion toxin to receptor sites associated with voltage-sensitive sodium channels in synaptic nerve ending particles. *J. Biol. Chem.*, **253**, 7307–7313.

Richie, J.M. and Rogart, R.B. (1977) The binding of saxitoxin and tetrodotoxin to excitable tissue. *Rev. Physiol. Biochem. Pharmacol.*, **79**, 1–50.

Rogart, R.B. (1986) High-STX-affinity vs. low-STX-affinity Na^{+} channel subtypes in nerve, heart, and skeletal muscle. *Ann. NY Acad. Sci.*, **479**, 402–430.

Rogart, R.B. Cribbs, L.L., Muglia, L.K., Kephart, D.D. and Kaiser, M.W. (1989) Molecular cloning of a putative tetrodotoxin-resistant rat heart sodium channel isoform. *Proc. Natl. Acad. Sci. USA*, **86**, 8170–8174.

Ruben, P.C., Starkus, J.G. and Rayner, M.D. (1990) Holding potential affects the apparent voltage sensitivity of sodium channel activation in crayfish giant axons. *Biophys. J.*, **58**, 1169–1181.

Salgado, V.L., Yeh, J.Z. and Narahashi, T. (1986) Use- and voltage-dependent block of the sodium channel by saxitoxin. *Ann. NY Acad. Sci.*, **479**, 84–95.

Satin, J., Kyle, J.W., Chen, M., Bell, P., Cribbs, L.L., Fozzard, H.A. and Rogart, R.B. (1992) A mutant of TTX-resistant cardiac sodium channels with TTX-sensitive properties. *Science*, **256**, 1202–1205.

Sato, K., Ishida, Y., Wakamatsu, K.R., Honda, H., Ohizumi, Y., Nakamura, H., Ohya, M., Lancelin, J.M., Kohda, D. and Inagaki, F. (1991) Active site of μ-conotoxin GIIIA, a peptide blocker of muscle sodium channels. *J. Biol. Chem.*, **266**, 16989–16991.

Scanley, B.E., Hanck, D.A., Chay, T. and Fozzard, H.A. (1990) Kinetic analysis of single sodium channels from canine cardiac purkinje cells. *J. Gen. Physiol.*, **95**, 411–437.

Schantz, E.J. (1986) Chemistry and biology of saxitoxin and related toxins. *Ann. NY Acad. Sci.*, **479**, 15–23.

Schreibmayer, W., Kazerani, H. and Tritthart, H.A. (1987) A mechanistic interpretation of the action of toxin II from *Anemonia sulcata* on the cardiac sodium channel. *Biochim. Biophys. Acta*, **901**, 273–282.

Sharkey, R.G., Jover, E., Couraud, F., Baden, D.G. and Catterall, W.A. (1987) Allosteric modulation of neurotoxin binding to voltage-sensitive sodium channels by *Ptychodiscus brevis* toxin 2. *Mol. Pharmacol.*, **31**, 273–278.

Sherman, S.J. and Catterall., W.A. (1982) Biphasic regulation of development of the high affinity saxitoxin receptor by innervation in rat skeletal muscle. *J. Gen. Physiol.*, **80**, 753–768.

Shon, K., Grilley, M.M., Marsh, M., Yoshikami, D., Hall, A.R., Kurz, B., Gray, W.R., Imperial, J.S., Hillyard, D.R. and Olivera, B.M. (1995) Purification, characterization. Synthesis and cloning of the lockjaw peptide from *Conus purpurascens* venom. *Biochemistry*, **34**, 4913–4918.

Shon, K., Hasson, A., Spira, M.E., Cruz, L.J., Gray, W.R. and Olivera, B.M. (1994) δ-Conotoxin GmVIA, a novel peptide from the venom of *Conus gloriamaris*. *Biochemistry*, **33**, 11420–11425.

Stankiewicz, M., Ben Khalifa, R., Grolleau, F., Tomaszewski, R., Kadziela, W. and Pelhate, M. (1996) Late sodium current induced by scorpion toxins and pyrethroid insecticides in insect neuronal membrane. In: *Physiology and ecotoxicology of insects and mechanisms of adaptation in vertebrates* (Z. Bargiel, Edt.). Scientific Society of Torun, Torun, Poland. In press.

Stephan, M.M., Potts, J.F. and Agnew, W.S. (1994) The μ1 skeletal muscle sodium channel: Mutation E403Q eliminates sensitivity to tetrodotoxin but not to μ-conotoxin GIIIA and GIIIB. *J. Membrane Biol.*, **137**, 1–8.

Stocker, M. Pongs, O., Hoth, M., Heinemann, S.H., Stuhmer, W., Schroter, K.H. and Ruppersberg, J.P. (1991) Swapping of functional domains in voltage-gated K channels. *Proc. Royal Soci. London Series B. Biological Sciences*, **245**, 101–107.

Strichartz, G. (1984) 'Structural determinants of the affinity of saxitoxin for neuronal sodium channels. *J. Gen. Physiol.*, **84**, 281–305.

Strichartz, G., Rando, T. and Wang, G.K. (1987) An integrated view of the molecular toxicology of sodium channel gating in excitable cells. *Ann. Rev. Neurosci.*, **10**, 237–267.

Strichartz, G., Rando, T., Hall, S., Gischier, J., Magnani, B. and Bay, C.H. (1986) On the mechanism by which saxitoxin binds to and blocks sodium channels. *Ann. NY Acad. Sci.*, **479**, 96–112.

Stuhmer, W. (1991) Structure-function studies of voltage-gated ion channels. *Annu. Rev. Biophys. Chem.*, **20**, 65–78.

Stuhmer, W., Conti, F., Suzuki, H., Wang, X., Noda, M., Yahagi, N., Kubo, H. and Numa, S. (1989) Structural parts involved in activation and inactivation of the sodium channel. *Nature*, **339**, 597–603.

Tanguy, J. and Yeh, J.Z. (1988) Batrachotoxin uncouples charge immobilization from fast inactivation in squid giant axon. *Biophys. J.*, **54**, 719–730.

Tejedor, F. and Catterall, W.A.. (1990) Photoaffinity labeling of the receptor site for α-scorpion toxins on purified and reconstituted sodium channels by a new toxin derivative. *Cellu. Mollec. Neurobiol.*, **10**, 257–265.

Tejedor, F., McHugh, E. and Catterall, W.A. (1988) Stabilization of sodium channel state with high affinity for saxitoxin by intramolecular cross-linking. Evidence for allosteric effects of saxitoxin binding. *Biochemistry*, **27**, 2389–2397.

Terlau, H., Heinemann, S.H., Stuhmer, W., Pusch, M., Conti, F., Imoto, K. and Numa, S. (1991) Mapping the site of block by tetrodotoxin and saxitoxin of sodium channel II. *FEBS Lett.*, **293**, 93–96.

Thomsen, W.J. and Catterall, W.A. (1989) Localization of the receptor site for α-scorpion toxins by antibody mapping implications for sodium channel topology. *Proc. Natl. Acad. Sci. USA*, **86**, 10161–10165.

Trainer, V.L., Baden, D.G. and Catterall, W.A. (1994) Identification of peptide components of the brevetoxin receptor site of rat brain sodium channels. *J. Biol. Chem.*, **269**, 19904–19909.

Trainer, V.L., Moreau, E., Guedin, D., Baden, D.G. and Catterall, W.A. (1993) Neurotoxin binding and allosteric modulation at receptor sites 2 and 5 on purified and reconstituted rat brain sodium channels. *J. Biol. Chem.*, **268**, 17114–17119.

Trainer, V.L., Thomsen, W.J., Catterall, W.A. and Baden, D.G. (1991) Photoaffinity labeling of the brevetoxin receptor on sodium channels in rat brain synaptosomes. *Mol. Pharmacol.*, **40**, 988–994.

Trimmer, J.S. and Agnew, W.S. (1989) Molecular diversity of voltage-sensitive Na channels. *Annu. Rev. Physiol.*, **51**, 401–418.

Trimmer, J.S., Cooperman, S.S., Tomiko, S.A., Zhou, J. Crean, S.M., Boyle, M.B., Kallen, R.G., Sheng, Z., Barchi, R.L., Sigworth, F.J., Goodman, R.H., Agnew, W.S. and Mandel, G. (1989) Primary structure and functional expression of a mammalian skeletal muscle sodium channel. *Neuron*, **3**, 33–49.

Vandenberg, C. and Benzanilla, F. (1991) A model of sodium channel gating based on single channel, macroscopic ionic, and gating currents in the squid giant axon. *Biophys. J.*, **60**, 1511–1533.

Vassilev, P., Scheuer, T. and Catterall, W.A. (1988) Identification of as intracellular peptide segment involved in sodium channel inactivation. *Science*, **241**, 1658–1661.

Vassilev, P., Scheuer, T. and Catterall, W.A. (1989) Inhibition of inactivation of single sodium channels by a site-directed antibody. *Proc. Natl. Acad. Sci. USA*, **86**, 8147–8151.

Vijverberg, H.P., Pauron, D. and Lazdunski, M. (1984) The effect of *Tityus serrulatus* scorpion toxin γ on sodium channel in neuroblastoma cells. *Pflugers Arch.*, **40**, 297–303.

West, J.W., Patton, D.E., Scheuer, T., Wang, W., Goldin, A.L. and Catterall, W.A. (1992) A cluster of hydrophobic amino acid residues required fast sodium channel inactivation. *Proc. Natl. Acad. Sci. USA*, **89**, 10910–10914.

Wheeler, K.P., Watt, D.D. and Lazdunski, M. (1983) Classification of sodium channel receptors specific for various scorpion toxins. *Pflugers Arch.* **397**, 164–165.

Willow, M. and Catterall, W.A. (1982) Inhibition of binding of [^{3}H]Batrachotoxinin A 20-α-benzoate to sodium channels by anticonvulsant drugs diphenyhydantoin and carbamazepine. *Mol. Pharmacol.* **22**, 627–635.

Wu, C.H. and Narahashi, T. (1987) Mechanism of action of novel marine neurotoxins on ion channels. *Annu. Rev. Pharmacol. Toxicol.* **28**, 141–161.

Yang, L. and Kao, C.Y. (1992) Action of chiriquitoxin on frog skeletal muscle fibers and implications for the tetrodotoxin/saxitoxin receptor. *J. Gen. Physiol.*, **100**, 609–622.

Yatani, A., Kirsch, G.E., Possani, L.D. and Brown, A.M. (1988) Effects of a new world scorpion toxins on single-channel and whole cell cardiac sodium currents. *Am. J. Physiol.*, **254**, H443–H451.

Zabrodovskaya, L.D. and Khodorov, B.I. (1985) Effect of tityus γ toxin on the activation process in sodium channels of frog myelinated nerve. *Gen. Physiol. Biophys.*, **4**, 101–104.

Zlotkin, E., Kadouri, D., Gordon, D., Pelhate, M., Martin, M.F. and Rochat, H. (1985) An excitatory and a depressant insect selective toxin from scorpion venom both affect sodium conductance and possess a common binding site. *Arch. Biochem. Biophys.*, **240**, 877–887.

8. ω-TOXINS, CALCIUM CHANNELS AND NEUROSECRETION

ANTONIO G. GARCÍA, ALMUDENA ALBILLOS,
LUIS GANDÍA, MANUELA G. LÓPEZ, PEDRO MICHELENA
and CARMEN MONTIEL

*Departamento de Farmacología, Facultad de Medicina, Universidad Autónoma de Madrid,
Arzobispo Morcillo, 4, 28029 Madrid, Spain*

INTRODUCTION

The combination of patch-clamp techniques, ω-toxins and molecular genetic strategies have revealed a great heterogeneity of voltage-dependent Ca^{2+} channels in neurones. Peptide toxins derived from the venoms of marine snails *Conus geographus* (ω-conotoxin GVIA), *Conus magus* (ω-conotoxins MVIIA, MVIIC and MVIID), as well as from *Agelenopsis aperta* spider venom (FTX, ω-agatoxin IVA) are powerful diagnostic pharmacological tools to discriminate between different subtypes of neuronal Ca^{2+} channels. Thus, so-called high-threshold activated (HVA) Ca^{2+} channels are selectively recognized by ω-conotoxin GVIA and MVIIA (N-type), by low concentrations (nanomolar) of ω-agatoxin IVA (P-type), or by high concentrations of ω-agatoxin IVA (micromolar) or the ω-conotoxins MVIIC and MVIID (Q-type). L-type HVA Ca^{2+} channels present in neurones, cardiovascular tissues, skeletal and smooth muscle, and in endocrine cells are targetted by so-called organic Ca^{2+} antagonists such as the 1,4-dihydropyridines (DHP) nifedipine or Bay K 8644, the benzylalkylamine verapamil or the benzothiazepine diltiazem; they are also specifically blocked by snake toxins calciseptine and calcicludine. Wide-spectrum ω-toxins (ω-conotoxin MVIIC, ω-agatoxin IA, IIA and IIIA) and organic compounds (flunarizine, dotarizine, cinnarizine, fluspirilene, R56865) can block several classes of HVA Ca^{2+} channels, including the L-type. A neuronal R-type HVA channel seems to be resistant to all known toxins. Low-voltage-activated (LVA) channels are blocked by 1-octanol and amiloride, and are more sensitive to Ni^{2+} than to Cd^{2+}; no toxins are known that recognize these channels.

It is interesting that a single cell can express different subtypes of HVA Ca^{2+} channels and that the quantitative expression of each channel subtype differs with the animal species. The example of adrenal medulla chromaffin cells is illustrative. In the bovine, PQ-type (45%) and N-type (35%) are predominant; the L-type Ca^{2+} channel carries a minor component of the whole-cell current (20%). In the rat, the L-type predominates (50%), together with the N-type (35%), while the PQ family accounts for a minor component (15%). A pattern similar to the rat was seen in cat chromaffin cells.

Ca^{2+} channels consist basically of a multiple subunit protein complex with a central pore-forming α_1 subunit and several regulatory and/or auxiliary subunits, which include a β subunit and the disulfide-linked α_2/δ subunits. Additionally, a fifth subunit has been reported in some tissues, such as the skeletal muscle γ subunit or the neural P95 subunit. The α_1 subunit contains the Ca^{2+} conductance pore, the essential gating machinery and the receptor sites for the most prominent pharmacological agents. The mammalian family of Ca^{2+} channel α_1 genes may be grouped into two subfamilies. The L-subfamily consists of three genes designed as α_{1S} (cloned from skeletal muscle), α_{1C} (cardiac, smooth muscle, neuronal), and α_{1D} (neuronal, endocrine). The DHP-resistant non-L subfamily have another three genes that are expressed almost exclusively in neuronal tissues. The α_{1B} gene encodes an ω-conotoxin GVIA-sensitive N-type Ca^{2+} channel. The α_{1A} subunit is likely related to the P-Q-family of Ca^{2+} channels. The α_{1E} subunit seems to be related to the R channel expressed in rat cerebellar granule neurones.

Ca^{2+} channels are known to be regulated by a large number of neurotransmitters which cause their phosphorylation by protein kinase A, by protein kinase C and by others kinases and messengers. Many neurotransmitters (noradrenaline, GABA, opiates, adenosine, ATP, serotonin, dopamine) inhibit the neuronal Ca^{2+} channels in a voltage-dependent manner. This modulatory effect is exerted through G-protein mediated mechanisms. The use of ω-conotoxins to isolate different Ca^{2+} channel subtypes has facilitated the study of this modulatory mechanisms, which mostly affect the non-L subtypes of HVA channels. This modulatory mechanism works through a feed-back autocrine loop which is activated by released neurotransmitters acting on presynaptic receptors.

N-type Ca^{2+} channels are highly involved in the control of noradrenaline release from sympathetic neurones, as well as acetylcholine release from the electric fish muscle end-plate, the myenteric plexus and detrusor muscle. Also N-channels partially control the non-adrenergic non-cholinergic neurotransmission in smooth muscle, GABA release in cerebellar neurons, glycine release in dorsal horn neurons of the spinal cord, adrenaline release from the dog adrenal, dynorphin release in dentate girus and the synaptic neurotransmission in retinal ganglion neurons and the hippocampus. P channels dominate the release of GABA from deep cerebellar neurons, glycine from dorsal horn neurons of the spinal cord and acetylcholine from the mammalian neuromuscular juction. They also seem to participate partially in the control of the release of other neurotransmitters. Up to now, Q channels have been implicated in the control of neurotransmission in the hippocampus and in the release of catecholamines from bovine chromaffin cells. L-type Ca^{2+} channels dominate the release of catecholamines in rat and cat chromaffin cells, and partially control the secretory process in bovine chromaffin cells.

Today, there is a growing consensus that Ca^{2+} channels are segregated to specialized areas of the plasmalemma were exocytosis occurs ("active sites"). In fact, there is evidence that N-type Ca^{2+} channels form a complex with the proteins of the exocytotic machinery neurexin, synaptotagmin and syntaxin, the "synaptosecretosome". The $[Ca^{2+}]_i$ gradients generated by the opening of such highly localized Ca^{2+} channels are not perceived by techniques measuring average changes of

$[Ca^{2+}]_i$. Thus, different indirect experimental and theoretical approaches predicts that local $[Ca^{2+}]_i$ can reach as much as 100 μM at specific exocytotic micro-domains, nearby the inner mouth of Ca^{2+} channels.

A critical question is why a neurosecretory cell expresses several Ca^{2+} channel sub-types. In bovine adrenal chromaffin cells L, N, P and Q channels have been found, yet only the L- and Q-types seem to be involved in the control of catecholamine release induced by depolarizing stimuli. Therefore, it seems that those "exocytotic channels" (L and Q) must be located nearby the secretory surface of the cell; however, it is uncertain what might be the functions of Ca^{2+} channels (i.e. N and P) that do not participate in the immediate control of the secretory process. Many other questions remain unanswered. For instance, the nature of the R type channel and whether a toxin can be found to recognize this channel. A third question relates to the number of Ca^{2+} channels yet unrecognized. The functions of the Ca^{2+} channels not related to exocytosis (i.e. the neuronal L-type channels) are a mistery. Finally, it is important to stress the need of finding non-peptide molecules to target specifically different channel subtypes; these compounds should cross the blood brain barrier and thus serve as therapeutic drugs to treat different brain diseases.

To communicate between them, cells use a language essentially based in electrical and chemical signals. This language is specially elaborate in the central nervous system, where communication between neurons is established at synapses. There, electrical signals (action potentials arriving to the presynaptic nerve termi-nal) and chemical signals (neurotransmitters pre-stored at presynaptic vesicles are released in milliseconds time to the synaptic gap, to act on pre- and postsynaptic receptors) alternate in a dynamic interconverting process implying the transforma-tion of the chemical into an electrical signal, and viceversa. Changes in the cytosolic concentration of ionized calcium ($[Ca^{2+}]_i$) play a critical role in the control of both, the electrical activity of neurones (i.e. through the regulation of their rate of firing by Ca^{2+}-dependent K^+ channels, the inactivation of Ca^{2+} channels, or the generation of Ca^{2+}-dependent action potentials) and the release of neurotransmitters (where Ca^{2+} serves as the coupling agent between the stimulus and the activation of the secretory machinery; Douglas, 1969; Katz, 1969). Both processes are controlled by the $[Ca^{2+}]_i$ at specific cell microdomains.

Various subcellular organelles (smooth endoplasmic reticulum, mitochondria, synaptic vesicles), Ca^{2+}-binding proteins and Ca^{2+}-transport proteins are responsible for the sequestration of Ca^{2+} and for maintaining low (around 100 nM) the $[Ca^{2+}]_i$. Plasmalemmal receptor-gated and voltage-sensitive Ca^{2+} channels are responsible for the fast transients of the $[Ca^{2+}]_i$ during cell activation. To characterize the voltage-gated Ca^{2+} channels in neurones, toxins from various marine species of genus *Conus* and from spiders have been invaluable tools to clarify the role of Ca^{2+} channels subtypes in controlling transmitter release. Without these so-called ω-toxins (Adams *et al.*, 1993; Olivera *et al.*, 1994) we could not have the insight on three aspects of neuronal communication that we will review in this article: (i) An ever increasing diversity of voltage-dependent Ca^{2+} channels being identified in neurones; (ii) their relevance to the control of neurosecretion; (iii) the functional specialization and strategic localization of Ca^{2+} channel subtypes to generate $[Ca^{2+}]_i$

microdomains nearby the secretory machinery. Some recent reviews analyze different aspects related to the use of ω-toxins and of molecular genetic approaches to define the diversity of Ca^{2+} channels in excitable cells (Olivera *et al.*, 1994; Gandía *et al.*, 1994; Hofmann *et al.*, 1994).

MULTIPLE SUBTYPES OF CALCIUM CHANNELS

Two approaches are mainly responsible for the discovery of the reach diversity of voltage-dependent Ca^{2+} channels. On the one hand, the improvement of the patch-clamp techniques (Hamill *et al.*, 1981) that allow the characterization of the biophysical properties of Ca^{2+} channels both at the single-channel and at the whole-cell level (kinetics of activation, inactivation and deactivation, voltage-range for activation, conductance). On the other hand, the isolation, purification and synthesis of small peptides from *Agelenopsis aperta* spider venom (ω-agatoxins) and from the venom of the piscivorous *Conus* marine snails (ω-conotoxins), have provided ligands with remarkable discrimination for different subtypes of high-threshold Ca^{2+} channels (Olivera *et al.*, 1994).

For the N-type Ca^{2+} channel, there are two *Conus* peptides which have become widely used as diagnostic pharmacological agents: ω-conotoxin GVIA, originally isolated from the venom of *Conus geographus* and ω-conotoxin MVIIA, originally characterized from the venom of *Conus magus*. For P-type Ca^{2+} channels, the toxin fraction (FTX) of the *Agelenopsis aperta* spider venom was initially used. Further purification of the venom led to the isolation of the ω-agatoxins. One of these, ω-agatoxin IVA, specifically blocks P-type Ca^{2+} channels. Finally, other ω-conotoxins have broader Ca^{2+} channel blocking properties than ω-conotoxin GVIA and ω-conotoxin MVIIA. cDNA clones encoding previously unknown ω-conotoxins were identified from a cDNA library made from the venom duct of *Conus magus* (Hillyard *et al.*, 1992). The predicted peptides (ω-conotoxin MVIIC and ω-conotoxin MVIID) were chemically synthetized and characterized. Both peptides inhibit Ca^{2+} channels resistant to 1,4-dihydropyridines (DHP), ω-conotoxin GVIA and ω-agatoxin IVA; thus these new conotoxins are particularly promising ligands for investigating the neuronal Ca^{2+} channel subtypes for which no other defined pharmacological agents are available.

With the combination of the patch-clamp techniques and these pharmacological probes at least six subtypes of voltage-dependent Ca^{2+} channels have been described up to now: T, L, N, P, Q and R (Table 8.1). These channels can be classified according to their range of activation in two main groups: one with a low threshold for activation (low-voltage-activated: LVA) and other with a high threshold for activation (high voltage-activated: HVA).

Low Voltage-Activated Channels: T-type Ca^{2+} Channels

LVA Ca^{2+} channels (Carbone and Lux, 1984) are characterized by a low threshold for activation and a similar permeability for Ca^{2+} and Ba^{2+} (Fox *et al.*, 1987a, b). A

Table 8.1 Ca^{2+} channel subtypes

Channel subtype	Activation	Conductance	Inactivation kinetics	Pharmacology
LOW THRESHOLD (LVA)				
T	> −60 mV	6–9 pS	τ = 5–10 ms Voltage-dependent	Ni^{2+} Amiloride
HIGH THRESHOLD (HVA)				
L	> −50 mV	18–28 pS	τ > 200 ms Calcium-dependent	Co^{2+}, Mn^{2+}, Cd^{2+} Dihydropyridines
N	> −50 mV	7–14 pS	τ = 50 ms Calcium-dependent	Co^{2+}, Mn^{2+}, Cd^{2+} ω-conotoxin GVIA ω-conotoxin MVIIA ω-conotoxin MVIIC ω-conotoxin MVIID
P	> −45 mV	9,14,19 pS	τ = 1 s	Co^{2+}, Mn^{2+}, Cd^{2+} FTX; sFTX ω-agatoxin IVA (nM) ω-conotoxin MVIIC
Q	—	—	—	ω-conotoxin MVIIC ω-conotoxin MVIID ω-agatoxin IVA (μM)
R	—	—	—	Ni^{2+} > Cd^{2+}

single subtype of Ca^{2+} channel has been identified in this group (Carbone and Lux, 1984) and was termed T (for "Transient" or "Tiny"). The main characteristics of T-type Ca^{2+} channels are their fast inactivation, which generates a transient current, and their inactivation when the holding potential is fixed between −60 and −50 mV. The single-channel conductance has been estimated to be around 8 pS.

Pharmacologically, T-type channels can be distinguished from other subtypes because they are more sensitive to blockade by the inorganic Ca^{2+} channel blocker Ni^{2+} than for Cd^{2+} (Fox *et al.*, 1987a, b). It has also been described that two organic compounds, 1-octanol and amiloride can selectively block T-type channels. Although some DHP derivatives as Bay K 8644 and nifedipine have been described to block T-type Ca^{2+} channels in glomerulosa cells of rat adrenal glands (Durroux *et al.*, 1988) and hypothalamic neurones (Akaike *et al.*, 1989), there is a general agreement that DHPs do not block T-type channels and are selective blockers for L-type channels when used at low concentrations, as described below.

High Voltage-Activated Ca^{2+} Channels

In contrast to LVA channels, HVA channels are characterized by their activation by strong depolarizing steps (Fox *et al.*, 1987a,b), a higher permeability to Ba^{2+} than to Ca^{2+} and a higher sensitivity to Cd^{2+} than to Ni^{2+}. Up to now, four major subtypes (L, N, P and Q) of HVA channels have been identified; less characterized is a fifth subtype named R. The major differences between these subtypes are related to their inactivation kinetics and their pharmacological properties.

L-type Ca^{2+} channels

L-type (for "long lasting") Ca^{2+} channels are kinetically characterized by showing little inactivation during depolarizing steps (τ_{inact} >500 ms) and their lower sensitivity to depolarized holding potentials. Single channel conductance was estimated to be around 18–28 pS. This subtype of Ca^{2+} channel seems to be present in all excitable cells and in many non-excitable cells, and they constitute the main pathway for Ca^{2+} entry in heart and smooth muscle, serving also to control hormone and transmitter release from endocrine cells and some neuronal preparations. Pharmacologically, L-type Ca^{2+} channels are highly sensitive to DHPs, both agonists (i.e. Bay K 8644) and antagonists (i.e. nifedipine, nimodipine, furnidipine). DHP agonists effects are characterized by the prolongation of the mean time for channel opening (Nowycky *et al.*, 1985; Gandía *et al.*, 1995), typically observed in whole-cell electrophysiological recordings as a prolongation of tail currents (Plummer *et al.*, 1989).

Other organic compounds have been described to effectively block L-type Ca^{2+} channels (Fleckenstein, 1983; Spedding, 1985): the arylalkylamines (i.e. verapamil) and benzothiazepines (i.e. diltiazem) are particularly useful in cardiac and smooth muscle cells, where they exert negative inotropic effects. Some piperazine derivatives (cinnarizine, flunarizine, dotarizine, R56865) also block L-type Ca^{2+} channels, but they block other subtypes of Ca^{2+} channels as well and thus, have been proposed as "wide-spectrum" Ca^{2+} channel blockers (Garcez do Carmo *et al.*, 1993; Villarroya *et al.*, 1995).

Some toxins have also been shown to block L-type Ca^{2+} channels. Of these, two toxins isolated from snake venoms, i.e. calciseptine (De Weille *et al.*, 1991; Kuroda *et al.*, 1992) and calcicludine (Schweitz *et al.*, 1994) seem to specifically block L-type Ca^{2+} channels, while ω-toxins from spider *Agelenopsis aperta* venom as ω-agatoxin IA (Scott *et al.*, 1990a; Adams *et al.*, 1990), ω-agatoxin IIA (Adams *et al.*, 1990; Venema *et al.*, 1992) and ω-agatoxin IIIA (Adams *et al.*, 1990; Mintz *et al.*, 1991; Cohen *et al.*, 1992; Mintz, 1994; Ertel *et al.*, 1994) are less specific and can block several subtypes of voltage-dependent Ca^{2+} channels.

Taicatoxin, an oligomeric toxin isolated from the *Oxyuranus s. scutellatus* snake venom, has been claimed to selectively block L-type Ca^{2+} channels, in a reversible and voltage-dependent manner (Possani *et al.*, 1992); however, we have not observed any significant effect of this toxin in bovine chromaffin cells (unpublished data).

N-type Ca²⁺ channels

N-type Ca^{2+} channels are distinguishable from L-type for their faster inactivation kinetics (τ_{inact} 50–80 ms). Although their relative fast inactivation can lead to their inactivation when maintaining a depolarizing holding potential, in some preparations N-type Ca^{2+} channels can contain a non-inactivating component, even at the end of long depolarizations, for instance in bovine chromaffin cells, in which N-type channels have been described as "non classical N-type" (Artalejo *et al.*, 1992). Also, in cat chromaffin cells the N-component of the whole-cell current through Ca^{2+} channels does not seem to inactivate (Albillos *et al.*, 1994). Thus, separation of current components based only on the activation and inactivation kinetics may lead to erroneous conclusions. Single-channel conductance of N-type channels has been estimated to be around 13 pS.

Pharmacologically, N-type Ca^{2+} channels are characterized by the irreversible blockade induced by the *Conus geographus* toxin ω-conotoxin GVIA (Nowycky *et al.*, 1985; Kasai *et al.*, 1987; Olivera *et al.*, 1994) and the reversible blockade induced by the *Conus magus* toxin ω-conotoxin MVIIA (Valentino *et al.*, 1993; Vega *et al.*, 1995). Other wide-spectrum toxins isolated from the venom of *Conus magus* snails as ω-conotoxin MVIIC and ω-conotoxin MVIID (Hillyard *et al.*, 1992; Monje *et al.*, 1993) can also block N-type Ca^{2+} channels in a non-selective manner. As described above, several ω-agatoxins (i.e. ω-agatoxin IIA and ω-agatoxin IIIA) can also block N-type Ca^{2+} channels.

A new toxin isolated from tarantula *Grammostola spatulata* venom, termed ω-grammotoxin SIA also blocks N-type Ca^{2+} channels in a non selective manner. This toxin also blocks P- and Q-, but not L-type channels (Lampe *et al.*, 1993; Piser *et al.*, 1994; 1995; Turner *et al.*, 1995).

P-type Ca²⁺ channels

P-type Ca^{2+} channels were first described by Llinás *et al.* (1989) in cerebellar Purkinje cells, in which Ca^{2+} currents were resistant to blockade by DHPs and ω-conotoxin GVIA. The toxin fraction from the venom of the funnel web spider *Agelenopsis aperta* (FTX) was found to effectively block this resistant current. These results led these authors to suggest the existence of a new subtype of HVA Ca^{2+} channel, which was termed P (for "Purkinje"). P-type Ca^{2+} channels are relatively insensitive to changes in the holding potential, and do not inactivate during depolarizing steps (Regan, 1991; Mintz *et al.*, 1992a,b). Multiple single channel conductances have been described for P-type Ca^{2+} channels (Usowicz *et al.*, 1992; Umemiya and Berger, 1995).

Pharmacologically, although P-type Ca^{2+} channels were first described by their blockade by FTX and its synthetic analog sFTX, these two toxins were found to block not only P-type Ca^{2+} channels but also other voltage- and ligand-activated whole-cell currents (Scott *et al.*, 1992). These results led to a further purification of the venom and three types of toxins were found (acylpolyamines, ω-agatoxins and µ-conotoxins). The ω-agatoxins (Adams *et al.*, 1990) have been shown to specifically block different subtypes of Ca^{2+} channels (see above). The fraction identified as ω-agatoxin IVA selectively blocks P-type Ca^{2+} channels, with a K_D of 1–2 nM (Mintz *et al.*, 1992a,b; Mintz and Bean, 1993a) and thus, this toxin in the nanomolar range (<30–100 nM) is actually accepted to be the probe to identify the presence of P-type Ca^{2+} channels. At higher concentrations (>100 nM), ω-agatoxin IVA seems to block not only P-type channels but also Q-type Ca^{2+} channels (see below). Recently, a new fraction (ω-agatoxin IVB) has been purified from the venom of *Agelenopsis aperta*. This new toxin has been shown to be a potent (K_D approximately 3 nM) blocker of P-type Ca^{2+} channels (Adams *et al.*, 1994), being 10 times more abundant in the venom than ω-agatoxin IVA.

P-type Ca^{2+} channels can be also blocked in a non-selective manner by the *Conus magus* snail toxins with wider spectrum of action ω-conotoxin MVIIC (Hillyard *et al.*, 1992; Monje *et al.*, 1993) and ω-conotoxin MVIID and by the tarantula *Grammostola spatulata* venom ω-grammotoxin SVIA (Lampe *et al.*, 1993; Piser *et al.*, 1994; 1995; Turner *et al.*, 1995).

Q-type Ca^{2+} channels

In many neuronal preparations, a significant component of the whole-cell current through Ca^{2+} channels is resistant to blockade with DHPs, ω-conotoxin GVIA and ω-agatoxin IVA (<100 nM), suggesting the presence of a new subtype of Ca^{2+} channel different from L-, N- and P-types. The synthesis of the toxin from the marine snail *Conus magus* ω-conotoxin MVIIC (Hillyard *et al.*, 1992; Monje *et al.*, 1993) led to the identification and characterization of a new subtype of HVA channel termed Q (Randall *et al.*, 1993; Wheeler *et al.*, 1994). Q-type current exhibits a prominent inactivation (30–35%) during a 105 ms test pulse to 0 mV (Randall *et al.*, 1994).

Characterization of Q-type Ca^{2+} channels is mostly based on pharmacological criteria. As described, Q-type channels are resistant to blockade by DHPs, ω-conotoxin GVIA and low doses (<100 nM) of ω-agatoxin IVA, but are sensitive to ω-conotoxin MVIIC (1–3 μM). Increasing concentrations of ω-agatoxin IVA (up to 2 μM) can also block Q-type Ca^{2+} channels (Wheeler *et al.*, 1994). It should be noted that these toxins used to identify Q-type channels are not selective for this subtype of channels and they also block, in a non-selective manner, N- and P-types.

Other toxins which have also been described as Q-type Ca^{2+} channel blockers include the *Conus magus* snail toxin ω-conotoxin MVIID (Monje *et al.*, 1993) and the *Grammostola spatulata* tarantula toxin ω-grammotoxin SIA (Lampe *et al.*, 1993; Piser *et al.*, 1994; 1995; Turner *et al.*, 1995).

R-type Ca^{2+} channels

In neuronal tissues, a residual Ca^{2+} current, characterized by its insensitivity to blockade by DHPs, ω-conotoxin GVIA, ω-agatoxin IVA and ω-conotoxin MVIIC has also been described and termed R-type (for "Resistant"; Randall *et al.*, 1993; Randall and Tsien, 1995). This new subtype of Ca^{2+} channel belongs to the HVA group, is rapidly inactivating ($\tau = 22$ ms) and more sensitive to blockade by Ni^{2+} ($IC_{50} = 66$ μM) than to Cd^{2+}.

Species Differences

Drastic species differences have been found in the set of Ca^{2+} channels present in a given cell or tissue. For instance, chick brain synaptosomes take up Ca^{2+} by a route highly sensitive to ω-conotoxin GVIA, thus suggesting that N-type Ca^{2+} channels dominate in this preparation (Bowman *et al.*, 1993). The opposite occurs in rat cortical synaptosomes, where K^+-evoked Ca^{2+} entry was insensitive to ω-conotoxin GVIA and sensitive to the P-type Ca^{2+} channel blocker ω-agatoxin IVA (Bowman *et al.*, 1993). Indirect studies on neurotransmitter release indicate that in motor nerve terminals from fish and amphibians N-type Ca^{2+} channels dominate, while at the mammalian muscle end-plate Ca^{2+} entry into presynaptic nerve terminals to trigger acetylcholine release takes place through P-type Ca^{2+} channels (Wessler *et al.*, 1990). Detailed direct electrophysiological studies on species differences have only been performed in primary cultures of adrenal medullary chromaffin cells. These are very close, embriologically and functionally, to sympathetic neurons and offer the advantage of the facility of culturing, even if they come from adult animals.

By combining electrophysiological patch-clamp techniques (Hamill *et al.*, 1981) and different drugs and toxins, we and others have investigated the presence of various subtypes of Ca^{2+} channels in chromaffin cells from three different species: rat, cat and bovine (Figure 8.1). These three species were chosen because different ω-toxins and organic Ca^{2+} channel blockers inhibited differently the catecholamine secretory responses to depolarizing stimuli (Ceña *et al.*, 1983; García *et al.*, 1984; Gandía *et al.*, 1990; López *et al.*, 1992; Jiménez *et al.*, 1993). Those earlier pharmacological data on secretion predicted that cells from different species could be endowed with different proportions of Ca^{2+} channel subtypes.

In bovine chromaffin cells, the presence of both, DHP-sensitive and DHP-resistant Ca^{2+} channels have been described by whole-cell (Ceña *et al.*, 1989; Artalejo *et al.*, 1991a; Bossu *et al.*, 1991a) and single-channel recordings (Artalejo *et al.*, 1991b; Bossu *et al.*, 1991b). DHP-insensitive but ω-conotoxin GVIA-sensitive components have also been found in the whole-cell currents through Ca^{2+} channels (Hans *et al.*, 1990; Artalejo *et al.*, 1992; Gandía *et al.*, 1993a; Albillos *et al.*, 1993). Furthermore, a large third component, DHP- and ω-conotoxin GVIA-resistant was always present. Experiments using both sFTX (Gandía *et al.*, 1993) and ω-agatoxin IVA (Albillos *et al.*, 1993) suggested the presence of a large fraction of P-like Ca^{2+} channels in bovine chromaffin cells.

Similar experiments were also carried out in cat chromaffin cells, in which the secretion of catecholamines seems to be fully controlled through DHP-sensitive, L-type Ca^{2+} channels (García *et al.*, 1984; Gandía *et al.*, 1987; Gandía *et al.*, 1990). Recordings of whole-cell currents through Ca^{2+} channels from cat chromaffin cells superfused with a solution containing nisoldipine (Albillos *et al.*, 1994) or furnidipine (López *et al.*, 1994a) showed a 45% blockade of I_{Ba}. When the cells were superfused with a solution containing ω-conotoxin GVIA (1 μM), also a significant blockade of I_{Ba} (about 45%) was observed. In the presence of both DHP and ω-conotoxin GVIA, a small component of the whole-cell currents remained unblocked, suggesting the presence of a subpopulation of P-type Ca^{2+} channels in these cells. This possibility has been confirmed by experiments in which cat chromaffin cells have been superfused with a solution containing ω-agatoxin IVA (50–200 nM). This treatment induced a 20% blockade of I_{Ba}, even in the presence of DHP and ω-conotoxin GVIA.

In the rat, both acetylcholine and K^+-evoked catecholamine secretory responses from perfused adrenal glands are effectively modulated by DHP Ca^{2+} channel antagonists and agonists. However, a DHP-resistant component of the secretory process in response to splanchnic nerve stimulation has also been described (López *et al.*, 1992). This could be attributed to the existence of at least two types of Ca^{2+} channels in the membrane of these cells. This possibility has been further investigated by using the whole-cell configuration of the patch-clamp technique. As in bovine and cat chromaffin cells, only HVA Ca^{2+} channels were found. No indications of the presence of low-voltage-activated T-type Ca^{2+} channels were observed. In rat cells, HVA currents showed variable sensitivity to DHP Ca^{2+} channel agonists and antagonists. Application of ω-conotoxin GVIA (1 μM) blocked 31% of the HVA Ba^{2+} currents, suggesting the existence of N-type Ca^{2+} channels. Superfusion of rat chromaffin cells with a solution containing ω-agatoxin IVA (300 nM) led to a 15% blockade of HVA Ba^{2+} currents. These data suggest the existence of a small proportion of PQ- family of Ca^{2+} channels in rat chromaffin cells (Gandía *et al.*, 1995).

In summary, three subtypes of Ca^{2+} channels have been characterized in chromaffin cells by using electrophysiological techniques and pharmacological tools. The differences between the bovine, cat and rat chromaffin cells lye in the different proportion of these subtypes of Ca^{2+} channels (Figure 8.1). Thus, in

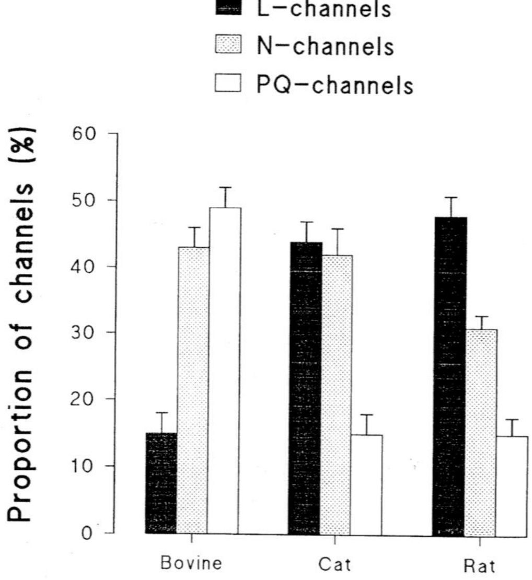

Figure 8.1 Relative densities of different Ca^{2+} channel subtypes in chromaffin cells from bovine, rat and cat adrenal medullary tissues.

the bovine, both PQ-type (45%) and N-type (35%) are predominant, being the L-type Ca^{2+} channel a minor component of the whole-cell current (20%). In the rat, the L-type is the predominant (50%), together with the N-type (35%), while the PQ-type Ca^{2+} channel seems to be a minor component (15%). Recently the use of ω-conotoxin MVIIC allowed the finding of a Q-like component in the whole-cell I_{Ba} of bovine chromaffin cells (López *et al.*, 1994b). A similar pattern to that of the rat is observed in the cat. These drastic differences might have profound physiological implications for the relative contribution of each channel subtype to the regulation of catecholamine release in different animal species (discussed below).

MOLECULAR STRUCTURE OF CALCIUM CHANNELS

Ca^{2+} channels consist basically of a multiple subunit protein complex with a central pore-forming α_1 subunit and several regulatory and/or auxiliary subunits, which include a β subunit and the disulfide-linked α_2/δ subunits. Additionally, a fifth subunit has been reported in some tissues, such as the skeletal muscle γ subunit or the neural p95 subunit. In Figure 8.2, the hypothetical subunit arrangement for the skeletal muscle high-voltage activated L-type Ca^{2+} channel is represented. Probably a combination of factors could explain the functional Ca^{2+} channel diversity found, among others, the existence of different classes of α_1 and β subunits encoded by multiple genes, alternative splicing from a single gene and multiple possible combinations among the subunits which make up the channel complex.

Structural Diversity of Ca^{2+} Channel α_1 Subunits

Expression studies have shown that basic HVA Ca^{2+} channel function, which is typical for the L- (skeletal muscle, cardiac muscle, and neuroendocrine tissue), N-,

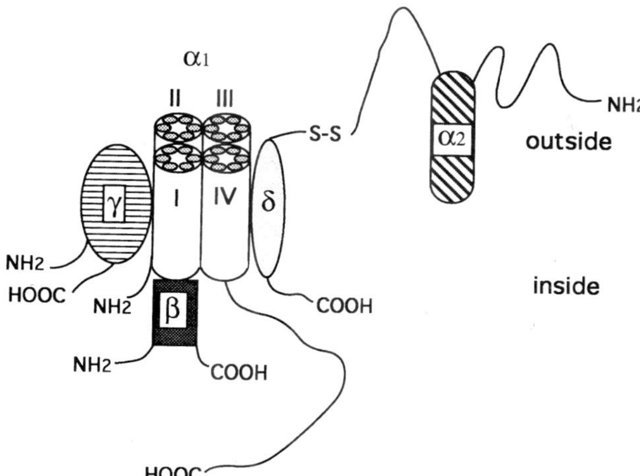

Figure 8.2 Subunit structure arrangement of the L-type Ca^{2+} channel from skeletal muscle.

P-, Q- and R type channels is carried by the corresponding α_1 subunit. This subunit contains the Ca^{2+}-conductance pore, the essential gating machinery and the receptor sites for the most prominent pharmacological agents. The mammalian Ca^{2+} channel α_1 subunits that have been isolated, sequenced and expressed can be grouped into six classes (S, C, D, B, A and E). These classes correspond to the six Ca^{2+} channel α_1 genes identified so far. Alternative splicing from single gene adds a large structural diversity to the multitude of Ca^{2+} channel α_1 gene subproducts.

Functional expression studies using the complementary DNAs of the different α_1 genes reveal properties that fall into the categories of L, N, P/Q and R channels. Table 8.2 shows the most recent agreed nomenclature for the family of the Ca^{2+} channel α_1 genes (Birnbaumer *et al.*, 1994). The table also compares the name of cloned α_1 genes, their major sites of expression, and their correlation with the functional known Ca^{2+} channels.

Subfamilies of α_1 Subunit Genes

The mammalian family of Ca^{2+} channel α_1 genes may be grouped into two subfamilies. The first subfamily includes the three genes designed as α_{1S} (cloned from skeletal muscle), α_{1C} (cardiac, smooth muscle, neuronal) and α_{1D} (neuronal, endocrine) (Tanabe *et al.*, 1987; Mikami *et al.*, 1989; Williams *et al.*, 1992b). Channels expressed from members of this group exhibit typical functional and pharmacological characteristics of L-type Ca^{2+} channels (i.e. sensitivity to DHPs). Thus, the group is termed the L-subfamily and its members display at least 60–70% amino acid identity among them.

The second subfamily of α_1 subunits can be named as non-L-type or DHP-resistant. This group is of great interest because its members (α_{1A}, α_{1B} and α_{1E}) are nearly exclusively distributed in neuronal tissues and they encode HVA Ca^{2+} chan-

Table 8.2   High-voltage activated Ca^{2+} channel α_1 subunit genes

Gene product	*Sites of Expression*	*Functional correlates (channel type)*
α_{1S}	Skeletal muscle	L
α_{1C}	Heart, brain, aorta, lung fibroblast, kidney, PC12 and GH3 cells	L
α_{1D}	Brain, pancreas, PC12 and GH3 cells	L
α_{1B}	Brain, peripheral neurons and PC12 cells	N
α_{1A}	Brain, cerebellum, Purkinje and granule cells, kidney, PC12 cells.	P? Q?
α_{1E}	Brain, heart	R?

nels that are insensitive to DHPs. Members of this subfamily show only about a 40% amino acid identity with members of the L subfamily, but >60% homology with each other. This high homology among the members of the non-L-type subfamily suggests that, in addition to the common lack of sensitivity to DHPs, they might share some other common pharmacological or functional features. Regarding the first, no agent has been found so far to influence the members of this subfamily with the effectiveness of DHPs acting on the L-type subfamily. In relation with a common functional property, two interesting possibilities are the general susceptibility to modulation by pertussis toxin-sensitive G-proteins, or an interaction with molecular components involved in neurotransmitter release (see later).

The α_{1B} gene encodes an ω-conotoxin GVIA-sensitive, DHP-resistant N-type Ca^{2+} channel and its expression is restricted to the nervous system and neuronally derived cell lines (Williams *et al.*, 1992a; Dubel *et al.*, 1992).

The α_{1A} subunit was the first α_1 subunit isolated from nervous tissue and it is highly expressed in mammalian brain, particularly in cerebellar Purkinje and granule cells (Mori *et al.*, 1991; Starr *et al.*, 1991; Fujita *et al.*, 1993). When the α_{1A} subunit was expressed in *Xenopus* oocytes, the current obtained was blocked by the crude venom from the funnel web spider *Agelenopsis aperta*. Since the first description of functional P-type channels was done in cerebellar Purkinje neuron cells using a toxin fraction (FTX) contained in the *Agelenopsis aperta* venom (Llinás *et al.*, 1989), it was suggested that the α_{1A} subunit represented the subunit of the P-type Ca^{2+} channel. However, several pharmacological and kinetic differences between the P-type Ca^{2+} channel (Mintz *et al.*, 1992a,b; Hillyard *et al.*, 1992) and the α_{1A} channel expressed in the *Xenopus* oocytes have been found (Sather *et al.*, 1993).

As discussed before, P-type channels hardly inactivate at all during depolarizations lasting up to 1 s and are potently blocked by ω-agatoxin IVA (IC_{50} 1–2 nM). However α_{1A} current expressed in oocytes decays with a time constant of 50–100 ms and the expressed Ca^{2+} channels show 100 times less sensitivity to ω-agatoxin IVA (IC_{50} around 200 nM). The opposite sensitivity is obtained using the toxin ω-conotoxin MVIIC, which shows a high selectivity to block the channel expressed in oocytes (IC_{50} <150 nM) while the concentration required for half-block of cerebellar P-type channels is between 1–10 μM. These results show that the α_{1A} subunit is able to support a Ca^{2+} channel activity quite different to that described for the native P-type channel. Therefore α_{1A} Ca^{2+} channels expressed in oocytes were termed Q channels. However, the above results do not exclude the possibility that P-type channels could be generated by any of the other α_{1A} subunit splice variants which have been reported in brain (Mori *et al.*, 1991). Alternatively, the properties of identical α_{1A} subunits may be altered by differences in the composition of the other subunits making up the channel complex. This seems not to be the case since the blocking effect of ω-agatoxin IVA on the α_{1A} currents expressed in oocytes was not modified upon the co-expression of the α_{1A} subunit with different classes of β subunits (Sather *et al.*, 1993).

Since the Q current was first described in α_{1A} injected frog oocytes, it was important to prove that this expressed current had a native real current counterpart. Evidence that it was not as artefact of the post-translation system of the oocyte comes from studies in cerebeller granule neurones. In those cells, a Q-type current with biophysical and pharmacological properties similar to the α_{1A} current expressed in oocytes, has been described. (Randall and Tsien, 1995).

The Q-type Ca^{2+} current in cerebellar granule neurons represents the bulk (35%) of the total current in these cells and coexists with other high-voltage-activated Ca^{2+} currents (L, N and P). However a component of the current, with the faster inactivation among all the currents present in granule cells, was resistant to the combination of selective blockers for the other four Ca^{2+} channels. This resistant current was blocked by low concentrations of Ni^{2+}, and the Ca^{2+} channel that gives rise to it has been termed R-type channel. The R current represents about 19% of the total current in cerebellar granule cells (Randall and Tsien, 1995). The electrical and pharmacological properties of the R-type current are similar to the current resulting from the expression of the α_{1E} subunit gene in *Xenopus* oocytes or in HEK293 cells (Soong *et al.*, 1993; Williams *et al.*, 1994). In light of the overall similarities and the presence of α_{1E} transcripts in rat cerebellar granule neurons (Soong *et al.*, 1993) it seems reasonable to attribute the R-type current to the α_{1E} subunit. The functional roles for R-type channels are not known and would greatly benefit from the availability of a selective blocker for this Ca^{2+} channel which is still lacking.

In spite of the divergence among different α_1 genes, the deduced amino acid sequences of the translated Ca^{2+} channel α_1 subunits show a simple generalized secondary structure which resembles the transmembrane topology described for the voltage-gated Na^+ and K^+ channels (Figure 8.3). This structure consists of four internal repeated motifs (I-IV), each motif being comprised of six α-helical transmembrane segments (S_1–S_6) including one (S_4) that is positively charged and is thought to be part of the voltage sensor. The short N-terminal and the long C-terminal domains of the protein, as well as several protein kinase A and numerous protein kinase C consensus sites have intracellular locations.

Other Ca^{2+} Channel Subunit Genes

The intracellular β subunit of Ca^{2+} channels is also represented by a multigene family and, up to now, a minimum of four genes have been established with alternative splic-

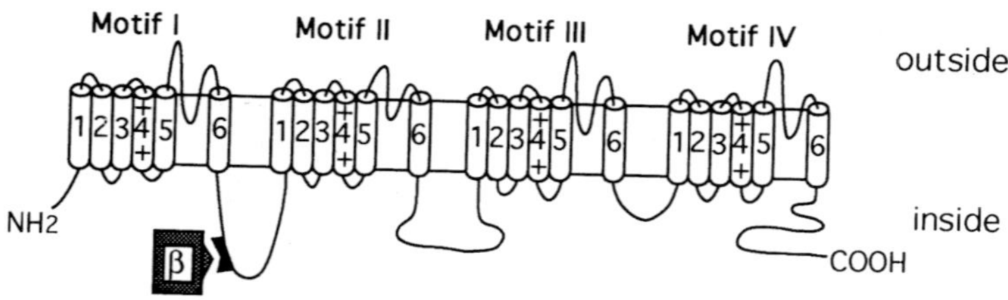

Figure 8.3   Proposed transmembrane folding model for the Ca^{2+} channel α_1 subunit, and interaction site between the α_1 and β subunits.

Table 8.3 High-voltage activated Ca^{2+} channel β subunits genes

Gene product	Splice variant	Sites of Expression	Component of
β_1	β_{1a}	Skeletal muscle	DHP receptor
	β_{1b}	Brain, heart	?
	β_{1c}	Brain, heart	?
	β_{1d}	n.d.	?
β_2	β_{2a}	Brain, heart	?
	β_{2b}	Brain, heart	?
	β_{2c}	Brain, heart	?
	β_{2d}	n.d.	?
β_3	β_{3a}	Brain, heart	ω-CTx receptor
	β_{3b}	n.d	?
β_4	β_{4a}	Brain	?
	β_{4b}	n.d.	?

n.d.- not determined.

ing (Ruth *et al.*, 1989; Pérez-Reyes *et al.*, 1992; Castellano *et al.*, 1993a, b; Hofmann *et al.*, 1994; Isom *et al.*, 1994). Therefore, the combination of 12 mature messages can coexist in various tissues. Table 8.3 summarizes the β subunit genes described so far according to the more recent consensus nomenclature (Birnbaumer *et al.*, 1994). The table also shows the known splice variants from each gene and some, but not all expression sites of mature β messenger RNAs.

Only one gene has been identified for the α_2/δ subunit with alternative splicing that gives rise to the existence of five messenger RNA species (Brust *et al.*, 1993). Also one gene has been described for the transmembrane γ subunit, apparently without the existence of any splice variants (Jay *et al.*, 1990; Hofmann *et al.*, 1994).

Modulation of α_1 Activity by the Other Auxiliary Subunits

Although all classes of α_1 subunits direct the expression of functional Ca^{2+} channels in different expression systems (i.e. *Xenopus* oocytes, mouse fibroblast L cells, HEK293 cells), the contribution of auxiliary subunits to the native properties of the channel complex was unknown until the first α_2, β and γ cDNAs obtained from rabbit skeletal muscle became available (Ellis *et al.*, 1988; Ruth *et al.*, 1989; Jay *et al.*, 1990). In recent years, cloning of new subunit genes and splice variants have greatly increased the number of potential combinations among α_1 and other subunits.

The results obtained until now show that co-expression of several classes of β-subunits with different α_1 subunits alters the voltage-dependence, kinetics and magnitude of the Ca^{2+} channel current, with different effects depending on the particular β- and α_1-subunits used (Mori *et al.*, 1991; Singer *et al.*, 1991; Lacerda *et al.*, 1991; Varadi *et al.*, 1991; Ellinor *et al.*, 1993; De Waard and Campbell, 1995).

The most complete studies about the role of auxiliary subunits within the Ca^{2+} channel complex have been done with the neuronal α_{1A}, and the cardiac α_{1C} subunits. Therefore, the co-expression of four different β subunit genes (β_{1b}, β_{2a}, β_3

and β_4) with the α_{1A} subunit, dramatically increase the current amplitude expressed in oocytes. All β subunits trigger significant changes in the voltage dependence of both activation and inactivation. The α_2/δ_b subunit does not modify the properties of α_{1A}-Ca^{2+} channel in the absence of β subunits. However it increases the β-induced stimulation of current amplitude and also regulates the β-induced change in inactivation kinetics. Both α_2/δ_b and β_{1b} slightly modified the sensitivity of the α_{1A} subunit to ω-conotoxin MVIIC (De Waard and Campbell, 1995).

It has also been demonstrated that β subunits increase the peak current induced by the expression of cardiac α_{1C} and this effect can be correlated with an increase in the number of DHP binding sites (Pérez-Reyes *et al.*, 1992; Castellano *et al.*, 1993b). Since DHPs bind to the α_{1C} subunit, the increase both in the current and in number of DHP binding sites would suggest that β subunits act by increasing the number of channels present at the plasma membrane. However, gating charge measurements (Neely *et al.*, 1993) and immunoblot analysis (Nishimura *et al.*, 1993) demonstrate that the number of α_{1C} subunits expressed at the plasma membrane was not altered by the β subunits. In order to reconcile these apparently contradictory results it was hypothesized that β subunits could induce important conformational changes on the expressed α_{1C} subunit which in turn, would increase not only the opening probability of the channel, but also the accessibility of the drug to its binding site.

Elegant experiments done by Pragnell *et al.* (1994) have shown the existence of a specific region in all the cloned α_1 subunits which contains the binding site for the β subunits. This region is highly conserved in all classes of α_1 subunits and it is located in the cytoplasmic connecting loop between motif I and II, 24 amino acids away from the sixth transmembrane segment of the motif I (IS$_6$, Figure 8.3).

In spite of the numerous data available about the functional role of ancillary subunits on the Ca^{2+} channel complex, it was necessary to be cautious with the co-expression results obtained, since many of the subunit combinations assayed are not co-expressed in the same cell or tissue *in vivo*. In fact, although the core subunit composition for the Ca^{2+} channel is clear (α_1, β and α_2/δ subunits), the specific gene and the splice variant of each of the subunits, as well as the presence of additional unknown subunits in the protein complex *in vivo*, is known for only a few Ca^{2+} channels, as the skeletal muscle DHP-sensitive L-type (α_{1S}, α_2/δ_a, β_{1a}, γ; Takahashi *et al.*, 1987) and the brain ω-conotoxin-sensitive N-type (α_{1B}, α_2/δ, β_3 and a 95 kDa protein; Witcher *et al.*, 1993).

MODULATION OF Ca^{2+} CHANNELS

As shown before, Ca^{2+} channels can inactivate in a voltage- or Ca^{2+}-dependent manner. Several neurotransmitters have been shown to influence Ca^{2+} channel currents either in their kinetics or their size. The phosphorylation of Ca^{2+} channels or their coupling with different G-proteins are possibly the two main transduction pathways regulating the kinetics of different subtypes of Ca^{2+} channels. We will briefly refer to these aspects which have been possible to elucidate because individual Ca^{2+} channel subtypes could be pharmacologically isolated by using the ω-toxins.

Phosphorylation by Protein Kinase A

Noradrenaline was the first neurotransmitter reported to enhance L-type Ca^{2+} current (I_{Ca}) in cardiac myocytes. This effect was exerted via activation of a β-adrenoceptor (Cachelin *et al.*, 1983; Bean *et al.*, 1984) that activates adenylyl cyclase through the GTP-binding protein G_s. The increase in cAMP causes activation of protein kinase A and phosphorylation of components of the Ca^{2+} channel.

Histamine and glucagon are also adenylate cyclase stimulating agents in cardiac cells. In contrast, acetylcholine, adenosine and atrial natriuretic factor decrease the level of cytosolic cAMP. These agents are thought to affect I_{Ca} via the same intracellular cascade, as proposed for β-adrenergic stimulation. In cardiac preparations, histamine enhances the upstroke velocity of slow action potentials, elevates the plateau and increases I_{Ca} (Eckel *et al.*, 1982). The effect of histamine on I_{Ca} is not additive to that of β-adrenergic agonists (Hescheler *et al.*, 1987b). In mammalian ventricular preparations, acetylcholine and adenosine decreased I_{Ca} when the current was enhanced by stimulation of the adenylate cyclase (Belardinelli and Isenberg, 1983; Hescheler *et al.*, 1986).

There is less evidence for a significant regulation of neuronal voltage-dependent Ca^{2+} channels by changes in intracellular cAMP. Several reports show that neuronal Ca^{2+} currents exhibit little or no sensitivity to cAMP, despite the presence of a DHP-sensitive component of the macroscopic current (McFadzean *et al.*, 1989; Wanke *et al.*, 1994). However, noradrenaline and β-adrenoceptor agonists increase Ca^{2+} currents in hippocampal neurons by a mechanism involving protein kinase A (Gray and Johnson, 1987). In addition, the α_1-like subunit of a ω-conotoxin GVIA-sensitive brain Ca^{2+} channel may be phosphorylated by cAMP-dependent protein kinase (Ahlijanian *et al.*, 1991). Expression studies have shown that cAMP increases Ca^{2+} currents which have the properties of P-type currents expressed in oocytes after injection of cerebellar mRNA (Fournier *et al.*, 1993).

Phosphorylation by Protein Kinase C

Phorbol esters increase single Ca^{2+} channel activity in cardiac cells and sympathetic neurons (Lacerda *et al.*, 1988; Lipscombe *et al.*, 1988). In addition to this, L-type cardiac Ca^{2+} channels expressed in *Xenopus* oocytes are both enhanced and decreased by PKC activators (Bourinet *et al.*, 1992; Singer *et al.*, 1992). There is also evidence that phorbol esters induce the appearance of new Ca^{2+} channels in *Aplysia* neurons (Strong *et al.*, 1987).

Protein kinase C has been reported to reduce the inhibitory effects of G-proteins on Ca^{2+} channels, possibly by disrupting the coupling of G proteins to the channels (Swartz, 1993). In hippocampal CA3 and cortical pyramidal neurons, activation of protein kinase C enhances current through N-type Ca^{2+} channels and, in addition, dramatically reduces the G protein-dependent inhibition of these same channels by the metabotropic glutamate receptor. In fast excitatory transmission at corticostriatal synapses, protein kinase C activators were also found to reduce the inhibitory effect produced by stimulation of the metabotropic glutamate receptor (Swartz *et al.*, 1993). However, in frog sympathetic neurons, protein kinase C enhances both N- and L- currents (Yang and Tsien, 1993). In contrast, the inhibition of Ca^{2+} channels by noradrenaline in sensory neurons is blocked by a specific protein kinase C inhibitor (Rane *et al.*, 1987). This inhibitory modulation involves N-type Ca^{2+} channels (Cox and Dunlap, 1992).

Other Kinases and Second Messengers

Other second messengers and kinases are also involved in the regulation of Ca^{2+} currents in neurons. Nitric oxide has been shown to modulate Ca^{2+} currents in superior cervical ganglion neurons (Chen and Schofield, 1993). Low oxygen tension inhibits L-type Ca^{2+} channels in arterial myocytes by a voltage-dependent mechanism (Franco-Obregón *et al.*, 1995). Some diffusible second messenger different to intracellular Ca^{2+}, cGMP, cAMP or PKC is mediating the oxotremorine-induced inhibition of N- and L-type Ca^{2+} channels in rat sympathetic neurons (Mathie *et al.*, 1992).

Several actions have been reported for the cyclic GMP-dependent protein kinase. It might phosphorylate the class C Ca^{2+} channel α_1-subunit (Hell *et al.*, 1993) on a site different from the C-terminal tail and modulate the inhibition induced by somatostatin on neuronal Ca^{2+} channels (Meriney *et al.*, 1994).

Dephosphorylation of Ca^{2+} Channels

Okadaic acid, which inhibits phosphatases 1 and 2A, and a specific peptide inhibitor of phosphatase 1, both enhanced Ca^{2+} currents in cerebellar granule neurons (Leighton *et al.*, 1994). The peptide inhibitor of phophatase 1 enhanced Ca^{2+} currents in sensory neurons too (Dolphin, 1991). The Ca^{2+}-activated phosphatase 2B (calcineurin) is likely to mediate some aspects of Ca^{2+}-dependent current inhibition (Armstrong, 1989; Kostyuk and Lukyanetz, 1993).

Modulation by G proteins of High-Voltage-Activated Ca^{2+} Channels

Dunlap and Fishbach (1978; 1981) were first to report the inhibition of neuronal Ca^{2+} channels by micromolar concentrations of noradrenaline, γ-aminobutyric acid and serotonin in chick sensory neurons. This mechanism of presynaptic inhibition was confirmed and extended to rat sympathetic neurons (Galvan and Adams, 1982) and suggested a marked selectivity for the high-threshold Ca^{2+} channels of chick sensory neurons (Deisz and Lux, 1985). This neurotransmitter-mediated inhibition of Ca^{2+} currents was modulated by voltage, since dopamine caused a marked slow-down of HVA channel activation at low membrane potentials that was partially relieved at higher membrane depolarizations (Marchetti *et al.*, 1986).

Similar effects were reported for many neurotransmitters in different types of neurosecretory cells. Thus, somatostatin (Luini *et al.*, 1986; Tsunoo *et al.*, 1986; Ikeda *et al.*, 1987; Sah, 1990), leucine-enkephalin (Tsunoo *et al.*, 1986; Albillos *et al.*, 1996a), acetylcholine (Wanke *et al.*, 1987), $GABA_B$ (Grassi and Lux, 1989), adenosine (Kasai and Aosaki, 1989), dynorphin (Bean, 1989; Sah, 1990), LHRH (Elmslie *et al.*, 1990) or ATP (Gandía *et al.*, 1993b), among other neurotransmitters, were reported to down-modulate HVA Ca^{2+} channels. This modulation is usually accompanied by a partial depression of Ca^{2+} currents and a slow phase of activation attributed to a receptor-mediated reaction coupled to G proteins (Lewis *et al.*, 1986; Holz *et al.*, 1986; Dolphin and Scott, 1987; Hescheler *et al.*, 1987a; Wanke *et al.*, 1987; Ikeda, 1991; Gandía *et al.*, 1993b).

With the recognition of distinct Ca^{2+} channels it has become important to ask which channels are targets of modulation by neurotransmitters. Transmitter regulation of ω-conotoxin GVIA-sensitive N-type Ca^{2+} channels has been extensively documented. Many transmitters inhibit Ca^{2+} channels through a pathway which is G protein-mediated and voltage-dependent (Dolphin and Scott, 1987; Lipscombe *et al.*, 1989; Aicardi *et al.*, 1991; Pollo *et al.*, 1992; Golard and Siegelbaum, 1993; Foucart *et al.*, 1993; Boland and Bean, 1993).

There is also evidence that ω-agatoxin IVA-sensitive P-type Ca^{2+} channels present in cerebellar Purkinje cells and spinal cord neurons are modulated by $GABA_B$ receptors. This modulation involves changes in voltage-dependence and kinetics (Mintz and Bean, 1993b). The Q component of current in cerebellar granule neurons is modulated by agonists of $GABA_B$, glutamate and adenosine receptors (Randall *et al.*, 1993). L-type Ca^{2+} channels appear to be also modulated in non-neuronal secretory cells (Luini *et al.*, 1986; Schmidt *et al.*, 1991; Pollo *et al.*, 1993) and chromaffin cells (Albillos *et al.*, 1996a).

In neurons, however, voltage-dependent modulation is mostly confined to non-L type Ca^{2+} channels (Lipscombe *et al.*, 1989; Aicardi *et al.*, 1991; Plummer *et al.*, 1991; Pollo *et al.*, 1992; Cox and Dunlap, 1992; Boland and Bean, 1993; Golard and Siegelbaum, 1993; Mintz and Bean, 1993b). Voltage-dependent inhibition is usually partial, leaving a component of voltage-independent depression, which have been reported to be associated to N-type channels (Luebke and Dunlap, 1994), P-type channels (Swartz, 1993), but mostly to L-type channels in neuro-secretory cells (Pollo *et al.*, 1993; Albillos *et al.*, 1996a), and peripheral and central neurons (Bley and Tsien, 1990; Amico *et al.*, 1995).

In bovine chromaffin cells, for instance, L- and non-L Ca^{2+} channels, including N-, P-, and Q-types Ca^{2+} channels have been reported to be modulated by opioids through different mechanisms (Albillos *et al.*, 1996a). While L-type channels are modulated by a G protein in a voltage-independent manner, non L-type channels appear to be inhibited through a G protein, voltage-dependent mechanism. The methionine-enkephalin-inhibited HVA Ca^{2+} current is reversed by depolarizing prepulses, recovering its initial size and shape (facilitation). The facilitated current completely dissapears when ω-conotoxin MVIIC is perfused in the presence of methionine-enkephalin, suggesting that N, P, and Q-like channels are involved in the voltage-dependent modulation. The slow phase of activation in the presence of methionine-enkephalin plus ω-conotoxin MVIIC is absent too. In contrast, when nifedipine is applied in the presence of methionine-enkephalin, neither the facilitated current nor the kinetics of inhibition are modified, indicating that L-type Ca^{2+} channels are modulated through a voltage-independent pathway (Figure 8.4C and D). As in neurons, voltage-dependent modulation is confined to non-L type Ca^{2+} channels in bovine chromaffin cells (Albillos *et al.*, 1996a).

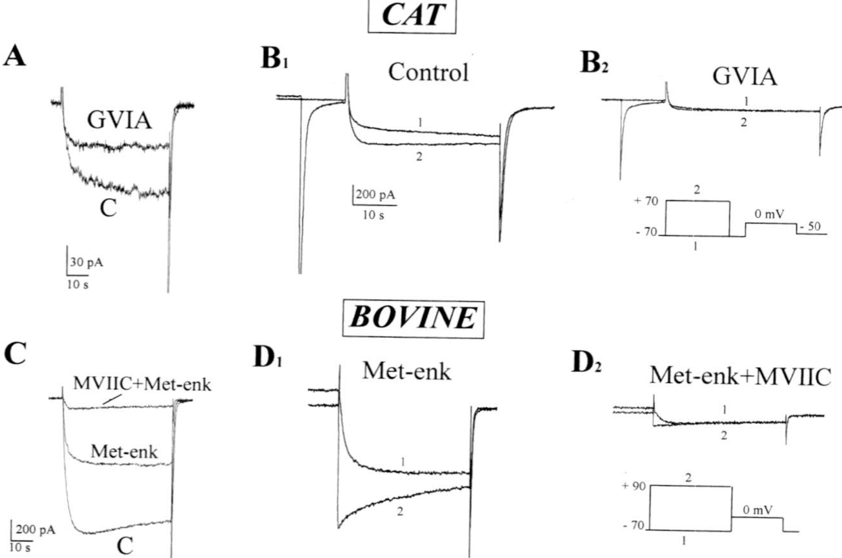

Figure 8.4 Non-L type Ca^{2+} channels are modulated by a voltage-dependent mechanism in cat and bovine chromaffin cells. Cat chromaffin cells: panels A, B_1 and B_2. Panel A, a step depolarization from −70 mV to 0 mV was applied to a voltage-clamped chromaffin cell. Notice that the slow activation kinetics in control conditions disappears afer perfusing the cell with 2 μM ω-conotoxin GVIA (GVIA). B_1, B_2: the double-pulse protocol illustrated in the inset is applied to a different cell. The facilitated current recruited in control conditions is absent in the presence of GVIA. Bovine chromaffin cells: panels C, D_1 and D_2. Panel C, same protocol as panel A. The slow activation of the current in the presence of 10 μM methionine-enkephalin (met-enk) disappears when 3 μM ω-conotoxin MVIIC (MVIIC) is added to the extracellular solution. The facilitated current elicited by the double-pulse protocol illustrated in the inset is absent when MVIIC plus met-enk are perfused to the same bovine chromaffin cell.

The same is true for cat chromaffin cells. These cells exhibit a higher degree of tonic inhibition by neurotransmitters or endogenous GTP, so that there is facilitated current in control conditions (Albillos *et al.*, 1994). The slow activation kinetics of the control currents, reflecting some tonic inhibition similar to that described for other neurotransmitters, and the corresponding facilitation recruited by prepulses, disappear after perfusing ω-conotoxin GVIA (Figure 8.4A and B). This does not occur after application of any DHP. Given that N and L type Ca^{2+} channels constitute most of the total current in cat chromaffin cells, we conclude that non-L type Ca^{2+} channels are the channels responsible for the facilitated current, as well as for the voltage-dependent modulation in these cells.

Modulation of Low-Voltage-Activated Ca^{2+} Channels

T-type Ca^{2+} channels are very rarely modulated by neurotransmitters or by other mechanisms. In chick sensory and sympathetic neurons, dopamine and noradrenaline were reported to block T-type Ca^{2+} channels by decreasing their open probability. (−)Baclofen and acetylcholine also inhibit T-type currents in rat sensory neurons. However, LVA Ca^{2+} channels have been reported to be enhanced or inhibited depending on the level of G protein activation (Scott *et al.*, 1990b). In neonatal rat ventricular myocytes, endothelin-1 enhances T-type Ca^{2+} channels through a mechanism involving protein kinase C (Furukawa *et al.*, 1992).

An Autocrine Loop for the Modulation of Ca^{2+} Channels

Albillos *et al.* (1996b) have recently provided direct evidence for the presence of a voltage-dependent autocrine loop capable of regulating neuronal HVA Ca^{2+} channels and hence, neurotransmitter release. They perfused a soluble vesicle lysate obtained from an extract of bovine adrenal medulla on whole-cell clamped bovine chromaffin cells. The soluble vesicle lysate inhibited Ca^{2+} or Ba^{2+} HVA currents in a concentration- and voltage-dependent manner, with the same slow kinetics of activation to that described for many neurotransmitters. This inhibition, which was mediated through a G-protein pathway, was mainly due to the ATP and opioids contained in the vesicles. In addition to this, they demonstrated that the facilitated current recruited by depolarizing prepulses has its origin in the suppression of the tonic inhibition of Ca^{2+} channels by the secreted material during the cellular stimulation. Ever since Fenwick *et al.* (1982) showed that Ca^{2+} currents in bovine chromaffin cells can be facilitated by strong depolarizing pre-pulses, various attempts have been made to unreveal its underlying mechanism. Either more Ca^{2+} entry through the same channel (Hoshi *et al.*, 1984; Hoshi and Smith, 1987), selective recruitment of otherwise silent L-type channels (Artalejo *et al.*, 1991a; 1994) or removal of tonic inhibition on Ca^{2+} channels (Callewaert *et al.*, 1991; Gandía *et al.*, 1993b; Doupnik and Pun, 1994; Albillos *et al.*, 1994; 1996b) have been proposed.

In their studies Doupnik and Pun (1994) and Albillos *et al.* (1996b) observed that the time course and size of Ba^{2+} current in bovine chromaffin cells depend critically on the cell superfusion conditions and are likely to derive from the G protein-mediated coupling between ATP- or opioid-autoreceptors and the high-threshold Ca^{2+} channels controlling secretion. Cell activity under stop-flow conditions (unperfused cell) favor the local rise of secreted products outside the plasmalemma, the subsequent activation of membrane autoreceptors and the rapid inhibiton of spatially localized Ca^{2+} channels. This tonic inhibition, induced by low molecular weight compounds of the vesicle content released during cell stimulation, is minimal under cell superfusion and can be partially reversed (facilitated) by strong depolarizations. Under these conditions, the Ba^{2+} current accelerates and acquires its fast activation time course and maximal amplitude. This was also corroborated by the fact that inhibition of Ba^{2+} currents under stop-flow conditions was prevented by the dialysis of tetanus toxin into the cell. This toxin blocks the secretory process, and therefore, no accumulation of modulatory products takes place nearby the cellular surface. Superfusion and stop-flow conditions may thus account for the extreme variability of the time course and voltage-dependent facilitation of Ca^{2+} channel currents reported either using static incubation systems (Artalejo *et al.*, 1991a; 1992; 1994; Fenwick *et al.*, 1982; Hoshi *et al.*, 1984; Hoshi and

Smith 1987) or different superfusion systems (Gandía *et al.*, 1993b; Doupnik and Pun, 1994; Callewaert *et al.*, 1991; Albillos *et al.*, 1994; Albillos *et al.*, 1996a,b). So, earlier interpretations that pre-pulse facilitation of I_{Ba} was due to recruitment of L-type Ca^{2+} channels (Artalejo *et al.*, 1991a; 1992; 1994) seem to derive from a mis-understanding of the importance of superfusion conditions, and the ignorance of this autocrine loop to modulate the chromaffin cell Ca^{2+} channels.

CALCIUM CHANNELS AND NEUROSECRETION

It has been long demonstrated that Ca^{2+} is essential for neurotransmitter release. The existence of multiple types of Ca^{2+} channels and the fact that several of them can coexist in the same cell type has raised questions about which channel (or channels) contribute to the control of the delivery of the Ca^{2+} necessary to trigger the secretory signal in a particular synapse. We will therefore review throughout this section how the different Ca^{2+} channel subtypes control the release of neuro-transmitters depending on the synapse, the neurotransmitter, the tissue and the animal species (see Tables 8.4, 8.5 and 8.6).

Brain Synaptosomes

Transmitter release from brain synaptosomes is controlled by different Ca^{2+} chan-nels and is greatly dependent on the animal species studied. In chick brain synapto-somes, inositol phosphate production together with noradrenaline release is highly sensitive to ω-conotoxin GVIA (Hofmann and Haberman, 1990). Ca^{2+} transients measured in chick brain synaptosomes loaded with the Ca^{2+}-sensitive fluorescent dye fura-2 demonstrated that increases in the $[Ca^{2+}]_i$ induced by high K^+ was almost completely supressed by ω-conotoxin GVIA (Bowman *et al.*, 1993). On the contrary, in rat brain synaptosomes the production of inositol phosphate and secretion of noradrenaline was insensitive to ω-conotoxin GVIA, but sensitive to ω-agatoxin IVA (Turner *et al.*, 1993). Glutamate release from rat brain synaptosomes is blocked 56% by ω-agatoxin IVA and 23% by ω-agatoxin IIIA, a L-N-P-type Ca^{2+} channel blocker (Turner *et al.*, 1992). In rat cortical synaptosomes Ca^{2+} transients induced by high K^+ were also inhibited 60% by ω-agatoxin IVA (Bowman *et al.*, 1993).

These results indicate that in chick brain synaptosomes Ca^{2+} entry and therefore neurotransmitter release is predominantly controlled via a N-type Ca^{2+} channel, whereas in rat brain synaptosomes a P-type Ca^{2+} channel exerts such control par-tially. Since L- and N-type Ca^{2+} channel blockers do not modify $[Ca^{2+}]_i$ levels or transmitter release, another Ca^{2+} entry pathway besides P-type Ca^{2+} channels, and distinct from L- and N-type Ca^{2+} channels, seems to be involved in the control of neurotransmitter release in rat brain synaptosomes.

Sympathetic Neurons

In rat sympathetic neurons, whole cell recordings have provided evidence for two subtypes of Ca^{2+} channels, the N and the L-type (Nowycky *et al.*, 1985), although

Table 8.4 Control of neurotransmitter release by N-type Ca^{2+} channels

Neurotransmitter	Preparation	ω-conotoxin GVIA [μM]	% Inhibition	Reference
Glutamate	Hippocampal CA1 pyramidal cells (Rat)	3	80	Takahashi and Momiyama (1993)
	Hippocampal synaptosomes (Rat)	1	16	Luebke *et al.* (1993)
GABA	Deep cerebellar neurons (Rat)	0.1	50	Takahashi and Momiyama (1993)
Glycine	Dorsal horn neurons of the spinal cord (Rat)	3	50	Takahashi and Momiyama (1993)
Noradrenaline	Brain synaptosomes (Rat)	1	>90	Hoffman and Habermann(1990)
	Neocortex (Rabbit)	0.005	46	Wessler *et al.* (1990)
	Neocortex (Human)	$IC_{50} = 14$ nM		Feuerstein *et al.* (1990)
	Sympathetic neurons (Rat)	0.1	92	Przywara *et al.* (1993)
	Vas deferens (Rat)	1	100	Maggi *et al.* (1988)
	Anococcygeus (Guinea pig)	0.01	98	Maggi *et al.* (1988)
	Vas deferens (Guinea pig)	0.01	97	Lundy and Frew (1988)
	Atria (Guinea pig)	$IC_{50} = 0.20$ μM		Vega *et al.* (1995)
	Atria (Guinea pig)	0.1	80	Haass *et al.* (1991)
	Atria (Guinea pig)	$IC_{50} = 0.42$ μM		Hong and Chang (1995b)
	Chromaffin cells (Dog)	0.4 μg/min	32	Kimura *et al.*, (1994)
	Urethra (Rabbit)	0.1	77	Zygmunt *et al.* (1993)
Adrenaline	Chromaffin cells (Dog)	0.4 μg/min	33	Kimura *et al.* (1995)
Dopamine	Striatum (Rabbit)	0.005	39	Dooley *et al.* (1987)
	Striatum (Rat)	1	38	Herdon and Nahorsky (1989)
Catecholamines	Chromaffin cells (Bovine)	5	17	Artalejo *et al.* (1994)
	Chromaffin cells (Cat)	1	20	López *et al.* (1994a)
	Chromaffin cells (Rat)	1	42	Kim *et al.* (1995)

Table 8.4 *Continued*

Neurotransmitter	Preparation	ω-conotoxin GVIA [μM]	% Inhibition	Reference
Acetylcholine	Neocortex (Human)	IC_{50} = 3 nM		Feuerstein *et al.* (1990)
	Electric fish (*torpedo marmorata*)	5	30	Fariñas *et al.* (1992)
	Electric fish (*gymnotus carapo*)	2.5	>95	Sierra *et al.* (1995)
	Phrenic nerve (Rat)	0.1	47	Wessler *et al.* (1990)
	Myenteric plexus (Guinea pig)	0.01	92	Lundy and Frew (1994)
	Myenteric plexus (Rat)	0.1	70	Wessler *et al.* (1990)
	Detrusor (Guinea pig)	0.1	88	Zygmunt *et al.* (1993)
	Urinary bladder (Guinea pig)	1	71	Maggi *et al.* (1988)
	Urinary bladder (Rat)	1	25	Maggi *et al.* (1988)
	Urinary bladder (Rat)	0.3	54	Frew and Lundy (1995)
	Atria (Guinea pig)	10	49	Hong and Chang (1995b)
NANC	Detrusor (Rabbit)	0.1	85	Zygmunt *et al.* (1993)
	Anococcygeus (Guinea pig)	0.05	64	Maggi *et al.* (1988)
	Urinary bladder (Guinea pig)	0.1	58	Maggi *et al.* (1988)
	Urethra (Rabbit)	0.1	47	Zygmunt *et al.* (1993)
	Jejunum (Guinea pig)	0.1	33	Lundy and Frew (1994)
	Taenia caecum (Guinea pig)	0.05	20	Lundy and Frew (1994)
EPSCs	Retinal ganglion neurons (Rat)	5	67	Tashenberger and Grantyn (1995)
EPSP	Hippocampal synaptic transmission (Rat)	1	46	Wheeler *et al.* (1994)
Dynorphin	Dentate gyrus dendrite (Guinea pig)	1	79	Simmons *et al.* (1995)
	Dentate gyrus axon (Guinea pig)	1	50	Simmons *et al.* (1995)

NANC, non-adrenergic-non-cholinergic neurotransmission; EPSPs, excitatory postsynaptic currents; EPSP, excitatory postsynaptic potentials.

Table 8.5 Control of neurotransmitter release by P-type Ca^{2+} channels

Neurotransmitter	Preparation	ω-Agatoxin IVA [μM]	% Inhibition	Reference
Glutamate	Brain synaptosomes (Rat)	0.2	56	Turner *et al.* (1992)
	Hippocampal synaptosomes (Rat)	0.2	40	Luebke *et al.* (1993)
	Hippocampal CA1 pyramidal cells (Rat)	0.2	25	Takahashi and Momiyama (1993)
GABA	Deep cerebellar neurons (Rat)	0.2	98	Takahashi and Momiyama (1993)
Glycine	Dorsal horn neurons of the spinal cord (Rat)	0.2	98	Takahashi and Momiyama (1993)
Acetylcholine	Phrenic nerve (Guinea pig)	0.02	>95	Hong and Chang (1995a)
Catecholamines	Chromaffin cells (Bovine)	0.1	35	Artalejo *et al.* (1994)

Table 8.6 Control of neurotransmitter release by Q-type Ca^{2+} channels

A. Blockade of Q channels by ω-conotoxin MVIIC:

Neurotransmitter	Preparation	ω-conotoxin MVIIC [μM]	% Inhibition	Reference
Acetylcholine	Urinary bladder (Rat)	3	54	Frew and Lundy (1995)
	Atria(Guinea pig)	$IC_{50} = 0.28\ \mu M$		Hong and Chang (1995b)
Noradrenaline	Atria(Guinea pig)	0.5	100	Hong and Chang (1995b)
	Atria(Guinea pig)	$IC_{50} = 0.19\ \mu M$		Vega *et al.* (1995)
Catecholamines	Chromaffin cells (Bovine)	3	50	López *et al.* (1994b)
EPSPs	Hippocampal synaptic transmission (Rat)	5	100	Wheeler *et al.* (1994)

EPSPs, excitatory postsynaptic potentials.

B. Blockade of Q channels by ω-agatoxin IVA:

Neurotransmitter	Preparation	ω-Agatoxin IVA [μM]	% Inhibition	Reference
Acetylcholine	Urinary bladder (Rat)	3	46	Frew and Lundy (1995)
	Atria (Guinea pig)	3	37	Hong and Chang(1995b)
Noradrenaline	Atria (Guinea pig)	3	21	Hong and Chang (1995b)

noradrenaline release is predominantly blocked by the N-channel blocker ω-conotoxin GVIA (Hirning *et al.*, 1988; Przywara *et al.*, 1993). In the sympathetic nerve endings of the iris, ω-conotoxin GVIA (1 μM) blocked over 80% the noradrenaline synthesis induced by high K^+, while nicardipine had no effect, indicating that Ca^{2+} entry through N-type Ca^{2+} channels play a major role in noradrenaline synthesis (Rittenhouse and Zigmond, 1991) and noradrenaline release (Hirning *et al.*, 1988; Przywara *et al.*, 1993) from sympathetic neurons. It is puzzling however, that although DHP antagonists do not affect secretion, the DHP agonist Bay K 8644 did enhance the release of noradrenaline evoked by K^+ or electrical stimulation in the rat was deferens (Ceña *et al.*, 1985). This paradox could be explained if L-type Ca^{2+} channels are located at presynaptic sites away from exocytotic sites. Ca^{2+} overflow through those channels in the presence of Bay K 8644, would reach secretory sites to enhance the release of noradrenaline, but DHP blockade of such channels will not affect secretion (Miller, 1987).

Striatum

In the striatum, neurotransmitter release is controlled by different Ca^{2+} channels. Thus, dopamine release induced by K^+ is blocked around 30% by ω-conotoxin GVIA (Herdon and Nahorski 1989; Turner *et al.*, 1993) although dopamine release evoked by electrical stimulation was almost completely inhibited by ω-conotoxin GVIA (Herdon and Nahorski 1989). Turner and cowokers (1993), using subsecond measurements of glutamate and dopamine released from rat striatal synaptosomes, showed that P-type Ca^{2+} channels, which are sensitive to ω-agatoxin IVA, control the release of both neurotransmitters; however, dopamine release (but not glutamate) was also partially blocked by ω-conotoxin GVIA. Another interesting observation by these authors is that the blockade of neurotransmitter release was voltage-dependent. With strong depolarizations (60 mM K^+), neither ω-agatoxin IVA nor ω-conotoxin GVIA were effective alone, although a combination of both produced a synergistic inhibition of 60–80% of Ca^{2+} dependent dopamine release. With milder depolarizations (30 mM K^+), ω-agatoxin IVA (200 nM) blocked over 80% dopamine and glutamate release, while ω-conotoxin GVIA (1 μM) blocked dopamine release by 25% and left unaffected glutamate release.

Hippocampus

In the hippocampi of rabbits, Dooley *et al.* (1987) demonstrated that electrically-induced release of dopamine, 5-hydroxitryptamine and acetylcholine was similarly blocked (around 40%) by nanomolar concentrations of ω-conotoxin GVIA. Under the same experimental conditions, dopamine release from the corpus striatum and noradrenaline release from the neocortex was also blocked 40% by ω-conotoxin GVIA (5 nM). Using a superfusion system with subsecond temporal resolution Luebke *et al.* (1993) studied the effects of ω-conotoxin GVIA and ω-agatoxin IVA on glutamate release from rat hippocampal synaptosomes. The glutamate release induced by very high K^+ concentrations was inhibited 16% by ω-conotoxin GVIA and 40% by ω-agatoxin IVA; such blockade was increased when lower concentrations of K^+ were employed to induce secretion. The amplitude of excitatory postsynaptic potentials in CA1 pyramidal neurons was reduced by ω-conotoxin GVIA and ω-agatoxin IVA, although ω-agatoxin IVA was more rapid and efficacious (Luebke *et al.*, 1993). Thus, at least two Ca^{2+} channels seem to control glutamate release from hippocampal neurons, but P-type channels seem to play a major role.

Synaptic transmission between hippocampal CA3 and CA1 neurons is controlled by N-type Ca^{2+} channels together with Ca^{2+} channels whose pharmacology differs from L and P-type channels, but resembles that of Q-type Ca^{2+} channels encoded by the α_{1A} subunit gene. Using rat hippocampal slices, Wheeler *et al.* (1994) showed that ω-conotoxin GVIA blocked EPSP by 46%, while P- and L-type Ca^{2+} channel antagonists had no effect. In contrast ω-conotoxin MVIIC (N-P-Q Ca^{2+} channel blocker) inhibited 100% of the EPSP. These results suggest that hippocampal synaptic transmitter release is regulated by N- and Q-subtypes of Ca^{2+} channels. Measuring excitatory postsynaptic currents (EPSCs) from hippocampal CA1 pyramidal neurons, Takahashi and Momiyama (1993) demonstrated that synaptic transmission at this level is predominantly controlled by N-type Ca^{2+} channels (80% block of EPSPs by ω-conotoxin GVIA) and to a lesser extent by P-type Ca^{2+} channels (25% inhibition by ω-agatoxin IVA).

The release of the neuropeptide dynorphin from hippocampal granule cells is controlled by different Ca^{2+} channels, depending on the release site (dentrite or axon). L-type Ca^{2+} channels mediate dynorphin release from dentrites, and N-type Ca^{2+} channels mediate dynorphin release from the axons of hippocampal granule cells (Simmons *et al.*, 1995).

Cerebellum

Inhibitory postsynaptic currents (IPSCs) evoked in neurons of the deep cerebellar nuclei by stimulating presumptive Purkinje cell axons were reversibly abolished by bicuculline, indicating that the responses were mediated by GABA (γ-aminobutiric acid). The application of ω-agatoxin IVA (200 nM) blocked IPSCs amplitude by 50%, while the L-type Ca^{2+} channel blocker, nicardipine had no effect (Takahashi and Momiyama, 1993) indicating that GABA release from Purkinje cell axons is mediated via Ca^{2+} entry through P-type Ca^{2+} channels.

Retinal Ganglion Neurons

Glutamatergic synaptic responses in rat retinal ganglion neurons is partially sensitive to ω-conotoxin GVIA (30% block) and insentive to low concentrations of ω-agatoxin IVA (Tashenberger and Grantyn, 1995). These results indicate that the major part of synaptic glutamate release in retinal ganglion neurons is governed by a novel toxin-resistant Ca^{2+} channel that could possibly be of the Q or R type.

Spinal Cord and Dorsal Root Ganglion Neurons

Synaptic transmission has been studied *in vitro* in co-cultures of fetal neurons from the ventral half of the spinal cord (VH neurons) and from the dorsal root ganglion (DRG neurons). In 70% of the VH-VH connections, and in 50% of the DGR-VH connections, Bay K 8644 failed to affect transmitter release. ω-Conotoxin GVIA produced no consistent effects on EPSPs or IPSPs elicited by stimulation of nearby neurons. VH neurons EPSPs elicited by stimulation of nearby DGR neurons were reduced 50% by ω-conotoxin GVIA. Thus, N and L channels may be involved in transmitter release in approximately 30% of these neurons (Yu *et al.*, 1992).

In rat dorsal horn neurons of the spinal cord, release of glycine induces IPSCs. The IPSCs were almost completely blocked by ω-conotoxin GVIA (>95%) and partially inhibited by ω-agatoxin IVA (50%), while nicardipine had no effect. Thus, N channels seems to dominate the control of glycine release from these neurons.

Intestinal Tract

Electrically-evoked release of acetylcholine is predominantly controlled through N-type Ca^{2+} channels at the myenteric plexus (Lundy and Frew, 1988; 1994; Wessler *et al.*, 1990). ω-Conotoxin GVIA markedly reduced (70%) the evoked release of [^{3}H]-acetylcholine from the myenteric plexus of the small intestine, with an IC_{50} of 0.7 nmol/l; the blocking potency was similar at stimulation frequencies of 3 and 10 Hz. An increase in the extracellular Ca^{2+} concentration attenuated the inhibitory effect of ω-conotoxin GVIA (Wessler *et al.*, 1990). No species difference was observed as to the channel controlling Ca^{2+} entry for transmitter release at the myenteric plexus.

In the guinea-pig jejunum, ω-conotoxin GVIA blocked only partially (33%) the inhibitory non adrenergic non cholinergic (NANC) transmission upon electrical stimulation. This was also the case at the taenia caecum (20% inhibition) (Lundy and Frew, 1994). In the proximal duodenum the NANC trasmission was insensitive to ω-conotoxin GVIA (Maggi *et al.*, 1988).

Therefore, cholinergic transmission at the intestinal tract seems to be regulated by Ca^{2+} entering through N-type Ca^{2+} channels while NANC transmission is regulated by another Ca^{2+} entry pathway besides N channels.

Lower Urinary Tract

In the rat or guinea-pig isolated bladder, ω-conotoxin GVIA produced a concentration and time dependent inhibition of twitch reponses to electrical field stimulation without affecting the response to exogenous acetylcholine. In the rat bladder the maximal effect did not exceed 25% inhibition, while a much larger fraction of the response (70%) was inhibited in the guinea-pig bladder. In the rat bladder the effects of ω-conotoxin GVIA were frequency-dependent; maximal effects of the toxin were observed at 2–5 Hz. Frew and Lundy (1995) have demonstrated that neurotransmission in the rat urinary bladder is supported by both N- and Q-type

Ca^{2+} channels. In their experiments, the resistant portion (non-N-non-P) was sensitive to ω-conotoxin MVIIC, which in addition to N and P also blocks Q channels.

In the rabbit urethra and detrusor muscle, Zygmunt *et al.* (1993) have studied the effects of ω-conotoxin GVIA on adrenergic, cholinergic and NANC responses induced by electrical stimulation. The adrenergic contraction (at 25 Hz) and NANC relaxation (at 10 Hz) in the urethra, and the cholinergic and NANC contractions (at 10 Hz) in the detrusor, were inhibited in a concentration-dependent manner by ω-conotoxin GVIA. The adrenergic contraction of the urethra was 10 times, and the cholinergic contraction in the detrusor was 3 times more sensitive to ω-conotoxin GVIA than the NANC responses. These results suggest that NANC transmission is less sensitive to ω-conotoxin GVIA than transmission mediated by adrenergic and cholinergic nerves in the rabbit lower urinary tract.

Vas Deferens

In the isolated rat or guinea-pig vas deferens, ω-conotoxin GVIA (1 nM–1 μM) produced concentration and time dependent inhibition of the response to electrical field stimulation, while the responses to K^{+}, noradrenaline or ATP were unaffected. A toxin concentration as low as 1 nM produced almost complete inhibition of twitches, but this effect took about 1 h to be completed. With higher concentrations, the time-course of the inhibition was much faster (Maggi *et al.*, 1988). In any case, full blockade was achieved with ω-conotoxin GVIA (Maggi *et al.*, 1988; Lundy and Frew, 1994), indicating that trasmission at this level is fully controlled through Ca^{2+} entering via N-type channels.

Other experiments show that in guinea-pig electrically-stimulated vas deferens, the ATP-mediated component of the biphasic contraction was found to be more susceptible to ω-conotoxin GVIA than the adrenergic component (Hata *et al.*, 1992).

Heart

The innervation in mammalian atria is both sympathetic and parasympathetic, which regulate the heart rate and the contractile strength. The subtypes of Ca^{2+} channels involved in neurotransmitter release have been studied by various investigators. Vega *et al.* (1995) have shown that electrically-stimulated left guinea-pig atria are sensitive to N-type Ca^{2+} channel blockers. Thus, ω-conotoxin GVIA and ω-conotoxin MVIIA blocked the inotropic response in a concentration-dependent fashion with IC_{50} values of 0.20 and 0.044 μM respectively. The N-P-Q channel blocker, ω-conotoxin MVIIC showed an IC_{50} of 0.19 μM; ω-agatoxin IVA had no effect. Hong and Chang (1995b) have studied the Ca^{2+} channel subtypes mediating the cholinergic and adrenergic neurotransmission in the guinea-pig atria. In left atria paced at 2–4 Hz, the negative inotropic effect induced by electrical field stimulation of parasympathetic nerves (in the presence of propranolol) was abolished by ω-conotoxin MVIIC. On the other hand, the inotropic response resulting from electrical field stimulation of the sympathetic nerves (in the presence of atropine) was abolished by ω-conotoxin GVIA and ω-conotoxin MVIIC. None of the peptide toxins affected the chronotropic and the inotropic responses evoked by carbachol, isoprenaline or noradrenaline (Hong and Chang, 1995b; Vega *et al.*, 1995).

These results suggest that under physiological conditions, the release of acetylcholine from parasympathetic nerves in the heart is dominated by a PQ subfamily of Ca^{2+} channels, while noradrenaline release from sympathetic nerves is controlled by an N-type Ca^{2+} channel.

Motor Nerve Terminals

Neurotransmitter release at this level is controlled by different types of Ca^{2+} channels depending on the species. The electroplax of marine electric fish is highly rich in motor nerve endings; this is the reason why it has been so widely used as a model to study transmitter release in motor nerve endings. ω-Conotoxin GVIA blocks the release of acetylcholine and Ca^{2+} uptake induced by depolarization in electric organ nerve terminals of the ray; the IC_{50} values were 3 μM for blocking transmitter release and 2 μM for blocking Ca^{2+} entry (Ahmad and Miljanich, 1988). Sierra *et al.* (1995) have also shown that N-type Ca^{2+} channels mediate transmitter release at the electromotoneuron-electrocyte synapses of the weakly electric fish *Gymnotus carapo;* ω-conotoxin GVIA (2.5 μM) blocked over 95% of the EPP while ω-agatoxin IVA and nifedipine had no effect. In contrast to these data, in *torpedo* synaptosomes Fariñas *et al.* (1992) found that ω-conotcxin GVIA (10^{-8} to 5×10^{-5} M) showed a differential effect on acetylcholine and ATP release: nucleotide release was inhibited 90% at the highest concentration tested while acetylcholine release was only moderately reduced (30%). In the frog neuromuscular junction Jahromi *et al.* (1992) have demonstrated that synaptic transmission is also governed by Ca^{2+} entry through an N-type Ca^{2+} channel.

In contrast to fish and amphibians where N-type Ca^{2+} channels control neurotransmitter release, in mammalian motor nerve terminals, Ca^{2+} entry serving to discharge acetylcholine release seems to be ruled by a P-type Ca^{2+} channel. So in the rat phrenic nerve, $[^{3}H]$-acetylcholine release was only partially inhibited by ω-conotoxin GVIA (Wessler *et al.*, 1990). In the mouse, EPP were almost completely abolished (>95%) with 200 nM ω-agatoxin IVA, indicating that a P-type channel regulates neurotransmitter release at this level (Hong and Chang, 1995a).

Chromaffin Cells

As described above, chromaffin cells contain different subtypes of Ca^{2+} channels at their plasmalemmal membrane. The presence and proportion of the various subtypes of Ca^{2+} channels depends on the animal species (Figure 8.1). Therefore, secretion of catecholamines from cells of different species will presumably be controlled in a different manner according to the Ca^{2+} channels it contains. In this section we will revise how catecholamine secretion is controlled in four different animal species (cat, cow, rat and dog) and how some subtypes of Ca^{2+} channels but not others, are more directly implicated in the control of exocytosis.

Cat chromaffin cells

K[+]-evoked catecholamine secretion was markedly potentiated by the L-type Ca^{2+} channel activator Bay K 8644 and blocked by nifedipine (García *et al.*, 1984). Ca^{2+} evoked secretion was equally potentiated by Bay K 8644 (Montiel *et al.*, 1984). Secretion was blocked by various DHPs in a concentration dependent manner, as well as by other drugs acting on L-type Ca^{2+} channels like verapamil and diltiazem (Gandía *et al.*, 1987). Measuring differential secretion of adrenaline and noradrenaline, Cárdenas *et al.* (1988) demonstrated that secretion of both amines were completely blocked when secretion is induced either by high K[+] or the nicotinic agonist dimethylphenylpiperazinium (DMPP). All these data indicated that secretion in these cells was controlled by a L-type channel. But Albillos *et al.* (1994) showed that cat chromaffin cells also contained ω-conotoxin GVIA sensitive channels besides L-type. It was then demonstrated that both HVA Ca^{2+} channels in cat chromaffin cells were present in an approximate proportion of 50–50%, and that the increase of the $[Ca^{2+}]_i$ induced by short (10 s) depolarizing pulses (70 mM K[+]) was also partially reduced by furnidipine (44%) and by ω-conotoxin GVIA (43%). In the perfused adrenal gland or in isolated cat chromaffin cells catecholamine release induced by 10 s pulses with 70 mM K[+] was blocked over 95% by furnidipine and around 25% by ω-conotoxin GVIA. These results show that though Ca^{2+} entry through both channels (N- and L-type) lead to similar increments of the average $[Ca^{2+}]_i$, the control of catecholamine release is dominated by Ca^{2+} entering through L-type Ca^{2+} channels (López *et al.*, 1994a).

Bovine chromaffin cells

Catecholamine secretion from bovine chromaffin cells is greatly potentiated in the presence of Bay K 8644 (Jiménez *et al.*, 1993); this increase in secretion went in parallel with an increase in $^{45}Ca^{2+}$ uptake (García *et al.*, 1984). Ceña *et al.* (1983) showed complete blockade of catecholamine release by nitrendipine in bovine chromaffin cells stimulated with high K[+]. These results disagreed with those obtained by other authors who found that blockade of secretion in bovine chromaffin cells by DHPs did not exceed further than 40–50% (Boarder *et al.*, 1987; Owen *et al.*, 1989; Gandía *et al.*, 1990; Jiménez *et al.*, 1993; López *et al.*, 1994b).

When toxins were available to block selectively other subtypes of Ca^{2+} channels, it was demonstrated that these cells contained other Ca^{2+} channel subtypes besides L, i.e. N, P and Q-type (Ballesta *et al.*, 1989, Rosario *et al.*, 1989; Bossu *et al.*, 1991a, b; Artalejo *et al.*, 1991a, b; Albillos *et al.*, 1993; Gandía *et al.*, 1993). ω-Conotoxin GVIA showed to be ineffective or very little effective in blocking catecholamine secretion (Owen *et al.*, 1989; Duarte *et al.*, 1993; Jiménez *et al.*, 1993; López *et al.*, 1994b; Artalejo *et al.*, 1994; Fernández *et al.*, 1995).

As to the contribution of P-type Ca^{2+} channels to secretion of catecholamines, we find different results in the literature. Granja *et al.* (1995) found that catecholamine secretion induced by high-K[+] remains unaffected by ω-agatoxin IVA (100 nM); nevertheless, when secretion was activated by nicotine, catecholamine release was significantly blocked by 50%; thus, Granja *et al.* (1995) concluded that ω-agatoxin IVA could also be acting on the nicotinic receptor. On the other hand, Duarte *et al.* (1993) showed that FTX decreased the K$^+$-evoked noradrenaline release by 75% and adrenaline release by 60%; the combination of FTX plus nitrendipine further decreased noradrenaline and adrenaline release by 90 and 75%, respectively. By measuring cell capacitance changes, Artalejo *et al.* (1994) found a significant participation of P-type Ca^{2+} channels on secretion from bovine chromaffin cell that could be blocked by ω-agatoxin IVA (100 nM); however, we failed to see any effect of this same toxin on secretion induced by K$^+$ depolarization (López *et al.*, 1994b).

The L-N-P-insensitive portion of catecholamine release seems to be ω-conotoxin MVIIC- sensitive. López *et al.* (1994b) have shown that catecholamine release from superfused bovine chromaffin cells (stimuli: 70 mM K$^+$ for 10 s) was inhibited 50% by furnidipine (3 μM). ω-Conotoxin MVIIC (3 μM) also reduced the secretory response by 50%. The combination of furnidipine and ω-conotoxin MVIIC completely abolished secretion. These authors also demonstrate that ω-conotoxin GVIA and ω-agatoxin IVA had no effect on secretion. These results strongly suggest that in bovine chromaffin cells, secretion is predominantly controlled by Ca^{2+} entering through L- and Q-type Ca^{2+} channels.

Rat chromaffin cells

DHPs block secretion in rat adrenal glands in a concentration dependent manner. The magnitude of such blockade is related to the type of stimuli employed to induce secretion. The DHP isradipine can fully block secretion when the stimuli used are K$^+$ or nicotine. In contrast, when electrical field stimulation is used only partial blockade is obtained with DHPs, being such inhibition frequency-dependent (López *et al.*, 1992). Measuring Ca^{2+} currents and capacitance Kim *et al.* (1995) have shown that ω-conotoxin GVIA (1 μM) blocks 40% and nicardipine around 60% of the total capacitance increase in rat chromaffin cells. Therefore, in these cells secretion would be controlled by L as well as by N-type Ca^{2+} channels.

Dog chromaffin cells

Kimura *et al.* (1994) have studied the effects of ω-conotoxin GVIA and L-type blockers (nifedipine and verapamil) on catecholamine release in anesthetized dogs. Catecholamine release into the blood stream was induced either by electrical stimulation of the splanchnic nerve or by intraarterial injection of acetylcholine. Administration of 0.4 μg/ml of ω-conotoxin GVIA reduced catecholamine secretion by 30% when electrical stimulation was used; under these experimetal conditions nifedipine (1 μg/ml) or verapamil (10 μg/ml) had no effect. When catecholamine release was induced by acetylcholine, ω-conotoxin GVIA blocked secretion around 50% and nifedipine also reduced it by 50%. These results suggest that N- and L-type Ca^{2+} channels located in the adrenal medullary cells may contribute to the release of adrenal catecholamines in the dog adrenal.

CALCIUM AND EXOCYTOSIS

The simple picture that Ca^{2+} channels open to produce a rise of averaged $[Ca^{2+}]_i$ to trigger secretion, is no more so simple. Today we know that steep Ca^{2+} gradients are produced, that elevations of $[Ca^{2+}]_i$ are highly inhomogeneous, that different steps of exocytosis might require different $[Ca^{2+}]_i$ and that Ca^{2+} microdomains are produced at specialized exocytotic plasmalemmal sites where specific Ca^{2+} channel subtypes are likely segregated and concentrated. We will discuss next all these interesting aspects that relates the Ca^{2+} channel subtypes to the neurosecretory process.

Calcium Requirements for Secretion

Two different approaches have been used to determine the Ca^{2+} requirements for secretion. Studies on permeabilized chromafin cells have shown that secretion is triggered by relatively low $[Ca^{2+}]_i$, in the order of 1 μM (Knight and Baker, 1982; Knight, 1988). Consistent with this view, direct measurements of $[Ca^{2+}]_i$ in intact cells indicate levels during secretion in the order of 1 μM (Knight and Kesteven, 1983; O'Sullivan *et al.*, 1989). On the other hand, theoretical predictions of the spatiotemporal changes of the $[Ca^{2+}]$ in the vicinity of release sites predict that $[Ca^{2+}]_i$ at release sites can reach concentrations in the order of 10–100 μM, due to steep spatial gradients undetectable with the available methods for measuring averaged global changes of $[Ca^{2+}]_i$ (Simon and Llinás, 1985; Fogelson and Zucker, 1985; Smith and Augustine, 1988). One possible explanation for these divergent views is to propose that while secretion can occur at the low levels of $[Ca^{2+}]$ typically used to evoke secretion from permeabilized cells (i.e. around 1 μM), the kinetics of secretion may be accelerated by higher $[Ca^{2+}]_i$ (Zimmemberg *et al.*, 1985). These concentrations cannot be detected with the techniques used in the direct measurements of averaged $[Ca^{2+}]_i$ due to their low spatiotemporal resolution. Therefore, it is possible that cells that produce rapid secretion in response to physiological stimuli, such as neurons (Llinás *et al.*, 1981), chromaffin cells (Neher and Marty, 1982) and anterior pituitary cells (Thomas *et al.*, 1990), do so by elevating $[Ca^{2+}]_i$ to levels substantially higher than those used in experiments on permeabilized cells.

Augustine and Neher (1992) studied the Ca^{2+} dependence of secretion in single adrenal chromaffin cells using simultaneous measurements of membrane capacitance, as an assay of secretion, and $[Ca^{2+}]_i$. They found that the intracellular dialysis of Ca^{2+}, through a patch pipette, promoted secretion. Its rate increased monotonically as $[Ca^{2+}]_i$ was elevated over a range above 0.2 μM, and saturated at concentrations greater than 10 μM, if at all. The total amount of secretion reached a maximum at 1.5 μM Ca^{2+} and declined at higher $[Ca^{2+}]_i$. Secretion appears to have a Hill coefficient for Ca^{2+} of about 2. Brief depolarizing pulses produced transient rises in both the $[Ca^{2+}]_i$ and the rate of secretion. Comparison of the rates of secretion measured during depolarization with those produced by Ca^{2+} dialysis suggests that $[Ca^{2+}]_i$ at secretory sites can exceed 10 μM during depolarization. The spatially averaged measurements of $[Ca^{2+}]_i$ indicate much smaller levels of $[Ca^{2+}]_i$; thus, there must be pronounced spatial gradients of $[Ca^{2+}]_i$ during depolarization. By comparing the rate and kinetics of secretion at different $[Ca^{2+}]$ in electroporated and intact bovine chromaffin cells superfused at high speed, Michelena *et al.* (1995) came to the conclusion that the rapidly releasable vesicle pool required local $[Ca^{2+}]_i$ of about 100 μM.

Similar conclusions were obtained from a different methodological approach (Michelena *et al.*, 1993). $[Ca^{2+}]_i$ changes and catecholamine release were measured in chromaffin cells stimulated with a depolarizing solution (100 mM K$^+$). Once depolarized, external Ca^{2+} or Ba^{2+} ions were offered to cells either as a single 2.5 mM step or as a ramp that linearly increased the concentration from 0 to 2.5 mM over a 10-min period. A clear separation between the changes of the $[Ca^{2+}]_i$ and the time course of secretion was observed. Specifically, secretion and $[Ca^{2+}]_i$ rose in parallel when a Ca^{2+} step was used to reach a peak in few seconds; however, while secretion declined to the basal level, $[Ca^{2+}]_i$ remained elevated at a plateau of 400 nM. With a Ca^{2+} ramp, only a transient small peak of secretion was observed, yet the $[Ca^{2+}]_i$ remained elevated throughout the 10-min stimulation period. The separation between secretion and $[Ca^{2+}]_i$ was observed even when L-type Ca^{2+} channels were expected to remain open (mild depolarization in the presence of 1 μM Bay K 8644). By using Ba^{2+} steps or ramps, sustained non-inactivating secretory responses were obtained. These results suggest that the rate and extent of secretion are not a simple function of the $[Ca^{2+}]_i$ at a given time; they are compatible with the following conclusions: (i) A steep extracellular-to-cytosolic Ca^{2+} gradient is required to produce a sharp increase in the $[Ca^{2+}]_i$ (tenths micromolar) at exocytotic sites capable of evoking a fast but transient secretory response. (ii) As a result of Ca_i^{2+}-dependent inactivation of Ca^{2+} channels, those high $[Ca^{2+}]_i$ are possible only at early times after cell depolarization.

The rate of secretion achieved with dialysis and perfusion of Ca^{2+} buffers is substantially lower for a given $[Ca^{2+}]$ than that observed in response to brief depolarizations or to other means of rapidly elevating $[Ca^{2+}]_i$, even at very high $[Ca^{2+}]_i$ (10–100 μM). On the other hand, the total amount of secretion is always much larger with Ca^{2+} dialysis than with rapid elevation of $[Ca^{2+}]_i$. These results were interpreted to reflect the presence of two pools of vesicles, a relatively small "rapidly releasable pool" and a larger "reserve pool". Neher and Zucker (1993) used the caged Ca^{2+} compound DM-nitrophen to investigate the kinetics of Ca^{2+}-dependent secretion in bovine chromaffin cells. Flash photolysis elicited sharp 100 μM $[Ca^{2+}]_i$ steps that evoked intense secretion, lasting a few seconds, followed by a persistent, slow secretion. Distinct kinetic components may reflect secretion from pools that are differentially capable of release. Both secretion and movement of vesicles between pools appear to be $[Ca^{2+}]_i$ sensitive. Heinemann *et al.* (1993) have proposed a two-step model of secretion that simulates the dynamics of a pool of release-ready granules for different time courses of $[Ca^{2+}]_i$ changes. A variety of results from the literature can be simulated with this model. In particular, the model predicts the occurrence of secretory depression and augmentation under appropriate conditions. Thus, after intense stimuli the rapidly releasable pool decreases, causing a depression of the secretory response to subsequent stimuli. The recovery is accelerated when the concentration of cytosolic Ca^{2+} is moderately increased, probably as a result of a Ca^{2+}-dependent refilling of this pool. In this way the pool can be even over-filled and subsequent secretory responses would be augmented (Rüden and Neher, 1993). In conclusion, the immediate secretory response seems to require high $[Ca^{2+}]_i$ nearby exocytotic sites, while the transport of vesicles from a reserve pool to secretory active zones requires milder elevations of the $[Ca^{2+}]_i$.

Ca^{2+} Channels, Ca^{2+} Microdomains, Ca^{2+} gradients and Exocytosis

To trigger release of neurotransmitter, Ca^{2+} ions must diffuse in the cytoplasm from the open channel to a Ca^{2+} binding site that regulates exocytosis. Unfortunately, techniques with both temporal and spatial resolution required for measuring Ca^{2+} transients ocurring during the few millisecond of evoked secretion have been lacking. Thus, direct measurements of $[Ca^{2+}]_i$ have so far provided only limited information about the presynaptic Ca^{2+} concentration transients. Photomultiplier measurements of fura-2 fluorescence allow quantification of the presynaptic $[Ca^{2+}]_i$ changes, but they reflect a spatial average and not spatial gradients of $[Ca^{2+}]_i$. Thus, the measured Ca^{2+} signal will be relatively small and slow in comparison to the signal near the open channels, which is presumably the Ca^{2+} signal that triggers release.

A detailed view of Ca^{2+} movement in presynaptic cytoplasm is obtainable only by theoretical extrapolation from more readily measured parameters. Theoretical studies of diffusion of Ca^{2+} ions into presynaptic cytoplasm suggest that Ca^{2+} signals drop off steeply in the vicinity of open channels (Llinás *et al.*, 1981; Zucker and Stockbridge, 1983; Chad and Eckert, 1984; Simon and Llinás, 1985; Fogelson and Zucker, 1985; Zucker and Fogelson, 1986; Sala and Hernández-Cruz, 1990). Each model includes strong cytoplasmic Ca^{2+} buffering, so that diffusion of Ca^{2+} within the cytoplasm is greatly restricted compared to simple aqueous diffusion (Hodgkin and Keynes, 1957). Under these conditions, each study has found that individual Ca^{2+} domains may sometimes be small in comparison to estimated distances between active Ca^{2+} channels. The idea is that near an open channel, $[Ca^{2+}]_i$ should rapidly rise to high levels and then decline very rapidly when the channel closes. At some distance away from the channel, however, the change in $[Ca^{2+}]_i$ will be much smaller and much slower because this signal is low-filtered by diffusion throughout the terminal. Chad and Eckert (1984) have introduced the term "domain" to refer to this volume of cytoplasm over wich the $[Ca^{2+}]_i$ transient produced by a single Ca^{2+} channel exerts some regulatory action. The lack of overlap between individual $[Ca^{2+}]_i$ domains means that Ca^{2+} channels could act independently to trigger transmitter release or to regulate other functions that are dependent on the $[Ca^{2+}]_i$. Ca^{2+} diffusion might also serve to coordinate a multiplicity of spatially distributed functions associated with release, e.g. vesicle transport and recycling, energy and transmitter metabolism or gene expression.

As Ca^{2+} ions diffuse slowly in the cytosol, changes in intracellular Ca^{2+} concentrations are often restricted to relatively small and localized regions within cells. As a consequence of this compartmentalization, it is likely that Ca^{2+} ions are near the physiological effector molecules that they activate. This seems to be the case. Neurons contain specialized regions at the plasma membrane where exocytosis occurs. These zones are characterized by a high local concentration of secretory vesicles and specialized presynaptic and postsynaptic regions. These regions, called "active zones", were first observed by electron microscopy (Heuser *et al.*, 1979).

Results obtained with several different experimental approaches point towards the possibility that presynaptic Ca^{2+} channels are selectively clustered within the active zone. Voltage clamp measurements in the squid giant synapse indicate a brief time delay between presynaptic Ca^{2+} entry and the generation of postsynaptic responses (Llinás *et al.* 1981), implying a short distance between Ca^{2+} channels and active zones. The superior potency of BAPTA compared to EGTA in blocking exocytosis apparently results from the faster Ca^{2+}-binding kinetics of BAPTA, and suggests that the Ca^{2+} receptor that triggers secretion binds Ca^{2+} rapidly (less than 200 μs). This Ca^{2+} receptor appears to be located very close to Ca^{2+} channels because Ca^{2+} ions can diffuse no more than a few tens of nanometers in a few tens of μs (Adler *et al.*, 1991). Freeze fracture studies of presynaptic membranes also reveal aggregates of large (10 nm diameter) intramembranous particles that are present only at active zones (Heuser *et al.*, 1974); several correlations suggest that these particles may be Ca^{2+} channels (Pumplin *et al.*, 1981). Finally, histochemical studies illustrate that binding of a specific toxin to a Ca^{2+} channel is restricted to the active zones of neuromuscular junctions (Robitaille *et al.*, 1990; Cohen *et al.*, 1991). More recently, using fura-2, digital imaging microscopy and correlative electron microscopy to identify the location of Ca^{2+} accumulation within squid synaptic terminals, Smith *et al.* (1993) found that sites of Ca^{2+} accumulation are selectively concentrated in the presynaptic terminals and preferentially occur at active zones. Even more, a possible multi-channel organization has been suggested. Haydon *et al.* (1994) were able to localize single Ca^{2+} channels by atomic force microscopy (nanometer level of resolution) on the calyx-type nerve terminals of the chick ciliary ganglion. They found a constant interchannel spacing of about 40 nm, which could indicate an intermolecular linkage.

The existence of localized active zones associated to Ca^{2+} domains has also been demonstrated in paraneuronal secretory cells. Schröeder *et al.* (1994) showed that catecholamine release occurs from spatially localized restricted zones in chromaffin cells using carbon-fiber ultramicroelectrodes to monitor secretion from small areas. High-resolution confocal microscopy using anti-dopamine-β-hydroxylase antibodies to mark exocytotic sites confirm the existence of zones of release wich are of similar dimensions to those revealed by the electrochemical measurements (Robinson *et al.*, 1995). With the aid of a pulsed-laser imaging system these authors conclude that the sites of Ca^{2+} entry are colocalized with active zones of catecholamine release in bovine adrenal chromaffin cells.

Theoretical models predict a high $[Ca^{2+}]_i$ domain close to the open Ca^{2+} channel. There are some indirect experimental evidences that strengthen this idea. Adler *et al.* (1991) examined the ability of several Ca^{2+} buffers to reduce evoked transmitter release when injected into the presynaptic terminal of squid giant synapse.

Injection of EGTA was ineffective at reducing transmitter release. Conversely, the buffer BAPTA, which has an equilibrium affinity for Ca^{2+} similar to that of EGTA, produced a substancial reduction in transmitter release when injected presynaptically. Several BAPTA derivatives with different Ca^{2+} dissociation constants were also used. Consideration of the relative efficacy of the various chelators in attenuating release suggests that the presynaptic Ca^{2+} transient that triggers release may be large, perhaps hundreds of μM or more. Other estimates of $[Ca^{2+}]_i$ in the transmitter releasing region were obtained from the frog hair cells suggesting that Ca^{2+} levels may reach hundreds of μM or more during depolarization (Roberts *et al.*, 1990). As discussed above, it also appears that Ca^{2+} concentrations at secretory sites of chromaffin cells reaches similar high values (10–100 μM) during depolarization (Augustine and Neher, 1992).

More direct evidences of the existence of high $[Ca^{2+}]_i$ domains come from different experimental approaches. Localized zones of Ca^{2+} entry, termed hotspots, have been identified in neurons, both on dendrites and at nerve terminals. In the hippocampal dendritic spines, Ca^{2+} accumulates in single postsynaptic spines with presynaptic stimulation but not in the parent dendrite, showing an independent regulation of Ca^{2+} in dendritic spines (Muller and Connor, 1991; Guthrie *et al.*, 1991). In the presynaptic terminal of the giant squid synapse, $[Ca^{2+}]_i$ microdomains of the order of 200 to 300 μM occur against the cytoplasmic surface of the plasmalemma during transmitter secretion (Llinás *et al.*, 1992). A recent report (Monck *et al.*, 1994) describes the existence of hotspots of localized Ca^{2+} entry in patch-clamped bovine adrenal chromaffin cells following the transient opening of voltage-sensitive Ca^{2+} channels. The pulsed-laser imaging system used in these studies has an improved spatial an temporal resolution compared with those imaging systems that have previously been used in attempts to capture the early stages of Ca^{2+} influx into chromaffin cells: submicron spatial resolution and ms time resolution. It can be observed how Ca^{2+} gradients are initially restricted to discrete submembrane domains, before developing into complete rings of elevated Ca^{2+} concentration. Issa and Hudspeth (1994) used confocal microscopy to localize Ca^{2+} channels on the cellular surface in frog hair cells loaded with the Ca^{2+} indicator fluo-3. When a cell is depolarized, fluo-3 fluorescence due to local Ca^{2+} influx marks active zones. Measurement of currents through membrane patches at fluorescently labeled active zones demonstrates that Ca^{2+} channels occur there.

In conclusion, the Ca^{2+} channels opening causes an increase of $[Ca^{2+}]_i$ in a small cytosolic volume (microdomain) localized near the channels. As a consequence of this, different types of Ca^{2+} channels with different patterns of activation might act independently, to generate different cytosolic Ca^{2+} signals to perform specific functions.

Are Exocytotic Ca^{2+} Channels a Component of the Secretory Machinery?

Many experimental evidences support the idea that Ca^{2+} channels tightly interact with proteins involved in synaptic vesicle docking and fusion to plasma membrane.

As stated above, the biophysical analysis of regulated secretion and modeling of the Ca^{2+} dynamics in the vicinity of Ca^{2+} channels suggest that the high $[Ca^{2+}]_i$ required for triggering secretion could only occur in close proximity to a single Ca^{2+} channel.

Biochemical analysis of the constitutive secretory pathways in the yeast *Saccharomyces cerevisiae* and in animal cells have revealed a number of similarities with synaptic vesicle membrane trafficking (Bennet and Scheller, 1993; Söllner *et al.*, 1993). This suggests that common mechanisms may underlie most membrane-trafficking events. An additional level of regulation involving Ca^{2+} is required in the case of synaptic vesicle fusion, probably utilizing proteins that are not shared with constitutive trafficking steps. Because of the high level of conservation among different membrane-trafficking pathways, the regulatory role of Ca^{2+} in synaptic vesicle fusion may be to relieve an inhibition of the constitutive fusion machinery (Bennet and Scheller, 1993; Söllner *et al.*, 1993; Popov and Poo, 1993).

The molecular machinery involved in synaptic vesicle docking and fusion includes protein components localized to both the vesicle and plasma membrane, as well as soluble factors (Bennet and Scheller, 1994). Synaptic vesicle associated proteins are VAMP or synaptobrevin, Rab3A, a low-molecular-weight GTP-binding protein involved in the regulation of a wide range of membrane-trafficking pathways, synaptotagmin, proposed as a Ca^{2+} sensor that triggers exocytosis because of its apparent low affinity for Ca^{2+} and the multiple binding sites for Ca^{2+} (Brose *et al.*, 1992), and synaptophysin, a protein that can form transmembrane ion channels when reconstituted into artificial bilayers (Thomas *et al.*, 1988), suggesting a role in the formation of a fusion pore during exocytosis. Plasma membrane proteins are syntaxins, SNAP-25, GAP-43, and neurexins, identified as the α-latrotoxin receptor (Petrenko *et al.*, 1991) (α-latrotoxin is a component of black widow spider venom that induces a massive Ca^{2+}-independent release of neurotransmitters; Petrenko, 1993). The following proteins are soluble factors: SEC1, NSF (N-Ethylmaleimide-sensitive factor), and SNAPs (soluble NSF attachment proteins).

All these proteins display specific interactions between them, pointing to the existence of a supramolecular complex. Thus, addition of detergent-solubilized brain membranes to NSF and SNAPs results in their assembly into a particle that sediments at 20S. The components of the 20S particle include the synaptic vesicle protein VAMP, and the synaptic plasma membrane proteins syntaxin and SNAP-25 (Söllner *et al.*, 1993). These proteins have been termed SNAREs, for SNAP receptors. Direct interactions of synaptotagmin with syntaxin (Bennet *et al.*, 1992a) and neurexin (Hata *et al.*, 1993), and either direct or indirect interactions with N-type Ca^{2+} channels has been described (Leveque *et al.*, 1992, Bennet *et al.*, 1992b). Syntaxins also interact with ω-conotoxin-sensitive N-type Ca^{2+} channels (Bennet *et al.*, 1992b, Yoshida *et al.*, 1992).

Recent experiments (O'Connor *et al.*, 1993; Leveque *et al.*, 1994) have identified interactions between presynaptic and synaptic vesicle membrane proteins, that might be important in organizing the components of the fast neurotransmitter release. In addition to the α-latrotoxin receptor (neurexin) and synaptotagmin, syntaxin co-purifies on an α-latrotoxin affinity column. Furthermore, antibodies against syntaxin and the α-latrotoxin receptor immunoprecipitate with [125I] ω-conotoxin binding sites, indicating that Ca^{2+} channels are associated with this complex. Thus, neurexin, synaptotagmin, syntaxin, and Ca^{2+} channels can be found in a structure that O'Connor *et al.* (1993) propose to call the "synapto-secretosome".

The Cholino-Chromaffin Cell Synapse as an Example of the Geographic Specialization of Ca^{2+} Channels

The scheme of Figure 8.5 shows the likely sequence of steps involved in the stimulus-secretion coupling process at the cholino-chromaffin synapse of the adrenal medulla. When an action potential invades the splanchnic nerve terminal innervating a chromaffin cell, the release of acetylcholine into the synaptic gap occurs (Collier *et al.*, 1984). It is not yet known what type of Ca^{2+} channel control the release of acetylcholine at this synapse. The ensuing activation of postsynaptic nicotinic receptors on the surface of chromaffin cells causes local depolarization of these cells (Douglas *et al.*, 1967) and the firing of action potentials (Kidokoro

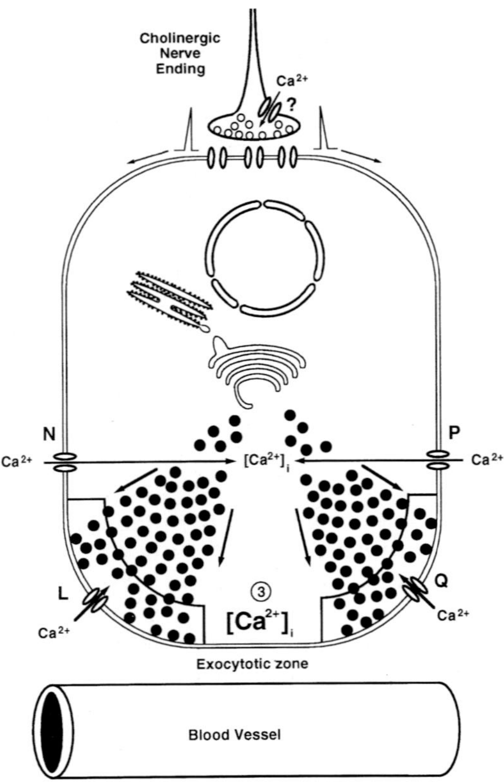

Figure 8.5   Scheme showing the arrangement of different Ca^{2+} channel subtypes to control the secretion of catecholamines in adrenal medullary chromafin cells (see text for details).

et al., 1982). The depolarizing wave carried by the action potentials recruit HVA Ca^{2+} channels of the L-type (García *et al.*, 1984), N-type (Hans *et al.*, 1990), P-type (Gandía *et al.*, 1993; Albillos *et al.*, 1993) and Q-type (López *et al.*, 1994b). External Ca^{2+} then gains the cell interior and increases the local $[Ca^{2+}]_i$ in a microdomain where exocytosis takes place.

We have recently suggested that depending on the Ca^{2+} channel pathway used by Ca^{2+}_o, the changes in $[Ca^{2+}]_i$ and their functional consequences might differ (López *et al.*, 1994a,b). Thus, Ca^{2+}_o entering through Ca^{2+} channels localized nearby exocytotic sytes (mostly Q- and L-subtypes) will increase the subplasmalemmal $[Ca^{2+}]_i$ at very high levels (10–100 μM) to activate exocytosis from a readily releasable vesicle pool. This pool contains about 200–400 vesicles per cell and is selectively depleted by tyramine (Michelena *et al.*, 1995). On the other hand, Ca^{2+} entering through N- and P-subtypes of Ca^{2+} channels, which are located far away from exocytotic active sites, will cause milder increases of $[Ca^{2+}]_i$ in cytosolic microdomains located deep in the cell. These Ca^{2+} transients could serve other slower cell functions requiring lower $[Ca^{2+}]_i$ changes, and serving pre-exocytotic steps such as the cytosolic transport of vesicles from the Golgi apparatus to subplasmalemmal sites.

PERSPECTIVES

In 10 years of ω-toxins use, at least 5 subtypes of HVA Ca^{2+} channels have been identified and characterized. New toxins are needed to target selectively the Q-type Ca^{2+} channel without affecting the N or P. The R or T-type channels also need new toxins to characterize their functions. The question of how many Ca^{2+} channel subtypes remain to be discovered is also relevant. In addition, differences among tissues and cell types for a given Ca^{2+} channel are emerging; L-type Ca^{2+} channels differ from skeletal, to cardiac, to smooth muscles and the brain. Are identical the Q channels from hippocampal and chromaffin cells? What about the N, P or R channels? Why different Ca^{2+} channels are required to control exocytosis of the same transmitter (i.e. acetylcholine, catecholamines) in the same cell type but from different animal species? Another important question relates to the expression of various channel subtypes in the same cell. Why exocytosis requires Ca^{2+} from different pathways? Is it a valve safety to secure the efficiency of the process? If the N channel is a part of the secretory machinery, what about the L, P or Q channels? How close are they from exocytotic active sites? And most interesting, are the channels of a paraneuronal cell such as the chromaffin cell equally organized than those of brain synapses? Why the release of noradrenaline is controlled by N channels in sympathetic neurones and by L or Q channels in chromaffin cells? Do action potentials recruit different Ca^{2+} channel subtypes in those two catecholaminergic cell types? Will a K^+ depolarizing stimulus recruit Ca^{2+} channels different to those recruited by action potentials in neurones, or by acetylcholine receptors in chromaffin cells? Is the electrical pattern of different excitable cells causing different secretion patterns by simply recruiting specific Ca^{2+} channels with particular gating and kinetic properties?

Another critical question relates to the development of a pharmacology for neuronal Ca^{2+} channels. While L-type Ca^{2+} channels have a rich pharmacology that have provided novel therapeutic approaches to treat cardiovascular diseases, non peptide molecules which block or inactivate the N, P, Q, T or R channels are lacking. Thus, a major goal for research in this field is the search for selective blockers of specific Ca^{2+} channel subtypes. The knowledge of the three-dimensional structure in solution of the different toxins is very important for studying the specificity of their interactions with Ca^{2+} channel subtypes, and to define active sites that can serve as models to design and synthetize non-peptide blockers. The ω-conotoxins are small peptides containing 24–29 amino acid residues. It is interesting that the amino acid sequence of ω-conotoxin MVIIA is much more similar to that of ω-conotoxin MVIIC than to ω-conotoxin GVIA; yet the pharmacology of ω-conotoxin MVIIA is much more closer to that of ω-conotoxin GVIA (blockade of N-type channels). Thus, it will be very important to define structural differences determining the toxin selectivity for N- or Q-type Ca^{2+} channels. The three-dimensional structures of ω-conotoxin GVIA (Sevilla *et al.*, 1993; Davis *et al.*, 1993; Pallaghy *et al.*, 1993; Skalicky *et al.*, 1993), ω-conotoxin MVIIA (Kohno *et al.*, 1995) and ω-conotoxin MVIIC (Nemoto *et al.*, 1995; Farr-Jones *et al.*, 1995) have been recently elucidated. Elucidation of the structures of ω-conotoxin MVIID, ω-agatoxin IVA and other new toxins will facilitate their comparisons and the definition of structural determinants for specific binding to Ca^{2+} channel subtypes.

Non-peptide blockers for neuronal Ca^{2+} channels are emerging, but they lack selectivity. For instance, the piperazine derivatives flunarizine, R56865 and dotarizine are "wide-spectrum" Ca^{2+} channel blockers (Garcez-do-Carmo *et al.*, 1993; Villarroya *et al.*, 1995). Fluspirilene, a member of the diphenylbutylpiperidine class of neuroleptic drugs (which also includes pimozide, clopimozide and penfluridol) has anti-schizophrenic actions and block N-type Ca^{2+} channels in PC12 cells (Grantham *et al.*, 1994). It may be that its neuroleptic properties are due, at least in part, to an inhibition of neuronal N-type Ca^{2+} channels. Thus, inhibition (or facilitation) of specific neurotransmitter release by selective blockers (or activators) of Ca^{2+} channels may have functional and therapeutic consequences. For instance, synthetic ω-conotoxin MVIIA protected hippocampal CA1 pyramidal neurons from damage caused by transient, global forebrain ischemia in the rat (Valentino *et al.*, 1993). Selective blockade of N-type Ca^{2+} channels may also be beneficial in treatment of specific pain syndromes (Olivera *et al.*, 1994). Thus, intrathecal administration of as little as 0.3 μg of ω-conotoxin MVIIA completely suppresed the nociceptive responses in the rat hindpaw formalin test. Tactile allodynia was also selectively abolished in a rat neuropathic pain model by intrathecal administration of ω-conotoxin MVIIA at doses that did not impair motor function; the toxin was found to be 100 times more potent than morphine. It seems clear that several other neurological or psychyatric diseases will benefit from the development of drugs that interfere selectively with different Ca^{2+} channel subtypes.

Many questions have been answered during a decade of intense research in the field of neuronal voltage-dependent Ca^{2+} channels. Most of the studies performed were possible only because, particularly the group of professor Baldomero Olivera, maked available to many groups of researchers potent ω-toxins to type and characterize those channels. The same is true for the patch-clamp techniques, which exploded at the same time than ω-toxins thanks to the efforts of the group of professor Erwin Neher. We would like to dedicate this review to these two scientists that with their original ideas and contributions helped many groups, including ours, to pose and resolve many questions related to Ca^{2+} ions, Ca^{2+} channels and neurosecretion. We thank them for this, but also because the answers to those questions opened up many new ones for our intellectual joy during the 10 coming years. We would finally like to apologize to those authors that were unvoluntarily cited in this review.

REFERENCES

Adams, M.E., Bindokas, V.P., Hasegawa, L. and Venema, V.J. (1990) ω-Agatoxins: novel calcium channel antagonists of two subtypes from Funnel web spider (Agelenopsis aperta) venom. *J. Biol. Chem.*, **265**, 861–867.

Adams, M.E., Mintz, I.M., Reily, M.D., Thanabal, V. and Bean, B.P. (1994) Structure and properties of ω-agatoxin IVB, a new antagonist of P-type calcium channels. *Mol. Pharmacol.*, **44**, 681–688.

Adams, M.E., Myers, R.A., Imperial, J.S. and Olivera, B.M. (1993) Toxityping rat brain Ca^{2+} channels with ω-toxins from spider and cone snail venoms. *Biochemistry*, **32**, 12566–12570.

Adler, E.M., Augustine, G.J., Duffy, S.N. and Charlton, M.P. (1991) Alien intracellular calcium chelators attenuate neurotransmitter release at the squid giant synapse. *J. Neurosci.*, **11**, 1496–1507.

Akaike, N., Kostyuk, P.G. and Osipchuk, Y.V. (1989) Dihydropyridine-sensitive low-threshold calcium channels in isolated rat hypothalamic neurones. *J. Physiol.*, **412**, 181–195.

Ahlijanian, M.K., Striessnig, J. and Catterall, W.A. (1991) Phosphorylation of an α_1-like subunit of an ω-conotoxin-sensitive brain Ca^{2+} channel by cAMP-dependent protein kinase and protein kinase C. *J. Biol. Chem.*, **266**, 20192–20197.

Ahmad, S.N. and Miljanich, G.P. (1988) The calcium antagonist, ω-conotoxin, and electric organ nerve terminals: binding and inhibition of transmitter release and calcium influx. *Brain Res.*, **453**, 247–256.

Aicardi, G., Pollo, A., Sher, E. and Carbone, E. (1991) Noradrenergic inhibition and voltage-dependent facilitation of ω-conotoxin sensitive Ca channels in insulin secreting RINm5F cells. *FEBS Lett.*, **281**, 201–204.

Albillos, A., Artalejo, A.R., López, M.G., Gandía, L., García, A.G. and Carbone, E. (1994) Ca^{2+} channel subtypes in cat chromaffin cells. *J. Physiol.*, **477**, 197–213.

Albillos, A., Carbone, E., Gandia, L., Garcia, A.G. and Pollo, A. (1996a) Opioid inhibition of Ca^{2+} channel subtypes in bovine chromaffin cells: selectively of action and voltage-dependence. *Eur. J. Neurosci.*, **8**, 1561–1570.

Albillos, A., García, A.G. and Gandía, L. (1993) ω-Agatoxin-IVA-sensitive calcium channels in bovine chromaffin cells. *FEBS Lett.*, **336**, 259–262.

Albillos, A., Gandia, L., Michelena, P., Gilabert, J-A., Valle, M. del, Carbone, E. and Garcia, A.G. (1996b) The mechanism of calcium channel facilitation in bovine chromaffin cells. *J. Physiol.*, **494**, 687–695.

Amico, C., Marchetti, C., Nobile, M. and Usai, C. (1995) Pharmacological types of Ca^{2+} channels and their modulation by baclofen in cerebellar granules. *J. Neurosci.*, **19**, 2839–2848.

Armstrong, D.L. (1989) Ca^{2+} channel regulation by calcineurin, a Ca^{2+}-activated phosphatase in mammalian brain. *Trends Neurosci.*, **12**, 117–122.

Artalejo, C.R., Adams, M.E. and Fox, A.P. (1994) Three types of Ca^{2+} channels trigger secretion with different efficacies in chromaffin cells. *Nature*, **367**, 72–76.

Artalejo, C.R., Dahmer, M.K., Perlman, R.L. and Fox, A.P. (1991a) Two types of Ca^{2+} currents are found in bovine chromaffin cells: facilitation is due to the recruitment of one type. *J. Physiol.*, **432**, 681–707.

Artalejo, C.R., Mogul, D.J., Perlman, R.L. and Fox, A.P. (1991b) Three types of bovine chromaffin cell Ca^{2+} channels: facilitation increases the open probability of a 27 pS channel. *J. Physiol.*, **444**, 213–240.

Artalejo, C.R., Perlman, R.L. and Fox, A.P. (1992) ω-Conotoxin GVIA blocks a Ca^{2+} current in chromaffin cells that is not of the "classic" N Type. *Neuron*, **8**, 85–95.

Augustine, G.J. and Neher, E. (1992) Calcium requirements for secretion in bovine chromaffin cells. *J. Physiol.*, **450**, 247–271.

Ballesta, J.J., Palmero, M., Hidalgo, M.J., Gutiérrez, L.M., Reig, J.A., Viniegra, S. and García, A.G. (1989) Separate binding and functional sites for ω-conotoxin and nitrendipine suggest two types of calcium channels in bovine chromaffin cells. *J. Neurochem.*, **53**, 1050–1056.

Bean, B.P. (1989) Neurotransmitter inhibition of neuronal Ca^{2+} currents by changes in channel voltage dependence. *Nature*, **340**, 153–156.

Bean, B.P., Nowycky, M.C. and Tsien, R.W. (1984) β-adrenergic modulation of Ca^{2+} channels in frog ventricular heart cells. *Nature*, **307**, 371–375.

Belardinelli, L. and Isenberg, G. (1983) Actions of adenosine and isoproterenol on isolated mammalian ventricular myocytes. *Circ. Res.*, **53**, 287–297.

Bennet, M.K., Calakos, N., Kreiner, T. and Scheller, R.H. (1992a) Synaptic vesicle membrane proteins interact to form a multimeric complex. *J. Cell. Biol.*, **116**, 761–775.

Bennet, M.K., Calakos, N. and Scheller, R.H. (1992b) Syntaxin: a synaptic protein implicated in docking of synaptic vesicles at presynaptic active zones. *Science*, **257**, 255–259.

Bennet, M.K. and Scheller, R.H. (1993) The molecular machinery for secretion is conserved from yeast to neurons. *Proc. Natl. Acad. Sci. USA*, **90**, 2559–2563.

Bennet, M.K. and Scheller, R.H. (1994) A molecular description of synaptic vesicle membrane trafficking. *Annu. Rev. Biochem.*, **63**, 63–100.

Birnbaumer, L., Campbell, K.P., Catterall, W.A., Harpold, M.M., Hofmann, F., Horne, W.A., Mori, Y., Schwartz, A., Snutch, T.P., Tanabe, T. and Tsien, R.W. (1994) The naming of voltage-gated calcium channels. *Neuron*, **13**, 505–506.

Bley, K.R. and Tsien, R.W. (1990) Inhibition of Ca^{2+} and K^+ channels in sympathetic neurons by neuropeptides and other ganglionic transmitters. *Neuron*, **4**, 379–391.

Boarder, M.R., Marriott, D. and Adams, M. (1987) Stimulus secretion coupling in cultured chromaffin cells. *Biochem. Pharmacol.*, **36**, 163–167.

Boland, L.M. and Bean, B.P. (1993) Modulation of N-type-Ca^{2+} channels in bullfrog sympathetic neurons by luteinizing hormone-releasing hormone: kinetics and voltage dependence. *J. Neurosci.*, **13**, 516–533.

Bossu J.L., De Waard M. and Feltz, A. (1991a) Inactivation characteristics reveal two calcium currents in adult bovine chromaffin cells. *J. Physiol.*, **437**, 603–620.

Bossu, J.L., De Waard, M. and Feltz, A. (1991b) Two types of calcium channels are expressed in adult bovine chromaffin cells. *J. Physiol.*, **437**, 621–634.

Bourinet, E., Fournier, F., Lory, P., Charnet, P. and Nargeot, J. (1992) Protein kinase C regulation of cardiac calcium channels expressed in *Xenopus*-oocytes. *Pflügers Arch. Eur. J. Physiol.*, **42**, 247–255.

Bowman, D., Alexander, S. and Lodge, D. (1993) Pharmacological characterisation of the calcium channels coupled to the plateau phase of KCl-induced intracellular free Ca^{2+} elevation in chicken and rat synaptosomes. *Neuropharmacology*, **32**, 1195–1202.

Brose, N., Petrenko, A.G., Südhof, T.C. and Jahn, R. (1992) Synaptotagmin: a calcium sensor on the synaptic vesicle surface. *Science*, **256**, 1021–1025.

Brust, P.F., Simerson, S., McCue, A.F., Deal, C.R., Schoonmaker, S., Williams, M.E., Veliçelebi, G., Johnson, E.C., Harpold, M.M. and Ellis, S.B. (1993) Human neuronal voltage-dependent calcium channels: studies on subunit structure and role in channel assembly. *Neuropharmacology*, **32**, 1089–1102.

Cachelin, A.B, De Peyer, J.E., Kokubun, S. and Reuter, H. (1983) Ca^{2+} channel modulation by 8-Bromocyclic AMP in cultured heart cells. *Nature*, **304**, 462–464.

Callewaert, G., Johnson, R.G. and Morad, M. (1991) Regulation of the secretory response in bovine chromaffin cells. *Am. J. Physiol.*, **260**, C851–C860.

Carbone, E. and Lux, H.D. (1984) A low voltage-activated, fully inactivating Ca channel in vertebrate sensory neurones. *Nature*, **310**, 501–502.

Cárdenas, A.M., Montiel, C., Esteban, C., Borges, R. and García, A.G. (1988) Secretion from adrenaline- and noradrenaline-storing adrenomedullary cells is regulated by a common dihydropyridine-sensitive calcium channel. *Brain Res.*, **456**, 364–366.

Castellano, A., Wei, X., Birnbaumer, L. and Perez-Reyes, E. (1993a) Cloning and expression of a third calcium channel β subunit. *J. Biol. Chem.*, **268**, 3450–3455.

Castellano, A., Wei, X., Birnbaumer, L. and Perez-Reyes, E. (1993b) Cloning and expression of a neuronal calcium channel β subunit. *J. Biol. Chem.*, **268**, 12359– 12366.

Ceña, V., García, A.G., Khoyi, M.A., Salaices, M. and Sánchez-García, P. (1985) Effect of the dihydropyridine Bay K 8644 on the release of [^{3}H]-noradrenaline from the rat isolated vas deferens. *Br. J. Pharmacol.*, **85**, 691–696.

Ceña, V., Nicolás, G.P., Sánchez-García, P., Kirpekar, S.M. and García, A.G. (1983) Pharmacological dissection of receptor-associated and voltage-sensitive ionic channels involved in catecholamine release. *Neuroscience*, **10**, 1455–1462.

Ceña, V., Stutzin, A. and Rojas, E. (1989) Effects of calcium and Bay K 8644 on calcium currents in adrenal medullary chromaffin cells. *J. Memb. Biol.*, **112**, 255–265.

Chad, J.E. and Eckert, R.O. (1984) Calcium domains associated with individual channels can account for anomalous voltage relations of Ca-dependent responses. *Biophys. J.*, **45**, 993–999.

Chen, C. and Schofield, G.G. (1993) Nitric oxide modulates Ca^{2+} currents of superior cervical ganglion neurons. *Soc. Neurosci. Abstr.*, **19**, 1129.

Cohen, C.J., Ertel, E.A., Smith, M.M., Venema, V.J., Adams, M.E. and Lebowitz, M.D. (1992) High affinity block of myocardial L-type calcium channels by the spider toxin ω-agatoxin IIIA: advantages over 1,4-dihydropyridines. *Mol. Pharmacol.*, **42**, 947–951.

Cohen, M.W., Jones, O.T. and Angelides, K.J. (1991) Distribution of Ca^{2+} channels on frog motor nerve terminals revealed by fluorescent ω-conotoxin. *J. Neurosci.*, **11**, 1032–1039.

Collier, B., Johnson, G., Kirpekar, S.M. and Prat, J.C. (1984) The release of acetylcholine and catecholamines from cat adrenal glands. *Neuroscience*, **13**, 957–964.

Cox D.H. and Dunlap, K. (1992) Pharmacological discrimination of N-type from L-type Ca^{2+} current and its selective modulation by transmitters. *J. Neurosci.*, **12**, 906–914.

Davis, J.H., Bradley, E.K., Miljanich, G.P., Nadasdi, L., Ramachandran, J. and Basus, V.J. (1993) Solution structure of omega-conotoxin GVIA using 2-D NMR spectroscopy and relaxation matrix analysis. *Biochemistry*, **32**, 7396–7405.

De Waard, M. and Campbell. K.P. (1995) Subunit regulation of neuronal α_{1A} Ca^{2+} channel expressed in *Xenopus* oocytes. *J. Physiol.*, **485**, 619–634.

De Weille, J.R., Schweitz, H., Maes, P., Tartar, A. and Lazdunski, M. (1991) Calciseptine, a peptide isolated from black mamba venom is a specific blocker of the L-type calcium channel. *Proc. Natl. Acad. Sci. USA*, **88**, 2437–2440.

Deisz, R.A. and Lux, H.D. (1985) γ-aminobutyric acid-induced depression of Ca^{2+} currents of chick sensory neurons. *Neurosci. Lett.*, **56**, 205–210.

Dolphin A.C. (1991) Regulation of Ca^{2+} channel activity by GTP binding proteins and second messengers. *Biochim. Biophys. Acta*, **1091**, 68–80.

Dolphin, A.C. and Scott, R.H. (1987) Ca^{2+} channel currents and their inhibition by (–)baclofen in rat sensory neurones, modulation by guanine nucleotides. *J. Physiol.*, **386**, 670–672.

Dooley, D.J., Lupp, A. and Herting, G. (1987) Inhibition of central neurotransmitter release by ω-conotoxin GVIA, a peptide modulator of the N-type voltage-sensitive calcium channel. *Naunyn-Schmiedeberg's Arch. Pharmacol.*, **336**, 467–470.

Douglas, W.W. (1969) Stimulus-secretion coupling: The concept and clues from chromaffin and other cells. *Br. J. Pharmacol.*, **34**, 451–474.

Douglas, W.W., Kanno, T. and Sampson, S.R. (1967) Influence of the ionic enviroment on the membrane potential of adrenal chromaffin cells and on the depolarizing effect of acetylcholine. *J. Physiol.*, **191**, 107–121.

Doupnik, C.A. and Pun, R.Y.K. (1994) G-protein activation mediates prepulse facilitation of Ca^{2+} channel currents in bovine chromaffin cells. *J. Memb. Biol.*, **140**, 47–56.

Duarte, C.B., Rosario, L.M., Sena, C.M. and Carvalho, A.P. (1993). A toxin fraction FTX from the funnel-web spider poison inhibits dihydropyridine-insensitive Ca^{2+} channels coupled to catecholamine release in bovine adrenal chromaffin cells. *J. Neurochem.*, **60**, 908–913.

Dubel, S.J., Starr, T.V.B., Hell, J., Ahlijanian, M.K., Enyeart, J.J., Catterall, W. A. and Snutch, T.P. (1992) Molecular cloning of the α_1 subunit of an ω-conotoxin-sensitive calcium channel. *Proc. Natl. Acad. Sci. USA*, **89**, 5058–5062.

Dunlap, K. and Fischbach, G.D. (1978) Neurotransmitters decrease the Ca^{2+} component of sensory neurone action potentials. *Nature*, **276**, 837–838.

Dunlap, K. and Fischbach, G.D. (1981) Neurotransmitters decrease the Ca^{2+} conductance activated by depolarization of embryonic chick sensory neurones. *J. Physiol.*, **317**, 519–535.

Durroux, T., Gallo-Payet, N. and Payet, M.D (1988) Three components of the calcium current in cultured glomerulous cells from rat adrenal gland. *J. Physiol.*, **404**, 713–729.

Eckel, L., Gristwood, R.W., Nawrath, H., Owen, D.A. and Satter, P. (1982) Inotropic and electrophysiological effects of histamine on human ventricular heart muscle. *J. Physiol.*, **330**, 111–123.

Ellinor, P.T., Zhang, J-F., Randall, A.D., Zhou, M., Schwarz, T.L., Tsien, R.W. and Horne, W.A. (1993) Functional expression of a rapidly inactivating neuronal calcium channel. *Nature*, **363**, 455–458.

Ellis, S.B., Williams, M.E., Way, N.R., Brenner, R., Sharp, A.H., Leung, A.T., Campell, K.P., Mckenna, E., Koch, W.J., Hui, A., Schwartz, A. and Harpold, M.M. (1988) Sequence and expression of mRNAs encoding the α_1 and α_2 subunits of a DHP-sensitive calcium channel. *Science*, **241**, 1661–1664.

Elmslie, K.S., Zhou, W. and Jones, S.W. (1990) LHRH and GTP-γ-S modify Ca^{2+} current activation in bullfrog sympathetic neurons. *Neuron*, **5**, 75–80.

Ertel, E.A., Warren, V.A., Adams, M.E., Griffin, P.R., Cohen, C.J. and Smith, M.M. (1994) Type III omega-agatoxins: a family of probes for similar binding sites on L- and N-type calcium channels. *Biochemistry*, **33**, 5098–5108.

Fariñas, I., Solsona, C. and Marsal, J. (1992) Omega-conotoxin differentially block acetylcholine and adenosine triphosphate releases from torpedo synaptosomes. *Neuroscience*, **47**, 641–648.

Farr-Jones, S., Miljanich, G.P., Nadasdi, L., Ramachandran, J. and Basus, V.J. (1995) Solution structure of omega-conotoxin MVIIC, a high affinity ligand of P-type calcium channels, using ^{1}H NMR spectroscopy and complete relaxation matrix analysis. *J. Memb. Biol.*, **248**, 106–124.

Fenwick, E.M., Marty, A. and Neher, E. (1982) Sodium and calcium channels in bovine chromaffin cells. *J. Physiol.*, **331**, 599–635.

Fernández, J.M., Granja, R., Izaguirre, V., González-García, C. and Ceña, V. (1995) ω-Conotoxin GVIA blocks nicotine-induced catecholamine secretion by blocking the nicotine receptor-activated inward currents in bovine chromaffin cells. *Neurosci. Lett.*, **191**, 1–4.

Feuerstein, T., Doodley, D.J. and Seeger, W. (1990) Inhibition of norepinephrine and acetylcholine release from human neocortex by ω-conotoxin GVIA. *J. Pharmacol. Exp. Ther.*, **252**, 778–795.

Fleckenstein, A. (1983) History of calcium antagonists. *Circ. Res.*, **52** (Suppl. I), 3–16.

Fogelson, A.L. and Zucker, R.S. (1985) Presynaptic calcium diffusion from various arrays of single channels. Implications for transmitter release and synaptic facilitation. *Biophys. J.*, **48**, 1003–1017.

Foucart, S., Bleakman, D., Bindokas, V.P. and Miller, R.J. (1993) Neuropeptide Y and pancreatic polypeptide reduce Ca^{2+} currents in acutely dissociated neurons from adult rat superior cervical ganglia. *J. Pharmacol. Exp. Ther.*, **265**, 903–909.

Fournier, F., Bourinet, E., Nargeot, J. and Charnet, P. (1993) Cyclic AMP-dependent regulation of P-type Ca^{2+} channels expressed in *Xenopus* oocytes. *Pflügers Arch. Eur. J. Physiol*, **423**, 173–180.

Fox, A.P., Nowycky, M.C. and Tsien, R.W. (1987a) Kinetic and pharmacological properties distinguishing three types of calcium currents in chick sensory neurones. *J. Physiol.*, **394**, 149–172.

Fox, A.P., Nowycky, M.C. and Tsien, R.W. (1987b) Single-channel recordings of three types of calcium channels in chick sensory neurones. *J. Physiol.*, **394**, 173–200.

Franco-Obregón, A., Ureña, J. and López-Barneo, J. (1995) Oxygen-sensitive Ca^{2+} channels in vascular smooth muscle and their possible role in hypoxic arterial relaxation. *Proc. Natl. Acad. Sci. USA*, **92**, 4715–4719.

Frew,R. and Lundy, P.M. (1995) A role for Q-type Ca^{2+} channels in neurotransmission in the rat urinary bladder. *Br. J. Pharmacol.*, **116**, 1595–1598.

Fujita, Y., Mynlieff, M., Dirksen, R.T., Kim M-S., Niidome, T., Nakai, J., Friedrich, I., Iwabe, N., Miyata, T., Furuichi, T., Furutama, D., Mikoshiba, K., Mori, Y. and Beam, K.G. (1993) Primary structure and functional expression of the ω-conotoxin-sensitive N-type channel from rabbit brain. *Neuron*, **10**, 585–598.

Furukawa, T., Ito, J., Nitta, J., Tsujino, M., Adachi, D., Hiroe, M., Marumo, F., Sawanonori, T. and Hiraoka, M. (1992) Endothelin-1 enhances Ca^{2+} entry through T-type Ca^{2+} channels in cultured neonatal rat ventricular myocytes. *Circ. Res.*, **71**, 1242–1253.

Galvan, M. and Adams, P.R. (1982) Control of Ca^{2+} current in rat sympathetic neurons by norepinephrine. *Brain Res.*, **244**, 135–144.

Gandía, L., Albillos, A. and García, A.G. (1993a) Bovine chromaffin cells possess FTX-sensitive calcium channels. *Biochem. Biophys. Res. Commun.*, **194**, 671–676.

Gandía, L., Albillos, A., López, M.G., Michelena, P. and García, A.G. (1994) Diversity of calcium channels: relevance to secretion. *Bio-Reguladores*, **3**, 71–80.

Gandía, L., Borges, R., Albillos, A. and García, A.G. (1995) Multiple types of calcium channels are present in rat chromaffin cells. *Pflügers Arch. Eur. J. Physiol.*, **430**, 55–63.

Gandía, L., Fonteríz, R.I., López, M.G., Artalejo, C.R. and García, A.G. (1987) Relative sensitivities of chromaffin cell calcium channels to organic and inorganic calcium antagonists. *Neurosci. Lett.*, **77**, 333–338.

Gandía, L., García, A.G. and Morad, M. (1993b) ATP modulation of calcium channels in bovine chromaffin cells. *J. Physiol.*, **470**, 55–72.

Gandía L., Michelena, P., De Pascual, R., López, M.G. and García, A.G. (1990) Different sensitivities to dihydropyridines of catecholamine release from cat and ox adrenals. *NeuroReport*, **1**, 119–122.

Garcez-do-Carmo, L., Albillos, A., Artalejo, A.R., De la Fuente, M.T., López, M.G., Gandía, L., Michelena, P. and García, A.G. (1993) R56865 inhibits catecholamine release from bovine chromaffin cells by blocking calcium channels. *Br. J. Pharmacol.*, **110**, 1149–1155.

García, A.G., Sala, F., Reig, J.A., Viniegra, S., Frías, J., Fonteríz, R.I. and Gandía, L. (1984) Dihydropyridine Bay K 8644 activates chromaffin cell calcium channels. *Nature*, **309**, 69–71.

Golard, A. and Siegelbaum, S.A. (1993) Kinetic basis for the voltage-dependent inhibition of N-type Ca^{2+} current by somatostatin and norepinephrine in chick sympathetic neurons. *J. Neurosci.*, **13**, 3884–3894.

Granja, R., Fernández-Fernández, J.M., Izaguirre, V., González-García, C. and Ceña, V. (1995) ω-Agatoxin IVA blocks nicotinic receptor channels in bovine chromaffin cells. *FEBS Lett.*, **362**, 15–18.

Grantham, C.J., Main, M.J. and Cannell, M.B. (1994) Fluspirilene block of N-type calcium current in NGF-differenciated PC12 cells. *Br. J. Pharmacol.*, **111**, 483–488.

Grassi, F. and Lux, H.D. (1989) Voltage-dependent GABA-induced modulation of Ca^{2+} currents in chick sensory neurons. *Neurosci. Lett.*, **105**, 113–119.

Gray, R. and Johnson, D. (1987) Noradrenaline and β-adrenoceptor agonists increase activity of voltage-dependent Ca^{2+} channels in hippocampal neurons. *Nature*, **327**, 620–622.

Guthrie, P.B., Segal, M. and Kater, S.B. (1991) Independent regulation of calcium revealed by imaging dendritic spines. *Nature*, **354**, 76–80.

Haass, M., Richardt, G., Brenn, T., Schömig, E. and Schömig, A. (1991). Nicotine induced release of noradrenaline and neuropeptide Y in guinea-pig heart: role of calcium channels and protein kinase C. *Naunyn-Schmiedeberg's Arch. Pharmacol.*, **344**, 527–531.

Hamill, O.P., Marty, A., Neher, E., Sakmann, B. and Sigworth, F.J. (1981) Improved patch-clamp techniques for high-resolution current recording from cells and cell-free membrane patches. *Pflügers Arch. Eur. J. Physiol.*, **391**, 85–100.

Hans, M., Illes, P. and Takeda, K. (1990) The blocking effects of ω-conotoxin on Ca current in bovine chromaffin cells. *Neurosci. Lett.*, **114**, 63–38.

Hata, Y., Davletov, B., Petrenko, A.G., Jahn, R. and Südhof, T.C. (1993) Interaction of synaptotagmin with the cytoplasmic domains of neurexins. *Neuron*, **10**, 307–315.

Hata, F., Fujita, A., Saeki, K., Kishi, I., Takeuchi, T. and Yagasaki, O. (1992) Selective inhibitory effects of calcium channel antagonists on the two components of the neurogenic responses of guinea-pig vas deferens. *J. Pharmacol. Exp. Ther.*, **263**, 214–220.

Haydon, P.G., Henderson, E. and Stanley, E.F. (1994) Localization of individual calcium channels at the release face of a presynaptic nerve terminal. *Neuron*, **13**, 1275–1280.

Heinemann, C., Rüden, L.v., Chow, R.H. and Neher, E. (1993) A two-step model of secretion control in neuroendocrine cells. *Pflügers Arch. Eur. J. Physiol.*, **424**, 105–112.

Hell, J.W., Yokoyama, C.T., Wong, S.T., Warner, C., Snutch, T.P. and Catterall, W.A. (1993) Differential phosphorylation of two size forms of the neuronal class C L-type Ca^{2+} channel α_1 subunit. *J. Biol. Chem.*, **268**, 19451–19457.

Herdon, H. and Nahorski, S.R. (1989) Investigations of the role of dihydropyridine and ω-conotoxin-sensitive calcium channels in mediating depolarization-evoked endogenous dopamine release from striatal slices. *Naunyn-Schmiedeberg's Arch. Pharmacol.,* **340**, 36–40.

Hescheler, J., Kameyama, M. and Trautwein, W. (1986) On the mechanism of muscarinic inhibition of the cardiac Ca current. *Pflügers Arch. Eur. J. Physiol.,* **407**, 182–189.

Hescheler, J., Rosenthal, W., Trautwein, W. and Schultz, G. (1987a) The GTP-binding protein, G_o, regulates neuronal Ca^{2+} channels. *Nature,* **325**, 445–447.

Hescheler, J., Tang, M., Jastorff, B. and Trautwein, W. (1987b) On the mechanism of histamine induced enhacement of the cardiac Ca^{2+} current. *Pflügers Arch. Eur. J. Physiol.,* **407**, 182–189.

Heuser, J.E., Reese, T.S., Dennis, M.J., Jan, Y., Jan, L. and Evans, L. (1979) Synaptic vesicle exocytosis captured by quick freezing and correlated with quantal transmitter release. *J. Cell. Biol.,* **81**, 275–300.

Heuser, J.S., Reese, T.S. and Landis, D.M.D. (1974) Functional changes in frog neuromuscular junctions studied with freeze-fracture. *J. Neurocytol.,* **3**, 109–131.

Hillyard, D.R., Monje, V.D., Mintz, M.I., Bean, B.P., Nadasdi, L., Ramachandran, J., Miljanich, G., Azimi-Zoonooz, A., McIntosh, J.M., Cruz, L.J., Imperial, J.S. and Olivera, B.M. (1992) A new conus peptide ligand for mammalian presynaptic Ca^{2+} channels. *Neuron,* **9**, 69–77.

Hirning, L.D., Fox, A.P., McClesky, E.W., Olivera, B.M., Thayer, S.A., Miller, R.J. and Tsien, R.W. (1988) Dominant role of N-type Ca^{2+} channels in evoked release of norepinephrine from sympathetic neurons. *Science,* **239**, 57–61.

Hodgkin, A.L. and Keynes, R.D. (1957) Movements of labelled calcium in squid giant axons. *J. Physiol.,* **138**, 253–281.

Hofmann, F., Biel, M. and Flockerzi, V. (1994) Molecular basis of Ca^{2+} channel diversity. *Annu. Rev. Neurosci.,* **17**, 399–418.

Hofmann, F. and Habermann, E. (1990) Role of ω-conotoxin-sensitive calcium channels in inositolphosphate production and noradrenaline release due to potassium depolarisation or stimulation with carbachol. *Naunyn-Schmiedeberg's Arch. Pharmacol.,* **341**, 200–205.

Holz, G.G., Rane, S.G. and Dunlap, K. (1986) GTP-binding protein mediate transmitter inhibition of voltage-dependent Ca^{2+} channels. *Nature,* **319**, 670–672.

Hong, S.J. and Chang, C.C. (1995a) Inhibition of acetylcholine release from mouse motor nerve by P-type calcium channel blocker, ω-agatoxin IVA. *J. Physiol.,* **482**, 283–290.

Hong, S.J. and Chang, C.C. (1995b) Calcium channel subtypes for the sympathetic and parasympathetic nerves of guinea pig atria. *Br. J. Pharmacol.,* **116**, 1577–1582.

Hoshi, T., Rothlein, J. and Smith, S.J. (1984) Facilitation of Ca^{2+} channel currents in bovine adrenal chromaffin cells. *Proc. Natl. Acad. Sci. USA,* **81**, 5871–5875.

Hoshi, T. and Smith, S.J. (1987) Large depolarization induces long openings of voltage-dependent calcium channels in adrenal chromaffin cells. *J. Neurosci.,* **7**, 571–580.

Ikeda, S.R. (1991) Double-pulse calcium current facilitation in adult rat sympathetic neurones. *J. Physiol.,* **439**, 181–214.

Ikeda, S.R., Schofield, G.G. and Weight, F.F. (1987) Somatostatin blocks a Ca^{2+} current in acutely isolated adult rat superior cervical ganglion neurons. *Neurosci. Lett.,* **81**, 123–128.

Isom, L.L., De Jongh, K.S. and Catterall, W.A. (1994) Auxiliary subunits of voltage gated ion channels. *Neuron,* **12**, 1183–1194.

Issa, N.P. and Hudspeth, A.J. (1994) Clustering of Ca^{2+} channels and Ca^{2+}-activated K^+ channels at fluorescently labeled presynaptic active zones of hair cells. *Proc. Natl. Acad. Sci. USA,* **91**, 7578–7582.

Jahromi, B.S., Robitaille, R. and Charlton, M.P. (1992) Transmitter release increases intracellular calcium in perisynaptic schwann cells in situ. *Neuron,* **8**, 1069–1077.

Jay, S.D., Ellis, S.B., McCue, A.F., Williams, M.E., Vedvick, T.S., Harpold, M.M. and Campell, K.P. (1990) Primary structure of the γ subunit of the DHP-sensitive calcium channel from skeletal muscle. *Science,* **248**, 490–492.

Jiménez, R.R., López, M.G., Sancho, C., Maroto, R. and García, A.G. (1993) A component of the catecholamine secretory response in the bovine adrenal gland is resistant to dihydropyridines and ω-conotoxin. *Biochem. Biophys. Res. Commun.,* **191**, 1278–1283.

Kasai, H. and Aosaki, T. (1989) Modulation of Ca-channel current by an adenosine analog mediated by a GTP-binding protein in chick sensory neurons. *Pflügers Arch. Eur. J. Physiol.,* **414**, 145–149.

Kasai, H., Aosaki, T. and Fukuda, J. (1987) Presynaptic Ca-antagonist ω-conotoxin irreversibly blocks N-type Ca channels in chick sensory neurons. *Neurosci. Res.*, **4**, 228–235.

Katz, B. (1969). The relevance of neural transmitter substances. The Sherrington Lectures X. Springfield, Ill. Thomas, pp. 1–60.

Kidokoro, Y., Miyazaki, S. and Ozawa, S. (1982) Acetylcholine-induced membrane depolarization and potential fluctuations in the rat adrenal chromaffin cell. *J. Physiol.*, **324**, 203–220.

Kim, S.J., Lim, W. and Kim, J. (1995). Contribution of L- and N-type calcium currents to exocytosis in rat adrenal medullary cells. *Brain Res.*, **675**, 289–296.

Kimura, T., Takeuchi, A. and Satoh, S. (1994). Inhibition by ω-conotoxin GVIA of adrenal catecholamine release in response to endogenous and exogenous acetylcholine in anesthetized dogs. *Eur. J. Pharmacol.* **264**, 169–175.

Knight, D.E. (1988) Intracellular factors controlling exocytosis. In *Molecular Mechanisms in Secretion*, Alfred Benzon Symposium 25, ed. Thorn, N.A., Treiman, M. and Peterson, O.H., pp 192–206. Munksgaard, Copenhagen.

Knight, D.E. and Baker, P.F. (1982) Calcium-dependence of catecholamine release from bovine adrenal medullary cells after exposure to intense electric fields. *J. Memb. Biol.*, **68**, 107–140.

Knight, D.E. and Kesteven, N.T. (1983) Evoked transient intracellular free Ca^{2+} changes and secretion in isolated bovine adrenal medullary cells. *Proc. Royal Soc.*, **B218**, 177–199.

Kohno, T., Kim, J.j., Kobayashi, K., Kodera, Y., Maeda, T. and Sato, K. (1995) Three-dimensional structure in solution of the calcium channel blocker ω-conotoxin MVIIA. *Biochemistry*, **34**, 10256–10265.

Kostyuk, P.G. and Lukyanetz, E.A. (1993) Mechanisms of antagonistic action of internal Ca^{2+} on serotonin-induced potentiation of Ca^{2+} currents in *Helix* neurons. *Pflügers Arch. Eur. J. Physiol.*, **424**, 73–83.

Kuroda, H., Chen, Y.N., Watanabe, T.X., Kimura, T. and Sakakibara, S. (1992) Solution synthesis of calciseptine, an L-type specific calcium channel blocker. *Pept. Res.*, **5**, 265–268.

Lacerda, A.E., Kim, H.S., Ruth, P., Perez-Reyes, E., Flockerzi, V., Hofmann, F., Birnbaumer, L. and Brown, A.M. (1991) Normalization of current kinetics by interaction between the α_1 and β subunits of the skeletal muscle dihydropyridine-sensitive. *Nature*, **352**, 527–530.

Lacerda, A.E., Rampe, D. and Brown, A.M. (1988) Effects of protein kinase C activators on cardiac Ca^{2+} channels. *Nature*, **335**, 249–251.

Lampe, R.A., Defeo, P.A., Davison, M.D., Young, J., Herman, J.L., Spreen, R.C., Horn, M.B., Mangano, T.J. and Keith, R.A. (1993) Isolation and pharmacological characterization of omega-grammotoxin SIA, a novel peptide inhibitor of neuronal voltage-sensitive calcium channel responses. *Mol. Pharmacol.*, **44**, 451–460.

Leighton, C., Mathie, A. and Dolphin, A.C. (1994) Modulation by phosphorylation and neurotransmitters of voltage-dependent Ca^{2+} channel currents in cultured rat cerebellar granule neurons. *J. Physiol.*, **477**, 89P.

Leveque, C., Hoshino, T., David, P., Shoji-Kasai, Y. Leys, K., Omori, A., Lang, B., El-Far, O., Sato, K., Martin-Moutot, N., Newson-Davis, J., Takahashi, M. and Seagar, M.J. (1992) The synaptic vesicle protein synaptotagmin associates with calcium channels and is a putative Lambert-Eaton myasthenic syndrome antigen. *Proc. Natl. Acad. Sci. USA*, **89**, 3625–3629.

Leveque, C., El-Far, O., Martin-Moutot, N., Sato, K., Kato, R., Takahashi, M. and Seagar, M. J. (1994) Purification of the N-type calcium channel associated with syntaxin and synaptotagmin. A complex implicated in synaptic vesicle exocytosis. *J. Biol. Chem.*, **269**, 6306–6312.

Lewis, D., Weight, F.F and Luini, A. (1986) A guanine nucleotide-binding protein mediates the inhibition of voltage dependent Ca^{2+} current by somatostatin in a pituitary cell line. *Proc. Natl. Acad. Sci. USA*, **83**, 9035–9039.

Lipscombe, D., Bley, K. and Tsien, R.W. (1988) Modulation of neuronal Ca channels by cAMP and phorbol esters. *Soc. Neurosci. Abstr.*, **14**, 153.

Lipscombe, D., Kongsamut, S. and Tsien, R.W. (1989) α-adrenergic inhibition of sympathetic neurotransmitter release mediated by modulation of N-type Ca^{2+}-channel gating. *Nature*, **340**, 639–642.

Llinás, R., Steinberg, I.Z. and Walton, K. (1981) Relationship between presynaptic calcium current and postsynaptic potencial in squid giant synapse. *Biophys. J.*, **33**, 323–352.

Llinás, R., Sugimori, M., Lin, J.-W. and Cherksey, B. (1989) Blocking and isolation of a calcium channel from neurons in mammals and cephalopods utilizing a toxin fraction (FTX) from funnel-web spider poison. *Proc. Natl. Acad. Sci. USA*, **86**, 1689–1693.

Llinás, R., Sugimori, M. and Silver, R.B. (1992) Microdomains of high calcium concentration in a presynaptic terminal. *Science*, **256**, 677–679.

López, M.G., Albillos, A., De la Fuente, M., Borges, R., Gandía, L., Carbone, E., García, A.G. and Artalejo, A.R. (1994a) Localized L-type calcium channels control exocytosis in cat chromaffin cells. *Pflügers Arch. Eur. J. Physiol.*, **427**, 348–354.

López, M.G., Shukla, R., García, A.G. and Wakade, A.R. (1992) A dihydropyridine-resistant component in the rat adrenal secretory response to splanchnic nerve stimulation. *J. Neurochem.*, **58**, 2139–2144.

López, M.G., Villarroya, M., Lara, B., Martínez-Sierra, R., García, A.G. and Gandía, L. (1994b) Q- and L-type Ca^{2+} channels dominate the control of secretion in bovine chromaffin cells. *FEBS Lett.*, **349**, 331–337.

Luebke, J.I. and Dunlap, K. (1994) Sensory neuron N-type Ca^{2+} currents are inhibited by both voltage-dependent and -independent mechanisms. *Pflügers Arch. Eur. J. Physiol.*, **365**, 258–262.

Luebke, J.I., Dunlap, K. and Turner, T.J. (1993) Multiple calcium channel types control glutamatergic synaptic transmission in the hippocampus. *Neuron*, **11**, 895–908.

Luini, A., Lewis, D., Guild, S., Schofield, G. and Weight, F. (1986) Somatostatin, an inhibitor of ACTH secretion, decreases cytosolic free Ca^{2+} and voltage-dependent Ca^{2+} current in a pituitary cell line. *J. Neurosci.*, **6**, 3128–3132.

Lundy, P.M. and Frew, R. (1988) Evidence of ω-conotoxin GVIA-sensitive Ca^{2+} channels in mammalian peripheral nerve terminals. *Eur. J. Pharmacol.*, **156**, 325–330.

Lundy, P.M. and Frew, R. (1994) Effect of ω-agatoxin-IVA on autonomic neurotransmission. *Eur. J. Pharmacol.*, **261**, 79–84.

Maggi, C.A., Patachini, R., Santiciol, P., Lippe, I.T., Guiliano, S., Geppetti, P., Del Blanco, E., Selleri, S. and Meli, A. (1988) The effect of omega-conotoxin GVIA, a peptide modulator of the N-type voltage sensitive calcium channels on motor responses produced by activation of efferent sensory nerves in mammalian smooth muscle. *Naunyn-Schmiedeberg's Arch. Pharmacol.*, **338**, 107–113.

Marchetti, C., Carbone, E. and Lux, H.D. (1986) Effects of dopamine and noradrenaline on Ca^{2+} channels of cultured sensory and sympathetic neurons of chick. *Pflügers Arch. Eur. J. Physiol.*, **406**, 104–111.

Mathie, A., Bernheim, L. and Hille, B. (1992) Inhibition of N- and L-type Ca^{2+} channels by muscarinic in rat sympathetic neurons. *Neuron*, **8**, 907–914.

McFadzean, I., Mullaney, I., Brown, D.A. and Milligan, G. (1989) Antibodies to the GTP binding protein, G_o, antagonize noradrenaline-induced Ca^{2+} current inhibition in NG108-15 hybrid cell. *Neuron*, **3**, 177–182.

Meriney, S.D., Gray, D.B. and Pilar, G.R. (1994) Somatostatin-induced inhibition of neuronal Ca^{2+} current modulated by cGMP-dependent protein kinase. *Nature*, **369**, 336–339.

Michelena, P., García-Pérez, L.E., Artalejo, A.R. and García, A.G. (1993) Dissociation between cytosolic calcium and secretion in chromaffin cells superfused with calcium ramps. *Proc. Natl. Acad. Sci. USA*, **90**, 3284–3288.

Michelena, P., Vega, T., Montiel, C., López, M.G., García-Pérez, L.-E., Gandía, L. and García, A.G. (1995) Effects of tyramine and calcium on the kinetics of secretion in intact and electroporated chromaffin cells superfused at high speed. *Pflügers Arch. Eur. J. Physiol.* (in press).

Mikami, A., Imoto, K., Tanabe, T., Niidome, T., Mori, Y., Takashima, H., Narumiya, S. and Numa, S. (1989) Primary structure and functional expression of the cardiac dihydropyridine-sensitive calcium channel. *Nature*, **340**, 230–233.

Miller, R. (1987) Multiple calcium channels and neuronal function. *Science*, **235**, 46–51.

Mintz, I.M. (1994) Block of Ca channels in rat central neurons by the spider toxin omega-Aga-IIIA. *J. Neurosci.*, **14**, 2844–2853.

Mintz, I.M., Adams, M.E. and Bean, B.P. (1992a) P-type calcium channels in central and peripheral neurons. *Neuron*, **9**, 1–20.

Mintz, I.M., Venema, V.J., Swidereck, K., Lee, T., Beau, B.P. and Adams, M.E. (1992b) P-type calcium channels blocked by the spider toxin ω-Aga-IVA. *Nature*, **355**, 827–829.

Mintz, I.M. and Bean, B.P. (1993a) Block of calcium channels in rat neurons by synthetic ω-Aga-IVA. *Neuropharmacology*, **32**, 1161–1169.

Mintz, I.M. and Bean, B.P. (1993b) $GABA_B$ receptor inhibition of P-type Ca^{2+} channels in central neurons. *Neuron*, **10**, 889–989.

Mintz, I.M., Venema, V.J., Adams, M.E. and Bean, B.P. (1991) Inhibition of N- and L-type Ca^{2+} channels by the spider venom toxin ω-agatoxin-IIIA. *Proc. Natl. Acad. Sci. USA*, **88**, 6628–6631.

Mintz, I.M., Venema, V.J., Swiderek, K., Lee, T., Bean, B.P. and Adams, M.E. (1992b) P-type calcium channels blocked by the spider toxin ω-Aga-IVA. *Nature*, **355**, 827–829.

Monck, J.R., Robinson, I.M., Escobar, A., Vergara, J. and Fernández, J.M. (1994) Pulsed laser imaging of rapid Ca^{2+} gradients in excitable cells. *Biophys. J.*, **67**, 505–514.

Monje, V.D., Haack, J.A., Naisbitt, S.R., Miljanich, G., Ramachandran, J., Nadasdi, L., Olivera, B.M., Hillyard, D.R. and Gray, W.R. (1993) A new conus peptide ligand for Ca channel subtypes. *Neuropharmacology*, **32**, 1141–1149.

Montiel, C., Artalejo, A.R. and García, A.G. (1984) Effects of the novel dihydropyridine Bay-K-8644 on adrenomedullary catecholamine release evoked by calcium reintroduction. *Biochem. Biophys. Res. Commun.*, **120**, 851–857.

Mori, Y., Friedrich, T., Kim, M-S., Mikami, A., Nakai, J., Ruth, P., Bosse, E., Hofmann, F., Flockerzi, V., Furuichi, T., Mikoshiba, K., Imoto, K., Tanabe, T. and Numa, S. (1991) Primary structure and functional expression from complementary DNA of a brain calcium channel. *Nature*, **350**, 398–402.

Muller, W. and Connor, J.A. (1991) Dendritic spines as individual neuronal compartments for synaptic Ca^{2+} responses. *Nature*, **354**, 73–76.

Neely, A., Wei, X., Olcese, R., Birnbaumer, L. and Stefani, E. (1993) Potentiation by the β subunit of the ratio of the ionic current to the charge movement in the cardiac calcium channel. *Science*, **262**, 575–578.

Neher, E. and Marty, A. (1982) Discrete changes of cell membrane capacitance observed under conditions of enhanced secretion in bovine adrenal chromaffin cells. *Proc. Natl. Acad. Sci. USA*, **79**, 6712–6716

Neher, E. and Zucker, R.S. (1993), Multiple calcium-dependent processes related to secretion in bovine chromaffin cells. *Neuron*, **10**, 21–30.

Nemoto, N., Kubo, S., Yoshida, T., Chino, N., Kimura, T., Sakakibara, S., Kyogoku, Y. and Kobayashi, Y. (1995) Solution structure of omega-conotoxin MVIIC detgermined by NMR. *Biochem. Biophys. Res. Commun.*, **207**, 695–700.

Nishimura, S., Takeshima, H., Hofmann, F., Flockerzi, V. and Imoto, K. (1993) Requirement of the calcium channel β subunit for functional conformation. *FEBS Lett.*, **324**, 283–286.

Nowycky, M.C., Fox, A.P. and Tsien, R.W. (1985) Three types of neuronal calcium channels with different calcium agonist sensitivity. *Nature*, **316**, 440–443.

O'Connor, V.M,. Shamotienko, O., Grishin, E. and Betz, H. (1993) On the structure of the 'synaptosecretosome'. Evidence for a neurexin/synaptotagmin/syntaxin/Ca^{2+} channel complex. *FEBS Lett.*, **326**, 255–260.

Olivera, B.M., Miljanich, G., Ramachandran, J. and Adams, M.E. (1994) Calcium channel diversity and neurotransmitter release: the ω-Conotoxins and ω-Agatoxins. *Annu. Rev. Biochem.*, **63**, 823–867.

O'Sullivan, A.J., Cheek, T.R., Moreton, R.B., Berridge, M.J. and Burgoyne, R.D. (1989) Localization and heterogeneity of agonist- induced changes in cytosolic concentrations in single bovine adrenal chromaffin cells from video imaging of Fura-2. *EMBO J.*, **8**, 401–411.

Owen, P.J., Marriot, D.B. and Boarder, M.R. (1989) Evidence for a dihydropyridine and conotoxin-insensitive release of noradrenaline and uptake of calcium in adrenal chromaffin cells. *Br. J. Pharmacol.*, **97**, 133–138.

Pallaghy, P.K., Duggan, B.M., Pennington, M.W. and Norton, R.S. (1993) Three-dimensional structure in solution of the calcium channel blocker omega-conotoxin. *J. Mol. Biol.*, **234**, 405–420.

Pérez-Reyes, E., Castellano, A., Kim, H.S., Bertrand, P., Baggstrom, E., Lacerda, A.E., Wei, X. and Birnbaumer, L. (1992) Cloning and expression of a cardiac/brain β subunit of the L-type calcium channel. *J. Biol. Chem.*, **267**, 1792–1797.

Petrenko, A.G. (1993) Alpha-Latrotoxin receptor. Implications in nerve terminal function. *FEBS Lett.*, **325**, 81–85.

Petrenko, A.G., Perin, M.S., Davletov, B.A., Ushkaryov, Y.A., Geppert, M. and Südhof, T.C. (1991) Binding of synaptotagmin to the alpha-latrotoxin receptor implicates both in synaptic vesicle exocytosis. *Nature*, **353**, 65–68.

Piser, T.M., Lampe, R.A., Keith, R.A. and Thayer, S.A. (1994) Omega-Grammotoxin blocks action-potential-induced Ca^{2+} influx and whole-cell Ca^{2+} current in rat dorsal-root ganglion neurons. *Pflügers Arch. Eur. J. Physiol.*, **426**, 214–220

Piser, T.M., Lampe, R.A., Keith, R.A. and Thayer, S.A. (1995) ω-Grammotoxin SIA blocks multiple, voltage-gated, Ca^{2+} channel subtypes in cultured rat hippocampal neurons. *Mol. Pharmacol.*, **48**, 131–139.

Plummer, M.R., Logothetis, D.E. and Hess, P. (1989) Elementary properties and pharmacological sensitivities of calcium channels in mammalian peripheral neurons. *Neuron*, **2**, 1453–1463.

Plummer, M.R., Rittenhouse, A., Kanevsky, M. and Hess, P. (1991) Neurotransmitter modulation of calcium channels in rat sympathetic neurones. *J. Neurosci.*, **11**, 2334–2348.

Pollo, A., Lovallo, M., Sher, E. and Carbone, E. (1992) Voltage-dependent noradrenergic modulation of ω-conotoxin-sensitive Ca^{2+} channels in human neuroblastoma IMR32 cells. *Pflügers Arch. Eur. J. Physiol.*, **422**, 75–83.

Pollo, A., Lovallo, M., Biancardi, E., Sher, E., Socci, C. and Carbone, E. (1993) Sensitivity to dihydropyridines, ω-conotoxin and noradrenaline reveals multiple high-voltage activated Ca^{2+} channels in rat insulinoma and human pancreatic β-cells. *Pflügers Arch. Eur. J. Physiol.*, **423**, 462–471.

Popov, S.V. and Poo, M-M. (1993) Synaptotagmin: a calcium-sensitive inhibitor of exocytosis?. *Cell*, **73**, 1247–1249.

Possani L.D., Martin, B.M., Yatani, A., Mochca-Morales, J., Zamudio, F.Z., Gurrola, G.B. and Brown, A.M. (1992) Isolation and physiological characterization of taicatoxin, a complex toxin with specific effects on calcium channels. *Toxicon*, **30**, 1343–1364.

Pragnell, M., De Waard, M., Mori, Y., Tanabe, T., Snutch, T.P. and Campbell, K.P. (1994) Calcium channel β-subunit binds to a conserved motif in the I-II cytoplasmic linker of the α_1-subunit. *Nature*, **368**, 67–70.

Przywara, D.A., Bhave, S.V., Chowdhury, P.S., Wakade, T.D. and Wakade, A.R. (1993) Sites of transmitter release and relation to intracellular Ca^{2+} in cultured sympathetic neurons. *Neuroscience*, **52**, 973–986.

Pumplin, D.W., Reese, T.S. and Llinás, R. (1981) Are the presynaptic membrane particles the calcium channel?. *Proc. Natl. Acad. Sci. USA*, **78**, 7210–7213.

Randall, A. and Tsien, R.W. (1995) Pharmacological dissection of multiple types of Ca^{2+} channel currents in rat cerebellar neurons. *J. Neurosci.*, **15**, 2995–3012.

Randall, A.D., Adams, M.E. and Tsien, R.W. (1994). P-, Q- and R-type Ca^{2+} channels in cerebellar granule cells and their dinstinctive responses to ω-aga-IVA and ω-aga-IIIA. *Soc. Neurosci. Abstr.*, **20**, 268–5.

Randall, A.D., Wendland, B., Schweizer, F., Miljanich, G., Adams M.E. and Tsien R.W. (1993) Five pharmacologically distinct high voltage activated Ca^{2+} channels in cerebellar granule cells. *Soc. Neurosci. Abstr.*, **19**, 1478.

Rane, S.G., Walsh, M.P. and Dunlap, K. (1987) Norepinephrine inhibition of sensory neuron Ca^{2+} current is blocked by a specific protein kinase C inhibitor. *Soci. Neurosci. Abstr.*, **13**, 793.

Regan, L.J. (1991) Voltage-dependent calcium currents in Purkinje cells from rat cerebellar vermis. *J. Neurosci.*, **11**, 2259–2269.

Reuter, H. (1995) Measurements of exocytosis from single presynaptic nerve terminals reveal heterogeneous inhibition by Ca^{2+} channel blockers. *Neuron*, **14**, 773–779.

Rittenhouse, A. and Zigmond, R.E. (1991) ω-Conotoxin inhibits the acute activation of tyrosine hydroxilase and the stimulation of norepinephrine release by potassium depolarisation and sympathetic nerve endings. *J. Neurochem.*, **56**, 615–622.

Roberts, W.M., Jacobs, R.A., Hudspeth, A.J. (1990) Colocalization of ion channels involved in frequency selectivity and synaptic transmission at presynaptic active zones of hair cells. *J. Neurosci.*, **10**, 3664–3684.

Robinson, I.M., Finnegan, J.M., Monck, J.R., Wightman, R.M. and Fernández, J.M. (1995) Colocalization of calcium entry and exocytosis release sites in adrenal chromaffin cells. *Proc. Natl. Acad. Sci. USA*, **92**, 2474–2478.

Robitaille, R., Adler, E.M. and Charlton, M.P. (1990) Strategic location of calcium channels at transmitter release sites of frog neuromuscular synapse. *Neuron*, **5**, 773–779.

Rosario, L.M., Soria, B., Feuerstein, G. and Pollard, H.B. (1989) Voltage-sensitive calcium flux into bovine chromaffin cells occurs through dihydropyridine-sensitive and dihydropyridine- and ω-conotoxin-insensitive pathways. *Neuroscience*, **29**, 735–747.

Rüden, L.v. and Neher, E. (1993), A Ca-dependent early step in the release of catecholamines from adrenal chromaffin cells. *Science*, **262**, 1061–1065.

Ruth, P., Röhrkasten, A., Biel, M., Bosse, E., Regulla, S., Meyer, H.E., Flockerzi, V. and Hofmann, F. (1989) Primary structure of the β subunit of the DHP-sensitive calcium channel from skeletal muscle. *Science*, **245**, 1115–1118.

Sah, D.W.Y. (1990) Neurotransmitter modulation of Ca^{2+} current in rat spinal cord neurons. *J. Neurosci.*, **10**, 136–141.

Sala, F. and Hernández-Cruz, A. (1990) Ca^{2+} diffusion modeling in a spherical neuron: relevance of buffering properties. *Biophys. J.*, **57**, 313–324.

Sather, W.A., Tanabe, T., Zhang, J.-F., Mori, Y., Adams, M.E. and Tsien, R.W. (1993) Distinctive biophysical and pharmacological properties of class A (BI) calcium channel α_1 subunits. *Neuron*, **11**, 291–303.

Schmidt, A., Hescheler, J., Offermanns, S., Spicher, K., Hinsch, K.D., Klinz, F.J., Codina, J., Birnbaumer, L., Gausepohl, H., Frank, R., Schultz, G. and Rosenthal, W. (1991) Involvement of pertussis toxin-sensitive G-proteins in the hormonal inhibition of dihydropyridine-sensitive Ca^{2+} currents in an insulin-secreting cell line (RINm5F). *J. Biol. Chem.*, **266**, 18025–18033.

Schröeder, T.J., Jankowski, J.A., Senyshyn, J., Holz, R.W. and Whightman, R.M. (1994) Zones of exocytotic release on bovine adrenal medullary cells in culture. *J. Biol. Chem.*, **269**, 17215–17220.

Schweitz, H., Heurteaux, C., Bois, P., Moinier, D., Romey, G. and Lazdunski, M. (1994) Calcicludine, a venom peptide of the Kunitz-type protease inhibitor family, is a potent blocker of high-threshold Ca^{2+} channels with a high affinity for L-type channels in cerebellar granule neurons. *Proc. Natl. Acad. Sci. USA*, **91**, 878–882.

Scott, R.H., Dolphin, A.C., Bindokas, V.P. and Adams, M.E. (1990a) Inhibition of neuronal Ca^{2+} channel currents by the funnel web spider toxin omega-Aga-IA. *Mol. Pharmacol.*, **38**, 711–718.

Scott, R.H., Sweeney, M.I., Kobrinsky, E.M., Pearson, H.A., Timms, G.H., Pullar, I.A., Wedley, S. and Dolphin, A.C. (1992) Actions of arginine polyamine on voltage and ligand-activated whole cell currents recorded from cultured neurones. *Br. J. Pharmacol.*, **106**, 199–207.

Scott, R.J., Wootton, J.F. and Dolphin, A.C. (1990b) Modulation of neuronal T-type Ca^{2+} channel currents by photoactivation of intracellular guanosine 5'-O(3-thio) triphosphate. *Neuroscience*, **38**, 285–294.

Sevilla, P., Bruix, M., Santoro, J., Gago, F., García, A.G. and Rico, M. (1993) Three-dimensional structure of ω-conotoxin GVIA determined by ^{1}H NMR. *Biochem. Biophys. Res. Commun.*, **192**, 1238–1244.

Sierra, F., Lorenzo, F., Macadar, O. and Buño, W. (1995) N-type Ca^{2+} channels mediate transmitter release at the electromotoneuron-electrocyte synapses of the weakly electric fish *Gymnotus carapo*. *Brain Res.*, **683**, 215–220.

Simmons, M.L., Terman, G., Gibbs, S.M. and Chaukin, C. (1995) L-type calcium channels mediate dynorphin neuropeptide release from dendrites but not axons of hippocampal granule cells. *Neuron*, **14**, 1265–1272.

Simon, S.M. and Llinás, R.R. (1985) Compartmentalization of the submembrane calcium activity during calcium influx and its significance in transmitter release. *Biophys. J.*, **48**, 485–498.

Singer, D., Biel, M., Lotan, I., Flockerzi, V., Hofmann, F. and Dascal, N. (1991) The roles of the subunits in the function of the calcium channel. *Science*, **253**, 1553–1557.

Singer, D., Gershon, E., Lotan, I., Hullin, R., Biel, M., Flockerzi, V., Hofmann, F. and Dascal, N. (1992) Modulation of cardiac Ca^{2+} channels in *Xenopus* oocytes by protein kinase C. *FEBS Lett.*, **306**, 113–118.

Skalicky, J.J., Metzler, W.J., Ciesla, D.J., Galdes, A. and Pardi, A. (1993) Solution structure of the calcium channel antagonist omega-conotoxin GVIA *Protein Sci.*, **2**, 1591–1603.

Smith, S.J. and Augustine, G.J. (1988) Calcium ions, active zones and synaptic transmitter release. *Trends Neurosci.*, **11**, 458–464.

Smith, S.J., Buchanan, J., Osses, L.R., Charlton, M.P. and Augustine, G.J. (1993) The spatial distribution of calcium signals in squid presynaptic terminals. *J. Physiol.*, **472**, 573–593.

Spedding, M. (1985) Calcium antagonist subgroups. *Trends Pharmacol. Sci.*, **6**, 109–114.

Söllner, T. Whiteheart, S.W., Brunner, M., Erdjument-Bromage, H., Geromanos, S., Tempst, P. and Rothman, J.E. (1993) SNAP receptors implicated in vesicle targeting and fusion. *Nature*, **362**, 318–324.

Soong, T.W., Stea, A., Hodson, C.D., Dubel, S.J., Vincent, S.R. and Snutch, T.P. (1993) Structure and functional expression of a member of the low voltage-activated calcium channel family. *Science*, **260**, 1133–1136.

Starr, T.V., Prystay, W. and Snutch, T.P. (1991) Primary structure of a calcium channel that is highly expressed in rat cerebellum. *Proc. Natl. Acad. Sci. USA*, **88**, 5621–5625.

Strong, J.A., Fox, A.P., Tsien, R.W. and Kaczmarek, L.K. (1987) Stimulation of protein kinase C recruits covert Ca^{2+} channels in *Aplysia* bag cell neurons. *Nature*, **325**, 714–717.

Swartz, K.J. (1993) Modulation of Ca^{2+} channels by protein kinase C in rat central and peripheral neurons: disruption of G-protein-mediated inhibition. *Neuron*, **11**, 305–320.

Swartz, K.J., Merritt, A., Bean, B.P. and Lovinger, D.M. (1993) Protein kinase C modulates glutamate receptor inhibition of Ca^{2+} channels and synaptic transmission. *Nature*, **361**, 165–168.

Takahashi, T. and Momiyama, A. (1993) Different types of calcium channels mediate central synaptic transmission. *Nature*, **366**, 156–158.

Takahashi, M., Seagar, M.J., Jones, J.F., Reber, B.F.X. and Catterall, W.A. (1987) Subunit structure of dihydropyridine sensitive calcium channels from skeletal muscle. *Proc. Natl. Acad. Sci. USA*, **84**, 5478–5482.

Tanabe, T., Takeshima, H., Mikami A., Flockerzi, V., Takahashi, H., Kangawa, K., Kojima, M., Matsuo, H., Hirose, T. and Numa, S. (1987) Primary structure of the receptor for calcium channel blockers from skeletal muscle. *Nature*, **328**, 313–318.

Tashenberger, H. and Grantyn, R. (1995) Several types of Ca^{2+} channels mediate glutamatergic synaptic responses to activation of single Thy-1-inmunolabeled rat retinal ganglion neurons. *J. Neurosci.*, **15**, 2240–2254.

Thomas, L., Hartung, K., Langosch, D., Rehm, H., Bamberg, E., Franke, W.W. and Betz, H. (1988) Identification of synaptophysin as a hexameric channel protein of the synaptic vesicle membrane. *Science*, **242**, 1050–1053.

Thomas, P, Suprenant, A. and Almers, W. (1990) Cytosolic Ca^{2+}, exocytosis, and endocytosis in single melanotrophs of the rat pituitary. *Neuron*, **5**, 723–733.

Tsunoo, A., Yoshi, M. and Narahashi, T. (1986) Block of Ca^{2+} channels by enkephalin and somatostatin in neuroblastoma glioma hybrid NG108-15 cells. *Proc. Natl. Acad. Sci. USA*, **83**, 9832–9836.

Turner, T.J., Adams, M.E. and Dunlap, K. (1992) Calcium channels coupled to glutamate release identified by ω-aga-IVA. *Science*, **258**, 310–313.

Turner, T.J., Adams, M.E. and Dunlap, K. (1993) Multiple Ca^{2+} channel types coexist to regulate synaptosomal neurotransmitter release. *Proc. Natl. Acad. Sci. USA*, **90**, 9518–9522.

Turner, T.J., Lampe, R.A. and Dunlap, K. (1995) Characterization of presynaptic calcium channels with omega-conotoxin MVIIC and omega-grammotoxin SIA: role for a resistant calcium channel type in neurosecretion. *Mol. Pharmacol.*, **47**, 348–353.

Umemiya, M. and Berger, A.J. (1995) Single-channel properties of four calcium channel types in rat motoneurones. *J. Neurosci.*, **15**, 2218–2224.

Usowicz, M.M., Sugimori, M., Cherksey, B. and Llinás, R. (1992) P-type calcium channels in the somata and dendrites of adult cerebellar Purkinje cells. *Neuron*, **9**, 1185–1199.

Valentino, K., Newcomb, R., Gadbois, T., Singh, T., Bowersox, S., Bitner, S., Justice, A., Yamashiro, D., Hoffmann, B.B., Ciaranello, R., Miljanich, G. and Ramachandran, J. (1993) A selective N-type calcium channel antagonist protects against neuronal loss after global cerebral ischemia. *Proc. Natl. Acad. Sci. USA*, **90**, 7894–7897.

Varadi, G., Lory, P., Schultz, D., Varadi, M. and Schwartz, A. (1991) Acceleration of activation and inactivation by the β subunit of the skeletal muscle calcium channel. *Nature*, **352**, 159–162.

Vega, T., De Pascual, R., Bulbena, O. and García, A.G. (1995) Effects of ω-conotoxins on noradrenergic neurotransmission in beating guinea-pig atria. *Eur. J. Pharmacol.*, **276**, 231–238.

Venema, V.J., Swiderek, K.M, Lee, T.D., Hathaway, G.M. and Adams, M.E. (1992) Antagonism of synaptosomal calcium channels by subtypes of ω-agatoxins. *J. Biol. Chem.*, **267**, 2610–2615.

Villarroya, M., Gandía, L., Lara, B., Albillos, A., López, M.G. and García, A.G. (1995) Dotarizine versus flunarizine as calcium antagonists in chromaffin cells. *Br. J. Pharmacol.*, **114**, 369–376.

Wanke, E., Ferroni, A., Malgaroli, A., Ambrosinii, A., Pozzan, T. and Meldolesi, J. (1987) Activation of a muscarinic receptor selectively inhibits a rapidly inactivated Ca^{2+} current in rat sympathetic neurons. *Proc. Natl. Acad. Sci. USA*, **84**, 4313–4317.

Wanke, E., Bianchi, L., Mantegazza, M., Guatteo, E., Mancinelli, E. and Ferroni, A. (1994) Muscarinic regulation of Ca^{2+} currents in rat sensory neurons: channel and receptor types, dose-response relationships and cross-talk pathways. *Eur. J. Neurosci.*, **6**, 381–391.

Wessler, I., Dooley, D.J., Werhand, J. and Schlemmer, F. (1990) Differential effects of calcium channels antagonists (ω-conotoxin GVIA, nifedipine, verapamil) on the electrically-evoked release of [³H]-acetylcholine from the myenteric plexus, phrenic nerve and neocortex of rats. *Naunyn-Schmiedeberg's Arch. Pharmacol.*, **341**, 288–294.

Wheeler, D.B., Randall, A. and Tsien, R.W. (1994) Roles of N-type and Q-type Ca^{2+} channels in supporting hippocampal synaptic transmission. *Science*, **264**, 107–111.

Williams, M.E., Brust, P.F., Feldman, D.H., Saraswathi, P., Simerson, S., Maroufi, A., McCue, A.F., Veliçelebi, G., Ellis, S.B. and Harpold, M.M. (1992a) Structure and functional expression of an ω-conotoxin-sensitive human N-type calcium channel. *Science*, **257**, 389–395.

Williams, M.E., Feldman, D.H., McCue, A.F., Brenner, R., Veliçelebi, G., Ellis, S.B. and Harpold, M.M. (1992b) Structure and functional expression of α_1, α_2, and β subunits of a novel human neuronal calcium channel subtype. *Neuron*, **8**, 71–84.

Williams, M.E., Marubio, L.M., Deal, C.R., Hans, M., Brust, P.F., Philipson, L.H., Miller, R.J., Johnson, E.C., Harpold, M.M. and Ellis, S.B. (1994) Structure and functional characterization of neuronal α_{1E} calcium channel subtypes. *J. Biol. Chem.*, **269**, 22347–22357.

Witcher, D.R., De Waard, M., Sakamoto, J., Franzini-Armstrong, C., Pragnell, M., Kahl, S.D. and Campbell K.P. (1993) Subunit identification and reconstitution of N-type Ca channel complex purified from brain. *Science*, **261**, 486–489.

Yang, J. and Tsien, R.W. (1993) Enhancement of N- and L-type Ca^{2+} channel currents by protein kinase C in frog sympathetic neurons. *Neuron*, **10**, 127–136.

Yoshida, A., Oho, C., Omori, A. Kuwahara, R., Iti, T. and Takahashi, M. (1992) HPC-1 is associated with synaptotagmin and omega-conotoxin receptor. *J. Biol. Chem.*, **267**, 24925–24928.

Yu, C., Lin, P.X., Fitzgerald, S. and Nelson, P. (1992). Heterogeneous calcium currents and transmitter release in cultured mouse spinal cord and dorsal root ganglion neurons. *J. Neurophysiol.*, **67**, 561–575.

Zimmerberg, J., Sardet, C. and Epel, D. (1985) Exocytosis of sea urchin egg cortical vesicles in vitro is retarded by hyperosmotic sucrose: kinetics of fusion monitored by quantitative light-scattering microscopy. *J. Cell Biol.*, **101**, 2398–2410.

Zucker, R.S and Fogelson, A.L. (1986) Relationship between transmitter release and presynaptic calcium influx when calcium enters through discrete channels. *Proc. Natl. Acad. Sci. USA*, **83**, 3032–3036.

Zucker, R.S. and Stockbridge, N. (1983) Presynaptic calcium diffusion and the time courses of transmitter release and synaptic facilitation at squid giant synapse. *J. Neurosci.*, **3**, 1263–1269.

Zygmunt, P.M., Zygmunt, P.K.E., Högestätt, E.D. and Andersson, K.E. (1993) Effects of ω-conotoxin on adrenergic, cholinergic and NANC neurotransmission in the rabbit. *Br. J. Pharmacol.*, **110**, 1285–1290.

9. IONOPHORE POLYPEPTIDE TOXINS AND SIGNAL TRANSDUCTION

E. BLOCH-SHILDERMAN, S. ABU-RAYA and P. LAZAROVICI

*Dept. of Pharmacology and Experimental Therapeutics,
School of Pharmacy, Faculty of Medicine, The Hebrew University of Jerusalem,
Jerusalem 91120, Israel*

INTRODUCTION

Ionophore toxins act directly on cell plasma membrane. They make pores in the cell membrane, disturb the ionic homeostasis and normal cell physiology and ultimately cause cell death (Thelestam and Mollby, 1979). Ionophore toxins from microbial plant and animal origin represent a very abundant, heterogeneous group of chemical structures with various mechanisms of actions (Harvey, 1990). Since the cellular mechanisms of ionphore toxins are largely unknown, their classification with respect signal transduction mechanisms is not trivial. In this chapter, we will focus on selected simple polypeptide, ionophore toxins which form discrete, transmembrane pores or channels (Bernheimer and Rudy, 1986). We will not consider other types of ionophore toxins, such as cytolytic toxins with phospholipase A_2, C and D activities, (Bernheimer and Rudy, 1986) cytolysins which activate cellular phospholipases (Dufton and Hider, 1988), or B subunits of binary, complex, microbial toxins (e.s. cholera, pertussis, shiga, heat-labile enterotoxin which produce membrane pores) (Zhang *et al.*, 1995; Fraser *et al.*, 1994; Stein *et al.*, 1994).

Eukaryotic cells plasma membranes have an important barrier role in compartmentalizing ions and macromolecules between the cytoplasm of the cell and the extracellular tissue fluid. The plasma membrane maintain different electrochemical potentials between the intracellular and extracellular enviroments and perform specialized biochemical reactions. Additionally, they contain a wide variety of different lipids, cations, enzymes, receptors and ionic channels that contribute to the specificity of the individual cell's plasma membrane. Therefore, it is not surprising that ionophore toxins have evolved with different lipid specificities and ionophore properties (Table 9.1).

The most extensively studied, major types of ionophore toxins, (see Table 9.1), are polypeptide toxins of different molecular weights and phospholipid selectivity, which aggregate different number of toxin monomers to assemble a functional pore in the cell plasma membrane. This produces a variety of pore sizes. According to the pore size and mechanism of cell membrane perforation we prefere to classify ionophore toxins in four groups. The first group, represented by the thiol-activated ionophore toxins from certain Gram-positive bacteria, require a thiol group for the binding of the toxins to the cholesterol in the eukaryotic cell's plasma membrane (Alouf, 1986). In general, these toxins form aggregates of 70–80 molecules to make large transmembrane pores with a ring or arc shaped structure as seen in electron micrographs (Niedermeyer, 1985). The estimated diameter of such pores is about 15 nm, therefore a single membrane lesion of such size will collapse voltage- and receptor-operated ionic channels active in signal transduction pathways. This results in cell death. The second group of ionophore toxins is represented by the single chain, low molecular weight polypeptides of 32 (streptolysin S) or 26 (Delta toxin, melitin) amino acid residues. They do not form visible, pore-like structures in electron micrographs of intoxicated cells, but cause leakage of intracellular low molecular weight markers, indicative of a size of about 4–5 nm plasma membrane lesions (Buckingham and Duncan, 1983; Thelestam and Mollby, 1979). These lesions formed by toxins olygomers do not have the typical pore/channel behaviour and may be characterized as a surfactant membranal effect, different from the detergent-like membrane solubilization process. The end result is a rapid loss of transmembrane ionic gradients, resulting in the collapse of the signal transduction pathaways, loss of ionic homeostasis and cell death. The third group of toxins represented by the α — toxin from *Staphylococcus aureus* and certain sea anemone toxins, are proteins that form 4–6 mer pores in the cell membrane which can be visualized using electron microscopy (Bhakdi and Tranum-Jensen, 1987). These toxins appear to bind selectively to cardiolipin and sphingomyelin. They also interact non-selectively with different lipids in artificial membranes (Menestrina, 1986). They form discrete, non-selective ion channels and small plasma membrane pores, 0.5–2 nm in diameter (Thelestam and Mollby, 1979). At high concentrations and/or long exposures this leads to cell lysis. A fourth group of ionophore toxins is the low molecular weight polypeptide antibiotics which do not appear to bind selectively to any particular phospholipid, which forms dimers (gramicidin) or octamers (alamethicin, pardaxin). These are small (>0.2 nm), ion permeable channels that may be voltage-dependent and cation-selective (Table 9.1). Despite the large number of ionophore toxins discovered and identified (Harvey, 1990), their poor chemical or pharmacological characterizations does not yet allow their recognition as tools with defined actions in signal transduction. In contrast, the biophysical and pharmacological progress which has been made during the last two decades using the ionophore toxins presented in Table 9.1, as well as their availability and simplicity, make them attractive tools to study signal transduction processes.

PARDAXIN, A REPRESENTATIVE MODEL FOR IONOPHORE TOXINS

The red sea flatfish, *Pardachirus marmoratus* (Figure 9.1a, b), produces a secretion toxic to marine organisms from more than 200 epithelial glands (Figure 9.1b) located along its dorsal and anal fins (Lazarovici *et al.*, 1990). The major toxic, ionophoric component of this secretion has been isolated (Figure 9.1c), character-

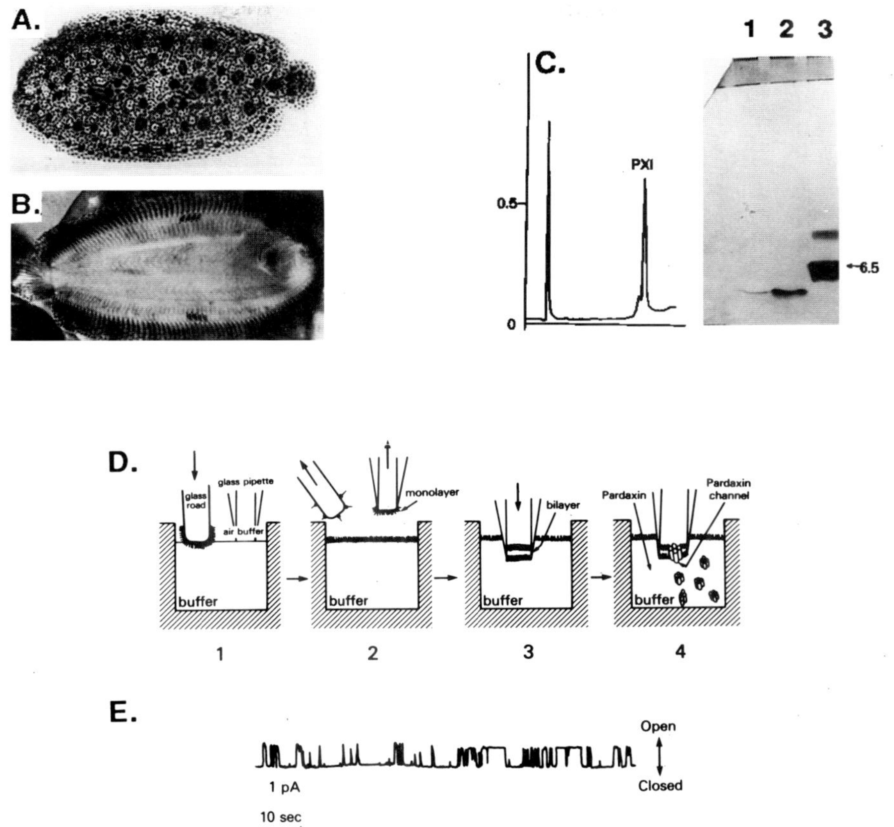

Figure 9.1 *Pardachirus marmoratus* fish (a, b), HPLC purification of pardaxin (c) and patch-clamp measurments of pardaxin single channels activity (d, e). a — dorsal view; b — venteral view; arrows point to the glands location at the basis of the fin. c — HPLC purification of pardaxin (Lazarovici *et al.*, 1986); Lane 1 and 2 — Silver staining of SDS-PAGE urea gels of 10 and 50 ng pardaxin (MW ~ 3.5 kDa). d — schematic representation of the process of bilayer formation from monolayers at the air-water interface at the tip of the patch pipet (Suarez-Isla *et al.*, 1983). Step 1 — a patch pipet is introduced into the solution and a monolayer is formed; step 2 — the pipet is removed from the solution; step 3 — the pipet is reimmersed into the solution, a new monolayer is attached, forming a bilayer; step 4 — addition of pardaxin to the bath and pore formation. e — single-channels currents of pardaxin pores; the recording was obtained at V = 30 mV and the seal resistance was 10 GΩ. An upward deflection corresponds to a channel-opening event. The single events of about 10 pS conductance appear in burst of channel activity followed by quiescent periods.

ized (Lazarovici *et al.*, 1986), sequenced and synthesized (Shai *et al.*, 1988). This compound, named pardaxin, is an acidic, hydrophobic, amphipathic, single chain polypeptide, composed of 33 amino acids and with a molecular weight of 3300 daltons (Lazarovici *et al.*, 1986). Five isotoxins forms of pardaxin have been characterized (Table 9.2). They differ slightly in certain amino acids at positions 4, 5, 14 and 31 from the amino terminal (Table 9.2). For comparative purposes, the sequences of several other ionophore toxins, mentioned in Table 9.1, are presented. In spite of the similarity in pore forming activity, examination of the ionophore toxin sequences does not reveal any obvious homology (Table 9.2). Also the primary structures, net charge and lengths are quite different. In liposomes (Loew *et al.*, 1985) and planar lipid bilayers (Lazarovici *et al.*, 1992; 1995), pardaxin

Table 9.1   The origin, classification and properties of some ionophore toxins

Toxin	Molecular Weight (kilodaltons)	Lipid Specificity	Pore Size (nm)	Number of Monomers Assembling the Pore	Mechanism of Action
Group 1					
Sreptolysin O[1]	67.0	cholesterol	~15	70-80	Thiol- activated
Cereolysin[2]	55.5	cholesterol	~15	ND	pores forming
Tetanolysin[3]	48.0	cholesterol	~15	ND	(large ring-like
Metridiolysin[4]	80.0	cholesterol	ND	ND	structure)
Group 2					
Streptolysin[5]	1.8	acidic and	4-5	ND	Pore and
Delta toxin[6]	3.0	neutral	4-5	aggregate	surfactant
Melittin[7]	2.8	phospholipids	4-5	4ers	activity
Group 3					
Alpha toxin[8]	33.0	cardiolipin phospholipids	0.5–2.0	5 or 6-ers	Pore activity
Cytolysin III[9]	17.0	sphingomyelin	0.5	4ers	Pore activity
Group 4					
Gramicidin[10]	1.1	phospholipids	0.2	2ers	Voltage-
Alamethicin[11]	1.2-1.6	phospholipids	0.2	8ers	dependent
Pardaxin[12]	3.3	phospholipids	0.2-1.0	8ers	cation channels

[1] *Streptococcus pyogens* bacteria, Alouf, 1986; [2] *Bacillus cereus* bacteria, Turnbull, 1986; [3] *Clostridium tetani* bacteria, Blumenthal and Habig, 1984; [4] *Metridium senile* sea anemone, Bernheimer *et al.*, 1979; [5] *Streptococcus pyogens* bacteria, Alouf, 1986, Buckingham and Duncan, 1983; [6] *Staphylococcus aureus* bacteria, Freer *et al.*, 1984; [7] *Apis mellifera* bee, Habermann, 1980; [8] *Staphylococcus aureus* bacteria, Thelstam and Blomquist, 1988, Walker and Bayley, 1995; [9] *Stichodactyla helianthus* sea anemone, Bernheimer and Avigad, 1976, Varanda and Finkelstein,1980, Belmonte *et al.*, 1993; [10] *Bacillus brevis* bacteria, Andersen, 1984; [11] *Trichoderma viridis* fungus, Hall *et al.*,1984; [12] *Pardachirus marmoratus* fish, Lazarovici *et al.*, 1992. ND-Not determined.

Table 9.2 Amino acid sequences of pardaxin isotoxins and related ionophore toxins

TOXIN	SEQUENCE						
	1 5	10	15	20	25	30	
Pardaxin I[1	NH-GFFAL	IPKII	SSPLF	KTLLS	AVGSA	LSSSG	GQE-COOH
Pardaxin II[1	NH-GFFFP-X						
Pardaxin P₁[2	NH-GFFAL	IPKII	SSPLF	KTLLS	AVGSA	LSSSG	EQE-COOH
Pardaxin P₂[2	NH-GFFAL	IPKII	SSPIF	KTLLS	AVGSA	LSSSG	GQE-COOH
Pardaxin P₃[3	NH-GFFAF	IPKII	SSPLF	KTLLS	AVGSA	LSSSG	EQE-COOH
Mellitin[3	NH-QQRKR	KIWSI	LAPLG	TTLVK	LVAGI	G-amidated	C terminal
Delta-lysin [4	NH-KKTFK	NVTDI	IWKVL	DGITS	IIDQA	M-COOH	

	1 5	10	15
	V	W	
Gramicidins[5	Formyl- GALA	VVNWL	FLWLW-ethanolamine
	I	Y	

	1 5	10	15
Alamethicin[6	Ac-UPUAU	AQUVU GLUPV	UUEQPhol

	5	10	15	20	25	30
Dermaseptin[7	NH₂-ALWKT	MLKKL	GTMAL HAGKA	ALGAA	ADTIS	QGTQ-COOH
Cecropin B[8	NH₂-RWKIF	KKIEK	MGRNI RDGIV	KAGPA	IEVLG	SAKAI-NH₂

[1]Shai *et al.*, 1988; Sequence of pardaxin I and partial amino acid sequence of pardaxin II isotoxins from the Red Sea Moses sole fish *Pardachirus marmoratus* (Lazarovici *et al.*, 1986); [2]Amino acid sequences of pardaxins P₁, P₂ and P₃ isotoxins isolated from the Pacific Ocean sole fish *Pardachirus pavoninus* (Thompson *et al.*, 1986); [3]Habermann, 1980; [4]Freer *et al.*, 1984; [5]Andersen, 1984; The amino terminal is either formyl Val or formyl Ile; amino acid 11 is either Trp, Phe or Tyr in gramicidins A, B or C respectively; [6]Hall and Vodyandy, 1984; Bruckner and Graf, 1983; A$_{ib}$ is α-aminoisobutyric acid; Phol is L-phenyl-alaninol; [7]Mor *et al.*, 1994; [8]Christensen *et al.*, 1988.
The code of one letter for amino acids has been used: A — alanine; C — cysteine; D — aspartic acid; E — glutamic acid; F — phenylalanine; G — glycine; H — histidine; I — isoleucine; K — lysine; L — leucine; M — methionine; N — asparagine; P — proline; Q — glutamine; R — arginine; S — serine; T — threonine; V — valine; W — tryptophan; Y — tyrosine.

formed voltage dependent pores (Figure 9.1e), with single-channel recordings of pardaxin pores forming in lipid bilayers at the tip of patch pipets (Suarez-Isla *et al.*, 1983). Similarly the properties of other ionophore toxins (Table 9.1) have been investigated by measuring both the leakage of intracellular compounds from intact cells (Thelstam and Mollby, 1979) as well as the ion movement across planar lipid membrane models (Menestrina, 1991; Braun and Focareta, 1991; Bernheimer and Rudy, 1986). Measurements of conductivity changes induced by ionophore toxins across lipid planar bilayers have been instructive in defining the basic pore properties (conductance, selectivity, rectification). To date, these measurements define the basic requirements to be fulfilled by the newly discovered ionophore toxins (Schonherr *et al.*, 1994; Belmonte *et al.*, 1993).

IONOPHORE TOXINS BINDING, OLIGOMERIZATION AND ASSEMBLY INTO AN ACTIVE MEMBRANAL PORE

The current mechanistic concepts of ionophore toxin action suggest that upon contact with the lipid bilayer phospholipids in artificial or plasma cell membrane, water soluble-ionophore toxins, bind to the membrane and subsequently reorganize in the lipid bilayer to form oligomeric, transmembrane pores. This concept is supported by extensive studies performed over the last two decades and with more recent reports using a variety of new, modern experimental approaches (Table 9.3). Binding of ionophore toxins to the lipid membrane has been characterized both qualitatively and quantitatively, using radiolabeled toxins, chemical cross-linking, gel filtration and analytical ultracentrifugation. Because of the lipid nature of the membrane, these studies indicate numerous unsaturated binding sites and enable estimation of the number of ionophore toxins monomers associated with the lipid membrane (Table 9.3). The development of a variety of polyclonal and monoclonal antibodies directed towards selective ionophore toxin epitopes provides useful tools in the examination of toxin conformational changes during membrane binding and ionophore organization (Table 9.3). Hydrophobic photolabeling using a radioactive apolar reagent has been used to establish the insertion of ionophore toxins in membranes, identify protein segments that interact with the membrane lipids and identify amino acid residues that line the pore (Table 9.3).

Finally, recent studies by mutagenesis, gel-shift electrophoresis and x-ray crystallography have established the oligomerization concept of the transmembrane pore in three-dimensional crystals and biological membranes (Table 9.3). They also provide accurate measurements of the number of *Staphylococcus aureus* α-toxin monomers assembling a pore and contributing valuable insights into the key amino acid residues important for the functioning of this toxin (Table 9.3). A working scheme for the membrane binding and oligomerization of the ionophore toxins mentioned in Table 9.1 is presented in Figure 9.2 Ionophore toxins can be found in physiological buffers either as monomers (Figure 9.2 (1)) or oligomers (Figure 9.2 (1a)) with the hydrophilic residues exposed to the water (Figure 9.2). In the first monomer category, we classify the large ionophore toxins belonging to groups 1 and 3 (Table 9.1), and in the second oligomer category we can include the small, amphipatic ionophore toxins from groups 2 and 4 (Table 9.1). Upon binding to cholesterol or sphingomyelin, the membrane-bound, ionophore toxin monomer diffuses laterally on the membrane plane and upon encountering one another they aggregate (3) to form supramolecular arc-and ring-shaped structures constituting large pores (4) that allow flux of ions and macromolecules (Figure 9.2). The small, amphipathic ionophore toxins are found as aggregates in aqueous solution. Upon binding to the membrane as an aggregate (Figure 9.2, 9.2b), or upon dissociation to the monomeric form (Figure 9.2, 9.2a), it also undergoes conformational changes resulting in formation of oligomers (Figure 9.2–3) which then organize to form an active pore (Figure 9.2–4). The ionophore toxins

Table 9.3 Different recent approaches to study ionophore toxins binding, oligomerization and pore assembly

Toxin	Methodology	Conclusions	Reference
Streptolysin O	Radioiodinated toxin, ELISA, Separation of oligomers by centrifugation.	Many non saturated binding sites, oligomers of 50-80 monomers were detected.	Walev *et al.*, 1995
Pseudomonas aeruginosa cytotoxin	Western blotting, Hemoglobin and carboxyfluorescein release.	A pentamer organization is a prerequisite for the pore activity.	Ohnishi *et al.*, 1994
Staphylococcus aureus α-toxin	Photolabeling with a hydrophobic probe.	Identification of membrane-inserted segments of toxin.	Lala *et al.*, 1995
	ELISA and neutralization assays using monoclonal structural and conformational antibodies.	Defined epitopes are involved in the different conformational steps required for pore formation.	Heveker *et al.*, 1994
	Cysteine scanning mutagenesis and targeted chemical modifications.	Systematic identification of the key residues important for binding, oligomerization and pore formation.	Walker and Bayley, 1995
	X-Ray diffraction, Chemical modification and Gel-shift electrophoresis.	The heptamer organization is the pore both in three dimensional crystals and on a biological membrane.	Gouaux *et al.*, 1994

belonging to group 2 (Table 9.1) form the pore only transiently, under certain experimental conditions. There is a high tendency for pore aggregation (Figure 9.2–5) resulting in a surfactant membrane disruption (Figure 9.2–6). In general, all the ionophore toxins presented in Table 9.1, at high concentrations and/or upon long times of exposures to membranes, and at different temperatures, may aggregate the pores (Palmer *et al.*, 1995), resulting in detergent-like membrane disruption (Lazarovici *et al.*, 1988).

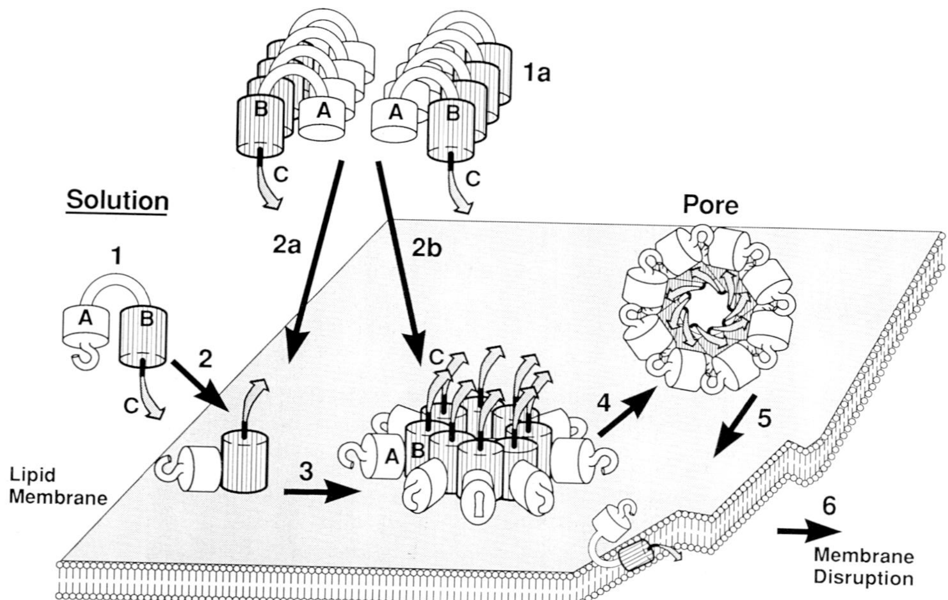

Figure 9.2   A schematic hypothesis of membrane binding, oligomerization and pore assembly of ionophore toxins. The toxin contains two α-helical domains and is found in solution either as a monomer (1) or an oligomer (1a). Upon binding to the membrane (2, 2b), an organization step occurs (3, 2b) followed by formation of an active pore (4). Alternatively, solution toxin olygomers first dissociate to monomers (2a) and then reorganize to form pores. At high toxin concentrations the pores may aggregate (5) resulting with surfactant lytic disruption of the plasma membrane (Adapted from Raghunathan *et al.*, 1990, Lazarovici *et al.*, 1992 and Bayley, 1994).

STRUCTURAL MODELS OF PARDAXIN PORES

The ionophore toxins (Table 9.1) have protein sequences consistent with the formation of an amphipathic α-helix (Guy and Raghunathan, 1989). An amphipathic α-helix is defined as an asymmetric distribution of polar and apolar residues between the opposite sides of the helix. Transmembrane portions of many proteins and in particular, ion channels and ionophore toxins, appear to be comprised primarily of amphipathic α-helices packed in a parallel or antiparallel manner and spanning a lipid bilayer membrane with the hydrophilic faces of the α-helix lining a central pore and the hydrophobic surfaces interacting with the hydrophobic phospholipid chains of the membrane (Raghunathan *et al.*, 1990; Guy and Conti, 1990; Kerr *et al.*, 1993). The pardaxins, 33 amino acids residues monomer sequence (Table 9.2), can be divided into three regions (Figure 9.3). The amino terminal region A (Gly[1]–Ser[12]), with the exception of Lys[8], is highly

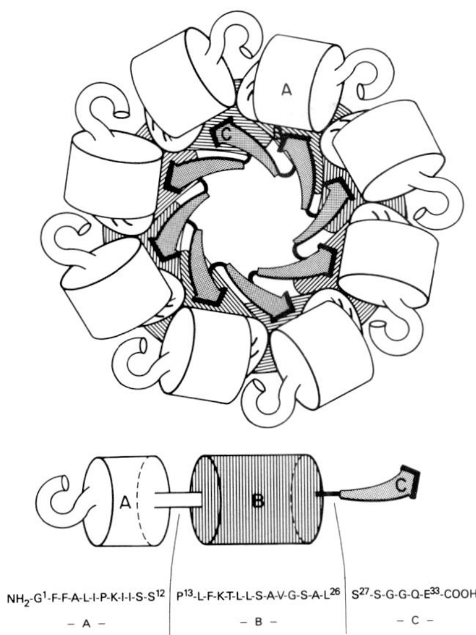

Figure 9.3 A schematic view of pardaxin secondary structure and pore. The pardaxin pore is composed of eight monomers. Each pardaxin monomer is composed of three regions: A — residues 1–12 from amino terminal containing one helix; B — residues 13–26 containing an amphipathic α-helix; C — residues 27–32 at the carboxy terminal involved in ion permeation.

hydrophobic and Lys^8–Pro^{13} segment has a helical nature (Lazarovici *et al.*, 1992; Barrow *et al.*, 1992; Shai *et al.*, 1990). Region B (Ile^{14}–Leu^{26}) is an amphipathic α-helix (Lazarovici *et al.*, 1992; Barrow *et al.*, 1992; Zagorsky *et al.*, 1991). The non polar residues (Ile^{14}, Phe^{15}, Leu^{18}, Leu^{19}, Ala^{21}, Val^{22}, Ala^{25}, Leu^{26}) reside on the opposite face to the polar or charged residues (Lys^{16}, Thr^{17}, Ser^{20}, Ser^{24}) (Zagorsky *et al.*, 1991). Pro^{13} causes a bend between the two helices (Lazarovici *et al.*, 1992; Zagorsky *et al.*, 1991). Carboxy terminal region C (Ser^{27}–Glu^{33}) appears to exist in an extended coil or β sheet conformation, as evident from NMR data (Zagorsky *et al.*, 1991) and molecular modeling (Lazarovici *et al.*, 1992). Using synthetic pardaxin analogues, the basic structural activity requirements for pore formation has been shown (Shi *et al.*, 1990; Shai *et al.*, 1991; Rapaport and Shai, 1992; Barrow *et al.*, 1992): 1. Pardaxin binds strongly to the membrane using the hydrophobic amino terminal region A. 2. Pardaxin penetrates into the hydrophobic phase of the membrane as detected by fluorescence energy transfer (Rapaport and Shai, 1992) or proteolysis and HPLC analysis (Lazarovici *et al.*, 1992). 3. Pardaxin self associates in its membrane-bound state with parallel orientation of the monomers within the aggregate (Rapaport and Shai, 1992). 4. Pardaxin B region contains the minimal requirements for amphipathicity and maintenance of the two α helix segments (Barrow *et al.*, 1992). 5. Pardaxin carboxy terminal C region is crucial for pore activity and is the pore binding segment (Shai *et al.*, 1990). 6. Upon association of pardaxin with the membrane a rearrangement of the secondary structure occurs resulting in an increase in the α-helicity (Lazarovici *et al.*, 1992). 7. An oligomer of eight pardaxin monoers is required for an active pore (Loew *et al.*, 1985). Based on these data, three dimensional models of pardaxin pores were developed (Figure 9.4), using computer graphics (Lazarovici *et al.*, 1992; Guy and

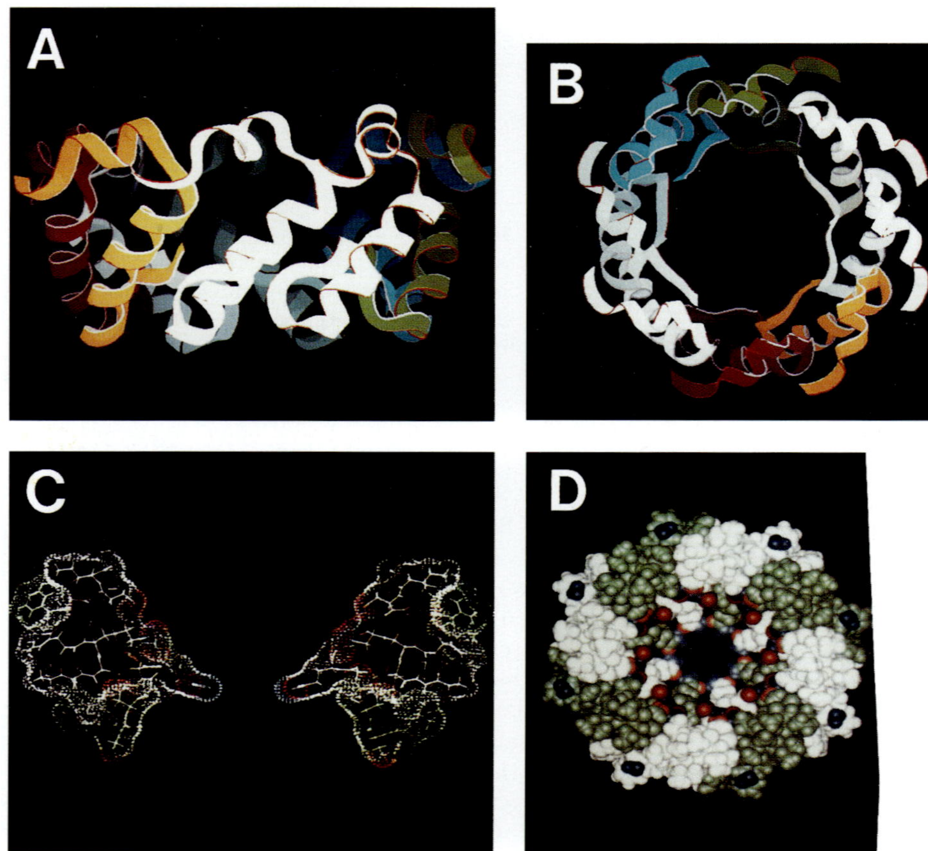

Figure 9.4 Computer graphic models of pardaxin pores. Side (A) and top (B) views of the ribbon representations of the pardaxin channel. Eight parallel monomers form the channel and each monomer is shown in a different colour. The β-barred forms part of the channel lining and is on the *cis* side of the membrane and is somewhat parallel to the membrane surface. (C) A cross-section of the channel showing the molecular surface. This shows the hourglass shape of the channel with the narrow region formed by Gln 32. (D) A space filled rendering of the pardaxin channel looking down the channel from the *cis* side. Charged oxygens are shown in red, uncharged oxygens in brown, charged nitrogens in blue and uncharged nitrogens in light blue. The ring of hydrogens bonds formed by Gln 32 at the narrow region and salt bridges formed between Gln 33 and Lys 8 and Lys 16 in the entrance of the channel are shown. Every other monomer is coloured green or white.

Raghunathan, 1989). The pardaxin pore was modeled as a cylinder of eight parallel monomers in which eight stranded β-barrel formed by residues 28–33 (region C, Figure 9.3) are surrounded by eight amphipathic α-helices formed by residues 13–26 (region B, Figure 9.3). Residues 1–10 (Region A, Figure 9.3) are postulated to form a ring of α-helices around the outer entrance of the pore (Figure 9.4, Lazarovici *et al.*, 1992). The shape of the pore is an hourglass with a narrow selectivity filter of 4.8A° diameter and free charges are absent inside the pore lining (Figure 9.4), which is consistent with observations that these channels transport both cations and anions (Shi *et al.*, 1995). This model is also consistent with structure activity studies (Shai, 1994) and is similar to other models developed for melittin, *Staphylococcus aureus* delta toxin, magainins and other ionophore toxins (Raghunathan *et al.*, 1990; Guy and Raghunathan, 1989). In the absence of crystal structures, the major value of these models is to suggest more informative, experiments (Shai, 1994). The relation between structure and function at the single amino acid residue level and the understanding of ion binding and movement through putative ionic channels or ionophore toxins pores is nonetheless still unavailable. Pardaxin and other ionophore pores should continue to serve as general models for transmembrane pores as their small size and availability make them ideal tools for biophysical investigations.

IONOPHORE TOXINS FORM NONSELECTIVE ION CHANNELS AS DETERMINED USING PARDAXIN PORES

Selectivity for different ions might be an important criteria for the classification of the ionophore toxins as several signal transduction mechanisms rely primarily on plasma membrane selective permeation to Na^+, K^+, Ca^{2+} and Cl^-. In contrast to the relative selectivity of the plasma membrane ionic channels (Hill, 1991), there is growing evidence for the partial or complete lack of ion selectivity of ionophore toxins (Menestrina, 1991). This poor selectivity is emphasized with pardaxin pores (Lazarovici *et al.*, 1992; Shi *et al.*, 1995). Single channel conductance of pardaxin pores is low (~10 pS) and the selectivity sequences calculated by comparison of the reversal potentials under ionic conditions in plasma lipid bilayer are: *Monovalent cations*: $Tl^+>Cs^+>Rb^+>K^+>NH_4^+>$methylamine$^+>Li^+>$dimethylamine$^+>Na^+$; *Divalent cations*: $Ba^{++}>Sr^{++}>Mn^{++}>Mg^{++}>$ethylendiamine; *Anions*: $I^->NO_3^->Br^->Cl^->ClO_4^->HCOO^->CH_3COO^-$. The permeability ratios of pardaxin pores of 0.78 Cl^- over K^+, 0.43 Na^+ over K^+ and 1.8 Cl^- over Na^+ were measured (Lazarovici *et al.*, 1992). Thus, the pardaxin pore shows only a modest selectivity for charge amongst these small anions and cations. Poorly selective channels or pores like pardaxin operating under physiological conditions likely could not select for a particular ion against the vast majority of Na^+, K^+, Ca^{2+}, Mg^{2+} or Cl^- in the tissue. This implies that ionophore toxins will collapse the cellular ionic gradients under physiological conditions. The poor selectivity of ionophore toxins is reminiscent of the newly discovered nonselective cation channels such as bacterial porins, cyclic nucleotide gated-channels, gap-junction channels and others of unknown function (Siemen and Hescheler, 1993). Although the ion selectivity can be calculated from the Goldman-Hodgkin-Katz relationship (Hille, 1991), there is still no correlate for a lack of ion selectivity at the molecular level. Recently, site-directed mutagenesis of single amino acids in the transmembranal α-helices switched ion selectivity from Na^+ to Ca^{2+} in sodium channels (Heinemann *et al.*, 1992) or from cation to anions in the nicotinic acetylcholine receptor (Galzi *et al.*, 1992). Therefore, it is tempting to suggest that the poor selectivity of the ionophore toxins is due to the absence of certain amino acid residues responsible for ion selectivity in their amphipathic α-helices.

SIGNAL TRANSDUCTION MECHANISMS OF IONOPHORE TOXINS

Binding experiments with ionophore toxins to eukaryotic cells indicated the absence of specific glycoprotein receptors and that pore formation is initiated by binding to different lipids (Table 9.1). The role of integral plasma membrane proteins in ionophore toxin binding and insertion remains to be determined, although in a few-recent studies it has been found that the *Pseudomonas aeruginosa* cytotoxin binds to the water channel CHIP28 protein in erythrocytes (Xiong *et al.*, 1994) and the *Clostridium perfringes* enterotoxin forms complex pores in the plasma membrane together with certain integral proteins (McClare, 1994). Although the precise phamacological mechanisms of ionophore toxins are largely unknown and many fundamental questions remained unanswered, the past several years have been productive in increasing our understanding of *S. aureus* a toxin and pardaxin actions on signal transduction pathways as summarized in Figure 9.5. Upon exposure of cells to these ionophore toxins, they bind and organize to form conductive pores within the cell membrane (step 1, Figure 9.5). Due to the poor selectivity, these pores will transport both monovalent and divalent cations into the cell (Fink *et al.*, 1989; Lazarovici and Lelkes, 1992; Nikodijevic *et al.*, 1992; Abu-Raya *et al.*, 1993; Dibas *et al.*, 1995). The increase in the intracellular Na^+ concentration will depolarize the cell which, in turn, will open voltage-sensitive calcium channels (VDCC, step 2, Figure 9.5), resulting in a small influx of extracellular calcium into the cell (Nikodijevic *et al.*, 1992). Toxin pores cause a rapid or gradual progressive increase in intracellular calcium (Abu-Raya *et al.*, 1993; Lazarovici and Lelkes, 1992). The influx of calcium from toxin pores and VDCC (Nikodijevic *et al.*, 1992) caused a sustained increase of high (μM) concentrations of intracellular calcium. A rapid consequence of the increase in intracellular calcium is the exocytotic secretion of a variety of chemical mediators, hormones or neurotransmitters, depending on the cell type or ionophore toxin (step 3, Figure 9.5). Pardaxin was found to induce exocytotic secretion of acetylcholine (Renner *et al.*, 1987; Arribas *et al.*, 1992), 5-HT and norepinephrine (Wang and Friedman, 1986), dopamine and ACTH (Bloch-Shilderman *et al.*, 1995). Gramicidin is a potent insulin secretagogue (Dibas *et al.*, 1995). As secretory granules and the cytoplasm of cells contain high ATP concentrations, ATP is also released by the action of ionophore toxins

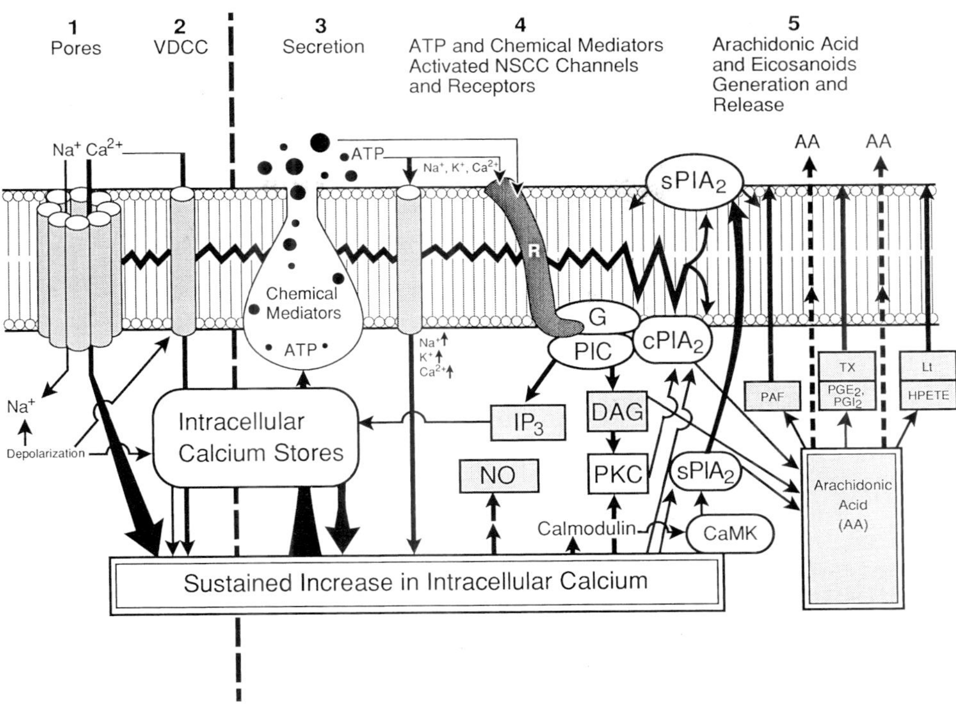

Figure 9.5 A schematic of ionophore toxins signal transduction pathways. (1) Pores formation in the plasma membrane. Influx of Na^+ and Ca^{2+}. (2) The depolarization will open voltage-dependent calcium channels (VDCC) resulting with Ca^{2+} influx and Ca^{2+} release from intracellular stores. (3) The elevation in intracellular calcium induces secretion of ATP and chemical mediators. (4) ATP and chemical mediators (such as bradykinin) activates non-selective cation channels (NSCC) and receptors (R) coupled by G proteins to phospholipase C (PLC) to increase intracellular calcium in a second wave by influx, depolarization and/or release from intracellular stores. The increase in intracellular calcium activates calmodulin, protein kinase C (PKC) and calcium-calmodulin (CaM) kinases. (5) Release of arachidonic acid (AA) from membrane phospholipids occurs by several pathways: I. activation of cytosolic PLA_2 ($cPLA_2$) by calcium, protein-kinases and/or receptors mediated signals; II. activation of secretory PLA_2 ($sPLA_2$) by calcium and/or protein phosphorylation signals; III. activation of PLC/DAG lipase pathway. After release, free AA may be incorporated into the membrane phospholipids or metabolized by one of the three pathways to generate platelet activating factor (PAF), or cycloxygenation to form prostaglandins (PG), prostacyclin (PGI_2) and thromboxane (TX), or lypooxygenases to form hydroperoxyeicosatetreanoic acids (HPETE) which undergo conversion into leukotrienes (Lt). The model is based on studies performed mainly with pardaxin or *Staphylococcus aureus* α-toxin using pheochromocytoma PC12 cells cultures. Na^+, Ca^{2+}, K^+ — sodium, calcium and potassium ions, respectively; NO nitric oxide; IP_3 — inositol (1,4,5) P_3; arrows indicate influx of ions or directional flow of activation.

(Lazarovici and Lelkes, 1992; Suttorp *et al.*, 1992) when added to the extracellular media (step 3, Figure 9.5). From this moment, either in a paracrine or autocrine fashion, the chemical mediators, hormones and/or ATP released, may activate their respective receptors and/or ionic channels (step 4, Figure 9.5), thus amplify and propagating the initial ionophore toxin signals. For example, external ATP triggers a rapid and short-lived increase of the internal calcium concentration by activating homeostatic channels permeable to Na^+, Ca^{2+} and K^+ (step 4, Figure 9.5; Fasolato *et al.*, 1990; deSouza *et al.*, 1995). Extracellular ATP will also activate purinergic receptors which will increase intracellular calcium through influx or release from IP_3 (inositol 1,4,5-trisphosphate) — sensitive intracellular calcium stores, through activation of phospholipase C (step 4, Figure 9.5, deSouza *et al.*, 1995). The picture might be more complex, since ATP can be hydrolyzed to adenosine which can than activate additional signal transduction pathways through adenosine A_2 receptors. Release of chemical mediators such as bradykinin will induce phospholipase C activation resulting in the formation of the second messengers, IP_3 and diacylglycerol (DAG), and subsequent mobilization of calcium from internal stores to increase cytoplasmic calcium (step 4, Figure 9.5). Indeed, *Staphylococcus aureus* α-toxin was found to stimulate phospholipase C (Fink *et al.*, 1989). In many cell types, receptor-mediated IP_3 formation and calcium release from intracellular stores is followed by a calcium influx across the plasma membrane (Putney *et al.*, 1989) required to refill depleted stores and provide enough calcium to produce important metabolic changes. These sequential, progressive four steps of ionophore toxins action (Figure 9.5) might cause a self-maintained, sustained increase in intracellular calcium due to influx from extracellular medium and release from intracellular stores, probably initiating in the short term a host array of cellular changes on metabolism, secretion, muscle contraction, gene expression, cell cycle and cross-talk between different signal transduction pathways. Most probably, chronic increase of intracellular calcium to milimolar concentrations may be the lethal insult causing necrotic cell death (Orrenius *et al.*, 1989). This causes the influx of water and extracellular ions, cell swelling and rupture, typical for ionophore toxins action (Shier, 1985; Lazarovici *et al.*, 1982). Physiologic effects of μM elevations in the intracellular free calcium concentration are transduced by many effector systems such as protein kinases and phosphatases. Calcium and DAG will stimulate protein kinase C (Figure 9.5), while the sustained increase in intracellular calcium will saturate calmodulin and strongly activate both dedicated calcium-calmodulin (CaM) kinases (Figure 9.5), such as myosin light chain kinase (MLCK), and multifunctional CaM kinases such as CaM kinase I, II etc. (Shulman, 1993). In the pheochromocytoma PC12 cells, multiple signals converge on CaM kinases, including ligand gated calcium influx via purinergic and nicotinic cholinergic receptors and IP_3 mediated release of intracellular calcium via muscarinic and bradykinin receptors (MacNicol and Schulman, 1992). PKC may cross-talk with CaM kinases and other calmodulin-dependent processes, finaly resulting with phosphorylation changes in multiple substrates localized in the nucleus, cytoskeleton, plasma membrane and cytosol (Figure 9.5). *Staphylococcus aureus* α-toxin induced reduction of epidermal growth factor affinity in

PC12 cells (Lazarovici and Chan, 1987), activation of PKC and PKA phosphorylation in myelin (Chan and Lazarovici, 1987), and increased c-fos expression in PC12 cells (unpublished data). The *E. coli* hemolysin and *Staphylococcus aureus* α-toxin also induce release of nitric oxide in endothelial cells (Suttorp *et al.*, 1993). These studies represent some of the physiological consequences of ionophore toxins increase in intracellular calcium and activation of the protein kinases phosphorylation cascades.

Another major cell signaling effect of ionophore toxins is the stimulation of the arachidonic acid cascade (Abu-Raya *et al.*, 1993; Suttorp *et al.*, 1985; 1992) initiated via activation of phospholipase A_2 (PLA$_2$, phosphtide 2-acylhydrolase) enzymes. PLA$_2$ action release polyunsaturated arachidonic fatty acid (AA) which is subsequently converted to a family of biologically active metabolites, collectively called "eicosanoids" (Figure 9.5; Piomelli, 1993). Some of the enzymatic pathways involved in AA release and metabolism are shown in Figure 9.5. In PC12 cells, streptolysin toxin, *Staphylococcus aureus* α and δ-toxins, *S. helianthus* toxin, allamethicin and pardaxin strongly stimulated both PLA$_2$ cyclooxygenase and lipoxygenase pathways resulting in formation of prostaglandin E_2 (PGE$_2$), thromboxane B_2 (TXB$_2$) and 5-HETE (Abu-Raya *et al.*, 1993; Fink *et al.*, 1989; Bloch-Shilderman *et al.*, 1995). It has also been shown that *Staphylococcus aureus* α-toxin strongly stimulated AA pathways with subsequent formation of prostacyclin PGI$_2$ in endothelial cells (Suttorp *et al.*, 1985), leukotriene B_4 in polymorphonuclear leukocytes (Suttorp *et al.*, 1987) and platelet-activating factor (PAF) in pulmonary artery endothelial cells (Suttorp *et al.*, 1992). The ionophore toxins generation of eicosanoids might explain a wide spectrum of pathological effects such as endotoxic shock, pulmonary hypertension, blood coagulation, inflammation and edema (Snyder, 1990; Bhakdi *et al.*, 1988; Seeger *et al.*, 1984). The initial step in eicosanoid generation is the hydrolysis of AA from the sn-2 position (where AA is most often esterified) of the glycerol backbone of plasma membrane phospholipids by the activities of two major groups of PLA$_2$: i. cytosolic, high molecular weight (70–110 kD) AA specific enzyme (cPLA$_2$); this enzyme translocates to plasma membrane in response to nanomolar raising levels of calcium concentrations upon receptor-stimulation, such as purinergic, adenosine A_1, dopamine D_2 receptors coupled to G-proteins and is regulated by PKC dependent phosphorylations (Piomelli, 1993). It is tempting to propose that cPLA$_2$s are activated by ionophore toxins to amplify the receptor-dependent hydrolysis of AA-containing phospholipids during step 4 (Figure 9.5); ii. Low molecular weight secreted and membrane-bound enzymes (sPLA$_2$, 14 kD), non-selectively hydrolyzing AA-containing phospholipids and regulated by glucocorticoids. These sPLA$_2$s have been involved in phospholipid remodeling, particularly important in preserving plasma membrane integrity and are secreted in large amount in inflammatory sites leading to a very marked increase in the release of AA and the generation of eicosanoids (Piomelli, 1993). Since ionophore toxins disturb cell membrane organization expressed in changes in the phase-transition properties (Mollat, 1976) and phospholipid aggregation (Lelkes and Lazarovici, 1988), it might recruit and activate sPLA$_2$s to repair the plasma membrane damaged by the toxins pores.

Alternatively AA release by ionophore toxins may be initiated by the activation of phosphoinositide-specific PLC (Fink *et al.*, 1989) producing 1,2-diacylglycerol (DAG). This intermediate has been shown to activate PKC which in turn was reported to activate PLA_2 (Parker *et al.*, 1987). In addition, DAG may be hydrolyzed by a DAG-lipase to yield AA and monoacylglycerol (Figure 9.5). This PLC-DAG lipase signal transduction pathway has been shown to be induced by bradykinin (Allen *et al.*, 1992) and is proposed as a second chemical mediator of ionophore toxins actions in certain cells (Figure 9.5). AA and eicosanoids may act both as intracellular second messengers to modify activities of phospholipases, protein-kinases, G-proteins, adenylate and guanylate cyclase as well as ion channels. Like local chemical mediators they may be released outside the intoxicated cells and act on neighbouring cells by binding to high-affinity membrane receptors triggering new cascades of cell signalling. The proposed working signal transduction model of ionophore toxins action (Figure 9.5) reflects complex, amplification, self-generating cascades of cellular signals propagated by intracellular calcium and the arachidonic acid cascade. Identification and characterization of ionophore toxins molecular mechanisms of action in different cells are fields of active research which are still faced with major unresolved questions. Understanding ionophore toxins action will lay down the foundation of efficient therapies and will provide new pharmacological concepts in cell signaling.

IONOPHORE TOXINS AS PHARMACOLOGICAL TOOLS

Certain ionophore toxins are used by cell biologists as tools for controlled permeabilization of cell plasma membranes (Booth and Fenwick, 1991; Bhakdi *et al.*, 1993). The most widely used are group 1 toxins such as streptolysin O (Table 9.1) and group 3 toxins such as *Staphylococcus aureus* α-toxin (Table 9.1). These toxins are preferable to detergents, digitonin or electroporation as the toxins are more easier to handle, stable, available commercially in a highly purified form and allow a controlled perforation of the cells. Also, the toxins do not require activation, they insert spontaneously to form transmembrane pores of a defined size (Table 9.1). Correct handling of ionophore toxins offer manipulations of the intracellular compartments in signal transduction studies by introducing small molecules such as second messengers, nucleotides, cations or even to incorporate very large molecule such as antibodies or other toxins (Bader *et al.*, 1986; Ahnert-Hilger *et al.*, 1985; 1988; 1992). Recently, protein engineering is being used to produce ionophore toxins conjugates and mutants which might have applications as components of ionic sensors in biotechnology, for drug delivery and for cell permeabilization (Bayley, 1994; Walker *et al.*, 1994). The knowledge provided by studies on the mechanism of oligomerization of *Staphylococcus aureus* α-toxin allowed the design of genetically and chemically modified toxin analogues, with pore-forming activities that can be switched on or off by chelating agents or divalent cations, respectively (Bayley, 1994). Immunolysins, such as *Staphylococcus aureus* α-toxin, genetically engineered mutants activated by proteolysis (Bayley, 1994) and pardaxin-transferrin conjugates (Welhoner, unpublished), might represent a new class of ionophore drugs to be used in chemotherapy. However, in as much as we are interested in progressing the use of ionophore toxins as pharmacological tools, it appears timely to point out that calcium homeostasis and arachidonic acid cascade might be affected at sub-cytotoxic permeabilizing concentrations (Abu-Raya *et al.*, 1993) and therefore permeabilized cells are not physiologically identical to untreated cells.

Although the precise relationship between a sustained rise in intracellular calcium concentration, eicosanoid production and cell death induced by ionophore toxins requires further investigation, these toxins represent a variety of pharmacological tools to explore the mechanism(s) of necrosis and apoptotic cell death (Duke *et al.*, 1994), to develop cell death model systems (Kubota *et al.*, 1995) and to search and develop cytoprotective drugs. The analysis of receptors signal transduction mechanisms will benefit from the use of ionophore toxins to study both calcium-dependent and calcium-independent pathways (Lazarovici, 1994).

CONCLUSIONS

Ionophore toxins belong to a large group of proteins of bacterial, plant and animal origin which alter the cells plasma membrane permeability. The importance of this group of toxins is witnessed by the fact that among the hundreds of protein toxins which have been studied, to date more than 200 are pore forming toxins. Although ionophore toxins appear to have little structure homology they share a common requirement — the need to insert and form pores in plasma cell membranes. The conversion from a water soluble toxin to a membrane pore necessitates a large change in conformation and organization, achieved by a new packaging of the hydrophobic, amphipathic α-helices and assembly in the plasma membrane in oligomeric, poorly selective channels. At sub-cytotoxic concentrations, these small lesions in the plasma membrane induce a sustained increase in intracellular calcium and activation of the arachidonic acid pathways. These crucial events are further amplified by cascades of cellular signaling due to the release of local mediators and/or hormones and modifications of intracellular enzymes, ion channels and receptors activities. The end result will be a lethal hit to the attacked cell signal transduction machinery resulting in necrosis. At high concentrations and/or chronic exposures of the cells to ionophore toxins, rapid cytotoxity will be achieved. However, subcytotoxic concentrations of ionophore toxins under defined experimental conditions, might be used as pharmacological tools to study exocytosis, to permeabilize cells, to change the intracellular milieu, to activate the arachidonic acid cascade, to develop immunolysins to be used in chemotherapy, for drug delivery and to investigate mechanisms of cell death and cytoprotection. The emerging diversity of signal transduction pathways on the one hand and the multiplicity of actions of ionophore toxins on these signaling cascades and cellular physiology might reveal potential sites of action for novel therapeutic agents.

REFERENCES

Abu-Raya, S., Trembovler, V., Shohami, E. and Lazarovici, P. (1993) Cytolysins increase intracellular calcium and induce eicosanoids release by pheochromocytoma PC12 cell cultures. *Natural Toxins*, **1**, 263–270.

Ahnert-Hilger, G., Bhakdi, S. and Gratzl, M. (1985) Minimal requirements for exocytosis: a study using PC12 cells permeabilized with Staphylococcal-toxin. *J. Biol. Chem.*, **260**, 12730–12734.

Ahnert-Hilger, G., Bader, M.F., Bhakdi, S. and Gratzl, M. (1988) Introduction of macromolecules into bovine adrenal medullary chromaffin cells and rat pheochromocytoma cells (PC12) by permeabilization with Streptolysin-O: inhibitory effect of tetanus toxin on catecholamine secretion. *J. Neurochem.*, **52**, 1751–1758.

Ahnert-Hilger, G., Wegenhorst, U., Stecher, B., Spicher, K., Rosenthal, W. and Gratz, M. (1992) Exocytosis from permeabilized bovine adrenal chromaffin cells is differently modulated by guanosine 5′-[8-thiol] triphosphate and guanosine 5′-[β8-imido] triphosphate. *Biochem. J.*, **284**, 321–326.

Allen, A.C., Gammon, C.M., Ousley, A.H., McCarthy, K.D. and Morell, P. (1992) Bradykinin stimulates arachidonic acid release through the sequential actions of an Sn-1 diacylglycerol lipase and a monocyl glycerol lipase. *J. Neurochem.*, **58**, 1130–1139.

Alouf, J.E. (1986) Streptococcal toxins (streptolysin O, streptolysin S, erythrogenic toxins) In: *Pharmacology of Bacterial Toxins* (Dorner, F. and Drews, J., eds.) pp. 199. *Int Enc. Pharmac. Therap.*, **199**, Pergamon Press, Oxford.

Andersen, O.S. (1984) Gramicidin channels *Ann. Rev. Physiol.*, **46**, 531–548.

Arribas, M., Blasi, J., Lazarovici, P. and Marsal, J. (1993) Calcium-dependent and independent acetylcholine release from electric organ synaptosomes by pardaxin: evidence of a biphasic action of an excitatory neurotoxin. *J. Neurochem.*, **60**, 552–558.

Bader, M.F., Thierse, D., Aunis, D., Ahnest-Hilger, G. and Gratzl, M. (1986) Characterization of hormone and protein release from α-toxin permeabilized chromaffin cells in primary culture. *J. Biol. Chem.*, **261**, 5777–5783.

Barrow, C.J., Nakanishi, K. and Tachibana, K. (1992) Structure and activity studies of pardaxin and analogues using model membranes of phosphatidylcholine. *Biochem. Biophys. Acta.*, **1112**, 235–240.

Bayley, H. (1994) Triggers and switches in a self-assembling pore-forming protein. *J. Cell Biochem.*, **56**, 177–182.

Belmonte, G., Pederzolli, C., Macek, P. and Menestrina, G. (1993) Pore formation by the sea anemone cytolysin equinatoxin II in the red blood cells and model lipid membranes. *J. Membrane Biol.*, **141**, 11–22.

Bernheimer, A.W. and Avigad, L.S. (1976) Properties of a toxin from the sea anemone *Stoichatus helianthus* inducing specific binding to sphingomyelin. *Proc. Natl. Acad. Sci. USA.*, **73**, 467–471.

Bernheimer, A.W. and Rudy, B. (1986) Interactions between membranes and cytolytic peptides. *Biochem. Biophys. Acta.*, **864**, 123–141.

Bernheimer, A.W., Avigad, L.S. and Kim, K.S.,(1979) Comparison of metridiolysin from the sea anemone *Metridium senile* with thiol activated cytolysins from bacteria. *Toxicon.*, **17**, 69–75.

Bhakdi, S. and Tranum-Jensen, J. (1987) Damage to mammalian cells by proteins that form transmembrane pores. *Rev. Physiol. Biochem. Pharmacol.*, **107**, 147–223.

Bhakdi, S., Muhly, M., Mannhardt, U., Hugo, F., Klapettek, K., Muller-Eckhardt, C. and Roka, L. (1988) Staphylococcal alpha-toxin promotes blood coagulation via attack on human platelets. *J. Exp. Med.*, **168**, 527–542.

Bhakdi, S., Weller, U., Walev, I., Martin, E., Jonas, D. and Palmer, M. (1993) A guide to the use of pore-forming toxins for controlled permeabilization of cell membranes. *Med. Microbiol. Immunol.*, **182**, 167–175.

Bloch-Shilderman, E., Abu-Raya, S., Rasouly, D., Furman, O., Trembovler, V., Shavit, D., Lelkes, P.I., Shohami, E., Gutman, Y. and Lazarovici, P. (1996) The role of calcium, protein kinase C, pertussis toxin substrates and eicosanoids on pardaxin-induced dopamine release from PC12 cells. In: *Biochemical Aspects of Marine Pharmacology* (Lazarovici, P., Spira, M. and Zlotkin, E., eds.) Alaken Inc., Ft Collins, Co. pp. 158–174.

Blumenthal, R. and Habig, W.H. (1984) Mechanism of tetanolysin — induced membrane damage: studies with black lipid membranes. *J. Bacteriol.*, **157**, 321–323.

Booth, P. and Fenwick, T. (1991) Permeabilization of cells using bacterial toxins. *Focus*, **13**, 54–56.

Braun, V. and Focareta, T. (1991) Pore-forming bacterial protein hemolysins (cytolysins). *Crit. Rev. Microbiol.*, **18**, 115–158.

Bruckner, H. and Graf, H. (1983) Parcelsin, a peptide antibiotic containing α-aminoisobutyric acid, isolated from *Trichoderma reesei Simmons*. *Experientia*, **39**, 528–530.

Buckingham, L. and Duncan, J.L. (1983) Approximate dimensions of membrane lesions produced by streptolysin S and streptolysin O. *Biochem. Biophys. Acta.*, **729**, 115–122.

Chan, K.F.J. and Lazarovici, P. (1987) *Staphylococcus aureus* α-toxin. Effect on protein phosphorylation in myelin. *Toxicon.*, **25**, 631–636.

Christensen, B., Fink, J., Merrifield, R.B. and Mauzerall, D. (1988) Channel-forming properties of cecropins and related model compounds incorporated into planar lipid membranes. *Proc. Natt. Acad. Sci. USA*, **85**, 5072–5076.

Dibas, A. I., Yorio, T. and Easom, R.A. (1995) Gramicidin A is a potent insulin secretagogue: dependence of extracellular calcium influx. *Biochem. Biophys. Res. Commun.*, **212**, 449–454.

Dufton, M.J. and Hider, R.C. (1988) The structure and pharmacology of elapid cytotoxins. *Pharmac. Therap.*, **36**, 1–40.

Duke, R.C., Witter, R.Z., Nash, P.B., Young, J.D.E. and Ojcius, D.M. (1994) Cytolysis mediated by ionophores and pore-forming agents: role of intracellular calcium in apoptosis. *FASEB J.*, **8**, 237–246.

Fasolato, C., Pizzo, P. and Pozan, T. (1990) Receptor mediated calcium influx in PC12 cells. *J. Biol. Chem.*, **265**, 20351–20355.

Fink, D., Contreras, M.L., Lelkes, P.I. and Lazarovici, P. (1989) *Staphylococcus aureus* α-toxin activates phospholipases and induces a Ca^{2+} influx in PC12 cells. *Cellular Signalling*, **4**, 387–393.

Fraser, M.E., Chernaia, M.M., Kozlov, Y.V. and James, M.N.G. (1994) Crystal structure of the holotoxin from *Shigella dyseteriae* at 2.5A° resolution. *Nature Struct. Biol.*, **1**, 59–64.

Freer, J.H., Birkbeck, T.H. and Bhakoo, M. (1984) Interaction of staphylococcal delta — lysin with phospholipid monolayers and bilayers — a short review. In: *Bacterial Protein Toxins*. (Alouf, J.E., Fehrenback, F.J., Freer, J.H. and Jeljaszewicz, J., eds) p. 181, Academic Press, London.

Galzi, J.L., Devillers-Thiery, A., Hussy, N., Bertrand, S., Changeux, J.P. and Bertrand, D. (1992) Mutations in the channel domain of a neuronal nicotinic receptor convert ion selectivity from cationic to anionic. *Nature*, **359**, 500–505.

Gouaux, J.E., Braha, O., Hobaugh, M.R., Song, L., Cheley, S., Shustak, C. and Bayley, H. (1994) Subunit stoichiometry of staphylococcal-hemolysin in crystals and in membranes: a heptameric transmembrane pore. *Proc. Natt. Acad. Sci. USA*, **91**, 12828–12831.

Guy, H.R. and Raghunathan, G. (1989) Structural models of membrane insertion and channel formation by antiparallel-helical membrane peptides. In: *Transport through membranes: carriers, channels and pumps.* (Pullman, E., ed.), pp. 369–379, Magness Found., Jerusalem.

Guy, H.R. and Conti, F. (1990) Pursuing the structure and function of voltage-gated channels. *TINS.*, **13**, 201–206.

Habermann, E. (1980) Mellittin — structure and activity. In: *Natural Toxins* (Eaker, D. and Wadstrom, T., eds) p. 173. Pergamon Press, Oxford.

Hall, J.E., Vodyanoy, I., Balasubramanian, T.M. and Marshall, G.R. (1984) Alamethicin. A rich model for channel behaviour. *Biophys. J.*, **45**, 233–247.

Harvey, A.L. (1990) Cytolytic toxins. In: *Handbook of Toxinology* (Shier, W.T. and Mebs, D. eds) pp. 1–66, Marcel Dekker, Inc, New York.

Heinemann, S.H., Sthumer, W., Imoto, K. and Numa, S. (1992) Calcium channel characteristics conferred on the sodium channel by single mutations. *Nature*, **356**, 441–443.

Hereker, N., Kiessing, S., Glaser, R., Hungerer, K.D. and Von-Baehr, R. (1994) Characterization of neutralizing monoclonal antibodies directed against *Staphylococcus aureus* alpha-toxin. *Hybridoma.*, **13**, 263–270.

Hille, B. (1991) *Ionic channels of excitable membranes.* Sinauer Associates Inc, Sunderland.

Kerr, I.D., Mark, S. and Sansom, P. (1993) Hydrophobic surface maps of channel-forming peptides:analysis of amphipathic helices. *Europ. Biophys. J.*, **22**, 269–277.

Kubota, M., Kataoka, A., Okuda, A., Bessho, R., Lin, Y.W., Wakazono, Y., Usami, I., Akiyama, Y. and Furusho, K. (1995) Selection and partial characterization of calcium ionophore resistant cells. *Biochem. Biophys. Res. Commun.*, **213**, 541–549.

Lala, A.K. and Raja, S.M. (1995) Photolabeling of a pore-forming toxin with the hydrophobic probe 2-[^{3}H]-diazofluorene. *J. Biol. Chem.*, **270**, 11348–11357.

Lazarovici, P., Menashe, M., Primor, N., Hochman, J. and Zlotkin, E. (1982) The cytotoxicity of a cobra snake phospholipase to S-49 mouse lymphoma cells. *Arch. Toxicol.*, **51**, 161–173.

Lazarovici, P., Primor, N. and Loew, L.M. (1986) Purification and pore forming activity of two hydrophobic polypeptides from the secretion of the Red Sea Moses sole (*Pardachirus marmoratus*). *J. Biol. Chem.*, **261**, 16704–16713.

Lazarovici, P. and Chan, K.F.J. (1987) *Staphylococcus aureus* α-toxin. Reduction of epidermal growth factor affinity in PC12 cells. *Toxicon*, **25**, 637–647.

Lazarovici, P., Primor, N., Caratsch, C.G., Munz, K., Lelkes, P., Loew, L.M., Shai, Y., McPhie, P., Louini, A., Contreras, M.L., Fox, J., Shih, Y.L. and Edwards, C. (1988) Action on artificial and neuronal membranes of pardaxin, a new presynaptic excitatory polypeptide neurotoxin with ionophore activity. In: *Neurotoxins in Neurochemistry*. (Dolly, O., ed.) pp. 219–240. Ellis Horwood, Chichester.

Lazarovici, P., Primor, N., Gennaro, J., Fox, J., Shai, Y., Lelkes, P.I., Caratsch, C.G., Raghunathan, G., Guy, H.R., Shih, Y.L. and Edwards, C. (1990) Origin, chemistry and mechanisms of action of a repellent, presynaptic excitatory, ionophore polypeptide. In: *Marine toxins: Origin, Structure and Molecular Pharmacology* (ACS Symposium Series) (Hall, S. and Strichartz, G., eds.) pp. 347–364. American Chemical Society, Washington, DC.

Lazarovici, P., Edwards, C., Raghunathan, G. and Guy, H.R. (1992) Secondary structure, permeability and molecular modeling of pardaxin pores. *J. Nat. Toxins.*, **1**, 1–15.

Lazarovici, P. and Lelkes, P.I. (1992) Pardaxin induces exocytosis in bovine adrenal medullary chromaffin cells independent of calcium. *J. Pharmacol. Exp. Therp.*, **263**, 1317–1326.

Lazarovici, P. (1994) Challenging catecholamine exocytosis with pardaxin, an excitatory ionophore fish toxin. *J. Toxicol. Toxin Reviews*, **13**, 45–63.

Lelkes, P.I. and Lazarovici, P. (1988) Pardaxin induces aggregation, but not fusion of phosphatidylserine vesicles. *FEBS Lett.*, **242**, 161–166.

Loew, L.M., Benson, L., Lazarovici, P. and Rosenberg, I. (1985) Fluorimetric analysis of transferable membrane pores. *Biochemistry*, **24**, 2101–2104.

MacNicol, M. and Shulman, H. (1992) Multiple Ca^2+ signaling pathways converge on CaM kinase in PC12 cells. *FEBS Lett.*, **304**, 237–240.

McClane, B.A. (1994) *Clostridium perfringens* enterotoxin acts by producing small molecules permeability alterations in plasma membranes. *Toxicology*, **87**, 43–67.

Menestrina, G. (1986) Ionic channels formed by *Staphylococcus aureus* alpha — toxin: voltage dependent inhibition by divalent and trivalent cations. *J. Memb. Biol.*, **90**, 177–190.

Menestrina, G. (1991) Electrophysiological methods for the study of toxin-membrane interaction. In: *Source book of bacterial protein toxins* (Alouf, J.E. and Freer, J.H., eds) pp 215–241. Academic Press, London.

Menestina, G. (1991) Pore-forming cytolysins studied with planar lipid membranes. *Period. Biol.*, **93**, 201–206.

Mollay, C. (1976) Effect of melittin and melittin fragments on the thermotropic phase transition of dipalmitoyllecithin and on the amount of lipid bound water. *FEBS Lett.*, **64**, 65–68.

Mor, A., Hani, K. and Nicolas, P. (1994) The vertebrate peptide antibiotics dermaseptins have overlapping structural features but target specific microorganism. *J. Biol. Chem.*, **269**, 31635–31641.

Niedermeyer, W. (1985) Interaction of streptolysin O with biomembranes: kinetic and morphological studies on erythrocyte membranes. *Toxicon*, **23**, 425–439.

Nikodijevic, B., Nikodijevic, D. and Lazarovici, P. (1992) Pardaxin stimulated calcium uptake in PC12 cells is blocked by cadmium and is not mediated by L-type calcium channels. *J. Basic Clin. Physiol. Pharmacol.*, **3**, 359–370.

Ohnishi, M., Hayashi, T., Tomita, T. and Terawaki, Y. (1994) Mechanism of the cytolytic action of *Pseudamonas aeruginosa* cytotoxin: oligomerizaton of the cytotoxin on target membranes. *FEBS Lett.*, **356**, 357–360.

Orrenius, S., McConkey, D.J., Bellomo, G. and Nicotera, P. (1989) Role of Ca^2+ in toxic cell killing. *Trends Pharmacol. Sci.*, **10**,281–285.

Palmer, M., Valeva, A., Kehoe, M. and Bhakdi, S. (1995) Kinetics of streptolysin O self-assembly. *Europ. J. Biochem.*, **231**, 388–395.

Parker, J., Daniel, L.W. and White, M. (1987) Evidence of protein kinase C involvement in phorbol diester-stimulated arachidonic acid release and prostaglandin synthesis. *J. Biol. Chem.*, **262**, 5385–5393.

Piomelli, D. (1993) Arachidonic acid in cell signaling. *Current Opin. Cell. Biol.*, **5**, 274–280.

Putney, J.W.Jr., Takemura, H., Hughes, A.R., Horstman, D.A. and Thastrup, O. (1989) How do inositol phosphate regulate calcium signaling? *FASEB*, **7**, 1899–1905.

Rapaport, D. and Shai, Y. (1992) Agregation and organization of pardaxin in phospholipid membranes. *J. Biol. Chem.*, **267**, 6502–6509.

Raghunathan, G., Seetharamulu, P., Brooks, B. and Guy, H.R. (1990) Models of delta-hemolysin membrane channels and crystal structures. *Proteins* : structure, function and Genetics. **8**, 213–225.

Renner, P., Caratsch, C.E., Waser, P.G., Lazarovici, P. and Primor, N. (1987) Presynaptic effects of the pardaxins, polypeptides isolated from the gland secretion of the flatfish *Pardachirus marmoratus*. *Neuroscience*, **23**, 319–325.

Schonherr, R., Hilger, M., Broer, S., Benz, R. and Braun, V. (1994) Interaction of *Serratia marcenscens* hemolysin (Sh1A) with artificial and erythrocyte membraness. *Europ. J. Biochem.*, **223**, 655–663.

Seeger, W., Bauer, M. and Bhakdi, S. (1984) Staphylococcal alpha-toxin elicits hypertension in isolated rabbit lungs due to stimulation of the arachidonic acid cascade. *J. Clin. Invest.*, **74**, 849–858.

Shai, Y., Fox, J., Caratsch, C., Shih, Y., Edwards, C. and Lazarovici, P. (1988) Sequencing and synthesis of pardaxin, a polypeptide from the Red Sea Moses sole with ionophore activity. *FEBS Lett.*, **242**, 161–166.

Shai, Y., Bach, D. and Yanovsky, A. (1990) Channel formation properties of synthetic pardaxin and analogues. *J. Biol. Chem.*, **265**, 20202–20209.

Shai, Y., Hadari, Y.R. and Finkels, A. (1991) pH-dependent pore formation properties of pardaxin analogues. *J. Biol. Chem.*, **266**, 22346–22354.

Shai, Y. (1994) Pardaxin: channel formation by shark repellant peptide from fish. *Toxicology*, **87**, 109–129.

Shi, Y., Edwards, C. and Lazarovici, P. (1995) Ion selectivity of the channels formed by pardaxin, an ionophore, in bilayer membranes. *Natural Toxins*, **3**, 151–155.

Shier, W.T. (1985) The final steps to toxic cell death. *J. Toxicol. Toxin Rev.*, **4**, 191–249.

Shulman, H. (1993) The multifunctional Ca^{2+}/Calmodulin-dependent protein kinases. *Current Opinion in Cell Biol.*, **5**, 247–253.

Shyder, F. (1990) Platelet-activating factor and related acetylated lipids as potent biologically active cellular mediators. *Am. J. Physiol.*, **259**, C697–C708.

Siemen, D. and Hescheler, J. (1993) *Nonselective cation channels: Pharmacology, Physiology and Biophysics.* BirVerlag, Basel, Switzerland.

Souza, L.R., Moore, H., Raha, S. and Reed, J. K. (1995) Purine and pyrimidine nucleotides activate distinct signalling pathways in PC12 cells. *J. Neurosci. Res.*, **41**, 753–763.

Stein, P.E., Boodhoo, A., Armstrong, G.D., Cockle, S.A., Klein, M.H. and Read, R.J. (1994) The crystal structure of pertussis toxin. *Structure*, **2**, 45–47.

Suarez-Isla, B.A., Wan, K., Lindstrom, J. and Montal, M. (1983) Single-channel recordings from purified acetylcholine receptors reconstituted in bilayers formed at the tip of patch pipets. *Biochemistry*, **22**, 2319–2323.

Suttorp, N., Seeger, W., Dewein, E., Bhakdi, S. and Roka, L. (1985) Staphylococcal alpha-toxin induced PGI_2 production in endothelial cells: role of calcium. *Amm. J. Physiol.*, **248**, C127–C134.

Suttorp, N., Seeger, W., Zucker-Raimann, J., Roka, L. and Bhakdi, S. (1987) Mechanism of leukotriene generation in polymorphonuclear leukocytes by staphylococccal alpha-toxin. *Infect. Immun.*, **55**, 104–110.

Suttorp, N., Buerke, M. and Tannert-Otto, S. (1992) Stimulation of PAF-synthesis in pulmonary artery endothelial cells by *Staphylococcus aureus* α-toxin. *Trombosis Res.*, **67**, 243–252.

Suttorp, N., Fuhrmann, M., Tannert-Otto, S., Grimminger, F. and Bhakdi, S. (1993) Pore-forming bacterial toxins potently induce release of nitric oxide in endothelial cells. *J. Exp. Med.*, **178**, 337–341.

Thelestam, M. and Mollby, R. (1979) Classification of microbial, plant and animal cytolysins based on their membrane damaging effects on human fibroblasts. *Biochem. Biophys. Acta.*, **557**, 156–169.

Thelestam, M. and Blomguist, L. (1988) Staphylococcal alpha toxin — recent advances. *Toxicon.*, **26**, 51–65.

Thompson, S.A., Tachibana, K., Nakanishi, K. and Kubota, I. (1986) Mellitin-like peptides from the shark-repelling defense secretions of the sole *Padachirus pavoninus. Science*, **233**, 341.

Turnbull, P.C.B. (1986) *Bacillus cereus* toxins In: *Pharmacology of Bacterial Toxins.* (Dorner, F. and Drews, J., eds) p 397, *Int. Enc. Pharmac. Therap.*, **119**. Pergamon Press, Oxford.

Varanda, W. and Finkelstein, A. (1980) Ion and nonelectrolyte permeability properties of channels formed in planar bilayer membranes by the cytolytic toxin from the sea anemone *Stoichatus helianthus. J. Memb. Biol.*, **55**, 203–211.

Walev, I., Palmer., M., Valeva, A., Weller, U. and Bhakdi, S. (1995) Binding, oligomerization and pore formation by streptolysin O in erythrocytes and fibroblast membranes: detection of non lytic polymers. *Infect. and Immun.*, **63**, 1188–1194.

Walker, B., Kasianowicz, J., Krishnasastry, M. and Bayley, H. (1994) A pore forming protein with a metal-activated switch. *Protein Engineering*, **7**, 655–662.

Walker, B. and Bayley, H. (1995) Key residues for membrane binding, oligomerization and pore forming activity of staphylococcal-hemolysin, identified by cysteine scanning mutagenesis and targeted chemical modification. *J. Biol. Chem.*, **270**, 23065–23071.

Wang, H.Y. and Friedman, E. (1986) Increased 5-hydroxytryptamine and norepinephrine release from rat brain slices by the Red Sea flatfish toxin pardaxin. *J. Neurochem.*, **47**, 656–658.

Xiong, G., Struckmeier, M. and Lutz, F. (1994) Pore-forming *Pseudomonas aeruginosa* cytotoxin. *Toxicology*, **87**, 69–83.

Zagorski, M.G., Norman, D.G., Barrow, C.J., Iwashita, T., Tachibana, K. and Patel, D.J. (1991) Solution structure of pardaxin P-2. *Biochemistry*, **30**, 8009–8017.

Zhang, R.G., Westbrook, M.L., Westbrook, E.M., Scott, D. L., Otwinowski, Z., Maulik, P.R., Reed, R.A. and Shipley, G.G. (1995) The 2.4A crystal structure of cholera toxin B subunit pentamer: choleragenoid. *J. Mol. Biol.*, **251**, 550–562.

10. ACTIVATION OF CHOLERA TOXIN BY ADP-RIBOSYLATION FACTORS (ARFs), 20 kDa GUANINE NUCLEOTIDE-BINDING PROTEIN COMPONENTS OF THE VESICULAR TRAFFICKING SYSTEM

WALTER A. PATTON, GUI-FENG ZHANG, JOEL MOSS
and MARTHA VAUGHAN

*Pulmonary/Critical Care Medicine Branch, National Heart, Lung, and Blood Institute,
National Institutes of Health, Building 10, Room 5N307, 10 Centre Drive,
Bethesda, MD 20892-1434*

INTRODUCTION

Vibrio cholerae synthesizes and secretes a proteinaceous toxin termed cholera toxin or CT that is the major causative agent of the diarrheal syndrome associated with cholera (Kaper *et al.*, 1995). The first step in cholera toxin action in intestinal epithelial cells is its interaction with a membrane-bound ganglioside, G_{M1} (Fishman *et al.*, 1993). Following endocytosis, CT catalyzes the ADP-ribosylation and thereby activation of $G_{s\alpha}$, a subunit of the heterotrimeric guanine nucleotide-binding protein (G protein) activator of adenylyl cyclase (Moss and Vaughan, 1988), which causes accelerated production and accumulation of high levels of cyclic AMP (cAMP). The action of CT in the intestine also results in the production of mediators such as 5-hydroxytryptamine (5-HT or serotonin) and prostaglandin E_2 (PGE_2) (Kaper *et al.*, 1995). *In vitro*, PGE_2 induces the abnormal fluid and electrolyte flux characteristic of cholera more effectively than does cAMP (*e.g.*, Peterson *et al.*, 1993), consistent with the notion that the increased levels of cAMP may not be directly responsible for the bulk of the secretory response induced by CT.

Vibrio cholerae secretes two additional toxins that apparently are not structurally related to cholera toxin and cause less severe diarrhea than CT. These are: Zot, which increases permeability of rabbit ileal mucosa by perturbing mucosal cell tight junctions (Fasano *et al.*, 1991) and Ace, which causes fluid secretion in rabbit ileal loops (Trucksis *et al.*, 1993). *Escherichia coli* produces an ADP-ribosylating toxin termed *E. coli* heat-labile enterotoxin-I (LT-I or LT) that is almost identical in activities and structure to CT (Spangler, 1992). Other toxins that specifically modify guanine nucleotide-binding proteins by catalyzing their ADP-ribosylation, include the toxins from *Bordetella pertussis* (pertussis toxin or PT), which modifies the heterotrimeric G protein α subunits of G_o, G_i and G_t, as well as diphtheria toxin (DT) from *Corynebacterium diphtheriae* and exotoxin A (ETA) from *Pseudomonas*, both of which modify EF2, a component of the protein synthetic machinery. These ADP-ribosylating toxins and those whose substrates are other than G proteins, are considered in several reviews in this volume as well as those of Rappuoli and Pizza (1991) and of Passador and Iglewski (1994).

In vitro studies have demonstrated enhancement of cholera toxin activity by monomeric 20-kDa guanine nucleotide-binding proteins, termed ADP-ribosylation factors or ARFs. ARFs, apart from their effects on CT *in vitro*, function in signaling-related events and vesicular trafficking. This review will focus first on cholera toxin structure and function, followed by the activation of toxin ADP-ribosyltransferase by ARF and then turn to the involvement of ARF in vesicular trafficking and phospholipase D activation.

CHOLERA TOXIN STRUCTURE AND FUNCTION

Mode of Cholera Toxin Entry into Cells

Cholera toxin (CT) is composed of one A subunit of 29 kDa (CTA), composed of CTA_1 and CTA_2 proteins linked by a single disulfide bond, and a B pentamer (CTB), the subunits of which are 11.6 kDa each (Ohtomo *et al.*, 1976). The mechanism by which CTB binding to a cell surface triggers holotoxin uptake is not fully understood. It is clear, however, from photolabeling (Wisnieski and Bramhall, 1981) and crystallization data (Ribi *et al.*, 1988) that CTB binding to the cell surface ganglioside G_{M1} results in the intimate association of CT with a lipid membrane, which is prerequisite to toxin action in the cells.

CTB pentamer binds to ganglioside G_{M1} (galactosyl-N-acetylgalactosaminyl-[N-acetylneuraminyl]-galactosylglucosylceramide) (Cuatrecasas, 1973; King and van Heyningen, 1973) on the surface of intestinal epithelial cells through five single-affinity sites (reviewed by Fishman, 1982) that act in a cooperative manner (Schafer and Thakur, 1982). After holotoxin binding to G_{M1} on the apical surface of the cell, there is a delay before CT-induced Cl⁻ secretion occurs. It is believed that the bound holotoxin is rapidly endocytosed and conveyed via a brefeldin A- (BFA) sensitive vesicular transport event to a compartment where CTA is reduced to produce CTA_1. There follows a temperature-sensitive transcytosis of the toxin and/or ADP-ribosylated $G_{s\alpha}$ to the basolateral membrane for activation of adenylyl cyclase (Lencer *et al.*, 1993). Orlandi *et al.* (1993) reported that BFA did not inhibit CT internalization, but did inhibit generation of CTA_1 and CT-induced increases in cAMP content. Likewise, Nambiar *et al.* (1993) found that a BFA-sensitive Golgi was required for BFA protection against CT cytotoxicity. Consistent with a Golgi requirement were earlier observations on CT endocytosis into Golgi-endoplasmic reticulum-lysosome (GERL) structures in neuronal cells (Joseph *et al.*, 1978) and more recent studies of internalization of rhodamine-labeled CT to the Golgi, also in neuronal cells (Sofer and Futerman, 1995). Together these observations are consistent with the trafficking of CT through a portion of the Golgi, presumably prior to movement to an acidic (Janicot *et al.*, 1991) or nonacidic (Lencer *et al.*, 1995) compartment where it appears that CTA_1 is generated for subsequent release to the cytosol.

Several studies suggest that CTB binding results in the perturbation of the interaction of G_{M1} with surrounding lipids. Following CTB binding, changes in acrylamide-quenching of a fluorescent analogue of G_{M1}, in G_{M1} micelles or in phospholipid micelles containing G_{M1}, suggested differences in the packing of membrane lipids (Picking, 1993; Picking *et al.*, 1995). It is possible that the changes result from G_{M1}, which is randomly distributed in model phosphatidylcholine bilayers (Thompson *et al.*, 1985), becoming ordered by the binding of CTB. Previous work had shown that CTB binding was increased in cells in which the fatty acid moiety of G_{M1} had been replaced with an acetyl group (Fishman *et al.*, 1980); the putative ordering of G_{M1} may have been facilitated by a less bulky substituent. Together these data support the idea that toxin binding to G_{M1} affects the orientation or structure of surrounding lipids and in doing so may be involved in triggering endocytosis.

An additional effect of CTB binding, the implications of which are unknown, is a net increase in calcium influx that occurs upon binding of CTB to cells (Knoop and Thomas, 1984; Maenz and Forsyth, 1986; Dixon *et al.*, 1987).

Enzymology of Cholera Toxin A Subunit

The CTA subunit is synthesized as a single polypeptide that is proteolytically nicked by the *Vibrio* after amino acid 192 to generate the 23 kDa CTA_1 and the 5 kDa CTA_2 proteins, which remain covalently linked by a single disulfide bond between Cys187 and Cys199 (reviewed by Spangler, 1992). Both proteolytic cleavage and disulfide bond reduction are necessary for the generation of catalytically active CTA_1 in a cell-free assay (Tomasi *et al.*, 1979; Mekalanos *et al.*, 1979a). In intact cells, CT is active only after a delay that correlates with production of the CTA_1 protein (Kassis *et al.*, 1982). For diphtheria toxin, reduction that occurs during the lag phase is the rate-limiting step in translocation of the toxin to the cytosol (Papini *et al.*, 1993), where it exerts its action. On the other hand, LT Arg192Gly (a mutant that cannot be cleaved by trypsin-like proteases, but can be reduced) elicited morphological changes similar to those produced by wild-type LT, but had lower NAD:agmatine ADP-ribosyltransferase activity and effects on cell cAMP content (Grant *et al.*, 1994). In separate experiments, effects of the same mutation in LT were tested in a cell toxicity assay with similar results (Pizza *et al.*, 1994a).

Investigation of the mechanism of CT action led to the recognition that NAD was one of several factors required for the cholera toxin-catalyzed activation of adenylyl cyclase in pigeon erythrocyte lysates (Gill, 1975). It was subsequently determined that CTA_1 catalyzes several reactions involving NAD (with K_m values in the millimolar range) including: 1) the hydrolysis of NAD to ADP-ribose and nicotinamide (NAD glycohydrolase activity) (Moss *et al.*, 1976; 1978; 1979); 2) the transfer of ADP-ribose from NAD to free arginine or other simple guanidino compounds (*e.g.*, agmatine) (ADP-ribosyltransferase activity) (Moss and Vaughan, 1977a; Mekalanos *et al.*, 1979b); guanidino compounds with larger hydrophobic domains were better substrates (Tait and Nassau, 1984); 3) the transfer of ADP-ribose from NAD to arginine in proteins (ADP-ribosyltransferase activity) (Abood *et al.*, 1982; Van Dop *et al.*, 1984); 4) the transfer of ADP-ribose from NAD to CTA_1 (auto-ADP-ribosylation) (Trepel *et al.*, 1977), primarily at Arg146 (Lai *et al.*, 1983; Xia *et al.*, 1984) and to additional positions in multi-(ADP-ribosylated) CTA_1 (Moss *et al.*, 1980a) and 5) the transfer of ADP-ribose from NAD to the heterotrimeric G protein subunit, $G_{s\alpha}$ (ADP-ribosyltransferase activity) (Gill and Meren, 1978; Johnson *et al.*, 1978; Kaslow *et al.*, 1980; Northup *et al.*, 1980). (Assays of CT activity using these five reactions have been reviewed by Moss *et al.*, 1994a.) ADP-ribosylation reactions involving the modification of guanidino groups appeared to be of the S_N2-type, with β-NAD as substrate and resulting in an α-anomeric product (Oppenheimer, 1978). Kinetically, the NAD:arginine ADP-ribosyltransferase reaction catalyzed by CT has been described as a rapid equilibrum, random sequential reaction (Osborne *et al.*, 1985).

Activation of Adenylyl Cyclase by Cholera Toxin

As an ADP-ribosyltransferase, CT transfers the ADP-ribose portion of NAD to Arg201 (Robishaw *et al.*, 1986) in $G_{s\alpha}$, a conserved amino acid equivalent to the ADP-ribosylation site (Arg174) in the α-subunit of the transducing G protein complex from rod outer segments, transducin ($G_{t\alpha}$) (Abood *et al.*, 1982; Van Dop *et al.*, 1984; Lochrie *et al.*, 1985; Tanabe *et al.*, 1985; Medynski *et al.*, 1985; Yatsunami and Khorana, 1985). GDP-bound $G_{s\alpha}$ resides on the plasma membrane as part of a heterotrimeric complex with G protein β and γ subunits that, as a tightly coupled dimer, function as a membrane anchor for $G_{s\alpha}$. Heterotrimeric G protein complexes ($\alpha\beta\gamma$) are signal-transducing elements in all eukaryotic cells, the α subunit of which is normally activated by binding of GTP in response to an extracellular stimulus acting on a cell surface receptor (reviewed by Spiegel, 1992; Emala *et al.*, 1994; Neer, 1995). Studies with transducin suggest that the substrate for cholera toxin is a transient complex of GTP-bound transducin $\alpha\beta\gamma$ and its activating receptor, rhodopsin (Navon and Fung, 1984). Following ADP-ribosylation, $G_{s\alpha}$ dissociates from the $\beta\gamma$ complex (Kahn and Gilman, 1984a). In crude membrane preparations, ADP-ribosylation by CT is believed to inhibit GTP hydrolysis by $G_{s\alpha}$ (Cassel and Selinger, 1977; Johnson and Bourne, 1977; Navon and Fung, 1984) and also appears to accelerate release of GDP (Burns *et al.*, 1982). After GDP release and GTP binding, $G_{s\alpha}$ is persistently activated and, in turn, activates adenylyl cyclase (Enomoto and Gill, 1980), which frequently results in accumulation of high levels of cAMP. As downstream effectors of heterotrimeric G proteins, adenylyl cyclases are a family of related membrane-bound proteins that catalyze the formation of intracellular cAMP from ATP (reviewed by Taussig and Gilman, 1995).

Although cAMP generation is most often mentioned as a consequence of cholera toxin activity, cAMP may not be the primary effector of ion flux. It was reported that cholera toxin caused release of prostaglandin E_2 (PGE_2) into the lumen of intestinal loops *in vitro* (Peterson and Ochoa, 1989), secondary to an effect on arachidonic acid formation (Reitmeyer and Peterson, 1990; Peterson *et al.*, 1990a).

CT also caused the release of 5-HT from enterochromaffin cells (Nilsson *et al.*, 1983), which could then stimulate PGE_2 synthesis (Beubler *et al.*, 1989). Direct application of PGE_2 to rabbit intestinal loops increased fluid flux (Peterson and Ochoa, 1989). The effects of cAMP and derivatives on PGE_2 levels and fluid accumulation appeared to differ depending on how the intestinal segment was prepared (Peterson *et al.*, 1994), thus leaving unclear the effects of cAMP on PGE_2. It was clear, however, that inhibition of protein synthesis by incubation with cycloheximide before addition of CT, inhibited fluid accumulation in rabbit intestine and arachidonic acid release in CHO cells, but did not inhibit the accumulation of cAMP in rabbit intestine (Peterson *et al.*, 1987) and CHO cells (Peterson *et al.*, 1991). Likewise, chloroquine, an inhibitor of phospholipase C and phospholipase A_2, inhibited arachidonic acid release, but only minimally affected cAMP concentrations (Liang *et al.*, 1990); the generation of cAMP was apparently independent of PGE_2. Together these data suggest that arachidonic acid metabolites play a role in fluid and ion flux, and cAMP may not be primarily responsible for the secretory effects of CT.

Structural Information from the Crystallization of CT and LT

The crystallization of isoelectrically homogeneous CT has been reported (Spangler and Westbrook, 1989). Considerable information concerning the three-dimensional structure of both toxins has come from X-ray crystallographic analysis of LT (Sixma *et al.*, 1991; 1993; Merritt *et al.*, 1994a; 1994b; 1994c; 1995) and more recently of CT (Zhang *et al.*, 1995a; 1995b). Early indications of similarity between LT and CT came from immunological data and the demonstration of identical catalytic activity (reviewed by Spangler, 1992) and the fact that the amino acid sequences of the A and B subunits of LT are almost identical to those of CT (Dallas and Falkow, 1980; Gill *et al.*, 1981; Spicer *et al.*, 1981). The two toxins were postulated (based on the sequence similarity) to have evolved from a common ancestor by a series of mutations that did not disrupt ADP-ribosyltransferase function (Yamamoto *et al.*, 1984), or the ability to interact with ARF (Lee *et al.*, 1991).

The LT crystal structure has been solved to 2.3Å resolution (Sixma *et al.*, 1991) and to 1.95Å resolution (Sixma *et al.*, 1993). (These structures were reviewed by Verlinde *et al.* (1994), who also emphasized their potential for structure-based drug design.) Crystals of LT holotoxin demonstrate the characteristics of AB_5 toxin structure. The five identical B subunits are assembled, noncovalently, in a ring-like structure with a highly charged central pore lined with five long amphiphilic helices, each contributed by an individual subunit. Each B subunit is in contact with two adjacent subunits through a total of approximately 30 hydrogen bonds and six salt bridges, forming a stable oligomer. The carboxy-terminal portion of the LTA_2 subunit serves as a link between the A and B subunits by extending into the B pentamer pore along with at least 66 water molecules. Most of the specific contacts of A_2 with the B pentamer are at the entrance to the pore, whereas contacts inside the pore, with the exception of two salt bridges, are water-mediated and hydrophilic in nature. The amino terminus of the LTA_2 protein

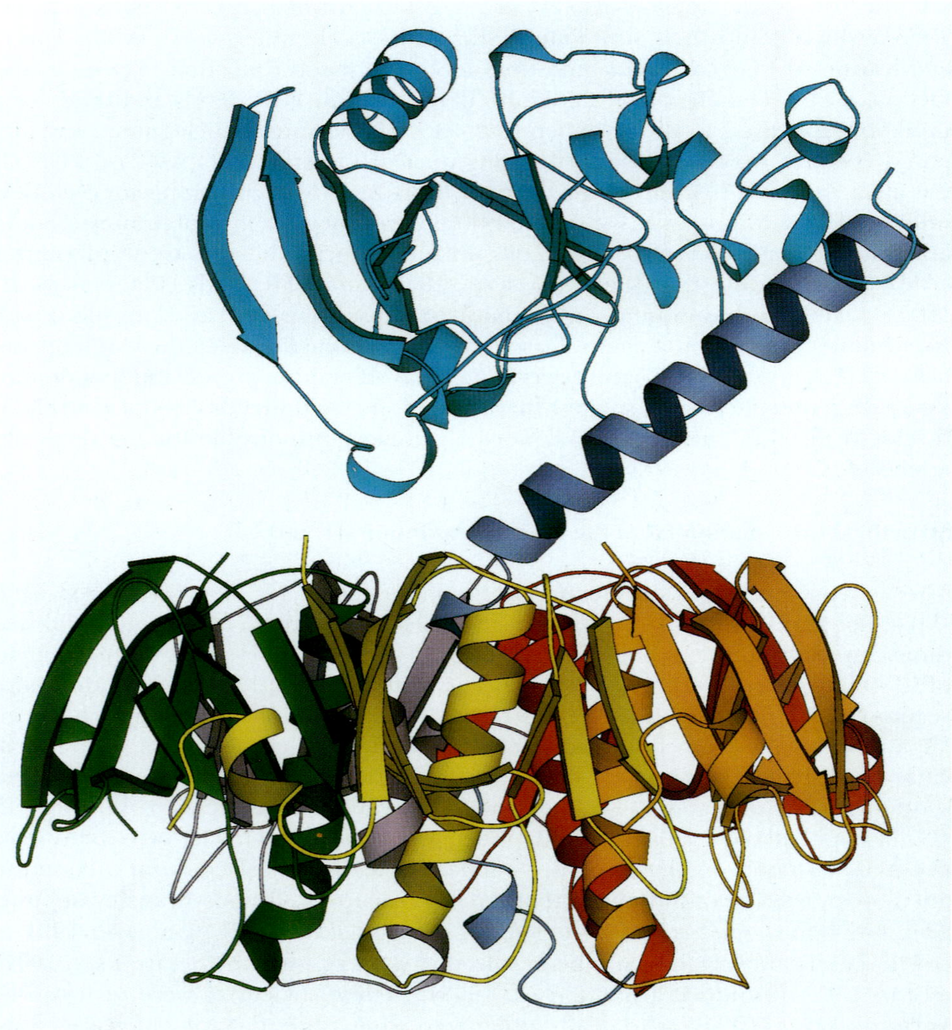

Figure 10.1   Structure of Native LT AB$_5$ Holotoxin. Note that the A$_1$ subunit (light blue) is anchored to the B subunit complex through the A$_2$ subunit (blue-grey). The amino-terminal end of the A$_2$ subunit does not enter the pore of the B pentamer and is primarily α-helical in structure. The carboxy-terminal end of the A$_2$ subunit penetrates the B pentamer and a portion is exposed at the bottom of the pore. The pore of the B pentamer is lined by α-helices from the five individual B subunits. The surface of the B subunit opposite to that exposed to the A$_1$ subunit, binds to G$_{M1}$ ganglioside in the plasma membrane (Merritt *et al.* 1994c). The figure is a reproduction of that published by Verlinde *et al.* (1994) and was a generous gift from Dr. Wim Hol, University of Washington, Seattle.

forms a long α-helix that is in close contact with the surface of the LTA$_1$ protein, except for the region close to the trypsinization site. Residues between Cys187 of LTA$_1$ and Cys199 of LTA$_2$ were reportedly difficult to visualize, consistent with the flexibility required for protease accessibility (Figure 10.1). The authors of the recently published CT holotoxin structure noted that the most striking difference between the LT and CT structures occurs in the A$_2$ subunit (Zhang *et al.*, 1995a); the portion of LTA$_2$ that extends through the B pentamer is mostly an extended chain, whereas, CTA$_2$ appears to be a continuous α-helix throughout its length.

The LTA subunit contains a potential NAD$^+$-binding site, located within an elongated crevice at the interface of β-sheet 1 (β1) and β-sheet 3 (β3) (Sixma *et al.*, 1993). The putative catalytic residue glutamate-112 (Glu112) (discussed below) lies close to this crevice. Glu112 forms a salt bridge with Arg54 to form a link between two loops, whose interaction would otherwise be mostly hydrophobic. The interacting loops lie at right angles to the elongated crevice and are believed to be involved in control of substrate access to the catalytic site or in translocation of the A$_1$ protein through the membrane. At the bottom of the crevice, the functionally important Arg7 (mentioned below) makes a salt bridge with Asp9 and hydrogen bonds with Val53, Arg54 and Ser61. The structure of a partially activated (trypsinized, but not reduced) LT (Merritt *et al.*, 1994a) had no significant differences from the structure discussed above, consistent with previous observations that trypsinization alone was not sufficient to produce active cholera toxin (Moss *et al.*, 1976; Tumasi *et al.*, 1979; Mekalanos *et al.*, 1979a).

Recently, computer modeling of the NAD-binding site of CT (Domenighini, *et al.*, 1994) was used to determine its structural features. The NAD-binding site was described as a cleft "formed by an α-helix bent over a β-strand and flanked on both sides with a β-strand". NAD was depicted aligned in the cleft, such that the nicotinamide could be stabilized by interaction with Arg7; the critical Glu112 of both CT and LT is positioned nearby for interaction with the nicotinamide moiety of NAD as well (Domenighini *et al.*, 1994).

Crystal structures (2.2Å resolution) of galactose-bound LT holotoxin (Merritt *et al.*, 1994b) and of G$_{M1}$-bound CTB pentamer (Merritt *et al.*, 1994c) show that the galactose-binding interactions are mostly within a single B subunit and that the majority of the interactions involve residues 51–60 of the B subunit. Residues 51–60 of the CTB monomer form a flexible loop that contains several tightly bound water molecules. Upon G$_{M1}$ binding, these residues become ordered by the formation of hydrogen bonds to the saccharide and to the tightly bound water. In the G$_{M1}$-bound CTB pentamer structure, the majority of interactions between G$_{M1}$ and B pentamer (which the authors described as "a two-fingered grip") involve the terminal galactose and sialic acid of G$_{M1}$, with minor interactions of the N-acetylgalactosamine (see abbreviations for G$_{M1}$ structure). The saccharide-binding cleft is composed of three loops from one subunit (residues 10–14 connecting α_1 to β_1, 51–58 connecting β_4 to the central helix, α_2, and 89–93 connecting β_5 to β_6) and one loop (residues 31–36 connecting β_2 to β_3) from an adjacent subunit.

Mutant Constructs and Toxin Function

Based on corresponding positions in related toxins and recently on positions predicted by computer modeling of the structure of LT, site-specific mutagenesis of CT and the almost identical LT has been used to determine which amino acids are critical to the function of the A_1 subunit. Perhaps the most studied amino acid in these toxins is Glu112. Positions equivalent to Glu112 in CT and LT are Glu148 in DT (Carroll and Collier, 1984), Glu553 in ETA (Carroll and Collier, 1987) and Glu129 in PT (Barbieri *et al.*, 1989). All of these glutamate residues had been shown to be located at or near the active site of the respective toxins by photo-crosslinking of NAD and by mutagenesis.

The importance of Glu112 in the activity of LT was shown directly by random mutation of LT DNA with hydroxylamine, giving rise to Glu112Lys (Tsuji *et al.*, 1990; 1991) and by direct replacement using recombinant techniques to produce Glu112Asp (Lobet *et al.*, 1991). These mutant LTs lacked or had greatly reduced ADP-ribosyltransferase activity, respectively. Glu112Lys LT retained the ability to interact with ADP-ribosylation factor (discussed below), but only after it was trypsinized and reduced (Moss *et al.*, 1993).

Based on the fact that the mutation Arg9Lys in PT reduced toxin activity (Barbieri and Cortina, 1988; Cieplak *et al.*, 1988; Pizza *et al.*, 1988; Burnette *et al.*, 1989; Lobet *et al.*, 1989; Locht *et al.*, 1989), mutation at the equivalent position of LT (Arg7) was used to produce Arg7Lys LT (Lobet *et al.*, 1991) or CT (Burnette *et al.*, 1991). Both constructs lacked demonstrable ADP-ribosyltransferase activity with $G_{s\alpha}$ as the ADP-ribose acceptor. Recombinant CT with both Arg7Lys and Glu112Gln mutations, also predictably lacked ADP-ribosyltransferase activity and did not activate adenylyl cyclase (Häse *et al.*, 1994). Another mutant LT (Ser61Phe), generated randomly using hydroxylamine, was also shown to be inactive in a fluid flux assay (Harford *et al.*, 1989).

In a comprehensive survey of the amino acids in and around the NAD-binding and catalytic sites of LT, computer modeling led to a prediction of residues involved in functional activity (Domenighini *et al.*, 1994a). Specific amino acid replacements resulted in loss of enzymatic activity, altered structural assembly, or no effect on enzymatic activity (Pizza *et al.*, 1994a). Mutations that caused loss or reduction of enzymatic activity were Arg7Lys, Val53Asp, Val53Glu, Ser63Lys, Val97Lys, Tyr104Lys, Tyr104Asp, Tyr104Ser, His107Glu, Glu110Ser, Glu112Ala, Ser114Lys and Ser114Glu. Loss of activity with mutations at these sites correlated for the most part with loss or introduction of hydrogen bonding or charge interactions; for example, Val97Lys, through these effects and/or steric constraints may significantly reduce the volume of the active site or interfere with a conformational change (Merritt *et al.*, 1995). Mutations that altered assembly of the holotoxin were Leu41Phe, Ala45Tyr, Ala45Asp, Val53Tyr, Val60Gly, Ser68Pro, His70Pro, Val97Tyr and Ser114Tyr. The loss of structure upon mutation at these sites resulted mostly from disruption of structure in the core of the protein. The mutations of Arg54Lys, Arg54Ala, Tyr59Met, Ser68Lys, Ala72Arg, Ala72His, Ala72Glu and Arg192Asn, which were designed to confirm or rule out the involvement of these residues, had no effect on CT. It was notable that the Arg192Asn mutant retained activity, although it should have been resistant to proteolysis (similar to Grant *et al.*, 1994). Thus, proteases other than trypsin may also act in this region to activate toxin.

A survey of important amino acids in CT by computer modeling (Fontana *et al.*, 1995) led to the mutations Val53Asp, Ser63Lys, Val97Lys, Tyr104Lys, Pro106Ser and a double mutant of Val53Asp and Ser63Lys, each of which assembled into AB_5 toxin, but was nontoxic in a cell-based assay. Mutations Arg54His and His107Asn had no effect on toxin assembly or cell toxicity, whereas a Ser114Glu mutant was unable to assemble into AB_5 toxin and only produced B pentamer that was not toxic to cells.

Cholera Toxin as a Molecular Tool

CT has been useful in identifying components of signaling pathways by labeling or modifying the activity of certain G protein α subunits *in vitro*. CT has also been used as a targeting molecule, and thus may have both diagnostic and therapeutic applications. Covalent attachment of CTB to liposomes via the reaction of free sulfhydryl groups in CTB with maleimide on the surface of phosphatidylethanolamine liposomes enhanced (~200-fold over native CTB) its ability to compete with RCA_{120} agglutinin for binding to brush border surfaces (Uwiera *et al.*, 1992). The authors speculated that such a modification could be useful for directing liposomes to cell surfaces that contain G_{M1}. Another potential use of CT as a molecular tool was shown by generating fusion proteins of cholera toxin and a protein of interest. A peptide of 15 amino acids was attached to the amino terminus of CTB without affecting CTB pentamer formation or its binding to G_{M1} (Dertzbaugh *et al.*, 1990). Jobling and Holmes (1992) exploited the CTB-G_{M1} interaction, as well as the interaction of CTA_2 with CTB, by preparing fusion proteins of CTA_2 with bacterial alkaline phosphatase, maltose-binding protein, or β-lactamase. These fusion proteins bound to membranes containing G_{M1}, although only the alkaline phosphatase and β-lactamase constructs retained enzymatic activity.

Several approaches (*i.e.*, administration of the nontoxic B subunit, killed *Vibrio cholerae*, attenuated live *Vibrio cholerae* and/or chemically modified CT) have been utilized in cholera toxin vaccine development (reviewed by Mekalanos and Sadoff, 1994). Another very promising approach is the administration of an inactive mutant (deletion or point) or chemically modified CT to elicit an immune response. Among the first mutant molecules to be considered were those produced by *Vibrio* strains in which a portion or all of the CTA gene was deleted (Mekalanos *et al.*, 1983). Administration of immunogen produced by these strains, however, caused some degree of cholera-like symptoms (Mekalanos and Sadoff, 1994). Recently, point mutations that inactivate CT or LT ADP-ribosyltransferase activity have been prepared for use in vaccine development (Burnette *et al.*, 1994). For example, LT Ser63Lys (Pizza *et al.*, 1994b) and LT Arg7Lys (Douce *et al.*, 1995) were reported to be effective in generating significant titers of anti-toxin antibodies. Fragments of CTB sequence, not large enough to bind G_{M1} have also been used for immunization. In one example, CTB peptides chemically linked in the form of a branched polymer, when administered to mice, either orally or intraperitoneally, elicited an antibody response stronger than that elicited by the non-polymeric peptide (Halimi and Rivaille, 1993).

Endogenous ADP-ribosyltransferases and ADP-ribosylarginine Hydrolases

Mono-ADP-ribosylation catalyzed by an endogenous eukaryotic, intracellular ADP-ribosyltransferase was first reported by Moss and Vaughan (1978). This work led to the identification of several arginine-specific ADP-ribosyltransferases from the cytosolic, membrane and nuclear fractions of avian tissues, including those from turkey erythrocytes (Moss *et al.*, 1980b; Moss and Stanley, 1981a; 1981b; Yost and Moss, 1983; West and Moss, 1986) and hen liver (Tanigawa *et al.*, 1984). All of the avian ADP-ribosyltransferases are ~28 kDa proteins, but differ in physical, regulatory and kinetic properties (reviewed by Zolkiewska *et al.*, 1994). The function(s) of these intracellular transferases remains to be determined.

Potential reversibility of the ADP-ribosylarginine modification in animal cells was established with the demonstration (Chang *et al.*, 1986; Moss *et al.*, 1985; Smith *et al.*, 1985) and purification (Moss *et al.*, 1988) of a turkey erythrocyte ADP-ribosylarginine hydrolase that catalyzed the removal of ADP-ribose from modified proteins (reviewed by Takada *et al.*, 1994). Comparison of rat (Moss *et al.*, 1992) and human (Takada *et al.*, 1993) hydrolase sequences revealed the presence of an extra cysteine in the rat sequence at postion 108 (equivalent to Ser103 in human). Site-specific mutagenesis demonstrated that the additional cysteine in the rat sequence was responsible for the DTT-dependence of the activity, which the human enzyme lacked (Takada *et al.*, 1993). Although a function for intracellular ADP-ribosylation has not been shown, the existence of transferase and hydrolase enzymes could reflect the presence of an ADP-ribosylation cycle, with a regulatory function.

A different type of arginine-specific ADP-ribosyltransferase, partially purified (Peterson *et al.*, 1990b) and cloned (Zolkiewska *et al.*, 1992) from rabbit skeletal muscle, contained hydrophobic amino and carboxy signal sequences consistent with those found in glycosylphosphatidylinositol (GPI)-anchored proteins. The extracellular location of this protein was confirmed by its release from the surface of cultured mouse myotubes (Zolkiewska *et al.*, 1992) and NMU cells expressing recombinant ADP-ribosyltransferase (Okazaki *et al.*, 1994) by incubation with phosphatidylinositol-specific phospholipase C. Identification of the endogenous substrate for the extracellular GPI-anchored muscle ADP-ribosyltransferase as integrin $\alpha 7$ (Zolkiewska *et al.*, 1993), suggested that this extracellular ADP-ribosyltransferase might play a role in the regulation of cell adhesion and related signaling events.

THE ROLE OF ARF IN CHOLERA TOXIN ACTION

Identification of ARF

GTP or GTP analogues (Moss and Vaughan, 1977b; Lin *et al.*, 1978; Enomoto and Gill, 1979; Nakaya *et al.*, 1980; Watkins *et al.*, 1980) and tissue factors (Enomoto and Gill, 1979; 1980; Schleifer *et al.*, 1982; Pinkett and Anderson, 1982) are required for cholera toxin activation of adenylyl cyclase. The nature of the nucleotide and tissue factor requirements was revealed after purification and characterization of an ADP-ribosylation factor (ARF) from rabbit liver membranes that migrated as a doublet of ~21.5 kDa on SDS-PAGE (Kahn and Gilman, 1984b). GTP, but not GDP, supported the ARF-stimulated, CT-catalyzed ADP-ribosylation of $G_{s\alpha}$ (Kahn and Gilman, 1986). Although the purified ARF contained tightly bound GDP, no intrinsic GTPase activity was detected (Kahn and Gilman, 1984b). Purified ARF bound GTP, GDP and non-hydrolyzable analogues of both most effectively in the presence of 3 mM DMPC and 2 mM Mg^{2+}, whereas adenine nucleotides did not bind under these conditions. Additionally, two soluble ARFs, sARFI and sARFII (Tsai *et al.*, 1988), as well as a membrane-bound ARF (Tsai *et al.* 1987) were purified from bovine brain; sARFI and sARFII were later identified as ARF1 and ARF3, respectively (Tsai *et al.*, 1992).

Mechanism of ARF Action and Biochemical Characterization of ARF

ARF apparently acts as an allosteric activator of CTA_1 (Tsai *et al.*, 1987; 1988; Noda *et al.*, 1990). In the presence of GTP, sARFII lowered the K_m for NAD and agmatine (Noda *et al.*, 1990); in the presence of 0.003% SDS, a non-denaturing concentration, sARFII caused a greater decrease in the K_m and an increase in V_{max}. Lipids and detergents appear to enhance ARF action, at least partially, by influencing GTP binding. The presence of 3 mM DMPC/0.2% cholate permitted (apparently maximal) high affinity binding of GTP with a K_d of 70 nM (Bobak *et al.*, 1990). For the stimulation of CTA_1 NAD:agmatine ADP-ribosyltransferase by sARFII in 0.003% SDS, the EC_{50} for GTP was 5 μM, whereas in DMPC/cholate it was 50 nM. Thus, both high and low affinity interactions of sARFII with GTP promoted activation of CTA_1.

Using gel filtration, the existence of stable complexes of sARFII and CTA was observed in the presence of 100 μM GTPγS and 0.003% SDS, but not in 3 mM DMPC/ 0.2% cholate or with 100 μM GDPβS (Tsai *et al.*, 1991a). In the presence of DMPC/cholate, sARFII alone formed GTPγS-dependent aggregates, although these did not form a complex with CTA that was demonstrable by gel filtration. These aggregates stimulated ADP-ribosylation of agmatine to the same extent as did non-aggregated sARFII, but were substantially more effective in enhancing the auto-ADP-ribosylation of CTA_1. Franco *et al.* (1993) also reported the formation of an ARF-CTA_1 complex, in the presence of azolectin/cholate, as well as the formation of ARF aggregates. Data from both groups suggest that ARF activation of CTA does not require a high affinity interaction.

The ability of ARFs to activate CT was affected differently by different lipids (Price *et al.*, 1992). Activities of recombinant ARFs 2, 3 and 5 were enhanced by 3 mM DMPC/0.2% cholate, whereas, rARF6 stimulated CTA activity significantly without added detergent or lipid. Lipids, detergents and ARF appear to play a role in promoting CT activity at physiological temperature. The optimal temperature for the ADP-ribosylation of agmatine by CTA or the auto-ADP-ribosylation of CTA, without lipid or ARF, was 25°–30°C, with considerably lower activity at 37°C (Murayama *et al.*, 1993). In the presence of sARFII and 3 mM DMPC/0.2% cholate, there was a shift in the optimal temperature of the reaction towards 37°C and significant enhancement of activity at 37°C (Murayama *et al.*, 1993). In these experiments, cardiolipin enhanced the maximal stimulation by sARFII at least twice as much as did 3 mM DMPC/0.2 % cholate or 0.003% SDS, by decreasing the apparent K_a for GTPγS (Murayama *et al.*, 1993).

The binding of GTP (or a non-hydrolyzable analogue) to ARF is necessary for all known ARF activities. ARF-GDP is inactive, whereas ARF-GTP is the active species. GTP activation cannot be mimicked by GDP•AlF$_4^-$, as is possible with heterotrimeric G proteins (Kahn, 1991). Non-hydrolyzable analogues of GTP, but not GDP, promote the association of native, soluble ARF with phospholipids and membranes (Kahn, 1991; Regazzi *et al.*, 1991; Walker *et al.*, 1992). Phospholipid composition also influenced the ability of ARF to bind to membrane vesicles (Walker *et al.*, 1992). Some cytosolic ARF bound to phosphatidylinositol or cardiolipin vesicles (to varying degrees) without added guanine nucleotide or analogue, whereas binding to phosphatidylserine was strictly dependent on nonhydrolyzable GTP analogues (Walker *et al.*, 1992). With the nucleotide binding to ARF being influenced by lipids to which ARF subsequently binds, it is possible that ARF is "loosely" associated with the membrane prior to GTP binding. In support of this are experiments by Franco *et al.* (1995), which indicated that myristoylated ARF1-GDP binds to membranes.

The amino-terminal glycine of ARF1 is myristoylated *in vivo* after the removal of the initiating methionine (Kahn *et al.*, 1988; Kunz *et al.*, 1993). Expression of a mutant ARF that was incapable of being myristoylated, demonstrated that the covalently attached lipid was required for ARF function in cells (Kahn *et al.*, 1995). For the purification of large quantities of myristoylated ARF, recombinant ARFs can be myristoylated in *E. coli* that have been cotransformed with a plasmid containing the yeast *N*-myristoyltransferase gene (reviewed by Moss *et al.*, 1994b), based on the procedure described by Duronio *et al.* (1990). The percentage of ARF that is myristoylated with this method is variable, but has been reported to be as high as 85% (Franco *et al.*, 1995).

Myristoylation of recombinant hARF5 promoted its association with Golgi membranes (Haun *et al.*, 1993a). In contrast, myristoylation was apparently not necessary for GTP-dependent membrane association of ARF1 *in vitro* (Franco *et al.*, 1993), but rather appeared to facilitate GDP dissociation and thus increase the rate of GTP binding (Franco *et al.*, 1995). The myristoylated amino terminus has been postulated to be a phospholipid- and GTP-sensitive switch required for the interaction of ARF with some of its effectors and for the increase in GTPγS affinity in the presence of phospholipids (Randazzo *et al.*, 1995). The fatty acid chain of myristoylated GDP-bound ARF is believed to be tightly associated with ARF. When GTP is bound, the myristate is believed to change position relative to the ARF protein, such that the fatty acid chain is available for interaction with membrane lipids, as described for other proteins (Johnson *et al.*, 1994).

ARF Structure

Mammalian ARFs are members of a multigene family with a high percentage of identity. Deduced amino acid sequences of ARFs from lower eukaryotes exhibit significant identity to mammalian ARFs (Figure 10.2). To date, six mammalian ARF cDNAs have been identified from numerous sources. In addition, three ARFs have been cloned from *Drosophila melanogaster*, three from *Saccharomyces cerevisiae* and two from *Giardia lamblia*. A deduced amino acid sequence from *Caenorhabditis elegans* indicates the presence of ARF in that species as well (References are found in Table 10.1). Plant and fungal ARF deduced amino acid sequences have also been identified, but a description of all known ARFs or deduced amino sequences of ARFs is beyond the scope of the review.

Mammalian ARFs are grouped into three classes based on amino acid sequence, size, phylogenetic analysis and gene structure. Class I ARFs, human ARFs 1 and 3 (hARF1 and hARF3) and bovine ARF2 (bARF2), contain 181 amino acids and are 95–96% identical in sequence, with differences primarily near the amino and carboxy-terminal. Class II ARFs, human ARFs 4 and 5 (hARF4 and hARF5), with 180 amino acids are 90% identical to each other; differences hARF4 and hARF5 are mostly in the carboxy-terminal half the proteins and both are approximately 79–80% identical to class I ARFs. Human ARF6 (hARF6), the only known class III ARF with 175 amino acids, is ~68% identical to class I and ~64% identical to class II ARFs. Differences between ARF6 and the others are spread throughout the protein, with most differences occurring near the amino terminus and in the carboxy half of the protein. *Drosophila* ARFs 1, 2 and 3 (dARF1, dARF2 and dARF3) are most similar to class I, II and III mammalian ARFs, respectively. Yeast ARF1 (yARF1), yeast ARF2 (yARF2), *C. elegans* ARF1 (cARF1) and *Giardia* ARF1 (gARF1) appear to be divergent from all mammalian classes. Yeast ARF3 (yARF3), however, is most similar to mammalian ARF6 (Class III). It is possible that sequence differences between ARFs from lower species and mammalian ARF classes, are functionally significant and could thus help in delineating the function of specific segments of ARF sequence.

Among regions of conserved amino acid sequence in ARFs, are several consensus sequences, similar to those found in the heterotrimeric G proteins and the Ras family of monomeric GTP-binding proteins, that are thought to be responsible for guanine nucleotide binding and hydrolysis. The first of these, GXXXXGK (the putative GTP-hydrolysis/phosphate-binding domain), is present in all mammalian and *Drosophila* ARFs as GLDAAGK. Other consensus sequences include the proposed Mg^{2+}- and guanine nucleotide β-phosphate-binding sequence, DXXG, represented in all known ARFs (except *C. elegans*) as DVGG and the sequence that putatively interacts with the purine ring of guanine nucleotides, NKXD, present in all ARFs as NKQD. Interestingly, the DVGG sequence in ARFs is also found in the heterotrimeric G protein α subunits, but not in other monomeric GTP-binding proteins (Botstein *et al.*, 1988).

Despite the presence and similarity of the consensus sequences in ARF to Ras, ARFs do not have intrinsic GTPase activity. The GXXXXGK region in most Ras

```
hARF1      ..NIFAN..K G........ .......... .......... .......... 50
bARF2      ..NVFE...K S........ .......... .......... .......... 50
hARF3      ..NIFGN.LK S.I...... .......... .......... .......... 50
hARF4      ..LTISS..S R.....Q... .......... .......... .......... 50
hARF5      ..LTVSA..S RI....Q... .......... .......... .......... 50
hARF6      ..----.VLS KI..N..... ..L....... .......... QS......V. 46
dARF1      ..NVFAN..K G........ .......... .......... .......... 50
dARF2      ..LTISS.LT R.....Q... .......... .......... .......... 50
dARF3      ..----..LS KI..N..... ..L....... .......... QS....G.V. 46
yARF1      ..LFAS...S N...N..... ......G... ..V....... .VI...... 50
yARF2      ..LYAS...S N...N..... ......G... ..V....... .VI...... 50
yARF3      ..NSIS.VLG K...S...K. ..L...K... .........N K.K.ST..V. 50
cARF1      ..LIMA...Q SWWIG.KYK. IV....N... .....NYVTK DQ.E.K.... 50
gARF1      ..QGAS.I.G K..S...V.. .......... ......M.. .V...V.... 50

Consensus  MG....KLF. .LFGKKEMRI LMVGLDAAGK TTILYKLKLG EIVTTIPTIG 50


hARF1      .......... .S........ .......... .......... .......E.. 98
bARF2      .......... .S........ .......... .......... .......E.. 98
hARF3      .......... .S........ .......... .......... .......E.. 98
hARF4      .......... .C........ ..R.....K. .......... .......E.. 98
hARF5      .......... .C........ .......... .......... .......E.. 98
hARF6      ......T... VK.N...... .......... .YTG...... ...CA..D.. 94
dARF1      .......... .S........ .......... .......... .......E.. 98
dARF2      .......... .C........ .......... .......... .......D.. 98
dARF3      ......T... VK.N...... .......... .YTG...... ...CA..D.. 94
yARF1      ......Q... .S........ ..R..S.... .YR..E.V.. .......S.. 98
yARF2      ......Q... .S........ ..R..S.... .YR..E.V.. .I....S.. 98
yARF3      ......T... VK.NM..... .QRL...... ..PA.TA... .I..SA.N.. 98
cARF1      S...E.S.R. LD.VI..I.. .ESL.KS.ST .YVQ.DVV.V .I..S.TT.. 98
gARF1      .......... .N........ ..S....... .Y...DA..Y .I..A.LEPK 100

Consensus  FNVETVEYKN I.FTVWDVGG QDKIRPLWRH YFQNTQGLIF VVDSNDR.-- 100


hARF1      .VN......M ...A...... .......... ...N..NA.. .......HS. 147
bARF2      .VN......T ...A...... ......V... ...N..NA.. .......HS. 147
hARF3      .VN......M ...A...... .......... ...N..NA.. .......HS. 147
hARF4      ..Q.VAD..Q K..LV..... ....L..... ...N..AIS. M......QS. 147
hARF5      .VQ.SAD..Q K..Q...... .......... .M.N..PVS. L......QH. 147
hARF6      ..D...Q..H .IINDR.M.. .II.I..... ...D..KPH. .QE....TR. 143
dARF1      ..G......M ...A...... ....I..... ...N..NA.. .......HS. 147
dARF2      ..T..ER..Q N..Q...... .......... ...N..TA.. L....R.NQ. 147
dARF3      ..D...T..H .IINDR.M.. .II.I..... ...D..KPH. .QE....TR. 143
yARF1      ..G....VMQ ...N.....N .AW....... ...E..SA.. ..E....HS. 147
yARF2      ..G....VMQ ...N.....N ..W....... ...E..SA.. ..E....HS. 147
yARF3      .ME..K...Y SIIG.K.MEN V....W.... ..KD..KPQ. VS.F.E.EKN 148
cARF1      ..PIMK.Q.H N..QHED.AR .HI..L.... ...G..NP.. VSTQ...QTL 148
gARF1      ..ED.KN..H TL.G...... .A........ ...K..STTD L.ER...QE. 149

Consensus  RI.EAREEL. RML.EDELRD AVLLVFANKQ DLP.AM..AE ITDKLGL..- 150


hARF1      ..H.N....A ......D... .........Q. RNQK...... ..   181
bARF2      ..Q.N....A ......D... .........Q. KNQK...... ..   181
hARF3      ..H.N....A ......D... .......A.Q. KNKK...... ..   181
hARF4      ..N.T..V.A ....Q.T... .........E. SKR-...... ..   180
hARF5      ..S.T..V.A ....Q.T... D......HE. SKR-...... ..   180
hARF6      I.D.N..V.P S.....D... ...T..TSNY KS--...... ..   175
dARF1      ..N.N....A ......D... .........Q. KNANRP.... ..   183
dARF2      ..N.H.F..S ....Q.H... .......AE. AKK-...... ..   180
dARF3      I.D.N..V.P S.....D..S ...I..TSNH KL--...... ..   175
yARF1      I.N.P.F..A ......E... ...E....S. KNST...... ..   181
yARF2      I.N.P.F..S ......E... ...E....N. KNQS...... ..   181
yARF3      .KNQP.CVIG SN.L..Q..V ...S.I..NT NVPKK..... ..   183
cARF1      RGA.K.Q.NG C..VK.E..P .A.E.IA.N. ----...... ..   178
gARF1      .KK.D....P ...R..D... Q......DYI FDKKNKKKGK KR   191

Consensus  LR.R.WYIQ. TCATSG.GLY EGLDWLSN.L ....------ --   192
```

Figure 10.2   Alignment of ARF Deduced Amino Acid Sequences. ARFs were aligned using the GeneWorks program for the Macintosh. Residues identical to the consensus sequence are indicated with a period and gaps with a dash. References and database accession numbers for these sequences are in Table 10.1.

Table 10.1 ARF and ARL Identification

Protein	Source	Reference	DB/Accession #
hARF1	Human	Bobak *et al.*, 1989 Peng *et al.*, 1989	SP/P32889
bARF2	Bovine	Price *et al.*, 1988	SP/P16500
hARF3	Human	Bobak *et al.*, 1989	SP/P16587
hARF4	Human	Monaco *et al.*, 1990	SP/P18085
hARF5	Human	Tsuchiya *et al.*, 1991	SP/P26437
hARF6	Human	Tsuchiya *et al.*, 1991	SP/P26438
dARF1	*Drosophila*	Murtagh *et al.*, 1993	SP/P35676
dARF2	*Drosophila*	Lee *et al.*, 1994a	GB/L25062
dARF3	*Drosophila*	Lee *et al.*, 1994a	GB/L25064
yARF1	*Saccharomyces*	Sewell and Kahn, 1988	SP/P11076
yARF2	*Saccharomyces*	Stearns *et al.*, 1990a	SP/P19146
yARF3	*Saccharomyces*	Lee *et al.*, 1993	SP/P40994
cARF1	*C. elegans*	Wilson *et al.*, 1994	SP/P34212
gARF1	*Giardia*	Murtagh *et al.*, 1992	SP/26991
hARL1	Human	Zhang *et al.*, 1995	GB/L28997
rARL1	Rat	Schürman *et al.*, 1994	SP/P41276
dARL1	*Drosophila*	Tamkun *et al.*, 1991	SP/P25160
hARL2	Human	Clark *et al.*, 1993	SP/P36404
dARL2	*Drosophila*	Clark *et al.*, 1993	GB/L14923
hARL3	Human	Cavenagh *et al.*, 1994	GB/U07151
rARL3	Rat	Cavenagh *et al.*, 1994	SP/P37996
rARL4	Rat	Schürman *et al.*, 1994	SP/P41275

SP — SwissProt Database
GB — GenBank Database

molecules is [10]GA<u>GG</u>VGK[16]. Replacement of Gly12 with other amino acids, except for proline resulted in a loss of GTPase activity (Barbacid, 1987). With the recognition that Gly12 of Ras corresponds to Asp26 of ARF, it had been speculated that a lack of intrinsic GTPase activity in ARF may be due to (in part) this amino acid difference. Comparison of hARF1 and human ARL1 (hARL1), an ARF-like protein that does not activate CTA, but does display intrinsic GTPase activity, reveals that hARL1 contains an Asp26. Gln61 of Ras is also important for its GTPase activity (Barbacid, 1987). Mutation of the equivalent residue in ARF (Gln71) also resulted in a protein that exists in an activated, GTP-bound form (Tanigawa *et al.*, 1993; Teal *et al.*, 1994; Zhang *et al.*, 1994; Kahn *et al.*, 1995), due to a loss of ability to hydrolyze GTP. The position of Gln71 in the ARF-GDP crystal differed from that in Ras (Amor *et al.*, 1994) and, in part, may also account for a lack of intrinsic GTPase activity (discussed below).

Although the Ras crystal structure had often been thought of as a model for ARF because of similarities in amino acid sequences, the crystal structure of nonmyristoylated, GDP-bound ARF (2.0Å resolution) revealed several differences between ARF and Ras (Amor *et al.*, 1994). It was most notable that, unlike Ras or the heterotrimeric α subunits of G proteins, ARF crystallized as a dimer, with an additional β-sheet (β_2), not present in Ras, as the interface between the two molecules. Greasley *et al.* (1994) described both dimeric and monomeric crystals, although it was noted that the monomeric form was not "reproducibly grown". The region of L4-$\beta2$ (approximately amino acids 46–54) corresponds to the effector region of Ras and thus the substantial difference in the structure of this segment (a β-sheet structure in ARF vs. a loop structure in Ras) led the authors to postulate that it may be the ARF effector region. Another prominent feature of the structure, that is not seen in Ras, is the amphipathic amino-terminal helix (that lies in a groove on the protein) followed by a flexible loop. (Ras lacks sequence corresponding to the amino terminus of ARF and thus neither the helix nor the loop is seen in the Ras structure.) A portion of the amino-terminal α-helix appears to form a positively-charged patch that, along with the amino terminal myristoyl moiety, may facilitate the interaction of ARF with phospholipids. Lastly, Gln71, a residue believed critical for GTP hydrolysis (because of its correspondence in the linear sequence to Gln61 in Ras, where it has that function, and the activity of a point mutation Q71L, discussed below), is located ~5Å farther away from the putative γ-phosphate binding site, than it is in Ras. This distance could account for the lack of intrinsic GTPase activity of ARF (Amor *et al.*, 1994).

Functions of specific segments of ARF have been investigated using deletion and point mutants, chimeras with ARF-like proteins (ARLs) that do not activate CTA, and peptides corresponding to regions of the ARF sequence. The amino terminus has received the most attention, based on the fact that in an amino acid sequence alignment of ARF with Ras, ARF appeared to have an amino-terminal extension of 14–16 amino acids. It was, therefore, proposed that the "extra" amino acids might account for the activation of CTA by ARF (Kahn *et al.*, 1992). A deletion mutant lacking the amino-terminal 17 amino acids, ($\triangle$1–17)mARFp, was constructed. It appeared devoid of ARF activity (assayed as stimulation of ADP-ribosylation of $G_{s\alpha}$), but was able to bind nucleotide with high affinity (Kahn *et al.*, 1992). To investigate further the importance of the amino terminus, a chimeric protein was constructed with the first 17 amino acids from human ARF1 (hARF1) and the remaining 164 from *Drosophila* ARL1 (dARL1), an ARF-like protein that lacks ARF activity (see below). This construct was reported to be active in stimulating the ADP-ribosylation of $G_{s\alpha}$ (Kahn *et al.*, 1992). To determine whether the 2–17 sequence was the ARF segment critical for CTA activation, a peptide with the sequence of ARF amino acids 2 to 17, which was predicted to form an amphipathic alpha helix (based on circular dichroism of the peptide in solution and amino acid sequence), was prepared and was found to inhibit ADP-ribosylation of $G_{s\alpha}$ by CTA as well as protein transport through the Golgi (Kahn *et al.*, 1992), where ARF functions in vesicular trafficking (see below). However, the interpretation of experiments with the ARF amino-terminal peptide has more recently been questioned. This peptide and peptides of similar structure appeared to inhibit nonspecifically the activities of PLC and PLD isoforms (Fensome *et al.*, 1994), as well as to damage lipid membranes irreversibly (Fensome *et al.*, 1994; Weidman and Winter, 1994), perhaps by mechanisms related to their amphipathic character.

The function of the ARF amino terminus was also investigated by Hong *et al.* (1994) using an amino-terminal deletion mutant lacking amino acids 2–13 (r△13 ARF1). When assayed by stimulation of CTA-catalyzed ADP-ribosylation of agmatine, r△13ARF1 (synthesized in *E. coli*) was fully active in the absence of added nucleotide; HPLC analysis showed that r△13ARF1 was isolated with GTP and GDP bound (Hong *et al.*, 1994; Randazzo *et al.*, 1994a). Surprisingly, r△13ARF1 from which bound nucleotide had been removed by prolonged dialysis against 7 M urea also activated CTA. In addition ITP was more effective in replacing GTP with r△13ARF1 than it was with intact rARF1 (Hong *et al.*, 1995). The ARF amino terminus is clearly not required for stimulation of CTA NAD:agmatine ADP-ribosyltransferase activity, although it appears to influence the specificity and affinity of nucleotide binding. A role of the amino terminus in nucleotide binding is also consistent with studies on ARF6. ARF6, which compared to other mammalian ARFs lacks four amino acids within its amino terminus, had a higher apparent affinity for GTP than did ARF2, a class I ARF (Price *et al.*, 1992). When synthesized in *E. coli* as an amino-terminal fusion protein with maltose-binding protein (MBP), rARF6 was isolated in the GTP-bound form, with activity independent of exogenous GTP (Welsh *et al.*, 1992; 1994).

Comparisons of r△13ARF1 and [△17]ARF1, an amino-terminal 17 amino acid deletion mutant, in CTA NAD:$G_{s\alpha}$ ADP-ribosyltransferase assays, rather than the NAD:agmatine ADP-ribosyltransferase assay used by Hong *et al.* (1994), led to the conclusion that the amino terminal 13 amino acids are required for ADP-ribosylation of $G_{s\alpha}$ (Randazzo *et al.*, 1994a). Stimulation of CT-catalyzed ADP-ribosylation of $G_{s\alpha}$ by rARF1 was inhibited by r△13ARF1, but not [△17]ARF1. In addition, the GTPase activity of r△13ARF1 was slightly activated by ARF-GTPase-activating protein (ARF-GAP, discussed below), which had no effect on [△17]ARF1 (Randazzo *et al.*, 1994a). It was, therefore, also postulated that amino acids 14–17 interact with cholera toxin and GAP (Randazzo *et al.*, 1994a).

Chimeric proteins based on sequences of hARF1 and hARL1, have proven very useful in defining function and domains. Two chimeras ARF-73-ARL and ARL-73-ARF, separating the cDNA sequences at the end of the ^{67}DVGGQDK^{73} sequence, were synthesized (Zhang *et al.*, 1995c). ARF-73-ARL was ineffective in stimulating CTA-catalyzed NAD:agmatine ADP-ribosyltransferase activity, whereas the opposite construct, ARL-73-ARF, activated CTA. Thus, the carboxy-terminal 60% of ARF contains a domain responsible for the activation of CTA. These constructs were also used to define the phospholipase D activation domain, as described below.

Point mutations have also led to a better understanding of regions important for activity in ARF. Mutations in the regions of yARF1 involved in nucleotide binding and hydrolysis were based on effects of analogous mutations in Ras (Kahn *et al.*, 1995). Mutation of Asp26 (^{24}GL<u>D</u>AAGK^{30} region) to Ala (D26A), Gly (D26G) or Val (D26V) slowed the rate of release of GDP, thus lowering the rate of GTPγS binding. Mutation of Gln71 (^{66}WDVGG<u>Q</u>^{71} region) to Leu (Q71L) (also reported by Tanigawa *et al.*, 1993, Teal *et al.*, 1994 and Zhang *et al.*, 1994), yielded an ARF that bound GTP as readily as native yARF1 and was effectively activated, as evidenced by the isolation of GTP-bound Q71L from transformed yeast. Mutation of Asn126 (^{126}<u>N</u>KQD^{129} region) to Ile (N126I), was lethal in yeast, whereas mutation of Trp66 (^{66}<u>W</u>DVGGQ^{71}

region) to Arg (W66R) resulted in an ARF with less than 1% of the ability to stimulate the ADP-ribosylation of G$_{s\alpha}$, despite having GTPγS bound. (Trp66 has been postulated to be one of the tryptophans responsible for the change in ARF intrinsic fluorescence upon GTP binding; Kahn and Gilman, 1986). Mutation of Gly2 to Ala (G2A) at the amino terminus of ARF resulted in a construct that was not myristoylated, or functional in complementing *arf⁻* gene deletions in yeast (Kahn *et al.*, 1995).

ARF Gene Structure

The six mammalian ARFs, which fall into three classes, are the products of six distinct genes (Tsuchiya *et al.*, 1991). ARF 1, 3, 4, 5 and 6 gene products appear to be ubiquitously expressed in mammalian systems, with quantitative differences related to time of development and tissue source (Tsai *et al.*, 1991b). ARF 2 has been found thus far only in bovine, rabbit, rat and mouse tissues (Bobak *et al.*, 1989; Monaco *et al.*, 1990; Tsuchiya *et al.*, 1989; 1991).

Analysis of class I gene sequences by Lee *et al.* (1992; human ARF1, hARF1), Serventi *et al.* (1993; bovine ARF2, bARF2) and Tsai *et al.* (1991c; human ARF3, hARF3) revealed a conserved arrangement of five exons and four introns, in which each of the consensus GTP-binding domains of ARF is contained in a separate exon except that the codons for the NKQD sequence are divided between exon four and five. The first exon contains only untranslated sequence; the translation initiation sequence is contained in exon two, which also encodes the GTP-hydrolysis (phosphate-binding) sequence, GLDAAGK. Exon three encodes the DVGG sequence and exon five contains the carboxy-terminal coding sequence as well as a long 3′ untranslated region with two polyadenylation sites (Figure 10.3).

The promoter regions of class I ARFs have characteristics similar to those of housekeeping genes (Haun *et al.*, 1993b). The hARF1 promoter, which lacks TATA and CAAT boxes, contains multiple transcription initiation sites and several poten-

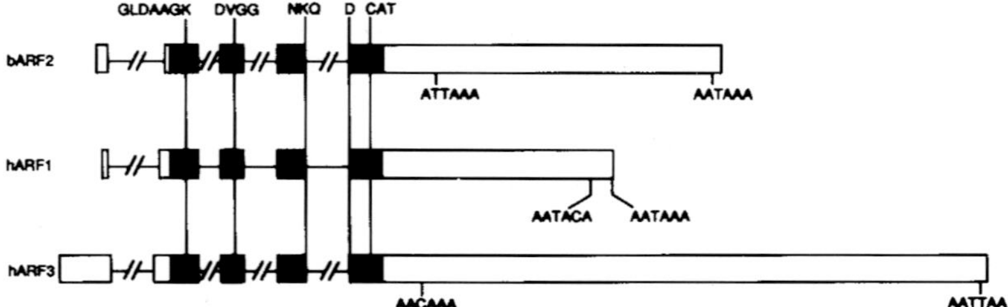

Figure 10.3 Mammalian ARF Class I Gene Structure. Bovine ARF2 (bARF2; Serventi *et al.*, 1993), human ARF1 (hARF1; Lee *et al.*, 1992) and human ARF3 (hARF3; Tsai *et al.*, 1991c) are aligned schematically to show structural features. Untranslated regions are represented by open boxes, coding regions by shaded boxes and introns by horizontal lines. Positions of consensus sequences for GTP binding and hydrolysis are shown by vertical lines. Sequences of potential polyadenylation signals are noted below each gene. The figure is from Serventi *et al.* (1993).

tial binding regions for Sp1 (Lee *et al.*, 1992). The bARF2 promoter differs, in that it contains six inverted CCAATT motifs (Serventi *et al.*, 1993). The hARF3 promoter lacks TATA and CAAT boxes, but has several GC boxes, multiple transcription initiation sites and, in addition, contains a palindromic sequence, shown by mutational analysis to be important for transcriptional regulation, that binds a protein from nuclear extracts (Haun *et al.*, 1993b).

The dARF1 gene structure was very similar, including the placement of introns, to those of mammalian class I ARF genes (Murtagh *et al.*, 1993); the dARF1 gene was mapped to the left arm of chromosome III. The yARF1 and yARF2 genes are located on chromosome IV (Stearns *et al.*, 1990a), whereas yARF3 is located on chromosome XV (Lee *et al.*, 1994a). *Giardia* ARF1 appears to be a single copy gene, lacking introns, with a short 5'-untranslated region of six nucleotides (Murtagh *et al.*, 1992).

Regulation of ARF Gene Expression

Investigations into the regulation of ARF gene expression have focused on regulation during development and regulation in response to exogenous stimuli. As class I ARF levels are generally higher in brain than in non-neural tissues, rat brain ARF1 and ARF3 mRNA and proteins were evaluated during postnatal development (Tsai *et al.*, 1991b). ARF 1 and ARF 3 proteins (sARFI and sARFII, respectively) were present in similar amounts on the second postnatal day, whereas by the 10th postnatal day ARF3 predominated. Over a period of 27 days, the level of ARF3 mRNA increased, while ARF2 and ARF4 levels decreased and ARF1, ARF5 and ARF6 mRNA levels were unchanged. At least two distinct ARF4 mRNAs, one considerably shorter than the other, were found in rat testis, but not in other tissues (Mishima *et al.*, 1992). The presence of short hARF4 mRNA was consistent with the presence of multiple polyadenylation sites and was postulated to occur in the late stages of spermatogenesis coincident with the cessation of transcription.

Duman *et al.* (1990) assessed the expression of ARF mRNA and protein levels in rat cerebral cortex in response to steroid administration. Corticosterone increased levels of ARF1 and ARF3 mRNA and protein, whereas adrenalectomy lowered the levels of ARF1 and ARF3 mRNA, suggesting that rat brain ARF1 and ARF3 expression (*in vivo*) is under the control of endogenous glucocorticoids. ARFs in lower eukaryotes may also be influenced by exogenous stimuli; levels of yARF3 were controlled by a glucose-repressible promoter (Lee *et al.*, 1994b)

ARF-related Proteins (ARLs and ARD)

ARF-related proteins are either 40–50 percent identical in sequence to ARFs or contain sequence that is very similar to that of ARFs plus additional protein sequence(s) peripheral to the ARF domain. ARF-like proteins (ARLs) were first identified in *Drosophila* by cDNA cloning (Tamkun *et al.*, 1991). The deduced amino acid sequence of dARL1 was ~55% identical to those of mammalian and yeast ARFs, but the protein did not stimulate CT-catalyzed ADP-ribosylation of $G_{s\alpha}$, despite binding GTP. These characteristics, along with the inability of the proteins to rescue the yeast *arf1⁻arf2⁻* lethal mutant, have become the criteria for defining an ARL. ARL was shown to be an essential protein in *Drosophila* as its deletion proved lethal (Tamkun *et al.*, 1991). Human ARL2 and *Drosophila* ARL2 were identified by Clark *et al.* (1993), who reported their seemingly ubiquitous distribution in human and fly tissues, respectively, with no apparent differences in dARL2 levels during *Drosophila* development. Human ARL3 and rat ARL3 were identified by Cavenagh *et al.* (1994); interestingly hARL3 protein was detected in several cell lines (derived from solid tumors) by immunoblotting. Rat ARL4 was identified by Schürmann *et al.* (1994), who noted that rat ARL4 mRNA was detectable in mouse 3T3-L1 fibroblasts on day 6 after the start of differentiation to the adipocyte phenotype. In normal rats, ARL4 mRNA was detected in adipose tissue, heart, skeletal muscle and brain. The known ARL sequences are aligned in Figure 10.4 and although a class system similar to that for mammalian ARFs has not been described for the ARLs, there appear to be at present four distinct types of ARLs.

A non-ARL, ARF-related protein identified by cDNA cloning was termed an <u>ARF</u><u>domain</u> protein, or ARD (Mishima *et al.*, 1993; GenBank Accession # L04510). Based on deduced amino acid sequence, human ARD1 (hARD1) is a 64-kDa protein with 18 kDa of the carboxy-terminus corresponding closely to a hARF1 that

```
hARL1      .G.FFS..FS S.-.FGT... M...I....G .......Y.. QVGEVV.TI.   47
rARL1      .G.FFS..FS S.-.FGT... M...I....G .......Y.. QVGEVV.TI.   47
dARL1      .G.V...YFR G.-.LGS... M...I....G .......Y.. QVGEVV.TI.   46
hARL2      .-.L..T.LK KMN.QKE... L.L.M....N .......KKF NGEDID.IS.   46
dARL2      .-.F..TVLK KMR.QKE... M...L....N .......K.F NGEPID.IS.   46
hARL3      .-.L....LR K.KSAPD.Q. V...L....N .....L.KQ. ASEDISHIT.   47
rARL3      .-.L....LR K.KSAPD.Q. V...L....N .....L.KQ. ASEDISHIT.   47
rARL4      .GNG.SDQTS I.SSLPSFQS FH.VI....C .....V.Y.. QFNEFVNTV.   50

Consensus  M.G.L-SI.. .L.-...-RE .RIL.LGLD. AGKTTIL.RL ......T..P   50


hARL1      .I...VET.. ....TYKNL. FQ...L...T .......C.Y S...AV....   92
rARL1      .I...VET.. ....TYKNL. FQ...L...T .......C.Y S...AV....   92
dARL1      .I...VEQ.. ....TYKNL. FQ...L...T .......C.Y S...AI....   91
hARL2      .L...IKTL. ....EHRGF. LNI..V...K .L.S...N.F ES..GL.W..   91
dARL2      .L...IKTL. ....EHNGYT LNM..V...K .L.S...N.F ES..GLVW..   91
hARL3      .Q...IKS.. ....QSQGF. LN...I...R K.....KN.F E...IL...I   92
rARL3      .Q...IKS.. ....QSQGF. LN...I...R K......S.F E...IL...I   92
rARL4      .K...TEKIK VTLGNSKTVT FHF..V...E KL..L.KS.T RC..GIVF..  100

Consensus  T.GFN...V- ----.....K ..VWD.GGQ. SIRPYWR.Y. .NTD..IYVV  100


hARL1      ..C..D.IGI SKS..VAM.E ..E.RK.I.V .......MEQ .MTSS.MANS  142
rARL1      ..C..D.IGI SKS..VAM.E ..E.RK.I.V .......MEQ .MTPS.MANA  142
dARL1      .....D.IGI SKD..LYM.R ..E.AG.I.V .L.....MDG CMTVA.VHHA  141
hARL2      .....Q.MQD CQR..QSL.V ..R.AG.T.. I.......PG .LSSNA.REA  141
dARL2      .....M.LES CGQ..QVL.Q ..R.AG.T.. .LC.....PG .LSSN..KEI  141
hARL3      .....K.FEE TGQ..AEL.E ..K.SCVPV. I.......LT .APAS..AEG  142
rARL3      .....K.FEE TGQ..TEL.E ..K.SCVPV. ........LT .APAA..AEG  142
rARL4      ..V.VE.MEE AKT..HKITR ISENQGVPV. IV......RN SLSLS..EKL  150

Consensus  DSADR.R... ...EL...L. EE.L..A.LL VFANKQDL.. A....EI...  150


hARL1      .G.PALKD.K ....FKT..T K.T..DEA.E ..VETLKS.Q .........--  181
rARL1      .G.PALKD.K ....FKT..T K.T..DEA.E ..VETLKS.Q .........--  181
dARL1      .G.ENLKN.T .F..FKT..T K....DQA.D ..SNTLQS.K .........--  180
hARL2      .E.DSIRSHH ..C.QGC..V T..N.LPGID ..LDDISS.- .....IFTAD  184
dARL2      .H.EDITTHH ..LVAGV..V T..K.LSS.D ..IADIAK.- .....IFTLD  184
hARL3      .N.HTIRD.V ....QSC..L T...VQDG.N .VCKNVNAK- ........KK  182
rARL3      .N.HTIRD.V ....QSC..L T...VQDG.N .VCKNVNAK- ........KK  182
rARL4      .AMGELSSST P.HLQPTC.I I.D..KEGLE K.HDMIIK.R KMLRQQKKKR  200

Consensus  L.L.....R. -WQI...SA. .GEGL...M. WL......R. ---------..  200
```

Figure 10.4   Alignment of ARL Deduced Amino Acid Sequences. ARLs were aligned as described for Figure 10.2. References and database accession numbers are found in Table 10.1.

lacks amino acids 1 through 15. Recombinant hARD1 stimulated CTA-catalyzed ADP-ribosylation of agmatine, in a GTP-dependent manner. Stimulation was observed in the presence of Tween-20, but not with DMPC/cholate or other lipids often used in the CTA assay. ARD1 mRNA was detected in rat skeletal muscle, testis, spleen, kidney, liver, lung, heart and brain as well as human fibro-blasts and neuro-blastoma cells, but not in HL-60 cells (Mishima *et al.*, 1993). It is not known whether ARD1 protein is present in any of the cells in which the mRNA was detected.

THE ROLE OF ARF IN VESICULAR TRAFFICKING

Association of ARF With Intracellular Structures

To determine the function of ARF, a yeast ARF1 *null* mutant strain was generated by Stearns *et al.* (1990a; 1990b). The phenotype of the mutant yeast suggested a defect in the secretory pathway and incomplete glycosylation of secreted invertase was later shown, as well as limited accumulation of invertase in the cells. Immunofluorescence and immunoelectron microscopy of NIH-3T3 cells showed that ARF1 was partly cytosolic and partly localized to the Golgi, primarily on the cytoplasmic surface of the *cis*-Golgi (Stearns *et al.*, 1990b). Localization of ARF1 to Golgi was GTPγS-dependent (Donaldson *et al.*, 1991; Regazzi *et al.*, 1991).

Brefeldin A (BFA), a fungal metabolite previously shown to inhibit intracellular transport of secretory proteins (Misumi *et al.*, 1986) and cause disintegration of the Golgi as well as enlargement of the ER (Lippincott-Schwartz *et al.*, 1989; 1990), was used by Donaldson *et al.* (1991) to assess the involvement of ARF1 in protein secretion. Prior treatment of isolated Golgi membranes with BFA decreased the amount of ARF1 bound. Removal of BFA from normal rat kidney cells allowed reassociation of ARF1 with the Golgi, thus BFA inhibition was reversible (Donaldson *et al.*, 1991). In related studies, two distinct pools of Golgi membrane-bound ARF were demonstrated using phosphatidylcholine (PC) liposomes to extract ARF from Golgi membranes (Helms *et al.*, 1993). Approximately 60% of bound ARF was extracted into the liposomes, whereas the remaining, more tightly associated ARF appeared to be in a saturable pool, perhaps reflecting a specific membrane-binding site (Helms *et al.*, 1993).

The association of different classes of ARF with intracellular structures has been investigated by several groups and is consistent with the idea that different classes of ARF have different functions. Binding of ARFs 1, 3 and 5 to Golgi and other membranes was investigated using isolated Golgi membranes and other subcellular fractions (Tsai *et al.*, 1992). Differential interaction of native ARFs 1, 3 and 5 with membrane fractions enriched in specific organelles revealed that ARF1 associated (mostly) with microsomes and Golgi, whereas ARF3 appeared to be less specifically located and was found in fractions containing Golgi, microsomes, mitochondria and plasma membrane; ARF5 also bound predominantly to Golgi fractions. Interaction of ARFs 1 and 3, but not ARF5 with Golgi was inhibited by BFA (Tsai *et al.*, 1993). In separate experiments, a requirement was demonstrated for myristoylation of recombinant hARF5 in its GTP-dependent association with Golgi (Haun *et al.*, 1993a), a requirement currently presumed for the *in vivo* activities of all ARFs.

ARF6 is the only mammalian ARF for which the native protein has not been detected, despite the presence of the mRNA. Overexpressed recombinant hARF6 localized to the cell periphery, specifically, the plasma membrane and some internal punctate structures (Peters *et al.*, 1995; D'Souza-Schorey *et al.*, 1995). Peters *et al.* (1995) noted that the tubulovesicular internal structures labeled by ARF6 were characteristic of early endosomes. Unlike ARF1 association with the Golgi, the membrane localization of wild-type ARF6 was not affected by BFA (10 μg/ml) after 1 h (Peters *et al.*, 1995). Both Peters *et al.* (1995) and D'Souza-Schorey *et al.* (1995) used mutant ARF6(Q67L), a mutation analogous to ARF1(Q71L) used by Tanigawa *et al.* (1993), Teal *et al.* (1994), Zhang *et al.* (1994) and Kahn *et al.* (1995), which prevents hydrolysis of bound GTP and ARF6(T27N), a mutation that decreases GTP binding, to study aspects of ARF6 function. Peters *et al.* (1995) noted that expression of the Q67L mutant resulted in increased plasma membrane surface area more than the slight increase seen with expression of wild-type ARF6, with the mutant protein diffusely distributed at the plasma membrane.

Effects of overexpression of ARF6 and ARF6 mutants on transferrin receptor localization (D'Souza-Schorey *et al.*, 1995; Peters *et al.*, 1995) and cell-surface transferrin binding and uptake (D'Souza-Schorey *et al.*, 1995) were also examined. ARF6(Q67L) and wild-type ARF6 increased the number of cell surface transferrin receptors and the amount of cell surface-bound transferrin. Overexpression of ARF6(T27N) resulted in the localization of the ARF to internal structures, in both large aggregates and punctate structures. Cell surface-bound transferrin was decreased and ARF6(T27N) and transferrin receptors were redistributed to internal structures (D'Souza-Schorey *et al.*, 1995; Peters *et al.*, 1995). Both groups concluded that their data are consistent with the involvement of ARF6 in an endocytic pathway, where hydrolysis of GTP bound to ARF 6 may facilitate membrane budding in early endosomes.

Exchange Factor and GTPase-activating Protein

The dissociation of GDP from and the subsequent binding of GTP to, ARF is necessary for ARF function. In cells, the exchange of GDP for GTP is postulated to occur via the action of a guanine nucleotide-exchange factor or protein (GEF or GEP). (Lipids, especially acidic phospholipids, also play a role in nucleotide exchange *in vitro* (Terui *et al.*, 1994).) Activated ARF-GTP may interact with a GTPase-activating protein (GAP) that promotes hydrolysis of bound GTP to GDP, thus producing inactive ARF-GDP. The GDP-GTP cycle is analogous to that of other guanine nucleotide-binding proteins such as Ras and has been discussed in numerous reviews (*e.g.*, Boguski and McCormick, 1993) (Figure 10.5).

Two groups reported that Golgi membranes catalyzed GTP binding to ARF and that BFA blocked GTP for GDP exchange and subsequent binding of ARF to Golgi

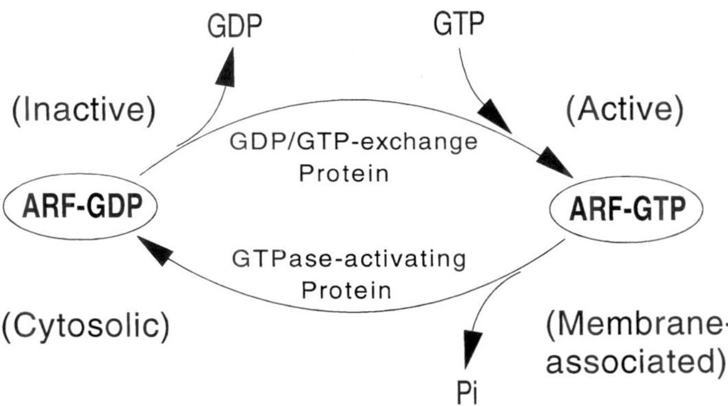

Figure 10.5   Model for the Regulation of Nucleotide Binding to ARF *In Vivo*. ARF-GDP, the inactive form, normally found in the cytosol, is acted upon by an ARF-specific guanine nucleotide exchange protein to form the active ARF-GTP, which becomes associated with membrane lipids. Action of an ARF GTPase-activating protein causes ARF to hydrolyze GTP to GDP and inactive ARF-GDP is released back to the cytosol.

membranes (Donaldson *et al.*, 1992a; Helms and Rothman, 1992). The Golgi-associated exchange factor was protease-sensitive; amino-terminal myristoylation of ARF was not required for activation by the Golgi factor, but was required for association of the active ARF with membrane (Randazzo *et al.*, 1993). Protease-insensitive activation of ARF was also observed in Golgi preparations and was presumed to result from the activation of ARF by lipids (Randazzo *et al.*, 1993).

More recently, a soluble ARF-GEP was identified Tsai *et al.* (1994). The partially purified protein increased the rate of nucleotide exchange on ARFs 1 and 3, but did not affect GTP affinity. Crude GEP behaved as a protein of ~700 kDa and was sensitive to BFA. The partially purified GEP was apparently an ~60 kDa protein and was insensitive to the effects of BFA, consistent with the notion that BFA does not act on GEP itself.

ARF activation was also facilitated by phosphatidylinositol 4,5-bisphosphate (PIP$_2$) and other acidic phospholipids (Terui *et al.*, 1994). PIP$_2$ (400 μM) enhanced GDP-dissociation and thus appeared to stabilize the nucleotide-free form of ARF. At the same concentration, it also inhibited GTP association (Terui *et al.*, 1994). The amino-terminal deletion mutant [$\triangle$1-17]ARF1 (described above) was also sensitive to the effects of PIP$_2$ and thus it was concluded that the amino terminus, which was required for several ARF activities, is not required for interaction with PIP$_2$ (Terui *et al.*, 1994).

ARF-GAP activity identified in and partially purified from a bovine brain particulate fraction, was activated by PIP$_2$ and to a lesser extent by phosphatidic acid (PA), but was inhibited by phosphatidylcholine (PC) (Randazzo and Kahn, 1994). Based on experiments with amino-terminal deletion mutants, the ARF amino terminus including amino acids 14–17 was postulated to be essential for the interaction of ARF1 with GAP (Randazzo *et al.*, 1994). Further characterization of a purified ARF-GAP was reported by Makler *et al.* (1995); the partially-purified ARF-GAP of ~49 kDa, was stimulated by PIP$_2$ and inhibited by PC, similar to the findings of Randazzo and Kahn (1994).

ARFs as Regulators of Vesicle Formation

ARF was implicated in the control of vesicular trafficking (in the ER and Golgi) when disruption of the yARF1 gene resulted in defective invertase secretion and accumulation in the cell of incompletely glycosylated invertase (Stearns *et al.*, 1990b); deletion of both yARF1 and yARF2 was lethal (Stearns *et al.*, 1990a). Deletion of yARF3 did not affect viability and thus yARF3 was postulated not to be involved in protein trafficking in the Golgi (Lee *et al.*, 1994b). ARF1 has also been implicated in regulation of nuclear vesicle dynamics (Boman *et al.*, 1992) and endosome-endosome fusion (Lenhard *et al.*, 1992; Spiro *et al.*, 1995). In *Saccharomyces cerevisiae*, ARF1 was localized to the Golgi (in agreement with Stearns *et al.*, 1990b; Donaldson *et al.*, 1991; Regazzi *et al.*, 1991) and shown to be required for an early step in ER to Golgi transport (Balch *et al.*, 1992). Subsequently, ARF was found associated with coat proteins (COPs; originally described by Serafini *et al.*, 1991a and Waters *et al.*, 1991) in the coatomer complex on the surface of non-clathrin coated vesicles derived from the Golgi (Serafini *et al.*, 1991b). ARF-GTP was required for the binding of β-COP to Golgi-derived vesicles (Donaldson *et al.*, 1992b; Palmer *et al.*, 1993). ARF and coatomer were the only cytosolic proteins required for budding and formation of Golgi-derived COP-coated vesicles (Orci *et al.*, 1993). The constitutively activated mutant ARF1(Q71L) prevented uncoating of COP-coated vesicles, made vesicle formation resistant to BFA (analogous to GTPγS, Donaldson *et al.*, 1991) and inhibited Golgi membrane traffic (Tanigawa *et al.*, 1993; Teal *et al.*, 1994; Zhang *et al.*, 1994). In the presence of GTP, AlF$_4^-$ also promoted the stable ARF-dependent association of coatomer with Golgi, similar to GTPγS or activating mutants, but only for a sub-fraction of membrane-associated ARF (Finazzi *et al.*, 1994). This was especially interesting because ARF (like other ~20-kDa guanine nucleotide-binding proteins) is not activated by AlF$_4^-$ (Kahn, 1991).

The model for ARF1 action in vesicular trafficking at the Golgi, originally proposed by Serafini *et al.* (1991b) suggested that ARF-GTP is formed and associates with a donor membrane, after which COPs can bind to initiate vesicle formation. COP recruitment continues, a vesicle eventually buds and finally (in a process that was later shown to require the addition of fatty acyl-CoA) membrane fusion occurs at the base of the bud to form a vesicle. After translocation to a target membrane, the vesicle becomes uncoated perhaps following GTP hydrolysis by ARF and release of ARF-GDP and can then fuse with the membrane. Data from many sources support this model of the role of ARF1 in ER and Golgi vesicular transport. Since the initial proposal of this model, the involvement of numerous additional molecules, crucial to vesicular formation, release, targeting and docking, is continuing to be recognized (reviewed by Donaldson and Klausner, 1994; Rothman, 1994; Nuoffer and Balch, 1994).

THE ROLE OF ARF IN THE ACTIVATION OF PLD

Phosphatidylcholine-specific phospholipase D hydrolyzes phosphatidylcholine to produce choline and phosphatidic acid (PA). PA demonstrates mitogenic activity; its product, diacylglycerol (DAG), is an activator of protein kinase C (PKC) (reviewed by Exton, 1994). Partially pure PLD and PLD in permeabilized cells has been shown to be regulated by a number of factors (Cockcroft, 1992); among the first of many described were guanine nucleotides and divalent cations (Anthes *et al.*, 1989). An early clue to the nature of GTPγS activation of PLD was provided by Anthes *et al.* (1991) who showed that a cytosolic factor was required for GTP γS-dependent activation of PLD. Attempts to characterize the cytosolic factor responsible for the stimulation of membrane-based PLD activity, resulted in the identification of ARF1 (Brown *et al.*, 1993) and ARFs 1 and 3 (Cockcroft *et al.*, 1994) as activators of PLD. Native ARF1 and recombinant hARF1 (rARF1) (Cockcroft. *et al.*, 1994), as well as myristoylated recombinant hARF1 (mARF1) (Brown *et al.*, 1993) stimulated PLD activity, with mARF1 being more effective than rARF1 (Brown *et al.*, 1993). Other studies demonstrated that members of all three classes of ARF stimulated a partially-purified rat brain PLD and the myristoylated proteins were more effective than the non-myristoylated (Massenburg *et al.*, 1994). Studies using partially-purified PLD (Brown *et al.*, 1993, Cockcroft *et al.*, 1994, Massenburg *et al.*, 1994) and others using permeabilized cells (Liscovitch *et al.*, 1994; Pertile *et al.*, 1995), noted that PLD required PIP$_2$.

Attempts to purify and characterize ARF-sensitive PLD led to its separation from an oleate-sensitive PLD during chromatography on heparin-Sepharose or heparin 5PW HPLC (Massenburg *et al.*, 1994), confirming the existence of different isoforms of mammalian PLD. The BFA-sensitivity of Golgi membrane-derived PLD (Ktistakis *et al.*, 1995) is consistent with ARF-sensitivity as well as with a role for PLD in vesicular trafficking. The action of PLD in the Golgi could contribute to the asymmetric distribution of individual phospholipids in the bilayer membrane, which, in turn, might influence vesicle formation or fusion as discussed by Devaux (1991) and Eastman *et al.* (1992). Interestingly, Liscovitch *et al.* (1994) have proposed a positive feedback loop involving PIP$_2$ and PLD control of vesicular trafficking in the Golgi. In this model, PLD resides on the acceptor membrane and is activated by ARF after docking of the transport vesicle. Activation of PLD results in the production of PA, which in turn, stimulates phosphatidylinositol-4-phosphate 5-kinase (present on the transport vesicle) to produce PIP$_2$ from PIP. The increase in PIP$_2$ further activates PLD and results in a rapid change in membrane lipid composition. The change in membrane lipid composition and the potential activation of ARF-GAP by the increased levels of PIP$_2$ and PA, could accelerate release of ARF-GDP followed by coatomer proteins and, overall, could contribute to vesicle fusion at the acceptor membrane. Recently a partially purified 95-kDa ARF-sensitive PLD, which retained the requirement for PIP$_2$, was identified in porcine brain (Brown *et al.*, 1995), suggesting that the requirement is a property of the PLD itself and is not due to a factor that loosely associates with PLD. Further characterization of this PLD isoform revealed that it was activated by members of all three mammalian ARF classes and was partially activated by yARFs (but only when myristoylated) and was not activated by the ARF-like proteins, ARL2 and ARL3 (Brown *et al.*, 1995).

Several reports indicate that ARF is not the only stimulator of ARF-sensitive PLD and that cellular factors such as Rho A or cdc42 (Bowman *et al.*, 1993; Ohguchi *et al.*, 1995; Singer *et al.*, 1995; Siddiqi *et al.*, 1995) synergistically activate ARF-sensitive PLD activity. Siddiqi *et al.* (1995) showed that ARF-sensitive PLD may be a cytosolic protein, whereas a membrane-associated PLD is sensitive to rho A and cdc42 (Siddiqi *et al.*, 1995). An additional synergistic activator of PLD, with a molecular weight of 50,000 (Bourgoin *et al.*, 1995; Lambeth *et al.*, 1995) supports evidence for the interaction of ARF with a number of proteins that "sensitize" PLD to the action of ARF or vice versa. At least one other (uncharacterized) cellular factor also appears to stimulate the same PLD, but in a GTP- or ARF-independent manner (Singer *et al.*, 1995). With the identification of other activators of mammalian ARF-sensitive PLD (PMA and Mg-ATP — Whatmore *et al.*, 1994) and of uncharacterized mammalian PLDs (*e.g.*, EGF — Yeo and Exton, 1995; PMA — Olson *et al.*, 1991; Ras — Jiang *et al.*, 1995) it is likely that ARF, ARF activation and ARF activation of PLD are under the control of several systems within cells.

Localization of an ARF domain responsible for the activation of ARF-sensitive PLD was accomplished through the use of ARF-ARL and ARL-ARF chimeric molecules (that were used also to delineate the structural requirements for ARF activation of CTA). The ARF-73-ARL construct, which did not activate CTA, did activate PLD in a GTP-dependent manner, whereas, ARL-73-ARF activated CTA, but not PLD (Zhang *et al.*, 1995c). These data are consistent with the hypothesis that a domain required for PLD activation lies amino terminal to position 73 in ARF. Data from studies of r△13ARF1 (in this laboratory) also indicated that the amino terminus is involved in the activation of PLD (unpublished data). More precise localization of the PLD interaction domain of ARF should be possible using similar chimeric constructs and deletion mutants.

SUMMARY

It is well known that the ADP-ribosyltransferase activity of cholera toxin in intestinal epithelium results in the ADP-ribosylation of $G_{s\alpha}$, the stimulation adenylyl cyclase and abnormalities in fluid and electrolyte flux. The search for a cholera toxin counterpart in eukaryotic systems led to the discovery of mammalian ADP-ribosyltransfersases and with it the possibility that mono-(ADP-ribosylation) may play a role in normal cell physiology. The further discovery of a family of evolutionarily conserved cholera toxin activators, ARFs, has sparked a whole new realm of research on cellular processes that are clearly unrelated to ADP-ribosylation and increases in cAMP concentrations. ARFs appear not to be activators of the endogenous transferases, but instead have been shown to function in processes that all seemingly relate to the regulation of membrane vesicle formation and fusion at numerous locations within the cell. The existence of different classes of ARF proteins apparently reflects the regulation of ARF-controlled membrane traffic on several levels (*i.e.*, the amount of specific ARF and thus specific ARF effector domain) present at a given location, as well as the activation state of that ARF. It will be important to learn whether (systematic/detailed) comparisons of several ARF structures can yield clues to the functions of specific ARFs at defined locations, whether cholera toxin exploits active ARF in cells and whether the cholera toxin interaction site has a physiologic function in cells.

REFERENCES

Abood, M.E., Hurley, J.B., Pappone, M-C., Bourne, H.R. and Stryer, L (1982) Functional homology between signal-coupling proteins. *J. Biol. Chem.*, **257**, 10540–10543.

Amor, J., Harrison, D., Kahn, R. and Ringe, D. (1994) Structure of the human ADP-ribosylation factor 1 complexed with GDP. *Nature*, **372**, 704–708.

Anthes, J.C., Eckel, S., Siegel, M.I., Egan, R.W. and Billah, M.M. (1989) Phospholipase D in homogenates from HL-60 granulocytes: Implications of calcium and G protein control. *Biochem. Biophys. Res. Comm.*, **163**, 657–664.

Anthes, J.C., Wang, P., Siegel, M.I., Egan, R.W. and Billah, M.M. (1991) Granulocyte phospholipase D is activated by a guanine nucleotide-dependent protein factor. *Biochem. Biophys. Res. Comm.*, **175**, 236–243.

Balch, W.E., Kahn, R.A. and Schwaninger, R. (1992) ADP-ribosylation factor is required for vesicular trafficking between the endoplasmic reticulum and the cis-Golgi compartment. *J. Biol. Chem.*, **267**, 13053–13061.

Barbacid, M. (1987) *ras* Genes. *Ann. Rev. Biochem.*, **56**, 779–827.

Barbieri, J.T. and Cortina, G. (1988) ADP-ribosyltransferase mutations in the catalytic S-1 subunit of pertussis toxin. *Infect. Immun.*, **56**, 1934–1941.

Barbieri, J.T., Mende-Mueller, L.M., Rappuoli, R. and Collier, R.J (1989) Photolabeling of Glu-129 of the S-1 subunit of pertusis toxin with NAD. *Infect. Immun.*, **57**, 3549–3554.

Beubler, E., Kollar, G, Saria, A, Bukhave, K and Rask-Madsen, J (1989) Involvement of 5-hydroxytryptamine, prostaglandin E_2 and cyclic adenosine monophosphate in cholera toxin-induced fluid secretion in the small intestine of the rat *in vivo*. *Gasteroenterology*, **96**, 368–376.

Bobak, D.A., Bliziotes, M.M., Noda, M., Tsai, S-C., Adamik, R. and Moss, J. (1990) Mechanism of activation of cholera toxin by ADP-ribosylation factor (ARF): Both low- and high-affinity interactions of ARF with guanine nucleotides promote toxin activation. *Biochemistry*, **29**, 855–861.

Bobak, D.A., Nightingale, M.S., Murtagh, J.J., Price, S.R., Moss, J. and Vaughan, M. (1989) Molecular cloning, characterization and expression of human ADP-ribosylation factors: Two guanine nucleotide-dependent activators of cholera toxin. *Proc. Natl. Acad. Sci. USA*, **86**, 6101–6105.

Boguski, M.S. and McCormick, F. (1993) Proteins regulating Ras and its relatives. *Nature*, **366**, 643–654.

Boman, A.L., Taylor, T.C., Melançon, P. and Wilson, K.L. (1992) A role for ADP-ribsoylation factor in nuclear vesicle dynamics. *Nature*, **358**, 512–514.

Botstein, D., Segev, N., Stearns, T., Hoyt, M.A., Holden, J. and Kahn, R.A. (1988) Diverse biological functions of small GTP-binding proteins in yeast. *Cold Spring Harbor Symposia on Quantitative Biology*, **53 (pt2)**, 629–636.

Bourgoin, S., Harbour, D., Desmarais, Y., Takai, Y. and Beaulieu, A. (1995) Low molecular weight GTP-binding proteins in HL-60 granulocytes: Assessment of the role of ARF and of a 50-kDa cytosolic protein in phospholipase D activation. *J. Biol. Chem.*, **270**, 3172–3178.

Bowman, E.P., Uhlinger, D.J. and Lambeth, J.D. (1993) Neutrophil phospholipase D is activated by a membrane-associated Rho family small molecular weight GTP-binding protein. *J. Biol. Chem.*, **268**, 21509–21512.

Brown, H.A., Gutowski, S., Kahn, R.A. and Sternweis, P.C. (1995) Partial purfication and characterization of ARF-sensitive phospholipase D from porcine brain. *J. Biol. Chem.*, **270**, 14935–14943.

Brown, H.A., Gutowski, S., Moomaw, C.R., Slaughter, C. and Sternweis, P.C. (1993) ADP-ribosylation factor, a small GTP-dependent regulatory protein, stimulates phospholipase D activity. *Cell*, **75**, 1137–1144.

Burnette, W.N., Cieplak, W. Jr., Kaslow, H.R., Rappuoli, R., Tuomanen, E.I. (1994) Recombinant microbial ADP-ribosylating toxins of *Bordetella pertussis*, *Vibrio cholerae*, and enterotoxigenic *Escherichia coli*: Structure, function and toxoid vaccine development. *Bioprocess Tech.*, **19**, 185–203.

Burnette, W.N., Mar, V.L., Platler, B.W., Schlotterbeck, J.D., McGinley, M.D., Stoney, K.S., Rohde, M.F. and Kaslow, H.R. (1991) Site-specific mutagenesis of the catalytic subunit of cholera toxin: Substituting lysine for arginine 7 causes loss of activity. *Infect. Immun.*, **59**, 4266–4270.

Burnette, J.T., Mende-Mueller, L.M., Rappouli, R. and Collier, R.J. (1989) Pertussis toxin S1 mutant with reduced enzyme activity and a conserved protective epitope. *Science*, **242**, 72–74.

Burns, D.L., Moss, J. and Vaughan, M. (1982) Choleragen-stimulated release of guanyl nucleotides from turkey erythrocyte membranes. *J. Biol. Chem.*, **257**, 32–34.

Carroll, S.F. and Collier, R.J. (1984) NAD binding site of diptheria toxin: Identification of a residue within the nicotinamide subsite by photochemical modification with NAD. *Proc. Natl. Acad. Sci. USA*, **81**, 3307–3311.

Carroll, S.F. and Collier, R.J. (1987) Active site of *Pseudomonas aeruginosa* exotoxin A. *J. Biol. Chem.*, **262**, 8707–8711.

Cassel, D. and Selinger, Z. (1977) Mechanism of adenylate cyclase activation by cholera toxin: Inhibition of GTP hydrolysis at the regulatory site. *Proc. Natl. Acad. Sci. USA*, **74**, 3307–3311.

Cavenagh, M., Breiner, M., Schürmann, A., Rosenwald, A., Terui, T., Zhang, C-J., Randazzo, P., Adams, M., Joost, H. and Kahn, R. (1994) ADP-ribosylation factor (ARF)-like 3, a new member of the ARF family of GTP-binding proteins cloned from human and rat tissues. *J. Biol. Chem.*, **269**, 18937–18942.

Chang, Y-C., Soman, G. and Graves, D.J. (1986) Identification of an enzymatic activity that hydrolyzes protein-bound ADP-ribose in skeletal muscle. *Biochem. Biophys. Res. Comm.*, **139**, 932–939.

Cieplak, Jr, W., Burnette, W.N., Mar, V.L., Kaljot, K.T., Morris, C.F., Chen, K.K., Sato, H. and Keith, J.M. (1988) Identification of a region in the S1 subunit of pertussis toxin that is required for enzymatic activity and that contributes to the formation of a neutralizing antigenic determinant. *Proc. Natl. Acad. Sci. USA*, **85**, 4667–4671.

Clark, J., Moore, L., Kraninskas, A., Way, J., Battey, J., Tamkun, J. and Kahn, R.A. (1993) Selective amplification of additional members of the ADP-ribosylation factor (ARF) family: Cloning of additional human and *Drosophila* ARF-like genes. *Proc. Natl. Acad. Sci. USA*, **90**, 8952–8956.

Cockcroft, S. (1992) G-protein-regulated phospholipases C, D and A_2-mediated signalling in neutrophils. *Biochim. Biophys. Acta*, **1113**, 135–160.

Cockcroft, S., Thomas, G.M.H., Fensome, A., Geny, B., Cunningham, E., Gout, I., Hiles, I., Totty, N.F., Truong, O. and Hsuan, J.J. (1994) Phospholipase D: A downstream effector of ARF in granulocytes. *Science*, **263**, 523–526.

Cuatrecasas, P. (1973) Interaction of *Vibrio cholerae* enterotoxin with cell membranes. *Biochemistry*, **12**, 3547–3558.

Dallas, W.S. and Falkow, S. (1980) Amino acid sequence homology bewteen cholera toxin and *Escherichia coli* heat-labile toxin. *Nature*, **288**, 499–501.

Dertzbaugh, M.T., Peterson, D.L., Macrina, F. (1990) Cholera toxin B-subunit gene fusion: Structural and functional analysis of the chimeric protein. *Infect. Immun.*, **58**, 70–79.

Devaux, P. (1991) Static and dynamic lipid asymmetry in cell membranes. *Biochemistry*, **30**, 1163–1173.

Dixon, S.J., Stewart, D., Grinstein, S. and Spiegel, S. (1987) Transmembrane signalling by the B subunit of cholera toxin: Increased cytoplasmic free calcium in rat lymphocytes. *J. Cell Biol.*, **105**, 1153–1161.

Domenighini, M., Magagnoli, C., Pizza, M. and Rappuoli, R. (1994) Common features of the NAD-binding and catalytic site of ADP-ribosylating toxins. *Mol. Microbiol.*, **14**, 41–50.

Donaldson, J.G. and Klausner, R.D. (1994) ARF: a key regulatory switch in membrane traffic and organelle structure. *Curr. Op. Cell Biol.*, **6**, 527–532.

Donaldson, J.G., Cassel, D., Kahn, R.A. and Klausner, R.D. (1992b) ADP-ribosylation factor, a small GTP-binding protein, is required for binding of the coatomer protein beta-COP to Golgi membranes. *Proc. Natl. Acad. Sci. USA*, **89**, 6408–6412.

Donaldson, J.G., Finazzi, D. and Klausner, R.D. (1992a) Brefeldin A inhibits Golgi-membrane catalyzed exchange of guanine nucleotide onto ARF protein. *Nature*, **360**, 350–352.

Donaldson, J.G., Kahn, R.A., Lippincott-Schwartz, J. and Klausner, R.D. (1991) Binding of ARF and β-COP to Golgi membranes: Possible regulation by a trimeric G protein. *Science*, **254**, 1197–1199.

Douce, G., Turcotte, C., Cropley, I., Roberts, M., Pizza, M., Domenghini, M., Rappuoli, R., Dougan, G. (1995) Mutants of *Escherichia coli* heat-labile toxin lacking ADP-ribosyltransferase activity act as nontoxic, mucosal adjuvants. *Proc. Natl. Acad. Sci. USA*, **92**, 1644–1648.

D'Souza-Schorey, C., Li, G., Colombo, M.I. and Stahl, P.D. (1995) A regulatory role for ARF6 in receptor-mediated endocytosis. *Science*, **267**, 1175–1178.

Duman, R.S., Winston, S.M., Clark, J.A. and Nestler, E.J. (1990) Corticosterone regulates the expression of ADP-ribosylation factor messenger RNA and protein in rat cerebral cortex. *J. Neurochem.*, **55**, 1813–1816.

Duronio, R.J., Jackson-Machelski, E., Heuckeroth, R.O., Olins, P.O., Devine, C.S., Yonemoto, W., Slice, L.W., Taylor, S.S. and Gordon, J.I. (1990) Protein *N*-myristoylation in *Escherichia coli:* Reconstitution of a eukaryotic protein modification in bacteria. *Proc. Natl. Acad. Sci. USA*, **87**, 1506–1510.

Eastman, S.J., Hope, M.J., Wong, K.F. and Cullis, P.R. (1992) Influence of phospholipid asymmetry on fusion between large unilamellar vesicles. *Biochemistry*, **31**, 4262–4268.

Emala, C.W., Schwindinger, W.F., Wand, G.S. and Levine, M.A. (1994) Signal-transducing G proteins: Basic and clinical implications. *Prog. Nucleic Acid Res. and Mol. Biol.*, **47**, 81–111.

Enomoto, K. and Gill, D.M. (1979) Requirement for guanosine triphosphate in the activation of adenylate cyclase by cholera toxin. *J. Supramol. Struct.*, **10**, 51–60.

Enomoto, K. and Gill, D.M. (1980) Cholera toxin activation of adenylate cyclase. *J. Biol. Chem.*, **255**, 1252–1258.

Exton, J.H. (1994) Phosphatidylcholine breakdown and signal transduction. *Biochim. Biophys.*, *Acta* **1212**, 26–42.

Fasano, A., Baudry, B., Pumplin, D.W., Wasserman, S.S., Tall, B.D., Ketley, J.M. and Kaper, J.B. (1991) *Vibrio cholerae* produces a second enterotoxin, which affects intestinal tight junctions. *Proc. Natl. Acad. Sci. USA*, **88**, 5242–5246.

Fensome, A., Cunningham, E., Troung, O. and Cockcroft, S. (1994) ARF(2-17) does not specifically interact with ARF1-dependent pathways. Inhibition by peptide of phospholipases Cβ,D and exocytosis in HL60 cells. *FEBS Let.*, **349**, 34–38.

Finazzi, D., Cassel, D., Donaldson, J.G. and Klausner, R.D. (1994) Aluminum fluoride acts on the reversibility of ARF1–dependent coat protein binding to Golgi membranes. *J. Biol. Chem.*, **269**, 13325–13330.

Fishman, P.H. (1982) Role of membrane gangliosides in the binding and action of bacterial toxins. *J. Membrane Biol.*, **69**, 85–97.

Fishman, P.H., Pacuszka, T., Hom, B. and Moss, J. (1980) Modification of ganglioside GM1. *J. Biol. Chem.*, **255**, 7657–7664.

Fishman, P.H., Pacuszka, T and Orlandi, P.A. (1993) Gangliosides as receptors for bacterial enterotoxins. *Advan. Lipid Res.*, **25**, 165–187.

Fontana, M.R., Manetti, R., Giannelli, V., Magagnoli, C., Marchini, A., Olivieri, R., Domenighini, M, Rappuoli, R. and Pizza, M. (1995) Construction of nontoxic derivatives of cholera toxin and characterization of the immunological response against the A subunit. *Infect. Immun.*, **63**, 2356–2360.

Franco, M., Chardin, P., Chabre, M. and Paris, S. (1993) Myristoylation is not required for GTP–dependent binding of ADP-ribosylation factor ARF1 to phospholipids. *J. Biol. Chem.*, **268**, 24531–24534.

Franco, M., Chardin, P., Chabre, M. and Paris, S. (1995) Myristoylation of ADP-ribosylation factor 1 facilitates nucleotide exchange at physiological Mg^{2+} levels. *J. Biol. Chem.*, **270**, 1337–1341.

Gill, D.M. (1975) Involvement of nicotinamide adenine dinucleotide in the action of cholera toxin *in vitro*. *Proc. Natl. Acad. Sci. USA*, **72**, 2064–2068.

Gill, D.M., Clements, J.D., Robertson, D.C. and Finkelstein, R.A. (1981) Subunit number and arrangement in *Escherichia coli* heat-labile enterotoxin. *Infect. Immun.*, **33**, 677–682.

Gill, D.M. and Meren, R. (1978) ADP–ribosylation of membrane proteins catalyzed by cholera toxin: Basis of the activation of adenylate cyclase. *Proc. Natl. Acad. Sci. USA*, **75**, 3050–3054.

Grant, C.C.R., Messner, R.J. and Cieplak, Jr., W. (1994) Role of trypsin-like cleavage at Arginine 192 in the enzymatic and cytotonic activities of *Escherichia coli* heat-labile enterotoxin. *Infect. Immun.*, **62**, 4270–4278.

Greasley, S., Jhoti, H., Fensome, A., Cockcroft, S., Thomas, G. and Bax, B. (1994) Crystallization and preliminary X–ray diffraction studies on ADP–ribosylation Factor 1. *J. Mol. Biol.*, **244**, 65 1–653.

Halimi, H. and Rivaille, P. (1993) Immune response related to the molecular structure of a peptide from cholera toxin B subunit. *Vaccine*, **11**, 1233–1239.

Harford, S., Dykes, C.W., Hobden, A., Read, M.J. and Halliday, I.J. (1989) Inactivation of the *Escherichia coli* heat-labile enterotoxin by *in vitro* mutagenesis of the A-subunit gene. *Eur. J. Biochem.*, **183**, 311–316.

Häse, C., Thai, L.S., Boesman-Finkelstein, M., Mar, V.L., Burnette, W.N., Kaslow, H.R., Stevens, L.A., Moss, J. and Finkelstein, R.A. (1994) Construction and characterization of recombinant *Vibrio cholerae* strains producing inactive cholera toxin analogs. *Infect. Immun.*, **62**, 3051–3057.

Haun, R.S., Moss, J. and Vaughan, M. (1993b) Characterization of the human ADP-ribosylation factor 3 promoter. *J. Biol. Chem.*, **268**, 8793–8800.

Haun, R., Tsai, S-C., Adamik, R., Moss, J. and Vaughan, M. (1993a) Effect of myristoylation on GTP-dependent binding of ADP-ribosylation factor to Golgi. *J. Biol. Chem.*, **268**, 7064–7068.

Helms, J.B., Palmer, D.J. and Rothman, J.E. (1993) Two distinct populations of ARF bound to Golgi membranes. *J. Cell Biol.*, **121**, 751–760.

Helms, J.B. and Rothman, J.E. (1992) Inhibition by brefeldin A of a Golgi membrane enzyme that catalyzes exchange of guanine nucleotide bound to ARF. *Nature*, **360**, 352–354.

Hong, J-X., Haun, R.S., Tsai, S-C., Moss, J. and Vaughan, M. (1994) Effect of ADP-ribosylation factor amino-terminal deletions on its GTP-dependent stimulation of cholera toxin activity. *J. Biol. Chem.*, **269**, 9743–9745.

Hong, J-X., Zhang, X., Moss, J. and Vaughan, M. (1995) Isolation of an amino-terminal deleted recombinant ADP-ribosylation factor 1 in an activated nucleotide-free state. *Proc. Natl. Acad. Sci. USA*, **92**, 3056–3059.

Janicot, M., Fouque, F. and Desbuquois, B. (1991) Activation of rat liver adenylate cyclase by cholera toxin requires toxin internalization and processing in endosomes. *J. Biol. Chem.*, **266**, 12858–12865.

Jiang, H., Lu, Z., Luo, Jing-Q., Wolfman, A. and Foster, D.A. (1995) Ras mediates the activation of phospholipase D by v-Src. *J. Biol. Chem.*, **270**, 6006–6009.

Jobling, M.G. and Holmes, R.K. (1992) Fusion proteins containing the A2 domain of cholera toxin assemble with B polypeptides of cholera toxin to form immunoreactive and functional holotoxin-like chimeras. *Infect. Immun.*, **60**, 4915–4924.

Johnson, G.L. and Bourne, H.R. (1977) Influence of cholera toxin on the regulation of adenylate cyclase by GTP. *Biochem. Biophys. Res. Commun.*, **78**, 792–798.

Johnson, D.R., Bhatnagar, R.S., Knoll, L.J. and Gordon, J.I. (1994) Genetic and biochemical studies of protein *N*-myristoylation. *Ann. Rev. Biochem.*, **63**, 869–914.

Johnson, G.L., Kaslow, H.R. and Bourne, H.R. (1978) Genetic evidence that cholera toxin substrates are regulatory components of adenylate cyclase. *J. Biol. Chem.*, **253**, 7120–7123.

Joseph, C.K., Kim, U.S., Stieber, A. and Gonatas, N.K. (1978) Endocytosis of cholera toxin into neuronal GERL. *Proc. Natl. Acad. Sci. USA*, **75**, 2815–2819.

Kahn, R.A. (1991) Fluoride is not an activator of the smaller (20–25-kDa) GTP-binding proteins. *J. Biol. Chem.*, **266**, 15595–15597.

Kahn, R.A. and Gilman, A.G. (1984a) ADP-ribosylation of Gs promotes the dissociation of its α and β subunits. *J. Biol. Chem.*, **259**, 6235–6240.

Kahn, R.A. and Gilman, A.G. (1984b) Purification of a protein cofactor required for ADP-ribosylation of the stimulatory regulatory component of adenylate cyclase by cholera toxin. *J. Biol. Chem.*, **259**, 6228–6234.

Kahn, R.A. and Gilman, A.G. (1986) The protein cofactor necessary for ADP-ribosylation of Gs by cholera toxin is itself a GTP-binding protein. *J. Biol. Chem.*, **261**, 7906–7911.

Kahn, R.A., Clark, J., Rulka, C., Stearns, T., Zhang, C-J., Randazzo, P.A., Terui, T. and Cavenagh, M. (1995) Mutational analysis of *Saccharomyces cerevisiae* ARF1. *J. Biol. Chem.*, **270**, 143–150.

Kahn, R.A., Goddard, C. and Newkirk, M. (1988) Chemical and immunological characterization of the 21-kDa ADP-ribosylation factor of adenylate cyclase. *J. Biol. Chem.*, **263**, 8282–8287.

Kahn, R.A., Randazzo, P., Serafini, T., Weiss, O., Rulka, C., Clark, J., Amherdt, M., Roller, P., Orci, L. and Rothman, J.E. (1992) The amino terminus of ADP-ribosylation factor (ARF) is a critical determinant of ARF activities and is a potent and specific inhibitor of protein transport. *J. Biol. Chem.*, **267**, 13039–13046.

Kaper, J.B., Morris, J.G. Jr., Levine, M.M. (1995) Cholera. *Clin. Microbiol. Rev.*, **8**, 48–86.

Kaslow, H.R., Johnson, G.L., Brothers, V.M. and Bourne, H.R. (1980) A regulatory component of adenylate cyclase from human erythrocyte membranes. *J. Biol. Chem.*, **255**, 3736–3741.

Kassis, S., Hagmann, J., Fishman, P.H., Chang, P.P. and Moss, J. (1982) Mechanism of action of cholera toxin on intact cells. Generation of A_1 peptide and activation of adenylate cyclase. *J. Biol. Chem.*, **257**, 12148–12152.

King, C.A. and van Heyningen, W.E. (1973) Deactivation of cholera toxin by a sialidase-resistant monosialosylganglioside. *J. Infect. Dis.*, **127**, 639–647.

Knoop, F.C. and Thomas, D.D. (1984) Effect of cholera enterotoxin on calcium uptake and cyclic-AMP accumulation in rat basophilic leukemia cells. *Int. J. Biochem.*, **16**, 275–280.

Ktistakis, N.T., Brown, H.A., Sternweis, P.C. and Roth, M.G. (1995) Phospholipase D is present on Golgi-enriched membranes and its activation by ADP-ribosylation factor is sensitive to brefeldin A. *J. Biol. Chem.*, **92**, 4952–4956.

Kunz, B.C., Muczynski, K.A., Welsh, C.F., Stanley, S.J., Tsai, S-C., Adamik, R., Chang, P.P., Moss, J. and Vaughan, M. (1993) Characterization of recombinant and endogenous ADP-ribosylation factors synthesized in Sf9 insect cells. *Biochem.*, **32**, 6643–6648.

Lai, C-Y., Xia, Q-C. and Salotra, P.T. (1983) Location and amino acid sequence around the ADP-ribosylation site in the cholera toxin active subunit A1. *Biochem. Biophys. Res. Comm.*, **116**, 341–348.

Lambeth, J.D., Kwak, J-Y., Bowman, E.P., Perry, D., Uhlinger, D.J. and Lopez, I. (1995) ADP-ribosylation factor functions synergistically with a 50-kDa cytosolic factor in cell-free activation of human neutrophil phospholipase D. *J. Biol. Chem.*, **270**, 2431–2434.

Larew, J.S.A., Peterson, J.E., Graves, D.J. (1991) Determination of the kinetic mechanism of arginine-specific ADP-ribosyltransferases using a high performance liquid chromatographic assay. *J. Biol. Chem.*, **266**, 52–57.

Lee, C-M., Chang, P.P., Tsai, S-C., Adamik, R., Price, S.R., Kunz, B.C., Moss, J. Twiddy, E.M. and Holmes, R.K. (1991) Activation of *Escherichia coli* heat-labile enterotoxins by native and recombinant adenosine diphosphate-ribosylation factors, 20-kD guanine nucleotide-binding proteins. *J. Clin. Invest.*, **87**, 1780–1786.

Lee, C-M., Haun, R.S., Tsai, S-C., Moss, J. and Vaughan, M. (1992) Characterization of the human gene encoding ADP-ribosylation factor 1, a guanine nucleotide-binding activator of cholera toxin. *J. Biol. Chem.*, **267**, 9028–9034.

Lee, F-J.S., Stevens, L.A., Hall, L.M., Murtagh, J.J., Kao, Y.L., Moss, J. and Vaughan, M. (1994a) Characterization of class II and class III ADP-ribosylation factor genes and proteins in *Drosophila melanogaster*. *J. Biol. Chem.*, **269**, 21555–21560.

Lee, F-J.S., Stevens, L.A., Kao, Y.L., Moss, J. and Vaughan, M. (1994b) Characterization of a glucose-repressible ADP-ribosylation factor 3 (ARF3) from *Saccharomyces cerevisiae*. *J. Biol. Chem.*, **269**, 20931–20937.

Lencer, W.I., de Almeida, J.B., Moe, S., Stow, J.L., Ausiello, D.A. and Madara, J.L. (1993) Entry of cholera toxin into polarized human intestinal epithelial cells: Identification of an early brefeldin A sensitive event required for A1-peptide generation. *J. Clin. Invest.*, **92**, 2941–2951.

Lencer, W.I., Strohmeier, S., Moe, S., Carlson, S.L., Constable, C.T. and Madara, J.L. (1995) Signal transduction by cholera toxin: processing in vesicular compartments does not require acidification. *Am. J. Physiology*, **269**, G548–G557.

Lenhard, J.M., Kahn, R.A. and Stahl, P.D. (1992) Evidence for ADP-ribosylation factor (ARF) as a regulator of *in vitro* endosome-endosome fusion. *J. Biol. Chem.*, **267**, 13047–13052.

Liang, Y-F., Peterson, J.W., Jackson, C.A. and Reitmeyer, J.C. (1990) Chloroquine inhibition of cholera toxin. *FEBS Lett.*, **275**, 143–145.

Lin, M.C., Welton, A.F. and Berman, M.F. (1978) Essential role of GTP in the expression of adenylate cyclase activity after cholera toxin treatment. *J. Cycl. Nuc. Res.*, **4**, 159–168.

Lippincott-Schwartz, J., Donaldson, J.G., Schweizer, A., Berger, E.G., Hauri, H.P., Yuan, C. and Klausner, R.D. (1990) Microtubule-dependent retrograde transport of proteins into the ER in the presence of brefeldin A suggests an ER recycling pathway. *Cell,* **60**, 821–836.

Lippincott-Schwartz, J., Yuan, L.C., Bonifacino, J.S. and Klausner, R.D. (1989) Rapid redistribution of Golgi proteins into the ER in cells treated with brefeldin A: Evidence for membrane cycling from the Golgi to the ER. *Cell,* **56**, 801–813.

Liscovitch, M., Chalifa, V., Pertile, P., Chen, C-S. and Cantley, L.C. (1994) Novel function of phosphatidylinositol 4,5-bisphosphate as a cofactor for brain membrane phospholipase D. *J. Biol. Chem.,* **269**, 21403–21406.

Lobet, Y., Cieplak, Jr., W, Smith, S.G. and Keith, J.M. (1989) Effects of mutations on enzymatic activity and immunoreactivity of the S1 subunit of pertussis toxin. *Infect. Immun.,* **57**, 3660–3662.

Lobet, Y., Cluff, C.W. and Cieplak, Jr., W (1991) Effect of site-directed mutagenic alterations on ADP-ribosyltransferase activity of the A subunit of *Escherichia coli* heat-labile enterotoxin. *Infect. Immun.,* **59**, 2870–2879.

Lochrie, M.A., Hurley, J.B. and Simon, M.I. (1985) Sequence of the alpha subunit of photoreceptor G protein: Homologies between transducin, Ras and elongation factors. *Science,* **228**, 96–99.

Locht, C., Capiau, C. and Feron, C. (1989) Identification of amino acid residues essential for the enzymatic activities of pertussis toxin. *Proc. Natl. Acad. Sci. USA,* **86**, 3075–3079.

Maenz, D.D. and Forsyth, G.W. (1986) Cholera toxin facilitates calcium transport in jejunal brush border vesicles. *Can. J. Physiol. Pharmacol.,* **64**, 568–574.

Makler, V., Cukierman, E., Rotman, M., Admon, A. and Cassel, D. (1995) ADP-ribosylation-directed GTPase-activating protein: Purification and partial characterization. *J. Biol. Chem.,* **270**, 5232–5237.

Massenburg, D., Han, J-S., Liyanage, M., Patton, W.A., Rhee, S.G., Moss, J. and Vaughan, M. (1994) Activation of rat brain phospholipase D by ADP-ribosylation factors 1,5 and 6: Separation of ADP-ribosylation factor-dependent and oleate-dependent enzymes. *Proc. Natl. Acad. Sci. USA,* **91**, 11718–11722.

Medynski, D.C., Sullivan, K., Smith, D., Van Dop, C., Chang, F-H., Fung, B.K-K., Seeburg, P.H. and Bourne, H.R. (1985) Amino acid sequence of the alpha subunit of transducin deduced from the cDNA sequence. *Proc. Natl. Acad. Sci. USA,* **82**, 4311–4315.

Mekalanos, J.J. and Sadoff, J.C. (1994) Cholera vaccines: Fighting and ancient scourge. *Science,* **265**, 1387–1389.

Mekalanos, J.J., Collier, R.J. and Romig, W.R. (1979a) Enzymatic activity of cholera toxin: II. Relationships to proteolytic processing, disulfide bond reduction and subunit composition. *J. Biol. Chem,.* **254**, 5855–5861.

Mekalanos, J.J., Collier, R.J. and Romig, W.R. (1979b) Enzymatic activity of cholera toxin: I. New method of assay and the mechanism of ADP-ribosyl transfer. *J. Biol. Chem.,* **254**, 5849–5854.

Mekalanos, J.J., Swartz, D.J., Pearson, G.D.N., Harford, N., Groyne, F. and de Wilde, M. (1983) Cholera toxin genes: Nucleotide sequence, deletion analysis and vaccine development. *Nature,* **306**, 551–557.

Merritt, E.A., Pronk, S.E., Sixma, T.K., Kalk, K.H., van Zanten, B.A.M. and Hol, W.G.J. (1994a) Structure of partially-activated *E. coli* heat-labile enterotoxin (LT) at 2.6Å resolution. *FEBS Let.,* **337**, 88–92.

Merritt, E.A., Sarfaty, S., Pizza, M., Rappuoli, R. and Hol, W.G.J. (1995) Mutation of a buried residue causes loss of activity, but no conformational change in the heat-labile enterotoxin of *Escherichia coli. Struct. Biol.,* **2**, 269–272.

Merritt, E.A., Sarfaty, S., Van Den Akker, F., L'Hoir, C., Martial, J.A. and Hol, W.G.J. (1994c) Crystal sturcture of cholera toxin B-pentamer bound to receptor G_{M1} pentasaccharide. *Prot. Sci.,* **3**, 166–175.

Merritt, E.A., Sixma, T.K., Kalk, K.H., van Zanten, B.A.M. and Hol, W.G.J. (1994b) Galactose-binding site in *Escherichia coli* heat-labile enterotoxin (LT) and cholera toxin (CT). *Mol. Microbiol.,* **13**, 745–753.

Mishima, K., Price, S.R., Nightingale, M.S., Kousvelar, E., Moss, J. and Vaughan, M. (1992) Regulation of ADP-ribosylation factor (ARF) expression: Cross species conservation of the developmental and tissue-specific alternative polyadenylation of ARF 4 mRNA. *J. Biol. Chem.,* **267**, 24109–24116.

Mishima, K., Tsuchiya, M., Nightingale, M.S., Moss, J. and Vaughan, M. (1993) ARD1, a 64-kDa guanine nucleotide-binding protein with a carboxy-terminal ADP-ribosylation factor domain. *J. Biol. Chem.,* **268**, 8801–8807.

Misumi, Y., Miki, A., Takatsuki, A., Tamura, G. and Ikehara, Y. (1986) Novel blockade by brefeldin A on intracellular transport of secretory proteins in cultured rat hepatocytes. *J. Biol. Chem.*, **261**, 11398–11403.

Monaco, L., Murtagh, J.J., Newman, K.B., Tsai, S-C., Moss, J. and Vaughan, M. (1990) Selective amplification of an mRNA and related pseudogene for a human ADP-ribosylation factor, a guanine nucleotide-dependent protein activator of cholera toxin. *Proc. Natl. Acad. Sci. USA*, **87**, 2206–2210.

Moss, J. and Stanley, S.J. (1981a) Amino acid-specific ADP-ribosylation. *J. Biol. Chem.*, **256**, 7830–7833.

Moss, J. and Stanley, S.J. (1981b) Histone-dependent and histone-independent forms of an ADP-ribosyltransferase from human and turkey erythrocytes. *Proc. Natl. Acad. Sci. USA*, **78**, 4809–4812.

Moss, J. and Vaughan, M. (1977a) Mechanism of action of choleragen: Evidence for ADP-ribosyltransferase activity with arginine as an acceptor. *J. Biol. Chem.*, **252**, 2455–2457.

Moss, J. and Vaughan, M. (1977b) Choleragen activation of solubilized adenylate cyclase: Requirement for GTP and protein activator for demonstration of enzymatic activity. *Proc. Natl. Acad. Sci. USA*, **74**, 4396–4400.

Moss, J. and Vaughan, M. (1978) Isolation of an avian erythrocyte protein possessing ADP-ribosyltransferase activity and capable of activating adenylate cyclase. *Proc. Natl. Acad. Sci. USA*, **75**, 3621–3624.

Moss, J. and Vaughan, M. (1988) Cholera toxin and *E. coli* enterotoxins and their mechanisms of action. *Handbook of Natural Toxins, Vol. 4: Bacterial Toxins* (Hardegree, M.C. and Tu, A.T., eds.) Marcell Dekker, New York, 39–87.

Moss, J., Haun, R.S., Tsai, S-C., Welsh, C.F., Lee, F-J.S., Price, S.R. and Vaughan, M. (1994b) Activation of cholera toxin by ADP-ribosylation factors: 20-kDa guanine nucleotide-binding proteins. *Meth. Enzym.*, **237**, 44–63.

Moss, J., Jacobson, M.K. and Stanley, S.J. (1985) Reversibility of arginine-specific mono-(ADP-ribosyl)ation: Identification in erythrocytes of an ADP-ribose-L-arginine cleavage enzyme. *Proc. Natl. Acad. Sci. USA*, **82**, 5603–5607.

Moss, J., Manganiello, V.C. and Vaughan, M. (1976) Hydrolysis of nicotinamide adenine dinucleotide by choleragen and its A protomer: Possible role in the activation of adenylate cyclase. *Proc. Natl. Acad. Sci. USA*, **73**, 4424–4427.

Moss, J., Ross, P.S., Agosto, G., Birken, S., Canfield, R.E. and Vaughan, M. (1978) Mechanism of action of choleragen and the glycopeptide hormones: Is the nicotinamide adenine dinucleotide glycohydrolase activity observed in purified hormone preparations instrinsic to the hormone? *Endocrinology*, **102**, 415–419.

Moss, J., Stanley, S.J. and Lin, M.C. (1979) NAD-glycohydrolase and ADP-ribosyltransferase activities are intrinsic to the A_1 peptide of choleragen. *J. Biol. Chem.*, **254**, 11993–11996.

Moss, J., Stanley, S.J., Nightingale, M.S., Murtagh, J.J. Jr., Monaco, L., Mishima, K., Chen, H-C., Williamson, K.C. and Tsai, S-C. (1992) Molecular and immunological characterization of ADP-ribosylarginine hydrolases. *J. Biol. Chem.*, **267**, 10481–10488.

Moss, J., Stanley, S.J., Vaughan, M. and Tsuji, T. (1993) Interaction of ADP-ribosylation factor with *Escherichia coli* enterotoxin that contains an inactivating Lysine-112 substitution. *J. Biol. Chem.*, **268**, 6383–6387.

Moss, J., Stanley, S.J. and Watkins, P.A. (1980b) Isolation and properties of an NAD- and guanidine-dependent ADP-ribosyltransferase from turkey erythrocytes. *J. Biol. Chem.*, **255**, 5838–5840.

Moss, J., Stanley, S.J., Watkins, P.A. and Vaughan, M. (1980a) ADP-ribosyltransferase activity of mono- and multi-(ADP-ribosylated) choleragen. *J. Biol. Chem.*, **255**, 7835–7837.

Moss, J., Tsai, S-C., Adamik, R., Chen, H-C. and Stanley, S.J. (1988) Purificiation and characterization of ADP-ribosylarginine hydrolase from turkey erythrocytes. *Biochemistry*, **27**, 5819–5823.

Moss, J., Tsai, S-C. and Vaughan, M. (1994a) Activation of cholera toxin by ADP-ribosylation factors. *Meth. Enzym.*, **235**, 640–647.

Murayama, T., Tsai, S-C., Adamik, R., Moss, J. and Vaughan, M. (1993) Effects of temperature on ADP-ribosylation factor stimulation of cholera toxin activity. *Biochemistry*, **32**, 561–566.

Murtagh, J., Lee, F-J., Deak, P., Hall, L., Monaco, L., Lee, C-M., Stevens, L., Moss, J. and Vaughan, M. (1993) Molecular characterization of a conserved guanine nucleotide-dependent ADP-ribosylation factor in *Drosophila melanogaster*. *Biochemistry*, **32**, 6011–6018.

Murtagh, J., Mowatt, M., Lee, C-M., Lee, F-J., Mishima, K., Nash, T., Moss, J. and Vaughan, M. (1992) Guanine nucleotide-binding proteins in the intestinal parasite *Giardia lamblia*. *J. Biol. Chem.*, **267**, 9654–9662.

Nambiar, M.P., Oda, T., Chen, C., Kuwazuru, Y and Wu, H.C. (1993) Involvement of the Golgi region in the intracellular trafficking of cholera toxin. *J. Cell. Physiol.*, **154**, 222–228.

Nakaya, S., Moss, J. and Vaughan, M. (1980) Effects of nucleoside triphosphates on choleragen-activated brain adenylate cyclase. *Biochemistry*, **19**, 4871–4874.

Navon, S.E. and Fung, B.K-K. (1984) Characterization of transducin from bovine retinal rod outer segments. Mechanism and effects of cholera toxin-catalyzed ADP-ribosylation. *J. Biol. Chem.*, **259**, 6686–6693.

Neer, E.J. (1995) Heterotrimeric G proteins: Organizers of transmembrane signals. *Cell*, **80**, 249–257.

Nilsson, O., Cassuto, J., Larsson, L-I., Jodal, M., Lidberg, P, Ahlman, H., Dahlström, A. and Lundgren, O. (1983) 5-hydroxytryptamine and cholera secretion: A histochemical and physiological study in cats. *Gut*, **24**, 542–548.

Noda, M., Tsai, S-C., Adamik, R., Moss, J. and Vaughan, M. (1990) Mechanism of cholera toxin activation by a guanine nucleotide-dependent 19-kDa protein. *Biochim. Biophys. Acta*, **1034**, 195–199.

Northup, J.K., Sternweis, P.C., Smigel, M.D., Schleifer, L.S., Ross, E.M. and Gilman, A.G. (1980) Purification of the regulatory component of adenylate cyclase. *Proc. Natl. Acad. Sci. USA*, **77**, 6516–6520.

Nuoffer, C. and Balch, W.E. (1994) GTPases: Multifunctional molecular switches regulating vesicular traffic. *Ann. Rev. Biochem.*, **63**, 949–990.

Ohguchi, K., Banno, Y., Nakashima, S. and Nozawa, Y. (1995) Activation of membrane-bound phospholipase D by protein kinase C in HL60 cells: Synergistic action of a small GTP-binding protein RhoA. *Biochem. Biophys. Res. Comm.*, **211**, 306–311.

Ohtomo, N., Muraoka, T., Tashiro, A., Zinnaka, Y. and Amako, K. (1976) Size and structure of the cholera toxin molecule and its subunits. *J. Infect. Dis.*, **133**, S31–S40.

Okazaki, I.J., Zolkiewska, A., Nightingale, M.S. and Moss, J. (1994) Immunological and structural conservation of mammalian skeletal muscle glycosylphosphatidylinositol-linked ADP-ribosyltransferases. *Biochemistry*, **33**, 12828–12836.

Olson, S.C., Bowman, E.P. and Lambeth, J.D. (1991) Phospholipase D activation in a cell-free system from human neutrophils by phorbol 12-myristate 13-acetate and guanosine 5′-O-(3-thiotriphosphate). *J. Biol. Chem.*, **266**, 17236–17242.

Oppenheimer, N.J. (1978) Structural determination and sterospecificity of the choleragen-catalyzed reaction of NAD⁺ with guanidines. *J. Biol. Chem.*, **253**, 4907–4910.

Orci, L., Palmer, D.J., Amherdt, M. and Rothman, J.E. (1993) Coated vesicle assembly in the Golgi requires only coatomer and ARF proteins from the cytosol. *Nature*, **364**, 732–734.

Orlandi, P.A., Curran, P.K. and Fishman, P.H. (1993) Brefeldin A blocks the response of cultured cells to cholera toxin. *J. Biol. Chem.*, **268**, 12010–12016.

Osborne, J.C. Jr., Stanley, S.J., Moss, J. (1985) Kinetic mechanisms of two NAD:arginine ADP-ribosyltransferases: The soluble, salt-stimulated transferase from turkey erythrocytes and choleragen, a toxin from *Vibrio cholerae*. *Biochemistry*, **24**, 5235–5240.

Palmer, D.J., Helms, J.B., Beckers, C.J.M., Orci, L. and Rothman, J.E. (1993) Binding of coatomer to Golgi membranes requires ADP-ribosylation factor. *J. Biol. Chem.*, **268**, 12083–12089.

Papini, E., Rappuoli, R., Murgia, M. and Montecucco, C. (1993) Cell penetration of diphtheria toxin. Reduction of the interchain disulfide bridge is the rate-limiting step of translocation in the cytosol. *J. Biol. Chem.*, **268**, 1567–1574.

Passador, L. and Iglewski, W. (1994) ADP-ribosylating toxins. *Meth. Enzym.* **235**, 617–631.

Peng, Z., Calvert, I., Clark, J., Helman, L., Kahn, R. and Kung, H. (1989) Molecular cloning, sequence analysis and mRNA expression of human ADP-ribosylation factor. *BioFactors*, **2**, 45–49.

Pertile, P., Liscovitch, M., Chalifa, V. and Cantley, L.C. (1995) Phosphatidylinositol 4,5-bisphosphate synthesis is required for activation of phospholipase D in U937 cells. *J. Biol. Chem.*, **270**, 5130–5135.

Peters, P.J., Hsu, V.W., Ooi, C.E., Finazzi, D., Teal, S.B., Oorschot, V., Donaldson, J.G. and Klausner, R.D. (1995) Overexpression of wild-type and mutant ARF1 and ARF6: Distinct perturbations of nonoverlapping membrane compartments. *J. Cell Biol.*, **128**, 1003–1017.

Peterson, J.W. and Ochoa, L.G. (1989) Role of prostaglandins and cAMP in the secretory effects of cholera toxin. *Science*, **245**, 857–859.

Peterson, J.W., Berg, W.D., Jr. and Coppenhaver, D.H. (1987) Synthesis of protein in intestinal cells exposed to cholera toxin (42599). *Proc. Soc. Exp. Biol. Med.*, **186**, 174–182.

Peterson, J.W., Cantu, J., Duncan, S. and Chopra, A.K. (1993) Molecular mediators formed in the small intestine in response to cholera toxin. *J. Diarrhoeal Dis. Res.*, **11**, 227–234.

Peterson, J.W., Jackson, C.A. and Reitmeyer, J.C. (1990a) Synthesis of prostaglandins in cholera toxin-treated Chinese hamster ovary cells. *Microbiol. Pathogen.*, **9**, 345–353.

Peterson, J.E., Larew, J.S-A. and Graves, D.J. (1990b) Purification and partial characterization of arginine-specific ADP-ribosyltransferase from skeletal muscle microsomal membranes. *J. Biol. Chem.*, **265**, 17062–17069.

Peterson, J.W., Lu, Y., Duncan, S., Cantu, J. and Chopra, A.K. (1994) Interactions of intestinal mediators in the mode of action of cholera toxin. *J. Med. Microbiol.*, **41**, 3–9.

Peterson, J.W., Reitmeyer, J.C., Jackson, C.A. and Ansari, G.A.S. (1991) Protein synthesis is required for cholera toxin-induced stimulation of arachidonic acid metabolism. *Biochim. Biophys. Acta*, **1092**, 79–84.

Picking, W.D. (1993) Interaction of pyrene-labeled monosialoganglioside G_{M1} micelles with cholera toxin. *Biochem. Biophys. Res. Comm.*, **195**, 1153–1158.

Picking, W.L., Moon, H., Wu, H. and Picking, W.D. (1995) Fluorescence analysis of the interaction between ganglioside G_{M1}-containing phospholipid vesicles and the B subunit of cholera toxin. *Biochim. Biophys. Acta*, **1247**, 65–73.

Pinkett, M.O. and Anderson, W.B. (1982) Plasma membrane-associated component(s) that confer(s) cholera toxin sensitivity to adenylate cyclase. *Biochim. Biophys. Acta*, **714**, 337–343.

Pizza, M., Bartoloni, A., Prugnola, A., Silvestri, S. and Rappuoli, R. (1988) Subunit S1 of pertussis toxin: Mapping the regions essential for ADP-ribosyltransferase activity. *Proc. Natl. Acad. Sci. USA*, **85**, 7521–7525.

Pizza, M., Domenighini, M., Hol, W., Giannelli, V., Fontana, M.R., Giuliani, M.M., Magagnoli, C., Peppoloni, S., Manetti, R. and Rappuoli, R. (1994a) Probing the structure-activity relationship of *Escherichia coli* LT-A by site-directed mutagenesis. *Mol. Microbiol.*, **14**, 51–60.

Pizza, M., Fontana, M.R., Giuliani, M.M., Domenighini, M., Magagnoli, C., Giannelli, V., Nucci, D., Hol, W., Manetti, R., Rappuoli, R. (1994b) A genetically detoxified derivative of heat-labile *Escherichia coli* enterotoxin induces neutralizing antibodies against the A subunit. *J. Exp. Med.*, **180**, 2147–2153.

Price, S.R., Nightingale, M., Tsai, S-C., Williamson, K.C., Adamik, R., Chen, H-C., Moss, J. and Vaughan, M. (1988) Guanine nucleotide-binding proteins that enhance choleragen ADP-ribosyltransferase activity: Nucleotide and deduced amino acid sequence of an ADP-ribosylation factor cDNA. *Proc. Natl. Acad. Sci. USA*, **85**, 5488–5491.

Price, S.R., Welsh, C.F., Haun, R.S., Stanley, S.J., Moss, J. and Vaughan, M. (1992) Effects of phospholipid and GTP on recombinant ADP-ribosylation factors (ARFs): Molecular basis for differences in requirements for activity of mammalian ARFs. *J. Biol. Chem.*, **267**, 17766–17772.

Randazzo, P.A. and Kahn, R.A. (1994) GTP hydrolysis by ADP-ribosylation factor is dependent on both an ADP-ribosylation factor GTPase-activating protein and acid phospholipids. *J. Biol. Chem.*, **269**, 10758–10763.

Randazzo, P.A., Terui, T., Sturch, S., Fales, H.M., Ferrige, A.G. and Kahn, R.A. (1995) The myristoylated amino terminus of ADP-ribosylation factor 1 is a phospholipid- and GTP-sensitive switch. *J. Biol. Chem.*, **270**, 14809–14815.

Randazzo, P.A., Terui, T., Sturch, S. and Kahn, R.A. (1994) The amino terminus of ADP-ribosylation factor (ARF) 1 is essential for the interaction of **Gs** and ARF GTPase-activating protein. *J. Biol. Chem.*, **269**, 29490–29494.

Randazzo, P.A., Yang, Y.C., Rulka, C. and Kahn, R.A. (1993) Activation of ADP-ribosylation factor by Golgi membranes. *J. Biol. Chem.*, **268**, 9555–9563.

Rappuoli, R. and Pizza, M. (1991) Structure and evolutionary aspects of ADP-ribosylating toxins. *Sourcebook of Bacterial Protein Toxins* (Alouf, J.E.and Freer, J.H., eds) *Academic Press, San Diego.*, 1–21.

Regazzi, R., Ullrich, S., Kahn, R.A. and Wollheim, C.B. (1991) Redistribution of ADP-ribosylation factor during stimulation of permeabilized cells with GTP analogues. *Biochem. J.*, **275**, 639–644.

Reitmeyer, J.C. and Peterson, J.W. (1990) Stimulatory effects of cholera toxin on arachidonic acid metabolism in Chinese hamster ovary cells (43022). *Proc. Soc. Exper. Biol. Med.*, **193**, 181–184.

Ribi, H.O., Ludwig, D.S., Mercer, K.L., Schoolnik, G.K. and Kornberg, R.D. (1988) Three-dimensional structure of cholera toxin penetrating a lipid membrane. *Science*, **239**, 1272–1276.

Robishaw, J.D., Russell, D.W., Harris, B.A., Smigel, M.D. and Gilman, A.G. (1986) Deduced primary structure of the α subunit of the GTP-binding stimulatory protein of adenylate cyclase. *Proc. Natl. Acad. Sci. USA*, **83**, 1251–1255.

Rothman, J.E. (1994) Mechanisms of intracellular protein transport. *Nature*, **372**, 55–63.

Schafer, D.E. and Thakur, A.K. (1982) Quantitative description of the binding of G_{M1} oligosaccharide by cholera enterotoxin. *Cell Biophysics*, **4**, 25–40.

Schleifer, L.S., Kahn, R.A., Hanski, E., Northup, J.K., Sternweis, P.C. and Gilman, A.G. (1982) Requirements for cholera toxin-dependent ADP-ribosylation of the purified regulatory component of adenylate cyclase. *J. Biol. Chem.*, **257**, 20–23.

Schürmann, A., Breiner, M., Becker, W., Huppertz, C., Kainulainen, H., Kentrup, H. and Joost, H-G. (1994) Cloning of two novel ADP-ribosylation factor-like proteins and characterization of their differential expression in 3T3-L1 cells. *J. Biol. Chem.*, **269**, 15683–15688.

Serafini, T., Orci, L., Amherdt, M., Brunner, M., Kahn, R.A. and Rothman, J.E. (1991b) ADP-ribosylation factor is a subunit of the coat of Golgi-derived COP-coated vesicles: A novel role for a GTP-binding protein. *Cell*, **67**, 239–253.

Serafini, T., Stenbeck, G., Brecht, A., Lottspeich, F., Orci, L., Rothman, J.E. and Wieland, F.T. (1991a) A coat subunit of Golgi-derived non-clathrin-coated vesicles with homology to the clathrin-coated vesicle coat protein β-adaptin. *Nature*, **349**, 215–220.

Serventi, I.M., Cavanaugh, E., Moss, J. and Vaughan, M. (1993) Characterization of the gene for ADP-ribosylation factor (ARF) 2, a developmentally regulated, selectively expressed member of the ARF family of ~20-kDa guanine nucleotide-binding proteins. *J. Biol. Chem.*, **268**, 4863–4872.

Sewell, J.L. and Kahn, R.A. (1988) Sequences of the bovine and yeast ADP-ribosylation factor and comparison to other GTP-binding proteins. *Proc. Natl. Acad. Sci. USA*, **85**, 4620–4624.

Siddiqi, A.R., Smith, J.L., Ross, A.H., Qiu, R-G., Symons, M. and Exton, J.H. (1995) Regulation of phospholipase D in HL 60 cells. *J. Biol. Chem.*, **270**, 8466–8473.

Singer, W.D., Brown, H.A., Bokoch, G.M. and Sternweis, P.C. (1995) Resolved phospholipase D activity is modulated by cytosolic factors other than ARF. *J. Biol. Chem.*, **270**, 14944–14950.

Sixma, T.K., Kalk, K.H., van Zanten, B.A.M., Dauter, Z., Kingma, J., Witholt, B. and Hol, W.G.J. (1993) Refined structure of *Escherichia coli* heat-labile enterotoxin, a close relative of cholera toxin. *J. Mol. Biol.*, **230**, 890–918.

Sixma, T.K., Pronk, S.E., Kalk, K.H., Wartna, E.S., van Zanten, B.A.M., Witholt, B. and Hol, W.G.J. (1991) Crystal structure of a cholera toxin-related heat-labile enterotoxin from *E. coli*. *Nature*, **351**, 371–377.

Smith, K.P., Benjamin, R.C., Moss, J. and Jacobson, M.K. (1985) Identification of enzymatic activities which process protein-bound mono-(ADP-ribose). *Biochemistry*, **126**, 136–142.

Sofer, A. and Futerman, A.H. (1995) Cationic amphiphilic drugs inhibit the internalization of cholera toxin to the Golgi apparatus and the subsequent elevation of cyclic AMP. *J. Biol. Chem.*, **270**, 12117–12122.

Spangler, B.D. and Westbrook, E.M. (1989) Crystallization of isoelectrically homogeneous cholera toxin. *Biochemistry*, **28**, 1333–1340.

Spangler, B.D. (1992) Structure and function of cholera toxin and the related *Escherichia coli* heat-labile enterotoxin. *Microbiol. Rev.*, **56**, 622–647.

Spicer, E.K., Kavanaugh, W.M., Dallas, W.S., Falkow, S., Konigsberg, W.H. and Schafer, D.E. (1981) Sequence homologies between A subunits of *Escherichia coli* and *Vibrio cholerae* enterotoxins. *Proc. Natl. Acad. Sci. USA* **78**, 50–54.

Spiegel, A.M. (1992) G proteins in cellular control. *Curr. Op. Cell Biol.*, **4**, 203–211.

Spiro, D.J., Taylor, T.C., Melançon, P. and Wessling-Resnick, M. (1995) Cytosolic ADP-ribosylation factors are not required for endosome-endosome fusion, but are necesary for GTPγS inhibition of fusion. *J. Biol. Chem.*, **270**, 13693–13697.

Stearns, T., Kahn, R.A., Botstein, D. and Hoyt, M.A. (1990a) ADP-ribosylation factor is an essential protein in *Saccharomyces cerevisiae* and is encoded by two genes. *Mol. and Cell. Biol.*, **10**, 6690–6699.

Stearns, T., Willingham, M.C., Botstein, D. and Kahn, R.A. (1990b) ADP-ribosylation factor is functionally and physically associated with the Golgi complex. *Proc. Natl. Acad. Sci. USA*, **87**, 1238–1242.

Tait, R.M. and Nassau, P.M. (1984) Artifical low-molecular-mass substrates of cholera toxin. *Eur. J. Biochem.*, **143**, 213–219.

Takada, T., Iida, K. and Moss, J. (1993) Cloning and site-directed mutagenesis of human ADP-ribosylarginine hydrolyase. *J. Biol., Chem.* **268**, 17837–17843.

Takada, T., Okasaki, I.J. and Moss, J. (1994) ADP-ribosylarginine hydrolases. *Mol. Cell. Biochem.* **138**, 119–122.

Tamkun, J.W., Kahn, R.A., Kissinger, M., Brizuela, B.J., Rulka, C., Scott, M.P. and Kennison, J.A. (1991) The arflike gene encodes an essential GTP-binding protein in *Drosophila*. *Proc. Natl. Acad. Sci. USA*, **88**, 3120–3124.

Tanabe, T., Nukada, T., Nishikawa, Y., Sugimoto, K., Suzuki, K., Takahashi, H., Noda, M., Haga, T., Ichiyama, A., Kangawa, K., Minamino, N., Matsuo, H. and Numa, S. (1985) Primary structure of the alpha-subunit of transducin and its relationship to Ras proteins. *Nature*, **315**, 242–245.

Tanigawa, G., Orci, L., Amherdt, M., Ravazzola, M., Helms, J. and Rothman, J. (1993) Hydrolysis of bound GTP by ARF protein triggers uncoating of Golgi-derived COP-coated vesicles. *J. Cell Biol.*, **123**, 1365–1371.

Tanigawa, Y., Tsuchiya, M., Imai, Y. and Shimoyama, M. (1984) ADP-ribosyltransferase from hen liver nuclei. *J. Biol. Chem.*, **259**, 2022–2029.

Taussig, R. and Gilman, A.G. (1995) Mammalian membrane-bound adenylyl cyclases. *J. Biol. Chem.*, **270**, 1–4.

Teal, S., Hsu, V., Peters, P., Klausner, R. and Donaldson, J. (1994) An activating mutation in ARF1 stabilizes coatomer binding to Golgi membranes. *J. Biol. Chem.*, **269**, 3135–3138.

Terui, T., Kahn, R. and Randazzo, P. (1994) Effects of acid phospholipids on nucleotide exchange properties of ADP-ribosylation factor 1. *J. Biol. Chem.*, **269**, 28130–28135.

Thompson, T.E., Allietta, M., Brown, R.E., Johnson, M.L. and Tillack, T.W. (1985) Organization of ganglioside G_{M1} in phosphatidylcholine bilayers. *Biochim. Biophys. Acta*, **817**, 229–237.

Tomasi, M., Battistini, A., Araco, A., Roda, G. and D'Agnolo, G. (1979) The role of the reactive disulfide bond in the interaction of cholera toxin functional regions. *Eur. J. Biochem.*, **93**, 621–627.

Trepel, J.B., Chuang, D-M. and Neff, N.H. (1977) Transfer of ADP-ribose from NAD to choleragen: A subunit acts as a catalyst and acceptor protein. *Proc. Natl. Acad. Sci. USA*, **74**, 5440–5442.

Trucksis, M., Galen, J.E., Michalski, J., Fasano, A. and Kaper, J.B. (1993) Accessory cholera enterotoxin (Ace), the third toxin of a *Vibrio cholerae* virulence cassette. *Proc. Natl. Acad. Sci. USA*, **90**, 5267–5271.

Tsai, S-C., Adamik, R., Haun, R.S., Moss, J. and Vaughan, M. (1992) Differential interaction of ADP-ribosylation factors 1,3 and 5 with rat brain Golgi membranes. *Proc. Natl. Acad. Sci. USA*, **89**, 9272–9276.

Tsai, S-C., Adamik, R., Haun, R.S., Moss, J. and Vaughan, M. (1993) Effects of brefeldin A and accessory proteins on association of ADP-ribosylation factors 1, 3 and 5 with Golgi. *J. Biol. Chem.*, **268**, 10820–10825.

Tsai, S-C., Adamik, R., Moss, J. and Vaughan, M. (1991a) Guanine nucleotide dependent formation of a complex between choleragen (cholera toxin) A subunit and bovine brain ADP-ribosylation factor. *Biochemistry*, **30**, 3697–3703.

Tsai, S-C., Adamik, R., Moss, J. and Vaughan, M. (1994) Indentification of a brefeldin A-insensitive guanine nucleotide-exchange protein for ADP-ribosylation factor in bovine brain. *Proc. Natl. Acad. Sci. USA*, **91**, 3063–3066.

Tsai, S-C., Adamik, R., Tsuchiya, M., Chang, P.P., Moss, J. and Vaughan, M. (1991b) Differential expression during development of ADP-ribosylation factors, 20-kDa guanine nucleotide-binding protein activators of cholera toxin. *J. Biol. Chem.* **266**, 8213–8219.

Tsai, S-C., Haun, R.S., Tsuchiya, M., Moss, J. and Vaughan, M. (1991c) Isolation and characterization of the human gene for ADP-ribosylation factor 3, a 20-kDa guanine nucleotide-binding protein activator of cholera toxin. *J. Biol. Chem.*, **266**, 23053–23059.

Tsai, S-C., Noda, M., Adamik, R., Moss, J. and Vaughan, M. (1987) Enhancement of choleragen ADP-ribosyltransferase activities by guanyl nucleotides and a 19-kDa membrane protein. *Proc. Natl. Acad. Sci. USA*, **84**, 5139–5142.

Tsai, S-C., Noda, M., Adamik, R., Chang, P.P., Chen, H-C., Moss, J. and Vaughan, M. (1988) Stimulation of choleragen enzymatic activities by GTP and two soluble proteins purified from bovine brain. *J. Biol. Chem.*, **263**, 1768–1772.

Tsuchiya, M., Price, S.R., Nightingale, M.S., Moss, J. and Vaughan, M. (1989) Tissue and species distribution of mRNA encoding two ADP-ribosylation factors, 20-kDa guanine nucleotide binding proteins. *Biochemistry*, **28**, 9668–9673.

Tsuchiya, M., Price, S.R., Tsai, S-C., Moss, J. and Vaughan, M. (1991) Molecular identification of ADP-ribosylation factor mRNAs and their expression in mammalian cells. *J. Biol. Chem.*, **266**, 2772–2777.

Tsuji, T., Inoue, T., Miyama, A. and Noda, M. (1991) Glutamic acid-112 of the A subunit of heat-labile enterotoxin from enterotoxigenic *Escherichia coli* is important for ADP-ribosyltransferase activity. *FEBS Let.*, **291**, 319–321.

Tsuji, T., Inoue, T., Miyama, A., Okamoto, K., Honda, T. and Miwatani, T. (1990) A single amino acid substitution in the A subunit of *Escherichia coli* enterotoxin results in a loss of its toxic activity. *J. Biol. Chem.*, **265**, 22520–22525.

Uwiera, R.E., Romancyia, D.A., Wong, J.P. and Forsyth, G.W. (1992) Effect of covalent modification on the binding of cholera toxin B subunit to ileal brush border surfaces. *Anal. Biochem.*, **204**, 244–249.

Van Dop, C., Tsubokawa, M., Bourne, H.R. and Ramachandran, J. (1984) Amino acid sequence of retinal transducin at the site ADP-ribosylated by cholera toxin. *J. Biol. Chem.*, **259**, 696–698.

Verlinde, C.L.M.J., Merritt, E.A., Van den Akker, F., Kim, H., Feil, I., Delboni, L.F., Mande, S., C, Sarfaty, S., Petra, P.H. and Hol, W.G.J. (1994) Protein crystallography and infectious diseases. *Prot. Sci.*, **3**, 1670–1686.

Walker, M.W., Bobak, D.A., Tsai, S-C., Moss, J. and Vaughan, M. (1992) GTP but not GDP analogues promote association of ADP-ribosylation factors, 20-kDa protein activators of cholera toxin, with phospholipids and PC-12 cell membranes. *J. Biol. Chem.*, **267**, 3230–3235.

Waters, M.G., Serafini, T. and Rothman, J.E. (1991) 'Coatomer': A cytosolic protein complex containing subunits of non-clathrin-coated Golgi transport vesicles. *Nature*, **349**, 248–251.

Watkins, P.A., Moss, J. and Vaughan, M. (1980) Effects of GTP on choleragen-catalyzed ADP-ribosylation of membrane and soluble proteins. *J. Biol. Chem.*, **255**, 3959–3963.

Weidman, P. and Winter, W. (1994) The G protein-activating peptide, Mastoparan and the synthetic amino-terminal ARF peptide, ARFp13, inhibit *in vitro* Golgi transport by irreversibly damaging membranes. *J. Cell Biol.*, **127**, 1815–1827.

Welsh, C.F., Moss, J. and Vaughan, M. (1992) ADP-ribosylation factor, a guanine nucleotide-binding protein activator of cholera toxin, is isolated in an activated state when expressed as a fusion protein in *Escherichia coli. Trans. Assoc. Amer. Phys.*, **CV**, 172–176.

Welsh, C.F., Moss, J. and Vaughan, M. (1994) Isolation of recombinant ADP-ribosylation factor 6, a ~20-kDa guanine nucleotide-binding protein, in an activated GTP-bound state. *J. Biol. Chem.*, **269**, 15583–15587.

West, R.E., Jr. and Moss, J. (1986) Amino acid-specific ADP-ribosylation: Specific NAD:arginine mono-ADP-ribosyltransferases associated with turkey erythrocyte nuclei and plasma membranes. *Biochemistry*, **25**, 8057–8062.

Whatmore, J., Cronin, P. and Cockcroft, S. (1994) ARF1-regulated phospholipase D in human neutrophils is enhanced by PMA and MgATP. *FEBS Let.*, **352**, 113–117.

Wilson, R., Ainscough, R., Anderson, K., Baynes, C., Berks, M., Bonfield, J., Burton, J., Connell, M., Copsey, T., Cooper, J., Coulson, A., Craxton, M., Dear, S., Du, Z., Durbin, R., Favello, A., Fraser, A., Fulton, L., Gardner, A., Green, P., Hawkins, T., Hillier, L., Jier, M., Johnston, L., Jones, M., Kershaw, J., Kirsten, J., Laisster, N., Latreille, P., Lightning, J., Lloyd, C., Mortimore, B., O'Callaghan, M., Parsons, J., Percy, C., Rifken, L., Roopra, A., Saunders, D., Shownkeen, R., Sims, M., Smaldon, N., Smith, A., Smith, M., Sonnhammer, E., Staden, R., Sulston, J., Thierry-Mieg, J., Thomas, K., Vaudin, M., Vaughan, K., Waterson, R., Watson, A., Weinstock, L., Wilkinson-Sproat, J. and Wohldman, P. (1994) 2.2 Mb of contiguous nucleotide sequence from chromosome III of *C. elegans. Nature*, **368**, 32–38.

Wisnieski, B.J. and Bramhall, J.S. (1981) Photolabeling of cholera toxin subunits during membrane penetration. *Nature*, **289**, 319–321.

Xia, Q-C., Chang, D., Blacher, R. and Lai, C-Y. (1984) The primary structure of the COOH-terminal half of cholera toxin subunit A$_1$ containing the ADP-ribosylation site. *Arch. Biochem. Biophys.*, **234**, 363–370.

Yamamoto, T., Nakazawa, T., Miyata, T., Kaji, A. and Yokota, T. (1984) Evolution and structure of two ADP-ribosylation enterotoxins, *Escherichia coli* heat-labile toxin and cholera toxin. *FEBS*, **169**, 241–246.

Yatsunami, K. and Khorana, H.G. (1985) GTPase of bovine rod outer segments: The amino acid sequence of the alpha subunit as derived from the cDNA sequence. *Proc. Natl. Acad. Sci. USA*, **82**, 4316–4320.

Yeo, E-J. and Exton, J.H. (1995) Stimulation of phospholipase D by epidermal growth factor requires protein kinase C activation in swiss 3T3 cells. *J. Biol. Chem.*, **270**, 3980–3988.

Yost, D.A. and Moss, J. (1983) Amino acid-specific ADP-ribosylation. Evidence for two distinct NAD:arginine ADP-ribosyltransferases in turkey erythrocytes. *J. Biol. Chem.*, **258**, 4926–2929.

Zhang, C., Rosenwald, A.G., Willingham, M.C., Skuntz, S., Clark, J. and Kahn, R. (1994) Expression of a dominant allele of human ARF1 inhibits membrane traffic *in vivo*. *J. Cell Biol.*, **124**, 289–300.

Zhang, G-F., Patton, W.A., Lee, F-J.S., Liyanage, M., Han, J-S., Rhee, S.G., Moss, J. and Vaughan, M. (1995c) Different ARF domains are required for the activation of cholera toxin and phospholipase D. *J. Biol. Chem.*, **270**, 21–24.

Zhang, R-G., Scott, D.L., Westbrook, M.L., Nance, S., Spangler, B.D., Shipley, G.G. and Westbrook, E.M. (1995a) The three-dimensional crystal structure of cholera toxin. *J. Mol. Biol.*, **251**, 563–573.

Zhang, R-G., Westbrook, M.L., Westbrook, E.M., Scott, D.L., Otwinowski, Z., Maulik, P.R., Reed, R.A. and Shipley, G.G. (1995b) The 2.4Å crystal structure of cholera toxin B subunit pentamer: Choleragenoid. *J. Mol. Biol.*, **251**, 550–562.

Zolkiewska, A. and Moss, J. (1993) Integrin α7 as substrate for a glycosylphosphatidylinositol-anchored ADP-ribosyltransferase on the surface of skeletal muscle cells. *J. Biol. Chem.*, **268**, 25273–25276.

Zolkiewska, A., Nightingale, M.S. and Moss, J. (1992) Characterization of NAD:arginine ADP-ribosyltransferase from rabbit skeletal muscle. *Proc. Natl. Acad. Sci. USA*, **89**, 11352–11356.

Zolkiewska, A., Okazaki, I.J. and Moss, J. (1994) Vertebrate mono-ADP-ribosyltransferases. *Mol. Cell. Biochem.*, **138**, 107–112.

11. RHO SMALL GTP-BINDING PROTEINS: SUBSTRATES OF CLOSTRIDIUM BOTULINUM C3-LIKE ADP-RIBOSYLTRANSFERASES

GERTRUD KOCH and KLAUS AKTORIES

*Pharmakologisches Institut der Albert-Ludwigs-Universität,
Hermann-Herder-Str. 5, D-79104 Freiburg, FRG*

INTRODUCTION

So far more than 100 different low molecular mass GTP-binding proteins have been identified (Bourne, 1993). Based on their structural and functional differences these proteins have been classified into various families which share conserved domains that are essential for GTP-binding and hydrolysis. Like the signal transducing heterotrimeric G proteins, these low molecular mass GTP-binding proteins are regulated by a GTPase cycle. They are inactive in the GDP-bound state and activated by GDP/GTP exchange, which allows the GTP-bound form to interact with its specific effectors. The active state of the GTP-binding protein and the activation of the effector system are terminated by hydrolysis of bound GTP by the endogenous GTPase activity. Thus, nucleotide binding and hydrolysis functions as a binary switch to turn on or off distinct molecular signal-effector pathways. (Figure 11.1a).

Among the low molecular mass GTP-binding proteins, Rho proteins are unique with respect to specific ADP-ribosylation by various bacterial ADP-ribosyltransferases, i.e. the C3-like transferases (Figure 11.1b). Moreover, recent findings indicate that Rho proteins are also substrate for a different family of bacterial toxins (*Clostridium difficile* toxins) that selectively modify these GTP-binding proteins by mono-glucosylation (Just *et al.*, 1995b; Just *et al.*, 1995c). Here, the structure and function of Rho proteins and of C3-like ADP-ribosyltransferases are described and studies are reviewed where C3-like ADP-ribosyltransferases have been used to elucidate the Rho-signaling pathway. Several reviews on structural, functional and regulatory aspects of small GTP-binding proteins (Boguski and McCormick, 1993), particularly on Rho protein function (Hall, 1994; Nobes and Hall, 1994; Hall *et al.*, 1993; Narumiya and Morii, 1993; Lamarche and Hall, 1994; Takai *et al.*, 1995) and on C3 ADP-ribosyltransferases (Aktories and Koch, 1994; Aktories, 1994; Aktories *et al.*, 1992; Aktories, 1992; Aktories *et al.*, 1990) have been published recently.

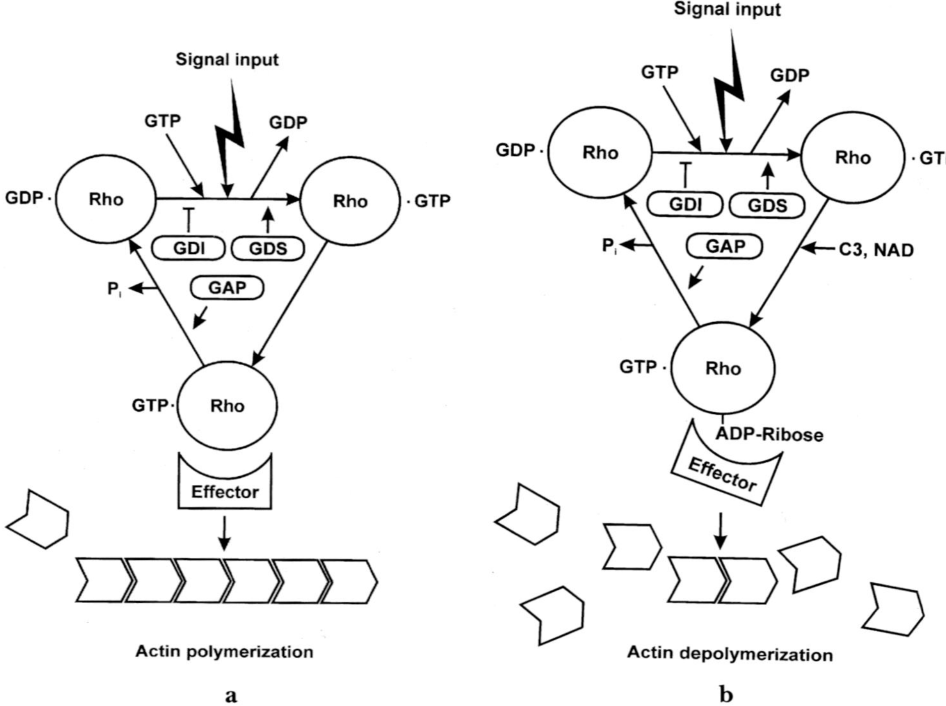

Figure 11.1 Schematic representation of the guanine nucleotide binding and hydrolysing cycle of Rho proteins. a) Normal sequence of events in Rho protein activation and inactivation (cf. text for details). GDI: guanine nucleotide dissociation inhibitor, GDS: guanine nucleotide dissociation stimulator, GAP: GTPase activating protein. b) ADP-ribosylation of Rho proteins prevents interaction with the effector and leads to actin depolymerization.

SEVERAL FAMILIES OF SMALL GTP-BINDING PROTEINS

According to function and sequence homologies, the group of small GTP-binding proteins has been divided into several subtype families. Among these are the Ras, Rab, Arf, Ran and Rho subtype families.

Ras proteins regulate cell proliferation and differentiation (Satoh *et al.*, 1992; Prendergast and Gibbs, 1993). The signal pathway involves activation of growth factor receptors, followed by Ras activation via adaptor molecules (Grb2) and nucleotide exchange factors (e.g. Sos). Ras subsequently activates via its effector Raf kinase the MAP kinase cascade which results in changes of nuclear transactivation (Vojtek *et al.*, 1993; Olivier *et al.*, 1993; Egan *et al.*, 1993; Rozakis-Adcock *et al.*, 1993). The elucidation of this pathway was a milestone in understanding of signal pathways regulated by low molecular mass GTP-binding proteins.

Rab and Arf proteins play a role in vesicle trafficking through the endocytic and exocytic compartment (Pfeffer, 1994; Novick and Brennwald, 1993; Balch, 1990; D'Souza-Schorey *et al.*, 1995; Lenhard *et al.*, 1992). At least 30 different Rab proteins have been identified and studies on their precise role are in progress. This will certainly increase our knowledge about intracellular trafficking. Arf proteins have also been recognized quite early as essential cofactors for cholera toxin induced ADP-ribosylation (Kahn *et al.*, 1988; Serventi *et al.*, 1992; Welsh *et al.*, 1994). Recently, Arf proteins were shown to regulate phospholipase D signal transduction (Cockcroft *et al.*, 1994). The members of the Ran subtype family are involved in nuclear transport and regulation of the cell cycle (Moore and Blobel, 1994). They play a role in DNA synthesis, RNA processing and trafficking of RNA and proteins between nucleus and cytosol (Ren *et al.*, 1993). Finally, the Rho protein family is involved in the organization of the actin cytoskeleton and apparently participates in regulation of various signal transduction pathways.

THE RHO PROTEIN FAMILY

The Rho (Rho = <u>R</u>as-<u>ho</u>molog) subtype family consists of RhoA, B, C, Rac1,2, Cdc42Hs (G25K), RhoG and TC10. Rho genes were first identified in a cDNA library of the marine snail *Aplysia californica* (Madaule and Axel, 1985). Subsequently, human homologs were recognized and termed RhoA, B, C (Chardin *et al.*, 1988; Yeramian *et al.*, 1987; Ogorochi *et al.*, 1989). On the amino acid level the various Rho proteins are about 85% identical and share about 35% homology with Ras. Rho proteins appear to be ubiquitously expressed. In the yeast *Saccharomyces cerevisiae*, four Rho protein homologs *Rho1-4p* have been found. *Rho1p* and *Rho2p* share 53% identity with each other. *Rho3p* and *Rho4p* display 35% homology to each other, and show 40% homology to other Rho proteins (Madaule *et al.*, 1987; Matsui and Toh-e, 1992; Yamochi *et al.*, 1994).

Rac1 and 2 (<u>R</u>as-related <u>C</u>3 botulinum toxin substrate) are 92% identical among each other and share 58% of the amino acids with RhoA, B and C (Didsbury *et al.*, 1989). They form a functional part of oxygen radical producing NADPH oxidase complex in neutrophils (Knaus *et al.*, 1992; Bokoch and Knaus, 1994) and macrophages (Abo *et al.*, 1991; Pick *et al.*, 1993). Rac proteins control membrane ruffling and lamellipodia formation in mouse fibroblasts (Ridley and Hall, 1992; Nobes and Hall, 1995) and regulate invasion of lymphocytes (Michiels *et al.*, 1995). Recently, Rac proteins were shown to participate in phospholipase A_2 regulation via growth factors (Peppelenbosch *et al.*, 1995). They appear to be essential for Ras-induced transformation (Qiu *et al.*, 1995) and are involved in activation of c-Jun N-terminal kinases (JNKs, also known as stress-activated protein kinases, SAPKs) (Minden *et al.*, 1995). A serine/threonine protein kinase (p65[PAK]) has been suggested to be an effector of Rac (Manser *et al.*, 1994).

Cdc42p was initially identified as a small GTP-binding protein in *S. cerevisiae* that regulates early stages of bud formation. The mammalian homolog Cdc42Hs was first recognized as an excellent substrate for tyrosine phosphorylation by epidermal growth factor receptor. It has a sequence identity of 80% with yeast *Cdc42p* (Shinjo *et al.*, 1990). Recently, the role of Cdc42Hs in reorganization of the cortical cytoskeleton has further been characterized indicating that Cdc42Hs is involved in formation of microspikes and filopodia (Kozma *et al.*, 1995; Nobes and

Hall, 1995). According to these studies, it has been suggested that all these Rho subtype proteins (Cdc42, Rac and Rho) regulate the actin cytoskeleton in a cascade like manner from Cdc42 protein via Rac to Rho proteins.

RhoG accumulates in late G1 of the cell cycle and has higher homologies with Rac than with Rho (Vincent *et al.*, 1992). Like Rac it is not or only a very poor substrate for C3 ADP-ribosylation (unpublished data). TC10 was identified in a human teratocarcinoma cell line (Drivas *et al.*, 1990). Most likely, TC10 is not a substrate for C3 ADP-ribosyltransferase, because it has no asparagine residue in the position equivalent to Asn41 of Rho which is modified by C3.

THE STRUCTURE OF RHO PROTEINS

The α-subunits of heterotrimeric G proteins, elongation factors and small GTP-binding proteins share four sequence motifs involved in guanine nucleotide binding and hydrolysis (Bourne *et al.*, 1990). In RhoA they are located between Gly12 and Thr19 (G-1 region), around the highly conserved Thr37 (G-2 region), between Asp59 and Gly62 (G-3 region) and between Asn117 and Asp120 (G-4 region). By sequence comparison with Ras, region G-2 is also a site for interaction with the effector (Pai *et al.*, 1989). As shown in Figure 11.2, inactivating modifications of Rho by bacterial toxins take place in this putative effector region (see below).

Rho proteins are posttranslationally modified by polyisoprenylation at Cys190 (RhoA and C) or Cys193 (RhoB) of the COOH-terminal CAAX-box (Figure 11.2). C represents cysteine, A an aliphatic amino acid and X any amino acid. RhoA and RhoC contain a C20 geranylgeranyl group, whereas RhoB is modified either by a C15 farnesyl group or a C20 geranylgeranyl group (Adamson *et al.*, 1992). After

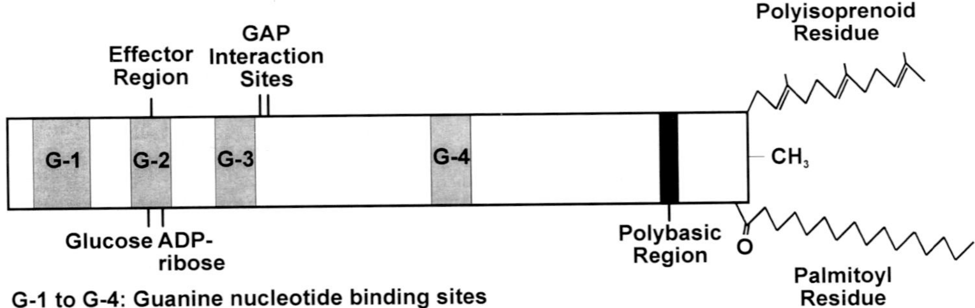

Figure 11.2 Structure of Rho proteins. Grey boxes indicate regions G-1 to G-4 of Rho protein involved in guanine nucleotide binding and hydrolysis. The black box shows localization of the membrane binding polybasic region of RhoA and RhoC. COOH-terminal modifications by methylation (RhoA, B, C), polyisoprenylation and palmitoylation (only for RhoB) are shown. In RhoA and C the polyisoprene residue is a geranylgeranyl residue and in RhoB it is a geranylgeranyl or a farnesyl residue. Positions of ADP-ribosylation and mono-glucosylation are shown.

covalent linkage of polyisoprenoid residues, the three COOH-terminal amino acids are cleaved and the COOH-terminus is carboxymethylated. In addition, a palmitoyl residue is suggested to be covalently linked to Cys189 and Cys192 of RhoB. RhoA and C contain polybasic regions (Arg182-Lys187). Palmitoylation, polyisoprenylation and the COOH-terminal polybasic residues play a role in membrane binding of Rho proteins and are important for the interaction of the GTP-binding proteins with regulatory proteins (see below), (Hori *et al.*, 1991; Glomset and Farnsworth, 1994).

THE GTPASE CYCLE OF RHO PROTEINS

Figure 11.1a shows a model of Rho protein activation and inactivation through GTP binding and hydrolysis. Upon signal input, GDP bound to Rho protein is exchanged for GTP which is abundantly present in the cytosol. GTP-bound Rho protein activates the effector which mediates down-stream transmission of the signal. Subsequently, Rho protein inactivates itself by GTP hydrolysis. Thereafter, it is again available for a new activation cycle.

Several proteins have been described which interfere with the GTPase cycle of Rho proteins (Figure 11.1a) (Hall, 1990). Guanine nucleotide dissociation inhibitors (GDI) for Rho, Rac and Cdc42 prevent dissociation of GDP and subsequent binding of GTP (Ueda *et al.*, 1990; Masuda *et al.*, 1994; Lelias *et al.*, 1993; Hiraoka *et al.*, 1992; Leonard *et al.*, 1992). GDIs are also able to extract Rho proteins from the cell membrane and keep them in their inactive forms in the cytosol.

Guanine nucleotide dissociation stimulators (GDS or GEF = guanine nucleotide exchange factor) increase nucleotide exchange at small GTP-binding proteins. Meanwhile a whole group of GDS proteins has been described (Feig, 1994). S*mg* GDS stimulates dissociation of GDP and subsequent association of GTP at RhoA, Rac1, Cdc42Hs, Rap1B, K-Ras and Smg p21 (Hiraoka *et al.*, 1992; Yaku *et al.*, 1994). Thus, this protein is rather unspecific. *Smg*GDS shows weak homology with Ras protein guanine nucleotide dissociation stimulators but not with Dbl (Kaibuchi *et al.*, 1991). The Dbl oncogene product acts specifically at members of the Rho protein family and stimulates guanine nucleotide exchange at RhoA, Rac1 and Cdc42Hs (Hart *et al.*, 1991; Yaku *et al.*, 1994). Dbl homology regions have been detected in numerous other proteins, e.g. Tiam1 (Habets *et al.*, 1994), Ost (Horii *et al.*, 1994), ect2 (Miki *et al.*, 1993), Cdc24 (Whitehead *et al.*, 1995), Bcr (Ridley *et al.*, 1993), Vav (Gulbins *et al.*, 1994; Adams *et al.*, 1992), Ras-GRF (Wei *et al.*, 1994), ABR (Heisterkamp *et al.*, 1993), FGD1 (Pasteris *et al.*, 1994) and Dbs (Whitehead *et al.*, 1995), TIM (Chan *et al.*, 1995) and Lbc (Zheng *et al.*, 1995). For most of these proteins a physiological role in activation of Rho subtype proteins is still unclear. Recently, it has been reported that Tiam1 which is able to turn non-invasive lymphoma cells into invasive cells activates Rac1 and Cdc42Hs and to a lesser extent RhoA (Habets *et al.*, 1994). On the other hand, Lbc activates RhoA, B, C but not Rac or Cdc42 (Zheng *et al.*, 1995). Ost, a focus forming protein isolated from an osteosarcoma cDNA expression library, catalyzes guanine nucleotide exchange at RhoA and Cdc42 but has low or no activity at other Rho subtype proteins (Horii *et al.*,1994).

A whole group of Rho GTPase-activating proteins (GAP) has been identified which stimulate GTP hydrolysis by Rho proteins (Lamarche and Hall, 1994). RhoGAP (Morii *et al.*, 1991; Lancaster *et al.*, 1994), Bcr (Diekmann *et al.*, 1991; Ridley *et al.*, 1993), chimaerin (Diekmann *et al.*, 1991; Ridley *et al.*, 1993; Leung *et al.*, 1993), Ras-GAP-binding protein p190 (Settleman *et al.*, 1992), myosin subtype Myr5 (Reinhard *et al.*, 1995) and phospholipase Cδ1-binding p122-RhoGAP (Homma and Emori, 1995) display GAP activity. In addition, the p85 subunit of PI 3-kinase, Abr, Bem3, rotund and 3BP-1 contain regions of RhoGAP homology (Lamarche and Hall, 1994). So far, however, a role of these proteins in regulation of Rho proteins has not been demonstrated.

C3-LIKE RHO PROTEIN ADP-RIBOSYLTRANSFERASES

ADP-ribosylation of nucleotide-binding proteins is a well-recognized mechanism by which bacterial protein toxins affect eukaryotic cells (Aktories and Just, 1993; Aktories, 1994). These toxins use NAD as cosubstrate and attach ADP-ribose onto eukaryotic proteins. In general, attachment of the bulky ADP-ribose group causes dramatic changes in protein function. For example, diphtheria toxin and *Pseudomonas aeruginosa* exotoxin A ADP-ribosylate elongation factor II to inhibit protein synthesis (Perentesis *et al.*, 1992). Pertussis toxin (Ui, 1990), cholera toxin and *E. coli* heat-labile enterotoxins (Spangler, 1992) ADP-ribosylate heterotrimeric G proteins and interfere with receptor mediated signal transduction. Furthermore, toxins produced by *Clostridium botulinum* (*C. botulinum* C2 toxin) (Aktories *et al.*, 1986), *C. perfringens* (iota toxin), (Schering *et al.*, 1988) and *C. spiroforme* (Simpson *et al.*, 1989) disrupt the cytoskeleton by ADP-ribosylation of actin (Aktories, 1994).

During screening for high producer strains of *C. botulinum* C2 toxin, we serendipitously detected *C. botulinum* ADP-ribosyltransferase C3 (Aktories *et al.*, 1987; Aktories *et al.*, 1988). In *C. botulinum* culture supernatant, apart from ADP-ribosylation of 42 kDa actin (Aktories *et al.*, 1986), we detected ADP-ribosylation of about 20 kDa human platelet membrane proteins which later were identified as small GTP-binding proteins of the Rho protein family. This novel ADP-ribosyltransferase proved to be distinct from *C. botulinum* neurotoxin C1 and actin ADP-ribosyltransferase C2 toxin and was, therefore, termed *C. botulinum* ADP-ribosyltransferase C3 (Aktories and Frevert, 1987).

STRUCTURE-FUNCTION RELATIONSHIPS OF C3-LIKE TRANSFERASES

The C3 ADP-ribosyltransferase gene is encoded by a DNA bacteriophage (Rubin *et al.*, 1988), which also carries the genes for neurotoxin C1 and D (Popoff *et al.*, 1991) and for hemagglutinin (Tsuzuki *et al.*, 1990). Cloning and sequencing of C3 ADP-ribosyltransferase genes from various *C. botulinum* strains showed that they were not identical. They rather encoded a whole group of isoenzymes (Figure 11.3a). C3-like ADP-ribosyltransferases are also produced by *Clostridium limosum* (Just *et al.*,

1992a), *Bacillus cereus* (Just *et al.*, 1992b) and *Staphylococcus aureus* (Inoue *et al.*, 1991). They are basic proteins (pI 9-10) and have molecular masses between 25 and 28 kDa. In Figure 11.3b their pairwise identity on the amino acid level and in Figure 11.3 a dendrogram with their hypothetical phylogenetic relationship is shown. Thus, C3 ADP-ribosyltransferases sequenced by Moriishi *et al.* (1991) and Nemoto *et al.* (1991) have an identity of 98%, whereas identity between C3-like exoenzymes from *C. botulinum* and *S. aureus* (EDIN = epidermal differentiation inhibitor) is as low as 30%.

During the ADP-ribosylation reaction, C3-like ADP-ribosyltransferases bind NAD, cleave NAD to ADP-ribose and nicotinamide and transfer the ADP-ribose moiety to Asn41 of Rho proteins (Figure 11.4). An NAD-interacting site of *C. limosum* exoenzyme was identified in our laboratory by UV-irradiation catalyzed linkage of NAD to the exoenzyme. Subsequently, in proteolytic digests of the exoenzyme, the amino acid with covalently bound NAD was found to be glutamic acid at position 174 of the transferase (Jung *et al.*, 1993). Because mutations of *C. limosum* exoenzyme at position 174 (E174Q and E174D) showed no major change in affinity for NAD but a more than 1000-fold reduced ADP-ribosyltransferase and NAD-glycohydrolase activity, Glu174 appears to be part of the active site of the transferase (Aktories *et al.*, 1995). In agreement with this hypothesis, a homolog glutamic acid is found at the same position in all C3-like transferases studied so far. Glutamic acid residues participating in NAD catalysis have also been identified in bacterial ADP-ribosyl-transferases like diphtheria toxin (Carroll and Collier, 1984), *Pseudomonas aeruginosa* exotoxin A (Carroll and Collier, 1987) and pertussis toxin (Barbieri *et al.*, 1989) and appear to be present in all bacterial ADP-ribosyltransferases.

By amino acid sequence comparison of various transferases three regions were suggested to be responsible for NAD-binding and hydrolysis. Region I contains histidine or arginine. About 50–75 amino acids downstream, region II contains aromatic amino acids (tyrosine or phenylalanine) or hydrophobic amino acids. About 130–170 amino acids downstream of region I, region III contains strictly conserved glutamic acid residues (Takada *et al.*, 1995). Arginine residues and glutamic acid residues corresponding to Arg51 (region I) and Glu174 (region III) of *C. limosum* exoenzyme are strictly conserved in all C3-like ADP-ribosyltransferases together with flanking amino acids (Figure 11.3a). As postulated above, distance between these two regions is about 130 amino acids. It is speculated that these arginine residues like glutamic acid play a role in NAD catalysis. Another cluster of well conserved amino acids can be found at positions corresponding to amino acids 130 to 137 (region II) of *C. limosum* exoenzyme. Again, as predicted above, the distance from region I is about 75 amino acids. This region contains conserved tyrosine residues as aromatic amino acids and an STS-motif which is also found in the family of ADP-ribosyltransferases consisting of cholera toxin, pertussis toxin and *E. coli* heat-labile enterotoxins 1 and 2 (Domenighini *et al.*, 1994). In conclusion, the amino acid composition and the distances of the three regions conserved within the group of C3-like ADP-ribosyltransferases correspond to the three regions identified in other bacterial and eukaryotic ADP-ribosyltransferases responsible for NAD-binding and catalysis.

```
        LIM    PYADSFKEFTNIDEARAWGDKQFAKYKLSSSEKNALTIYTRNAARINGPL    50

        C3M    SYADTFTEFTNVEEAKKWGNAQYKKYGLSKPEQEAIKFYTRDASKINGPL    50

        C3N    SYADTFTEFTNVEEAKKWGNAQYKKYGLSKPEQEAIKFYTRDASKINGPL    50

        C3P    AYSNTYQEFTNIDQAKAWGNAQYKKYGLSKSEKEAIVSYTKSASEINGKL    50

        EDIN   A---DVKNFTDLDEATKWGNKLIKQAKYSSDDKIALYEYTKDSSKINGPL    47

                .     .**...*  **.       *. .. *.  **.... *** *

        Region I
        LIM    RANQGNTNGLPADIRKEVEQIDKSFTKMQTPENIILFRGDDPGYLGP--D    98

        C3M    RANQGNENGLSSDILQKVKLIDQSFSKMKMPQNIILFRGDDPAYLGT--E    98

        C3N    RANQGNENGLPADILQKVKLIDQSFSKMKMPQNIILFRGDDPAYLGP--E    98

        C3P    RQNKGVINGFPSNLIKQVELLDKSFNKMKTPENIMLFRGDDPAYLGT--E    98

        EDIN   RLAGGDINKLDSTTQDKVRRLDSSISKSTTPESVYVYRLLNLDYLTSIVG   97

                *  . *  *  . ..   ..*  .* *..*   *... ..*  . .**..  .

                                            Region II
        LIM    FENTILNRDGTINKAVFEQVKLRFKGK--DRKEY---GYISTSLVNGSAF   143

        C3M    FQDKILNKDGTINRDVFEQVKAKFLKK--DRTEY---GYISTSLMSA-QF   142

        C3N    FQDKILNKDGTINKTVFEQVKAKFLKK--DRTEY---GYISTSLMSA-QF   142

        C3P    FQNTLLNSNGTINKTAFEKAKAKFLNK--DRLEY---GYISTSLMNVSQF   143

        EDIN   FTNEDLYKLQQTNNGQYDENLVRKLNNVMNSRIYREDGYSSTQLVSGAAV   147

                *  .. * .   .*.. ...    .    .  .. *   ** ** *..   .

                                            Region III
        LIM    AGRPIITKFKVLDGSKAGYIEP--ISTFKGQLEVLLPRSSTYTISDMQIA   191

        C3M    GGRPIVTKFKVTNGSKGGYIDP--ISYFPGQLEVLLPRNNSYYISDMQIS   190

        C3N    GGRPIVTKFKVTNGSKGGYIDP--ISYFPGQLEVLLPRNNSYYISDMQIS   190

        C3P    AGRPIITKFKVAKGSKAGYIDP--ISAFAGQLEMLLPRHSTYHIDDMRLS   191

        EDIN   GGRPIELRLELPKGTKAAYLNSKDLTAYYGQQEVLLPRGTEYAVGSVELS   197

                .****   ....  .*.*..*...  .. . ** *.**** ..*  .... ..


        LIM    PNNKQIIITALLK------R    205

        C3M    PNNRQIMITAMIF------K    204

        C3N    PNNRQIMITAMIF------K    204

        C3P    SDGKQIIITATMMGTAINPK    211

        EDIN   NDKKKIIITAIVF-----KK    212
```
a   . ..*.*** .   .

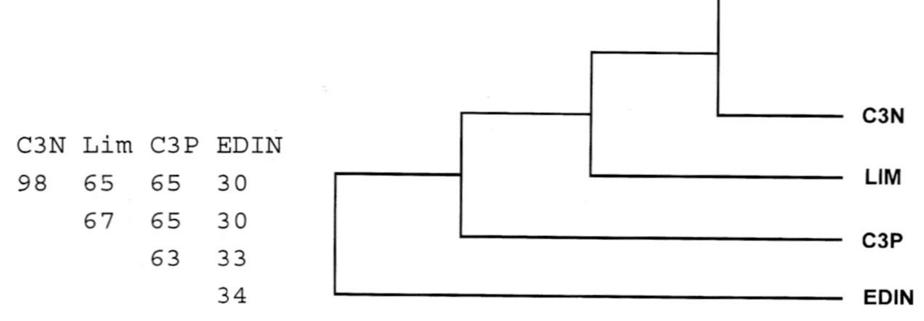

b   **c**

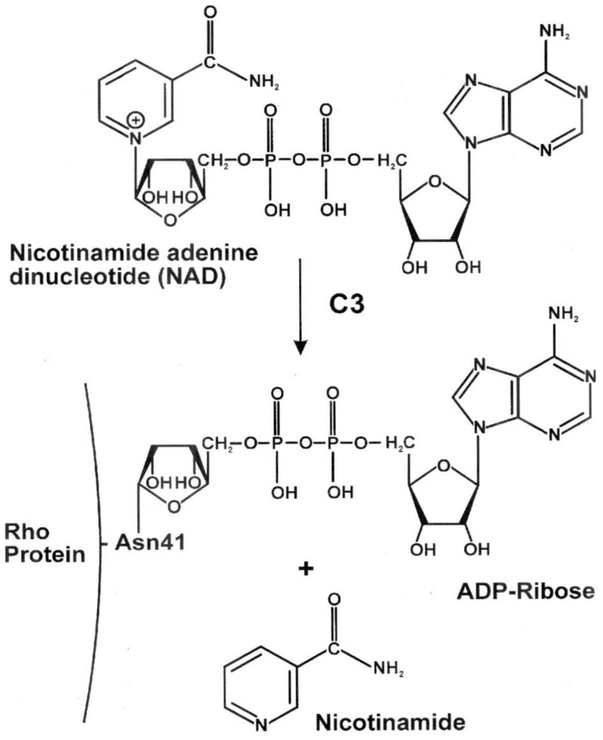

Figure 11.4   Schematic representation of the ADP-ribosyltransferase reaction catalyzed by C3-like ADP-ribosyltransferases. Nicotinamide adenine dinucleotide (NAD) from the cytosol is hydrolysed by C3-like ADP-ribosyltransferases to ADP-ribose and nicotinamide. ADP-ribose is covalently bound to Asn41 of Rho proteins.

◄ **Figure 11.3**   Comparison of amino acid sequences of C3-like ADP-ribosyltransferases sequenced by different groups (C3M: *C. botulinum* ADP-ribosyltransferase C3, Moriishi *et al.*, 1991; C3N: *C. botulinum* ADP-ribosyltransferase C3, Nemoto *et al.*, 1991; C3P: *C. botulinum* ADP-ribosyltransferase C3, Popoff *et al.*, 1991; EDIN: epidermal differentiation inhibitor from *S. aureus*, Inoue *et al.*, 1991; LIM: *C. limosum* exoenzyme, Just *et al.*, 1992a). a) Multiple amino acid sequence comparison of C3-like ADP-ribosyltransferases was performed with the computer program PCGENE, IntelliGenetics Inc., Switzerland. Asteriks show that an amino acid position in the alignment is perfectly conserved and dots show positions of conservative amino acid substitutions. Regions I-III assumed to bind NAD and catalyze ADP-ribosylation are indicated. b) Pairwise percentage determination of identity (upper panel) at the amino acid level of the C3-like ADP-ribosyltransferases. Similar amino acids are: A, S, T; D, E; N, Q; R, K; I, L, M, V and F, Y, W. c) Hypothetical phylogenetic tree obtained from pairwise similarity scores using the computer program PCGENE.

IN VITRO STUDIES OF THE C3 ADP-RIBOSYLTRANSFERASE REACTION

Substrates for ADP-ribosylation by C3-like transferases are RhoA, B, C which are ADP-ribosylated in Asn41 (Sekine *et al.,* 1989). Other Rho subtype proteins including Rac are poor or not at all *in vitro* substrates for ADP-ribosylation by C3. Accordingly, only RhoA, B, C proteins have been identified as substrates for C3 in intact cells.

The ADP-ribosyltransferase reaction catalyzed by C3-like ADP-ribosyltransferases is a mono-ADP-ribosylation and not a poly-ADP-ribosylation reaction. Phosphodiesterase releases 5′-AMP from ADP-ribosylated Rho and not phosphoribosyl-AMP, a cleavage product of poly(ADP-ribose)(Aktories *et al.,* 1988; Rubin *et al.,* 1988). Accordingly, the ADP-ribosyltransferase reaction is not blocked by thymidine, an inhibitor of poly(ADP-ribose)polymerase.

The K_m for NAD of C3 ADP-ribosyltransferase and *B. cereus* exoenzyme were calculated to be about 0.5 μM in cell lysates (Just *et al.,* 1992a; Just *et al.,* 1995a). However, the K_m (20 μM) is considerably higher with purified recombinant Rho proteins and recombinant *C. limosum* transferase. It appears that additional cellular factors decrease the K_m for NAD. ADP-ribosylation of Rho proteins by C3 is reversible in the presence of high concentrations of nicotinamide (10 mM) and at low pH (pH < 7) (Habermann *et al.,* 1991). As it is known for other ADP-ribosyltransferases, C3-like exoenzymes exhibit NAD hydrolase activity (Aktories *et al.,* 1988). This enzyme activity is at least 100 fold lower than the transferase activity and most likely without physiological significance.

Various detergents increase (e.g. sodium cholate, sodium dodecylsulfate at low concentrations) or decrease (e.g. CHAPS, Lubrol-PX, sodium dodecyl sulfate at higher concentrations) ADP-ribosylation of Rho proteins by C3-like transferases (Maehama *et al.,* 1990; Just *et al.,* 1993). Enhancement of ADP-ribosylation by detergents may partly be attributed to detergent disruption of the complex formed between Rho and Rho guanine nucleotide dissociation inhibitor (GDI) because GDI-bound Rho is not susceptible to ADP-ribosylation (Kikuchi *et al.,* 1992). Recently it has been shown that gangliosides like G_{T1b} and G_{q1b} inhibit ADP-ribosylation of Rho proteins by C3 (Hara-Yokoyama *et al.,* 1995). So far the significance of this finding is not clear.

Rho proteins are guanine nucleotide-binding as well as Mg^{2+}-binding proteins. At low magnesium concentrations ADP-ribosylation is usually increased by guanine nucleotides. This is probably due to the fact that without magnesium, guanine nucleotides are easily lost from Rho proteins leading to destabilization of the protein (Habermann *et al.,* 1991). Consequently, binding of guanine nucleotides to Rho proteins prevents proteolytic degradation and thereby ADP-ribosylation is increased. At higher magnesium concentrations ADP-ribosylation of membrane-bound Rho protein was higher with GDP than with GTP (Habermann *et al.,* 1991). In contrast, ADP-ribosylation of cytosolic Rho proteins was enhanced by GTP or GTPγS in comparison to GDP (Williamson *et al.,* 1990). This effect is most likely due to GTP or GTPγS-induced dissociation of Rho from the Rho-GDI complex.

C3 exoenzymes ADP-ribosylate RhoA in Asn41. This amino acid is located in the G-2 region which is also known as the effector region of the protein (Figure 11.2).

ADP-ribosylation does not change basal, i.e. single cycle GTPase activity of Rho very much (Paterson *et al.*, 1990). Steady-state GTPase activity, comprising several cycles of guanine nucleotide exchange, is increased by ADP-ribosylation by about 50% (Mohr *et al.*, 1992).

C3-LIKE TRANSFERASES AS TOOLS TO STUDY CELLULAR FUNCTIONS OF RHO PROTEINS

Evidence that C3 exoenzyme acts on the cytoskeleton was first reported by Rubin *et al.* (1991). Because C3 is difficult to introduce into cells, it was applied by osmotic shock. In subsequent studies, C3 was microinjected (Paterson *et al.*, 1990) or introduced by electroporation (Koch *et al.*, 1994) or pore-forming agents (Norman *et al.*, 1994). In some cases, C3 was used at rather high concentrations (50–100 μg/ml) to enable cell entry (Wiegers *et al.*, 1991). After microinjection, effects on cell morphology are visible usually within 30 min. In contrast, treatment of cultured cells (even with high concentrations of C3) requires 12 to 48 h to reveal C3 effects on the cytoskeleton. This latency is due to the slow uptake of the transferase which occurs most likely by unspecific endocytosis.

Typically C3 causes rounding up and retraction of the cell body. These morphological changes are accompanied by a decrease or loss of F-actin staining by FITC-phalloidin (Wiegers *et al.*, 1991; Paterson *et al.*, 1990; Chardin *et al.*, 1989). It has also been shown that intermediate filaments redistribute after treatment with C3, however, this effect occurs subsequently to depolymerization of F-actin (Wiegers *et al.*, 1991). In contrast, the microtubular network is much less affected (Chardin *et al.*, 1989; Wiegers *et al.*, 1991) indicating that C3-like transferase activity is specific. Induction of actin stress fiber formation and formation of focal adhesions in mouse fibroblast Swiss 3T3 cells by either dominant active Rho (Val14Rho), lysophosphatic acid (LPA) or bombesin which may act via G protein-coupled receptors is antagonized by C3 ADP-ribosylation (Paterson *et al.*, 1990; Ridley and Hall, 1992). Also, PDGF-induced formation of stress fibers is inhibited by C3. In contrast, membrane ruffling induced by PDGF remains unaffected because it is regulated by Rac (Ridley and Hall, 1992; Ridley *et al.*, 1992). In the same system C3-induced inhibition of focal adhesion formation could be bypassed by microinjection of constitutively activated Rac1 (Nobes and Hall, 1995). These findings support the view that in intact cells Rac is not a substrate for the C3 ADP-ribosyltransferase.

Moreover, ADP-ribosylation of Rho by C3 was used to study cellular functions which are dependent on the cytoskeleton. It was shown that C3 exoenzyme inhibits basal and FMLP-induced neutrophil motility (Stasia *et al.*, 1991). In fertilized *Xenopus laevis* oocytes, cytoplasmic but not nuclear division is inhibited by exoenzyme C3 (Kishi *et al.*, 1993). Also, ADP-ribosylation of Rho decreases bovine sperm motility (Hinsch *et al.*, 1993) and Swiss 3T3 cell phagokinesis (Takaishi *et al.*, 1993). Hepatocyte growth factor induced motility of cultured mouse keratinocytes is also diminished by ADP-ribosylation of Rho (Takaishi *et al.*, 1994). Further, lymphocyte cytotoxicity (Lang *et al.*, 1992) and platelet aggregation (Morii *et al.*, 1992) are inhibited by ADP-ribosylation. In rat peritoneal mast cells, the effect of C3 was studied on compound 48/80-induced actin polymerization (Norman *et al.*, 1994). It was shown that only certain actin compartments are affected by C3-induced ADP-ribosylation but not the complete actin cytoskeleton (Norman *et al.*, 1994). Moreover, in human myeloid leukaemic cells (HL60) polymerization of the actin cytoskeleton induced by N-formyl peptides was even increased after C3 treatment of cells (Koch *et al.*, 1994) supporting the view that Rho-dependent regulation of the actin cytoskeleton is one of several ways to control actin polymerization.

Direct influence of RhoA protein on smooth muscle contraction has also been shown. In rabbit arterial smooth muscle GTPγS enhanced calcium-dependent contraction. This increase in calcium sensitivity was abolished by EDIN and C3 transferase (Hirata *et al.*, 1992).

Recent findings demonstrate that Rho not only participates in organization of the actin cytoskeleton but also functions in cellular events which appear to be largely independent of the actin cytoskeleton. For example it has been shown that C3 inhibits lymphocyte aggregation independent of actin polymerization (Tominaga *et al.*, 1993). Recently, it was demonstrated that Rho participates in constitutive endocytosis in *Xenopus laevis* oocytes, also in an actin independent manner (Schmalzing *et al.*, 1995). Inactivation of Rho protein by microinjection of C3 ADP-ribosyltransferase led to an increase of cell surface area, membrane capacitance and the number of surface sodium pumps. Electron micrographs showed elongated and convoluted microvilli. Kinetic studies revealed that a decrease of constitutive endocytosis rather than an increase of exocytosis caused this enlargement of surface area. These C3 effects were also observed after microinjection of ADP-ribosylated RhoA. Moreover, C3 inhibited phorbol ester-induced fluid phase endocytosis indicating that protein kinase C acts upstream of Rho proteins to stimulate endocytosis.

INFLUENCE OF C3-LIKE ADP-RIBOSYLTRANSFERASES ON SIGNAL TRANSDUCTION PATHWAYS

In contrast to the Ras signaling pathway, a complete pathway for Rho signaling cannot be described until now. Proteins involved in signal transduction from activated Rho protein to actin are still obscure. Rho neither directly interferes with polymerization of fluorescently labeled actin *in vitro* (unpublished data) nor is it located in focal adhesions where actin stress fibers are formed (Adamson *et al.*, 1992). Exoenzyme C3 was used by various groups to examine whether Rho is a signal transmitter in established signaling pathways.

A possible involvement of Rho in signaling pathways comprising heterotrimeric G proteins and their respective receptors was shown rather early. For example, photoexcitation of rhodopsin in a membrane preparation from bovine retina diminished Rho ADP-ribosylation by exoenzyme C3 (Wieland *et al.*, 1990a; Wieland *et al.*, 1990b).

Recent studies indicate that Rho proteins are somehow involved in regulation of PI 3-kinase activity. PI 3-kinase consists of a p85 regulatory subunit and a p110 catalytic subunit and associates with tyrosine kinase as well as non-tyrosine kinase receptors (Stephens *et al.*, 1993). This enzyme attaches a phosphate group to the inositol ring of phosphatidyl inositol (PI) or its derivatives $PI(4)P$, $PI(3,4)P_2$ and $PI(3,4,5)P_3$. The regulatory p85 subunit has been shown to contain a RhoGAP homology region (Otsu *et al.*, 1991). This points to a functional link with the Rho signaling pathway.

C3 ADP-ribosyltransferase was applied as a tool to study a possible involvement of Rho proteins in the PI 3-kinase signaling pathway. By treatment of Swiss 3T3 cells with the transferase, PI 3-kinase activity was found to be inhibited in immunoprecipitates of C3-treated cells (Kumagai *et al.*, 1993). Also, in human platelet supernatant, activation of PI 3-kinase with GTPγS is inhibited by C3 ADP-ribosyltransferase (Zhang *et al.*, 1993). Moreover, EDIN, a C3-like transferase from *Staphylococcus aureus*, was shown to inhibit GTPγS-stimulated PI 3-kinase activity immunoprecipitated from platelet cytosol (Zhang *et al.*, 1995). So far, however, a direct interaction of Rho with PI 3-kinase is unclear, because PI 3-kinase activity in p85 immunoprecipitates from PC12 neuronal cells is stimulated by GTPγS-Rac1 and GTPγS-Cdc42, but not by GTPγS-RhoA (Zheng *et al.*, 1994). RhoA has also been suggested to control PI 4,5-kinase in a C3 dependent manner (Chong *et al.*, 1994).

As mentioned above, in Swiss 3T3 cells, bombesin or lysophosphatidic acid (LPA) rapidly stimulate formation of actin stress fibers and focal adhesions. Concomitantly, p125 focal adhesion kinase (p125 FAK) and paxillin which form part of focal adhesion plaques are phosphorylated (Sinnett-Smith *et al.*, 1993). C3 exoenzyme decreased bombesin and endothelin-stimulated tyrosine phosphorylation of p125 FAK and paxillin in Swiss 3T3 cells (Rankin *et al.*, 1994). In addition, C3 treatment of Swiss 3T3 cells inhibited LPA-stimulated tyrosine phosphorylation of p125 FAK, p43 mitogen-activated protein kinase and of two phosphorylated 72 and 88 kDa proteins (Seckl *et al.*, 1995; Kumagai *et al.*, 1993). The data suggest that Rho protein may be inserted in a signaling pathway between LPA or neuropeptide receptors and focal adhesions.

Phospholipase D plays a role in receptor- and protein kinase C-linked signal transduction (Ohguchi *et al.*, 1995; Malcolm *et al.*, 1994). It generates the biologically active molecule phosphatidic acid from phosphatidylcholine. In rat liver membranes activity of phospholipase D can be stimulated by GTPγS. Extraction of Rho proteins with RhoGDI from membranes diminished GTPγS-stimulated phospholipase D activity. This activity could be reconstituted by addition of RhoA. Surprisingly, however, this effect was not antagonized by prior C3 ADP-ribosylation of RhoA (Malcolm *et al.*, 1994). In contrast, in Rat 1 fibroblasts, C3 caused inhibition of Rho-mediated agonist stimulation of phospholipase D (Malcolm *et al.* 1996).

In summary, within these signaling pathways Rho protein obviously acts upstream of transphosphorylating enzymes analogous to Ras proteins which act upstream of MAP kinases.

OTHER TOXINS MODIFYING RHO SUBTYPE PROTEINS

Apart from ADP-ribosylating toxins, other Rho protein-modifying bacterial toxins have been identified recently. Most exciting is the finding that Rho subtype proteins are glucosylated by *C. difficile* toxin A and B (Figure 11.5). Toxin A and B are produced by certain strains of *Clostridium difficile* and are postulated to be the major virulence factors for inducing antibiotic-associated diarrhea and pseudomembranous colitis (Bartlett, 1990; Lyerly *et al.*, 1988; Kelly *et al.*, 1994). These toxins have a molecular mass of 308 and 270 kDa, respectively. Both toxins are cytotoxic, cause rounding up as well as arborization of cultivated cells by disruption of the actin cytoskeleton. The striking similarity to effects of C3-like ADP-ribosyltransferases on cell morphology has led to investigating the influence of these toxins on Rho proteins. Initially, a decrease in ADP-ribosylation by C3 was observed in toxin B treated cells (Just *et al.*, 1994). Subsequently, it was shown that toxin B mono-glucosylates Rho subtype proteins e.g. RhoA, Rac1 and Cdc42Hs but not Ras by using UDP-glucose as a cosubstrate (Just *et al.*, 1995b; Just *et al.*, 1995c). Interestingly, modification of Rho proteins occurs in Thr37. This amino acid residue is conserved between all small GTP-binding proteins and is involved in coordination of the magnesium ion and binding of the nucleotide. The area around Thr37 is also called switch I region because it undergoes major conformational changes upon GDP or GTP binding. Modification of Rho proteins in Thr37 inhibits the regulatory function of the GTP-binding proteins in a manner similar to C3-induced ADP-ribosylation.

Cytotoxic necrotizing factor 1 and 2 (CNF1 and 2) are synthesized by pathogenic strains of *E. coli* and are assumed to contribute to pathogenesis of *E. coli* diarrhea and extraintestinal infections. CNF1 and CNF2 have a molecular mass of about 110 kDa and an 85% amino acid identity. In contrast to *C. difficile* toxin B which disrupts the actin cytoskeleton, CNF1 and CNF2 increase actin stress fiber formation in cultured HEp2 cells (Oswald *et al.*, 1994) similar to constitutively activated Val14RhoA (Paterson *et al.*, 1990). Oswald *et al.* (1994) subsequently demonstrated that CNF2 treatment of Swiss 3T3 cells increased the molecular mass of

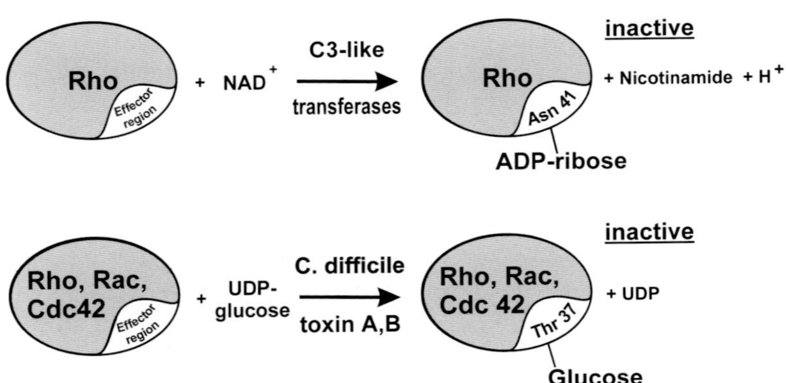

Figure 11.5   Comparison of the ADP-ribosyltransferase reaction catalyzed by C3-like exoenzymes with the glucosylation reaction catalyzed by *C. difficile* toxin A and B. NAD or UDP-glucose are substrates for C3-like exoenzymes or for toxin A and B, respecitively. Rho protein is modified and thereby inactivated at the effector region either at Asn41 by ADP-ribosylation or at Thr37 by glucosylation.

Rho proteins labeled by ADP-ribosylation. This increase in molecular mass may be attributed to an unknown posttranslational modification which leads to activation of Rho proteins and hence to stress fiber formation.

CONCLUDING REMARKS

Over the last years, C3-like exoenzymes have been widely used as specific inhibitors of Rho protein in various cellular and signal transducing systems. Thus, the transferases are established pharmacological and cell biological tools to study cellular functions of Rho. At least two pitfalls in application of C3 transferases have to be mentioned. Major disadvantage of C3 exoenzymes is their poor cell accessibility. This problem can be solved by microinjection of C3 or by using chimeric toxins consisting of the binding component of diphtheria toxin and the C3 transferase (Aullo *et al.*, 1993). In this respect, the *C. difficile* toxins may be of great advantage because these toxins are taken up by cells via membrane receptors. However, *C. difficle* toxins are less specific than C3 transferases and modify all Rho subtype proteins. In earlier studies C3-induced [^{32}P]ADP-ribosylation was applied not only to identify but also to quantify Rho proteins. However, as the ADP-ribosylation is modified by the nucleotide bound or by complexation of Rho with GDI, it is not possible to take the labeling by ADP-ribosylation as a measure for Rho protein concentrations. With these precautions in mind, C3-like transferases are extremely useful tools to study the functional role of Rho proteins and will help to further unravel the signal cascade involving Rho proteins.

REFERENCES

Abo, A., Pick, E., Hall, A., Totty, N., Teahan, C.G. and Segal, A.W. (1991) Activation of the NADPH oxidase involves the small GTP-binding protein p21rac. *Nature*, **353**, 668–670.

Adams, J.M., Houston, H., Allen, J., Lints, T. and Harvey, R. (1992) The hematopoietically expressed *vav* proto-oncogene shares homology with the *dbl* GDP-GTP exchange factor, the *bcr* gene and a yeast gene (*CDC24*) involved in cytoskeletal organization. *Oncogene*, **7**, 611–618.

Adamson, P., Marshall, C.J., Hall, A. and Tilbrook, P.A. (1992) Post-translational modifications of p21[rho] proteins. *J. Biol. Chem.*, **267**, 20033–20038.

Adamson, P., Paterson, H.F. and Hall, A. (1992) Intracellular localization of the P21[rho] proteins. *J. Cell Biol.*, **119**, 617–627.

Aktories, K. (1992) Clostridium botulinum C2 toxin and C. botulinum C3 ADP-ribosyltransferase. In: *Handbook of Experimental Pharmacology, Vol. 102, Selective Neurotoxicity.* H. Herken and F. Hucho, eds. Springer-Verlag, Berlin-Heidelberg, pp. 841–854.

Aktories, K. (1994) Clostridial ADP-ribosylating toxins: effects on ATP and GTP-binding proteins. *Mol. Cell. Biochem.*, **138**, 167–176.

Aktories, K., Bärmann, M., Ohishi, I., Tsuyama, S., Jakobs, K.H. and Habermann, E. (1986) Botulinum C2 toxin ADP-ribosylates actin. *Nature*, **322**, 390–392.

Aktories, K., Braun, U., Habermann, B. and Rösener, S. (1990) Botulinum ADP-ribosyltransferase C3. In: *ADP-ribosylating Toxins and G Proteins.* J. Moss and M. Vaughan, eds. American Society for Microbiology, Washington, pp. 97–115.

Aktories, K. and Frevert, J. (1987) ADP-ribosylation of a 21-24 kDa eukaryotic protein(s) by C3, a novel botulinum ADP-ribosyltransferase, is regulated by guanine nucleotide. *Biochem. J.*, **247**, 363–368.

Aktories, K., Jung, M., Böhmer, J., Fritz, G., Vandekerckhove, J. and Just, I. (1995) Studies on the active site structure of C3-like exoenzymes: Involvement of glutamic acid in catalysis of ADP-ribosylation. *Biochimie*, **77**, 326–332.

Aktories, K. and Just, I. (1993) GTPases and actin as targets for bacterial toxins. In: *GTPases in biology I.* B.F. Dickey and L. Birnbaumer, eds. Springer-Verlag, Berlin-Heidelberg, pp. 87–112.

Aktories, K. and Koch, G. (1994) Modification of actin and of Rho proteins by clostridial ADP-ribosylating toxins. In: *Microbial Toxins and Virulence Factors in Desease.* B. Iglewski, J. Moss, A.T. Tu and M. Vaughan, eds. Marcel Dekker, New York, Basal, pp. 491–520.

Aktories, K., Mohr, C. and Koch, G. (1992) Clostridium botulinum C3 ADP-ribosyltransferase. *Curr. Top. Microbiol. Immunol.*, **175**, 115–131.

Aktories, K., Rösener, S., Blaschke, U. and Chhatwal, G.S. (1988) Botulinum ADP-ribosyltransferase C3. Purification of the enzyme and characterization of the ADP-ribosylation reaction in platelet membranes. *Eur. J. Biochem.*, **172**, 445–450.

Aktories, K., Weller, U. and Chhatwal, G.S. (1987) Clostridium botulinum type C produces a novel ADP-ribosyltransferase distinct from botulinum C2 toxin. *FEBS Lett.*, **212**, 109–113.

Aullo, P., Giry, M., Olsnes, S., Popoff, M.R., Kocks, C. and Boquet, P. (1993) A chimeric toxin to study the role of the 21 kDa GTP binding protein rho in the control of actin microfilament assembly. *EMBO J.*, **12**, 921–931.

Balch, W.E. (1990) Small GTP-binding proteins in vesicular transport. *Trends Biochem. Sci.*, 473–477.

Barbieri, J.T., Mende-Mueller, M., Rappuoli, R. and Collier, R.J. (1989) Photolabeling of Glu-129 of the S-1 subunit of pertussis toxin with NAD. *Infect. Immun.*, **57**, 3549–3554.

Bartlett, J.G. (1990) Clostridium difficile: clinical considerations. *Rev. Infect. Dis.*, **12(suppl. 2)**, S243–S251.

Boguski, M.S. and McCormick, F. (1993) Proteins regulating Ras and its relatives. *Nature*, **366**, 643–654.

Bokoch, G.M. and Knaus, U.G. (1994) The role of small GTP-binding proteins in leukocyte function. *Curr. Opin. Immunol.*, **6**, 98–105.

Bourne, H.R. (1993) GTPases: A turn-on and a surprise. *Nature*, **366**, 628–629.

Bourne, H.R., Sanders, D.A. and McCormick, F. (1990) The GTPase superfamily: a conserved switch for diverse cell functions. *Nature*, **348**, 125–132.

Carroll, S.F. and Collier, R.J. (1984) NAD binding site of diphtheria toxin: identification of a residue within the nicotinamide subsite by photochemical modification with NAD. *Proc. Natl. Acad. Sci. USA*, **81**, 3307–3311.

Carroll, S.F. and Collier, R.J. (1987) Active site of Pseudomonas aeruginosa exotoxin A. Glutamic acid 553 is photolabeled by NAD and shows functional homology with glutamic acid 148 of diphtheria toxin. *J. Biol. Chem.*, **262**, 8707–8711.

Chan, A.M-L., McGovern, E.S., Catalano, G., Fleming, T.P. and Miki, T. (1995) Expression cDNA cloning of a novel oncogene with sequence similarity to regulators of small GTP-binding proteins. *Oncogene*, **9**, 1057–1063.

Chardin, P., Boquet, P., Madaule, P., Popoff, M.R., Rubin, E.J. and Gill, D.M. (1989) The mammalian G protein rho C is ADP-ribosylated by Clostridium botulinum exoenzyme C3 and affects actin microfilament in Vero cells. *EMBO J.*, **8**, 1087–1092.

Chardin, P., Madaule, P. and Tavitian, A. (1988) Coding sequence of human rho cDNAs clone 6 and clone 9. *Nucl. Acids Res.*, **16**, 2717.

Chong, L.D., Traynor-Kaplan, A., Bokoch, G.M. and Schwartz, M.A. (1994) The small GTP-binding protein Rho regulates a phosphatidylinositol 4-phosphate 5-kinase in mammalian cells. *Cell*, **79**, 507–513.

Cockcroft, S., Thomas, G.M.H., Fensome, A., Geny, B., Cunningham, E., Gout, I., Hiles, I., Totty, N.F., Truong, O. and Hsuan, J.J. (1994) Phospholipase D: a downstream effector of ARF in granulocytes. *Science*, **263**, 523–526.

D'Souza-Schorey, C., Li, G., Colombo, M.I. and Stahl, P.D. (1995) A regulatory role for ARF6 in receptor-mediated endocytosis. *Science*, **267**, 1175–1178.

Didsbury, J., Weber, R.F., Bokoch, G.M., Evans, T. and Snyderman, R. (1989) Rac, a novel ras-related family of proteins that are botulinum toxin substrates. *J. Biol. Chem.*, **264**, 16378–16382.

Diekmann, D., Brill, S., Garrett, M.D., Totty, N., Hsuan, J., Monfries, C., Hall, C., Lim, L. and Hall, A. (1991) *Bcr* encodes a GTPase-activating protein for p21^{rac}. *Nature*, **351**, 400–402.

Domenighini, M., Magagnoli, C., Pizza, M. and Rappuoli, R. (1994) Common features of the NAD-binding and catalytic site of ADP-ribosylating toxins. *Mol. Microbiol.*, **14**, 41–50.

Drivas, G.T., Shih, A., Coutavas, E., Rush, M.G. and D'Eustachio, P. (1990) Characterization of four novel *ras*-like genes expressed in a human teratocarcinoma cell line. *Mol. Cell. Biol.*, **10**, 1793–1798.

Egan, S.E., Giddings, B.W., Brooks, M.W., Buday, L., Sizeland, A.M. and Weinberg, R.A. (1993) Association of Sos Ras exchange protein with Grb2 is implicated in tyrosine kinase signal transduction and transformation. *Nature*, **363**, 45–51.

Feig, L.A. (1994) Guanine-nucleotide exchange factors: A family of positive regulators of Ras and related GTPases. *Curr. Opin. Cell Biol.*, **6**, 204–211.

Glomset, J.A. and Farnsworth, C.C. (1994) Role of protein modification reactions in programming interactions between Ras-related GTPases and cell membranes. *Annu. Rev. Cell Biol.*, **10**, 181–205.

Gulbins, E., Coggeshall, K.M., Baier, G., Telford, D., Langlet, C., Baier-Bitterlich, G., Bonnefoy-Berard, N., Burn, P., Wittinghofer, A. and Altman, A. (1994) Direct stimulation of Vav guanine nucleotide exchange activity for Ras by phorbol esters and diglycerides. *Mol. Cell. Biol.*, **14**, 4749–4758.

Habermann, B., Mohr, C., Just, I. and Aktories, K. (1991) ADP-ribosylation and de-ADP-ribosylation of the *rho* protein by *Clostridium botulinum* exoenzyme C3. Regulation by EDTA, guanine nucleotides and pH. *Biochim. Biophys. Acta*, **1077**, 253–258.

Habets, G.G.M., Scholtes, E.H.M., Zuydgeest, D., Van der Kammen, R.A., Stam, J.C., Berns, A. and Collard, J.G. (1994) Identification of an invasion-inducing gene, *Tiam-1*, that encodes a protein with homology to GDP-GTP exchangers for Rho-like proteins. *Cell*, **77**, 537–549.

Hall, A. (1990) *Ras* and GAP - who's controlling whom? *Cell*, **61**, 921–923.

Hall, A. (1994) Small GTP-binding proteins and the regulation of the actin cytoskeleton. *Annu. Rev. Cell Biol.*, **10**, 31–54.

Hall, A., Paterson, H.F., Adamson, P. and Ridley, A.J. (1993) Cellular responses regulated by rho-related small GTP-binding proteins. *Philos. Trans. R. Soc. Lond. [Biol.]*, **340**, 267–271.

Hara-Yokoyama, M., Hirabayashi, Y., Irie, F., Syuto, B., Moriishi, K., Sugiya, H. and Furuyama, S. (1995) Identification of gangliosides as inhibitors of ADP-ribosyltransferases of pertussis toxin and exoenzyme C3 from *Clostridium botulinum*. *J. Biol. Chem.*, **270**, 8115–8121.

Hart, M.J., Eva, A., Evans, T., Aaronson, S.A. and Cerione, R.A. (1991) Catalysis of guanine nucleotide exchange on the CDC42Hs protein by the *dbl* oncogene product. *Nature*, **354**, 311–314.

Heisterkamp, N., Kaartinen, V., Van Soest, S., Bokoch, G.M. and Groffen, J. (1993) Human ABR encodes a protein with GAP^{rac} activity and homology to the DBL nucleotide exchange factor domain. *J. Biol. Chem.*, **268**, 16903–16906.

Hinsch, K-D., Habermann, B., Just, I., Hinsch, E., Pfisterer, S., Schill, W-B. and Aktories, K. (1993) ADP-ribosylation of Rho proteins inhibits sperm motility. *FEBS Lett.*, **334**, 32–36.

Hiraoka, K., Kaibuchi, K., Ando, S., Musha, T., Takaishi, K., Mizuno, T., Asada, M., Ménard, L., Tomhave, E., Didsbury, J., Snyderman, R. and Takai, Y. (1992) Both stimulatory and inhibitory GDP/GTP exchange proteins, *smg* GDS and *rho* GDI, are active on multiple small GTP-binding proteins. *Biochem. Biophys. Res. Commun.*, **182**, 921–930.

Hirata, K-I., Kikuchi, A., Sasaki, T., Kuroda, S., Kaibuchi, K., Matsuura, Y., Seki, H., Saida, K. and Takai, Y. (1992) Involvement of *rho* p21 in the GTP-enhanced calcium ion sensitivity of smooth muscle contraction. *J. Biol. Chem.*, **267**, 8719–8722.

Homma, Y. and Emori, Y. (1995) A dual functional signal mediator showing RhoGAP and phospholipase C-δ stimulating activities. *EMBO J.*, **14**, 286–291.

Hori, Y., Kikuchi, A., Isomura, M., Katayama, M., Miura, Y., Fujioka, H., Kaibuchi, K. and Takai, Y. (1991) Post-translational modifications of the C-terminal region of the *rho* protein are important for its interaction with membranes and the stimulatory and inhibitory GDP/GTP exchange proteins. *Oncogene*, **6**, 515–522.

Horii, Y., Beeler, J.F., Sakaguchi, K., Tachibana, M. and Miki, T. (1994) A novel oncogene, *ost*, encodes a guanine nucleotide exchange factor that potentially links Rho and Rac signaling pathways. *EMBO J.*, **13**, 4776–4786.

Inoue, S., Sugai, M., Murooka, Y., Paik, S-Y., Hong, Y-M., Ohgai, H. and Suginaka, H. (1991) Molecular cloning and sequencing of the epidermal cell differentiation inhibitor gene from Staphylococcus aureus. *Biochem. Biophys. Res. Commun.*, **174**, 459–464.

Jung, M., Just, I., van Damme, J., Vandekerckhove, J. and Aktories, K. (1993) NAD-binding site of the C3-like ADP-ribosyltransferase from *Clostridium limosum. J. Biol. Chem.*, **268**, 23215–23218.

Just, I., Fritz, G., Aktories, K., Giry, M., Popoff, M.R., Boquet, P., Hegenbarth, S. and Von Eichel-Streiber, C. (1994) *Clostridium difficile* toxin B acts on the GTP-binding protein Rho. *J. Biol. Chem.*, **269**, 10706–10712.

Just, I., Mohr, C., Habermann, B., Koch, G. and Aktories, K. (1993) Enhancement of Clostridium botulinum C3-catalyzed ADP-ribosylation of recombinant rhoA by sodium dodecyl sulfate. *Biochem. Pharmacol.*, **45**, 1409–1416.

Just, I., Mohr, C., Schallehn, G., Menard, L., Didsbury, J.R., Vandekerckhove, J., van Damme, J. and Aktories, K. (1992a) Purification and characterization of an ADP-ribosyltransferase produced by *Clostridium limosum. J. Biol. Chem.*, **267**, 10274–10280.

Just, I., Schallehn, G. and Aktories, K. (1992b) ADP-ribosylation of small GTP-binding proteins by *Bacillus cereus. Biochem. Biophys. Res. Commun.*, **183**, 931–936.

Just, I., Selzer, J., Jung, M., van Damme, J., Vandekerckhove, J. and Aktories, K. (1995a) Rho-ADP-ribosylating exoenzyme from *Bacillus cereus* — purification, characterization and identification of the NAD-binding site. *Biochemistry,* **34**, 334–340.

Just, I., Selzer, J., Wilm, M., Von Eichel-Streiber, C., Mann, M. and Aktories, K. (1995b) Glucosylation of Rho proteins by *Clostridium difficile* toxin B. *Nature,* **375**, 500–503.

Just, I., Wilm, M., Selzer, J., Rex, G., Von Eichel-Streiber, C., Mann, M. and Aktories, K. (1995c) The enterotoxin from *Clostridium difficile* (ToxA) monoglucosylates the Rho proteins. *J. Biol. Chem.*, **270**, 13932–13936.

Kahn, R.A., Goddard, C. and Newkirk, M. (1988) Chemical and immunological characterization of the 21 kDa ADP-ribosylation factor (ARF) of adenylate cyclase. *J. Biol. Chem.*, **263**, 8282–8287.

Kaibuchi, K., Mizuno, T., Fujioka, H., Yamamoto, T., Kishi, K., Fukumoto, Y., Hori, Y. and Takai, Y. (1991) Molecular cloning of the cDNA for stimulatory GDP/GTP exchange protein for *smg* p21s (*ras* p21-like small GTP-binding proteins) and characterization of stimulatory GDP/GTP exchange protein. *Mol. Cell. Biol.*, **11**, No. 5, 2873–2880.

Kelly, C.P., Pothoulakis, C. and LaMont, J.T. (1994) Current concepts: *Clostridium difficile* colitis. *N. Engl. J. Med.*, **330**, 257–262.

Kikuchi, A., Kuroda, S., Sasaki, T., Kotani, K., Hirata, K., Katayama, M. and Takai, Y. (1992) Functional interactions of stimulatory and inhibitory GDP/GTP exchange proteins and their common substrate small GTP-binding protein. *J. Biol. Chem.*, **267**, 14611–14615.

Kishi, K., Sasaki, T., Kuroda, S., Itoh, T. and Takai, Y. (1993) Regulation of cytoplasmic division of *Xenopus* embryo by *rho* p21 and its inhibitory GDP/GTP exchange protein (*rho* GDI). *J. Cell Biol.*, **120**, 1187–1195.

Knaus, U.G., Heyworth, P.G., Kinsella, B.T., Curnutte, J.T. and Bokoch, G.M. (1992) Purification and characterization of Rac2. A cytosolic GTP-binding protein that regulates human neutrophil NADPH oxidase. *J. Biol. Chem.*, **267**, 23575–23582.

Koch, G., Norgauer, J. and Aktories, K. (1994) ADP-ribosylation of Rho by *Clostridium limosum* exoenzyme affects basal but not N-formyl-peptide-stimulated actin polymerization in human myeloid leukaemic (HL60) cells. *Biochem. J.*, **299**, 775–779.

Kozma, R., Ahmed, S., Best, A. and Lim, L. (1995) The Ras-related protein Cdc42Hs and bradykinin promote formation of peripheral actin microspikes and filopodia in Swiss 3T3 fibroblasts. *Mol. Cell. Biol.*, **15**, 1942–1952.

Kumagai, N., Morii, N., Fujisawa, K., Nemoto, Y. and Narumiya, S. (1993) ADP-ribosylation of *rho* p21 inhibits lysophosphatidic acid-induced protein tyrosine phosphorylation and phosphatidylinositol 3-kinase activation in cultured Swiss 3T3 cells. *J. Biol. Chem.*, **268**, 24535–24538.

Lamarche, N. and Hall, A. (1994) GAPs for rho-related GTPases. *Trends Genet.*, **10**, 436–440.

Lancaster, C.A., Taylor-Harris, P.M., Self, A.J., Brill, S., Van Erp, H.E. and Hall, A. (1994) Characterization of rhoGAP. A GTPase-activating protein for rho-related small GTPases. *J. Biol. Chem.*, **269**, 1137–1142.

Lang, P., Guizani, L., Vitté-Mony, I., Stancou, R., Dorseuil, O., Gacon, G. and Bertoglio, J. (1992) ADP-ribosylation of the *ras*-related, GTP-binding protein RhoA inhibits lymphocyte-mediated cytotoxicity. *J. Biol. Chem.*, **267**, 11677–11680.

Lelias, J-M., Adra, C.N., Wulf, G.M., Guillemot, J-C., Khagad, M., Caput, D. and Lim, B. (1993) cDNA cloning of a human mRNA preferentially expressed in hematopoietic cells and with homology to a

GDP-dissociation inhibitor for the rho GTP-binding proteins. *Proc. Natl. Acad. Sci. USA*, **90**, 1479–1483.

Lenhard, J.M., Kahn, R.A. and Stahl, P.D. (1992) Evidence for ADP-ribosylation factor (ARF) as a regulator of *in vitro* endosome-endosome fusion. *J. Biol. Chem.*, **267**, 13047–13052.

Leonard, D., Hart, M.J., Platko, J.V., Eva, A., Henzel, W., Evans, T. and Cerione, R.A. (1992) The identification and characterization of a GDP-dissociation inhibitor (GDI) for the CDC42Hs protein. *J. Biol. Chem.*, **267**, 22860–22868.

Leung, T., How, B-E., Manser, E. and Lim, L. (1993) Germ cell b-chimaerin, a new GTPase-activating protein for p21*rac*, is specifically expressed during the acrosomal assembly stage in rat testis. *J. Biol. Chem.*, **268**, 3813–3816.

Lyerly, D.M., Krivan, H.C. and Wilkins, T.D. (1988) Clostridium difficile: its disease and toxins. *Clin. Microbiol. Rev.*, **1**, 1–18.

Madaule, P. and Axel, R. (1985) A novel ras-related gene family. *Cell*, **41**, 31–40.

Madaule, P., Axel, R. and Myers, A.M. (1987) Characterization of two members of the rho gene family from the yeast Saccharomyces cerevisiae. *Proc. Natl. Acad. Sci. USA*, **84**, 779–783.

Maehama, T., Ohoka, Y., Ohtsuka, T., Takahashi, K., Nagata, K., Nozawa, Y., Ueno, K., Ui, M. and Katada, T. (1990) Botulinum ADP-ribosyltransferase activity as affected by detergents and phospholipids. *FEBS Lett.*, **263**, 376–380.

Malcolm, K.C., Ross, A.H., Qiu, R-G., Symons, M. and Exton, J.H. (1994) Activation of rat liver phospholipase D by the small GTP-binding protein RhoA. *J. Biol. Chem.*, **269**, 25951–25954.

Malcolm, K.C., Elliott, C.M. and Exton, J.H. (1996) Evidence for Rho-mediated agonist stimulation of phospholipase D in Rat 1 fibroblasts. *J. Biol. Chem.*, **271**, 13135–13139.

Manser, E., Leung, T., Salihuddin, H., Zhao, Z. and Lim, L. (1994) A brain serine/threonine protein kinase activated by Cdc42 and Rac1. *Nature*, **367**, 40–46.

Masuda, T., Tanaka, K., Nonaka, H., Yamochi, W., Maeda, A. and Takai, Y. (1994) Molecular cloning and characterization of yeast *rho* GDP dissociation inhibitor. *J. Biol. Chem.*, **269**, 19713–10718.

Matsui, Y. and Toh-e, A. (1992) Isolation and characterization of two novel *ras* superfamily genes in *Saccharomyces cerevisiae. Gene*, **114**, 43–49.

Michiels, F., Habets, G.G.M., Stam, J.C., Van der Kammen, R.A. and Collard, J.G. (1995) A role for rac in Tiam1-induced membrane ruffling and invasion. *Nature*, **375**, 338–340.

Miki, T., Smith, C.L., Long, J.E., Eva, A. and Fleming, T.P. (1993) Oncogene *ect2* is related to regulators of small GTP-binding proteins. *Nature*, **362**, 462–465.

Minden, A., Lin, A., Claret, F-X., Abo, A. and Karin, M. (1995) Selective activation of the JNK signaling cascade and c-Jun transcriptional activity by the small GTPases Rac and Cdc42Hs. *Cell*, **81**, 1147–1157.

Mohr, C., Koch, G., Just, I. and Aktories, K. (1992) ADP-ribosylation by Clostridium botulinum C3 exoenzyme increases steady state GTPase activities of recombinant rhoA and rhoB proteins. *FEBS Lett.*, **297**, 95–99.

Moore, M.S. and Blobel, G. (1994) A G protein involved in nucleocytoplasmic transport: The role of Ran. *Trends Biochem. Sci.*, **19**, 211–216.

Morii, N., Kawano, K., Sekine, A., Yamada, T. and Narumiya, S. (1991) Purification of GTPase-activating protein specific for the *rho* gene. *J. Biol. Chem.*, **266**, 7646–7650.

Morii, N., Teru-uchi, T., Tominaga, T., Kumagai, N., Kozaki, S., Ushikubi, F. and Narumiya, S. (1992) A *rho* gene product in human blood platelets. II. Effects of the ADP-ribosylation by botulinum C3 ADP-ribosyltransferase on platelet aggregation. *J. Biol. Chem.*, **267**, 20921–20926.

Moriishi, K., Syuto, B., Yokosawa, N., Oguma, K. and Saito, M. (1991) Purification and characterization of ADP-ribosyltransferases (exoenzyme C3) of Clostridium botulinum type C and D strains. *J. Bacteriol.*, **173**, 6025–6029.

Narumiya, S. and Morii, N. (1993) *rho* gene products, botulinum C3 exoenzyme and cell adhesion. *Cell. Signal.*, **5**, 9–19.

Nemoto, Y., Namba, T., Kozaki, S. and Narumiya, S. (1991) Clostridium botulinum C3 ADP-ribosyltransferase gene. *J. Biol. Chem.*, **266**, 19312–19319.

Nobes, C. and Hall, A. (1994) Regulation and function of the Rho subfamily of small GTPases. *Curr. Opin. Genet. Dev.*, **4**, 77–81.

Nobes, C.D. and Hall, A. (1995) Rho, Rac and Cdc42 GTPases regulate the assembly of multimolecular focal complexes associated with actin stress fibers, lamellipodia and filopodia. *Cell*, **81**, 53–62.

Norman, J.C., Price, L.S., Ridley, A.J., Hall, A. and Koffer, A. (1994) Actin filament organization in activated mast cells is regulated by heterotrimeric and small GTP-binding proteins. *J. Cell Biol.*, **126**, 1005–1016.

Novick, P. and Brennwald, P. (1993) Friends and family: The role of the Rab GTPases in vesicular traffic. *Cell*, **75**, 597–601.

Ogorochi, T., Nemoto, Y., Nakajima, M., Nakamura, E., Fujiwara, M. and Narumiya, S. (1989) cDNA cloning of Gb, the substrate for botulinum ADP- ribosyltransferase from bovine adrenal gland and its identification as a *rho* gene product. *Biochem. Biophys. Res. Commun.*, **163**, 1175–1181.

Ohguchi, K., Banno, Y., Nakashima, S. and Nozawa, Y. (1995) Activation of membrane-bound phospholipase D by protein kinase C in HL60 cells: Synergistic action of a small GTP-binding protein RhoA. *Biochem. Biophys. Res. Commun.*, **211**, 306–311.

Olivier, J.P., Raabe, T., Henkemeyer, M., Dickson, B., Mbamalu, G., Margolis, B., Schlessinger, J., Hafen, E. and Pawson, T. (1993) A Drosophila SH2-SH3 adaptor protein implicated in coupling the sevenless tyrosine kinase to an activator of Ras guanine nucleotide exchange, Sos. *Cell*, **73**, 179–191.

Oswald, E., Sugai, M., Labigne, A., Wu, H.C., Fiorentini, C., Boquet, P. and O'Brien, A.D. (1994) Cytotoxic necrotizing factor type 2 produced by virulent *Escherichia coli* modifies the small GTP-binding proteins Rho involved in assembly of actin stress fibers. *Proc. Natl. Acad. Sci. USA*, **91**, 3814–3818.

Otsu, M., Hiles, I., Gout, I., Fry, M., Ruiz-Larrea, F., Panayotou, G., Thompson, A., Dhand, R., Hsuan, J., Totty, N., Smith, A.D., Morgan, S.J., Courtneidge, S.A., Parker, P.J. and Waterfield, M.D. (1991) Characterization of two 85 kd proteins that associate with receptor tyrosine kinases, middle-T/pp60^{c-src} complexes and PI3-kinase. *Cell*, **65**, 91–104.

Pai, E.F., Kabsch, W., Krengel, U., Holmes, K.C., John, J. and Wittinghofer, A. (1989) Structure of the guanine-nucleotide-binding domain of the Ha-ras oncogene product p21 in the triphosphate conformation. *Nature*, **341**, 209–214.

Pasteris, N.G., Cadle, A., Logie, L.J., Porteous, M.E.M., Schwartz, C.E., Stevenson, R.E., Glover, T.W., Wilroy, R.S. and Gorski, J.L. (1994) Isolation and characterization of the faciogenital dysplasia (Aarskog-Scott syndrome) gene: A putative Rho/Rac guanine nucleotide exchange factor. *Cell*, **79**, 669–678.

Paterson, H.F., Self, A.J., Garrett, M.D., Just, I., Aktories, K. and Hall, A. (1990) Microinjection of recombinant p21^{rho} induces rapid changes in cell morphology. *J. Cell Biol.*, **111**, 1001–1007.

Peppelenbosch, M.P., Qiu, R-G., De Vries-Smits, A.M.M., Tertoolen, L.G.J., de Laat, S.W., McCormick, F., Hall, A., Symons, M.H. and Bos, J.L. (1995) Rac mediates growth factor-induced arachidonic acid release. *Cell*, **81**, 849–856.

Perentesis, J.P., Miller, S.P. and Bodley, J.W. (1992) Protein toxin inhibitors of protein synthesis. *BioFactors*, **3**, 173–184.

Pfeffer, S.R. (1994) Rab GTPases: Master regulators of membrane trafficking. *Curr. Opin. Cell Biol.*, **6**, 522–526.

Pick, E., Gorzalczany, Y. and Engel, S. (1993) Role of the *rac*l p21-GDP-dissociation inhibitor for *rho* heterodimer in the activation of the superoxide-forming NADPH oxidase of macrophages. *Eur. J. Biochem.*, **217**, 441–455.

Popoff, M.R., Hauser, D., Boquet, P., Eklund, M.W. and Gill, D.M. (1991) Characterization of the C3 gene of Clostridium Botulinum types C and D and its expression in Escherichia coli. *Infect. Immun.*, **59**, 3673–3679.

Prendergast, G.C. and Gibbs, J.B. (1993) Pathways of Ras function: Connections to the actin cytoskeleton. *Adv. Cancer Res.*, **62**, 19–64.

Qiu, R-G., Chen, J., Kirn, D., McCormick, F. and Symons, M. (1995) An essential role for Rac in Ras transformation. *Nature*, **374**, 457–459.

Rankin, S., Morii, N., Narumiya, S. and Rozengurt, E. (1994) Botulinum C3 exoenzyme blocks the tyrosine phosphorylation of p125^{FAK} and paxillin induced by bombesin and endothelin. *FEBS Lett.*, **354**, 315–319.

Reinhard, J., Scheel, A.A., Diekmann, D., Hall, A., Ruppert, C. and Bähler, M. (1995) A novel type of myosin implicated in signalling by rho family GTPases. *EMBO J.*, **14**, 697–704.

Ren, M., Drivas, G., D'Eustachio, P. and Rush, M.G. (1993) Ran/TC4: A small nuclear GTP-binding protein that regulates DNA synthesis. *J. Cell Biol.*, **120**, 313–323.

Ridley, A.J. and Hall, A. (1992) The small GTP-binding protein rho regulates the assembly of focal adhesions and actin stress fibers in response to growth factors. *Cell*, **70**, 389–399.

Ridley, A.J., Paterson, H.F., Johnston, C.L., Diekmann, D. and Hall, A. (1992) The small GTP-binding protein rac regulates growth factor-induced membrane ruffling. *Cell,* **70**, 401–410.

Ridley, A.J., Self, A.J., Kasmi, F., Paterson, H.F., Hall, A., Marshall, C.J. and Ellis, C. (1993) rho Family GTPase activating proteins p190, bcr and rhoGAP show distinct specificities *in vitro* and *in vivo*. *EMBO J.,* **12**, 5151–5160.

Rozakis-Adcock, M., Fernley, R., Wade, J., Pawson, T. and Bowtell, D. (1993) The SH2 and SH3 domains of mammalian Grb2 couple the EGF receptor to the Ras activator mSos1. *Nature,* **363**, 83–85.

Rubin, E.J., Gill, D.M., Boquet, P. and Popoff, M.R. (1988) Functional modification of a 21-Kilodalton G protein when ADP-ribosylated by exoenzyme C3 of Clostridium botulinum. *Mol. Cell. Biol.,* **8**, 418–426.

Satoh, T., Nakafuku, M. and Kaziro, Y. (1992) Function of Ras as a molecular switch in signal transduction. *J. Biol. Chem.,* **267**, 24149–24152.

Schering, B., Bärmann, M., Chhatwal, G.S., Geipel, U. and Aktories, K. (1988) ADP-ribosylation of skeletal muscle and non-muscle actin by *Clostridium perfringens* iota toxin. *Eur. J. Biochem.,* **171**, 225–229.

Schmalzing, G., Richter, H-P., Hansen, H. and Aktories, K. (1995) Involvement of the GTP-binding protein Rho in endocytosis of Xenopus oocytes. *J. Cell Biol.,* **130**, 1319–1332.

Seckl, M.J., Morii, N., Narumiya, S. and Rozengurt, E. (1995) Guanosine 5'-3-O-(thio)triphosphate stimulates tyrosine phosphorylation of p125[FAK] and paxillin in permeabilized Swiss 3T3 cells. Role of p21[rho]. *J. Biol. Chem.,* **270**, 6984–6990.

Sekine, A., Fujiwara, M. and Narumiya, S. (1989) Asparagine residue in the rho gene product is the modification site for botulinum ADP-ribosyltransferase. *J. Biol. Chem.,* **264**, 8602–8605.

Serventi, I.M., Moss, J. and Vaughan, M. (1992) Enhancement of cholera toxin-catalyzed ADP-ribosylation by guanine nucleotide-binding proteins. *Curr. Top. Microbiol. Immunol.,* **175**, 43–68.

Settleman, J., Albright, C.F., Foster, L.C. and Weinberg, R.A. (1992) Association between GTPase activators for Rho and Ras families. *Nature,* **359**, 153–154.

Shinjo, K., Koland, J.G., Hart, M.J., Narasimhan, V., Johnson, D.I., Evans, T. and Cerione, R.A. (1990) Molecular cloning of the gene for the human placental GTP-binding protein Gp (G25K): Identification of this GTP-binding protein as the human homolog of the yeast cell-division-cycle protein CDC42. *Proc. Natl. Acad. Sci. USA,* **87**, 9853–9857.

Simpson, L.L., Stiles, B.G., Zepeda, H. and Wilkins, T.D. (1989) Production by Clostridium spiroforme of an iotalike toxin that possesses mono(ADP-ribosyl)transferase activity: Identification of a novel class of ADP- ribosyltransferases. *Infect. Immun.,* **57**, 255–261.

Sinnett-Smith, J., Zachary, I., Valverde, A.M. and Rozengurt, E. (1993) Bombesin stimulation of p125 focal adhesion kinase tyrosine phosphorylation. Role of protein kinase C, Ca^{2+} mobilization and the actin cytoskeleton. *J. Biol. Chem.,* **268**, 14261–14268.

Spangler, B.D. (1992) Structure and function of cholera toxin and the related *Escherichia coli* heat-labile enterotoxin. *Microbiol. Rev.,* **56**, 622–647.

Stasia, M-J., Jouan, A., Bourmeyster, N., Boquet, P. and Vignais, P.V. (1991) ADP-ribosylation of a small size GTP-binding protein in bovine neutrophils by the C3 exoenzyme of Clostridium botulinum and effect on the cell motility. *Biochem. Biophys. Res. Commun.,* **180**, 615–622.

Stephens, L.R., Jackson, T.R. and Hawkins, P.T. (1993) Agonist-stimulated synthesis of phosphatidylinositol(3,4,5)-trisphosphate: A new intracellular signalling system. *Biochim. Biophys. Acta Mol. Cell Res.,* **1179**, 27–75.

Takada, T., Iida, K. and Moss, J. (1995) Conservation of a common motif in enzymes catalyzing ADP-ribose transfer. *J. Biol. Chem.,* **270**, 541–544.

Takai, A., Sasaki, T., Tanaka, K. and Nakanishi, H. (1995) Rho as a regulator of the cytoskeleton. *Trends Biochem. Sci.,* **20**, 227–231.

Takaishi, K., Kikuchi, A., Kuroda, S., Kotani, K., Sasaki, T. and Takai, Y. (1993) Involvement of *rho* p21 and its inhibitory GDP/GTP exchange protein (*rho* GDI) in cell motility. *Mol. Cell. Biol.,* **13**, 72–79.

Takaishi, K., Sasaki, T., Kato, M., Yamochi, W., Kuroda, S., Nakamura, T., Takeichi, M. and Takai, Y. (1994) Involvement of *Rho* p21 small GTP-binding protein and its regulator in the HGF-induced cell motility. *Oncogene,* **9**, 273–279.

Tominaga, T., Sugie, K., Hirata, M., Morii, N., Fukata, J., Uchida, A., Imura, H. and Narumiya, S. (1993) Inhibition of PMA-induced, LFA-1-dependent Lymphocyte aggregation by ADP-ribosylation of the small molecular weight GTP binding protein, *rho*. *J. Cell Biol.,* **120**, 1529–1537.

Tsuzuki, K., Kimura, K., Fujii, N., Yokosawa, N., Indoh, T., Murakami, T. and Oguma, K. (1990) Cloning and complete nucleotide sequence of the gene for the main component of Hemagglutinin produced by *Clostridium botulinum* Type C. *Infect. Immun.*, **58**, 3173–3177.

Ueda, T., Kikuchi, A., Ohga, N., Yamamoto, J. and Takai, Y. (1990) Purification and characterization from bovine brain cytosol of a novel regulatory protein inhibiting the dissociation of GDP from and the subsequent binding of GTP to *rho*B p20, a *ras* p21-like GTP-binding protein. *J. Biol. Chem.*, **265**, 9373–9380.

Ui, M. (1990) Pertussis toxin as a valuable probe for G-protein involvement in signal transduction. In: *ADP-ribosylating toxins and G proteins*. J. Moss and M. Vaughan, eds. American Society for Microbiology, Washington,D.C. Vol.1st, pp. 45–77.

Vincent, S., Jeanteur, P. and Fort, P. (1992) Growth-regulated expression of *rhoG*, a new member of the *ras* homolog gene family. *Mol. Cell. Biol.*, **12**, 3138–3148.

Vojtek, A.B., Hollenberg, S.M. and Cooper, J.A. (1993) Mammalian Ras interacts directly with the serine/threonine kinase Raf. *Cell*, **74**, 205–214.

Wei, W., Das, B., Park, W. and Broek, D. (1994) Cloning and analysis of human cDNAs encoding a 140 kDa brain guanine nucleotide-exchange factor, Cdc25[GEF], which regulates the function of Ras. *Gene*, **151**, 279–284.

Welsh, C.F., Moss, J. and Vaughan, M. (1994) ADP-ribosylation factors: A family of ~20 kDa guanine nucleotide-binding proteins that activate cholera toxin. *Mol. Cell. Biochem.*, **138**, 157–166.

Whitehead, I., Kirk, H. and Kay, R. (1995a) Retroviral transduction and oncogenic selection of a cDNA encoding Dbs, a homolog of the Dbl guanine nucleotide exchange factor. *Oncogene,* **10**, 713–721.

Whitehead, I., Kirk, H. and Kay, R. (1995b) Expression cloning of oncogenes by retroviral transfer of cDNA libraries. *Mol. Cell. Biol.*, 704–710.

Wiegers, W., Just, I., Müller, H., Hellwig, A., Traub, P. and Aktories, K. (1991) Alteration of the cytoskeleton of mammalian cells cultured *in vitro* by Clostridium botulinum C2 toxin and C3 ADP-ribosyltransferase. *Eur. J. Cell Biol.*, **54**, 237–245.

Wieland, T., Ulibarri, I., Aktories, K., Gierschik, P. and Jakobs, K.H. (1990a) Interaction of small G proteins with photoexcited rhodopsin. *FEBS Lett.*, **263**, 195–198.

Wieland, T., Ulibarri, I., Gierschik, P., Hall, A., Aktories, K. and Jakobs, K.H. (1990b) Interaction of recombinant rho A GTP-binding proteins with photoexcited rhodopsin. *FEBS Lett.*, **274**, 111–114.

Williamson, K.C., Smith, L.A., Moss, J. and Vaughan, M. (1990) Guanine nucleotide-dependent ADP-ribosylation of soluble rho catalyzed by *Clostridium botulinum* C3 ADP-ribosyltransferase. *J. Biol. Chem.*, **265**, 20807–20812.

Yaku, H., Sasaki, T. and Takai, Y. (1994) The *Dbl* oncogene product as a GDP/GTP exchange protein for the *Rho* family: Its properties in comparison with those of *Smg* GDS. *Biochem. and Biophys. Res. Commun.*, **198**, 811–817.

Yamochi, W., Tanaka, K., Nonaka, H., Maeda, A., Musha, T. and Takai, Y. (1994) Growth site localization of Rho1 small GTP-binding protein and its involvement in bud formation in *Saccharomyces cerevisiae. J. Cell Biol.*, **125**, 1077–1093.

Yeramian, P., Chardin, P., Madaule, P. and Tavitian, A. (1987) Nucleotide sequence of human rho cDNA clone 12. *Nucl. Acids Res.*, **15**, 1869

Zhang, J., Benovic, J.L., Sugai, M., Wetzker, R., Gout, I. and Rittenhouse, S.E. (1995) Sequestration of a G-protein βγsubunit or ADP-ribosylation of Rho can inhibit thrombin-induced activation of platelet phosphoinositide 3-kinases. *J. Biol. Chem.*, **270**, 6589–6594.

Zhang, J., King, W.G., Dillon, S., Hall, A., Feig, L. and Rittenhouse, S.E. (1993) Activation of platelet phosphatidylinositide 3-kinase requires the small GTP-binding protein Rho. *J. Biol. Chem.*, **268**, 22251–22254.

Zheng, Y., Bagrodia, S. and Cerione, R.A. (1994) Activation of phosphoinositide 3-kinase activity by Cdc42Hs binding to p85. *J. Biol. Chem.*, **269**, 18727–18730.

Zheng, Y., Olson, M.F., Hall, A., Cerione, R.A. and Toksoz, D. (1995) Direct involvement of the small GTP-binding protein Rho in *lbc* oncogene function. *J. Biol. Chem.*, **270**, 9031–9034.

12. DIPHTHERIA TOXIN MUTATIONS: THEIR ROLE IN STRUCTURE — FUNCTION STUDIES

VIRGINIA GRAY JOHNSON

*Laboratory of Bacterial Toxins, Center for Biologics Evaluation and Research,
Food and Drug Administration, Bethesda, MD 20892*

Diphtheria toxin (DT) is the main virulence factor of *Corynebacterium diphtheriae.* The structural gene for the toxin, *tox*, is carried by certain temperate bacteriophages that lysogenize *C. diphtheriae* converting it from the nontoxinogenic to the toxinogenic form. Expression of the *tox* gene is regulated at least in part by the bacterial host and occurs only under conditions of limited iron (Collier, 1975; Pappenheimer, 1977). The toxin is secreted by the bacterium as a single polypeptide chain (535 amino acids, M_r 58,342). A 25-residue leader peptide precedes the mature toxin and aids in the secretion of the toxin from the bacterium (Greenfield *et al.*, 1983; Kaczorek *et al.*, 1983; Ratti *et al.*, 1983). An arginine-rich region located within one of the two disulfide loops is proteolytically cleaved or "nicked", yielding an A (DTA, M_r 21,167) and a B (DTB, M_r 37,195) fragment linked by a disulfide bond. Proteolytic nicking and reduction of this disulfide bond are required for toxicity (Gill and Pappenheimer, 1971; Drazin *et al.*, 1971; Collier and Kandel, 1971; Papini *et al.*, 1993).

DTA contains the enzymatic function of the toxin. It catalyzes the transfer of ADP-ribose from NAD to elongation factor 2 (EF-2) (Honjo *et al.*, 1968; Gill *et al.*, 1969) according to the following reaction:

$$EF\text{-}2 + NAD \rightarrow ADP\text{-}ribose\text{-}EF\text{-}2 + nicotinamide + H+$$

ADP-ribose is covalently attached to a modified histidine residue called diphthamide, found only in EF-2 (van Ness *et al.*, 1980). EF-2 is required for the transfer of peptidyl-tRNA from the aminoacyl site to the peptidyl site of the 60S ribosome. ADP-ribosylation of EF-2 inactivates it thus inhibiting protein synthesis and ultimately resulting in the death of the cell. DTB has two functions: it is responsible for receptor binding as well as facilitating the transloacation of DTA across the membrane and into the cytosol where it is enzymatically active.

The intoxication of cells by DT is a multistep process, involving both DTA and DTB (Figure 12.1). The toxin first binds to its receptor, recently identified as the heparin-binding epidermal growth factor-like precursor (pro HB-EGF) (Naglich and Eidels, 1990; Naglich *et al.*, 1992). It is then endocytosed and routed to an acid-containing endosome. Exposure of the toxin to the acid environment of the

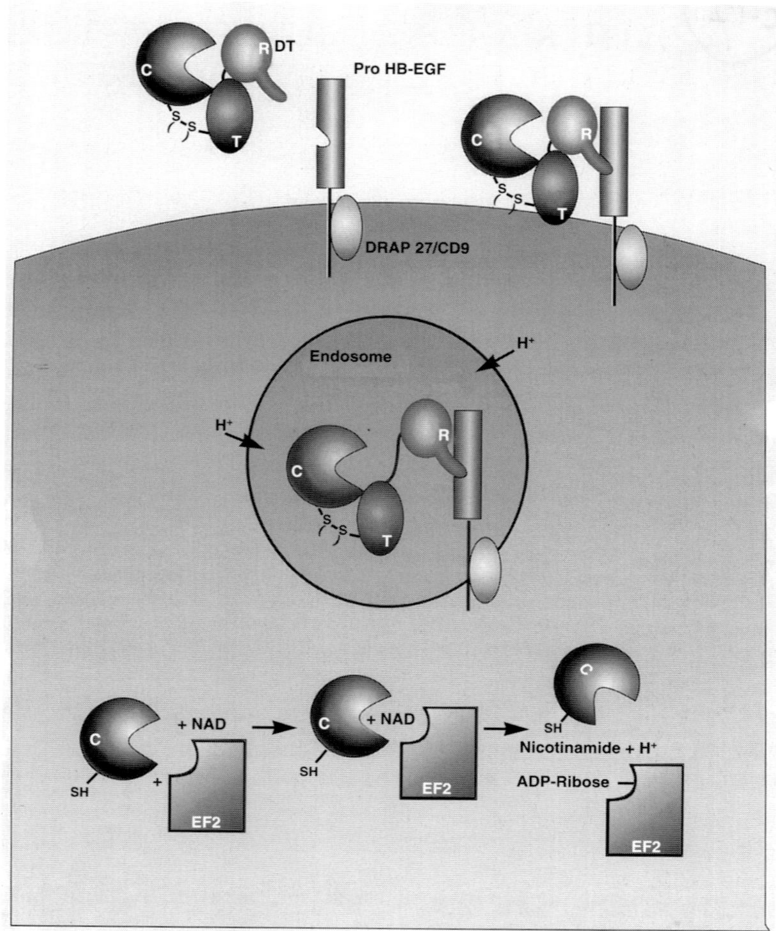

Figure 12.1 Schematic representation of the intoxication pathway for DT. The three domains of the toxin are shown as C (catalytic, residues 1–190), T (transmembrane, residues 191–378) and R (receptor-binding, residues 379–535). DT is proteolytically cleaved in an arginine-rich loop connecting the C and T domain. These domains continue to remain associated through a disulfide bond in the same loop (residues 186–201). The DT receptor is a complex consisting of a heparin-binding epidermal growth factor-like precursor (pro HB-EGF) and an associated transmembrane protein (DRAP 27/CD9) (panel 1). DT is believed to bind to its receptor via a surface exposed loop extending from the R domain (panel 2). The toxin-receptor complex is internalized by receptor-mediated endocytosis and routed via coated pits to acid-containing endosomes (panel 3). In response to the acidic environment of the endosome, the toxin undergoes a conformational change resulting in the penetration of portions of the T domain through the endosomal membrane. Upon reduction of the disulfide bond connecting the C and T domain, the C domain translocates through the endosomal membrane and is released into the cytosol. Here it catalyzes the transfer of ADP-ribose from NAD to elongation factor 2 (EF-2), resulting in the inhibition of protein synthesis and ultimately in the death of the cell (panel 4).

endosome results in a conformational change in the toxin molecule and leads to the translocation of DTA across the endosomal membrane and into the cytosol. Once in the cytosol, DTA catalyzes the ADP-ribosylation of EF-2, thereby inactivating protein synthesis resulting in cell death.

This chapter will focus on the structure-function relationships of the toxin, specifically how mutagenesis has facilitated our understanding of this topic. It will also describes how these mutants are utilized for a variety of pharmacological purposes.

CRYSTAL STRUCTURE OF DIPHTHERIA TOXIN

The crystallographic structure of DT, determined in 1992 by Choe *et al.* has revolutionized our understanding of the structure-function relationships of the toxin. The crystal reveals three distinct domains corresponding to the three distinct functions that the toxin performs (Figure 12.2). The catalytic (C) domain extends from residue 1 to 190 and comprises the A chain of the toxin. The B chain of the toxin is composed of two domains corresponding to its two functions: the transmembrane (T) domain (residues 191–378) and the receptor-binding (R) domain (residues 379–535). The three domains are arranged in the shape of a Y, extending approximately 90Å high, 30Å thick, and 50Å across, with the T domain forming the base of the Y and the C and R domains forming the two arms. An extended, hydrophilic 14-base loop (residues 187–200) connects the C and T domains. This loop contains the arginine-rich site that is the target of nicking by trypsin, furin and other proteases.

The crystal structure suggests why intact DT is unable to catalyze the ADP-ribosylation of EF-2. The active site is formed at the interface between the C domain and the R domain and entry of the bulky EF-2 molecule (Mr ~100 k) into the active site is blocked by the R domain in the intact toxin. Nicking and reduction of the loop connecting the C and T domains results in the separation of the toxin into A and B subunits, thus removing restrictions imposed by the R domain and allowing EF-2 free access to the active site. In contrast to this, NAD is much smaller (Mr ~700) and is able to access the active site even in the intact toxin. In addition to ADP-ribosyltransferase (ADPRT) activity, DTA manifests weak ADP-glycohydrolase activity, the physiologic significance of which is unknown. The fact that NAD-glycohydrolase activity is identical in the intact toxin and DTA demonstrates that NAD has free access to the active site even in the presence of DTB (Kandel *et al.*, 1974).

DT was initially crystalized as a dimer at 2.5Å resolution. Since then, the resolution of dimeric DT has been extended to 2.0Å (Bennett *et al.*, 1994a) and the structure of monomeric DT has been determined at 2.3Å resolution (Bennett and Eisenberg, 1994). The refined model reveals that each molecule in dimeric DT has a more open structure than that of monomeric toxin; the individual units of the dimer are called "open monomers". Diphtheria toxin forms dimers when it is exposed to freezing temperatures in the presence of phosphate buffer. Characteristically, phosphate buffer decreases in pH from 7 to 3.6 upon slow

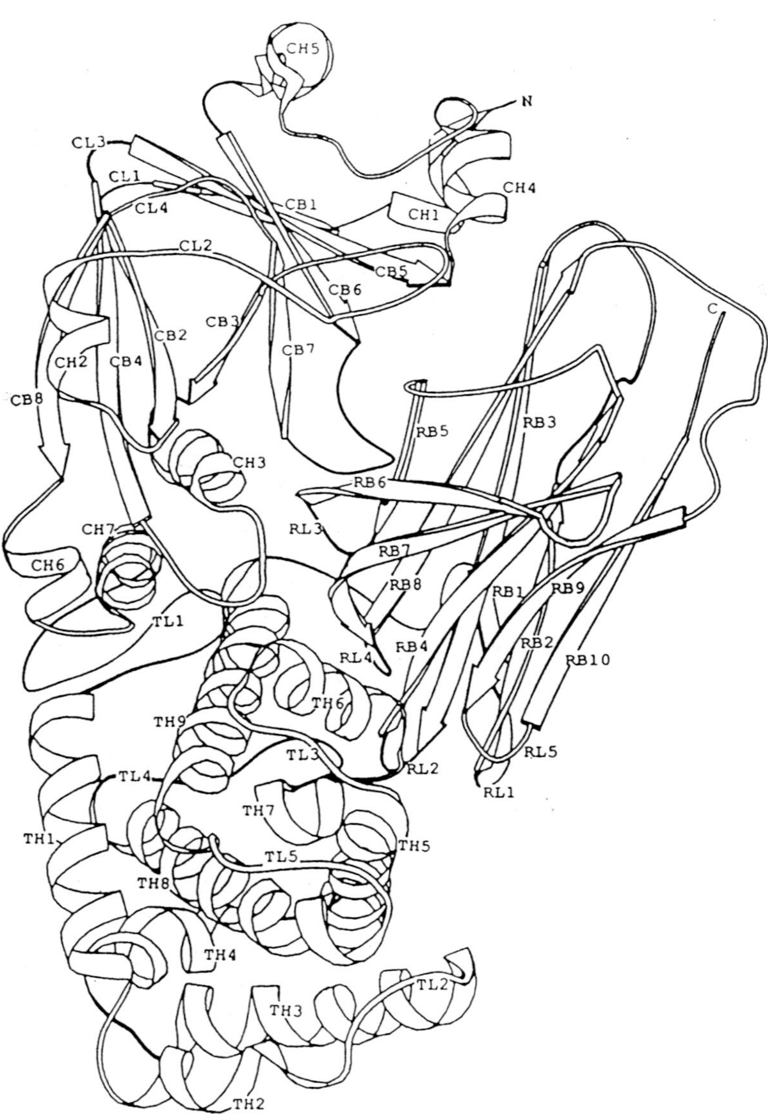

Figure 12.2 A ribbon diagram of DT. Secondary structures are labeled. The first letter denotes the domain (C for catalytic, T for transmembrane, and R for receptor-binding domains). The second letter denotes the type of secondary structure (H for helix, B for β-strand, and L for loop). The third number represents the sequential order from the N terminus of each domain. This ribbon diagram was kindly provided by Dr. David Eisenberg and reprinted with permission from Choe *et al.*, 1992.

freezing. Exposure to these low pH conditions appears to be a prerequisite for dimerization since DT does not dimerize if frozen in buffers that lack this property (Carroll *et al.*, 1988). At low pH the salt bridges that are believed to stabilize the interface between the R and the C domains are disrupted due to protonation of acidic residues. This converts monomeric DT into an open form which can then dimerize by "domain swapping" of the two R domains to form a compact, globular, dimeric structure that is long-lived and resistant to denaturation (Bennett *et al.*, 1994b). The toxin dimer is nontoxic; it is unable to bind to the DT receptor, presumably because dimer interaction sterically blocks the R domains of each monomer. In contrast, the C domain of the dimer appears to be identical to that of the monomer with both conformations exhibiting comparable ADPRT and NAD-glycohydrolase activities (Carroll *et al.*, 1986).

The "open monomer" subunit of the dimer may have structural features similar to the toxin monomer that has undergone conformation changes required for membrane translocation. Both forms are triggered by exposure to low pH. Both forms also have a more accessible phosphate-binding site (P-site) located within the R domain of the toxin (Lory *et al.*, 1980). Additionally, both open forms have a more accessible β-hairpin loop (residues 514–525) that is believed to be involved in the actual molecular recognition of the receptor by the toxin (Bennett and Eisenberg, 1994). In the compact monomeric structure, this β-hairpin loop lies very close to the T domain and as a result, membrane insertion of the T domain could be sterically hindered by the close proximity of the R domain and possibly by the DT cellular receptor (Bennett *et al.*, 1994a). Details of these sites are discussed in later sections.

TOXIN RECEPTOR

The existence of specific cell-surface receptors for DT was first demonstrated by Ittelson and Gill (1973). Subsequently, Middlebrook and coworkers (1978) showed that certain African green monkey kidney cell lines, such as Vero, express extremely high levels of binding sites per cell ($1 - 2 \times 10^5$/cell). The toxin-receptor interaction on Vero cells is specific, saturable and exhibited a Kd of 1×10^{-9} M. These cells have greatly facilitated the study of DT-receptor interactions and have become the standard cell line for the investigation of DT.

Eidels and coworkers (Naglich and Eidels, 1990; Naglich *et al.*, 1992) cloned and identified the toxin receptor by taking advantage of differences in toxin susceptibility of different mammalian cell lines. Mouse cell lines, which lack the receptor and are therefore resistant to DT, were transfected with cDNA derived from Vero cells. The DT receptor was subsequently cloned and identified as a heparin-binding epidermal growth factor-like precursor (pro HB-EGF). The precursor form of HB-EGF is anchored in the membrane and is cleaved to yield a biologically-active mature growth factor (HB-EGF) that exerts its mitogenic effect by binding to the EGF receptor. A considerable amount of pro HB-EGF remains uncleaved on the cell surface and provides mitogenic stimulation to neighboring cells in a juxtacrine manner (Nakamura *et al.*, 1995). DT opportunistically uses the pro HB-EGF as its receptor.

DT binds to the extracellular, EGF-like domain of the human pro HB-EGF. Mouse pro HB-EGF differs within the EGF-like domain and is therefore unable to bind DT and serve as a functional DT receptor (Hooper and Eidels, 1995; Mitamura *et al.*, 1995). HB-EGF is mitogenic via the binding of its EGF-like domain to the EGF receptor. The binding of DT to the EGF-like domain of human HB-EGF inhibits its mitogenic activity presumably by masking or competing for the binding sites on the EGF receptors (Mitamura *et al.*, 1995).

Toxin bound to the receptor complex, enters the cell by receptor-mediated endocytosis. The toxin receptor lacks a rapid internalization signal sequence in its cytoplasmic domain and, as a result, the rate of internalization of the DT-receptor complex is much slower (1–2%/min) than that of classic endocytic receptors (10–15%/min for the transferrin receptor) (Almond and Eidels, 1994). Instead, the toxin-receptor complexes are believed to be internalized by entrapment in clathrin-coated pits as part of bulk-phase turnover of cell surface proteins. The extreme potency of the toxin results in the success of the intoxication process in spite of this slow rate of internalization. Just one molecule of DTA in the cytosol is sufficient to produce cell death (Yamaizumi *et al.*, 1978).

Mekada and coworkers (Iwamoto *et al.*, 1991) identified a 27 kDa protein that closely associates with pro HB-EGF in Vero cells. This 27 kDa protein, called DRAP 27 (diphtheria toxin receptor-associated protein), is the monkey equivalent of human CD9, and although it does not actually bind DT, it serves to upregulate the number of functional receptors and thus increase DT binding (Mitamura *et al.*, 1992; Brown *et al.*, 1993; Iwamoto *et al.*, 1994). DRAP 27/CD9 is a tetra membrane-spanning protein and has also been found to upregulate the juxtacrine mitogenic activity of pro HB-EGF (Higashiyama *et al.*, 1995; Nakamura *et al.*, 1995).

RECEPTOR-BINDING DOMAIN

The R domain of DT extends from residue 379 to 535 and is composed of 10 β strands (RB1–RB10) (Figure 12.2) (Choe *et al.*, 1992). These β-strands are arranged into 2 β-sheets: RB2, RB3, RB5 and RB8 comprise a four stranded β-sheet that faces a five-stranded β-sheet composed of RB4, RB6, RB7, RB9 and RB10. The two β-sheets have a flattened-barrel topology with a jelly-roll fold similar to that found in immunoglobulin variable domains, tumor necrosis factor and the receptor-binding domain of Pseudomonas exotoxin A. A β-hairpin loop forms the connection between the last 2 β-strands in the R domain (residues 514–525) and is proposed to play an important role in receptor recognition. This protruding loop corresponds with CDR3 in the immunoglobulin variable domain, consistent with its role in molecular recognition (Bennett and Eisenberg, 1994).

The location of the receptor-binding site in the toxin was originally identified using mutant toxins known as CRMs (cross-reacting materials). CRM 45 (M_r 45,000) is a deletion mutant produced by nitrosoguanidine mutagenesis (Uchida *et al.*, 1973). It contains an early termination signal, resulting in the deletion of the C-terminal 17 kDa portion comprising the last 149 amino acids of DTB (Giannini *et al.*, 1984). CRM 45 has full enzymatic activity, but is incapable of binding cell surface receptors (Boquet and Pappenheimer, 1976) and as a result it was inferred that the receptor-binding domain of the toxin was located within this C-terminal fragment. Myers and Villemez (1988) further defined the location of the receptor-binding site. They cleaved the asparaginyl-glycyl bonds at positions 453/454 and 481/482 using hydroxylamine and eliminated the binding of the resulting toxins. The actual binding site was therefore localized to the C-terminal 7 kDa segment of the B chain. Eidels and coworkers (Rolf *et al.*, 1990; Rolf and Eidels, 1993) extended these findings by demonstrating that the peptide fragment generated by hydroxylamine cleavage could protect cells by blocking the binding of toxin to cell surface receptors. This peptide, consisting of the last 54 amino acids of DT, is composed of RB8, RB9 and RB10 together with the intervening loops (Figure 12.2).

Full length CRMs, containing mutations at individual amino acids, have played an important role in delineating not only the receptor-binding region in general but more specifically what amino acids are actually involved in receptor recognition. CRM 103 and 107 were isolated by Laird and Groman (1976) using nitrosoguanidine mutagenesis. These CRMs were found to have full enzymatic activity but receptor-binding activity was reduced 8,000-fold (CRM 107) or 100 fold (CRM 103) (Greenfield *et al.*, 1987). Sequencing revealed that both CRM 103 and CRM 107 have single mutations. The serine at position 508 was changed to a phenylalanine (S508F) in CRM 103. CRM 107 also has a serine to phenylalanine switch but at position 525 (S525F) (Johnson and Youle, 1991). CRM 9, an independently isolated mutant, shares this identical mutation (S525F) (Hu and Holmes, 1987). The mutations in the R domain of these three CRM's fall within the last two β-loops (RB9, RB10 and interconnecting loop RL15), proposed to be the region that is involved in actual receptor recognition (Choe *et al.*, 1992).

Both these mutations, S508F and S525F, involve the substitution of a small polar residue with a bulky apolar amino acid. Because of this, it was not known if the amino acids at these locations played a direct role in receptor recognition or if the inhibition of binding was due to a more general steric or conformational change. To address this question, Collier and colleagues (Shen *et al.*, 1994) substituted alanine for each of the 12 solvent-exposed residues in the loop region (residues 516–530) that extends from the molecule and includes parts of RB9 and RB10 together with intervening RL5 (Choe *et al.*, 1992). They found that Lys-516 and Phe-530 played a strong role in receptor recognition and that Tyr-514, Val-523, Asn-524 and Lys-526 played lesser but significant roles. The CD spectra of mutants containing K516A or F530A were identical to that of the wild type toxin suggesting that the decreased binding observed in these mutants was not due to conformational changes, but rather than these residues were actually involved in receptor recognition. They also substituted serine at position 525 with either alanine or with phenylalanine and found that substitution with alanine produced little inhibition of receptor binding, while phenylalanine substitution significantly reduced binding. Since Ser-525 is located within 4Å of Lys-516, it is proposed that the bulky side chain found in S525F mutant (CRM 107, CRM 9) indirectly blocks binding by interfering with Lys-516. Similarly, the Ser-508 side chain lies within 4Å of Phe-530. This suggests that the reduction of binding observed in CRM 103, CRM 107 and CRM 9 may be due to steric problems asociated with the substitution of an apolar bulky residue in place of a small, polar one. These bulky residues interfere with Lys-516 and Phe-530, believed to be instrumental in receptor recognition (Shen *et al.*, 1994).

In addition to the receptor-binding site, the R domain of DT contains a second site, known as the "P-site" or polyphosphate-binding domain (Lory and Collier, 1980; Proia *et al.*, 1979). Middlebrook and coworkers (Middlebrook *et al.*, 1978; Middlebrook and Dorland, 1979) were the first to demonstrate that ATP and other phosphate-containing compounds inhibit toxicity presumably by interfering with receptor binding. The P-site actually interacts with two regions of the toxin: the receptor-binding site of the R domain and the NAD-binding site of the C domain. Evidence for this interaction comes from the observation that occupancy of the P-site by compounds containing multiple phosphate residues inhibits both the binding of the toxin to the cell surface receptor and the binding of NAD or the endogeneous dinucleotide, adenylyl-(3′,5′)-uridine 3′-monophosphate (ApUp) to the dinucleotide site (Lory *et al.*, 1980). This implies a close proximity of the nucleotide binding site of the A fragment with the P-site and the receptor-binding site of the B fragment.

Initial evidence for the localization of the P-site came from CRM 45 which, because it was unable to bind ATP, suggested that the deleted C-terminal 17 kDa portion of the B fragment of CRM 45 contained the P-site (Lory *et al.*, 1980). Eidels and coworkers further localized the P-site to the 7 kDa cyanogen bromide fragment located at the C-terminal end of the B fragment (Proia *et al.*, 1980). Barbieri *et al.* (1986) demonstrated that receptor-bound DT was still able to bind ApUp and therefore concluded that the P-site and the receptor-binding site, while both located in the receptor-binding domain, were distinct and non-overlaping sites. Further proof for this comes from the observation that dimeric toxin does not bind cell receptors but does bind ApUp with an affinity similar to that of monomeric toxin (Carroll *et al.*, 1986). Furthermore, CRM 197, while incapable of binding ApUp, is fully capable of binding to cell surface receptors (Mekada and Uchida, 1985). Rolf and Eidels (1993) definitively demonstrated that the receptor-recognition and phosphate-binding functions were located on different peptides.

Examination of the sequence of the toxin (Greenfield *et al.*, 1983; Ratti *et al.*, 1983; Kaczorek *et al.*, 1983) reveals a concentration of basic residues within the carboxyl-terminal portion of DTB. Affinity labelling with periodate-cleaved ATP followed by sodium cyanoborohydride reduction, resulted in labeling Lys-474 which lies within this grouping of basic residues in DTB. The crystallographic structure of DT shows that these basic residues (Lys-447, -456 and -474; Arg-455, -458, -460, -462 and 472; His-449) are positioned on the side of the R domain that faces the enzymatic cleft of the C domain (Choe *et al.*, 1992). It is believed that this concentration of basic residues constitutes or contains the P-site. The crystal structure also reveals that Arg-458 is positioned so that a salt bridge can be formed with the 3′ terminal phosphate of ApUp positioned in the enzymatic cleft thus accounting for the high affinity between DT and ApUp ($K_d \sim 0.3$ nM).

THE TRANSMEMBRANE DOMAIN

How a soluble, hydrophilic protein such as diphtheria toxin is able to cross the hydrophobic lipid bilayer has attracted much attention over the last several decades. In spite of this interest, it remains perhaps the least understood of all the steps in the intoxication pathway. Exposure of the toxin to acidic conditions is a well established prerequisite for translocation (Kim and Groman, 1965). Low pH conditions (pH 5.0–5.5) are encountered within the cell in endosomal compartments or can be artifically induced by exposing surface-bound toxin to a brief acidic "pulse". This causes the toxin to translocate directly across the plasma membrane thus bypassing the usual intracellular routing pathway for DT (Sandvig and Olsnes, 1980; Draper and Simon, 1980).

Exposure to low pH brings about a conformational change in the toxin and exposes regions of the toxin that are normally hidden at neutral pH. This is demonstrated by the increase in trypsin sensitivity of the toxin and changes in its CD spectrum (Hu and Holmes, 1984). The conformational changes also expose hydrophobic regions of the toxin illustrated by the ability of the toxin to bind Triton X-100 (Boquet et al., 1976; Sandvig and Olsnes, 1981) and by the ability to insert into artificial lipid membranes or liposomes (Hu and Holmes, 1984; Zalman and Wisnieske, 1984; Montecucco et al., 1985; Papini et al., 1987; Dumont and Richards, 1988). Confirmation of the ability of DT to insert into lipid bilayers also comes from observations that DT can form ion channels, both in planar lipid membranes (Donovan et al., 1981; Kagan et al., 1981) and in living cells (Sandvig and Olsnes, 1988; Eriksen et al., 1994).

The conformational change in DT is believed to represent a partial unfolding of the toxin which disrupts the tertiary structure but which leaves the secondary structure largely intact (Blewitt et al., 1984; Zhao and London, 1986). This is characteristic of a molten globule-like structure, a conformation that has been observed in a number of proteins and is recognized to play an important role in general protein folding and unfolding phenomena including membrane translocation (Ptitsyn, 1987; Goto and Fink, 1989). Low pH-induced unfolding likely involves alterations of electrostatic forces due to protonation of Asp, Glu and His residues. Protonation could result in the formation of buried charges, disruption of carboxyl-dependent salt bridges or the induction of repulsions due to increased positive charges, all of which would lead to a partial unfolding of the toxin (London, 1992a; 1992b).

The crystallographic structure reveals that the transmembrane domain (T domain) is entirely α-helical, and that some of these α-helices are extremely hydrophobic in nature, similar to known transmembrane proteins (Choe et al., 1992). There are a total of nine α-helices (TH1–TH9) arranged in three layers, each layer consisting of two or more antiparallel helices (Figure 12.2). The two C-terminal helices, TH8 and TH9, are the most hydrophobic. They are arranged in a helical hairpin conformation and constitute the innermost layer of the transmembrane domain. The next layer, composed of helices, TH5, TH6 and TH7, is only slightly less hydrophobic and is arranged as a kinked hairpin that wraps around the TH8–TH9 helices. The outermost layer consists of TH1, TH2 and TH3, which are extremely hydrophilic. The α-helices are connected by flexible loops, two of which (TL3 between helices TH5 and TH6, and loop TL5 between helices TH8 and TH9) are acidic in nature containing a total of six Asp and Glu residues. At neutral pH, these loops are highly charged and water-soluble.

This overall structure led Choe *et al.* (1992) to propose a model for membrane insertion. Upon exposure to low pH conditions, the acidic residues in the inter-helix loops become at least partially protonated and therefore more neutral and membrane-soluble. This effect is believed to be particularly strong at the surface of the endosomal membrane where there is a higher concentration of protons due to surface potential. The low pH of the endosome would therefore convert water-soluble loops into membrane-soluble "daggers" capable of inserting into the lipid bilayer. Penetration of these loops into the membrane would be followed by inser-tion of the two connected apolar helix pairs. If the acidic loops penetrate through to the cytosolic side of the membrane, reexposure to the neutral pH of the cytosol would lead to reionization and this could effectively stabilize the hairpins in a transmembrane configuration (Choe *et al.*, 1992).

The results of a number of experiments are consistent with this model. Zhan *et al.* (1994) cloned and expressed the T domain as a discrete protein in *E. coli*. To investigate the relative movements of the three layers of α-helices, the T domain was mutagenized and artificial disulfide bonds were introduced between the TH8–TH9 helical hairpin and the surrounding layers. Disulfides inhibited the exposure of apolar regions at low pH and blocked channel formation in artificial bilayers. Reduction of the disulfide bonds abolished these effects. These results suggest that TH8–TH9 hairpin separates from the rest of the T domain and that this separation is triggered by contact of the domain with the membrane under acidic conditions. To extend this observation, a 61-aa stretch of the T domain containing TH8, TH9 and the TL5 loop between them, was found to form chan-nels that are indistinguishable in their conductance, ion selectivity, and pH dependence from channels formed by the entire T domain (Silverman *et al.*, 1994a).

Ionizable residues in the TH8 and TH9 helical hairpin were mutagenized and switched with amino acids of the opposite charge in order to determine the role of acidic residues in the insertion of helical hairpins. Two of these charge-reversal mutations, D349K and D352K, both located in the TL5 loop of the TH8/TH9 helical hairpin, demonstrated a 100-fold reduction in cytotoxicity. These mutants also lost their ability to form channels in artificial lipid bilayers and to permeabi-lize Vero cells to $^{86}Rb^+$ after exposure to acidic conditions (O'Keefe *et al.*, 1992; Silverman *et al.*, 1994b). Cabiaux *et al.* (1993) demonstrated that another muta-tion, I364K, found within TH9, produces a similar reduction in cytotoxicity, translocation and channel formation. These mutations did not alter the pH-dependent conformational changes in the toxins but instead appeared to alter only the toxin's ability to insert into the membrane. Since the insertion of a positively charged residue would be energetically unfavourable, the loss of activity exhibited by these mutants supports the proposal that the insertion of the TH8/TH9 helical hairpin is a critical step in the translocation process. Moreover, when wild-type DT was inserted into planar lipid bilayers from the cis compartment, D352 and the TL5 loop in which it resides, were localized to a site at or near the trans face of the bilayer near the channel entrance (Mindell *et al.*, 1992; 1994).

Mutagenesis of Pro-345 also altered the ability of the toxin to translocate (Johnsons *et al.*, 1993). Pro-345 occupies a strategic location at the end of the TH8 α-helix at the point where it connects with the TL5 loop. Proline residues, besides being classic α-helix breakers, have the ability to undergo a cis-trans isomerization reaction and produce a concomitant conformational change. This reaction is catalyzed by a family of enzymes called prolyl-peptidyl cis-trans isomerases (PPIases), ubiquitous throughout the cell. It is proposed that the presence of PPIases in endosomes could have important implications for DT translocation. Cis-trans isomerization of the bond at Pro-345 could alter the orientation of the TH8–TH9 helical hairpin which, in turn, could improve accessibililty of the protonated TL5 loop to the lipid bilayer. Substitution of Pro-345 with either Glu or Gly resulted in a 99% reduction in toxicity. Since substitution with both Glu, an α-helix breaker, or Gly, an α-helix former, both resulted in the same reduction in toxicity, it suggested that Pro-345 was not functioning simply as an α-helix breaker but instead may be playing a more specific role in membrane translocation, perhaps involving cis-trans isomerization (Johnson *et al.*, 1993).

Although the evidence described above for insertion of the TH8–TH9 helical hairpin into membranes in response to acidic conditions is strong, the data for the insertion of the TH5–TH7 hairpin is less conclusive. Three acidic residues exist in the TL3 interconnecting loop. However, charge reversal mutagenesis of two of these residues, D290K and E292K, produced no alteration in cytotoxicity or in channel formation relative to wild-type toxin. Mutagenesis of the third, D295K, produced reduced toxicity although the effect was much smaller than that observed with similar mutations in the TL5 loop (Silverman *et al.*, 1994b). Therefore, decisive evidence for the insertion of the TH5-TH7 hairpin into the membrane during the translocation process is lacking.

The TH1–TH4 helices are highly charged and hydrophilic in nature and form the outermost layer of the T domain. In the acidic environment of the endosome, the charged residues in these helices may interact with the phospholipid head groups of the lipid bilayer serving as the initial membrane contact for the translocation process (Collier, 1995). TH1 is amphipathic in nature, containing a hydrophobic face and a charged hydrophilic face. The amphipathic nature of this helix may be important for the orientation of the C domain for translocation or for the actual translocation process (vandereSpek *et al.*, 1993; 1994; Madshus, 1994). There is controversy over the exact role that these helices play in the translocation process. More exact details of how the C domain actually translocates through the endosomal membrane, its interaction with the channel formed by the T domain or the other helices of the T domain are not well understood.

THE CATALYTIC DOMAIN

The crystallographic structure of the C domain reveals that it is composed of a mixture of α and β structures (Choe *et al.*, 1992). The β-strands are arranged in two β-sheet subdomains which form the enzymatic cleft (Figure 12.2). One subdomain consists of β-strands CB2,CB4 and CB8 surrounded by α-helices CH2, CH3, CH6 and CH7. The second subdomain consists of CB1, CB3, CB5, CB6 and CB7 surrounded by α-helices CH1, CH4 and CH5. The two subdomains are connected by four extended loops, CL1–CL4, which endow this domain with a high degree of flexibility allowing it to open and unfold during membrane translocation. It is necessary for the C domain to refold in the cytosol after it translocates through the endosomal membrane and is released from the T domain. The fact that the crystal structure of recombinant DTA is virtually identical to that of the C domain in the intact toxin, demonstrates that the C domain does not require the remainder of the toxin for refolding (Weiss *et al.*, 1995).

The catalytic cleft of the C domain is occupied by either the endogenous dinucleotide, ApUp, or by the natural substrate, NAD. ApUp binds competitively with NAD but has a higher affinity (K_d 0.3 nM for ApUp; K_d 8 μM for NAD), presumably due to a greater number of contacts with residues in the C domain as well as salt bridges with the P site of the R domain. The crystallographic structure of DT was determined with ApUp bound in the enzymatic cleft. Although the exact conformation of NAD in the enzymatic cleft is unknown, certain assumptions can be made based on the sturctural similarities between ApUp and NAD. Both ApUp and NAD contain adenosine and since it is unlikely that DT would have two distinct adenosine-binding sites, it is assumed that both adenosines are bound in a similar way. The uridine ring of ApUp is looked on as the structural analogue of the nicotinamide ring of NAD. Based on these comparisons, it is assumed that NAD is bound roughly perpendicular to the orientation of ApUp (Bennett *et al.*, 1994a).

The first residues determined to be critical for ADPRT activity were identified in CRMs generated by nitrosoguanidine mutagenesis. Three CRMs generated by this method, CRM 176, 228 and 197, demonstrated a deficit in catalytic activity. CRM 176 retains 8 to 10% of its enzymatic activity (Uchida *et al.*, 1973). The substitution of glycine at position 128 with aspartic acid (G128D) was found to be responsible for this reduction in enzymatic activity (Maxwell *et al.*, 1987). CRM 228 had no detectible enzymatic activity (Uchida *et al.*, 1973). It has a total of five amino acid changes, two of which are in DTA (G79D, E162K) (Kaczorek *et al.*, 1983). The DTA sites were separately mutagenized and G79D was identified as the mutation that singularly accounts for the loss of enzymatic activity in CRM 228 (Johnson and Nicholls, 1994b). CRM 197 also had no detectible enzymatic activity (Uchida *et al.*, 1973). It has a single mutation; glycine at position 52 is changed to glutamic acid (G52E) (Ratti *et al.*, 1983). A number of amino acids inserted at this site were subsequently found to also reduce ADPRT activity to background levels (Johnson and Nicholls, 1995), indicating the importance of Gly-52 for ADPRT activity.

A variety of techniques have been used to identify sites that are critical for ADPRT activity or for NAD binding. Glu-148 is one of these residues. The first identification of Glu-148 as a putative active-site residue of DT came from experiments in which this residue was shown to be photolabeled in the presence of nicotinamide-radiolabeled NAD (Carroll and Collier, 1984; Carroll *et al.*, 1985). The entire nicotinamide moiety of NAD became linked to Glu-148. Site-directed mutagenesis has confirmed the importance of Glu-148 in ADPRT activity. Replacement of Glu-148 with either aspartic acid, glutamine or serine greatly reduced ADPRT activity (100- to 300-fold) yet produced only minimal changes in K_m or K_d for NAD or K_m for EF-2 (Tweten *et al.*, 1985; Barbieri and Collier, 1987; Wilson *et al.*, 1990). The retention of substrate affinites suggested that the mutations did not alter the overall conformation of the toxin but instead altered only the catalytic rate. The spatial distribution and the charge of the carboxyl side chain of Glu-148 appear to be crucial for ADPRT activity. Even relatively conservative changes such as the withdrawal of the carboxyl group by one methylene unit, as in the aspartate mutant, or replacement of the carboxyl group by an uncharged amide, as in the glutamine mutant, greatly reduce enzymatic activity (Wilson *et al.*, 1990).

A specific active-site Glu may be a universal feature of ADP-ribosyltransferases (Domenighini *et al.*, 1991; Domenighini *et al.*, 1995; Collier, 1995). In addition to Glu-148 in DT, it has been found that Glu-553 in Pseudomonas exotoxin A (Carroll and Collier, 1987), Glu-112 in *E. coli* heat-labile toxin (Sixma *et al.*, 1991), Glu-129 in pertussis toxin (Barbieri *et al.*, 1989), and Glu-998 in poly (ADP-ribose) polymerase (Marsischky *et al.*, 1995) all appear to play an important role in ADPRT activity. In DT, it is proposed that Glu-148 functions in catalysis by deprotonating the attacking diphthamide residue of EF-2. One may therefore infer that the homologous active-site Glu residues of the other ADP-ribosyltransferases similarily deprotonate the attacking nucleophilic residues of their respective target proteins.

His-21 has also been identified as being a critical residue for ADPRT activity. Analysis of the crystal structure of DT reveals that His-21 is located within the enzymatic cleft (Choe *et al.*, 1992). The region encompassing H21 also demonstrates sequence similarity with other toxins expressing ADPRT activity. This similarity is especially strong for Pseudomonas exotoxin A but there is also more limited similarity with pertussis toxin S1 subunit, cholera toxin, and *E. coli* heat labile toxin (Domenighini *et al.*, 1991; 1994). His-21 is the only histidine residue in DTA compared with 15 histidine residues in DTB. Chemical modification of DTA with diethyl pyrocarbonate selectively modified His-21 and resulted in inhibition of NAD binding and ADPRT activity (Papini *et al.*, 1989). His-21 was mutagenized and replaced with all alternative amino acids (Johnson and Nicholls, 1994a). The majority of the substitutions virtually abolish enzymatic activity, the exception being a mutant in which His-21 was replaced with asparagine (H21N). This mutant demonstrated only a slight increase in K_m and relatively small decreases in both reaction rate (k_{cat}) and catalytic efficiency (k_{cat}/K_m). Asparagine is a sterically conserved substitution for histidine, but its side-chain is unable to replace the imidazole group of histidine in general acid-base mechanisms or to participate in electrostatic interactions. The effective substitution of asparagine for His-21 therefore suggests that it is not the acid-base properties of histidine but rather its hydrogen-bond-forming ability that is important for ADPRT activity. His-21 is believed to form hydrogen bonds with the carboxamide moiety of the nicotinamide ring of NAD and thereby serves to orient it optimally within the active-cleft (Blanke *et al.*, 1994a).

The active site cleft of DTA has several prominent aromatic amino acids, Trp-50, Trp-153, Tyr-54 and Tyr-65, which could potentially participate in hydrophobic stacking interactions with the aromatic rings of NAD. The intrinsic protein fluorescence of DTA is quenched by the binding of NAD (Kandel *et al.*, 1974). This effect is thought to be due to one or both of the tryptophan residues (these are the only tryptophans in DTA). Trp-50 and Trp-153 were mutagenized to phenylalanine and alanine in an attempt to determine the role that these tryptophans play in NAD binding (Wilson *et al.*, 1994). Substitution with phenylalanine (W50F and W153F) produced only minimal reduction in both ADPRT activity and NAD binding, suggesting that substitution of the phenyl ring for the indole ring still allowed normal substrate binding and catalysis. Similarly, the W153A substitution did not produce a major reduction in enzyme activity. In contrast, the W50A mutation reduced ADPRT activity by 10^5-fold and the ability to bind NAD to below detectable limits. These results implicate Trp-50 in the binding of NAD; more specificalaly, the binding may involve the aromatic stacking of the adenine ring of NAD with the indole ring of Trp-50 (Wilson *et al.*, 1994).

Substitution with either phenylalanine or alanine were also carried out in place of Tyr-65 (Blanke *et al.*, 1994b). This site was previously implicated in NAD binding based on the results of photoaffinity labeling studies (Papini *et al.*, 1991). Phenylalanine could effectively substitute for Tyr-65 (Y65F) but the alanine-containing mutant (Y65A) showed a major reduction in both ADPRT activity and NAD binding. These results support a model of ADP-ribosylation in which the phenolic ring of Tyr-65 interacts with the nicotinamide ring of NAD, orienting the N-glycosidic bond of NAD for attack by the incoming nucleophile in a direct displacement mechanism (Blanke *et al.*, 1994b).

The binding of NAD to the active-site cleft of DTA is therefore thought to involve two subsites, the adenine subsite and the nicotinamide subsite (Domenighini *et al.*, 1995). Binding in the adenine subsite is believed to involve the aromatic stacking of the adenine ring of NAD with the indole ring of Trp-50. The nicotinamide subsite of the toxin involves the participation of His-21, Tyr-65 and indirectly Glu-148. The imidazole ring of His-21 forms a hydrogen bond with the carboxamide moiety of the nicotinamide ring of NAD. Similarly, there is hydrophobic interaction between the phenolic side chain of Tyr-65 and the nicotinamide ring. Together these two residues orient the nicotinamide ring such that the N-glycosidic bond between the nicotinamide ring and the adjacent ribose ring is in the optimum position for nucleophilic attack by diphthamide. This N-glycosidic bond normally has considerable rotational freedom. It is proposed that His-21 and Tyr-65 limit the rotational and translational freedom of the N-glycosidic bond and provide for a more stable and favorable orientation for nucleophilic attack (Blanke *et al.*, 1994b). The large size of EF-2 may limit the entry of diphthamide into the nicotinamide subsite of the toxin. The ADPRT reaction is believed to proceed by a direct displacement reaction, with the π-imidazole nitrogen of diphthamide being activated by Glu-148 (Wilson *et al.*, 1994).

PHARMACOLOGICAL USES OF NATIVE AND MUTANT DIPHTHERIA TOXINS

Diphtheria toxin is perhaps the best characterized of all the bacterial toxins. The potent toxicity of DT together with the wealth of information available about the structure and function of the toxin has fostered the use of the toxin for a wide variety of purposes. From a public health perspective, the most important and widespread pharmacological use of diphtheria toxin is as a vaccine. Before a comprehensive immunization program was implemented, diphtheria was widespread. More than 1 in 20 inhabitants, especially children, suffered from clinical diphtheria during their lifetime and 5–10% of these died from the disease (Griffith, 1979). The major clinical manifestations of diphtheria are not due to an invasive bacterial infection but rather to the release of the toxin. It therefore follows that the disease can be prevented by the presence of neutralizing antibodies against the toxin. Active immunity against diphtheria may be acquired following a natural infection or may be elicited by immunization with an inactivated form of the toxin (toxoid). Formaldehyde treatment proved to be an effective method to chemically detoxify DT and facilitated the production of an effective vaccine (Ramon, 1924; Glenny and Hopkins, 1923). A widespread immunization program was initiated in the United States in the 1930's and has essentially lead to the virtual elimination of diphtheria in this country. However, the situation is very different in Eastern Europe. In the 1990's, diphtheria made a spectacular comeback in the New Independent States of the former Soviet Union (CDC, 1995). This resurgence is believed to be due to a decline in the immunization coverage in infants and children, a waning immunity in adults and the lack of available vaccines (Galazka and Robertson, 1995; Galazka *et al.*, 1995; Maurice, 1995). This epidemic emphasizes the need for effective immunization programs, not just in infancy and childhood but throughout adult life.

Diphtheria toxoid for vaccine use is traditionally prepared by formaldehyde inactivation of the toxin, using essentially the same method as that developed in the 1920's (Ramon, 1924; Glenny and Hopkins, 1923). While this method produces an extremely safe and efficacious vaccine, there is increasing interest in developing improved vaccines against diphtheria. Adverse reactions to the vaccines are rare and are usually limited to mild, local reactions. Some believe that a more highly purified diphtheria toxoid would be less reactogenic (Relyveld *et al.*,

1979; Rappuoli, 1990; Aggerbeck and Heron, 1992) especially in adults where the amount of diphtheria toxoid is reduced relative to the infant formulation in order to minimize reactogenicity (Edsall *et al.*, 1951; Epsen, 1954). Purification of the toxin prior to formaldehyde treatment would produce a more purified toxoid but this approach necessitates the handling of large quantities of highly concentrated toxic materials and represents a safety hazard. A genetically-inactivated toxoid offers an alternative approach to chemically inactivating the toxin. This approach would eliminate not only the hazard of handling toxic material but also abolish the need for formaldehyde treatment and facilitate the production of a more highly purified toxoid. A number of DTA mutants that are essentially devoid of enzymatic activity are currently being evaluated as alternative diphtheria vaccines (Wilson *et al.*, 1994; Johnson and Nicholls, 1995). These genetic mutants could also facilitate the production of combination vaccines. Protein carriers are currently used to improve the efficacy of polysaccaride-based conjugate vaccines (Anderson, 1982). CRM 197, an enzymatically-inactive mutant of DT, often serves as a carrier in such vaccines. However, CRM 197 does not have a native conformation, is more sensitive to cellular proteases, does not raise high titer neutralizing antibodies, and therefore is not an ideal immunogen for diphtheria (Pappenheimer *et al.*, 1972; Porro *et al.*, 1980). Other DT mutants may maintain their native toxin conformation and thus not only serve as an effective protein carrier for conjugate vaccines but also generate an effective anti-diphtheria response. Such responses are highly desirable in the development of combination vaccines.

Knowledge of the structure-function relationships of DT together with new advances in genetic engineering and vaccinology have provided alternative approaches for a diphtheria immunization. Inactive forms of the toxin can be delivered by live attenuated bacterial or viral vaccine strains, such as BCG, *Salmonella* sp., *Vibrio cholerae*, and vaccinia virus. The safety considerations for the delivery of a potentially lethal toxin by a live vaccine strain are considerable. Because it is possible for secondary mutations to restore toxicity, multiple mutations, each independently capable of inactivating the toxin would be required (Killeen *et al.*, 1992). Perhaps an even safer approach would be to include only the receptor-bind domain as the immunogen, since the majority of the neutralizing antibodies are directed against this domain (Pappenheimer *et al.*, 1972). A number of groups are currently investigating the potential of this domain as an immunogen, either incorporated into live vaccine strains or as a substitute for the more conventional vaccines (Fu *et al.*, 1993; Gomez-Duarte *et al.*, 1995).

DNA vaccines, involving the direct innoculation of purified DNA to raise immune responses (Wolff *et al.*, 1990), provides another alternative approach for diphtheria immunization. This technique utilizes eukaryotic expression vectors containing DNA encoding either inactivated mutants of DT or subunits of DT. The uptake and expression of the vaccine DNA by host cells initiates the immune response. This approach has produced favorable results for other immunogens (reviewed in Schodel *et al.*, 1994) and may represent a promising alternative method for diphtheria immunization in the future.

The potent toxicity of DT has been effectively targeted both at a protein and a genetic level. At a genetic level, expression of DTA under the control of tissue-specific transcriptional regulatory elements has lead to the ablation of selected cell populations. This has been used for the selective destruction of both cells in culture and in transgenic mice (Maxwell *et al.*, 1986; Palmiter *et al.*, 1987; Aguila *et al.*, 1995). However, the potency of DTA combined with the intrinsic leakiness of certain transcriptional regulatory sequences has lead to undesirable toxicity in nontargeted tissues. The use of attenuated toxins that require a substantially higher level of expression than their wild-type counterparts has helped to counter this problem (Breitman *et al.*, 1990; Fisher *et al.*, 1991). This approach allows target tissue ablation directed by regulatory elements that, while preferentially active in the target tissue, also retain significant basal activity in other tissues. Alternatively, the stringency of control over DTA expression can be improved thus reducing nontargeted expression and toxicity (Robinson and Maxwell, 1995). The targeted expression of DTA may thus provide a novel approach to cancer or AIDS therapy in the future.

The targeted delivery of DT as a protein has also attracted significant interest. Paul Ehrlich was the first to suggest the simple yet elegant concept of linking cell-specific antibodies with toxins to create "magic bullets". He envisioned that such immunotoxins could be used to selectively destroy a specifically targeted population of cells. In order to precisely target the potent activity of DT, the native receptor-binding function of the toxin must be replaced by a specific targeting moiety such as a monoclonal antibody, growth factor or cytokine. To avoid nontargeted toxicity, the native R-domain must be either removed or inactivated. Truncated toxins, containing deletions of either the entire R-domain or a portion of it (Strom *et al.*, 1992) or mutant full-length toxins with attenuated receptor binding ability, such as S525F (CRM 107 or CRM 9) (Greenfield *et al.*, 1987; Neville *et al.*, 1992) are used for this purpose. The T domain is required for the effective translocation of DTA into the cytosol and is therefore usually included to maximize the potency of the immunotoxin. A wide variety of immunotoxins have been constructed using an assortment of targeting moieties. These are either chemically crosslinked with the toxin (Johnson *et al.*, 1989; Neville *et al.*, 1992) or constructed genetically and expressed in *E. coli* (Strom *et al.*, 1991).

A potential problem for the use of DT-based immunotoxins is the presence of pre-existing anti-DT antibody in human blood in response to the widespread immunization program. Neutralizing antibody severely limits the therapeutic efficacy of these immunotoxins (Johnson *et al.*, 1989). Two approaches have been used to overcome the blocking effect of pre-existing anti-DT antibodies. One approach utilizes immunotoxins made with truncated forms of DT. Since the majority of the neutralizing antibodies appear to be directed against the receptor-binding domain of the toxin (Pappenheimer *et al.*, 1972; Zucker *et al.*, 1984), removal of this domain limits the inhibitory effect of anti-DT antibodies toward the immunotoxin (Thompson *et al.*, 1995). The second approach involves the use of the imunotoxin in an environment that excludes pre-existing anti-DT antibodies. The central nervous system is an example of such an immunologically-privileged site. Since cerebral spinal fluid contains only 0.2% of the IgG levels found in serum, immunotoxin inhibition by pre-existing antibodies is avoided (Johnson *et al.*, 1989). The treatment of tumors of the central nervous system with immunotoxins offers early clinical promise (Pastan *et al.*, 1995).

Diphtheria toxin was the first toxin for which an enzymatic function was discovered and it continues to be perhaps the best characterized of all the bacterial toxins. Detailed knowledge of the structure — function of the toxin has lead to the development of new pharmacologic uses for the toxin and produced improvements in existing uses. As more of the unanswered questions about diphtheria toxin become resolved and advances are made in the areas of protein engineering and targeted delivery systems, even more pharmacologic uses for the toxin are anticipated.

REFERENCES

Aggerbeck, H. and Heron, I. (1992) Detoxification of diphtheria and tetanus toxin with formaldehyde. Detection of protein conjugates. *Biologicals*, **20**, 109–115.

Aguila, H.L., Hershberger, R.J. and Weissman, I.L. (1995) Transgenic mice carrying the diphtheria toxin A chain gene under the control of the granzyme A promoter: Expected depletion of cytotoxic cells and unexpected depletion of CD8 T cells. *Proc. Natl. Acad. Sci. USA*, **92**, 10192–10196.

Almond, B. and Eidels, L. (1994) The cytoplasmic domain of the diphtheria toxin receptor (HB-EGF precursor) is not required for receptor-mediated endocytosis. *J. Biol. Chem.*, **269**, 26635–26641.

Anderson, P. (1982) Antibody responses to *Haemophilus influenzae* type b and diphtheria toxin induced by conjugates of oligosaccharides of the type b capsule with the nontoxic protein CRM 197. *Infect. Immun.*, **39**, 233–238.

Barbieri, J.T., Collins, C.M. and Collier, R.J. (1986) Diphtheria toxin can simultaneously bind to its receptor and adenylyl-(3',5')-uridine 3'-monophosphate. *Biochem.*, **25**, 6608–6611.

Barbieri, J.T. and Collier, R.J. (1987) Expression of a mutant, full-length form of diphtheria toxin in *Escherichia coli*. *Infect. Immun.*, **55**, 1647–1651.

Barbieri, J.T., Mende-Meuller, L.M., Rappuoli, R. and Collier, R.J. (1989) Photolabeling of Glu-129 of the S-1 subunit of pertussis toxin with NAD. *Infect. Immun.*, **57**, 3549–3554.

Bennett, M.J., Choe, S. and Eisenberg, D. (1994a) Refined structure of dimeric diphtheria toxin at 2.0Å resolution. *Protein Science*, **3**, 1444–1463.

Bennett, M.J., Choe, S., and Eisenberg, D. (1994b) Domain swapping: entangling alliances between proteins. *Proc. Natl. Acad. Sci. USA*, **91**, 3127–313.

Bennett, M.J. and Eisenberg, D. (1994) Refined structure of monomeric diphtheria toxin at 2.3Å resolution. *Protein Science*, **3**, 1464–1475.

Blanke, S.R., Huang, K., Wilson, B.A., Papini, E., Covacci, A. and Collier, R.J. (1994a) Active-site mutations of the diphtheria toxin catalytic domain: role of histidine-21 in nicotinamide adenine dinucleotide binding and ADP-ribosylation of elongation factor 2. *Biochem.*, **33**, 5155–5161.

Blanke, S.R., Huang, K. and Collier, R.J. (1994b) Active-site mutations of diphtheria toxin: role of tyrosine-65 in NAD binding and ADP-ribosylation. *Biochem.*, **33**, 15494–15500.

Blewitt, M., Zhao, J., McKeever, B., Sarma, R. and London, E. (1984) Fluorescence characterization of the low pH-induced change in DT conformation: effect of salt. *Biochem. Biophys. Res. Comm.*, **120**, 286–290.

Boquet, P. and Pappenheimer, A. (1976) Interaction of DT with mammalian cell membranes. *J. Biol. Chem.*, **251**, 5770–5778.

Boquet, P., Silverman, M., Pappenheimer, A. and Vernon, W. (1976) Binding of Triton X-100 to diphtheria toxin, CRM 45 and their fragments. *Proc. Natl. Acad. Sci. USA*, **73**, 4449–4453.

Breitman, M.L., Rombola, H., Maxwell, I.H., Klintworth, G.K. and Bernstein, A. (1990) Genetic ablation in transgenic mice with an attenuated diphtheria toxin A gene. *Mol. Cell. Biol.*, **10**, 474–479.

Brown, J.G., Almond, B.D., Naglich, J.G. and Eidels, L. (1993) Hypersensitivity to diphtheria toxin by mouse cells expressing both diphtheria toxin receptor and CD9. *Proc. Natl. Acad. Sci. USA*, **90**, 8184–8188.

Cabiaux, V., Mindell, J. and Collier, R.J. (1993) Membrane translocation and channel-forming activities of diphtheria toxin are blocked by replacing isoleucine 364 with lysine. *Infect. Immun.*, **61**, 2200–2202.

Carroll, A.F. and Collier, R.J. (1984) NAD binding site of diphtheria toxin: Identification of a residue within the nicotinamide subsite by photochemical modification with NAD. *Proc. Natl. Acad. Sci. USA*, **81**, 3307–3311.

Carroll, S.F., McCloskey, J.A., Crain, P.F., Oppenheimer, N.J., Marschner, T.M. and Collier, R.J. (1985) Photoaffinity labeling of diphtheria toxin fragment A with NAD: structure of the photoproduct at position 148. *Proc. Natl. Acad. Sci. USA*, **82**, 7237–7241.

Carroll, S.F., Barbieri, J.T. and Collier, R.J. (1986) Dimeric Form of diphtheria toxin: purification and characterization. *Biochem.*, **25**, 24225–24230.

Carroll, S.F. and Collier, R.J. (1987) Active site of Pseudomonas aeruginosa exotoxin A: glutamic acid 553 is photolabeled by NAD and shows functional homology with glutamic acid 148 of diphtheria toxin. *J. Biol. Chem.*, **262**, 8707–8711.

Carroll, S.F., Barbieri, J.T. and Collier, J.R. (1988) Diphtheria toxin: Purification and properties. *Methods Enzymol.*, **165**, 68–76.

CDC (1995) Diphtheria epidemic — New Independent States of the Former Soviet Union, 1990–1994. *Morbidity and Mortality Weekly Report*, **44**, 177–181.

Choe, S., Bennett, M.J., Fujii, G., Curmi, P.M., Kantardjieff, K.A., Collier, R.J. and Eisenberg, D. (1992) The crystal structure of diphtheria toxin. *Nature*, **357**, 216–222.

Collier, R.J. and Kandel, J. (1971) Structure and activity of DT. 1. Thiol-dependent dissociation of a fraction of toxin into enzymically active and inactive fragments. *J. Biol. Chem.*, **246**, 1496–1503.

Collier, R.J. (1975) Diphtheria toxin: Mode of action and structure. *Bact. Rev.*, **39**, 54–85.

Collier, R.J. (1995) Three-dimensional structure of diphtheria toxin. In Bacterial Toxins and Virulence Factors in Disease. In J. Moss, M. Vaughan, B. Iglewski, and A.T. Tu, (eds.), *Handbook of Bacterial Toxins, Vol. 8.* Marcel Dekker, Inc., N.Y., pp. 81–93.

Domenighini, M., Montecucco, C., Ripka, W.C. and Rappuoli, R. (1991) Computer modelling of the NAD-binding site of ADP-ribosylating toxins: active-site structure and mechanism of NAD binding. *Mol. Micro.*, **5**, 23–31.

Domenighini, M., Magagnoli, C., Pizza, M. and Rappuoli, R. (1994) Common features of the NAD-binding and catalytic site of ADP-ribosylating toxins. *Mol. Micro.*, **14**, 41–50.

Domenighini, M., Pizza, M. and Rappuoli, R. (1995) Bacterial ADP-ribosyltransferases. In J. Moss, M. Vaughan, B. Iglewski, and A.T. Tu, (eds.), *Handbook of Bacterial Toxins, Vol. 8.* Marcel Dekker, Inc., N.Y., pp. 59–80.

Donovan, J.J., Simon, M.I., Draper, R.K. and Montal, M. (1981) Diphtheria toxins forms transmembrane channels in planar lipid bilayers. *Proc. Natl. Acad. Sci. USA*, **78**, 172–176.

Draper, R. and Simon, M. (1980) The entry of DT into the mammalian cell cytoplasm: evidence for lysosomal involvement. *J. Cell Biol.*, **87**, 849–854.

Drazin, R., Kandel, J. and Collier, R.J. (1971) Structure and activity of DT. II. Attack by trypsin at a specific site within the intact toxin molecule. *J. Biol. Chem.*, **246**, 1504–1510.

Dumont, M. and Richards, F. (1988) The pH-dependent conformational change of DT. *J. Biol. Chem.*, **263**, 2087–2097.

Edsall, G., Banton, J.B. and Wheeler, R.E. (1951) The antigenicity of single, graded doses of purified diphtheria toxoid in man. *Am. J. Hyg.*, **53**, 283–295.

Eriksen, S., Olsnes, S., Sandvig, K. and Sand, O. (1994) Diphtheria toxin at low pH depolarizes the membrane, increases the membrane conductance and induces a new type of ion channel in Vero cells. *EMBO J.*, **13**, 4433–4439.

Fisher, K.S., Maxwell, I.H., Murphy, J.R., Collier, J. and Glode, L.M. (1991) Construction and expression of plasmids containing mutated diphtheria toxin A-chain coding sequences. *Infect. Immun.*, **59**, 3562–3565.

Fu, H., Shen, W.H. and Collier, R.J. (1993) Receptor-binding domain of diphtheria toxin as a potential immunogen. In R.M. Chanock, F. Brown, H.S. Ginsberg and E. Norrby, (eds.), *Vaccines '93 Molecular*

Approaches to the Control of Infectious Diseases, Cold Spring Harbor Press, Cold Spring Harbor, N.Y., pp. 379–383.

Galazka, A.M. and Robertson, S.E. (1995) Diphtheria: Changing patterns in the developing world and the industrialized world. *Eur. J. Epidemiology*, **11**, 107–117.

Galazka, A.M., Robertson, S.E. and Obapenko, G.P. (1995) Resurgence of diphtheria. *Eur. J. Epidemiology*, **11**, 95–105.

Giannini, G., Rappuoli, R. and Ratti, G. (1984) The amino acid sequence of two non-toxic mutants of DT: CRM 45 and CRM 197. *Nucleic Acids Res.*, **12**, 4063–4069.

Gill, M., Pappenheimer, A., Brown, R. and Kurnick, J. (1969) Studies on the mode of action of DT. VII. Toxin-stimulated hydrolysis of nicotinamide adenine dinucleotide in mammalian cell extracts. *J. Exp. Med.*, **129**, 1–21.

Gill, M. and Pappenheimer, A. (1971) Structure and activity of diphtheria toxin. *J. Biol. Chem.*, **246**, 1492–1495.

Glenny, A.T. and Hopkins, B.E. (1923) Diphtheria toxoid as an immunizing agent. *Br. J. Exp. Pathol.*, **4**, 283–288.

Gomez-Duarte, O.G., Galen, J., Chatfield, S.N., Rappuoli, R., Eidels, L. and Levine, M.M. (1995) Expression of fragment C of tetanus toxin fused to a carboxyl-terminal fragment of diphtheria toxin in Salmonella typhi SVD 908 vaccine strain. *Vaccine*, **13**, 1596–1602.

Goto, Y. and Fink, A.L. (1989) Conformational states of β-lactamase: molten-globule states at acidic and alkaline pH with high salt. *Biochem.*, **28**, 945–952.

Greenfield, L., Bjorn, M., Horn, G., Fong, D., Buck, G., Collier, R. and Kaplan, D. (1983) Nucleotide sequence of the structural gene for diphtheria toxin carried by corynebacteriophage β. *Proc. Natl. Acad. Sci. USA*, **80**, 6853–6857.

Greenfield, L., Johnson, V.G. and Youle, R. (1987) Mutations in diphtheria toxin separate binding from entry and amplify immunotoxin selectvity. *Science*, **238**, 538–539.

Griffith, A.H. (1979) The role of immunization in the control of diphtheria. *Dev. Biol. Stand.*, **43**, 3–13.

Higashiyama, S., Iwamoto, R., Goishi, K., Raab, G., Taniguchi, N., Klagsbrun, M. and Mekada, E. (1995) The membrane protein CD9/DRAP27 potentiates the juxtacrine growth factor activity of the membrane-anchored heparin-binding EGF-like growth factor. *J. Cell Biol.*, **128**, 929–938.

Honjo, T., Nishizuka, Y. and Hayaishi, O. (1968) DT-dependent adenosine diphosphate ribosylation of aminoacyl transferase II and inhibition of protein synthesis. *J. Biol. Chem.*, **243**, 3553–3555.

Hooper, K., and Eidels, L. (1995) Localization of a critical diphtheria toxin-binding domain to the C-terminus of the mature heparin-binding EGF-like growth factor region of the diphtheria toxin receptor. *Biochem. Biophys. Res. Comm.*, **206**, 710–717.

Hu, V. and Holmes, R. (1984) Evidence for direct insertion of fragments A and B of DT into model membranes. *J. Biol. Chem.*, **259**, 12226–12233.

Hu, V.W. and Holmes, R. (1987) Single mutation in the A domain of diphtheria toxin results in a protein with altered membrane insertion behavior. *Biochim. et Biophys. Acta*, **902**, 24–30.

Ipsen, J. (1954) Immunization of adults against diphtheria and tetanus. *N. Eng. J. Med.*, **251**, 459–466.

Ittelson, T.R. and Gill, D.M. (1973) Diphtheria toxin: specific competition for cell receptors. *Nature*, **242**, 330–332.

Iwamoto, R., Senoh, H., Okada, Y., Uchida, T. and Mekada, E. (1991) An antibody that inhibits the binding of diphtheria toxin to cells revealed the association of a 27 kDa membrane protein with the diphtheria toxin receptor. *J. Biol. Chem.*, **266**, 20463–20469.

Iwamoto, R., Higashiyama, S., Mitamura, T., Taniguchi, N., Klagsbrun, M. and Mekada, E. (1994) Heparin-binding EGF-like growth factor, which acts as the diphtheria toxin receptor, forms a complex with membrane protein DRAP 27/CD9, which up-regulates functional receptors and diphtheria toxin sensitivity. *EMBO J.*, **13**, 2322–2330.

Johnson, V.G., Wrobel, C., Wilson, D., Zovickian, J., Greenfield, L., Oldfield, E.H. and Youle, R. (1989) Improved tumor-specific immunotoxins in the treatment of CNS and leptomeningeal neoplasia. *J. Neurosurg.*, **70**, 240–248.

Johnson, V.G. and Youle, R.J. (1991) Intracellular routing and membrane translocation of diphtheria toxin and ricin. In C.S. Steere and J.A. Hanover (eds), *Intracellular Trafficking of Proteins*, Cambridge Univ. Press., N.Y., pp. 183–225.

Johnson, V.G., Nicholls, P.J., Habig, W.H. and Youle, R.J. (1993) The role of proline 345 in diphtheria toxin translocation. *J. Biol. Chem.*, **268**, 3514–3519.

Johnson, V.G. and Nicholls, P.J. (1994a) Histidine 21 does not play a major role in diphtheria toxin catalysis. *J. Biol. Chem.*, **269**, 4349–4354.

Johnson, V.G. and Nicholls, P.J. (1994b) Identification of a single amino acid substitution in the diphtheria toxin A chain of CRM 228 responsible for the loss of enzymatic activity. *J. Bacteriol.*, **176**, 4766–4769.

Johnson, V.G. and Nicholls, P.J. (1995) Mutagenesis of diphtheria toxin A chain to inactivate ADP-ribosyltransferase activity. In R.M. Chanock, F. Brown, H.S. Ginsberg, and E. Norrby (eds.), *Vaccines 95: Molecular Approaches to the Control of Infectious Diseases.*, Cold Spring Harbor Laboratory Press, N.Y., pp. 207–211.

Kaczorek, M., Delpeyroux, F., Chenciner, N., Streeck, R., Murphy, J., Boquet, P. and Tiollais, P. (1983) Nucleotide sequence and expression of the diphtheria tox 228 gene in *E. coli. Science*, **221**, 855–858.

Kagan, B.L., Kinkelstein, A. and Colombini, M. (1981) Diphtheria toxin fragment forms large pores in phospholipid bilayer membranes. *Proc. Natl. Acad. Sci. USA*, **78**, 4950–4954.

Kandel, J.C., Collier, R.J. and Chung, D.W. (1974) Interaction of fragment A from diphtheria toxin with Nicotinamide Adenine Dinucleotide. *J. Biol. Chem.*, **249**, 2088–2097.

Killeen, K.P., Escuyer, V., Mekalanos, J.J. and Collier, R.J. (1992) Reversion of recombinant toxoids: mutations in diphtheria toxin that partially compensate for active-site deletions. *Proc. Natl. Acad. Sci. USA*, **89**, 6207–6209.

Kim, K. and Groman, N. (1965) Mode of inhibition of DT by ammonium chloride. *J. Bact.*, **90**, 1557–1562.

Laird, W. and Groman, N. (1976) Isolation and characterization of tox mutants of Corynebacteriophage beta. *J. Virology*, **19**, 220–227.

London, E. (1992a) How bacterial protein toxins enter cells: the role of partial unfolding in membrane translocation. *Mol. Micro.*, **6**, 3277–3282.

London, E. (1992b) Diphtheria toxin: membrane interaction and membrane translocation. *Biochim. Biophys. Acta*, **1113**, 25–51.

Lory, S., Carroll, S. and Collier, R.J. (1980) Ligand interactions of DT. II Relationships between the NAD site and the P site. *J. Biol. Chem.*, **255**, 12016–12019.

Lory, S. and Collier, R.J. (1980) DT: nucleotide binding and toxin heterogeneity. *Proc. Natl. Acad. Sci. USA*, **77**, 267–271.

Madshus, I.H. (1994) The N-terminal α-helix of fragment B of diphtheria toxin promotes translocation of fragment A into the cytoplasm of eukaryotic cells. *J. Biol. Chem.*, **269**, 17723–17729.

Marsischky, G.T., Wilson, B.A. and Collier, R.J. (1995) Role of glutamic acid 988 of human poly-ADP-ribose polymerase in polymer formation. *J. Biol. Chem.*, **270**, 3247–3254.

Maurice, J. (1995) Belated attempts to contain Russia's diphtheria. *Lancet*, **345**, 715.

Maxwell, I.H., Maxwell, F. and Glode, L.M. (1986) Regulated expression of a diphtheria toxin A-chain gene transfected into human cells: possible strategy for inducing cancer cell suicide. *Cancer Res.*, **46**, 4660–4664.

Maxwell, F., Maxwell, I.H. and Glode, L.M. (1987) Cloning, sequence determination, and expression in transfected cells of the coding sequence for the tox 176 attenuated diphtheria toxin A chain. *Mol. Cell Biol.*, **7**, 1576–1579.

Mekada, E. and Uchida, T. (1985) Binding properties of diphtheria toxin to cells are altered by mutation in the fragment A domain. *J. Biol. Chem.*, **260**, 12148–12153.

Middlebrook, J.L., Dorland, R.B. and Leppla, S.H. (1978) Association of diphtheria toxin with Vero cells: demonstration of a receptor. *J. Biol. Chem.*, **253**, 7325–7330.

Middlebrook, J. and Dorland, R. (1979) Protection of mammalian cells from DT by exogenous nucleotides. *Can. J. Micro.*, **25**, 285–290.

Mindell, J.A., Silverman, J.A., Collier, R.J. and Finkelstein, A. (1992) Locating a residue in the diphtheria toxin channel. *Biophys. J.*, **62**, 41–44.

Mindell, J.A., Silverman, J.A., Collier, R.J. and Finkelstein, A. (1994) Structure function relationships in diphtheria toxin channels: II. A residue responsible for the channel's dependence on trans pH. *J. Membrane Biol.*, **137**, 29–44.

Mitamura, T., Iwamoto, R., Umata, T., Yomo, T., Urabe, I., Tsuneoka, M. and Mekada, E. (1992) The 27 kDa diphtheria toxin receptor-associated protein (DRAP 27) from Vero cells is the monkey homologue of human CD9 antigen: expression of DRAP 27 elevates the number of diphtheria toxin receptors on toxin-sensitive cells. *J. Cell Biol.*, **118**, 1389–1399.

Mitamura, T., Higashiyama, S., Taniguchi, N., Klagsbrun, M. and Mekada, E. (1995) Diphtheria toxin binds to the epidermal growth factor (EGF)-like domain of human heparin-binding EGF-like growth factor/diphtheria toxin receptor and inhibits specifically its mitogenic activity. *J. Biol. Chem.*, **270**, 1015–1019.

Montecucco, C., Schiavo, G. and Tomasi, M. (1985) pH-dependence of the phospholipid interaction of DT fragments. *Biochem. J.*, **231**, 123–128.

Myers, D. and Villemez, C.L. (1988) Specific chemical cleavage of diphtheria toxin with hydroxylamine: purification and characterization of the modified proteins. *J. Biol. Chem.*, **263**, 17122–17127.

Naglich, J.G. and Eidels, L. (1990) Isolation of diphtheria toxin-sensitive mouse cells from a toxin-resistant population transfected with monkey DNA. *Proc. Natl. Acad. Sci. USA*, **87**, 7250–7254.

Naglich, J.G., Metherall, J.E., Russell, D.W. and Eidels, L. (1992) Expression cloning of a diphtheria toxin receptor: identity with a heparin-binding EGF-like growth factor precursor. *Cell*, **69**, 1051–1061.

Nakamura, K., Iwamoto, R. and Mekada, E. (1995) Membrane-anchored heparin-binding EGF-like growth factor (HB-EGF) and diphtheria toxin receptor-associated protein (DRAP 27)/CD9 form a complex with integrin a3b1 at cell-cell contact sites. *J. Cell Biol.*, **129**, 1691–1705.

Neville, D.M., Scharff, J. and Srinivasachar, K. (1992) *In vivo* T-cell ablation by a holo-immunotoxin directed at human CD3. *Proc. Natl. Acad. Sci. USA*, **89**, 2585–2589.

O'Keefe, D.O., Cabiaux, V., Choe, S., Eisenberg, D. and Collier, R.J. (1992) pH-dependent insertion of proteins into membranes: B-chain mutation of diphtheria toxin that inhibits membrane translocation, Glu-349-Lys. *Proc. Natl. Acad. Sci. USA*, **89**, 6202–6206.

Palamiter, R.D., Behringer, R.R., Quaife, C.J., Maxwell, F., Maxwell, I.H. and Brinster, R.L. (1987) Cell lineage ablation in transgenic mice by cell-specific expression of a toxin gene. *Cell*, **50**, 435–443.

Papini, E., Schiavo, G., Tomasi, M., Colombatti, M., Rappuoli, R. and Montecucco, C. (1987) Lipid interaction of DT and mutants with altered fragment B. 2. Hydrophobic photolabelling and cell intoxication. *Eur. J. Biochem.*, **169**, 637–644.

Papini, E., Schiavo, G., Sandona, D., Rappuoli, R. and Montecucco, C. (1989) Histidine 21 is at the NAD+ binding site of diphtheria toxin. *J. Biol. Chem.*, **264**, 12385–12388.

Papini, E., Santucci, A., Schiavo, G., Domenighini, M., Neri, P., Rappuoli, R. and Montecucco, C. (1991) Tyrosine 65 is photolabeled by 8-azidoadenine and 8-azidoadenosine at the NAD binding site of diphtheria toxin. *J. Biol. Chem.*, **266**, 2494–2498.

Papini, E., Rappuoli, R., Murgia, M. and Montecucco, C (1993) Cell penetration of diphtheria toxin: reduction of the interchain disulfide bridge is the rate-limiting step of translocation in the cytosol. *J. Biol. Chem.*, **268**, 1567–1574.

Pappenheimer, A.M., Uchida, T. and Harper, A. (1972) An immunological study of the diphtheria toxin molecule. *Immunochem.*, **9**, 891–906.

Pappenheimer, A.M. (1977) Diphtheria toxin. *Ann. Rev. Biochem.*, **46**, 69–94.

Pastan, I.H., Archer, G.E., McLendon, R.E., Friedman, H.S., Fuchs, H.E., Wang, Q.C., Pai, L.H., Herndon, J. and Bigner, D.D. (1995) Intrathecal administration of single-chain immunotoxin, LMB-7 [B3(Fv)-PE38], produces cures of carcinomatous meningitis in a rat model. *Proc. Natl. Acad. Sci. USA*, **92**, 2765–2769.

Porro, M., Saletti, M., Nencioni, L., Tagliaferri, L. and Marsili, I. (1980) Immunogenic correlation between cross-reacting material (CRM 197) produced by a mutant of *Corynebacterium diphtheriae* and diphtheria toxoid. *J. Inf. Dis.*, **142**, 716–724.

Proia, R.L., Hart, D.A. and Eidels, L. (1979) Interaction of diphtheria toxin with phosphorylated molecules. *Infect. Immun.*, **26**, 942–948.

Proia, R.L., Wray, S.K., Hart, D.A. and Eidels, L. (1980) Characterization and affinity labeling of the cationic phosphate-binding (nucleotide-binding) peptide located in the receptor-binding region of the B-fragment of diphtheria toxin. *J. Biol. Chem.*, **255**, 12025–12033.

Ptitsyn, O.B. (1987) Protein folding: hypothesis and experiments. *J. Prot. Chem.*, **6**, 273–293.

Ramon, G. (1924) Sur la toxine et sur l'anatoxine diphtheriques. *Ann. Inst. Pasteur*, **38**, 1–106.

Rappuoli, R. (1990) New and improved vaccines against diphtheria and tetanus. In G.C. Woodrow and M.M. Levine, (eds.) *New Generation Vaccines*, Marcel Dekker, N.Y., pp. 251–268.

Ratti, G., Rappuoli, R. and Giannini, G. (1983) The complete nucleotide sequence of the gene coding for DT in the corynephage omega (tox+) genome. *Nucleic Acids Res.*, **11**, 6589–6595.

Relyveld, E.H., Henocq, E. and Bizzini, B. (1979) Studies on untoward reactions to diphtheria and tetanus toxoids. *Develop. Biol. Standard.*, **43**, 33–37.

Robinson, D.F. and Maxwell, I.H. (1995) Suppression of single and double nonsense mutations introduced into the diphtheria toxin A-chain gene: a potential binary system for toxin gene therapy. *Human Gene Therapy*, **6**, 137–143.

Rolf, J.M., Gaudin, H.M. and Eidels, L. (1990) Localization of the diphtheria toxin receptor-binding domain to the carboxyl-terminal $M_r \sim 6000$ region of the toxin. *J. Biol. Chem.*, **265**, 7331–7337.

Rolf, J.M. and Eidels, L. (1993) Characterization of the diphtheria toxin receptor-binding domain. *Mol. Micro.*, **7**, 585–591.

Sandvig, K. and Olsnes, S. (1980) DT entry into cells is facilitated by low pH. *J. Cell Biol.*, **87**, 828–832.

Sandvig, K. and Olsnes, S. (1981) Rapid entry of nicked DT into cells at low pH. *J. Biol. Chem.*, **256**, 9068–9076.

Sandvig, K. and Olsnes, S. (1988) Diphtheria toxin-induced channels in Vero cells selective for monovalent cations. *J. Biol. Chem.*, **263**, 12353–12359.

Schodel, F., Aguado, M.T. and Lambert, P.H. (1994) Introduction: Nucleic acid vaccines, WHO, Geneva, 17–18 May, 1994. *Vaccine*, **12**, 1491–1492.

Shen, W.H., Choe, S., Eisenberg, D. and Collier, R.J. (1994) Participation of lysisne 516 and phenylalanine 530 of diphtheria toxin in receptor recognition. *J. Biol. Chem.*, **269**, 29077–29084.

Silverman, J.A., Mindell, J.A., Zhan, H., Finkelstein, A. and Collier, R.J. (1994a) Structure-function relationships in diphtheria toxin channels: I. Determining a minimal channel-forming domain. *J. Memb. Biol.*, **137**, 17–28.

Silverman, J.A., Mindell, J.A., Finkelstein, A., Shen, W.H. and Collier, R.J. (1994b) Mutational analysis of the helical hairpin region of diphtheria toxin transmembrane domain. *J. Biol. Chem.*, **269**, 22524–22532.

Sixma, T.K., Pronk, S.E., Kalk, K.H., Wartna, E.S., vanZanten, B.A., Witholt, B. and Hol, W.G. (1991) Crystal structure of a cholera toxin-related heat-labile enterotoxin from *E. coli*. *Nature*, **351**, 371–377.

Strom, T.B., Anderson, P.L., Rubin-Kelly, V.E., Williams, D.P., Kiyokama, T. and Murphy, J.R. (1991) Immunotoxins and cytokine fusion proteins. *Ann. N.Y. Acad. Sci.*, **636**, 233–250.

Strom, T.B., Kelley, V.R., Woodworth, T.G. and Murphy, J.R. (1992) Interleukin-2 receptor-directed immunosuppressive therapies: antibody- or cytokine-based targeting molecules. *Immun. Reviews*, **129**, 131–163.

Thompson, J., Hu, H., Scharff, J. and Neville, D.M. (1995) An anti-CD3 single-chain immunotoxin with a truncated diphtheria toxin avoids inhibition by pre-existing antibodies in human blood. *Proc. Natl. Acad. Sci. USA*, **270**, 28037–28041.

Tweten, R.K., Barbieri, J.T. and Collier, R.J. (1985) Diphtheria toxin: effect of substituting aspartic acid for glutamic acid 148 on ADP-ribosyltransferase activity. *J. Biol. Chem.*, **260**, 10392–10394.

Uchida, T., Pappenheimer, A.M. and Greany, R. (1973) Diphtheria toxin and related proteins. I. Isolation and properties of mutant proteins serologically related to diphtheria toxin. *J. Biol. Chem.*, **248**, 3838–3844.

Van Ness, B.G., Howard, J.B. and Bodley, J.W. (1980) ADP-ribosylation of elongation factor 2 by diphtheria toxin: isolation and properties of the novel ribosyl-amino acid and its hydrolysis products. *J. Biol. Chem.*, **255**, 10710–10716.

vanderSpek, J.C., Mindell, J.A., Finkelstein, A. and Murphy, J.R. (1993) Structure/function analysis of the transmembrane domain of DAB389–Interleukin-2, an Interleukin-2 receptor-targeted fusion toxin. *J. Biol. Chem.*, **268**, 12077–12082.

vanderSpek, J.C., Howland, K., Friedman, T. and Murphy, J.R. (1994) Maintenance of the hydrophobic face of the diphtheria toxin amphipathic transmembrane helix 1 is essential for the efficient delivery of the catalytic domain to the cytosol of target cells. *Prot. Eng.*, **7**, 985–989.

Weiss, M.S., Blanke, S.R., Collier, R.J. and Eisenberg, D. (1995) Structure of the isolated catalytic domain of diphtheria toxin. *Biochem.*, **34**, 773–781.

Wilson, B.A., Reich, K.A., Weinstein, B.R. and Collier, R.J. (1990) Active-site mutations of diphtheria toxin: effects of replacing glutamic acid-148 with aspartic acid, glutamine, or serine. *Biochem.*, **29**, 8643–8651.

Wilson, B.A., Blanke, S.R., Reich, K.A. and Collier, R.J. (1994) Active-site mutations of diphtheria toxin. Tryptophan 50 is a major determinant of NAD affinity. *J. Biol. Chem.*, **269**, 23296–23301.

Wolff, J.A., Malone, R.W., Williams, P., Wang, C., Acsadi, G., Jani, A. and Felgner, P. (1990) Direct gene transfer into mouse muscle *in vivo. Science*, **247**, 1465–1468.

Yamaizumi, M., Mekada, E., Uchida, T. and Okada, Y. (1978) One molecule of DT fragment A introduced into a cell can kill the cell. *Cell*, **15**, 245–250.

Zalman, L. and Wisieski, B. (1984) Mechanism of insertion of DT; peptide entry and pore size determination. *Proc. Natl. Acad. Sci. USA*, **81**, 3341–3345.

Zhan, H., Choe, S., Nuynh, P.D., Finkelstein, A., Eisenberg, D. and Collier, R.J. (1994) Dynamic transitions of the transmembrane domain of diphtheria toxin: disulfide trapping and fluorescence proximity studies. *Biochem.*, **33**, 11254–11263.

Zhao, J.M. and London, E. (1986) Similarity of the conformation of diphtheria toxin at high temperature to that in the membrane-penetrating low pH state. *Proc. Natl. Acad. Sci. USA*, **83**, 2002–2006.

Zucker, D.R., Murphy, J.R. and Pappenheimer, A.M. (1984) Monoclonal antibody analysis of diphtheria toxin II. Inhibition of ADP-ribosyltransferase activity. *Mol. Immun.*, **21**, 795–800.

13. ADENYLATE CYCLASE TOXIN FROM *BORDETELLA PERTUSSIS*: MODES OF INTERACTION WITH CELLS

MICHAL BEJERANO, ISRAEL NISAN and EMANUEL HANSKI

*Department of Clinical Microbiology, The Hebrew University-Hadassah
Medical School, Jerusalem 91010, Israel*

INTRODUCTION

The Gram-negative coccobacillus *Bordetella pertussis* is the aetiologic agent of whooping cough, in humans. Whooping cough ceased to be the major cause of childhood morbidity and mortality when whole-cell vaccine came into widespread use in developed countries (Cherry *et al.*, 1984). Nonetheless, concerns over the adverse reactions associated with this vaccine have discouraged its use in infants and children, with a consequent increase in pertussis cases (Centers for Disease Control, 1984; Romanus *et al.*, 1987). This has stimulated efforts to develop a safer acellular vaccine, which in turn has stimulated the study of *B. pertussis* pathogenesis.

Like most infectious diseases, whooping cough proceeds through the following sequence of events: 1) entry of the pathogen into the host and adherence to specific tissues; 2) bacterial colonization and evasion of host defenses; 3) development of a local damage; and 4) dissemination of bacterial products, causing systemic disease (Weiss and Hewlett, 1986). To accomplish these steps, *B. pertussis* produces several unique biologically active compounds considered to be virulence factors. The first steps involving adherence and colonization of the mucosal respiratory tract are mediated by filamentous haemagglutinin (FHA), pertactin, pertussis toxin (PTX) and fimbrial proteins that mediate adherence of the pathogen to ciliated cells (Sandros and Tuomanen, 1993). Once *B. pertussis* has colonized, the bacterium releases compounds such as: dermonecrotic toxin, tracheal cytotoxin, lipopolysacharide, PTX and adenylate cyclase (AC) toxin. These factors cause local damage and systemic symptoms, which are characterized by paroxysmal coughing, lymphocytosis and loss of weight. The genes encoding some of these factors are positively and negatively regulated in coordinate fashion by a single locus, *bvg*, which modulates their expression in response to environmental signals (Gross *et al.*, 1989; Miller *et al.*, 1989; Melton and Weiss, 1993). The *bvg* locus encodes two trans-acting proteins, BvgA and BvgS, which display homology to the response regulator-histidine protein kinase system (Arico *et al.*, 1991; Miller *et al.*, 1989), that enables prokaryotes to sense their environment and respond to it (Stock *et al.*, 1989).

Early studies with transposon mutants of *B. pertussis* suggested that at least two factors, PTX and AC toxin, are crucial for virulence in animal models of pertussis disease (Weiss *et al.*, 1983; 1984). AC toxin has several unusual features; its location in the bacterium is extracytoplasmic (Hewlett *et al.*, 1976) and it is greatly stimulated by the eukaryotic Ca^{2+}-binding protein calmodulin (CaM), which is absent from *B. pertussis* (Wolf *et al.*, 1980). Following penetration into eukaryotic cells and activation by CaM, AC toxin utilizes the intracellular ATP pool to generate hyperphysiological levels of intracellular cAMP (Confer and Eaton, 1982; Hanski and Farfel, 1985; Shattuck and Storm, 1985). The production of such high levels of intracellular cAMP in immune effector cells leads to impairment of several functions including: chemotaxis, phagocytosis and superoxide generation, resulting in a debilitated host-defense response (Confer and Eaton, 1982; Weiss and Hewlett, 1986; Weiss *et al.*, 1986; Friedman *et al.*, 1987; Galgiani *et al.*, 1988).

Molecular analysis of the structural gene (*cya*A) encoding the AC toxin (CyaA) has shown that it belongs to the RTX (Repeats in ToXins) family, which comprises cytolytic, pore-forming toxins (Glaser *et al.*, 1988a,b). RTX toxins are produced by Gram-negative bacterial pathogens and have several features in common. These include: 1) series of glycine-rich repeat units in the toxin's C-terminus; 2) secretion of mature toxin into the extracellular medium without the formation of a periplasmic intermediate; 3) an operon that encodes the structural gene along with genes required for its secretion and post-translational modification; and 4) a dependence of toxin activity on extracellular Ca^{2+} (reviewed by: Welch, 1991; Coote, 1992; Menestrina *et al.*, 1994). Among toxins of the RTX family, CyaA is the only member that possesses a known enzymatic activity and its pathophysiological role is well documented.

In this chapter, we will discuss the structure of CyaA, mode of its interaction with target cells and the toxin's pathophysiological effects. We will focus on recent advances in understanding how the AC toxic activity is related to the haemolytic activity and how both activities are connected to pathogenesis.

TOXIN STRUCTURE

Early attempts to purify CyaA led to the identification of small (<60 kDa), large (~200 kDa) and very large (>600 kDa) forms, which were not fully resolved, from each other. Each form possessed independently AC enzymatic activity as determined *in vitro*, but was unable to generate cAMP in target cells. Since then, several laboratories using different purification protocols have unequivocally demonstrated that the AC enzymatic activity and the toxic properties of CyaA reside on the same macromolecule, which migrates in SDS-PAGE with an apparent molecular weight (M_r) of 190 to 215 kDa. The low molecular weight forms that were detected earlier, represent degradation products of the full-length AC toxin and the very large forms are oligomers of the toxin molecule that are formed in solution (reviewed by Hanski and Coote, 1991).

cyaA was cloned using two different strategies. The first strategy involved complementation of a Tn5 insertion mutant of *B. pertussis*, deficient both in AC and haemolytic activities (Brownlie *et al.*, 1986) and the second involved complementation of an *E. coli* (*cya*-deficient) strain, harbouring a plasmid expressing CaM (Glaser *et al.*, 1988a). *cyaA* encodes a polypeptide of 1706 amino acids, having a calculated M_r of 177 kDa (Glaser *et al.*, 1988a). The discrepancy between the calculated and observed M_r (177 vs. 190–215 kDa) is caused by anomalous migration of CyaA on SDS-PAGE (Gentile *et al.*, 1990). The first 400 N-terminal amino acids of CyaA contains the AC catalytic moiety, which has both enzymatic and CaM binding domains (Glaser *et al.*, 1988a,b; Ladant *et al.*, 1989). Tryptic cleavage of the full-length molecule in the presence of CaM generates a 43 kDa fully catalytically-active fragment (Bellalou *et al.*, 1990a), which upon further proteolysis releases two fragments corresponding to residues 1–224 (T25) and residues 225–339 (T18). The ATP-binding domain is located on T25 and the CaM-binding domain on T18 (Ladant, 1988; Munier *et al.*, 1991). The two fragments are 2 to 3 orders of magnitude less active with respect to catalysis of cAMP formation and CaM binding than the 43 kDa catalytically-active fragment, but they may reassociate in the presence of CaM to form a fully active T25/T18 complex. It was found that a stretch of 72 amino acids overlapping the C-terminus of T25 and the N-terminus of T18 contributes 90% of the binding energy needed to form the active complex. The AC catalytic domain, though it contributes only 10% to the total binding energy, profoundly effects CaM binding by inducing a conformational change within the CaM binding domain (Bouhss *et al.*, 1993). The C-terminal region of T18 also has an essential function in enabling the CaM binding domain to attain the conformation necessary for CaM binding (Munier *et al.*, 1993). Similarly, binding of CaM facilitates tight binding of ATP to the active site on the AC domain (Sarfati *et al.*, 1990) and deletion of the C-terminal region of T18 consequently affects cAMP generation (Munier *et al.*, 1993). By site-directed mutagenesis, it has been demonstrated that Lys[58], Lys[65] and His[63] are directly involved in catalysis (Glaser *et al.*, 1989; Xia and Storm, 1990; Munier *et al.*, 1992). Substitution of Trp[242] by Gly or Asp decreases the affinity of CyaA for CaM by three orders of magnitude. This residue is situated in a region displaying amphiphatic α-helical structure known to be important for CaM binding (Glaser *et al.*, 1989).

The C-terminal part of CyaA (residues 400–1706) is responsible for the haemolytic phenotype of *B. pertussis* (Glaser *et al.*, 1988b) and exhibits several features common to the RTX family of toxins (Welch, 1991; Coote, 1992). These include: 1) a hydrophobic domain in the form of four potential membrane spans, which reside between amino acids 530 and 780; 2) 40 to 45 glycine- and aspartate-rich nonapeptide repeats, located between residues 1015 and 1613, possessing the consensus sequence GGXG(N/D)DX(L/I/F)X; and 3) an uncleaved C-terminal secretion signal. Like all other members of the RTX family, CyaA requires posttranslational modification in the form of fatty-acylation, for its toxic and haemolytic activities. In CyaA, this modification is mediated by a separate gene product, CyaC, which is part of the *cya* locus (see below) (Barry *et al.*, 1991;

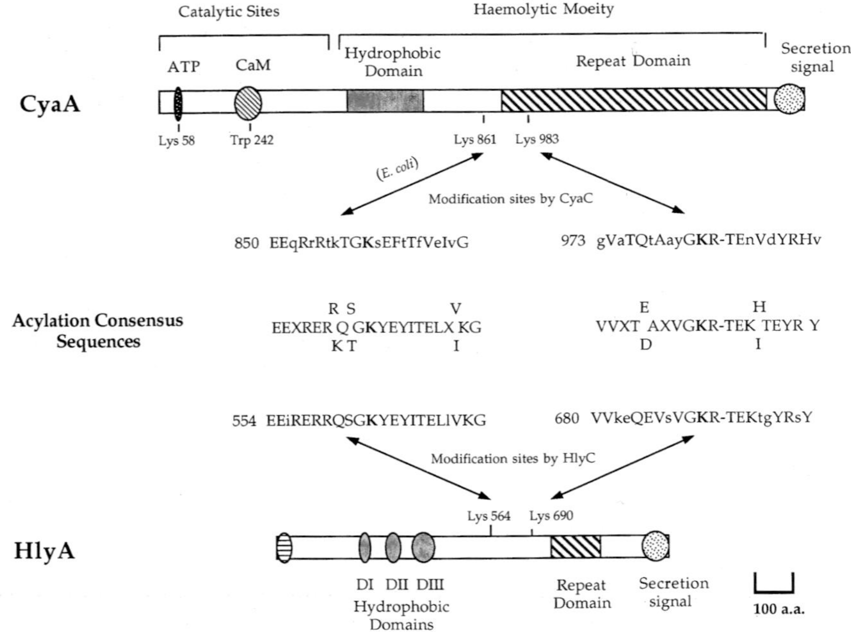

Figure 13.1 Structures of CyaA and HlyA. Regions involved in various functions are indicated for both proteins. Modification sites are shown and aligned with consensus sequences derived for both acylation sites (Stanley *et al.*, 1994). Non conserved amino acids are in lowercase and conserved amino acids are in upercase.

Hewlett *et al.*, 1993). Acylation of CyaA in *B. pertussis* occurs on Lys^{983}, which is palmitoylated (Hackett *et al.*, 1994). When *cya*A is expressed in *E. coli* together with *cya*C, both Lys^{983} and Lys^{860} are acylated (Hackett *et al.*, 1995), as was predicted by Stanley *et al.* (1994), on the basis of a consensus sequence for fatty-acyl modification derived from studies of *E. coli* α haemolysin (HlyA). The structural organization of both CyaA and HlyA are shown in Figure 13.1.

THE STRUCTURE OF CYA OPERON AND THE MECHANISM OF TOXIN SECRETION

The *cya* locus comprises five genes: *cya*C, A, B, D and E. *cya*C is located upstream from *cya*A and is transcribed from its own promoter in the opposite direction from that of *cya*A, B, D and E. (Barry *et al.*, 1991; Goyard and Ullmann, 1993). *cya*B, D and E, which are required for secretion of CyaA, are located downstream from *cya*A (Glaser *et al.*, 1988b). Transcription of *cya*A, B, D and E initiates at a single promoter located 115 bp up-stream from *cya*A. Most transcripts initiated at this promoter terminate within the *cya*A-*cya*B intergenic region. However, there is also a low level of read-through to produce a full length transcript encoding all four genes (Laoide and Ullmann, 1990). Transcription from this upstream promoter occurs only in virulent-(*bvg⁺*) strains and not in avirulent-(*bvg⁻*) strains, indicating that *bvg⁺* is necessary for transcription (Laoide and Ullmann, 1990). Nonetheless, when DNA containing the cyaA promoter is transformed into *E. coli*, in the presence of the *bvg⁺* locus, it is not expressed, indicating that *bvg⁺* is sufficient to regulate *cya*A transcription, but an additional factor may be required (Goyard and Ullmann, 1991). In addition to the promoter upstream of *cya*A, there is a second, weaker promoter, located downstream of *cya*A and upstream of *cya*B. Transcription from this second promoter is independent of the *bvg* locus. This promoter allows a constitutive low-level of expression of cyaB, D and E, ensuring the immediate and efficient secretion of CyaA when *B. pertussis* switches from the avirulent to the virulent state (Laoide and Ullmann, 1990). Among the RTX family of toxins, most are organized in operons similar to that of *cya* and contain genes homologous to *cya*C, B and D (Ludwig and Goebel, 1991; Welch, 1991; Coote, 1992).

HlyA serves as a prototype for studying the secretion of other toxins of the RTX family. Secretion of HlyA is mediated by two specific membrane-located secretory proteins, HlyB and HlyD and an outer membrane protein, TolC. HlyA, B and D are encoded within the *hly* operon, while TolC is not (Wandersman *et al.*, 1990; Koronakis *et al.*, 1992). On a basis of sequence and structural homology, HlyB belongs to the ATP-binding cassette (ABC) superfamily of proteins involved in transport of substrates across a membrane (Hyde *et al.*, 1990). Deletions within the C-terminal signal sequence of HlyA that disrupt its secretion, are suppressed by amino acid substitutions in HlyB, demonstrating a direct interaction between HlyA and HlyB (Sheps *et al.*, 1995). It is believed that HlyB, HlyD and TolC, form a transport complex, which spans the inner and the outer membranes and allows transport of HlyA directly into the medium, without forming a periplasmic intermediate (Koronakis *et al.*, 1992). Interestingly, the HlyA secretion machinery can export CyaA but at a low level (Masure *et al.*, 1990; Sebo and Ladant, 1993). Sequence similarity over several genes within the *cya* and *hly* operons (Glaser *et al.*, 1988b), along with the ability of HlyBD to secrete CyaA, suggest that CyaA is secreted by CyaB, D and E in a similar fashion (Ludwig and Goebel, 1991).

MODES OF AC TOXIN INTERACTION WITH CELLS

Intoxication of Cells by CyaA Occurs Directly From the Cell Surface and Not From Intracellular Vesicles

More than a decade ago, Confer and Eaton discovered that exposure of human polymorphonuclear cells or macrophages to an extract from *B. pertussis* caused an abrupt increase in the level of intracellular cAMP (Confer and Eaton, 1982). Today, it is well established that the burst of intracellular cAMP is caused by CyaA, which enters these cells, becomes activated by CaM and utilizes the cell's own ATP pool to generate cAMP.

Early studies on the mechanism of CyaA entry into eukaryotic cells demonstrated that intoxication occurs directly from the plasma membrane via an unusual mechanism. cAMP accumulated rapidly and without any noticeable lag period in human lymphocytes and erythrocytes exposed to CyaA, reaching a plateau within 10 to 20 min (Friedman *et al.*, 1987). Maintenance of the plateau level requires the continued presence of exogenous CyaA, since washing of cells (Farfel *et al.*, 1987) or addition of anti-cyaA antibody (Gentile *et al.*, 1988) leads to a rapid ($t_{1/2} = 10$–20 min) cessation of cAMP accumulation. It was therefore suggested that the plateau level of intracellular cAMP represents a steady state, in which the continuous penetration of CyaA into cells is balanced by its intracellular inactivation (Farfel *et al.*, 1987). This notion is supported by the following findings: 1) the accumulation of the AC catalytic activity within cells displays similar kinetics to that of cAMP accumulation, reaching a steady state level within 20 min (Friedman *et al.*, 1987); 2) more then 95% of the intracellular AC catalytic activity of CyaA is associated with plasma membrane fractions, regardless of how long cells had been exposed to the toxin (Farfel, *et al.*, 1987); and 3) upon removal of CyaA from the extracellular medium, intracellular AC catalytic activity is rapidly lost (Friedman *et al.*, 1987). More recently, we have demonstrated that CyaA is rapidly proteolysed by a fraction of human lymphocyte cytosol in an ATP-dependent manner, with a concomitant loss of catalytic activity (Gilboa-Ron *et al.*, 1989). Taken together, it appears that cAMP generation by CyaA, as well as its inactivation within cells, occurs while it is associated with the cells plasma membrane and accessible to the cell cytosol.

Support for an unusual mechanism of uptake of CyaA toxin is provided by the observation that compounds that interfere with receptor-mediated endocytosis and block intoxication of cells by a variety of toxins (Eidels *et al.*, 1983) have no effect on cAMP accumulation induced by CyaA (Gentile *et al.*, 1988; Gordon *et al.*, 1988; Gordon *et al.*, 1989). Particularly intriguing is the demonstration that the requirements for cAMP induction in Chinese hamster ovary (CHO) cells by *Bacillus antrachis* AC and CyaA are fundamentally different and that only the former enters CHO by receptor-mediated endocytosis (Gordon *et al.*, 1989). This finding, along with others, demonstrated that these two AC toxins, in spite of sharing amino acid homology within their catalytic domains and a common requirement for CaM, possess completely different mechanisms for penetration into cells (Mock and Ullmann, 1993).

No specific cell receptor has been identified for CyaA. Treatment of cells with trypsin or cycloheximide, does not affect CyaA penetration (Gordon *et al.*, 1989). The ganglioside G_{M1} serves as a receptor for cholera toxin (CT) (reviewed by Splanger, 1992). Interestingly, preincubation of CyaA with gangliosides inhibits cAMP accumulation in CHO cells, with the following ranking order of decreasing inhibition: $G_{T1b} > G_{M1} > G_{M3}$. This suggests that G_{T1b} may function as the cellular receptor for CyaA (Gordon *et al.*, 1989). However, the apparent affinity of CyaA for G_{T1b}, calculated from these inhibition studies is in the range of 50 μM, which is far lower than the affinity of CT for G_{M1} (Splanger, 1992). Therefore, the interaction of CyaA with cells may be more complex than that of CT, perhaps involving more than one cell-surface component. Such a notion is supported by further studies, in which

CyaA was mixed with liposomes containing a labeled marker. CyaA caused release of the marker and the release was maximal when negatively charged gangliosides were used (Gordon *et al.*, 1989). Furthermore, it was recently demonstrated that CyaA produces ion-permeable cation selective pores in bilayers made of negatively charged phospholipids (Szabo *et al.*, 1994). Thus, it appears that CyaA can interact directly with cell membrane phospholipids and this interaction is charge-dependent. The involvement of a charge-charge interaction between CyaA and its target cell is also supported by the observation that polycations and positively charged proteins, but not neutral proteins, inhibit penetration of CyaA into CHO cells (Raptis *et al.*, 1989).

Toxic and Haemolytic Activities of CyaA: Requirements for Extracellular Ca^{2+} and Fatty Acid Acylation

Early studies, using transposon mutagenesis of *B. pertussis*, indicated that the genes encoding AC and haemolytic activities reside in close proximity on the bacterial chromosome (Weiss *et al.*, 1983). On the basis of considerable homology between the C-terminal portion of CyaA and HlyA and similar organizations of *cya* and *hly* operons, Glaser and coworkers suggested that CyaA itself possesses both toxic and haemolytic activities and therefore named this toxin cyclolysin (Glaser *et al.*, 1988a,b). Subsequently, several laboratories demonstrated that purified CyaA is capable of lysing erythrocytes (Bellalou *et al.*, 1990b; Ehrmann *et al.*, 1991; Rogel *et al.*, 1991). However, initial characterization of CyaA-mediated haemolysis showed that its specific activity is much lower than that of HlyA (Bellalou *et al.*, 1990b; Ehrmann *et al.*, 1991; Rogel *et al.*, 1991). Furthermore, using sheep erythrocytes as target cells, it was demonstrated that generation of cAMP by CyaA occurs concomitantly with its penetration into cells, whereas haemolysis becomes apparent only after a lag time of about 40 min (Figure 13.2A). Moreover, haemolysis requires much higher concentrations of CyaA than does cAMP generation. In fact, amounts of CyaA sufficient to generate almost maximal intracellular cAMP production, do not induce haemolysis, even after a long incubation period (Figure 13.2B, see also Rogel *et al.*, 1991). Pertussis is a disease of man and CyaA can generate cAMP within human erythrocytes, however it fails to demonstrate any haemolytic activity (Rogel *et al.*, 1991). Taken together, these findings suggest that haemolytic activity of CyaA may not be an essential feature of its role in promoting disease. Indeed, *B. pertussis* strains expressing CyaA molecules that have lost their AC catalytic activity but retained haemolytic activitiy, are impaired in their ability to colonize the mouse respiratory tract (Khelef *et al.*, 1992) and fail to induce a consistent pulmonary lesion (Khelef *et al.*, 1995). Construction of *B. pertussis* mutants producing CyaA defective in the haemolytic activity only, will directly address the question of whether the haemolytic activity has any role in the disease process.

The observations described above regarding toxic and haemolytic activities of CyaA raise several fundamental questions concerning these two activities: 1) are these two processes connected; 2) what is the mechanism by which CyaA causes haemolysis; and 3) how is the catalytic moiety of CyaA delivered into the cell cytosol.

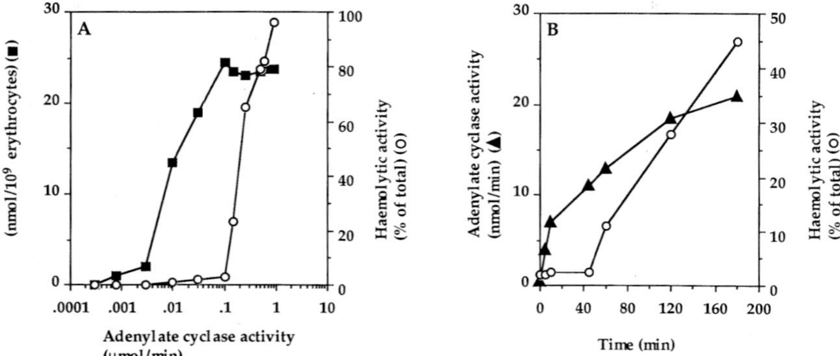

Figure 13.2   Dose response (A) and time dependence (B) of adenylate cyclase toxic and haemolytic activities in sheep erythrocytes.

By manipulating the composition of medium in which sheep erythrocytes are incubated, we were able to uncouple the toxic and the haemolytic activities of CyaA. CaM, when added to the culture medium, blocks CyaA-mediated cAMP generation but enhances haemolysis. Exposure of erythrocytes to CyaA in the absence of Ca^{2+}, eliminates the ability of CyaA to generate cAMP, but affects its haemolytic activity only slightly (Rogel *et al.*, 1991). We therefore proposed that these two activities, though mediated by the same molecule, may involve different domains and/or different conformations (Rogel *et al.*, 1991). Indeed it was previously demonstrated that the C-terminal 1300 residues of CyaA are absolutely required for binding and delivery of the AC catalytic moiety into cells (Rogel *et al.*, 1989; Hewlett *et al.*, 1989; Bellalou *et al.*, 1990b). Furthermore, a truncated AC toxin comprising of only the C-terminal 1300 residues, display haemolytic activity with features identical to those of the intact toxin (Sakamoto *et al.*, 1992; Ehrmann *et al.*, 1992).

The mechanisms of haemolysis induced by toxins of the RTX family has been studied most extensively for HlyA. Two different models have been suggested. The first model attributes cell disruption to the pore-forming activity of HlyA, which generates ion-permeable transmembrane channels, leading to osmotic lysis in hypotonic media (Menestrina *et al.*, 1994). The second model proposes that the toxin may evoke a perturbation in the membrane, even in the absence of channel formation, leading to cell lysis (Ostolaza *et al.*, 1993; Moayeri and Welch, 1994). There is also disagreement concerning how many molecules of HlyA are required to form an aqueous channel in the erythrocyte membrane. Some investigators have suggested that a single molecule is sufficient to form such a channel (Menestrina *et al.*, 1987; 1994; Menestrina 1988), while others proposed that oligomerization of the toxin is required for pore formation (Benz *et al.*, 1989; 1992; Ludwig *et al.*, 1993).

The size of the channel formed by CyaA was determined to be of 0.6–0.8 nm, on the basis of osmotic protection studies against CyaA-promoted lysis of sheep erythrocytes (Ehrmann *et al.*, 1991). This channel size is significantly smaller than

1–2 nm, which is the size of the channel formed by HlyA, as determined using similar methods (Menestrina *et al.*, 1994). The properties of the CyaA-induced channel deduced from studies in lipid bilayers are as follows: 1) the channel is selective for cations (Szabo *et al.*, 1994; Benz *et al.*, 1994); 2) it is absolutely Ca^{2+}-dependent and is polarity- and voltage-dependent (Szabo *et al.*, 1994); and, 3) it has a single-channel conductance of 27 pS, which is considerably smaller than the conductance of 1500 pS estimated for the channel produced by HlyA (Benz *et al.*, 1994).

As the concentration of CyaA is increased, its haemolytic activity increases more rapidly than its AC toxic activity, suggesting that a functional haemolytic unit may consist of more than one CyaA molecule (Betsou *et al.*, 1993). In addition, CyaA induces a conductance in a lipid bilayer, with a Hill coefficient greater than three (Szabo *et al.*, 1994). Furthermore, when *cya*A is expressed in *E. coli* together with *cya*C, it produces AC toxic activity identical to that of the native toxin expressed in *B. pertussis*, but its haemolytic activity is several-fold lower (Sebo *et al.*, 1991). These observations, taken together, suggest that the oligomerization of CyaA leading to pore-formation, constitutes an important part of the mechanism by which CyaA lyses erythrocytes.

There are several indications suggesting that the delivery of AC toxin into the cell cytosol could itself be independent of the pore-forming activity described above. Intoxication of cells by CyaA occurs as a linear function of its concentration (Hanski and Farfel, 1985; Rogel *et al.*, 1991; Betsou *et al.*, 1993: Hackett *et al*; 1995), suggesting a monomeric mechanism for delivery of the catalytic moiety, whereas haemolysis, as mentioned above, is likely to require the formation of oligomers. In addition, CyaA containing alternative lipid modifications has reduced ability to form pores in lipid bilayer (Benz *et al.*, 1994) and to lyse erythrocytes (Hackett *et al.*, 1995) in comparison to that of CyaA isolated from *B. pertussis*, but there is no difference in AC toxic activity (Hackett *et al.*, 1995). Furthermore, the entire CyaA molecule is required for delivery of the catalytic moiety of AC toxin into cells (Iwaki *et al.*, 1995). In contrast, CyaA lacking the AC catalytic moiety or a large portion of the Ca^{2+}-binding repeat domain, forms channels in lipid bilayers having properties that are identical to those formed by the full length toxin (Benz *et al.*, 1994).

Early studies found that the toxic and haemolytic activities of CyaA are strictly Ca^{2+} dependent (Confer *et al.*, 1984; Hanski and Farfel, 1985; Gentile *et al.*, 1990; Bellalou *et al.*, 1990b; Ehrmann *et al.*, 1991). However, when CyaA is purified in the presence of Ca^{2+} and subsequently exposed to EGTA, it retains the ability to bind and lyse erythrocytes, but its catalytic domain fails to penetrate into the cell cytosol and thus to generate cAMP (Rogel *et al.*, 1991). This suggests that Ca^{2+} is not required for cell binding or haemolysis, but it is absolutely required for delivery of the catalytic moiety of CyaA into cells (Rogel *et al.*, 1991). The apparent discrepancy between these and previous results were recently resolved by Rose *et al.* (1995) who demonstrated that CyaA possesses two distinct classes of Ca^{2+}-binding sites. The first class comprises of three to five high-affinity sites that bind Ca^{2+} so strongly that it cannot be removed by incubating CyaA with EGTA, while the second class of low affinity binding sites may be dissociated of Ca^{2+} in this manner.

The high affinity binding sites are required for binding of CyaA to erythrocytes and for its haemolytic activity, whereas the low affinity sites, which bind Ca^{2+} with a Km in the millimolar range, are required for the toxic activity of CyaA (Rose *et al.*, 1995). The location of the high affinity sites on CyaA has not yet been determined conclusively, but the low affinity sites have been localized to the Asp-Gly repeat domain in the C-terminal half of the molecule.

The structure of alkaline protease from *Pseudomonas aeruginosa*, which contains Asp-Gly repeats, analogous to those of CyaA, has recently been determined by X-ray crystallography (Baumann *et al.*, 1993). According to the deduced structure, the repeats are organized into parallel β-rolls, forming sites where Ca^{2+} is bound in an octahedral coordination. The first six residues of each Asp-Gly repeat form two half-sites for Ca^{2+} binding, thus each Ca^{2+} ion is bound between a pair of adjacent loops of the β-roll. On the basis of Ca^{2+} binding experiments, Rose *et al.* (1995) had detected the presence of 45 low-affinity Ca^{2+} binding sites, which is consistent with the number of Asp-Gly repeats.

CyaA-induced generation of intracellular cAMP, requires a concentration of free Ca^{2+} greater than 0.1 mM and reaches a maximum at 1–2 mM Ca^{2+} (Hanski and Farfel, 1985; Hewlett *et al.*, 1991). Within this range of Ca^{2+} concentrations, CyaA undergoes a conformational change from a "closed" to a more "open" configuration, detectable as an increase in mass and Stokes radius as the concentration of Ca^{2+} is increased (Masure *et al.*, 1988). CyaA also undergoes a similar change in the appearance as visualized by electron microscopy (Hewlett *et al.*, 1991). Rose *et al.* (1995) demonstrated by circular dichroism spectroscopy, that Ca^{2+}-binding to the low affinity sites induces a large conformational change in the Asp-Gly repeat region of CyaA. Since intoxication of cells by CyaA and the conformational change observed in the repeat domain exhibit similar Ca^{2+}-dependency (Rose *et al.*, 1995), it is conceivable that the conformational change plays an important role for the low-affinity Ca^{2+}-binding sites in the delivery of the catalytic moiety of CyaA into cells.

Binding and insertion of CyaA into cell membranes requires very low Ca^{2+} concentrations and occurs at temperatures lower than $10°C$, whereas translocation of AC toxin to the cell cytosol requires a Ca^{2+} concentration in the mM range and temperature higher than $20°C$. Thus it has been possible to uncouple these two steps experimentally (Rogel and Hanski, 1992). This uncoupling was performed on CyaA-treated sheep erythrocytes, in which the catalytic AC domain is internalized but not completely proteolysed. Instead, partial proteolysis occurs, releasing an active 43 kDa fragment containing the N-terminus of CyaA into the cell cytosol (Rogel and Hanski, 1992). This 43 kDa fragment is stable, allowing us to follow the accumulation of the AC catalytic activity of CyaA within cells in the absence of intracellular inactivation. In the first step of this procedure, CyaA is bound to cells in the absence of added exogenous Ca^{2+} or at low temperature. CyaA bound to cells under these conditions, is fully accessible to inactivation by extracellular treatment of cells with trypsin. In the second step, Ca^{2+} is added back to the medium of CyaA-bound cells or the temperature is raised to $36°C$. Following these treatments, more than 60% of AC catalytic activity that was initially bound to the cells surface, becomes resistant to the trypsin treatment. We therefore concluded that in the second step, the AC catalytic moiety of CyaA is translocated from the outer surface of the plasma membrane to its inner surface, where it is cleaved and (in sheep erythrocytes) released to the cytosol (Rogel and Hanski, 1992). Translocation occurs with very rapid kinetics, with a half time less than two minutes. In contrast accumulation of the N-terminal 43 kDa fragment in the cell cytosol proceeds at much slower rate, suggesting that translocation of CyaA is independent of its intracellular cleavage (Rogel and Hanski, 1992).

Acylation of CyaA is required for both its toxic and haemolytic activities (Barry *et al.*, 1991; Hewlett *et al.*, 1993). In the case of HlyA and other toxins of the RTX family, acylation is required for interaction with the cell membrane (Boehm *et al.*, 1990a,b; Cruz *et al.*, 1990; Ludwig *et al.*, 1992). The role of acylation in the case of CyaA is less clear. Hewlett *et al.* (1993) demonstrated that acylated and non-acylated CyaA bind to Jurkat cells at comparable levels, but only the acylated CyaA is able to intoxicate and lyse cells. Since acylation does not affect the Ca^{2+}-induced conformational change that is associated with AC toxin activity of CyaA, it may act to promote CyaA translocation across the membrane and/or formation of the haemolytic pore (Hewlett *et al.*, 1993). When CyaA is mixed with a 13 residues peptide derived from the portion of CyaA containing Lys^{983} (which is palmitoylated), the toxic activity of the full length molecule is inhibited in a manner dependent on the concentration of the peptide (Hackett *et al.*, 1994). This peptide is more efficient inhibitor when it is added to CyaA before mixing with cells than if it is added to target cells before addition of CyaA (Hackett *et al.*, 1994). This suggests that fatty-acyl modification is important for steps occurring prior to CyaA interaction with the cell membrane, such as formation of oligomers (Hackett *et al.*, 1994). In contrast to the observations of Hewlett and coworkers, we have observed that the membrane-binding properties of non-acylated CyaA are strikingly different from those of acylated CyaA when assayed with sheep erythrocytes. Non-acylated CyaA can be dissociated from cells by washing with EGTA or sodium carbonate, while the acylated CyaA withstands these treatments (Rogel and Hanski, unpublished data). Furthermore, non-acylated CyaA does not compete with acylated CyaA for binding to cells. Since non-acylated CyaA can partially complement the AC toxic activity of an acylated mutant lacking AC activity (Iwaki *et al.*, 1995), farther acylation does not appear to be essential for AC toxin penetration into cells. Rather, present evidence suggests that it is important for the initial interaction with the target cell membranes. Further studies are required to clarify the role of fatty acylation in the interaction of CyaA with cells.

Summary of the Findings and Postulated Mechanisms for Intoxication of Cells by CyaA

CyaA appears to be different from the bipartite A-B type bacterial toxins in its modes of interaction with the target cell surface and translocation across the cell membrane. Toxins such as diphtheria toxin, *Pseudomonas aeruginosa* exotoxin A, cholera toxin, shigella toxin, pertussis toxin, ricin and even *B. anthracis* AC toxin employ distinct domains for binding to specific cell-surface receptor (or intermediatory components) and for translocation across the membrane into the cell cytosol (Olsnes and Sandvig, 1985; Pastan and FitzGerald, 1989; London, 1992). In contrast, no target-cell binding domain, has been defined for CyaA and its penetration appears to begin upon a rather non-specific adsorption to the cell surface. Accordingly, CyaA is able to intoxicate all cell types tested to date. Structural integrity of the entire CyaA molecule and the cooperation of all the domains, is absolutely required for translocation of the AC catalytic moiety across the plasma membrane (Iwaki *et al.*, 1995).

We have suggested that intoxication of cells by CyaA is a monomolecular process, in which the C-terminal hydrophobic domain creates a protein channel in the membrane through which the catalytic moiety is translocated in a partially unfolded state (Rogel and Hanski, 1992). Alternatively, it has been suggested that a cross section of the haemolytic channel formed in lipid bilayers by CyaA is too small to allow the passage of a polypeptide, even one that is totally unfolded (Benz *et al.*, 1994). A model was therefore proposed in which the catalytic moiety undergoes a Ca^{2+}-induced conformational change, producing an amphipatic structure that interacts with the lipid bilayer and slides across it, perhaps in association with the channel formed by the hydrophobic domain (Benz *et al.*, 1994). However, as we described above, haemolysis is likely to be a distinct event, which may not be necessarily directly involved in the mechanism of AC-mediated cell intoxication. Furthermore, while the repeat domain of CyaA is not required for channel formation in lipid bilayers (Benz *et al.*, 1994), it is absolutely required for AC intoxication of cells (Rose *et al.*, 1995). Thus, the ability of the C-terminal portion of CyaA to form a protein channel in intact cells has not been completely ruled out by experiments performed in lipid bilayers.

There are still many points that must be clarified before the exact mechanism for CyaA translocation across the membrane can be elucidated. These include the following: 1) is intoxication of cells by CyaA a monomolecular phenomenon, or does it require oligomerization, as has been recently suggested (Iwaki *et al.*, 1995); 2) where are the high affinity Ca^{2+}-binding sites on CyaA located; 3) does the C-terminal hydrophobic domain of CyaA, which is essential for secretion and intoxication (Iwaki *et al.*, 1995), directly interact with the plasma membrane; and 4) what are the forces that drive the translocation of CyaA across the plasma membrane. Is it driven by: 1) the energy released upon the binding of the catalytic moiety of CyaA to CaM within the cell (Oldenburg *et al.*, 1992); 2) the difference in Ca^{2+} concentration between the exterior and interior of the cell that changes the conformation of the protein from an un-folded to a folded configuration (Rogel and Hanski, 1992); or, 3) the membrane potential, since membrane depolarization prevents cell intoxication by CyaA (Otero *et al.*, 1995). Further investigation of CyaA function according to these and other avenues should enhance our fragmentary knowledge of how intoxication occurs.

PATHOPHYSIOLOGICAL EFFECTS OF AC TOXIN AND ITS ROLE AS A PROTECTIVE ANTIGEN

Pathophysiological Effects of AC Toxin as Determined in Animal Models

The most widely used and best characterized model for whooping cough is intranasal challenge of infant mice which causes lethality at doses higher than 10^5 colony forming units (cfu) per infection (Pittman *et al.*, 1980). Using this model, Weiss *et al.* (1984; 1985) showed that *B. pertussis* Tn5 mutants deficient in production of PTX or AC toxin, had a diminished ability to cause a lethal infection, suggesting that these toxins are essential mediators of the disease. More recently, this observation was extended to include a broader collection of mutants deficient in other *bvg* regulated genes (Weiss and Goodwin, 1989). The most direct evidence for the involvement of CyaA in the pathogenesis of pertussis comes from experiments in which site-directed mutations were constructed in CyaA, followed by exchange of the mutated gene with the chromosomal allele by homologous recombination. *B. pertussis* mutants deficient in the AC toxic activity were much less virulent than the wild type (WT) strain (Gross *et al.*, 1992; Khelef *et al.*, 1992).

Challenging mice with a sublethal dose of *B. pertussis* leads to rapid proliferation of bacteria in the lung until they reach a peak, which is followed by gradual clearance. Inoculation with 10^3 cfu of a WT strain of *B. pertussis*, results in proliferation of bacteria within the lung to 10^7 cfu within 15 days post-inoculation, followed by clearance within 40 days post-inoculation (Goodwin and Weiss, 1990). Challenging mice with a similar dose of a *B. pertussis* AC mutant leads to a much lower proliferation and bacteria are cleared from the lungs within 10 days. When mice are challenged with a low dose of a PTX mutant, proliferation of the bacteria in the lungs occurs as efficiently as with the WT strain, however, the PTX mutant strain is cleared within 25 days (Goodwin and Weiss, 1990). On the basis of these and other results (Khelef *et al.*, 1992), it was concluded that CyaA is required to initiate infection and that PTX is important in subsequent steps of bacterial colonization. The importance of PTX and AC in virulence was also demonstrated by histological examination of mouse lung and bronchoalveolar lavage fluids, following infection with WT and mutant strains of *B. pertussis*. Both toxins are essential for lesion formation and inflammation of the lung (Khelef *et al.*, 1994).

Pathophysiological Effects of AC Toxin as Determined in Cell and Tissue Culture Models

AC toxin affects host immune cells through: 1) generation of cAMP, which disrupts signal transduction and other cell functions; and, 2) induction of apoptosis in macrophages. Exposure of neutrophils and macrophages to *B. pertussis* extract leads to impairment of chemotaxis, reduction in oxidative response and to decreased efficiency of killing (Confer and Eaton, 1982). Entry of CyaA to monocytes inhibits oxidative responses (Pearson *et al.*, 1987; Friedman *et al.*, 1987;

Galgiani *et al.*, 1988). When *B. pertussis* is mixed with monocyte-macrophage cell line, J774A.1, it causes apoptotic cell death, which is dependent on the expression of AC toxin (Khelef *et al.*, 1993). Subsequent work has demonstrated that apoptosis requires the AC toxic activity of CyaA and indeed can be induced by purified CyaA alone (Khelef and Guiso, 1995).

Although *B. pertussis* has traditionally been viewed as a non-invasive microorganism, it may enter and survive inside various eukaryotic cells such as HeLa cells (Ewanowich *et al.*, 1989), CHO cells (Mouallem *et al.*, 1990), polymorphonuclear leukocytes (Steed *et al.*, 1991) and macrophages (Friedman *et al.*, 1992). The involvement of known virulence factors in bacterial internalization and survival within the cell is not clear. Friedman *et al.* (1992) showed that both internalization and survival require the expression of virulence factors, including PTX, CyaA and FHA. Interestingly, Ewanowich *et al.* (1989) demonstrated that cAMP accumulation in HeLa cells causes a decrease in bacterial uptake, suggesting that CyaA expression could interfere with *B. pertussis* uptake. Indeed CyaA expression is down-regulated upon entry of the bacteria into macrophages, suggesting that the bacterium may switch from a virulent to an avirulent phenotype following internalization (Masure and Barret, 1991; Masure, 1992). One role of AC toxin, the induction of apoptosis in macrophages, may be important during early stages of infection, when bacteria escape from first line of immune defense. Penetration and survival within macrophages occur at later stages of infection, after a host immune response has developed. The ability of the bacteria to sense the stage of the infection and to switch between these two processes allows *B. pertussis* to successfully evade host defenses and survive within the host.

AC Toxin as a Protective Antigen

Prior vaccination of mice with CyaA protects against *B. pertussis* colonization of lung (Guiso *et al.*, 1989; 1991). Protection may be induced by the entire CyaA molecule or by the N-terminal catalytic portion of CyaA containing the first 400 residues, which is generated by partial digest of CyaA with trypsin. This early finding indicates that the major protective epitopes reside on the AC catalytic moiety of the toxin (Guiso *et al.*, 1991). However, it has been shown more recently that protection absolutely depends on post-translational modification of CyaA and is affected by the nature of the acylated fatty acid (Betsou *et al.*, 1993). By using a set of CyaA deletion mutants, investigators localized the protective region to the last 800 amino acids of the toxin molecule (Betsou *et al.*, 1995). This finding is supported by the observation that antibodies generated in *B. pertussis*-infected mice and humans, recognized the same C-terminal region (Betsou *et al.*, 1995). Interestingly, there is no correlation between the capacity of anti-CyaA antibodies, generated against different truncated forms of the toxin to neutralize AC toxic or haemolytic activities and their ability to protect mice against colonization by *B. pertussis*. The recent findings contrast with earlier observations of Guiso *et al.* (1991), which indicated that the N-terminal portion of the toxin may induce protection. To explain this apparent discrepancy, it was hypothesized that presentation of the N-terminal portion of AC toxin to the immune system induces the formation of protective antibodies, which are not formed when the entire CyaA molecule is processed (Betsou *et al.*, 1995). This hypothesis is supported by the fact that a monoclonal antibody directed against the part of the toxin located between amino acids 385 and 400, (in the N-terminal region), when given together with *B. pertussis* challenge, protects mice against pulmonary and intracerebral lesions (Guiso *et al.*, 1989).

Sera from patients recovering from pertussis (Arciniega *et al.*, 1991; Farfel *et al.*, 1990) and from children vaccinated with conventional diphtheria-tetanus-pertussis (DTP) vaccine (Farfel *et al.*, 1990) contain anti-CyaA antibodies, demonstrating that CyaA is immunogenic in humans. In agreement with the findings of Betsou *et al.* (1995) mentioned above, these sera do not neutralize the toxic activity of CyaA (Farfel *et al.*, 1990). A high proportion of neonates (Farfel *et al.*, 1990) and infants prior to DTP immunization (Arciniega *et al.*, 1991) possess anti-CyaA antibodies. This indicates that these antibodies are maternally derived, though it is unknown whether they help to protect young infants against pertussis (Arciniega *et al.*, 1993). Since a recombinant CyaA protein that lacks toxic and haemolytic activity is protective, it may be a good candidate for inclusion in a future acellular vaccine.

Acknowledgments

We are grateful to Carol Berkower for critical reading of this manuscript and constructive comments. Work in the authors' laboratory was supported by the German-Israeli Foundation (GIF).

REFERENCES

Arciniega, J.L., Hewlett, E.L., Johnson, F.D., Deforest, A., Wassilak, S.G.F., Onorato, I.M., Manclark, C.R. and Burns, D.L. (1991) Human serologic response to envelope-associated proteins and adenylate cyclase toxin of *Bordetella pertussis*. *J. Infect. Dis.*, **163**, 135–142.

Arciniega, J.L., Hewlett, E.L., Edwards, K.M. and Burns, D.L. (1993) Antibodies to *Bordetella pertussis* adenylate cyclase toxin in neonatal and maternal sera. *FEMS Immun. Med. Microbiol.*, **6**, 325–330.

Arico, B., Scarlato, V., Monack, D., Falkow, S. and Rappuoli, R. (1991) Structural and genetic analysis of the *bvg* locus in *Bordetella pertussis*. *Mol. Microbiol.*, **5**, 2481–2491.

Barry, E.M., Weiss, A.A., Ehrman, I.E., Gray, M.C., Hewlett, E.L. and Goodwin, M.S. (1991) *Bordetella pertussis* adenylate cyclase toxin and hemolytic activities require a second gene, *cyaC*, for activation. *J. Bacteriol.*, **173**, 720–726.

Baumann, U., Wu, S., Flaherty, K.M. and Mckay, D.B. (1993) Three-dimensional structure of the alkaline protease of *Pseudomonas aeroginosa*: a two-domain protein with a calcium binding parallel beta rol motif. *EMBO J.*, **12**, 3357–3364.

Bellalou, J., Ladant, D. and Sakamoto, H. (1990a) Synthesis and secretion of *Bordetella pertussis* adenylate cyclase as a 200 kilodalton protein. *Infect. Immun.*, **58**, 1195–1200.

Bellalou, J., Sakamoto, H., Ladant, D., Geoggroy, C. and Ullman, A. (1990b) Deletions affecting Hemolytic and toxin activities of *Bordetella pertussis* adenylate cyclase. *Infect. Immun.*, **58**, 3242–3247.

Benz, R., Schmid, A., Wagner, W. and Goebel, W. (1989) Pore formation by the *Eschrichia coli* haemolysin: evidence for an association-dissociation equilibrium of the pore-forming aggregates. *Infect. Immun.*, **57**, 667–895.

Benz, R., Maier, E., Ladant, D., Ullmann, A. and Sebo, P. (1994) Adenylate cyclase toxin (CyaA) of *Bordetella pertussis*. Evidence for the formation of small ion-permeable channels and comparison with HlyA of *Escherichia coli. J. Biol. Chem.*, **268**, 27231–27239.

Betsou, F., Sebo, P. and Guiso, N. (1993) CyaC-mediated activation is important not only for toxic but also for protective activities of *Bordetella pertussis* adenylate cyclase-hemolysin. *Infect. Immun.*, **61**, 3583–3589.

Betsou, F., Sebo, P. and Guiso, N. (1995) The C-terminal domain is essential for protective activity of the *Bordetella pertussis* adenylate cyclase-hemolysin. *Infect. Immun.*, **63**, 3309–3315.

Boehm, D.F., Welch, R.A. and Snyder, I.S. (1990a) Domains of *Escherichia coli* hemolysin (HlyA) involved in binding of calcium and erythrocytes membranes. *Infect. Immun.*, **58**, 1959–1964.

Boehm, D.F., Welch, R.A. and Snyder, I.S. (1990b) Calcium is required for binding of *Escherichia coli* hemolysin (HlyA) to erythrocyte membranes. *Infect. Immun.*, **58**, 1951–1958.

Bouhss, A., Krin, E., Munier, H., Gilles, A-M., Danchin, A., Glaser, P. and Barzu, O. (1993) Cooperative phenomena in binding and activation of *Bordetella pertussis* adenylate cyclase by calmodulin. *J. Biol. Chem.*, **268**, 1690–1694.

Brezin, C., Guiso, N., Ladant, D., Djavadi-Ohaniance, L., Megret, F., Onyeocha, I. and Alonso, J-M. (1987) Protective effects of anti-*Bordetella pertussis* adenylate cyclase antibodies against lethal respiratory infection of the mouse. *FEMS Microbiol. Lett.*, **42**, 75–80.

Brownlie, R.M., Coote, J.G., Parton, R., Schultz, J.E., Rogel, A. abd Hanski, E. (1988) Clonning of adenylate cyclase genetic determinant of *Bordetella pertussis* and its expression in *Escherichia coli* and *Bordetella pertussis*. *Microb. Pathogen.*, **4**, 335–344.

Centers for Disease Control (1984) Pertussis-Washington. *Morbid. Mortal. Weekly Rep.*, **34**, 390–400.

Cherry, J.D. (1984) The epidemiology of pertussis and pertussis immunization in the United Kingdom and the United States: A comparative study. *Curr. Prob. Pediatr.*, **14**, 1–78.

Confer. D.L. and Eaton. J.W. (1982) Phagocyte impotence caused by the invasive bacterial adenylate cyclase. *Science*, **217**, 948–950.

Confer. D.L., Slungaard, A.S., Graf, E., Panter, S.S. and Eaton. J.W. (1984) *Bordetella* adenylate cyclase toxin: entry of bacterial adenylate cyclase into mammalian cells. Adv. *Cyclic Nucleotide Res.*, **17**, 183–187.

Coote, J.G. (1991) Antigenic switching and pathogenicity: enviromental effects on virulence gene expression in *Bordetella pertussis. J. Gen. Microbiol.*, **137**, 2493–2503.

Coote, J.G. (1992) Structural and functional relationships among the RTX toxin determinants of Gram-negetive bacteria. *FEMS Microbiol.*, **88**, 137–162.

Cruz, W.T., Young, R., Chang, Y.F. and Struck, D.K. (1990) Deletion analysis resolves cell-binding and lytic domains of the Pastuerella leukotoxin. Mol. Microbiol., **4**, 1933

Ehrmann, I.E., Gray, M.C., Gordon, V.M., Gray, L.S. and Hewlett, E.L. (1991) Hemolytic activity of adenylate cyclase toxin from *Bordetella pertussis*. *FEBS Letters.*, **278**, 79–83.

Ehrmann, I.E., Weiss, A.A., Goodwin, M.S., Gray, M.C., Barry, E. and Hewlett, E.L. (1991) Enzymatic activity of adenylate cyclase toxin from *Bordetella pertussis* is not required for haemolysis. *FEBS Lett.*, **304**, 51–56.

Eidels, L., Proia, R.L. and Hart, D.A. (1983) Membrane receptors for bacterial toxins. *Microbiaol. Rev.*, **47**, 596–620.

Ewanowich, C.A., Melton, A.R., Weiss, A.A., Sherburne, R.K. and Peppier, M.S. (1989) Invasion of HeLa 229 cells by virulent *Bordetella pertussis*. *Infect. Immun.*, **57**, 2698–2704.

Farfel, Z., Friedman, E. and Hanski, E. (1987) The invasive adenylate cyclase of *Bordetella pertussis*. Intracellular localization and kinetics of penetration into various cells. *Biochem. J.*, **243**, 153–158.

Farfel, Z., Konen, S., Wiertz, E., Klapmuts, R., Addy, P.A-K. and Hanski, E. (1990) Antibodies to *Bordetella pertussis* adenylate cyclase are produced in man during pertussis infection and after vaccination. *J. Med. Microbiol.*, **32**, 173–177.

Friedman, R.L., Fierderlein, R.L., Glasser, L. and Galgiani, J.N. (1987) *Bordetella pertussis* adenylate cyclase: effects of affinity-purified adenylate cyclase on human polymorphonuclear leukocytes funcions. *Infect. Immun.*, **55**, 135–140.

Friedman, R.L., Nordensson, K., Wilson, L., Akporiaye, E.T. and Yocum, D.E. (1992) Uptake and intracellular survival of *Bordetella pertussis* in human macrophages. *Infect. Immun.*, **60**, 4578–4585.

Galgiani, J.N., Hewlett, E.L. and Friedman, R.L. (1988) Effects of adenylate cyclase toxin from *Bordetella pertussis* on human neutrophil interactions with *Coccidioides immitis* and *Staphylococcus aurus*. *Infect. Immun.*, **56**, 751–755.

Gilles, A-M., Munier, H., Rose, T., Glaser, P., Krin, E., Danchin, A., Pellecuer, C. and Barzu, O. (1990) Intrinsic fluorescence of a truncated *Bordetella pertussis* adenylate cyclase expressed in *Escherichia coli*. *Biochem.*, **29**, 8126–8130.

Gentile, F., Raptis, A., Knipling, L.G. and Wolff, J. (1988) *Bordetella pertussis* adenylate cyclase. Penetration into host cells. *Eur. J. Biochem.*, **175**, 447–453.

Gentile, F., Knipling, L.G., Sackett, D.L. and Wolff, J. (1990) Invasive adenylate cyclase of *Bordetella pertussis*: Physical, catalytic and toxic properties. *J. Biol. Chem.*, **265**, 10686–10692.

Gilboa-Ron, A., Rogel, A. and Hanski, E. (1989) *Bordetella pertussis* adenylate cyclase inactivation by the host cell. *Biochem. J.*, **262**, 25–31.

Glaser, P., Ladant, D., Sezer, O., Pichot, F., Ullmann, A. and Danchin, A. (1988a) The calmodulin-seneitive adenylate cyclase of *Bordetella pertussis*: cloning and expression in *Escherichia coli*. *Molec. Microbio.*, **2**, 19–30.

Glaser, P., Sakamoto, H., Bellalou, J., Ullmann, A. and Danchin, A. (1988b) Secretion of cyclolysin, the calmodulin-seneitive adenylate cyclase-hemolysin bifunctional protein of *Bordetella pertussis*. *EMBO J.*, **7**, 3997–4004.

Glaser, P., Elmaohlou-Lazaridou, A., Krin, E., Ladant, D., Barzu, O. and Danchin, A. (1989) Identification of residue essential for catalysis and binding of calmodulin in *Bordetella pertussis* adenylate cyclase by site directed mutagenesis. *EMBO J.*, **8**, 967–972.

Goodwin, M.S. and Weiss, A.A. (1990) Adenylate cyclase toxin is critical for colonization and pertussis toxin for lethal infection by *Bordetella pertussis* in infant mice. *Infect. Immun.*, **58**, 3445–3447.

Gordon, V.M., Leppla, S.H. and Hewlett, E.L. (1988) Inhibitors of receptor-mediated endocytocic block the entry of *Bacillus anthracis* adenylate cyclase toxin but not that of *Bordetella pertussis* adenylate cyclase toxin. *Infect. Immun.*, **56**, 1066–1069.

Gordon, V.M., Young, W.W.Jr., Lechler, S.M., Gray, M.C., Leppla, S.H. and Hewlett, E.L. (1989) Adenylate cyclase toxin from *Bacillus anthracis* and *Bordetella pertussis*: Different processes for interaction with and entry into target cells. *J. Biol. Chem.*, **264**, 14792–14796.

Goto, T., Herberman, R.B., Maluish, A. and Strong, D.M. (1983) Cyclic AMP as a mediator of prostaglandin E-induced suppression of human natural killer cell activity. *J. Immun.*, **130**, 1350–1355.

Goyard, S. and Ullmann, A. (1991) Analysis of *Bordetella pertussis* cya operon regulation by use of cya-lac fusions. *FEMS Microbiol. Lett.*, **77**, 251–256.

Goyard, S. and Ullmann, A. (1993) Functional analysis of the cya promoter of *Bordetella pertussis*. *Mol. Microbiol.*, **7**, 693–704.

Gross, R., Arico, B. and Rappuoli, R. (1989) Families of bacterial signal-transduction proteins *Molec. Microbiol.*, **3**, 1661–1667.

Gross. M.K., Douglas, C.Au., Smith, A.L. and Storm, D.R. (1992) Targeted mutations that ablate either the adenylate cyclase or hemolysin function of the bifunctional cyaA toxin of *Bordetella pertussis* abolish virulence. *Proc. Natl. Acad. Sci. USA*, **89**, 4898–4902.

Guiso, N., Rocancourt, M., Szatanik, M. and Alonso J.M. (1989) Bordetella adenylate cyclase is a virulence associated factor and an immunoprotective antigen. *Microb. Pathogen.*, **7**, 373–380.

Guiso, N., Szatanik, M. and Rocancourt, M. (1991) Protective activity of *Bordetella pertussis* adenylate cyclase-hemolysin against bacterial colonization. *Microb. Pathogen.*, **11**, 423–431.

Hackett, M., Guo, L., Shabanowitz, J., Hunt, D.F. and Hewlett, E.L. (1994) Internal Lysine Palmitoylation in adenylate cyclase toxin from *Bordetella pertussis*. *Science*, **266**, 433–435.

Hackett, M., Walker, C.B., Guo, L., Gray, M.C., Van Cuyk, S., Ullmann, A., Shabanowitz, J., Hunt, D.F., Hewlett, E.L and Sebo, P. (1995) Hemolytic, but not cell-invasive activity, of adenylate cyclase toxin is selectively affected by differential fatty-acylation in *Escherichua coli*. *J. Biol. Chem.*, **270**, 20250–20253.

Hanski, E. and Farfel, Z. (1985) *Bordetella pertussis* invasive adenylate cyclase: Partial resolution and properties of cellular penetration. *J. Biol. Chem.*, **260**, 5526–5532.

Hewlett, E.L., Urban, M.A., Manclark, C.R. and Wolff, J. (1976) Extracytoplasmic adenylate cyclase of *Bordetella pertussis*. *Proc. Natl. Acad. Sci. USA*, **73**, 1926–1930.

Hewlett, E.L., Gordon, V.M., McCaffery, J.D., Sutherland, W.M. and Gray, M.C. (1989) Adenylate cyclase toxin from *Bordetella pertussis*. Identification and purification of the holotoxin molecule. *J. Biol. Chem.*, **264**, 19379–19384.

Hewlett, E.L., Gray, L., Allieta, M., Ehrman, I.E., Gordon, V.M. and Gray, M.C. (1991) Adenylate cyclase toxin from *Bordetella pertussis*. Conformational change associated with toxin activity. *J. Biol. Chem.*, **266**, 17503–17508.

Hewlett, E.L., Gray, M.C., Ehrman, I.E., Maloney, N.J., Otero, A.S., Gray, L., Allieta, M., Szabo, G., Weiss, A.A. and Barry, E.M. (1993) Characterization of adenylate cyclase toxin from mutant of *Bordetella pertussis* defective in the activator gene, *cyaC. J. Biol. Chem.*, **268**, 7842–7848.

Hyde, S.C., Emsley, P., Hartshon, M.J., Mimmack, M., Pearce, S.R., Gallagher, M.P., Gill, D.R., Hubbard, R.E. and Higgins, C.F. (1990) Structural model of ATP-binding proteins associated with cystic fibrosis, multidrug resistance. *Nature*, **324**, 485–489.

Iwaki, M., Ullmann, A. and Sebo, P. (1995) Identification by *in vitro* complementation of regions required for cell-invasive activity of *Bordetella pertussis* adenylate cyclase toxin. *Mol. Microbiol.*, **17**, 1015–1024.

Johnson, K.W., Davis, B.H. and Smith, K.A. (1988) cAMP antagonizes interleukin 2-promoted T-cell cycle progression at a discrete point in early G1. *Proc. Natl. Acad. Sci. USA*, **85**, 6072–6076.

Khelef, N., Sakamoto, H. and Guiso, N. (1992) Both adenylate cyclase and hemolytic activities are required by *Bordetella pertussis* to initiate infection. *Microb. Pathogen.*, **12**, 227–235.

Khelef, N., Zychlinsky, A. and Guiso, N. (1993) *Bordetella pertussis* induces apoptosis in macrophages: role of adenylate cyclase-hemolysin. *Infect. Immun.*, **61**, 4064–4071.

Khelef, N., Bachelet, C.M., Vargaftig, B.B. and Guiso, N. (1994) Characterization of murine lung inflammation after infection with parental *Bordetella pertussis* and mutants deficient in adhesins or toxins. *Infect. Immun.*, **62**, 2893–2900.

Khelef, N. and Guiso, N. (1995) Induction of macrophage apoptosis by *Bordetella pertussis* adenylate cyclase-hemolysin. *FEMS Microbiol. letters*, **134**, 27–32.

Koronakis, V., Stanley, P., Koronakis, E. and Hughes, C. (1992) The HlyB/HlyD-dependent secretion of toxins by Gram-negative bacteria. *Fems Microbiol. Immun.*, **105**, 45–54.

Ladant, D. (1988) Interaction of *Bordetella pertussis* adenylate cyclase with calmodulin: identification of two seperated calmodulin-binding domains. *J. Biol. Chem.*, **263**, 2612–2618.

Ladant, D., Michelson, S., Sarfati, R., Gilles, A-M., Predeleanu, R. and Barzu, O. (1989) Characterization of the calmodulin-binding and the catalytic domains of *Bordetella pertussis* adenylate cyclase. *J. Biol. Chem.*, **264**, 4015–4020.

Laoide, B.M. and Ullmann, A. (1990) Virulence dependent and independent regulation of the *Bordetella pertussis cya* operon. *EMBO J.*, **9**, 999–1005.

London, E. (1992) How bacterial protein toxins enter cells: the role of parial unfolding in membrane translocation. *Mol. Microbiol.*, **6**, 3277–3282.

Ludwig, A.T. and Goebel, W. (1991) Genetic determinants of cytolytic toxins from gram negative bacteria. *Source book of bacterial protein toxins.*, 117–146.

Ludwig, A.T., Garcia, F., Benz, R., Jarchau, T., Orpeza-Wekerle, R.L., Hoppe, J. and Goebel, W. (1992) Structures essential for pore formation by *Escherichia coli* haemolysin. *Witholt B et al. (eds) Bacterial protein toxins.*, 450–460.

Ludwig, A.T., Benz, R. and Goebel, W. (1993) Oligomerization of *Escherichia coli* hemolysin (HlyA) is involved in pore formation. *Mol. Gen. Genet.*, **241**, 89–96.

Masure, H.R., Oldenburg, D.J., Donovan, M.G., Shattuck, R.L. and Storm, D.R. (1988) The interaction of Ca^{2+} with the calmodulin-sensitive adenylate cyclase from *Bordetella pertussis*. *J. Biol. Chem.*, **263**, 6933–6940.

Masure, H.R, Au, D.C., Gross, M.K. and Storm, D.R. (1990) Secretion of the *Bordetella pertussis* adenylate cyclase from *escherichia coli* containing the hemolysin operon. *Biochem.*, **29**, 140–145.

Masure, H.R. and Barret, J. (1991) *Abstracts of the annual meeting of the American society for microbiology.*, **B284**, p. 73.

Masure, H.R. (1992) Modulation of adenylate cyclase toxin production as *Bordetella pertussis* enters human macrophages. *Proc. Natl. Acad. Sci. USA*, **89**, 6521–6525.

Melton, A.R. and Weiss, A.A. (1993) Characterization of enviromental regulators of *Bordetella pertussis*. *Infect. Immun.*, **61**, 807–815.

Menestrina, G. (1988) *Escherichia coli* haemolysin permeabilizes small unilamellar vesicles loaded with calcein by a single-hit mechanism. *FEBS Letters.*, **232**, 217–220.

Menestrina, G., Mackman, N., Holland, I.B. and Bhakdi, S. (1987) *Escherichia coli* haemolysin forms voltage-dependent channels in lipid membranes. *Biochim. Biophys. Acta.*, **905**, 109–117.

Menestrina, G., Moser, C., Pellet, S. and Welch, R. (1994) Pore-formation by Escherichia coli hemolysin (HlyA) and other members of the RTX toxin family. *Toxicology*, **87**, 249–267.

Miller, J.F., Mekalanos, J.J. and Falkow, S. (1989) Coordinate regulation and sensory transduction in the control of bacterial virulence. *Science*, **243**, 916–922.

Moayeri, M. and Welch, R.A. (1994) Effects of temperature, time and toxin concentration on lesion formation by the *Escherichia coli* Hemolysin. *Infect. Immun.*, **62**, 4124–4134.

Mock, M. and Ullmann, A. (1993) Calmodulin-activated bacterial adenylate cyclases as virulanse factors. *Trends in Microbiol.*, **1**, 187–192.

Mouallem, M., Farfel, Z. and Hanski, E. (1990) *Bordetella pertussis* adenylate cyclase toxin: intoxication of host cells by bacterial invasion. *Infect. Immun.*, **58**, 3759–3764.

Munier, H., Gilles, A-M., Glaser, P., Krin, E., Danchin, A., Sarfati, S.R. and Barzu, O. (1991) Isolation and charcterization of catalytic and calmodulin-binding domains of *Bordetella pertussis* adenylate cyclase. *Eur. J. Biochem.*, **196**, 469–474.

Munier, H., Bouhss, A., Krin, E., Danchin, A., Gilles, A-M. Glaser, P. and Barzu, O. (1992) The role of histidine 63 in the catalytic mechanism of *Bordetella pertussis* adenylate cyclase. *J. Biol. Chem.*, **267**, 9816–9820.

Munier, H., Bouhss, A., Gilles, A-M., Krin, E., Glaser, P., Danchin, A. and Barzu, O. (1993) Structural flexibility of the calmodulin-binding locus in *Bordetella pertussis* adenylate cyclase. Reconstitution of catalytically active species from fragments or inactive forms of the enzyme. *Eur. J. Biochem.*, **217**, 581–586.

Murayama, T., Hewlett, E.L., Maloney, N.J., Justice, J.M. and Moss, J. (1994) Effect of temperature and host factors on the activities of pertussis toxin and *Bordetella* adenylate cyclase. *Biochem.*, **33**, 15293–15297.

Oldenburg, D.J., Gross, M.K., Wong, C.S. and Storm, D.R. (1992) High-affinity calmodulin binding is required for the rapid entry of *Bordetella pertussis* adenylyl cyclase into neuroblastoma cells. *Biochem.*, **31**, 8884–8891.

Olsnes, S. and Sandvig, K. (1985) Entry of polypeptide toxins into animal cells. *Endocytosis (Plenum Publishing Corporation, 1985).*, 195–234.

Ostolaza, H., Bartolome, B., de Zarate, I.O., de la Cruz, F. and Goni, F.M. (1993) Release of lipid vesicle contents by the bacterial protein toxin α-haemolysin. *Biochim. Biophys. Acta*, **1147**, 81–88.

Otero, A.S., Yi, X.B., Gray, M.C., Szabo, G and Hewlett, E.L. (1995) Membrane depolarization prevents cell invasion by *Bordetella pertussis* adenylate cyclase toxin. *J. Biol. Chem.*, **270**, 9695–9697.

Pastan, I. and FitzGerald, D. (1989) *Pseudomonas* exotoxin: chimeric toxins. *J. Biol. Chem.*, **264**, 15157–15160.

Pearson, R.D., Symes, P., Conboy, M., Weiss, A.A. and Hewlett, E.L. (1987) Inhibition of monocyte oxidative response by *Bordetella pertussis* adenylate cyclase. *J. Immun.*, **139**, 2749–2754.

Pitman, M., Furman, B.L. and Wardlaw, A.C. (1980) *Bordetella pertussis* respiratory tract infection in the mouse: pathophysiological responses. *J. Infect. Dis.*, **142**, 56–66.

Raptis, A., Knipling, L.G., Gentile, F. and Wolff, J. (1989) Modulation of invasiveness and catalytic activity of *Bordetella pertussis* adenylate cyclase by polycations. *Infect. Immun.*, **57**, 1066–1071.

Rogel, A., Schultz, J.E., Brownlie, R.M., Coote, J.G., Parton, R. and Hanski, E. (1989) *Bordetella pertussis* adenylate cyclase: Purification and characterization of the toxic form of the enzyme. *EMBO J.*, **8**, 2755–2760.

Rogel, A., Meller, R. and Hanski, E. (1991) Adenylate cyclase toxin from *Bordetella pertussis*. The relationship between induction of cAMP and hemolysis. *J. Biol. Chem.*, **266**, 3154–3161.

Rogel, A. and Hanski, E. (1992) Distinct steps in the penetration of adenylate cyclase toxin of *Bordetella pertussis* into sheep erythrocytes. *J. Biol. Chem.*, **267**, 22599–22605.

Romanus, V., Jonsell, R. and Bergquist, S.D. (1987) Pertussis in Sweden after the cessation of general immunization in 1979. *Pediatr. Infect. Dis. J.*, **6**, 364–371.

Rose, T., Sebo, P., Bellalou, J. and Ladant, D. (1995) Interaction of calcium with *Bordetella pertussis* adenylate cyclase toxin. Characterization of multiple calcium-binding sites and calcium-induced conformational changes. *J. Biol. Chem.*, **270**, 26370–26376.

Sakamoto, H., Bellalou, J., Sebo, P. and Ladant, D. (1992) *Bordetella pertussis* adenylate cyclase toxin: structural and functional independence of the catalytic and hemolytic activities. *J. Biol. Chem.*, **267**, 13598–13602.

Sandros, J. and Tuomanen, E. (1993) Attachment factors of *Bordetella pertussis*: mimicry of eukaryotic cell recognition molecules. *Trends in Microbiology*, **1**, 192–195.

Sarfati, R.S., Kansal, V.K., Munier, H., Glaser, P., Gilles, A-M., Labruyere, E., Mock, M., Danchin, A. and Barzu, O. (1990) Binding of 3′-anthraniloy-2′-deoxy-ATP to calmodulin-activated adenylate cyclase from *Bordetella pertussis* and *Bacillus anthracis. J. Biol. Chem.*, **265**, 18902–18906.

Sebo, P., Glaser, P., Sakamoto, H. and Ullmann, A. (1991) High level synthesis of active adenylate cyclase toxin of *Bordetella pertussis* in a reconstructed *Escherichia coli* system. *Gene*, **104**, 19–24.

Sebo, P. and Ladant, D. (1993) Repeat sequences in the *Bordetella pertussis* adenylate cyclase toxin can be recognized as alternative carboxy-proximal secretion signals by the *Escherichia coli* α-haemolysin translocator. *Mol. Microbiol.*, **9**, 999–1009.

Shattuck, R.L. and Storm, D.R. (1985) Calmodulin inhibits entry of *Bordetella pertussis* adenylate cyclase into animal cells. *Biochem.*, **24**, 6323–6328.

Sheps, J.A., Cheung, I. and Ling, V. (1995) Hemolysin transport in *Escherichia coli*. Point mutants in HlyB compensate for a deletion in the predicted amphiphilic helix region of the HlyA signal. *J. Biol. Chem.*, **24**, 14829–14834.

Splanger, B.D. (1992) Structure and function of cholera toxin and the related *Escherichia coli* heat-labile enterotoxin. *Micobiol. Rev.*, **56**, 622–647.

Stanley, P., Packman, L.C., Koronakis, V. and Hughes, C. (1994) Fatty acylation of two internal Lysine residues required for the toxic activity of *Escherichia coli* hemolysin. *Science*, **266**, 1992–1995.

Steed, L.L., Morey, S. and Friedman, R.L. (1991) Intracellular survival of virulent *Bordetella pertussis* in human polymorphonuclear leukocytes. *J. Leukocyte Biol.*, **50**, 321–330.

Stock, J.B., Ninfa, A.J. and Stock, A.M. (1989) Protein phosphorilation and regulation of adaptive responses in bacteria. *Microbiol. Rev.*, **53**, 450–490.

Szabo. G., Gray, M.C. and Hewlett, E.L. (1994) Adenylate cyclase toxin from *Bordetella pertussis* produces ion conductance across artificial lipid bilayers in a calcium- and polarity-dependent manner. *J. Biol. Chem.*, **269**, 22496–22499.

Wandersman, C. and Delepelaire, P. (1990) TolC, an *Escherichia coli* outer membrane protein required for hemolysin secretion. *Proc. Natl. Acad. Sci. USA*, **87**, 4776–4780.

Weiss, A.A., Hewlett, E.L., Myers, G.A. and Falkow, S. (1983) Tn5-induced mutations affecting virulence factors of *Bordetella pertussis. Infect. Immun.*, **42**, 33–41.

Weiss, A.A., Hewlett, E.L., Myers, G.A. and Falkow, S. (1984) Pertussis toxin and extracytoplasmic adenylate cyclase as virulence factors of *Bordetella pertussis. J. Infect. Dis.*, **150**, 219–222.

Weiss, A.A., Hewlett, E.L., Myers, G.A. and Falkow, S. (1985) Genetic studies of the molecular basis of whooping cough. Proceedings of the 4th International Symposium on Pertussis, Geneva, *Dev. Biol. Stand.*, **61**, 11–19.

Weiss, A.A. and Hewlett, E.L. (1986) Virulence factors of *Bordetella pertussis Ann. Rev. Microbiol.*, **40**, 661–686.

Weiss, A.A, Hewlett, E.L. and Cronin, M.J. (1986) Bacterial adenylate cyclase increases cyclic AMP and hormone release in pituitary tumor cells. *Biochemical and Biophysical research communications.*, **136**, 463–469.

Weiss, A.A. and Goodwin, M.S. (1989) Lethal infection by *Bordetella pertussis* mutants in the infant mouse model. *Infect. Immun.*, **57**, 3757–3764.

Welch, R.A. (1991) Pore-forming cytolysins of Gram-negative bacteria. *Mol. Microbiol.*, **5**, 521–528.

Wolff, J., Cook, G.H., Goldhammer, A.R. and Berkowitz, S.A. (1980) Calmodulin activates prokaryotic adenylate cyclase. *Proc. Natl. Acad. Sci. USA*, **77**, 3840–3844.

Xia, Z.G. and Storm, D.R. (1990) A-type ATP binding consensus sequences are critical for the catalytic activity of the calmodulin-sensitive adenylyl cyclase from *Bacillus anthracis. J. Biol. Chem.*, **265**, 6517–6520.

14. PHOSPHOINOSITIDE-SPECIFIC PHOSPHOLIPASE C AND TOXINS

MARGARITA L. CONTRERAS and DONALD W. FINK, JR.

*Department of Pharmacology and Toxicology, Michigan State University,
East Lansing, MI 48824
Laboratory of Cell Biology/Neurotrophic Factors, Division of Cytokine Biology,
Center for Biologics Evaluation and Research, Food and Drug Administration,
Rockville, MD 20852*

INTRODUCTION

Second messenger systems are critical for transduction of signals across the plasma membrane, from the surface receptors that bind the external stimuli to the intracellular compartment where the second messengers stimulate a cellular response. One second messenger system involves phospholipase C-mediated hydrolysis of inositol-containing phospholipids or phosphoinositides. It has been over 40 years since Hokin and Hokin (1953) first demonstrated a cholinergic agonist-induced incorporation of ^{32}P into phosphatidylinositol (PI) in pancreas and brain cortex slices. However, the consequence or significance of this rapid lipid turnover was not at first clear. It was Michell *et al.* (1981) and Nishizuka (1984) who proposed second messenger functions for the phosphoinositide metabolites, diacylglycerol (DAG) and 1,4,5-trisphosphate ($Ins(1,4,5)P_3$). Subsequently, the role of phosphatidylinositols as a second messenger system for a variety of receptors has been documented (Meldrum *et al.*, 1991b; Fisher *et al.*, 1992; Berridge, 1993; Tsunoda, 1993). It is now apparent that the phosphatidylinositol system is a common second messenger system involved in diverse cellular responses, such as mitogenesis (Carney *et al.*, 1985; La Morte *et al.*, 1993), contraction (Volpe *et al.*, 1986; Stull *et al.*, 1988), secretory processes (Blondel *et al.*, 1995) and sensory perception (Breer *et al.*, 1990; Hwang *et al.*, 1990; Kalinoski *et al.*, 1992).

The inositol-containing phospholipids are quantitatively minor membrane lipid components, constituting only 2 to 8% of the phospholipids in eukaryotic membranes (Majerus *et al.*, 1984; 1986). Over 80% of the phosphoinositides is in the form of PI, which is a unique lipid in that the polar *myo*-inositol headgroup can be phosphorylated to produce phosphatidylinositol 4-phosphate (PIP) and phosphatidylinositol 4,5-bisphosphate (PIP_2). Upon receptor activation, PIP_2 is hydrolyzed by an inositol lipid-specific phospholipase C (PLC) to produce two second messengers — $Ins(1,4,5)P_3$ and DAG (Fisher *et al.*, 1992; Michell, 1992). $Ins(1,4,5)P_3$ diffuses through the cytosol to specific $Ins(1,4,5)P_3$ receptors to promote the release of calcium from the endoplasmic reticulum and an influx of external calcium, sometimes in conjunction with inositol 1,3,4,5-tetrakisphosphate ($Ins(1,3,4,5)P_4$). DAG binds to and activates several isoforms of protein kinase C (Takai *et al.*, 1979;

Nishizuka, 1988; and Parker *et al.*, 1989). Generally, the activation of protein kinase C and the increase in intracellular Ca^{2+}, that is seen following hydrolysis of PIP_2, act in a synergistic fashion to produce the cellular responses (Nishizuka, 1985; Berridge, 1987). The DAG may also be further hydrolyzed to arachidonic acid, a precursor of icosanoid mediators (Majerus 1983; Majerus *et al.* 1984). Thus, the phosphatidylinositol system involves production of multiple second messengers to regulate calcium-dependent processes, protein kinase C-mediated protein phosphorylation reactions and arachidonic acid-involved pathways. In certain systems, other inositol phosphates derived from $Ins(1,4,5)P_3$, such as $Ins(1,3,4,5)P_4$, may also act as second messengers (Bansal and Majerus, 1990; Michell, 1992). Recent studies have also indicated the existence of a separate nuclear phosphoinositide cycle (Divecha *et al.*, 1993; Cocco *et al.*, 1994), which may provide insight into the relationship between inositides and the nuclear effects of various hormones, neurotransmitters and growth factors. The focus of this review, however, will be on the more familiar and better characterized plasma membrane phosphoinositide cycle.

FAMILY OF PHOSPHOLIPASE C ENZYMES

General Structural Features

A number of PLCs have been purified or molecularly cloned and sequenced from mammalian and nonmammalian tissues (Rhee *et al.*, 1989; Rhee and Choi, 1992a; Cockcroft and Thomas, 1992). Based on amino acid sequences and immunological cross-reactivities, the PLC enzymes have been categorized into three types or families, designated β, γ and δ. Each family consists of multiple subtypes, such as PLC-$\beta1$ and PLC-$\beta2$. The differences in the various members within a PLC family are not due to differences in species or tissue of origin. For example there is approximately 84% sequence conservation between bovine and rat PLC-$\delta1$ (Suh *et al.*, 1988b). Similarly, there is a better than 95% amino acid sequence identity between PLC-$\gamma1$ derived from rat, bovine and human brains and other mammalian tissues (Rhee and Choi, 1992a). Instead, each subtype is a product of a unique gene. In a few instances, a subtype has been shown to be present in multiple forms due to alternate splicing of the gene.

The three types of PLCs are single polypeptides with dissimilar amino acid sequences (Rhee *ei al.*, 1989; Rhee and Choi, 1992a; Cockcroft and Thomas, 1992). Due to the differences in amino acid sequences, there is a lack of immunological cross-reactivity between the three types of PLCs. Furthermore, the three types of PLCs have different molecular masses. In general, the molecular masses, as determined by SDS-PAGE, are 125 to 154 kd for PLC-βs, 145 to 148 kd for PLC-γs and 85 to 88 kd for PLC-δs. While the overall sequence homology between the three families are low, there is significant sequence homology in two domains, designated X and Y. The X domain, consisting of about 150 amino acids, is approximately 60% identical between the three types of PLC. The Y domain, composed of about 240 amino acids, is approximately 40% identical between the three PLC groups. The homology in these two regions suggests that the X and Y domains may comprise critically important regions for the catalytic activity of the enzymes, such as the catalytic domain or the recognition sites for phosphoinositides or Ca^{2+}. It has been suggested that the carboxyl-terminal half of the Y domain is a Ca^{2+}-dependent phospholipid-binding domain, since this region has an amino acid sequence that is homologous to the sequences of Ca^{2+}-dependent phospholipid-binding motifs that are found in cytosolic phospholipase A and protein kinase C (Clark *et al.*, 1991). Furthermore, in support of the idea that the X and Y domains are critical to the activity of the PLC enzymes, Emori *et al.* (1989) found that deletion of either the X or Y domain from the PLC-γ isozymes results in complete loss of biological activity. Similarly, partial deletion of either the X or Y domains in PLC-δ1 inactivates the enzyme (Ellis *et al.* 1993; Yagisawa *et al.*, 1994).

The region between the X and Y domains is variable in size and amino acid sequence. While there are approximately 50–70 amino acids separating the X and Y domains in PLC-β and δ isozymes, there are about 400 amino acids separating the X and Y domains in the PLC-γ isozymes (Stahl *et al.*, 1988: Suh *et al.*, 1988a). Interestingly, the region separating the X and Y domains in PLC-γ isozymes contain regions that are similar in amino acid sequence to the noncatalytic domain of the *src* oncogene product, which are designated SH2 and SH3 (src homology 2 and 3). The SH2 and SH3 domains in the PLC-γs are highly conserved regions, that have also been found in the regulatory domains of nonreceptor tyrosine kinases, such as is found in *fgr, lyn, abl* and *src* (Pawson 1988).

In almost all PLCs, there are about 300 amino acids at the amino-terminus preceding the X domain (Kriz *et al.*, 1990; Rhee and Choi, 1992b). In contrast, the carboxy-terminal region between the Y domain and the C-terminus is variable in size. The lack of a carboxy-terminal region after the Y domain readily distinguishes the PLC-δ type. In contrast, the PLC-β type generally have a carboxy terminal domain that is about 400 amino acids in size and contains an unusually high number of charged residues (about 40%) (Katan *et al.*, 1988; Kriz *et al.*, 1990). The carboxy-terminal region of the PLC-γ type is intermediate in size.

One possible reason for the multiplicity in PLC types is found in the different mechanisms by which the different PLCs are involved in signal transduction (Rhee and Choi, 1992a; Cockcroft and Thomas, 1992; Harden, 1992; Exton, 1993). The PLCγs, with SH2 and SH3 domains, are activated by tyrosine phosphorylation. In contrast, PLCs of the β family have been shown to be activated by G proteins. However, the mechanism for activation of the PLC-δs does not appear to involve either guanine nucleotide binding proteins or tyrosine phosphorylation. Thus, the mechanism by which PLC-δs mediate signal transduction still remains unclear.

Phospholipase-β Family

Several distinct members of the PLC-β family have been cloned from mammals, *Xenopus, Drosophila* and *Artemia*. The first member of this family, designated PLC-β1, was isolated from bovine brain (Ryu *et al.*, 1986) and rat brain (Suh *et al.*, 1988). More recently, in the rat brain, Bahk *et al.* (1994) has shown that there are really two forms of PLC-β1 that are generated by alternative splicing of a single gene. In addition to the PLC-β1s, other members of this family were detected by screening cDNAs from a variety of tissues, using a low stringency cross-hybridization probe made from the conserved X domain of PLC-β1. Thus, PLC-β2 was identified using a HL-60 library (Park *et al.*, 1992) and PLC-β3 was identified in a rat thyroid cell line and rat brain (Jhon *et al.*, 1993).

Two *Drosophila* genes, *norpA* (Bloomquist *et al.*, 1988) and *plc*-21 (Shortridge *et al.*, 1991) code for PLCs (PLC-norpA and PLC-21) that show a greater degree of homology with PLC-βs than with the PLCs of the δ or γ type. Further, the *Drosophila* PLCs resemble the PLC-β type with a carboxy-terminus of approximately 400 amino acids that contains a large number of charged residues and a small region separating the X and Y domains. Thus, norpA and PLC-21 are considered members of the PLC-β family. The gene encoding PLC-21 actually codes for two different enzymes that differ in only 7 amino acid residues in the carboxy terminal domain (Shortridge *et al.*, 1991). One PLC-21 mRNA is expressed only in the adult head, while the other mRNA is expressed throughout development in both the *Drosophila* head and body. It has been postulated that these two forms of PLC-21 may be differentially regulated since the 7 amino acid sequence in one form of the enzyme contains two serine residues, that may be sites for regulating enzyme activity through phosphorylation.

In invertebrates, phototransduction is dependent on hydrolysis of phosphoinositides. In *Drosophila*, PLC-norpA has been shown to be critically involved in phototransduction (Bloomquist *et al.*, 1988). In contrast, phototransduction is dependent on the activation of a cGMP phosphodiesterase in vertebrates (Stryer *et al.*, 1986). However, light-activated PLC activities have been reported in vertebrate rod outer segments (Jelsema, 1989; Ghalayini *et al.*, 1992). Thus, studies were initiated to find the vertebrate analog of PLC-norpA. A PLC, designated PLC-β4, was characterized from the bovine retina, which has an overall 54% sequence homology with PLC-norpA (Lee *et al.*, 1993). The degree of homology is greater in the X (77%) and Y (74%) domains. Since the homology is greatest with PLC-norpA than with any other member of the PLC-β family, PLC-β4 may be the mammalian analog of PLC-norpA. Although PLC-β4 is predominantly found in the retina, it is also expressed in the rat brain (Lee *et al.*, 1993) and bovine cerebellum (Kim *et al.*, 1993). In the bovine cerebellum, two PLC-β4s were found, suggesting that multiple forms of this PLC may exist. Similarly, Ferreira *et al.* (1993) found four types of mammalian norpA-like PLCs, that are derived by alternative processing of the mRNA.

The mammalian norpA-like PLCs are unique in that their amino acid sequences contain conserved GTPase sequence motifs. The G_1 motif, GX_4GKS, is one of the 5 motifs found in members of the GTPase superfamily that includes G proteins (Bourne *et al.*, 1991). The sequence recognizes the phosphoryl group of nucleotides. Lee *et al.* (1994) has demonstrated that PLC-β4 is unique in that the PLC activity is selectively inhibited by ribonucleotides with 5'-phosphate and 2'-hydroxyl groups of ribose. Thus monophosphate, diphosphate and triphosphate ribonucleotides inhibit PLC-β4 activity. Considering that PLC-β4 catalytic activity toward PIP_2 is 4 to 5 times higher than the specific activity of PLC-β1 or PLC-β3, the ribonucleotide-mediated inhibition of PLC-β4 activity may serve to maintain PLC-β4 activity low in the absence of hormonal stimulation. The G_1 motif is also found in PLC-norpA and PLC-21 (Smrcka *et al.*, 1991), indicating that the activity of these PLC isozymes would also be inhibited by ribonucleotides. No other members of the PLC-β family possess the GTPase sequence motifs.

Two additional members of the PLC-β family have been identified in nonmammalian tissues. In *Xenopus*, a PLC with approximately 80% homology with PLC-β3 has been identified (Ma *et al.*, 1993). This PLC was designated PLC-Xβ. More recently, a novel PLC-β was isolated from a *Artemia* brine shrimp cDNA library (Su *et al.*, 1994). The *Artemia* PLC, designated PLC-β_x, is novel in that it is highly truncated in the amino-terminal region. Unlike the other PLCs, this PLC does not have 300 amino acids in the region between the amino-terminus and the X-domain. Further, the X domain is truncated on the amino-terminal side, having only 14–25 amino acids instead of the usual approximately 150 amino acids that are found in the other PLCs of higher organisms. Thus, this PLC has a smaller mass (55 kDa) than the other known PLC-βs (125–154 kDa).

Phospholipase C-γ Family

The first member of this family, PLC-γ1, was isolated from bovine and rat brain (Suh *et al.*, 1998a). Using low stringency hybridization techniques, a full length cDNA for a PLC was isolated from HL-60 cells (Kriz *et al.*, 1990), rat muscle (Emori *et al.*, 1989) and human lymphocytes (Ohta *et al.*, 1988). This PLC was designated PLC-γ2, since it is similar to PLC-γ1 in amino acid sequence. There is a high percent of amino acid sequence identity in domains X (80%) and Y (62%) between the γ2 and γ1 forms of PLC. PLC-γ2 is a member of the PLC-γ family, since it has two SH2 and one SH3 motifs in the 400 amino acid region between the X and Y domains — a feature that is unique to the PLC-γs. There is approximately 70–76% amino acid sequence homology in the SH2 and SH3 regions in the two forms of PLC-γ, suggesting that the SH2 and SH3 domains are critical to the function of the PLC-γs. While the X and Y domains are critical for PLC phosphodiesterase activity (Emori *et al.*, 1989), the SH2 and SH3 domains are critical for receptor-activated PLC activity and interaction of PLC-γs with other proteins (Rhee and Choi, 1992a; Gergel *et al.*, 1994; Seedorf *et al.*, 1994; Koblan *et al.*, 1995).

Phospholipase C-δ Family

The cDNA encoding the first member of this family to be identified, PLC-δ1, was isolated from rat and bovine brain (Homma *et al.*, 1988; Suh *et al.*, 1988b). Subsequently, sequence analysis of a novel PLC led to the isolation of a cDNA that encodes a protein similar to PLC-δ1, which was designated PLC-δ2 (Meldrum *et al.*, 1991a). The sequence identity is greatest with PLC-δ1, especially in the X (69%) and Y (50%) domains, than with PLCs of the γ and β families. Furthermore, PLC-δ2 possesses the distinguishing feature of the PLC-δ family, in that it lacks a carboxy-terminal region after the X and Y domains. The cDNA for a third member of PLC-δ family was isolated from a fibroblast (WI-38) library and the amino acid sequence for PLC-δ3 was deduced (Kriz *et al.*, 1990). Based on the deduced amino acid sequence, it appears that PLC-δ3 has a high degree of sequence identity with PLC-δ1 in the X (69%) and Y (45%) domains, as well as in the amino-terminal region (43%). Similar to the other members of the PLC-δ family, PLC-δ3 also is missing a carboxy-terminal region beyond the Y domain.

As mentioned previously, the X and Y domains are thought to be critical for PLC activity. Studies by Cheng *et al.* (1995) have further delineated the amino acid residues in the X region important in the enzymatic activity of PLC-δ1. Using single point mutagenesis, the studies revealed that histidine[356] and glutamate[341] in the X-domain are critical for Ca^{2+}-dependent hydrolysis of PIP_2. However, the mutant PLCs could still bind PIP_2, suggesting that these two amino acid residues are required for either the binding of Ca^{2+} or for catalytic activity. Considering that these two residues are invariant in prokaryotic PLCs, whose activity does not depend on calcium, the two amino acid residues are probably involved in the catalytic activity of the enzyme.

Phospholipase C Catalytic Activity

The catalytic properties of the three families of PLCs are similar. All three types of PLCs can catalyze the hydrolysis of the three major inositol-containing phospholipids, PI, PIP and PIP_2. However, the phosphodiesterase activity depends on Ca^{2+}. Actually, elevating the level of Ca^{2+} can by itself result in activation of PLC. However, the specific rate of hydrolysis of each of the phosphoinositides have specific Ca^{2+} dose-dependent hydrolysis profiles. At low physiologically relevant concentrations of Ca^{2+}, PLC-β1 and β2 have been shown to have a catalytic preference for the polyphosphoinositides, with little activity towards PI (Kriz *et al.*, 1990; Katan and Parker, 1987). At high mM concentrations of Ca^{2+}, PI is hydrolyzed but still the polyphosphoinositides are the preferred substrate, being hydrolyzed several-fold better than PI. In the case of PLC-β1, it has been demonstrated that this selectivity is due to a higher Vmax value for the polyphosphoinositides (Katan and Parker, 1987).

The substrate selectivity of PLC-γ enzymes are different. At a physiological concentration of Ca^{2+}, PLC-γ1 has a similar catalytic activity towards PIP_2 and PI (Ryu *et al.*, 1987a). The catalytic profile is somewhat different for PLC-γ2. At an optimal concentration of Ca^{2+}, of approximately 30 mM, the phosphodiesterase hydrolyzes PI and PIP about eight-fold faster than PIP_2 (Kriz *et al.*, 1990). However, at a more physiologically relevant concentration of Ca^{2+} (0.01 mM), PIP_2 is the preferred substrate. Nevertheless, the selectivity of PIP_2 over PI is less with the PLC-γs than with the PLC-βs. Similar to the PLC-βs, PLC-δ1 and δ2 also show a substrate preference of polyphosphoinositides over PI at low physiologically relevant concentrations of Ca^{2+} (Ryu *et al.*, 1987b; Meldrum *et al.*, 1989). Only at high mM Ca^{2+} concentrations is PI efficiently hydrolyzed by the PLC-δs. Thus, the polyphosphoinositides are generally the preferred substrates of the PLC isozymes. This general selectivity of polyphosphoinositides as the preferred substrate has also been seen in *in vitro* studies, in which PIP_2 and/or PIP are preferably hydrolyzed to produce inositol phosphates and DAG (Berridge, 1983; Batty and Nahorski, 1989; Fisher *et al.*, 1990).

ACTIVATION OF PLC BY TYROSINE PHOSPHORYLATION

Growth Factor-Stimulated Tyrosine Phosphorylation of PLC-γ1

As mentioned previously, there are at least two general mechanisms through which activation of PLC hydrolytic activity is achieved. Stimulation of PLC-β isozymes PLC-β1, PLC-β2 and PLC-β3 is contingent upon a G protein-dependent coupling of the cytosolic enzyme to plasma membrane receptors. A distinct mechanism is involved for peptide growth factor-stimulated increases in the catalytic activity of PLC-γ isoforms, PLC-γ1 and PLC-γ2. In this instance, increased enzymatic activity occurs as a consequence of tyrosine phosphorylation of PLC-γ (Rhee, 1991).

Frequently, though not universally, enhanced phosphorylation of the PLC-γ1 isoform on aminotyrosine residues is mediated by growth factor receptors with intrinsic cytoplasmic tyrosine kinase activity that becomes stimulated upon binding of the receptor with its cognate ligand. It has been shown for a variety of peptide growth factors, eliciting biological actions through cell-surface transmembrane receptors possessing ligand-activated tyrosine kinase activity, that an early event in the signaling cascade for these growth factors is enhanced tyrosine phosphorylation of PLC-γ1. Included in this group of growth factors are epidermal growth factor [EGF] (Margolis *et al.*, 1989; Meisenhelder *et al.*, 1989; Nishibe *et al.*, 1989; Huckle *et al.*, 1990), platelet-derived growth factor [PDGF] (Downing *et al.*, 1989; Wahl *et al.*, 1989; Kumjian *et al.*, 1989; Morrison *et al.*, 1990; Kim *et al.*, 1991), nerve growth factor [NGF] (Kim *et al.*, 1991; Vetter *et al.*, 1991) and fibroblast growth factor [FGF] (Burgess *et al.*, 1990). In each instance it has been shown that growth factor stimulated tyrosine phosphorylation of PLC-γ1 is associated with activation of the enzyme's phospholipid hydrolytic activity (Nishibe *et al.*, 1990; Kim *et al.*, 1991; Kim *et al.*, 1991; Mohammadi *et al.*, 1992; Peters *et al.*, 1992).

Ligand-dependent stimulation of growth factor receptor tyrosine kinase activity results in tyrosine phosphorylation of PLC-γ1 at multiple sites. It is known, for example, that EGF and PDGF stimulate phosphorylation of PLC-γ1 on the same tyrosine residues both *in vitro* and within the intact cell. The specific phosphorylation sites have been identified as tyrosine residues 771, 783 and 1254 (Kim *et al.*, 1990; Wahl *et al.*, 1990). To determine the role of PLC-γ1 tyrosine phosphorylation in the mechanism of enzymatic activation, mutant enzymes were constructed that consisted of phenylalanine substitutions for each of the three tyrosine sites. The mutant constructs were then expressed in NIH 3T3 cells (Kim *et al.*, 1991). Substitution of phenylalanine for Tyr783 completely inhibited activation of PLC-γ1 by PDGF. The same substitution at Tyr1254 was also inhibitory, however, the inhibition was less complete resulting in a 40% reduction in the response. In contrast to the attenuating impact observed for these substitutions at Tyr783 and 1254, mutation of Tyr771 to phenylalanine facilitated responsiveness of PLC-γ1 to PDGF stimulation enhancing the response by 50%. Decreased responses of the PLC-γ1 mutants to PDGF do not occur as a consequence of enzymatic inactivation resulting from phenylalanine for tyrosine substitutions. Mutation at each of the residues, 771, 783, or 1254 did not affect catalytic activity of PLC-γ1 as assessed *in vitro* (Kim *et al.*, 1991).

In addition to inducing enzymatic activation, tyrosine phosphorylation of the PLC-γ isoform results in directional redistribution of the enzyme from a cytosolic to particulate cellular location. Translocation of PLC-γ from the cytosol to membrane has been observed as a consequence of growth factor treatment (Kim *et al.*, 1990; Todderud *et al.*, 1990) as well as incubation of cell cultures with an inhibitor of protein tyrosine phosphatase, orthovanadate (Atkinson *et al.*, 1993). The proposed effect of this translocation is to bring the putative catalytic domain regions X and Y into proximity with the cytoplasmic-facing surface of the cell membrane where the preferred PLC phospholipid substrate PIP_2 is located.

Tyrosine Phosphorylation of PLC-γ1 Mediated by Receptors Lacking Intrinsic Tyrosine Kinase Activity

There are several examples of PLC-γ1 tyrosine phosphorylation occurring as a result of the activation of cell membrane receptors lacking tyrosine kinase catalytic activity. Members of the intercellular adhesion molecule (ICAM) family (Kanner *et al.*, 1993) and proteins of the extracellular matrix such as collagen (Somogyi *et al.*, 1994) stimulate tyrosine phosphorylation and enzymatic activation of PLC-γ1. Enhanced tyrosine phosphorylation of PLC-γ1 elicited by ICAMs and collagen is mediated, in part, by integrin cell surface receptors (Richardson and Parsons, 1995). Integrins are heterodimeric, transmembrane cell adhesion receptors that transduce biologic signals across the membrane. Activation of integrins leads to tyrosine and serine phosphorylation of several proteins, including PLC-γ1. Coupling of integrins to additional signalling molecules with tyrosine kinase activity is essential as integrins are catalytically inactive. One likely candidate is the focal adhesion kinase, also referred to as pp125 FAK [FAK] (Schaller and Parsons, 1994). The FAK tyrosine kinase localizes to regions of focal contacts which are also characterized as sites of integrin clustering. Furthermore, FAK is capable of physically associating with integrins *in vitro*.

Additional examples whereby stimulation of receptors lacking protein tyrosine kinase activity result in tyrosine phosphorylation of PLC-γ1 include ligation of the T-cell receptor complex in T cells (Park *et al.*, 1991; Weiss *et al.*, 1991), cross-linking of high- or low-affinity Fc receptors for IgG in the human monocytic cell line U937 (Liao *et al.*, 1992) and cross-linking of the B cell antigen receptor (BCR) in B lymphocytes (Carter *et al.*, 1991). In each instance, activation of a protein tyrosine kinase(s) subsequent to an initiation event occurring at the cell membrane is presumed. In the particular case of BCR, a number of protein tyrosine kinases (PTKs) are known to associate with BCR including members of the src PTK family, $p53/p56^{lyn}$, $p59^{fyn}$ and $p55^{blk}$, as well as the non-src family PTK $p72^{syk}$. These associated PTKs become activated within seconds following BCR cross-linking. On the basis of evidence that indicates a physical association between PLC-γ1 and $p72^{syk}$ following BCR cross-linking, this non-src PTK has been implicated as the mediator of PLC-γ1 tyrosine phosphorylation (Sillman and Monroe, 1995).

Finally, it has been suggested that tyrosine phosphorylation and enzymatic activation of PLC-γ1 might occur through a mechanism involving G protein-receptor coupling. Transfection of Chinese hamster ovary cells with G protein-coupled m5 muscarinic receptors results in enhanced tyrosine phosphorylation of PLC-γ1 and activation of the enzyme upon stimulation with the muscarinic agonist carbachol (Gusovsky *et al.*, 1993). It has been suggested that muscarinic receptor-mediated tyrosine phosphorylation of PLC-γ1 occurs as a consequence of the activation of a calcium-sensitive tyrosine kinase(s). Perhaps more intriguing is the observation that pertussis toxin blocks EGF-induced tyrosine phosphorylation of PLC-γ1 in hepatocytes (Yang *et al.*, 1991) suggesting that a pertussis toxin-sensitive G protein, G_i, is involved in the signaling pathway that mediates EGF-induced tyrosine phosphorylation and activation of PLC-γ1 in hepatocytes.

The Z Region: SH2 and SH3 Domains in Phospholipase C-γ

Structural homology within the PLC family of isozymes is readily demonstrated by the significant degree of sequence identity in two domains designated X and Y. These structural regions are characteristically present within all members of the PLC family and are apparently responsible for the catalytic activation of the enzymes. As discussed previously, it has been shown that deletion of either the X or Y domain will result in complete loss of enzymatic activity (Emori *et al.*, 1989). A unique structural feature within the PLC-γ subtype is a stretch of approximately 400 amino acids separating the X and Y domains. This portion of the enzyme is referred to as the Z region and is comprised of amino acid sequences that are related to portions of the *src* oncogene product known as SH2 and SH3 (*src* homology 2 and 3) domains. Within the Z region of PLC-γ these sequences occur as two sequential SH2 domains followed by an SH3 domain (Rhee, 1991). SH2 and SH3 were initially identified as highly conserved amino acid sequences within the non-catalytic, regulatory regions of non-receptor tyrosine kinases including products of *src, abl and fps* (Pawson, 1988). In contrast to the X and Y domains, the SH2 and SH3 elements comprising the Z region are not required for catalytic activity of the enzyme. Deletion of the entire *src*-homology region from PLC-γ does not abolish enzymatic activation (Bristol *et al.*, 1988). It is now well established that the SH2 and SH3 domains of the Z region serve a regulatory function that facilitates direct protein-protein interaction between PLC-γ and a number of targets.

Protein-protein interactions mediated by the SH2 and SH3 domains of PLC-γ

The SH2 domains of intracellular signaling molecules, including PLC-γ, exhibit the capacity to bind sites containing phosphotyrosine on a number of growth factor receptors and cytosolic phosphoproteins (Koch *et al.*, 1991; Koch *et al.*, 1992). Upon stimulation by ligand binding, growth factor receptors become autophosphorylated on tyrosine residues residing outside the catalytic region of the receptor tyrosine kinase. Although common to SH2 domains is their binding to tyrosine phosphates, distinct proteins expressing SH2 motifs are able to discriminate different tyrosine phosphorylated sites through recognition of residues flanking the target phosphotyrosine (Cantley *et al.*, 1991). For example, trkA, the NGF receptor involved in high-affinity binding of NGF, possesses two tyrosine autophosphorylation sites at positions 490 and 785. It has been shown that tyrosine 785 is the site for interaction between trkA and PLC-γ1 (Obermeier *et al.*, 1993; Loeb *et al.*, 1994). The EGF receptor contains five sites distal to the kinase domain where tyrosine autophosphorylation occurs. These tyrosine residues are found at positions 992, 1068, 1086, 1148 and 1173 (Downward *et al.*, 1984; Hsuan *et al.*, 1989; Walton *et al.*, 1990). Mutational analysis has revealed that only tyrosine 992 is involved in the SH2-mediated interaction between PLC-γ1 and the EGF receptor (Vega *et al.*, 1992). Similarly, the PDGF receptor possesses a number of tyrosine autophosphorylation sites distributed throughout the intracellular portion of the receptor including the juxtamembrane region, the kinase insert and the carboxy-terminal tail. Tyrosine residues phosphorylated at these locations are capable of binding a variety of SH2-containing cytosolic proteins. Physical interaction between the SH2 domain of PLC-γ1 and the PDGF receptor occurs only at the phosphorylated tyrosine mapping to position 1021 which resides in the carboxy-terminal tail of the receptor (Rönnstrand *et al.*, 1992; Larose *et al.*, 1993; Kashishian and Cooper, 1993; Valius *et al.*, 1993). Specificity of this interaction has recently been demonstrated in studies that involve amino acid switching at the tyrosine 1021 site. Replacement of the amino acids leucine and proline, which are in carboxy-terminal positions +4 and +5, respectively, immediately adjacent to tyrosine 1021, with corresponding residues from a phosphatidylinositol 3′-kinase SH2 binding site abrogates binding of PLC-γ1 to the PDGF receptor (Larose *et al.*, 1995).

The *src* homology SH3 domain is also capable of promoting direct protein-protein interactions. SH3 domains bind to proline-rich amino acid sequence motifs within a target protein. At least one protein, the microtubule-activated GTPase dynamin, has been identified which appears to specifically associate with PLC-γ1 through its SH3 domain (Gout *et al.*, 1993; Seedorf *et al.*, 1994).

Regulatory function of PLC-γ1 SH2/SH3 domains

The SH2/SH3 domains of PLC-γ1, while not essential for promoting catalytic activation of the enzyme, do appear to serve important regulatory functions that facilitate interaction of PLC-γ1 with activated receptor tyrosine kinases and structural target proteins through direct protein-protein interaction. This, in turn, permits the tyrosine phosphorylation of PLC-γ1, subsequent activation and hydrolytic cleavage of the preferred phospholipid substrate. The significance of the direct association of PLC-γ1 with activated growth factor receptors autophosphorylated on tyrosine residues is underscored by studies in which this receptor-PLC-γ1 association is prevented through mutation of the receptor's SH2-target binding site.

The tyrosine kinase receptor for NGF, trkA, has a tyrosine in position 785 that is a major autophosphorylation site and also is part of a motif homologous to the PLC-γ1 SH2 binding site found in other tyrosine kinase growth factor receptors. Construction and expression of trkA receptors containing a tyrosine-to-phenylalanine point mutation at position 785 results in kinase active receptors that fail to bind and form complexes with PLC-γ1. Furthermore, trkA mutant receptors are incapable of tyrosine phosphorylating PLC-γ1 in response to stimulation by NGF (Loeb *et al.*, 1994). A similar loss of SH2-mediated PLC-γ1 binding is observed when the tyrosine autophosphorylation site at position 992 of a truncated EGF receptor is mutated to phenylalanine (Vega *et al.*, 1992). Interestingly, in contrast to the trkA mutant receptor, the modified EGF receptor retains the ability to phosphorylate PLC-γ1, however, this phosphorylation fails to result in activation of the enzyme. Although it appears, in general, that tyrosine phosphorylation of PLC-γ1 is necessary for enzymatic activation, this finding involving the mutated EGF receptor suggests that both tyrosine phosphorylation of PLC-γ1 as well as SH2-mediated binding of PLC-γ1 to sites of tyrosine autophosphorylation on the EGF receptor are necessary for activation of PLC-γ1.

In addition to mediating protein-protein interactions, SH2/SH3 domains may play a role in regulating PLC-γ1 biologic activity. Recombinant peptides generated from cDNA encoding the Z region (SH2 and SH3 domains) of PLC-γ1 have an inhibitory effect on the catalytic activity of not only PLC-γ1, but also on the other members of the PLC isoform family (Homma and Takenawa, 1992). Perhaps most striking is the observation that the SH3 domain of PLC-γ1, not the phospholipase activity of the enzyme, is required for mediating the mitogenic properties of PLC-γ1 (Huang *et al.*, 1995). Microinjection of a catalytically inactive PLC-γ1 into quiescent fibroblasts has been shown to elicit a full mitogenic response. After mapping the capacity to induce mitogenesis to the Z domain (SH2 and SH3 motifs) it was observed that inactivation of the tyrosine binding properties of both SH2 domains had no effect on the mitogenic activity elicited by a Z-domain peptide. However, deletion of the SH3 domain from the Z-domain construct resulted in a complete loss of induced mitogenic activity.

Proposed mechanism for activation of PLC-γ1 through tyrosine phosphorylation

A plausible scheme by which stimulation of receptor tyrosine kinases results in activation of PLC-γ1 has been suggested by Rhee (1991). Under basal conditions, a vast proportion (>95%) of PLC-γ1 resides in the cytosol. Stimulation of the tyrosine kinase receptor by ligand binding to the extracellular domain elicits autophosphorylation of the receptor on tyrosine residues located in the intracellular, carboxy-terminal region. The receptor phosphotyrosine sites are recognized by the SH2 domains of PLC-γ1 which results in the formation of a receptor-PLC-γ1 complex. Subsequently, phosphorylation of PLC-γ1 tyrosine residues 771, 783 and 1254 is mediated by the activated receptor tyrosine kinase. Following phosphorylation of the three tyrosine sites, PLC-γ1 is presumed to be released from the receptor. While direct evidence is lacking, it is proposed that at this point the two SH2 domains in the tyrosine-phosphorylated PLC-γ1 could interact intramolecularly with PLC-γ1 tyrosine phosphates. Based on findings indicating that full activation of PLC-γ1 in NIH3T3 cells by PDGF requires phosphorylation of PLC-γ1 at tyrosines 783 and 1254, but not tyrosine 771 (Kim *et al.*, 1991), tyrosines 783 and 1254 appear most likely to be involved in the intramolecular SH2 interaction. As a consequence of the intramolecular interaction, it is speculated that a conformational change is elicited which makes available the SH3 domain permitting binding of PLC-γ1 to the membrane cytoskeleton. This, in turn, would bring the X and Y catalytic domains to the cytoplasmic surface of the cell membrane where the PLC substrate PIP$_2$ is located.

ACTIVATION OF PLC BY G PROTEINS

G Proteins

A family of G proteins functions to couple membrane bound receptors to a variety of effector moieties, such as adenylate cyclase and PLC (Gilman, 1987; Simon *et al.*, 1991; Rens-Domiano and Hamm, 1995). The G proteins are heterotrimers, composed of α, β and γ subunits. The specificity of function primarily resides in the α subunit. The α subunit binds guanine nucleotides and has GTPase activity. In the basal state, GDP is bound to the α-subunit. Following ligand activation of a G-coupled receptor, interaction of the receptor with the G protein results in exchange of GDP for GTP. Subsequently, the G protein dissociates into α-GTP complex and βγ heterodimer. The α-GTP complex or the βγ heterodimer may then interact with and regulate the activity of effector proteins. In most cases, it is believed that the α subunit is responsible for regulating the effector activity, but in a few instances the βγ dimer regulates the activity of effector proteins. Hydrolysis of GTP to GDP by the intrinsic GTPase activity of the α subunit returns the α subunit to the inactive α-GDP complex, which then reassociates with the βγ heterodimer to return the system to the basal state. Thus, GTP analogues can be used to activate G proteins and nonhydrolyzable GTP analogues, such as GTPγS, produce a sustained activation of the G protein. Conversely, GDP analogues can inhibit activation of G proteins by GTP analogues. Fluoride is another tool used study G proteins. Fluoride in the presence of aluminum forms AlF$_4$, which mimics the γ-phosphate group of GTP in the GDP-bound α-subunit. Thus, AlF$_4$ activates G proteins. Consequently, the use of GTP analogues, GDP analogues and fluoride, in the presence of aluminum, is often used as an investigative tool to determine whether G proteins play a role in coupling a receptor to an effector, such as PLC.

The involvement of G proteins in receptor-activated PLC activity was implicated by the following phenomenological observations.

1) In various membrane preparations and permeabilized cells, stable GTP analogues stimulate PLC-mediated hydrolysis of polyphosphoinositides and receptor-stimulated PLC activity is dependent on the presence of GTP or a GTP analogue (Litosch *et al.*, 1985; Cockcroft and Gomperts, 1985; Cockcroft *et al.*, 1987; Martin, 1989; Harden, 1992). In the presence of GTP analogs, the rate of PLC catalytic activity displays hysteresis, which is reduced in the presence of receptor agonists, consistent with the idea of G-proteins regulating PLC activity. Furthermore, the effect of GTP and its analogues can be mimicked by fluoride, in the presence of aluminum (Hepler and Harden, 1986; Bigay *et al.*, 1987).

2) In radioligand binding studies, receptors linked to hydrolysis of polyphospho-inositides display a high-affinity binding complex, which is not present when GTP was included in the binding reaction, suggesting that the receptors inter-act with a G protein (Evans *et al.*, 1985a,b; Harden, 1992).

3) Receptors linked to PLC stimulate GTPase activity in a number of tissues (Grandt *et al.*, 1986; Houslay *et al.*, 1986).

4) Pertussis toxin (PT), which inactivates some G proteins, blocks receptor-activated hydrolysis of polyphosphoinositides in a variety of receptor systems (Nakamura and Ui, 1985; Smith *et al.*, 1985; Perney and Miller, 1989).

PT does not attenuate receptor-activated PLC activity in all receptor systems in which guanine nucleotides regulate PLC activity, suggesting that there are PT-sensitive and PT-insensitive G proteins involved in activation of PLCs (Hepler and Harden, 1986; Ashkenazi *et al.*, 1989; Harden, 1992). Furthermore, guanine nucleotides do not always influence receptor-stimulated PLC activity, particularly in situations where the receptors do not possess the requisite seven-membrane spanning receptor structure for interaction with G proteins (Harden, 1992; Cockcroft and Thomas, 1992). Thus, there are at least two pathways by which receptors are coupled to PLCs, of which only one pathway involves G proteins. Recent studies with purified G protein subunits and specific PLC types have pro-vided insight into two mechanisms by which G proteins couple receptors to PLCs.

Regulation of PLC by Gα Subunits

Identification of the Gα subunits that activate PLC activity

A number of investigations were instrumental in identifying the G proteins that mediate receptor-stimulated PLC activity. Using GTP-γS-treated bovine liver mem-branes, Taylor *et al.* (1990) isolated a unique 42 kd G protein α-subunit that acti-vated PLC. This 42 kd protein was not recognized by a battery of antibodies against known α-subunits, indicating that a novel α-subunit is involved in activating PLCs. This novel 42 kd protein contained sequences that were identical to those found in α_q and α_{11}, two α subunits from the G$_q$ family of G proteins (Pang and Sternweis, 1990). Proof that the α-subunit that stimulates PLC is a member of the G$_q$ family came from studies with antibodies developed to specifically recognize α_q and α_{11}. Taylor *et al.* (1991a; 1991b) found that the α-subunit preparation isolated from liver membranes could actually be resolved into 42 kd and 43 kd proteins. Using these specific antibodies, the 42 kd protein was identified as α_q and the 43 kd protein was determined to be α_{11}. Sequencing of the proteins confirmed these des-ignations. The α subunit-PLC interaction is specific for α_q and α_{11}, as α_t, α_z, α_{i2} and α_o subunits are ineffective in stimulating hydrolysis of PIP$_2$ (Wu *et al.*, 1992).

Both the GTP-γS-bound forms of α_q and α_{11} are equally capable of stimulating PLC activity (Taylor *et al.*, 1991b). This is not surprising considering that α_q and α_{11} are similar with 88% sequence identity (Strathmann and Simon, 1990). Activation of PLC by α_q/α_{11} is dependent on the presence of Mg^{2+} and Ca^{2+} and the order of guanine nucleotide specificity for activation of PLC is GTP-γ-S > GppNHp > GMPPCP (Blank *et al.*, 1991). Activation of PLC activity by GTP-γS can be blocked by high concentrations of GDP-βS and excess $\beta\gamma$ subunits. Furthermore, activation of PLC by α_q/α_{11} is unaffected by PT. Considering that the α-subunits of the G_q family lack the appropriate structure for modification by PT (Strathmann and Simon, 1990), the lack of an effect by PT is consistent with the idea that the α_q/α_{11} subunits mediate the PT-insensitive stimulation of PLC activity.

Further studies on the characteristics of the interaction between α_q/α_{11} and PLC have shown that the α subunit-mediated increase in PLC activity is due to an increase in the catalytic activity of the enzyme and an increase in Ca^{2+} sensitivity (Litosch, 1991; Smrcka *et al.*, 1991). Not only do the α_q/α_{11} subunits affect the activity of PLC, but PLC also affects the activity of the α subunits. A functional interaction between PLC-β1 and α_q/α_{11} has been demonstrated by the ability of PLC to act as a GTPase stimulating protein (Berstein *et al.*, 1992a). When purified M1 cholinergic receptors and G_q/G_{11} are reconstituted in lipid vesicles, the addition of PLC-β1 results in a further stimulation of the GTPase activity associated with G_q/G_{11} that is at least 50-fold greater in the presence of PLC-β1. The GTPase activating protein is PLC-β1, as antibodies against this PLC blocks the GTPase activation. The increased GTPase activity is also specifically associated with G_q/G_{11}, since PLC-β1 is ineffective in stimulating the GTPase activity of G_o and G_s.

While it was clear that the G_q/G_{11} proteins can regulate PLC activity, it was still necessary to prove that these G proteins also are coupled to receptors that regulate PLC activity, to prove that the G_q/G_{11} proteins couple the receptors to PLCs. A number of studies demonstrated that Ca^{2+}-mobilizing receptors do couple with G_q/G_{11} proteins, as indicated by Ca^{2+}-mobilizing agonist-dependent stimulated binding of radiolabeled GTP analogues to α_q/α_{11} subunits and increased GTPase activity (Wange *et al.*, 1991; Berstein *et al.*, 1992b). Furthermore, addition of anti-serum raised to a C-terminal peptide common to α_q and α_{11}, which is thought to be involved in receptor coupling, specifically inhibits Ca^{2+}-mobilizing agonist-stimulated GTPase activity (Shenker, 1991) and PLC activity (Gutowski *et al.*, 1991; Aragay *et al.*, 1992), indicating that α_q/α_{11} subunits mediate receptor-stimulated PLC activity.

Confirmation that α_q/α_{11} proteins alone can couple the receptors to the PLCs was provided by Berstein *et al.* (1992b), who demonstrated that only three components are needed to reconstitute carbachol-stimulated PLC activity in phospholipid vesicles — M1 muscarinic receptor, G_q/G_{11} proteins and PLC-β1. In this system, the receptor is equally efficacious in stimulating binding of ^{32}P-GTP-azidoanilide to α_q and α_{11}. The time course and dose response for carbachol-stimulated ^{35}S-GTPγS binding correlate with those associated with α_q- and α_{11}-stimulated hydrolysis of ^{3}H-PIP$_2$, as would be expected if G_q/G_{11} coupled the muscarinic receptor to PLC.

Other α subunits of the G_q family are also able to couple receptors to PLCs. Agonist-stimulated PLC activity can be mediated by α_{14}, α_{15} and α_{16} (Lee *et al.*, 1992; Jhon *et al.*, 1993; Jiang *et al.*, 1994; Offermanns and Simon, 1995). The α_{14} subunit is approximately 81% identical with α_q, but unlike α_q and α_{11} which are ubiquitously expressed, α_{14} has a more limited tissue distribution (Wilkie *et al.*,

1991). The human α_{16} and the murine cognate α_{15} are about 85% identical to each other and have the highest degree of homology with α_q than with α subunits from other G protein families and thus are considered members of the G_q family (Wilkie *et al.*, 1991; Armatruda *et al.*, 1991). The α_{15} and α_{16} subunits are restricted to a subset of hematopoietic cells.

Thus, there are five α subunits from the G_q family that can couple the receptors to PLCs. Studies with interleukin 8 and anaphylatoxin C5a receptors have suggested that there may be selectivity in the interaction of the receptors with specific α subunits that activate PLCs, as interleukin 8 and C5a receptors selectively interact with α_{16}, but not with α_q and α_{11} (Armatruda *et al.*, 1993; Buhl *et al.*, 1993; Wu *et al.*, 1993c). There is also selectivity in the interaction of these α subunits with specific PLCs, as discussed below.

Identification of the PLCs that are regulated by Gα subunits

In several of the above studies, the α_q/α_{11}-mediated receptor-stimulated PIP_2 hydrolysis was seen using PLC-β1 as the phosphodiesterase in the reconstitution or expression system (Berstein *et al.*, 1991a, b; Blank *et al.*, 1991; Taylor *et al.*, 1991a, b), demonstrating that α_q and α_{11} subunits activate this specific PLC isozyme. In contrast, PLCs of the γ or δ type are not responsive to activated α_q or α_{11} (Blank *et al.*, 1991; Hepler *et al.*, 1993; Taylor *et al.*, 1991a). The PLCs responsive to α_q or α_{11} are limited to the β family of PLCs. Other PLC-β isozymes, in addition to PLC-β1, are also stimulated by α_q and α_{11}. However, there are differences in efficacies of stimulating the different PLC-βs, indicating some specificity in the interaction of the α subunits with the different PLC-β isozymes. Activated α_q and α_{11} subunits are strong activators of PLC-β3 and β1 (Hepler, *et al.*, 1993; Jhon *et al.*, 1993; Smrcka and Sternweis, 1993). In contrast, activated α_q and α_{11} only stimulate PLC-β2 to a small or negligible extent.

Similarly, studies with α_{14} and α_{16} have demonstrated that these subunits specifically activate PLC-βs. PLC-β1 is activated by α_{14} and α_{16} (Lee *et al.*, 1992; Jhon *et al.*, 1993; Kozasa *et al.*, 1993). PLC-β2 is only stimulated by α_{14} to a small extent, similar to what is seen with α_q and α_{11}. Jhon *et al.* (1993) and Kozasa *et al.* (1993) have shown that PLC-β2 is only stimulated to a small extent by α_{16}, similar to the level produced by α_q. In disagreement, Lee *et al.* (1992) have shown that α_{16} is a moderate stimulator of PLC-β2, less that what is seen with PLC-β1 but greater than the level that is stimulated by α_q. The differences in the reports on the responsiveness of PLC-β2 to α_{16} may be due in part to whether saturating or subsaturating levels of enzyme or α subunits are used. Thus, the extent of interaction of α_{16} with PLC-β2 still needs to be clarified.

The more recently identified PLC-β3 and PLC-β4 are also responsive to activated α subunits of the G_q class of G proteins. PLC-β3 has been shown to be responsive to activated α_{16}, almost to the same degree as is seen with PLC-β1 (Jhon *et al.*, 1993; Kozasa *et al.*, 1993). PLC-β4 is efficiently activated by α_q, α_{11}, α_{14} and α_{15} (Jiang *et al.*, 1994). Considering that α_{16} is about 85% identical to α_{15}, it is a surprising result that α_{16} does not couple well to PLC-β4.

Research has been done to identify the region in the PLC enzyme that is involved in the interaction with α_q. Using variant forms of PLC-β1, Wu *et al.* (1993a) found that deletions in the carboxy terminus of PLC-β1 lead to loss of α_q-stimulated PLC activity but not loss of intrinsic activity, suggesting that the carboxy terminus is needed for interaction with the α subunit. The importance of the carboxy-terminal region in interaction of PLC-β with α_q was further supported by studies on proteolyzed forms of PLC-β1 (Park *et al.*, 1993b). Considering that the PLC-βs are distinguished from PLCs of the γ or δ type by a long carboxy region following the Y domain and that only members of the β type are responsive to α subunits, it is not surprising that this region is involved in the interaction with G proteins. The region in the carboxy terminus of PLC-β1 critical for activation by α_q has been further localized to the region between amino acid residues 1030 to 1042 (Wu *et al.*, 1993a).

Regulation of PLC Activity by $\beta\gamma$ Complexes

Identification of G protein $\beta\gamma$ complexes as regulators of PLC activity

A series of experiments have clearly demonstrated that G protein $\beta\gamma$ subunits can activate PLC activity. Camps *et al.* (1992) demonstrated that $\beta\gamma$ complexes stimulate PLC obtained from HL-60 human promyeolocytes and neutrophils. This stimulation is not due to a contaminating presence of α subunits, as the addition of GDP-α from G_t, a G protein involved in visual transduction, inhibits the response in a dose-dependent manner. Similarly, Blank *et al.* (1992) found that GTPγS activates PLC in a system composed of PLCs from bovine liver cytosol and G proteins from liver plasma membranes. Purification of the α and $\beta\gamma$ subunits clearly demonstrated that it is the $\beta\gamma$ subunits that are the active component. The α subunits from a variety of G proteins, such as G_t, G_o and G_{11} inhibit the response to $\beta\gamma$ (Blank *et al.*, 1992; Boyer *et al.*, 1992; Camps *et al.*, 1992). The increased PLC activity is due to an increase in the maximal activity of the enzyme without an apparent change in the sensitivity to calcium (Boyer *et al.*, 1992). The concentration of $\beta\gamma$ subunits needed to half-maximally activate the $\beta\gamma$-sensitive PLC is greater than the concentration of GTPγS-activated α_q or α_{11} that is needed to half-maximally activate PLC-β1, suggesting that the source of the $\beta\gamma$ subunits stimulating PLC activity is not G_q or G_{11} (Blank *et al.*, 1992; Boyer *et al.*, 1992). A more abundant source of $\beta\gamma$ may be needed, such as G_o or G_i. Thus, it is possible that receptors that are coupled to G_o or G_i may stimulate PLC activity via the $\beta\gamma$ complexes. However, this hypothesis still needs to be proven before a physiological role for $\beta\gamma$ dimers in regulating PLC activity can be assigned.

The regulation of PLC by $\beta\gamma$ complexes is different than the regulation of adenylate cyclase by $\beta\gamma$ complexes. The $\beta\gamma$ complexes stimulate certain adenylate cyclases (Types II and IV) to produce the second messenger cAMP, but this activation requires the presence of α_s, the α subunit that couples to and activates adenylate cyclase (Tang and Gilman, 1991). Thus, the $\beta\gamma$ subunits potentiate the α_s-stimulated adenylate cyclase activity. In contrast, $\beta\gamma$-stimulated PLC activity does not require the presence of an α subunit and actually is often inhibited by α subunits (Boyer *et al.*, 1992; Camps *et al.*, 1992). Smrcka and Sternweis (1993) have suggested that the α and $\beta\gamma$ subunits bind to different sites on the PLCs and occupation of one site influences the effectiveness of the other site.

Thus far, there are five β subunits, that are 50–80% identical and seven diverse γ subunits that have been identified as being involved in making up the $\beta\gamma$ complexes in trimeric G proteins (Clapham and Neer, 1993). There is specificity in the interaction between β and γ subunits to form $\beta\gamma$ complexes, as not all β and γ subunits can interact with each other (Schmidt *et al.*, 1992). Not all combinations of β and γ subunits have been tested, but the $\beta_1\gamma_1$ and $\beta_1\gamma_2$ dimers have been shown to be effective in stimulating PLC-β_2 (Katz *et al.*, 1992). More recent studies have also found that $\beta_1\gamma_5$ and $\beta_2\gamma_5$, but not $\beta_2\gamma_1$, activate PLC-β_2 (Wu *et al.*, 1993b), suggesting that there is specificity in the type of $\beta\gamma$ subunits that can interact with PLCs.

Identification of PLCs regulated by G protein $\beta\gamma$ complexes

As with the α subunit-mediated regulation of PLCs, the $\beta\gamma$ mediated regulation of PLCs is specific for PLCs of the β family (Blank *et al.*, 1992; Park *et al.*, 1993a). The $\beta\gamma$ complexes do not significantly stimulate PLCs of the γ or δ families. PLC-$\beta3$ is much more responsive to $\beta\gamma$ dimers than the $\beta1$ or $\beta2$ isozymes. The rank order of efficacies for $\beta\gamma$-stimulated PLC activity is PLC-$\beta3$ > PLC-$\beta2$ > PLC-$\beta1$ (Park *et al.*, 1993a). This is in contrast to the α-stimulated PLC activity, which has an order of efficacies of PLC-$\beta1 \geq$ PLC-$\beta3$ >> PLC-$\beta2$. The half-maximal concentration of $\beta\gamma$ dimers needed to stimulate PLC-$\beta3$ is approximately 25–50 fold higher than that of α_q subunits needed to activate PLC-$\beta1$. Considering that the level of $\beta\gamma$ dimers needed to activate PLCs is much higher than the level of G_q proteins in cells, it is likely that activation of the less abundant G_q, G_{11}, G_{14}, or G_{16} will result only in activation of PLC-$\beta1$ or PLC-$\beta3$. However, in tissues in which there is an abundance of G_o and G_i, activation of these G proteins will result in high enough levels of $\beta\gamma$ to activate PLC-$\beta3$ or PLC-$\beta2$. The exact PLC isozyme stimulated will depend on which isozymes are present.

More recently, the activation of PLC-$\beta4$ has been examined. Similar to PLC-$\beta1$, PLC-$\beta2$ and PLC-$\beta3$, PLC-$\beta4$ is also activated by α_q, α_{11}, α_{14} and α_{15}. However, PLC-$\beta4$ is not activated by $\beta_1\gamma_1$, $\beta_1\gamma_2$, $\beta_1\gamma_3$, or $\beta_2\gamma_2$ (Jiang *et al.*, 1994), suggesting that PLC-$\beta4$ is not activated by $\beta\gamma$ dimers. Thus the PLC-$\beta4$ isozyme is different from the $\beta1$, $\beta2$ and $\beta3$ isozymes in that it can only be activated by α subunits.

Activation by PLC activity by α_q and $\beta\gamma$ can be additive or synergistic, suggesting that the different G protein subunits interact with PLCs at different sites on the enzyme (Smrcka and Sternweis, 1993; Wu *et al.*, 1993b). While a region in the carboxy terminal end of PLC-βs is involved in the interaction with α subunits, the site critical for interaction of the PLC-βs with $\beta\gamma$ dimers is a region in the amino terminus (Wu *et al.*, 1993b). The $\beta\gamma$ binding domain has only been broadly mapped to the region from the amino terminus to the end of the Y domain. Within in the initial 100 amino acids of the region is a pleckstrin homology (PH) domain (Inglese *et al.*, 1995). The PH domain was named after the first protein shown to contain this domain, pleckstrin. The PH domain has been postulated to be involved in the $\beta\gamma$ binding domain of various proteins, since it is involved in the $\beta\gamma$ binding domain in the β-adrenergic receptor kinase. Thus, the PH domain may be involved in binding of $\beta\gamma$ complexes to PLC-βs.

REGULATION OF PLC ACTIVITY BY SECOND MESSENGERS

Cross-talk Between Second Messenger Signalling Cascades

There exists substantial experimental evidence from a variety of studies indicating that the phosphoinositide, adenylate cyclase and guanylate cyclase signalling pathways are biochemically interconnected. First, the formation of cyclic AMP (cAMP) is affected by both $Ins(1,4,5)P_3$ and diacylglycerol, second messenger molecules that are generated by the PLC-mediated hydrolysis of PIP_2. Calcium released from vesicular nonmitochondrial intracellular stores by the action of $Ins(1,4,5)P_3$ forms a complex with calmodulin which, in turn, binds to selected forms of adenylate cyclase resulting in their activation. Diacylglycerol generated by the hydrolytic activity of PLC serves as an endogenous activator of protein kinase C (PKC). Activation of PKC does not uniquely enhance or attenuate intracellular levels of cAMP that result from a hormone-mediated increase in receptor-coupled adenylate cyclase activity but rather may be either facilitating or inhibiting depending on the cell type. In S49 lymphoma cells, stimulation of PKC phosphorylating activity as a consequence of phorbol ester treatment results in a doubling of isoproterenol-stimulated cAMP accumulation (Bell *et al.*, 1985). In contrast, stimulation of PKC by phorbol ester resulted in a desensitization of gonadotrophin-responsive adenylate cyclase in the MLTC-1 murine Leydig tumor cell line (Rebois and Patel, 1985). Secondly, increasing cAMP concentrations within the cell resulting in activation of cAMP-dependent protein kinase (PKA) has a similar dichotomous effect on the generation of $Ins(1,4,5)P_3$. It has been shown in a variety of cell and tissue types including glomerulose cells (Gilman, 1984), insulin-secreting islets (Zawalich *et al.*, 1988), kidney (Neylon and Summers, 1988), lymphocytes (Granja *et al.*, 1991, Alava *et al.*, 1992), neurotumor cells (McAtee and Dawson, 1989, Kim *et al.*, 1989), neutrophils (Kato *et al.*, 1986), platelets (Watson *et al.*, 1984, Yada *et al.*, 1989) and smooth muscle (Madison and Brown, 1988) that cAMP-mediated activation of PKA exerts an inhibitory influence on agonist-stimulated formation of $Ins(1,4,5)P_3$. Antithetically, it has been demonstrated in a variety of other cell types such as Balb/c 3T3 cells (Pittner and Fain, 1989), hepatocytes (Blackmore and Exton, 1986) and rat parotid acinar cells (Horn *et al.*, 1988) that pretreatment with agents eliciting an increase in cAMP will potentiate receptor-mediated formation of $Ins(1,4,5)P_3$. Finally, it has been demonstrated that the short-lived, highly reactive radical, nitric oxide, is an inhibitory modulator of PLC activity (Clementi *et al.*, 1995). Nitric oxide has been reported to function as a neuronal signalling molecule involved in a number of physiologic processes including neurotransmitter release, long-term potentiation and gene transcription. Nitric oxide inhibits agonist-dependent mobilization of intracellular calcium in neurosecretory cells by attenuating PLC hydrolytic activity through a biochemical process that appears to involve nitric oxide-mediated activation of soluble guanylyl cyclase, elevation of cyclic GMP (cGMP) concentrations and subsequent stimulation of cGMP-dependent protein kinase I. Implicit in the interplay between second messenger cascades is the thesis that activation of kinases and subsequent protein phosphorylation results in cross-talk regulation of the PLC biochemical pathway (Cockcroft and Thomas, 1992, Rhee and Choi, 1992, Rhee *et al.*, 1993).

Modulation of PLC Enzymatic Activity by Protein Phosphorylation

It is widely recognized that allosteric modification of proteins by phosphorylation represents a generalized biochemical mechanism for regulating enzymatic activity within the intact cell. The effect of phosphorylation on PLC activity may result in either activation or inhibition depending upon the amino acid specificity of the kinase involved. As previously mentioned, agonist-dependent activation of receptors which possess intrinsic tyrosine kinase activity or are coupled to cytosolic tyrosine kinases promote enhanced PLC activity. In contrast, activation of PKC, PKA or cGMP-dependent protein kinase I which phosphorylate substrates on serine and/or threonine amino acid residues results in inhibition of agonist-stimulated PLC activity. The specific substrates for these serine/threonine kinases are largely unknown. Potential targets include tyrosine kinase growth factor receptors, G proteins, nonreceptor protein tyrosine kinases, protein-tyrosine phosphatases and isozymes of PLC.

Phosphorylation of PLC by PKA

It has been shown for a variety of cell types including rat glioma C6Bu1 cells (Kim *et al.*, 1989), 3T3 cells (Olashaw *et al.*, 1990) and the human leukemic T cell, Jurkat (Granja *et al.*, 1991, Park *et al.*, 1992) that activation of PKA resulting from treatment of intact cells with forskolin, dibutryl cyclic AMP or cholera toxin results in enhanced phosphorylation of PLC-γ1. The site of phosphorylation has been identified as serine 1248. This PKA-mediated serine phosphorylation is specific for the PLC-γ1 isoform as neither PLC-β1 or PLC-δ1 were similarly phosphorylated. Phosphorylation of PLC-γ1 *in vitro* by PKA has also been demonstrated (Rhee *et al.*, 1990). Although prior treatment with agents that provoke an increase in cAMP concentrations have been observed to attenuate agonist-mediated activation of PLC and are associated with an enhanced phosphorylation of PLC-γ1 on serine residues, it has not been demonstrated *in vitro* that serine phosphorylation of the enzyme affects a change in enzymatic activity (Olashaw *et al.*, 1990, Granja *et al.*, 1991). A decrease in ligand-stimulated tyrosine phosphorylation of PLC-γ1 and inhibition of PLC activity is associated with a PKA-mediated increase in PLC-γ1 serine phosphorylation (Park *et al.*, 1992).

Phosphorylation of PLC by PKC

Evidence obtained from studies conducted with a number of different cell types including human Jurkat T-cells (Park *et al.*, 1992), WB liver epithelial cells (Huckle *et al.*, 1990) and the neuronal PC12 cell line (Fink and Guroff, 1990) indicates that prior activation of PKC will attenuate agonist-mediated stimulation of PLC. Treatment of Jurkat cells with phorbol myristic acid (PMA), a phorbol ester that activates PKC, results in phosphorylation of PLC-γ1 at serine 1248, the same amino acid residue phosphorylated by PKA. As is the case for PKA, a prior increase in PLC-γ1 serine phosphorylation elicited by activation of PKC results in diminished agonist-stimulated tyrosine phosphorylation of PLC-γ1, the consequence of which is an inhibition of enzymatic activity. The most likely explanation for this phosphorylation-dependent inhibition of agonist-stimulated PLC activity is that enhanced phosphorylation of PLC-γ1 at serine 1248 elicited by either PKC or PKA alters the substrate character of the protein causing it to become less favorable for protein tyrosine kinases or more accessible to the action of a tyrosine phosphatase. It is important to recognize that PKC-dependent inhibition of PLC-γ1 is not universally observed (Ryu *et al.*, 1990, Ward and Cantrell, 1990). Additionally, PKC-stimulated phosphorylation of PLC-γ1 in Jurkat cells (Park *et al.*, 1992) may represent an exceptional case rather than the rule (Ryu *et al.*, 1990)

PKC, but not PKA, induces phosphorylation of the PLC-β1 isozyme. Furthermore, it has been shown that treatment of PC12, C6Bu1 and NIH 3T3 cells with PMA elicits a PKC-dependent phosphorylation of serine residue-887 in PLC-β1, whereas phosphorylation of PLC-δ1 and PLC-γ1 are not affected (Ryu *et al.*, 1990). The *in vitro* phosphorylation of purified PLC-β1 by PKC also results in incorporation of phosphate at a single serine site (serine-887) (Ryu *et al.*, 1990). It remains to be established whether phosphorylation of PLC-β1 has any impact on its catalytic function.

Regulation of PLC Activity by PKC-mediated Phosphorylation of Growth Factor Receptor

In addition to evidence presented suggesting that direct serine phosphorylation of PLC isozymes by PKA (and to some extent PKC) attenuates agonist-stimulated enzymatic activity, it has also been shown for the PLC-γ1 isoform that prior activation of PKC (Huckle *et al.*, 1990) or PKA (Park *et al.*, 1992) is associated with an attenuation in receptor tyrosine kinase-mediated tyrosine phosphorylation of PLC-γ1. The apparent mechanism by which PKC-mediated inhibition of epidermal growth factor (EGF) receptor tyrosine kinase activity occurs is a PKC-dependent phosphorylation of the growth factor receptor.

The EGF receptor contains several distinctive sites of serine and threonine phosphorylation, namely, threonine[654], threonine[669], serine[1046] and serine[1047]. Phosphorylation of the EGF receptor at these sites is increased by treatment with agents that activate PKC including EGF (Hunter *et al.*, 1985; Davis and Czech, 1984), platelet-derived growth factor (PDGF) and phorbol esters (Schlessinger, 1988). The identification of threonine[654] as the principal phosphorylation site involved in modulating EGF receptor function was accomplished by expressing EGF receptors with single point mutations or multiple mutations in Chinese hamster ovary cells (Countaway *et al.*, 1990) or NIH-3T3 cells (Decker, *et al.*, 1990). The major sites of serine and threonine phosphorylation were replaced with alanine residues. Inhibition of EGF receptor tyrosine kinase activity by PKC-mediated phosphorylation was not observed for any of the receptor mutants that lacked threonine[654]. Mutations at any of the other sites failed to alter phorbol ester-dependent inhibition of EGF receptor tyrosine phosphorylating activity. Receptors with alanine[654] substitutions remain functional as is evidenced by the fact that stimulation with EGF results in receptor tyrosine autophosphorylation as well as tyrosine phosphorylation of a number of additional substrates including PLC-γ1.

Details characterizing what appears to be an autoregulatory feedback mechanism have been provided for a PKC-mediated inhibition of EGF-stimulated tyrosine phosphorylation of PLC-γ1 and subsequent enzymatic activation. Addition of EGF to a diversity of cell types stimulates tyrosine phosphorylation of PLC-γ1 (Wahl *et al.*, 1989; Margolis *et al.*, 1989; Meisenhelder *et al.*, 1989) which is accompanied by enhanced phosphoinositide turnover (Pike and Eakes, 1987; Wahl *et al.*, 1987; Hepler *et al.*, 1987; Johnson and Garrison, 1987). Consequently, increases in levels of diacylglycerol and intracellular Ca^{2+} follow which, in turn, may result in activation of PKC. Down-regulation of phorbol ester-responsive PKC activity by prolonged treatment with PMA results in potentiation of both EGF-stimulated PLC-γ1 tyrosine phosphorylation and PI hydrolysis (Huckle *et al.*, 1990) further implicating PKC as a mediator of feedback inhibition.

TOXINS

In the process of trying to understand the mechanism of action of a variety of toxins, it has become apparent that numerous toxins exert some of their actions on the phosphoinositide second messenger system. Occasionally, when the action of a toxin on the phosphoinositide system has been elucidated, the toxin then has become a useful tool for examining receptor-coupled PIP_2 hydrolysis. For example, pertussis toxin, through its inhibitory effect on receptor-stimulated PIP_2 hydrolysis, was one tool used to indicate the involvement of PT-sensitive G proteins in coupling receptors to PLCs in certain systems. The actions of the toxins can be directly on the phosphoinositide signal transduction system, such as on the G proteins, or indirectly on the phosphoinositide system, such as on intracellular calcium levels.

The sources of these toxins are varied. Numerous toxins are secreted from bacteria and are referred to as exotoxins. Within the larger group of bacterial secreted toxins, there are toxins with inherent PLC activity or toxins that directly affect the activity of G proteins. In addition, there are a number of bacteria-derived toxins lacking inherent PLC activity that are capable of stimulating PLC-mediated PIP_2 hydrolysis in eukaryotic cells by creating ion permeable pores in the plasma membrane as well as inducing tyrosine phosphorylation mediated by receptor-coupled tyrosine kinases. Still other toxins eliciting activation of PLC have been isolated from marine dinoflagellates as well as from the venom of bees, wasps, spiders and snakes. As might be predicted from the variety of sources from which these toxins originate, the mechanisms by which activation of PLC is achieved are also quite variable. The breadth of physiologic effects resulting from the action of these toxins ranges from cytolysis and hemolysis to regulation of hormone release and modulation of neuronal function.

Bacterial Toxins That ADP-ribosylate G Proteins

Pertussis toxin

PT is a bacterial exotoxin produced by *Bordetella pertussis*, with a molecular mass of approximately 120 kDa (Ui, 1985; Sekura and Zhang, 1985; Ui, 1990; Yamane and Fung, 1993). The toxin is a hexameric protein composed of one S_1, one S_2, one S_3, two S_4 and one S_5 subunits. According to the A(Active)-B(Binding) model, the S_1 subunit is the A-protomer and subunits S_2 through S_5 comprise the B-oligomer. The A-protomer has ADP-ribosyltransferase activity and the B-oligomer functions to attach the toxin to glycoproteins on the cell surface. The minimal carbohydrate structure for the PT cell receptor is a branched mannose core with N-acetyl glucosamine attached. Attachment to this cell surface receptor is a prerequisite for entry of the A-protomer into the cell. Once inside the cell, the A-protomer undergoes an activation step that involves reduced glutathione and can be mimicked in the test tube with dithiothreitol. The activation process involves liberation of the A-protomer from the toxin and reductive cleavage of disulfide bonds in the A-subunit. The resulting A-protomer is then active and catalyzes the transfer ADP-ribose from endogenous NAD^+ to acceptor proteins.

Certain G protein α-subunits are ADP-ribosylated by PT at a cysteine residue four amino acid residues from the carboxy-terminus (West *et al.*, 1985; Bourne *et al.*, 1985; Ui, 1990). Early studies demonstrated that PT attenuates receptor-mediated inhibition of adenylate cyclase activity by ADP-ribosylating the α-subunit of G_i, the G protein involved in inhibition of adenylate cyclase activity (Bokoch *et al.*, 1983; Bourne *et al.*, 1985; Reisine, 1990). PT also ADP-ribosylates G_t or transducin, the G protein involved in visual transduction (Manning *et al.*, 1984, Bourne *et al.*, 1985) and G_o, a G protein that is abundant in the brain (Reisine, 1990; Yamane and Fung, 1993). The conformation of the G proteins that is preferentially covalently modified with ADP-ribose is the form of the G protein that is uncoupled from the R (Ui, 1984; Ribeiro-Neto *et al.*, 1985). Thus, the result is that PT via ADP-ribosylation of the α-subunits causes uncoupling of the G protein from the receptor, thereby interfering with signal transduction from the receptor to the effector proteins, such as adenylate cyclase. Hence, inhibition of signal transduction by PT is often used as an indication that a PT-sensitive G protein, such as G_o and G_i, is involved in the signal transduction mechanism.

Based on studies in which PT attenuated receptor-activated PLC activity, it was postulated that G_i or G_o was the PT-sensitive G protein involved in coupling the receptor to PLC (Nakamura and Ui, 1985; Smith *et al.*, 1985; Perney and Miller, 1989). Unambiguous identification of the G protein involved in PT-sensitive receptor-stimulated PLC activity has been difficult to obtain and thus only a few studies demonstrate that G_o or G_i couple the receptor to PLC. An early report suggested that addition of G_i and G_o to pertussis-toxin treated HL60 membranes restored PIP_2 hydrolysis in response to the chemotactic peptide f-met-leu-phe (FMLP) (Kikuchi *et al.* 1986). However, the level of PLC activities observed in this study was low and high levels of G proteins were required. Another study demonstrated that a muscarinic agonist-mediated IP_3-stimulated Cl⁻ current in *Xenopus* oocytes could be stimulated by injection of activated α_o (Moriarty *et al.*, 1990), suggesting that α_o has the potential to activate PLC, but it does not necessarily prove that α_o normally regulates PLC activity.

More recently, Quick *et al.* (1994) expressed serotonin 1c (5HT1c) and thyrotropin-releasing hormone (TRH) receptors in *Xenopus* oocytes and assessed functional coupling of receptor/G protein to PLC by again measuring an IP_3-mediated Cl⁻ current. The 5HT1c-stimulated current, which was inhibited by PT, was blocked 66% by injection of antisense oligonucleotides to the PT-sensitive α_o, whereas the TRH response, which was PT-insensitive, was only reduced 23%. In contrast, the TRH-stimulated current was inhibited approximately 60% by the injection of antisense oligonucleotides to the PT-insensitive α_q subunit, whereas the 5HT response was only reduced by approximately 28%. These studies indicate that the 5HT-1c and the TRH receptors were preferentially coupled to G_o and G_q respectively and the PT-sensitive response may be correlated with the involvement of G_o. Further studies were performed with coexpression of the receptors with different α-subunits. The 5HT- and TRH-stimulated Cl⁻ current was inhibited by PT, only when α_o was the α-subunit expressed, providing further evidence that the PT-sensitive response may involve the G_o type of G protein.

There are also a few studies implicating the involvement of the PT-sensitive G_i in coupling the receptor to PLC. For example, α_{i3} couples the m2 muscarinic receptors to PLC in a PT-sensitive manner (Hunt *et al.*, 1994). Replacement of cysteine[351], which is the site of PT-dependent ADP-ribosylation, with serine results in a α_{i3} subunit which is PT-insensitive. In the presence of this PT-insensitive α_{i3} subunit, carbachol stimulates hydrolysis of PIP_2 in a PT-insensitive manner, suggesting that G_{i3} mediates PT-sensitive muscarinic agonist-stimulated PLC activity.

Considering that the inhibitory effect of PT has been observed in systems where the receptor is thought to couple to G proteins that are not PT substrates (Shayman *et al.* 1987; Harden, 1992) and that the studies clearly indicating the involvement of specific PT-sensitive α-subunits in activation of PLCs are limited, an additional mechanism for the inhibitory effect of PT has been suggested (Harden, 1992; Exton, 1994). The effect of PT on PLC activity may be mediated by G protein $\beta\gamma$ complexes. In systems where the $\beta\gamma$-subunits activate PLC activity, the effect of PT may be to reduce the level of $\beta\gamma$-subunits. PT-mediated ADP-ribosylation of G proteins (G_i or G_o) results in increased association of $\beta\gamma$-subunits to the ADP-ribosylated α-subunits in the heterotrimeric form, thereby reducing the level of free $\beta\gamma$-subunits that would be available to stimulate PLC activity. In support of this hypothesis, is a study by Katz *et al.* (1992). In this study, M2 muscarinic receptors, transfected into COS-7 cells, did not stimulate PLC activity when the cells were only transfected with the α_i subunit. The β and γ subunits were also needed to stimulate PLC activity and this $\beta\gamma$-stimulated activity was inhibited by PT.

Cholera toxin

Cholera toxin (CT) is a bacterial exotoxin (M_r, 85,000) produced by *Vibrio cholera* (Ribeiro-Neto *et al.*, 1985; Fishman, 1990; Yamane and Fung, 1993). Similar to PT, CT is also a hexameric protein that can be divided into A and B protomers. The A protomer is composed of A_1 and A_2 polypeptide chains and the B protomer is comprised of 5 identical polypeptides. The A protomer, in the A_1 peptide, has ADP-ribosyltransferase activity and the B protomer binds the toxin to gangliosides (GM1) in the cell membrane and is required for penetration of the toxin into the cell. Once inside the cell, reduction and activation leads to release of the A_1 subunit from the holotoxin. The A_1 subunit can then catalyze ADP-ribosylation of proteins using NAD^+ as a cosubstrate.

In the presence of Mg^{2+} and GTP, CT ADP-ribosylates the α-subunits of G_s and G_t at an arginine residue in the carboxy terminus (Ribeiro-Neto *et al.*, 1985; Ribeiro-Neto *et al.*, 1987; Fishman, 1990; Yamane and Fung, 1993). Since the G protein in the presence of Mg_{2+} and GTP is the preferred substrate, the preferred substrate is the free active GTP-α subunit. The result of this covalent modification is inhibition of the intrinsic GTPase activity associated with the α-subunit, resulting in persistent activation of the G protein. In the case of G_s, which couples to adenylate cyclase, CT causes a persistent activation of adenylate cyclase and formation of cAMP.

A few reports have indicated an effect of CT on the phosphoinositide system. In certain cells, CT decreases agonist-stimulated inositol phosphate production (Kim *et al.*, 1990; Barnes and Conn, 1993). This effect can be mimicked by the addition of cAMP analogues or adenylate cyclase activators, suggesting that the effect of CT is due to increased formation of cAMP. The increase in cAMP levels would result in increased activation of PKA. Considering that PLC, in particular PLC-γ1, is a substrate for PKA (Kim *et al.*, 1989; Olashaw *et al.*, 1990; Granja *et al.*, 1991; Park *et al.*, 1992), it is likely that the inhibitory effect of CT on inositol phosphate formation is due to PKA-mediated phosphorylation of PLC. In fact, CT is an important tool that has been used to investigate PKA-mediated phosphorylation of PLC-γ1 in cells (Kim *et al.*, 1990; Olashaw *et al.*, 1990).

Bacterial Phospholipase C

The subject of bacterial phospholipase C enzymes has recently been reviewed in considerable detail (Titball, 1993). These enzymes have been isolated from a number of gram-positive and gram-negative bacteria. Classified on the basis of their characteristic enzymatic activity which is to mediate hydrolysis of the ester bond located in the head group of intact glycerophospholipids, bacterial PLCs are uniformly single polypeptide proteins. Analysis of these peptides reveals a structural relatedness. In instances where the deduced amino acid sequence is known they have been found to contain typical signal sequences facilitating their secretion. Although the normal function of bacterial PLCs may be to generate stores of phosphate, these peptides are also described by their toxic character. Cytolysis appears to represent an important mechanism by which toxic effects are elicited and which contributes to the pathogenesis of disease. In contrast to a number of bacterial toxins, PLCs do not require internalization in order to manifest toxicity which results from direct interaction of the peptides with membrane phospholipids.

Hydrolysis of phosphatidylinositol by bacterial phospholipase C

Within the greater bacterial phospholipase C family which exhibit broad substrate specificity for phospholipids, there is a subgroup of bacterial peptides capable of hydrolyzing phosphoinositides. These PLCs derived from *Bacillus cereus, Bacillus thuringiensis and Listeria monocytogenes* demonstrate considerable similarity in their deduced amino acid sequence, particularly in the N-terminal region (Henner *et al.*, 1988; Leimeister-Wachter *et al.*, 1991; Mengaud *et al.*, 1991). The molecular mass of these proteins ranges from 20,000 to 35,000 daltons. This subgroup of peptides, along with PLC from *Clostridium novyi*, are collectively designated as type I enzymes, capable of hydrolyzing phosphatidylinositol phosphates without being membrane associated (Camilli *et al.*, 1991; Mengaud *et al.*, 1991). In addition to their phosphoinositide hydrolytic activity, this group of bacterial PLCs is also capable of hydrolyzing phosphatidylinositol-glycan-ethanolamine (PI-glycan) which serves as an anchor for a number of eukaryotic membrane proteins (Camilli *et al.*, 1991; Kominami *et al.*, 1985; Mengaud *et al.*, 1991). In contrast to members of the bacterial phospholipase C family which hydrolyze lipids other than the phosphoinositides, the phosphoinositide-specific PLCs do not require divalent cations for activation (Kuppe *et al.*, 1989; Mengaud *et al.*, 1991). Aside from these limited characteristics, little else is known about the structure-function relationships in this group of enzymes.

The capacity for bacterial PLCs to elicit an increase in the levels of IP_3 within the intact eukaryotic cell has not been demonstrated, however, it is reasonable to assume that this could occur were the bacterial PI-PLC able to gain access to the cytoplasmic surface of the plasma membrane. Release of anchored cell membrane proteins resulting from bacterial PLC-mediated hydrolysis of PI-glycan appears to represent an important property of these enzymes. Cleavage of the PI-glycan anchor by PI-specific bacterial PLCs results in the liberation of a number of membrane-bound enzymes including alkaline phosphatase, 5′-nucleotidase, alkaline phosphodiesterase I and acetylcholinesterase (Ikezawa, 1986). An increase in blood alkaline phosphatase levels has been reported to occur following an intravenous administration of *Bacillus cereus* PLC indicating that members of this group of bacterial PLCs elicit comparable effects both *in vivo* and *in vitro*. The significance of bacterial PLC-mediated cleavage of PI-glycan and release of membrane-associated proteins remains to be established, however, it is worth mentioning in this instance that eukaryotic alkaline phosphatase released by the hydrolytic activity of bacterial PLCs could play a role in the phosphate-scavaging pathway of the bacterium. This is particularly important as the level of inorganic phosphate in the blood is insufficient to support the growth of many bacteria. In addition, it has been reported that the bacterial PLCs are capable of influencing the growth of cultured cells (Ikezawa, 1986).

The interaction of bacterial PLCs with membranes establishes these enzymes as an important investigative tool (Titball, 1993). Their enzymatic activity permits utilization of the bacterial PLCs as probes to examine membrane phospholipid composition. Furthermore, the phospholipid-hydrolyzing action of these enzymes defines the utility of bacterial PLCs as important reagents capable of mimicking the effects of eukaryotic PLCs on cell metabolism. Finally, accumulation of information detailing the precise mechanisms involved in bacterial PLC-membrane interactions could facilitate development of a strategy for the delivery of membrane active drugs. In contrast to a number of bacterial toxins, PLCs do not require internalization in order to exert their toxic effects. Consequently, it has been proposed that linking bacterial PLC with targeting antibodies might permit the development of an active cytotoxic agent that could be therapeutically efficacious.

Bacterial Toxins Lacking Intrinsic PLC Activity That Induce Phosphoinositide Metabolism

Selected staphylococcal exotoxins represent examples of secreted bacterial toxins capable of eliciting enhanced phosphatidylinositol metabolism in eukaryotic cells through mechanisms not dependent upon intrinsic PLC hydrolytic activity. Instead, alterations in membrane ion permeability as well as stimulation of tyrosine phosphorylation contribute to the activation of PLC.

Staphylococcus aureus α-toxin

Staphylococcus aureus α-toxin is a 33,000 dalton exotoxin secreted by pathogenic staphylococci (Rogolsky, 1979) that possesses dermanecrotic activity and is lethal in animals. The toxicity of this exotoxin is due, in large measure, to interaction of the toxin with the cell membrane resulting in perturbation of lipid bilayer structural organization resulting in cytolysis (Thelestam and Bloomquist, 1988). Subsequent to binding of *S. aureus* α-toxin to artificial or biological membranes, a self-oligomerization process occurs creating ring structures that represent transmembrane pores permeable to ions and small metabolites (Fussle *et al.*, 1981; Menestrina, 1988). It has been shown that addition of subcytotoxic concentrations of *S. aureus* α-toxin to undifferentiated PC12 cells results in an influx of calcium and activation of PI-specific PLC as indicated by enhanced phosphatidylinositol turnover (Fink *et al.*, 1989). It is likely that alteration of membrane structure, accompanied by formation of staphylococcal α-toxin pores, facilitate calcium influx resulting in activation of PLC.

Staphylococcal superantigens: toxic shock syndrome toxin-1

Staphylococcal superantigens define a family of bacterial exotoxins with the capacity to bind major histocompatibility complex (MHC) class II molecules resulting in T-cell activation (Fischer *et al.*, 1989; Fraser, 1989; Mollick *et al.*, 1989; Scholl *et al.*, 1989; Kappler *et al.*, 1989). These exotoxins are also able to act directly on monocytes inducing the release of interleukin-1 and tumor necrosis factor as a consequence of transcriptional activation of the corresponding genes (Parsonnet *et al.*, 1985; Parsonnet and Gillis, 1988; Trede *et al.*, 1991). Together, these observations suggest that biologic signals elicited by staphylococcal superantigens are initiated and transduced through MHC class II molecules. Addition of the staphylococcal superantigen toxic shock syndrome toxin-1 (TSST-1) to human peripheral blood monocytes elicits tyrosine phosphorylation of the PLC-γ1 isozyme which, in turn, is associated with enhanced phosphoinositide hydrolysis (Trede, *et al.*, 1994). Stimulation of PLC hydrolytic activity by TSST-1 involves biochemical mechanisms analogous to those utilized by peptide growth factors, namely ligand-induced activation of cognate growth factor receptor tyrosine kinase activity. Increased accumulation of inositol phosphates elicited by TSST-1 in monocytes can be blocked by prior exposure of the cells to inhibitors of protein tyrosine kinases. The signaling events initiated by interaction of TSST-1 with MHC class II molecules appear to depend on activation of receptor coupled, *src*-related protein tyrosine kinases *fgr* and *hck* (Morio *et al.*, 1994).

Non-bacterial Toxins That Regulate Phospholipase C Activity

There are a number of non-bacterial toxins obtained from a diversity of natural sources including marine protozoans, stinging insects such as wasps and bees, as well as venomous spiders and snakes that are capable of regulating the catalytic activity of phospholipase C. Not surprisingly, the putative mechanisms by which these toxins exert their effect on PLC activity are as diverse in their character as the sources from which the toxins are derived. Frequently, the availability of extracellular calcium is an necessary requirement for toxin-mediated stimulation of PLC activity.

Toxin-induced stimulation of PLC mediated through calcium channels

Maitotoxin, a water-soluble polyether terpenoid toxin derived from the marine protozoal dinoflagellate *Gambierdiscus toxicus* (Takahashi *et al.*, 1982), is a prototypic representative of the class of toxins capable of effecting PLC activity through activation of select calcium channels. This potent marine toxin elicits a number physiologic responses including stimulation of calcium uptake (Ohizumi and Yasumoto, 1983b; Freedman *et al.*, 1984; Login *et al.*, 1985; Lebrun *et al.*, 1987; Sladeczek *et al.*, 1988), enhanced neurotransmitter and hormone release (Takahashi *et al.*, 1982; Schettini *et al.*, 1984; Shalaby *et al.*, 1986; Lebrun *et al.*, 1987; Gusovsky *et al.*, 1988), evocation of smooth muscle and cardiac muscle contraction (Ohizumi and Yasumoto, 1983a; Ohizumi *et al.*, 1983; Legrand and Bagnis, 1984; Kobayashi *et al.*, 1986) and increased phosphoinositide breakdown (Gusovsky *et al.*, 1989; Gusovsky *et al.*, 1990; Choi *et al.*, 1990). The precise mechanism(s) by which maitotoxin exerts its biologic effects remain to be determined, however, it is certain that calcium plays a fundamental role in mediating these effects. In fact, every action of maitotoxin on cellular homeostasis including stimulation of PLC-mediated PIP_2 hydrolysis is dependent upon the presence of extracellular calcium. The biologic activities of maitotoxin are consistently associated with increased cytosolic free calcium levels. Stimulation of calcium influx by maitotoxin has been shown to be antagonized by a number of organic voltage-dependent L-type channel blockers (nifedipine, verapamil, diltiazem, nicardipine) and inorganic calcium channel blockers (cobalt, manganese, cadmium) suggesting that calcium channels are activated by maitotoxin (Choi *et al.*, 1990; Meucci *et al.*, 1992).

In view of the fact that enzymatic activity of PLC is calcium-sensitive, it seemed reasonable to propose that activation of PLC by maitotoxin resulted from stimulation of calcium channels. A preponderance of evidence suggests this is not the case. While such activation appears to be involved in maitotoxin-stimulated neurotransmitter and hormone release, as well as contraction of cardiac and smooth muscle, increased phosphoinositide breakdown stimulated by maitotoxin is not inhibited by blockade of voltage-dependent calcium channels. Additionally, the stimulatory effect of maitotoxin on PLC activity occurs at concentrations lower than those required for calcium channel activation (Berta *et al.*, 1986; Gusovsky *et al.*, 1988). Consequently, alternative mechanisms for maitotoxin-mediated increases in PLC activity have been suggested. Maitotoxin has not been shown to have a direct effect on PLC enzymatic activity in membrane preparations (Gusovsky *et al.*, 1987; Gusovsky *et al.*, 1990) nor has maitotoxin been demonstrated to possess pore-forming or ionophoric activity (Takahashi *et al.*, 1983). The highly polar character of the molecule, as well as its large molecular weight (3424 daltons), make it unlikely that maitotoxin even has the potential to penetrate the plasma membrane of cells. Additional evidence arguing against an ionophore-like action for maitotoxin derives from studies demonstrating that calcium ionophores such as ionomycin and A23187 fail to elicit stimulation of PLC hydrolytic activity to a level comparable to that observed for maitotoxin (Gusovsky *et al.*, 1989). It has been proposed that maitotoxin stimulation of PLC occurs as a consequence of direct activation of receptor-operated calcium channels (Choi *et al.*, 1990; Soergel *et al.*, 1992), however, this thesis has been recently challenged (Musgrave *et al.*, 1994).

Sarafotoxin and α-latrotoxin represent two additional examples of biologically active peptides that, like maitotoxin, regulate PLC activity through their effects on calcium channels. Sarafotoxin is a snake cardiotoxic, 21 amino acid peptide isolated from the venom of the burrowing asp, *Atractaspis engaddensis*. Cloned and sequenced, the nucleotide sequence encoding the mature peptide region exhibits considerable homology with members of the mammalian endothelin, vasoconstrictor peptide family (Takasaki *et al.*, 1992). As a consequence of their sequence identity, sarafotoxin and endothelin share significant functional and structural similarities (Takasaki *et al.*, 1988; Kloog *et al.*, 1988, Sokolovsky, 1992; Kochva *et al.*,

1993). Sarafotoxin binds with high affinity to the same homogeneous binding sites as members of the endothelin peptide family (Lin *et al.*, 1991). Receptor binding of sarafotoxin elicits hydrolysis of phosphoinositides in a variety of cell types including cultured tracheal smooth muscle cells (Yang *et al.*, 1994), pituitary cells (Lewy *et al.*, 1992), astrocytes, C-6 glioma and cerebellar granule cells (Lin *et al.*, 1989; 1990; 1991; MacCumber *et al.*, 1990; Marsault *et al.*, 1990). The mechanisms involved in mediating sarafotoxin-induced PLC activity are similar to those observed for maitotoxin-elicited activation of PLC. Sarafotoxin-stimulated phosphoinositide hydrolysis is dependent on the presence of extracellular calcium and is not inhibited by organic or inorganic calcium channel blockers (Lin *et al.*, 1991), suggesting that sarafotoxin activates receptor-operated calcium channels.

α-Latrotoxin is a high molecular weight protein isolated from black widow spider venom that possesses potent neurotoxic activity (Meldolesi *et al.*, 1986). In addition to inducing extensive neurotransmitter release, α-latrotoxin has been shown to markedly stimulate phosphoinositide hydrolysis in a manner completely dependent on the presence of extracellular calcium (Vicentini and Meldolesi, 1989; Rosenthal *et al.*, 1990). This dependence on extracellular calcium for toxin-induced enzymatic activation of PLC is reminiscent of the effects observed for both maitotoxin and sarafotoxin. The action of α-latrotoxin is initiated by binding to an oligomeric high affinity binding site that appears to be comprised of a family of related high molecular weight polypeptides complexed to a low molecular weight protein (Petrenko *et al.*, 1993). In contrast to the receptor-operated calcium channels that are stimulated by maitotoxin and sarafotoxin, binding of α-latrotoxin elicits the opening of a nonspecific cation channel resulting in profound calcium influx (Wanke *et al.*, 1986; Petrenko *et al.*, 1990). These cationic channels, which may be blocked by the trivalent cation, lanthanum, appear to be comprised of α-latrotoxin molecules associated into clusters that insert into the membrane following binding to high-affinity receptor sites (Filippov *et al.*, 1994). Once established, the persistently open channel permits calcium influx resulting in a sustained rise in intracellular calcium that is believed to be responsible for the activation of a calcium-sensitive PLC rather than a direct coupling of the α-latrotoxin receptor to the PLC enzyme (Rosenthal *et al.*, 1990).

Involvement of G proteins in toxin-mediated effects on PLC hydrolytic activity

Mastoparan is an amphiphilic hymenopterid tetradecapeptide isolated from wasp venom (*Vespula lewisi*). Originally recognized and named for its ability to stimulate release of histamine from rat peritoneal mast cells (Hirai *et al.*, 1979), mastoparan acts as a secretagogue on a variety of cells stimulating release of catecholamines from bovine adrenal chromaffin cells (Wilson, 1989), secretion of serotonin from platelets (Ozaki *et al.*, 1990), release of insulin and glucagon from rat pancreatic islets (Komatsu *et al.*, 1992) as well as secretion of prolactin from cultured rat anterior pituitary cells (Mau *et al.*, 1994). Mechanistically, mastoparan-induced secretion appears to involve, in part, activation of PLC and an associated hydrolysis of phosphoinositides. Mastoparan-stimulated phosphoinositide metabolism has been observed in a variety of cell types including rat pheochromocytoma cells (Choi *et al.*, 1992), human polymorphonuclear leukocytes (Perianin and Snyderman, 1989), HL-60 cells (Gusovsky *et al.*, 1991) and NIH3T3 fibroblasts (Shin *et al.*, 1994). It has been shown that mastoparan directly activates G proteins (Higashijima *et al.*, 1990) suggesting a possible mechanism for the activation of PLC. Mastoparan is capable of altering membrane structure and function by the formation of ion channels that facilitate the influx of calcium, however, in comparison with maitotoxin this effect is less robust (Choi, *et al.*, 1992). It is unlikely that channel-mediated calcium influx is responsible for PLC activation as maximal stimulation of the enzyme occurs at concentrations of mastoparan considerably below those eliciting maximal calcium influx.

Available experimental evidence suggests that stimulation of PLC by mastoparan is mediated primarily by guanine nucleotide-binding proteins (G proteins). It has been reported that mastoparan stimulates the binding of GTP to purified G_0 and/ or G_i in phospholipid vesicles (Higashijima *et al.*, 1988). Activation of G proteins results in hydrolysis of GTP, consequently, the increased GTPase activity is indicative of G protein activation. Mastoparan has been shown to enhance the GTPase activity and rate of nucleotide exchange for several purified G proteins including G_0 and G_i (Higashijima *et al.*, 1988; Rhoden and Douglas, 1994). Furthermore, pertussis toxin-mediated ADP-ribosylation of G proteins attenuates mastoparan-stimulated binding of GTP to G_0 (Higashijima *et al.*, 1988) and diminishes stimulation of phosphoinositide metabolism elicited by mastoparan in HL-60 cells (Gusovsky *et al.*, 1991). Involvement of a pertussis-toxin sensitive G-protein in the mastoparan-mediated increase in phosphoinositide hydrolysis observed in HL-60 cells may represent a relatively unique situation. In a variety of other cell types, including PC12 cells (Choi *et al.*, 1992) and cultured rat anterior pituitary cells (Mau *et al.*, 1994), pertussis toxin is without effect on the stimulation of phosphoinositide hydrolysis by mastoparan. In addition to directly activating G proteins, mastoparan also appears capable of stimulating G protein activity indirectly through activation of nucleoside diphosphate kinase. This enzyme, in conjunction with ATP, is reported to be involved in converting GDP-bound G proteins into the active GTP-bound form by catalyzing the conversion of GDP to GTP (Kikkawa *et al.*, 1990). Mastoparan elicits activation of nucleoside diphosphate kinase in reconstituted as well as intact cellular systems (Kikkawa *et al.*, 1992; Klinker *et al.*, 1994).

Activation of PLC hydrolytic activity by mastoparan is not an effect without exception. Treatment of human neuroblastoma and astrocytoma cells with mastoparan results in inhibition of phosphoinositide hydrolysis (Wojcikiewicz and Nahorski, 1989; Nakahata *et al.*, 1990; Nakahata *et al.*, 1994). The inhibitory influence of mastoparan on PLC activity is not diminished by pretreatment of pertussis toxin suggesting involvement of a pertussis toxin-insensitive G protein. This contradictory observation indicates that caution need be exercised when utilizing bioactive toxins to dissect biochemical pathways associated with activation and regulation of PLC. Tissue and cell-specific differences are likely to impact the type of biochemical effect elicited by a particular toxin.

CONCLUDING REMARKS

Efficient transduction of signals across the plasma membrane is essential for the appropriate response of cells to environmental physiologic stimuli. An important signal transducing function is subserved by PLC. Activation of PLC results in generation of two significant intracellular second messenger molecules, namely, $Ins(1,4,5)P_3$ and DAG. Specificity of the signal transduction mechanism from external stimuli to specific PLC isozymes is due to the specificity in the interaction between receptors and PLCs. Receptors, that have the capacity to activate G proteins, are coupled to certain types of PLC-βs by the G_q family of G proteins. There is further specificity in the type of G protein, from the G_q family, with which a specific receptor will interact. An additional coupling mechanism is the $\beta\gamma$ complex-mediated activation of certain PLC-β isozymes. Recent studies have suggested that the specificity of coupling is more extensive, in that only certain combinations of specific βs and γs will result in activation of PLC-βs. Receptors, with tyrosine kinase activity or the capability to activate intracellular tyrosine kinases, stimulate PLCs of the γ family. Thus, the specificity in the response to an extracellular ligand will by influenced by the receptor types, PLC isozymes and G protein subunits that are present in a cell. The regulation of the PLC-δs is unknown and clearly further investigation will be required to clarify the function and regulatory mechanisms for this family of PLCs.

Toxins derived from a diversity of sources including bacteria, marine organisms and venom of spiders, snakes and stinging insects have played a fundamental role in determining physicochemical properties associated with the regulation of PLC activity. The bacterial exotoxins, pertussis toxin and cholera toxin, have provided insight into the regulation of PLC activity by G proteins and the adenylate cyclase system. Recognition of the inherent phospholipid hydrolytic activity of secreted bacterial PLC enzymes have established these molecules as important investigative agents for probing membrane phospholipid composition and mimicking effects of eukaryotic PLCs on cell metabolism. It has been speculated that these bacterial PLCs might be useful in development of a strategy for delivering membrane-active drugs. Other bacteria-derived toxins have been shown to regulate PLC activity through a membrane pore-mediated influx of extracellular calcium as well as the induction of tyrosine kinase activity. Non-bacterial toxins also affect PLC activity through a variety of mechanisms, including the activation of G proteins and stimulation of calcium influx. However, the use of agents, such as mastoparan, as investigative tools should be used with caution since tissue and cell-specific differences may result in different toxin-mediated effects.

REFERENCES

Alava, M.A., DeBell, K.E., Conti, A., Hoffman, T. and Bonvini, E. (1992) Increased intracellular cyclic AMP inhibits inositol phospholipid hydrolysis induced by perturbation of the T cell receptor/CD3 complex but not by G-protein stimulation. *Biochemical Journal*, **284**, 189–199.

Aragay, A.M., Katz, A. and Simon, M.I. (1992) The $G\alpha_q$ and $G\alpha_{11}$ proteins couple the thyrotropin-releasing hormone receptor to phospholipase C in GH_3 rat pituitary cells. *Journal of Biological Chemistry*, **267**, 24983–24988.

Armatruda, T., Gerard, N.P. Gerard, C. and Simon, M.I. (1993) Specific interactions of chemoattractant factor receptors with G-proteins. *Journal of Biological Chemistry*, **268**, 10139–10144.

Armatruda, T., Steele, D., Slepak, V. and Simon, M.I. (1991) $G\alpha_{16}$, a G protein subunit specifically expressed in hematopoietic cells. *Proceedings of the National Academy of Science, USA*, **88**, 5587–5591.

Ashkenzai, A., Peralta, E.G., Winslow, J.W., Ramachandran, J. and Capon, D.J. (1989) Functionally distinct G proteins selectively couple different receptors to PI hydrolysis on the same cell. *Cell*, **56**, 487–493.

Atkinson, T.P., Lee, C-W., Rhee, S.G. and Hohman, R.J. (1993) Orthovanadate induces translocation of phospholipase C-γ1 and -γ2 in permeabilized mast cells. *Journal of Immunology*, **151**, 1448–1455.

Barnes, S.J. and Conn, P.M. (1993) Cholera toxin and dibutyryl cyclic adenosine $3',5'$-monophosphate sensitize gonadotropin-releasing hormone-stimulated inositol phosphate production to inhibition in protein kinase-C (PKC)-depleted cells: evidence for cross-talk between a cholera toxin-sensitive G protein and PKC. *Endocrinology*, **133**, 2756–2760.

Batty, I.H. and Nahorski, S.R. (1989) Rapid accumulation and sustained turnover of inositol phosphates in cerebral-cortex slices after muscarinic-receptor stimulation. *Biochemical Journal*, **260**, 237–241.

Bell, J.D., Buxton, I.L. and Brunton, L.L. (1985) Enhancement of adenylate cyclase activity in S49 lymphoma cells by phorbol esters. Putative effect of C kinase on alpha s-GTP-catalytic subunit interaction. *Journal of Biological Chemistry*, **260**, 2625–2628.

Berridge, M.J. (1983) Rapid accumulation of inositol trisphosphate reveals that agonists hydrolyse polyphosphoinositides instead of phosphatidylinositol. *Biochemical Journal*, **212**, 849–858.

Berridge, M.J. (1987) Inositol trisphosphate and diacylglycerol: two interacting second messengers. *Annual Review of Biochemistry*, **56**, 159–193.

Berridge, M.J. (1993) Inositol trisphosphate and calcium signalling. *Nature*, **361**, 315–325.

Berstein, G. Blank, J.L., Jhon, D-Y., Exton, J.H., Rhee, S.G. and Ross, E.M. (1992a) Phospholipase C-β1 is a GTPase-activating protein for $G_{q/11}$, its physiologic regulator. *Cell*, **70**, 411–418.

Berstein, G. Blank, J.L., Smrcka, A.V., Higashijima, T. and Sternweis, P.C. (1992b) Reconstitution of agonist-stimulated phosphatidylinositol 4,5-bisphosphate hydrolysis using purified m1 muscarinic receptor, Gq/11 and phospholipase C-β1. *Journal of Biological Chemistry*, **267**, 8081–8088.

Berta, P., Sladeczek, F., Derancourt, J., Durand, M., Travo, P. and Haiech, J. (1986) Maitotoxin stimulates the formation of inositol phosphates in rat aortic myocytes. *FEBS Letters*, **197**, 349–352.

Bigay, J., Deterra, P., Pfister, C. and Chabre, M. (1987) Fluoride complexes of aluminum or berylium act on G-proteins as reversibly bound analogues of the gamma phosphate of GTP. *EMBO Journal*, **6**, 2907–2913.

Blackmore, P.F. and Exton, J.H. (1986) Studies on the hepatic calcium-mobilizing activity of aluminum fluoride and glucagon. Modulation by cAMP and phorbol myristate acetate. *Journal of Biological Chemistry*, **261**, 11056–11063.

Blank, J.L., Brattain, K.A. and Exton, J.H. (1992) Activation of cytosolic phosphoinositide phospholipase C by G-protein $\beta\gamma$ subunits. *Journal of Biological Chemistry*, **267**, 23069–23075.

Blank, J.L., Ross, A.H. and Exton, J.H. (1991) Purification and characterization of two G-proteins which activate the β1 isozyme of phosphoinositide-specific phospholipase C. Identification as members of the G_q class. *Journal of Biological Chemistry*, **266**, 18206–18216.

Blondel, O., Graeme I.B. and Seino S. (1995) Inositol 1,4,5-trisphosphate receptors, secretory granules and secretion in endocrine and neuroendocrine cells. *Trends in Neuroscience*, **81**, 157–161.

Bloomquist, B.T., Shortridge, R.D., Schneuwly, S. Perdew, M., Montell, C., Steller, H. Rubin, G. and Pak, W.L. (1988) Isolation of a putative phospholipase C gene of *Drosophila*, norpA and its role in phototransduction. *Cell*, **54**, 723–733.

Bokoch, G.M., Katada, T., Northup, J.K., Hewlett, E.L. and Gilman, A.G. (1983) Identification of the predominant substrate for ADP-ribosylation by islet activating protein. *Journal of Biological Chemistry*, **258**, 2072–2075.

Bourne, H.R., Medynski, D., Van Dop, C., Sullivan, K. and Chang, F-H. (1985) Genetic and functional studies of pertussis toxin substrates. In *Pertussis Toxin* (Sekura, R.D., Moss, J. and Vaughan, M., eds.), Academic Press, Inc., London, 167–184.

Bourne, H.R., Sanders, D.A. and McCormick, F. (1991) The GTPase superfamily: conserved structure and molecular mechanism. *Nature*, **349**, 117–127.

Boyer, J.L., Waldo, G.L. and Harden, T.K. (1992) $\beta\gamma$-subunit activation of G-protein-regulated phospholipase C. *Journal of Biological Chemistry*, **267**, 25451–25456.

Breer H., Boekhoff, I. and Tareilus, E. (1990) Rapid kinetics of second messenger formation in olfactory transduction. *Nature*, **345**, 65–68.

Bristol, A., Hall S.M., Kriz, R.W., Stahl, M.L., Fan, Y.S., Byers, M.G., Eddy, R.L., Shows, T.B. and Knopf, J.L. (1988) Phospholipase C-148: chromosomal location and deletion mapping of functional domains. *Cold Spring Harbor Symposium Quantitation Biology*, **53**, 915–920.

Buhl, A.M., Eisfelder, B.J., Worthen, G.S., Johnson, G.L. and Russell, M. (1993) Selective coupling of the human anaphylatoxin C5a receptor and $\alpha16$ in human kidney 293 cells. *FEBS Letters*, **323**, 132–134.

Burgess, W.H., Dionne, C.A., Kaplow, J., Mudd, R., Friesel, R., Zilberstein, A., Schlessinger, J. and Jaye, M. (1990) Characterization and cDNA cloning of phospholipase C-γ, a major substrate for heparin-binding growth factor 1 (acidic fibroblast growth factor)-activated tyrosine kinase. *Molecular and Cellular Biology*, **10**, 4770–4777.

Camilli, A., Goldfine, H. and Portnoy, D.A. (1991) *Listeria monocytogenes* mutants lacking phosphatidylinositol-specific phospholipase C are avirulent. *Journal of Experimental Medicine*, **173**, 751–754.

Camps, M. Hou, C. Sidiropoulos, D. Stock, J.B., Jakobs, K.H. and Gierschik, P. (1992) Stimulation of phospholipase C by guanine-nucleotide-binding protein $\beta\gamma$ subunits. *European Journal of Biochemistry*, **206**, 821–831.

Cantley, L.C., Auger, K.R., Carpenter, C., Duckworth, B., Graziani, A., Kapeller, R. and Soltoff, S. (1991) Oncogenes and signal transduction. *Cell*, **65**, 281–302.

Carney, D.J., Scott, D.L., Gordon, E.A. and LaBelle, E.F. (1985) Phosphoinositides in mitogenesis: Neomycin inhibits thrombin-stimulated phosphoinositide turnover and initiation of cell proliferation. *Cell*, **42**,479–488.

Carter, R.H., Park, D.J., Rhee, S.G. and Fearon, D.T. (1991) Tyrosine phosphorylation of phospholipase C induced by membrane immunoglobulin in B lymphocytes. *Proceedings of the National Academy of Sciences, USA*, **88**, 2745–2749.

Cheng, H-F., Jiang, M-J., Chen, C-L., Liu, S-M., Wong, L-P., Lomasney, J.W. and King, K. (1995) Cloning and identification of amino acid residues of human phospholipase C$\delta1$ essential for catalysis. *Journal of Biological Chemistry*, **270**, 5495–5505.

Choi, O. H., Padgett, W.L. and Daly, J.W. (1992) Effects of the amphiphilic peptides melittin and mastoparan on calcium influx, phosphoinositide breakdown and arachidonic acid release in rat pheochromocytoma PC12 cells. *Journal of Pharmacology and Experimental Therapeutics*, **260**, 369–375.

Choi, O.H., Padgett, W.L., Nishizawa, Y., Gusovsky, F., Yasumoto, T. and Daly, J.W. (1990) Maitotoxin: effects on calcium channels, phosphoinositide breakdown and arachidonate release in pheochromocytoma PC12 cells. *Molecular Pharmacology*, **37**, 222–230.

Clapham, D.E. and Neer, E.J. (1993) New roles for G-protein $\beta\gamma$ dimers in transmembrane signalling. *Nature*, **365**, 403–406.

Clark, J.D., Lin, L-L., Kriz, R.W., Ramesha, C.S., Sultzman, L.A., Lin, A.Y., Milona, N. and Knopf, J.L. (1991) A novel arachidonic acid-selective cytosolic PLA_2 contains a Ca^{2+}-dependent translocation domain with homology to PKC and GAP. *Cell*, **65**, 1043–1051.

Cocco L., Martelli, A.M. and Gilmour, R.S. (1994) Inositol lipid cycle in the nucleus. *Cellular Signaling*, **6**, 481–485.

Cockcroft, S. and Gomperts, B.D. (1985) Role of guanine nucleotide binding protein in the activation of polyphosphoinositide phosphodiesterase. *Nature*, **314**, 534–536.

Cockcroft, S., Howell, T.W. and Gomperts, B.D. (1987) Two G-proteins act in series to control stimulus-secretion coupling in mast cells: use of neomycin to distinguish between G-proteins controlling polyphosphoinositide phosphodiesterase and exocytosis. *Journal of Cellular Biology*, **105**, 2745–2750.

Cockcroft, S. and Thomas, G.M.H. (1992) Inositol-lipid-specific phospholipase C isoenzymes and their differential regulation by receptors. *Biochemical Journal*, **288**, 1–14.

Countaway, J.L., McQuilkin, P., Gironés, N. and Davis, R.J. (1990) Multisite phosphorylation of the epidermal growth factor receptor. *Journal of Biological Chemistry*, **265**, 3407–3416.

Davis, R.J. and Czech, M.P. (1984) Tumor-promoting diesters mediate phosphorylation of the epidermal growth factor receptor. *Journal of Biological Chemistry*, **259**, 8545–8549.

Decker, S.J., Ellis, C., Pawson, T. and Velu, T. (1990) Effects of substitution of threonine 654 of the epidermal factor receptor on epidermal growth factor-mediated activation of phospholipase C. *Journal of Biological Chemistry*, **265**, 7009–7015.

Divecha N., Banfic, H. and Irvine, R.F. (1993) Inositides and the nucleus and inositides in the nucleus. *Cell*, **74**, 405–407.

Downing, J.R., Margolis, B.L., Zilberstein, A., Ashmun, R.A., Ullrich, A., Sherr, C.J. and Schlessinger, J. (1989) Phospholipase C-γ, a substrate for PDGF receptor kinase, is not phosphorylated on tyrosine during the mitogenic response to CSF-1. *EMBO Journal*, **8**, 3345–3350.

Downward, J., Parker, P. and Waterfield, M.D. (1984) Autophosphorylation sites on the epidermal growth factor receptor. *Nature*, **311**, 483–485.

Ellis, M.V., Carne, A. and Katan, M. (1993) Structural requirements of phosphatidylinositol-specific phospholipase C delta 1 for enzyme activity. *European Journal of Biochemistry*, **213**, 339–347.

Emori, Y, Homma, Y, Sorimachi, H., Kawasaki, H., Nakanishi, O., Suzuki, K. and Takenawa, T. (1989) A second type of rat phosphoinositide-specific phospholipase C containing a *src*-related sequence not essential for phosphoinositide-hydrolyzing activity. *Journal of Biological Chemistry*, **264**, 21885–21890.

Evans, T., Hepler, J.R., Masters, S.B., Brown, J.H. and Harden, T.K. (1985b) Guanine nucleotide regulation of agonist binding to muscarinic receptors: relation to efficacy of agonists for stimulation of phosphoinositide breakdown and Ca^{2+} mobilization. *Biochemical Journal*, **232**, 751–757.

Evans, T., Martin, M.W., Hughes, A.R. and Harden, T.K. (1985a) Guanine nucleotide-sensitive, high affinity binding of carbachol to muscarinic cholinergic receptors of 132N1 astrocytoma cells is insensitive to pertussis toxin. *Molecular Pharmacology*, **27**, 32–37.

Ferreira, P.A., Shortridge, R.D. and Pak, W.L. (1993) Distinctive subtypes of bovine phospholipase C that have preferential expression in the retina and high homology to the *norpA* gene product of *Drosophila*. *Proceedings of the National Academy of Science, USA*, **90**, 6042–6046.

Filippov, A.K., Tertishnikova, S.M., Alekseev, A.E., Tsurupa, G.P., Pashkov, V.N. and Grishin, E.V. (1994) Mechanism of α-latrotoxin action as revealed by patch-clamp experiments on *Xenopus* oocytes injected with rat brain messenger RNA. *Neuroscience*, **61**, 179–189.

Fink, D.W., Contreras, M.L., Lelkes, P.I. and Lazarovici, P. (1989) *Staphylococcus aureus* α-toxin activates phospholipases and induces a calcium influx in PC12 cells. *Cellular Signalling*, **1**, 387–393.

Fink, D.W. and Guroff, G. (1990) Nerve growth factor stimulation of arachidonic acid release from PC12 cells: Independence from phosphoinositide turnover. *Journal of Neurochemistry*, **55**, 1716–1726.

Fischer, H., Dohlsten, M., Lindvall, M., Sjogren, H-O. and Carlsson, R. (1989) Binding of staphylococcal enterotoxin A to HLA-DR on B cell lines. *Journal of Immunology*, **142**, 3151–3157.

Fisher, S.K., Heacock, A.M. and Agranoff, B.W. (1992) Inositol lipids and signal transduction in the nervous system: An update. *Journal of Neurochemistry*, **58**, 18–38.

Fisher, S.K., Heacock, A.M., Seguin, E.B. and Agranoff, B.W. (1990) Polyphosphoinositides are the major source of inositol phosphates in carbamoylcholine-stimulated SK-N-SH neuroblastoma cells. *Molecular Pharmacology*, **38**, 54–63.

Fishman, P.H. (1990) Mechanism of action of cholera toxin. In: *ADP-ribosylating toxins and G proteins: insights into signal transduction* (Moss, J. and Vaughan, M., eds.) American Society of Microbiology, Washington, D.C., 127–140.

Fraser, J.D. (1989) High-affinity binding of staphylococcal enterotoxins A and B to HLA-DR. *Nature*, **339**, 221–223.

Freedman, S.B., Miller, D.M., Miller, R.J. and Tindall, D.R. (1984) Interaction of maitotoxin with voltage-sensitive calcium channels in cultured neuronal cells. *Proceedings of the National Academy of Sciences, USA*, **81**, 4582–4585.

Fussle, R., Bhakdi, S., Sziegoleit, A., Tranum-Jensen, J., Kranz, T. and Wellensiek, H.J. (1981) On the mechanism of membrane damage by *Staphylococcus aureus alpha*-toxin. *Journal of Cell Biology*, **91**, 83–94.

Gergel, J.R., McNamara, D.J., Dobrusin, E.M., Zhu, G., Saltiel, A.R. and Miller, W.T. (1994) Identification of amino acids in the N-terminal SH2 domain of phospholipase Cγ1 important in the interaction with epidermal growth factor receptor. *Biochemistry*, **33**, 14671–14678.

Ghalayini, A.J. and Anderson, R.E. (1992) Activation of bovine rod outer segment phospholipase C by arrestin. *Journal of Biological Chemistry*, **267**, 17977–17982.

Gilman, A.G. (1984) G proteins and dual control of adenylate cyclase. *Cell*, **36**, 577–579.

Gilman, A.G. (1987) G proteins: transducers of receptor-generated signals. *Annual Review of Biochemistry*, **56**, 615–649.

Gout, I., Dhand, R., Hiles, I.D., Fry, M.J., Panayotou, G., Das, P., Truong, O., Totty, N.F., Hsuan, J., Booker, G.W., Campbell, I.D. and Waterfield, M.D. (1993) The GTPase dynamin binds to and is activated by a subset of SH3 domains. *Cell*, **75**, 25–36.

Grandt, R., Aktories, K. and Jacobs, K.H. (1986) Evidence for two GTPases activated by thrombin in membranes of human platelets. *FEBS Letters*, **196**, 279–283.

Granja, C., Lin, L.L., Yunis, E.J., Relias, V. and Dasgupta, J.D. (1991) PLC-γ1, a possible mediator of T-cell receptor function. *Journal of Biological Chemistry*, **266**, 16277–16280.

Gusovsky, F., Bitran, J., Yasumoto, T. and Daly, J.W. (1990) Mechanism of maitotoxin-stimulated phosphoinositide breakdown in HL-60 cells. *Journal of Pharmacology and Experimental Therapeutics*, **252**, 466–473.

Gusovsky, F., Daly, J.W., Yasumoto, T. and Rojas, E. (1988) Differential effects of maitotoxin on ATP secretion and on phosphoinositide breakdown in rat pheochromocytoma cells. *FEBS Letters*, **233**, 139–142.

Gusovsky, F., Lueders, J.E., Kohn, E.C. and Felder, C.C. (1993) Muscarinic receptor-mediator tyrosine phosphorylation of phospholipase C-γ. *Journal of Biological Chemistry*, **268**, 7768–7772.

Gusovsky, F., Soergel, D.G. and Daly, J.W. (1991) Effects of mastoparan and related peptides on phosphoinositide breakdown in HL-60 cells and cell-free preparations. *European Journal of Pharmacology*, **206**, 309–314.

Gusovsky, F., Yasumoto, T. and Daly, J.W. (1987) Maitotoxin stimulates phosphoinositide breakdown in neuroblastoma hybrid NCB-20 cells. *Cellular and Molecular Neurobiology*, **7**, 317–322.

Gusovsky, F., Yasumoto, T. and Daly, J.W. (1989) Maitotoxin, a potent, general activator of phosphoinositide breakdown. *FEBS Letters*, **243**, 307–312.

Harden, T.K. (1992) G-protein-regulated phospholipase C: Identification of component proteins. *Advances in Second Messengers and Phosphoprotein Research*, **28**, 65–72.

Henner, D.J., Yang, M., Chen, E., Hellmiss, R., Rodriquez, H. and Low, M.G. (1988) Sequence of the *Bacillus thuringiensis* phosphatidylinositol specific phospholipase C. *Nucleic Acids Research*, **16**, 10383.

Hepler, J.R. and Harden, T.K. (1986) Guanine-nucleotide-dependent pertussis-toxin-insensitive stimulation of inositol phosphate formation by carbachol in a membrane preparation from human astrocytoma cells. *Biochemical Journal*, **239**, 141–146.

Hepler, J.R., Kozasa, T., Smrcka, A.V., Simon, M.I., Rhee, S.G., Sternweis, P.C. and Gilman, A.G. (1993) Purification from Sf9 cells and characterization of recombinant $G_q\alpha$ and $G_{11}\alpha$: activation of purified phospholipase C isozymes by Gα subunits. *Journal of Biological Chemistry*, **268**, 14367–14375.

Hepler, J.R., Nakahata, N., Lovenberg, T.W., DiGuiseppi, J., Herman, B., Earp, H.S. and Harden, T.K. (1987) Epidermal growth factor stimulates the rapid accumulation of inositol (1,4,5)-trisphosphate and a rise in cytosolic calcium mobilized from intracellular stores in A431 cells. *Journal of Biological Chemistry*, **262**, 2951–2956.

Higashijima, T., Burnier, J. and Ross, E.M. (1990) Regulation of G_i and G_0 by mastoparan, related amphiphilic peptides and hydrophobic amines. *Journal of Biological Chemistry*, **265**, 14176–14186.

Higashijima, T., Uzu, S., Nakajima, T. and Ross, E.M. (1988) Mastoparan, a peptide toxin from wasp venom, mimics receptors by activating GTP-binding regulatory proteins (G proteins). *Journal of Biological Chemistry*, **263**, 6491–6494.

Hirai, Y., Yasuhara, Y., Yoshida, H., Nakajima, T., Fujino, M. and Kitada, C. (1979) A new mast cell degranulating peptide 'mastoparan' in the venom of *Vespula lewisii*. *Chemical Pharmacology Bulletin (Tokyo)*, **27**, 1942–1944.

Homma, Y., Imaki, J., Nakanishi, O. and Takenawa, T. (1988) Isolation and characterization of two different forms of inositol phospholipid-specific phospholipase C from rat brain. *Journal of Biological Chemistry*, **263**, 6592–6598.

       MARGARITA L. CONTRERAS and DONALD W. FINK, JR.

Homma, Y. and Takenawa, T. (1992) Inhibitory effect of *src* homology (SH)2/SH3 fragments of phospholipase C-γ on the catalytic activity of phospholipase C isoforms. *Journal of Biological Chemistry*, **267**, 21844–21849.

Horn, V.J., Baum, B.J. and Ambudkar, I.S. (1988) Beta-adrenergic receptor stimulation induces inositol trisphosphate production and Ca2+ mobilization in rat parotid acinar cells. *Journal of Biological Chemistry*, **263**, 12454–12460.

Houslay, M.D., Bojanic, D. and Wilson, A. (1986) Platelet activating factor and U44069 stimulate a GTPase activity in human platelets which is distinct from the guanine nucleotide regulatory proteins, N_s and N_i. *Biochemical Journal*, **234**, 737–740.

Hsuan, J.J., Totty, N. and Waterfield, M.D. (1989) Identification of a novel autophosphorylation site (P4) on the epidermal growth factor receptor. *Biochemical Journal*, **262**, 659–663.

Huang, P.S., Davis, L., Huber, H., Goodhart, P.J., Wegrzyn, R., Oliff, A. and Heimbrook, D. (1995) An SH3 domain is required for the mitogenic activity of microinjected phospholipase C-γ1. *FEBS Letters*, **358**, 287–292.

Huckle, W.R., Hepler, J.R., Rhee, S.G., Harden, T.K. and Earp, S.H. (1990) Protein kinase C inhibits epidermal growth factor-dependent tyrosine phosphorylation of phospholipase C-γ and activation of phosphoinositide hydrolysis. *Endocrinology*, **127**, 1697–1705.

Hunt, T.W., Carroll, R.C. and Peralta, E.G. (1994) Heterotrimeric G proteins containing $G\alpha_{i3}$ regulate multiple effector enzymes in the same cell: activation of phospholipases C and A_2 and inhibition of adenylyl cyclase. *Journal of Biological Chemistry*, **269**, 29565–29570.

Hunter, T., Ling, N. and Cooper, J.A. (1984) Protein kinase C phosphorylation of the EGF receptor at a threonine residue close to the cytoplasmic face of the plasma membrane. *Nature*, **311**, 480–483.

Hwang, P.M., Verma. A., Bredt, D.S. and Snyder, S.H. (1990) Localization of phosphatidylinositol signaling components in rat taste cells: Role in bitter taste transduction. *Proceedings of the National Academy of Sciences, USA*, **87**, 7395–7399.

Ikezawa, H. (1986) The physiological action of bacterial phosphatidylinositol-specific phospholipase C. The release of ectoenzymes and other effects. *Journal of Toxicology and Toxin Reviews*, **5**, 1–24.

Inglese, J., Kock, W.J., Kazushige, T. and Lefkowitz, R.J. (1995) Gβγ interactions with PH domains and Ras-MAPK signaling pathways. *Trends in Biochemical Sciences*, **20**, 151–156.

Jelsema, C.L. (1989) Regulation of phospholipase A2 and phospholipase C in rod outer segments of bovine retina involves a common GTP-binding protein but different mechanisms of action. *Annals of the New York Academy of Science*, **559**, 158–177.

Jhon, D-Y., Lee, H-H., Park, D., Lee, C-W., Lee, K-H., Yoo, O.J. and Rhee S.G. (1993) Cloning, sequencing, purification and G_q-dependent activation of phospholipase C-β3. *Journal of Biological Chemistry*, **268**, 6654–6661.

Jiang, H., Wu, D. and Simon, M.I. (1994) Activation of phospholipase C β4 by heterotrimeric GTP-binding proteins. *Journal of Biological Chemistry*, **269**, 7593–7596.

Johnson, R.M. and Garrison, J.C. (1987) Epidermal growth factor and angiotensin II stimulate formation of inositol 1,4,5- and 1,3,4-trisphosphate in hepatocytes. Differential inhibition by pertussis toxin and phorbol 12-myristate 13-acetate. *Journal of Biological Chemistry*, **262**, 17285–17293.

Kalinoski D.L., Aldinger, S.B., Boyle, A.G., Huque, T., Marececk, F., Prestwich, G.D. and Restrepo, D. (1992) Characterization of a novel inositol 1,4,5-trisphosphate receptor in isolated olfactory cilia. *Biochemical Journal*, **281**, 449–456.

Kanner, S.B., Grosmaire, L.S., Ledbetter, J.A. and Damle, N.K. (1993) Beta 2-integrin LFA-1 signaling through phospholipase C-γ1 activation. *Proceedings of the National Academy of Sciences, USA*, **90**, 7099–7103.

Kappler, J., Kotzin, B., Herron, L., Gelfand, E.W., Bigler, R.D., Boylston, A., Carrel, S., Posnett, D.N., Choi, Y. and Marrack, P. (1989) Vβ-specific stimulation of human T cells by staphylococcal toxins. *Science*, **244**, 811–813.

Kashishian, A. and Cooper, J.A. (1993) Phosphorylation sites at the C-terminus of the platelet-derived growth factor receptor bind phospholipase C-γ1. *Molecular Biology of the Cell*, **4**, 49–57.

Katan, M. and Parker, P.J. (1987) Purification of PI-specific PLC from a particulate fraction of bovine brain 154 kDa. *European Journal of Biochemistry*, **168**, 413–418.

Kato, H., Ishitoya, J. and Takenawa, T. (1986) Inhibition of inositol phospholipids metabolism and calcium mobilization by cyclic AMP-increasing agents and phorbol ester in neutrophils. *Biochemical and Biophysical Research Communications,* **139,** 1272–1278.

Katz, A., Wu, D. and Simon, M.I. (1992) Subunits $\beta\gamma$ of heterotrimeric G protein activate β2 isoform of phospholipase C. *Nature,* **360,** 686–689.

Kikkawa S., Takahashi, K., Takahashi, K., Shimada, N., Ui, M., Kimura, N. and Katada, T. (1990) Conversion of GDP into GTP by nucleoside diphosphate kinase on the GTP-binding proteins. *Journal of Biological Chemistry,* **265,** 21536–21540.

Kikkawa S., Takahashi, K., Takahashi, K., Shimada, N., Ui, M., Kimura, N. and Katada, T. (1992) Activation of nucleoside diphosphate kinase by mastoparan, a peptide isolated from wasp venom. *FEBS Letters,* **305,** 237–240.

Kikuchi, A., Kozawa, O., Kaibuchi, K., Katada, T. and Ui, M. (1986) Direct evidence for involvement of a guanine nucleotide-binding protein in chemotactic peptide-stimulated formation of inositol bisphosphate and trisphosphate in differentiated human leukemic (HL-60) cells. *Journal of Biological Chemistry,* **261,** 11558–11562.

Kim, H.K., Kim, J.W., Zilberstein, A., Margolis, B., Kim, J.G., Schlessinger, J. and Rhee, S.G. (1991) PDGF stimulation of inositol phospholipid hydrolysis requires PLC-γ1 phosphorylation on tyrosine residues 783 and 1254. *Cell,* **65,** 435–441.

Kim, M.J., Bahk, Y.Y., Min, D.S., Lee, S-J., Ryu, S.H. and Suh, P-G. (1993) Cloning of cDNA encoding rat phospholipase C-β4, a new member of the phospholipase C. *Biochemical Biophysical Research Communications,* **194,** 706–712.

Kim, U-H., Fink, D.W., Jr., Kim, H.S., Park, D.J., Contreras, M.L., Guroff, G. and Rhee, S.G. (1991) Nerve growth factor stimulates phosphorylation of phospholipase C-γ in PC12 cells. *Journal of Biological Chemistry,* **266,** 1359–1362.

Kim, U-H., Kim, H-S. and Rhee, S.G. (1990) Epidermal growth factor and platelet-derived growth factor promote translocation of phospholipase C-γ from cytosol to membrane. *FEBS Letters,* **270,** 33–36.

Kim, U-H., Kim, J.W. and Rhee, S.G. (1989) Phosphorylation of phospholipase C-γ by cAMP-dependent protein kinase. *Journal of Biological Chemistry,* **264,** 20167–20170.

Klinker, J.F., Hageluken, A., Grunbaum, L., Heilmann, I., Nurnberg, B., Harhammer, R., Offermanns, S., Schwaner, I., Ervens, J., Wenzel-Seifert, K., Muller, T. and Seifert, R. (1994) Mastoparan may activate GTP hydrolysis by G_i-proteins in HL-60 membranes indirectly through interaction with nucleoside diphosphate kinase. *Biochemical Journal,* **304,** 377–383.

Kloog, Y., Ambar, I., Sokolovsky, M., Kochva, E., Wollberg, Z. and Bdolah, A. (1988) Sarafotoxin, a novel vasoconstrictor peptide: Phosphoinositide hydrolysis in rat heart and brain. *Science,* **242,** 268–270.

Kobayashi, M., Kondo, S., Yasumoto, T. and Ohizumi, Y. (1986) Cardiotoxic effects of maitotoxin, a principal toxin of seafood poisoning, on guinea pig and rat cardiac muscle. *Journal of Pharmacology and Experimental Therapeutics,* **238,** 1077–1083.

Koblan, K.S., Schaber, M.D., Edwards, G., Gibbs, J.B. and Pompliano, D.L. (1995) *src*-Homology 2 (SH2) domain ligation as an allosteric regulator: Modulation of phosphoinositide-specific phospholipase Cγ1 structure and activity. *Biochemical Journal,* **305,** 745–751.

Koch, C.A., Anderson, D., Moran, M.F., Ellis, C. and Pawson, T. (1991) SH2 and SH3 domains: elements that control interactions of cytoplasmic signaling proteins. *Science,* **252,** 668–674.

Koch, C.A., Moran, M.F., Anderson, D., Liu, X.Q., Mbamalu, G. and Pawson, T. (1992) Multiple SH2-mediated interactions in v-*src*-transformed cells. *Molecular and Cellular Biology,* **12,** 1366–1374.

Kochva, E., Bdolah, A. and Wollberg, Z. (1993) Sarafotoxins and endothelins: evolution, structure and function. *Toxicon,* **31,** 541–568.

Komatsu, M., Aizawa, T., Yokokawa, N., Sato, Y., Okada, N., Takasu, N. and Yamada, T. (1992) Mastoparan-induced hormone release from rat pancreatic islets. *Endocrinology,* **130,** 221–228.

Kominami, T., Maki, A. and Ikehara, Y. (1985) Electrophoretic characterization of hepatic alkaline phosphatase released by phosphatidylinositol-specific phospholipase C. *Biochemical Journal,* **227,** 183–189.

Kozasa, T., Hepler, J.R., Smrcka, A.V., Simon, M.I., Rhee, S.G., Sternweis, P.C. and Gilman, A.G. (1993) Purification and characterization of recombinant $G_{16}\alpha$ from Sf9 cells: activation of purified phospholipase C isozymes by G-protein α subunits. *Proceedings of the National Academy of Science,* **90,** 9176–9180.

Kriz, R., Lin, L-L., Sultzman, L., Ellis, C., Heldin, C-H., Pawson, T. and Knopf, J.L. (1990) Phospholipase isozymes: structural and functional similarities. *Ciba Foundation Symposium*, **150**, 112–123.

Kumjian, D.A., Wahl, M.I., Rhee, S.G. and Daniel, T.O. (1989) Platelet-derived growth factor (PDGF) binding promotes physical association of PDGF receptor with phospholipase C. *Proceedings of the National Academy of Sciences, USA*, **86**, 8232–8236.

Kuppe, A., Evans, L.M., McMillen, D.A. and Griffith, O.H. (1989) Phosphatidylinositol-specific phospholipase C of *Bacillus cereus*: cloning, sequencing and relationship to other phospholipases. *Journal of Bacteriology*, **171**, 6077–6083.

LaMorte, V.J., Harootunian, A.T., Spiegel, A.M., Tsien, R.Y. and Feramisco, J.R. (1993) Mediation of growth factor induced DNA synthesis and calcium mobilization by Gq and Gi2. *Journal of Cell Biology*, **121**, 91–99.

Larose, L., Gish, G. and Pawson, T. (1995) Construction of an SH2 domain-binding site with mixed specificity. *Journal of Biological Chemistry*, **270**, 3858–3862.

Larose, L., Gish, G., Shoelson, S. and Pawson, T. (1993) Identification of residues in the beta platelet-derived growth factor receptor that confer specificity for binding to phospholipase C-γ1. *Oncogene*, **8**, 2493–2499.

Lebrun, P., Hermann, M., Yasumoto, T. and Herchuelz, A. (1987) Effects of maitotoxin on ionic and secretory events in rat pancreatic islets. *Biochemical and Biophysical Research Communications*, **144**, 172–177.

Lee, C.H., Park, D., Wu, D., Rhee, S.G. and Simon, M.I. (1992) Members of the G_q α subunit gene family activate phospholipase C β isozymes. *Journal of Biological Chemistry*, **267**, 16044–16047.

Lee, C-W., Lee, K-H., Lee, S.B, Park, D. and Rhee, S.G. (1994) Regulation of phospholipase C-β4 by ribonucleotides and the α subunit of Gq. *Journal of Biological Chemistry*, **269**, 25335–25338.

Lee, C-W., Park D.J., Lee, K-H., Kim C.G. and Rhee, S.G. (1993) Purification, molecular cloning and sequencing of phospholipase C-β4. *Journal of Biological Chemistry*, **268**, 21318–21327.

Legrand, A.M. and Bagnis, R. (1984) Effects of ciguatoxin and maitotoxin on isolated rat atria and rabbit duodenum. *Toxicon*, **22**, 471–475.

Leimeister-Wachter, M., Domann, E. and Chakraborty, T. (1991) Detection of a gene encoding a phosphatidylinositol-specific phospholipase C that is co-ordinately expressed with listeriolysin in *Listeria monocytogenes*. *Molecular Microbiology*, **5**, 361–366.

Lewy, H., Galron, R., Bdolah, A., Sokolovsky, M. and Naor, Z. (1992) Paradoxical signal transduction mechanism of endothelins and sarafotoxins in cultured pituitary cells: stimulation of phosphoinositide turnover and inhibition of prolactin release. *Molecular and Cellular Endocrinology*, **89**, 1–9.

Liao, F., Shin, H.S. and Rhee, S.G. (1992) Tyrosine phosphorylation of phospholipase C-γ1 induced by cross-linking of the high-affinity or low-affinity Fc receptor for IgG in U937 cells. *Proceedings of the National Academy of Sciences, USA*, **89**, 3659–3663.

Lin, W-W., Lee, C-Y. and Chuang, D-M. (1989) Cross-desensitization of endothelin-II and sarafotoxin-induced phosphoinositide turnover in neurons. *European Journal of Pharmacology*, **166**, 581–582.

Lin, W-W., Lee, C-Y. and Chuang, D-M. (1990) Comparative studies of phosphoinositide hydrolysis induced by endothelin-related peptides in cultured cerebellar astrocytes, C-6 glioma and cerebellar granule cells. *Biochemical and Biophysical Research Communications*, **168**, 512–519.

Lin, W-W., Lee, C-Y. and Chuang, D-M. (1991) Endothelin and sarafotoxin-induced phosphoinositide hydrolysis in cultured cerebellar granule cells: biochemical and pharmacological characterization. *Journal of Pharmacology and Experimental Therapeutics*, **257**, 1053–1061.

Litosch, I., Wallis, C. and Fain, J.N. (1985) 5-Hydroxytryptamine stimulates inositol phosphate production in a cell-free system from blowfly salivary glands. Evidence for a role of GTP in coupling receptor activation to phosphoinositide breakdown. *Journal of Biological Chemistry*, **260**, 5464–5471.

Loeb, D.M., Stephens, R.M., Copeland, T., Kaplan, D.R. and Greene, L.A. (1994) A Trk nerve growth factor (NGF) receptor point mutation affecting interaction with phospholipase C-γ1 abolishes NGF-promoted peripherin induction but not neurite outgrowth. *Journal of Biological Chemistry*, **269**, 8901–8910.

Login, I.S., Judd, A.M., Cronin, M.J., Koike, K., Schettini, G., Yasumoto, T. and Macleod, R.M. (1985) The effects of maitotoxin on $^{45}Ca^{2+}$ flux and hormone release in GH_3 rat pituitary cells. *Endocrinology*, **116**, 622–627.

Ma, H-W., Blitzer, R.D., Healy, E.C., Premont, R.T., Landau, E.M. and Iyengar, R. (1993) Receptor-evoked Cl⁻ in *Xenopus* oocytes is mediated through a β-type phospholipase C. *Journal of Biological Chemistry*, **268**, 19915–19918.

MacCumber, M.W., Ross, C.A. and Snyder, S.H. (1990) Endothelin in brain: receptors, mitogenesis and biosynthesis in glial cells. *Proceedings of the National Academy of Sciences, USA*, **87**, 2359–2363.

Madison, J.M. and Brown, J.K. (1988) Differential inhibitory effects of forskolin, isoproterenol and dibutyryl cyclic adenosine monophosphate on phosphoinositide hydrolysis in canine tracheal smooth muscle. *Journal of Clinical Investigations*, **82**, 1462–1465.

Majerus, P.W. (1983) Arachidonate metabolism in vascular disorders. *Journal of Clinical Investigations*, **72**, 1521–1525.

Majerus, P.W., Connolly, T.M., Deckmyn, H., Ross, T.S. and Bross, T.E., Ishii, H., Bansal, V.S. and Wilson D.B. (1986) The metabolism of phosphoinositide-derived messenger molecules. *Science*, **234**, 1519–1526.

Manning, D.R., Fraser, B.A., Kahn, R.A. and Gilman, A.G. (1984) ADP-ribosylation of transducin by islet activating protein. Identification of asparagine as the site of ADP-ribosylation. *Journal of Biological Chemistry*, **259**, 749–756.

Margolis, B., Rhee, S.G., Felder, S., Mervic, M., Lyall, R., Levitzki, A., Ullrich, A., Zilberstein, A. and Schlessinger, J. (1989) EGF induces tyrosine phosphorylation of phospholipase C-II: a potential mechanism for EGF receptor signaling. *Cell*, **57**, 1101–1107.

Marsault, R., Vigne, P., Breittmayer, J-P. and Frelin, C. (1990) Astrocytes are target cells for endothelins and sarafotoxin. *Journal of Neurochemistry*, **54**, 2142–2144.

Martin, T.F.J. (1989) Lipid hydrolysis by phosphoinositidase C. Enzymology and regulation by receptors and guanine nucleotides. In: *Inositol lipids in cell signalling* (Michell, R.H., Drummond, A.H., Downes, C.P., eds.) Academic Press, London, 81–107.

Mau, S.E., Witt, M.R. and Vilhardt, H. (1994) Mastoparan, a wasp venom peptide, stimulates release of prolactin from cultured rat anterior pituitary cells. *Journal of Endocrinology*, **142**, 9–18.

McAtee, P. and Dawson, G. (1989) Rapid dephosphorylation of protein kinase C substrates by protein kinase A activators results from inhibition of diacylglycerol release. *Journal of Biological Chemistry*, **264**, 11193–11199.

Meisenhelder, J., Suh, P-G., Rhee, S.G. and Hunter, T. (1989) Phospholipase C-γ is a substrate for the PDGF and EGF receptor protein-tyrosine kinases *in vivo* and *in vitro*. *Cell*, **57**, 1109–1122.

Meldolesi, J., Scheer, H., Madeddu, L. and Wanke, E. (1986) Mechanism of action of α-latrotoxin: the presynaptic stimulatory toxin of the black widow spider venom. *Trends in Pharmacological Science*, **7**, 151–155.

Meldrum, E., Katan, M. and Parker, P.J. (1989) A novel inositol-phospholipid-specific phospholipase C: Rapid purification and characterization. *European Journal of Biochemistry*, **182**, 673–677.

Meldrum, E., Kriz, R.W., Totty, N. and Parker, P.J. (1991a) A second gene product for the inositol-phospholipid-specific phospholipase Cδ subclass. *European Journal of Biochemistry*, **196**, 159–165.

Meldrum, E. Parker, P.J. and Carozzi, A. (1991b) The PtdIns-PLC superfamily and signal transduction. *Biochimica Biophysica Acta*, **1092**, 49–71.

Menestrina, G. (1986) Ionic channels formed by *Staphylococcus aureus alpha*-toxin: voltage-dependent –inhibition by divalent and trivalent cations. *Journal of Membrane Biology*, **90**, 177–190.

Mengaud, J., Barun-Breton, C. and Cossart, P. (1991) Identification of a phosphatidylinositol-specific phospholipase C in *Listeria monocytogenes*: a novel type of virulence factor? *Molecular Microbiology*, **5**, 367–372.

Meucci, O., Grimaldi, M., Scorziello, A., Govoni, S., Bergamaschi, S., Yasumoto, T. and Schettini, G. (1992) Maitotoxin-induced intracellular calcium rise in PC12 cells: involvement of dihydropyridine-sensitive and omega-conotoxin-sensitive calcium channels and phosphoinositide breakdown. *Journal of Neurochemistry*, **59**, 679–688.

Michell, R.H. (1992) Inositol lipids in cellular signalling mechanisms. *Trends in Biochemical Sciences*, **17**, 274–276.

Michell, R.H., Kirk, C.J., Jones, L.M., Downes, C.P. and Creba, J.A. (1981) The stimulation of inositol lipid metabolism that accompanies calcium mobilization in stimulated cells - defined characteristics and unanswered questions. *Philosophical Transcripts of Royal Society of London Series B*, **296**, 123–137.

Mohammadi, M., Dionne, C., Li, W., Li, N., Spivak, T., Honegger, A.M., Jaye, M. and Schlessinger, J. (1992) Point mutation in FGF receptor eliminates phosphatidylinositol hydrolysis without affecting mitogenesis. *Nature*, **358**, 681–684.

Mollick, J.A., Cook, R.G. and Rich, R.R. (1989) Class II MHC molecules are specific receptor for staphylococcus enterotoxin A. *Science*, **244**, 817–820.

Moriarty, T.M., Padrell, E., Carty, D.J., Omri, G. and Landau, M. (1990) G_o protein as a signal transducer in the pertussis toxin-sensitive phosphatidylinositol pathway. *Nature*, **343**, 79–82.

Morio, T., Geha, R.S. and Chatila, T.A. (1994) Engagement of MHC class II molecules by staphylococcal superantigens activates *src*-type protein tyrosine kinases. *European Journal of Immunology*, **24**, 651–658.

Morrison, D.K., Kaplan, D.R., Rhee, S.G. and Williams, L.T. (1990) Platelet-derived growth factor (PDGF)-dependent association of phospholipase C-γ with the PDGF receptor signaling complex. *Molecular and Cellular Biology*, **10**, 2359–2366.

Musgrave, I.F., Seifert, R. and Schultz, G. (1994) Maitotoxin activates cation channels distinct from the receptor-activated non-selective cation channels of HL-60 cells. *Biochemical Journal*, **301**, 437–441.

Nakahata, N., Abe, M.T., Matsuoka, I. and Nakanishi, H. (1990) Mastoparan inhibits phosphoinositide hydrolysis via pertussis toxin-insensitive G-protein in human astrocytoma cells. *FEBS Letters*, **260**, 91–94.

Nakahata, N., Ishimoto, H., Mizuno, K., Ohizumi, Y. and Nakanishi, H. (1994) Dual effects of mastoparan on intracellular free Ca^{2+} concentrations in human astrocytoma cells. *British Journal of Pharmacology*, **112**, 299–303.

Nakamura, T. and Ui, M. (1985) Simultaneous inhibitions of inositol phospholipid breakdown, arachidonic acid release and histamine secretion in mast cells by islet activating protein, pertussis toxin. *Journal of Biological Chemistry*, **260**, 3584–3593.

Neylon, C.B. and Summers, R.J. (1988) Inhibition by cAMP of the phosphoinositide response to alpha 1-adrenoceptor stimulation in rat kidney. *European Journal of Pharmacology*, **148**, 441–444.

Nishibe, S., Wahl, M.I., Hernández-Sotomayor, S.M.T., Tonks, N.K., Rhee, S.G. and Carpenter, G. (1990) Increase of the catalytic activity of phospholipase C-γ1 by tyrosine phosphorylation. *Science*, **250**, 1253–1256.

Nishibe, S., Wahl, M.I., Rhee, S.G. and Carpenter, G. (1989) Tyrosine phosphorylation of phospholipase C-II *in vitro* by the epidermal growth factor receptor. *Journal of Biological Chemistry*, **264**, 10335–10338.

Nishizuka, Y. (1984) The role of protein kinase in cell surface signal transduction and tumour promotion. *Nature*, **308**, 693–694.

Nishizuka, Y. (1986) Studies and perspectives of protein kinase C. *Science*, **233**, 305–312.

Nishizuka, Y. (1988) The molecular heterogeneity of protein kinase C and its implications for cellular regulation. *Nature*, **334**, 661–665.

Obermeier, A., Halfter, H., Wiesmüller, K-H., Jung, J., Schlessinger, J. and Ullrich, A. (1993). Tyrosine 785 is a major determinant of Trk-substrate interaction. *EMBO Journal*, **12**, 933–941.

Offermanns, S. and Simon, M.I. (1995) $G\alpha_{15}$ and $G\alpha_{16}$ couple a wide variety of receptors to phospholipase C. *Journal of Biological Chemistry*, **270**, 15175–15180.

Ohizumi, Y., Kajiwara, A. and Yasumoto, T. (1983) Excitatory effect of the most potent marine toxin, maitotoxin, on the guinea pig vas deferens. *Journal of Pharmacology and Experimental Therapeutics*, **227**, 199–204.

Ohizumi, Y. and Yasumoto, T. (1983a) Contractile response of the rabbit aorta to maitotoxin, the most potent marine toxin. *Journal of Physiology (London)*, **337**, 711–721.

Ohizumi, T. and Yasumoto, T. (1983b) Contraction and increase in tissue calcium content induced by maitotoxin, the most potent known marine toxin, in intestinal smooth muscle. *British Journal of Pharmacology*, **79**, 3–5.

Ohta, S., Matsui, A., Nazawa, Y. and Kagawa, Y. (1988) Complete cDNA encoding a putative phospholipase C from transformed human lymphocytes. *FEBS Letters*, **242**, 31–35.

Olashaw, N.E., Rhee, S.G. and Pledger, W.J. (1990) Cyclic AMP agonists induce the phosphorylation of a 76 kDa protein co-precipitated by anti-(phospholipase C-tau) monoclonal antibodies in BALB/c-3T3 cells. Relationship to inositol phosphate formation. *Biochemical Journal*, **272**, 297–303.

Ozaki, Y., Matsumoto, Y., Yatomi, Y., Higashihara, M., Kariva, T. and Kume, S. (1990) Mastoparan, a wasp venom, activates platelets via pertussis toxin-sensitive GTP-binding proteins. *Biochemical and Biophysical Research Communications*, **170**, 779–785.

Pang, I.H. and Sternweis, P.C. (1990) Purification of unique α-subunits of GTP-binding regulatory proteins by affinity chromatography with immobilized $\beta\gamma$-subunits. *Journal of Biological Chemistry*, **265**, 18707–18712.

Park, D., Jhon, D-Y., Kriz, R., Knopf, K.L. and Rhee, S.G. (1992) Cloning, sequencing, expression and G_q-independent activation of phospholipase C-β2. *Journal of Biological Chemistry*, **267**, 16048–16055.

Park, D., Jhon, D-Y., Lee, C-W., Lee, K-H. and Rhee, S.G. (1993a) Activation of phospholipase C isozymes by G protein $\beta\gamma$ subunits. *Journal of Biological Chemistry*, **268**, 4573–4576.

Park, D., Jhon, D-Y., Lee, C-W., Ryu, S.H. and Rhee, S.G. (1993b) Removal of the carboxyl-terminal region of phospholipase C-β1 by calpain abolishes activation by Gα_q. *Journal of Biological Chemistry*, **268**, 3710–3714.

Park, D.J., Min, H.K. and Rhee, S.G. (1992) Inhibition of CD3-linked phospholipase C by phorbol ester and by cAMP is associated with decreased phosphotyrosine and increased phosphoserine contents of PLC-γ1. *Journal of Biological Chemistry*, **267**, 2496–1501.

Park, D.J., Rho, H.W. and Rhee, S.G. (1991) CD3 stimulation causes phosphorylation of phospholipase C-γ1 on serine and tyrosine residues in a human T-cell line. *Proceedings of the National Academy of Sciences, USA*, **88**, 5453–5456.

Parker, P.J., Kour, G., Marais, R.M., Mitchell, F., Pears, C.J., Schaap, D., Stabel, S. and Webster, C. (1989) Protein kinase C - a family affair. *Molecular Cellular Endocrinology*, **65**, 1–11.

Parsonnet, J. and Gillis, Z.A. (1988) Production of tumor necrosis factor by human monocytes in response to toxic shock syndrome toxin-1. *Journal of Infectious Diseases*, **158**, 1026–1033.

Parsonnet, J., Hickman, R.K., Eardley, D.P. and Pier, G.B. (1985) Induction of human interleukin-1 by toxic shock syndrome toxin-1. *Journal of Infectious Diseases*, **151**, 514–522.

Pawson, T. (1988) Non-catalytic domains of cytoplasmic protein-tyrosine kinases: regulatory elements in signal transduction. *Oncogene*, **3**, 491–495.

Perianin, A. and Snyderman, R. (1989) Analysis of calcium homeostasis in activated human polymorphonuclear neutrophils: evidence for two distinct mechanisms for lowering calcium. *Journal of Biological Chemistry*, **264**, 1005–1009.

Perney, T.M. and Miller, R.J. (1989) Two different G-proteins mediate neuropeptide Y and bradykinin-stimulated phospholipid breakdown in cultured rat sensory neurons. *Journal of Biological Chemistry*, **264**, 7317–7327.

Peters, K.G., Marie, J., Wilson, E., Ives, H.E., Escobedo, J., Del Rosario, M., Mirda, D. and Williams, L.T. (1992) Point mutation of an FGF receptor abolishes phosphatidylinositol turnover and Ca^{2+} flux but not mitogenesis. *Nature*, **358**, 678–681.

Petrenko, A.G., Lazaryeva, V.D., Geppert, M., Tarasyuk, T.A., Moomaw, C., Khokhlatchev, A.V., Ushkaryov, Y.A., Slaughter, C., Nasimov, I.V. and Sudhof, T.C. (1993) Polypeptide composition of the α-latrotoxin receptor. High affinity binding protein consists of a family of related high molecular weight polypeptides complexed to a low molecular weight protein. *Journal of Biological Chemistry*, **138**, 91–102.

Petrenko, W.G., Kovalenko, V.A., Shamotienko, O.G., Surkova, I.N., Tarasyuk, T.A., Ushkaryov, T.A. and Grishin, E.V. (1990) Isolation and properties of the α-latrotoxin receptor. *EMBO Journal*, **6**, 2023–2027.

Pike, L.J. and Eakes, A.T. (1987) Epidermal growth factor stimulates the production of phosphatidylinositol monophosphate and the breakdown of polyphosphoinositides in A431 cells. *Journal of Biological Chemistry*, **262**, 1644–1651.

Pittner, R.A. and Fain, J.N. (1989) Exposure of cultured hepatocytes to cyclic AMP enhances the vasopressin-mediated stimulation of inositol phosphate production. *Biochemical Journal*, **257**, 455–460.

Quick, M.W., Simon, M.I., Davidson, N., Lester, H.A. and Aragay, A.M. (1994) Differential coupling of G protein α subunits to seven-helix receptors expressed in *Xenopus* oocytes. *Journal of Biological Chemistry*, **269**, 30164–30172.

Rebois, R.V. and Patel, J. (1985) Phorbol ester causes desensitization of gonadotropin-responsive adenylate cyclase in a murine Leydig tumor cell line. *Journal of Biological Chemistry*, **260**, 8026–8031.

Reisine, T. (1990) Pertussis toxin in the analysis of receptor mechanisms. *Biochemical Pharmacology*, **39**, 1499–1504.

Rens-Domino, S. and Hamm, H.E. (1995) Structural and functional relationships of heterotrimeric G-proteins. *FASEB Journal*, **91**, 1059–1066.

Rhee, S.G. (1991) Inositol phospholipid-specific phospholipase C: interaction of the γ_1 isoform with tyrosine kinase. *Trends in Biochemical Science*, **16**, 297–301.

Rhee, S.G. and Choi, K, D. (1992a) Multiple forms of phospholipase C isozymes and their activation mechanisms. *Advances in Second Messengers and Phosphoprotein Research*, **26**, 35–61.

Rhee, S.G. and Choi, K, D. (1992b) Regulation of inositol phospholipid-specific phospholipase C isozymes. *Journal of Biological Chemistry*, **267**, 12393–12396.

Rhee, S.G., Kim, U-H. and Kim, J.W. (1990) Cross-talk between cellular signalling cascade pathways suggested by cAMP-induced phosphorylation of phospholipase C-γ. *Advances in Second Messenger and Phosphoprotein Research*, **24**, 164–169.

Rhee, S.G., Lee, C-W. and Jhon, D-Y. (1993) Phospholipase C isozymes and modulation by cAMP-dependent protein kinase. *Advances in Second Messenger and Phosphoprotein Research*, **28**, 57–64.

Rhoden, K.J. and Douglas, J.S. (1994) Stimulation of GTP hydrolysis in guinea pig bronchial membranes by mastoparan. *Lung*, **172**, 355–363.

Richardson, A. and Parsons, J.T. (1995) Signal transduction through integrins: a central role for focal adhesion kinase? *Bioessays*, **17**, 229–236.

Rogolsky, M. (1979) Nonenteric toxins of *Staphylococcus aureus*. *Microbiology Reviews*, **43**, 320–360.

Rönnstrand, L., Mori, S., Arridsson, A.K., Eriksson, A., Wernstedt, C., Hellman, U., Claesson-Welsh, L. and Heldin, C.H. (1992) Identification of two C-terminal autophosphorylation sites in the PDGF beta-receptor: involvement in the interaction with phospholipase C-γ. *EMBO Journal*, **11**, 3911–3919.

Rosenthal, L., Zacchetti, D., Madeddu, L. and Meldolesi, J. (1990) Mode of action of α-latrotoxin: role of divalent cations in Ca^{2+}-dependent and Ca^{2+}-independent effects mediated by the toxin. *Molecular Pharmacology*, **38**, 917–923.

Rubeiro-Neto, F.A.P., Mattera, R., Grenet, D. Sekura, R.D., Birnbaumer, L. and Field, J.B. (1987) Adenosine diphosphate ribosylation of G proteins by pertussis and cholera toxin in isolated membranes. Different requirements for and effects of guanine nucleotides and Mg^{2+}. *Molecular Endocrinology*, **1**, 472–481.

Rubeiro-Neto, F.A.P., Mattera, R., Hildebrandt, J.D., Codina, J., Field, J.B., Birnbaumer, L. and Sekura, R.D. (1985) ADP-ribosylation of membrane components by pertussis and cholera toxin. *Methods in Enzymology*, **109**, 566–571.

Ryu, S.H., Cho, K.S., Lee, K-Y., Suh, P-G. and Rhee, S.G. (1987a) Purification and characterization of two immunologically distinct phosphoinositide-specific phospholipase C from bovine brain. *Journal of Biological Chemistry*, **262**, 12511–12518.

Ryu, S.H., Kim, U-H., Wahl, M.I., Brown, A.B., Carpenter, G., Huang, K.P. and Rhee, S.G. (1990) Feedback regulation of phospholipase C-beta by protein kinase C. *Journal of Biological Chemistry*, **265**, 17941–17945.

Ryu, S.H., Suh, P-G., Cho, K.S., Lee, K-Y. and Rhee, S.G. (1987b) Bovine brain cytosol contains three immunologically distinct forms of inositol phospholipid-specific phospholipase C. *Proceedings of the National Academy of Sciences, USA*, **84**, 6649–6653.

Schaller, M.D. and Parsons, J.T. (1994) Focal adhesion kinase and associated proteins. *Current Opinion in Cell Biology*, **6**, 705–710.

Schettini, G., Koike, K., Login, I.S., Judd, A.M., Cronin, M.J., Yasumoto, T. and MacLeod, R.M. (1984) Maitotoxin stimulates hormonal release and calcium flux in rat anterior pituitary cells *in vitro*. *American Journal of Physiology*, **247**, E520–E525.

Schlessinger, J. (1988) The epidermal growth factor as a multifunctional allosteric protein. *Biochemistry*, **27**, 3119–3123.

Schmidt, C.J., Thomas, T.C., Levine, M.A. and Neer, E.J. (1992) Specificity of G-protein β and γ subunit interaction. *Journal of Biological Chemistry*, **267**, 13807–13810.

Scholl, P., Diez, A., Mourad, W., Parsonnet, J., Geha, R.S. and Chatila, T. (1989) Toxic shock syndrome toxin-1 binds to class II major histocompatibility molecules. *Proceedings of the National Academy of Sciences, USA*, **86**, 4210–4214.

Seedorf K., Kostka, G., Lammers, R., Bashkin, P., Daly, R., Burgess, W.H., van der Bliek, A.M., Schlessinger, J. and Ullrich, A. (1994) Dynamin binds to SH3 domains of phospholipase Cγ and GRB-2. *Journal of Biological Chemistry*, **269**, 16009–16014.

Sekura, R.D. and Zhang, Y.-L. (1985) Pertussis toxin: Structural elements involved in the interaction with cells. In *Pertusis Toxin* (Sekura, R.D., Moss, J. and Vaughan M., eds.), Academic Press, Inc., London, 45–64

Shalaby, I.A., Kongsamut, S. and Miller, R.J. (1986) Maitotoxin-induced release of γ-[^{3}H]aminobutyric acid from cultures of striatal neurons. *Journal of Neurochemistry*, **46**, 1161–1165.

Shenker, A., Goldsmith, P. Unson, C.G. and Spiegel, A.M. (1991) The G protein coupled to the thromboxane A$_2$ receptor in human platelets is a member of the novel G$_q$ family. *Journal of Biological of Chemistry*, **266**, 9309–9313.

Shin, Y., Moni, R.W., Lueders, J.E., Daly, J.W. (1994) Effects of the amphiphilic peptides mastoparan and adenoregulin on receptor binding, G proteins, phosphoinositide breakdown, cyclic AMP generation and calcium influx. *Cellular and Molecular Neurobiology*, **14**, 133–157.

Shortridge, R.D., Yoon, H., Lending, C.R., Bloomquist, B.T., Perdew, M.H. and Pak, W.L. (1991) A *Drosophila* phospholipase C gene that is expressed in the central nervous system. *Journal of Biological Chemistry*, **266**, 12474–12480.

Sillman, A. and Monroe, J.G. (1995) Association of p72^{syk} with the src homology-2 (SH2) domains of PLC-γ1 in B lymphocytes. *Journal of Biological Chemistry*, **270**, 11806–11811.

Simon, M.I., Strathmann, M.P. and Gautam, N. (1991) Diversity of G proteins in signal transduction. *Science*, **252**, 802–808.

Sladeczek, F., Schmidt, B.H., Alonso, R., Vian, L., Tep, A., Yasumoto, T., Cory, R.N. and Bockaert, J. (1988) New insights into maitotoxin action. *European Journal of Biochemistry*, **174**, 663–670.

Smith, C.D., Lane, B.C., Kusaka, I., Verghese, M.W. and Snyderman, R. (1985) Chemoattractant receptor-induced hydrolysis of phosphatidylinositol 4,5-bisphosphate in human polymorphonuclear leukocyte membranes. *Journal of Biological Chemistry*, **260**, 5875–5878.

Smrcka, A.V., Hepler, J.R., Brown, K.O. and Sternweis, P.C. (1991) Regulation of polyphosphoinositide-specific phospholipase C activity by purified Gq. *Science*, **250**, 804–807.

Smrcka, A.V. and Sternweis, P.C. (1993) Regulation of purified subtypes of phosphatidylinositol-specific phospholipase Cβ by G-protein α and βγ subunits. *Journal of Biological Chemistry*, **268**, 9667–9674.

Soergel, D.G., Yasumoto, T., Daly, J.W. and Gusovsky, F. (1992) Maitotoxin effects are blocked by SK&F 96365, an inhibitor of receptor–mediated calcium entry. *Molecular Pharmacology*, **41**, 487–493.

Sokolovsky, M. (1992) Structure-function relationships of endothelins, sarafotoxins and their receptor subtypes. *Journal of Neurochemistry*, **59**, 809–821.

Somogyi, L., Lasic, Z., Vukicevic, S. and Banfic, H. (1994) Collagen type IV stimulates an increase intracellular Ca^{2+} in pancreatic acinar cells via activation of phospholipase C. *Biochemical Journal*, **299**, 603–611.

Stahl, M.L., Ferenz, C.R., Kelleher, K.L., Kriz, R.W. and Knopf, J.L. (1988) Sequence similarity of phospholipase C with the non-catalytic region of src. *Nature*, **332**, 269–272.

Strathmann, M. and Simon, M.I. (1990) G protein diversity: a distinct class of α subunits is present in vertebrates and invertebrates. *Proceedings of the National Academy of Sciences, USA*, **87**, 9113–9117.

Stryer, L.C. (1986) Cyclic GMP cascade of vision. *Annual Review of Neuroscience*, **9**, 87–119.

Stull, H.T., Kamm, K.E. and Taylor, D.A. (1988) Calcium control of smooth muscle contractility. *American Journal of Medical Sciences*, **296**, 241–245.

Su, X., Chen, F. and Hokin, L.E. (1994) Cloning and expression of a novel, highly truncated phosphoinositide-specific phospholipase C cDNA from embryos of the brine shrimp, *Artemia*. *Journal of Biological Chemistry*, **269**, 12925–12931.

Suh, P.G., Ryu, S.H., Moon, K.H., Suh, H.W. and Rhee, S.G. (1988a) Inositol phospholipid-specific phospholipase C: complete cDNA and protein sequences and sequence homology to tyrosine kinase-related oncogene products. *Proceedings of the National Academy of Sciences, USA*, **85**, 5419–5423.

Suh, P.G., Ryu, S.H., Moon, K.H., Suh, H.W. and Rhee, S.G. (1988b) Cloning and sequence of multiple forms of phospholipase C. *Cell*, **54**, 161–169.

Takahashi, M., Ohizumi, Y. and Yasumoto, T. (1982) Maitotoxin, a Ca^{2+} channel activator candidate. *Journal of Biological Chemistry*, **257**, 7287–7289.

Takahashi, M., Tatsumi, M., Ohizumi, Y. and Yasumoto, T. (1983) Ca^{2+} channel activating function of maitotoxin, the most potent marine toxin known, in clonal rat pheochromocytoma cells. *Journal of Biological Chemistry*, **258**, 10944–10949.

Takai, Y., Kishimoto, A., Kikkawa, U., Mori, T. and Nishizuka, Y. (1979) Unsaturated diacylglycerol as a possible messenger for the activation of calcium-activated, phospholipid-dependent protein kinase system. *Biochemical Biophysical Research Communications*, **91**, 1218–1224.

Takasaki, C., Itoh, Y., Onda, H. and Fujino, M. (1992) Cloning and sequence analysis of a snake, *Atractaspis engaddensis* gene encoding sarafotoxin S6c. *Biochemical and Biophysical Research Communications*, **189**, 1527–1533.

Takasaki, C., Tamiya, N., Bdolah, A., Wollberg, Z. and Kochva, E. (1988) Sarafotoxin S6: Several isotoxins from *Atractaspic engaddensis* (burrowing asp) venom that affect the heart. *Toxicon*, **26**, 543–548.

Tang, W-J. and Gilman, A.G. (1993) Type-specific regulation of adenylyl cyclase by G protein $\beta\gamma$ sub-units. *Science*, **254**, 1500–1503.

Taylor, S.J., Smith, J.A. and Exton, J.H. (1990) Purification from bovine liver membranes of a guanine nucleotide-dependent activator of phosphoinositide-specific phospholipase C. Immunologic identification as a novel G-protein α subunit. *Journal of Biological Chemistry*, **265**, 17150–17156.

Taylor, S.J., Chae, H.Z., Rhee, S.G. and Exton, J.H. (1991a) Activation of the β1 isozyme of phospholipase C by α subunits of the Gq class of G proteins. *Nature*, **350**, 516–518.

Taylor, S.J. and Exton, J.H. (1991b) Two α subunits of the Gq class of G proteins stimulate phospho-inositide phospholipase C-β1 activity. *FEBS Letters*, **286**, 214–216.

Thelestam, M. and Bloomquist, L. (1988) *Staphylococcal alpha* toxin- recent advances. *Toxicon*, **26**, 55–65.

Titball, R.W. (1993) Bacterial phospholipases C. *Microbiological Reviews*, **57**, 347–366.

Todderud, G., Wahl, M.I., Rhee, S.G. and Carpenter, G. (1990) Stimulation of phospholipase C-γ1 membrane association by epidermal growth factor. *Science*, **249**, 296–298.

Trede, N.S., Geha, R.S. and Chatila, T. (1991) Transcriptional activation of Il-1β and tumor necrosis factor-α genes by MHC class II ligands. *Journal of Immunology*, **146**, 2310–2315.

Trede, N.S., Morio, T., Scholl, P.R., Geha, R.S. and Chatila, T.A. (1994) Early activation events induced by the staphylococcal superantigen toxic shock syndrome toxin-1 in human peripheral blood mono-cytes. *Clinical Immunology and Immunopathology*, **70**, 137–144.

Tsunoda, Y. (1993) Receptor-operated Ca^{2+} signaling and crosstalk in stimulus secretion coupling. *Biochimica Biophysica Acta*, **1154**, 105–156.

Ui, M. (1985) Islet-activating protein, pertussis toxin: subunit structure and mechanism for its multiple biological actions. In *Pertusis Toxin* (Sekura, R.D., Moss, J. and Vaughan, M., eds.) Academic Press, Inc., London, 19–43

Ui, M. (1990) Pertussis toxin as a valuable probe for G-protein involvement in signal transduction. In: *ADP-ribosylating toxins and G proteins: insights into signal transduction* (Moss, J. and Vaughan, M., eds.) American Society of Microbiology, Washington, D.C., 45–77.

Valius, M., Bazenet, C. and Kazlauskas, A. (1993) Tyrosines 1021 and 1009 are phosphorylation sites in the carboxy terminus of the platelet-derived growth factor receptor beta subunit and are required for binding of phospholipase C-γ and a 64-kilodalton protein, respectively. *Molecular and Cellular Biology,*, **13**, 133–143.

Vega, Q.C., Cochet, C., Filhol, O., Chang, C-P., Rhee, S.G. and Gill, G. (1992) A site of tyrosine phos-phorylation in the C terminus of the epidermal growth factor receptor is required to activate phos-pholipase C. *Molecular and Cellular Biology*, **12**, 128–135.

Vetter, M.L., Martin-Zanca, D., Parada, L.F., Bishop, J.M. and Kaplan, D.R. (1991) Nerve growth factor rapidly stimulates tyrosine phosphorylation of phospholipase C-γ1 by a kinase associated with the product of the trk protooncogene. *Proceedings of the National Academy of Sciences, USA*, **88**, 5650–5654.

Vicentini, L. and Meldolesi, J. (1984) α-Latrotoxin of black widow spider venom binds to a specific receptor coupled to phosphoinositide breakdown in PC12 cells. *Biochemical and Biophysical Research Communications*, **121**, 538–544.

Volpe, P., Virgillo, F.D., Pozzan, T. and Salvaitai, G. (1986) Role of inositol 1,4,5-trisphosphate in excita-tion-contraction coupling in skeletal muscle. *FEBS Letters*, **1974**, 1–4.

Wahl, M.I., Nishibe, S., Kim, J.W., Kim, H., Rhee, S.G. and Carpenter, G. (1990) Identification of two epidermal growth factor-sensitive tyrosine phosphorylation sites of phospholipase C-γ in intact HSC-1 cells. *Journal of Biological Chemistry,* **265,** 3944–3948.

Wahl, M.I., Nishibe, S., Suh, P-G., Rhee, S.G. and Carpenter, G. (1989) Epidermal growth factor stimulates tyrosine phosphorylation of phospholipase C-II independently of receptor internalization and extracellular calcium. *Proceedings of the National Academy of Sciences, USA,* **86,** 1568–1572.

Wahl, M.I., Olashaw, N.E., Nishibe, S., Rhee, S.G., Pledger, W.J. and Carpenter, G. (1989) Platelet-derived growth factor induces rapid and sustained tyrosine phosphorylation of phospholipase C-γ in quiescent BALB/c 3T3 cells. *Molecular and Cellular Biology,* **9,** 2934–2943.

Wahl, M.I., Sweatt, J.D. and Carpenter, G. (1987) Epidermal growth factor (EGF) stimulates inositol trisphosphate formation in cells which overexpress the EGF receptor. *Biochemical and Biophysical Research Communications,* **142,** 688–695.

Walton, B.M., Chen, W.S., Rosenfeld, M.G. and Gill, G.N. (1990) Analysis of deletions of the carboxyl terminus of the epidermal growth factor receptor reveals self- phosphorylation at tyrosine 992 and enhanced *in vivo* tyrosine phosphorylation of cell substrates. *Journal of Biological Chemistry,* **265,** 1750–1754.

Wange, R.L., Smrcka, A.V. Sternweis, P.C. and Exton, J.H. (1991) Photoaffinity labeling of two rat liver plasma membrane proteins with [32P]γ-azidoanilide GTP in response to vasopressin. *Journal of Biological Chemistry,* **266,** 11409–11412.

Wanke, E., Ferroni, A., Gattanini, P. and Meldolesi, J. (1986) α-Latrotoxin of the black widow spider venom opens a small, non-closing cation channel. *Biochemical and Biophysical Research Communications,* **134,** 320–325.

Ward, S.G. and Cantrell, D.A. (1990) Heterogeneity of the regulation of phospholipase C by phorbol esters in T lymphocytes. *Journal of Immunology,* **144,** 3523–3528.

Watson, S.P., McConnell, R.T. and Lapetina, E.G. (1984) The rapid formation of inositol phosphates in human platelets by thrombin is inhibited by prostacyclin. *Journal of Biological Chemistry,* **259,** 13199–13203.

Weiss, A., Koretzky, G., Schatzman, R.C. and Kadlecek, T. (1991) Functional activation of the T-cell antigen receptor induces tyrosine phosphorylation of phospholipase C-γ1. *Proceedings of the National Academy of Sciences, USA,* **88,** 5484–5488.

West, R.E., Moss, J., Vaughan, M., Liu, T. and Liu, T-Y. (1985) Pertussis toxin-catalyzed ADP-ribosylation of transducin. Cysteine 347 is the ADP ribose acceptor site. *Journal of Biological Chemistry,* **260,** 14428–14430.

Wilkie, T.M., Scherle, P.A., Strathmann, M.P., Slepak, V.Z. and Simon, M.I. (1991) Characterization of G protein α-subunits in the G_q class: expression in murine tissues and in stromal and hematopoietic cell lines. *Proceedings of the National Academy of Science, USA,* **88,** 10049–10053.

Wilson, S.P. (1989) Effects of mastoparan on catecholamine release from chromaffin cells. *FEBS Letters,* **247,** 239–241.

Wojcikiewicz, R.J.H. and Nahorski, S.R. (1989) Phosphoinositide hydrolysis in permeabilized SH-SY5Y human neuroblastoma cells is inhibited by mastoparan. *FEBS Letters,* **247,** 341–344.

Wu, D., Jiang, H., Katz, A. and Simon, M.I. (1993a) Identification of critical regions on phospholipase C-β1 required for activation by G-proteins. *Journal of Biological Chemistry,* **268,** 3704–3709.

Wu, D., Katz, A. and Simon, M.I (1993b) Activation of phospholipase C β2 by the α and βγ subunits of trimeric GTP-binding protein. *Proceedings of the National Academy of Science, USA,* **90,** 5297–5301.

Wu, D., LaRosa, G.J. and Simon, M.I. (1993c) G protein-coupled signal transduction pathways for interleukin-8. *Science,* **261,** 101–103.

Wu, D. Lee, C.H., Rhee, S.G. and Simon, M.I. (1992) Activation of phospholipase C by the α subunits of the Gq and G11 proteins in transfected COS-7 cell. *Journal of Biological Chemistry,* **267,** 1811–1817.

Yada, Y., Nagao, S., Okano, Y. and Nozawa, Y. (1989) Inhibition by cyclic AMP of guanine nucleotide-induced activation of phosphoinositide-specific phospholipase C in human platelets. *FEBS Letters,* **242,** 368–372.

Yagisawa, H., Hirata, M., Kanematsu, T., Watanabe, Y., Ozaki, S., Sakuma, K., Tanaka, H., Yabuta, N., Kamata, H., Hirata, H. and Nojima, H. (1994) Expression and characterization of an inositol 1,4,5-trisphosphate binding domain of phosphatidylinositol-specific phospholipase C-δ1. *Journal of Biological Chemistry,* **269,** 20179–20188.

Yamane, H.K. and Fung, K-K. (1993) Covalent modifications of G-proteins. *Annual Review of Pharmacology and Toxicology*, **32**, 201–241.

Yang, C.M., Yo, Y.L., Ong, R. and Hsieh, J.T. (1994) Endothelin- and sarafotoxin-induced phosphoinositide hydrolysis in cultured canine tracheal smooth muscle cells. *Journal of Neurochemistry*, **62**, 1440–1448.

Yang, L., Baffy, G., Rhee, S.G., Manning, D., Hansen, C.A. and Williamson, J.R. (1991) Pertussis toxin-sensitive G_i protein involvement in epidermal growth factor-induced activation of phospholipase C-γ in rat hepatocytes. *Journal of Biological Chemistry*, **266**, 22451–22458.

Zawalich, W.S., Diaz, V.A. and Zawalich, K.C. (1988) Influence of cAMP and calcium on [^{3}H]inositol efflux, inositol phosphate accumulation and insulin release from isolated rat islets. *Diabetes*, **37**, 1478–1483.

15. RYANODINE RECEPTOR-SPECIFIC SCORPION TOXINS: NEW PROBES FOR THE ANALYSIS OF CHANNEL STRUCTURE AND FUNCTION

JEFFERY MORRISSETTE and ROBERTO CORONADO

Department of Physiology, University of Wisconsin School of Medicine, Madison, WI. 53706, USA

INTRODUCTION

Ryanodine receptors are intracellular Ca^{2+} channels which in muscle, are ultimately responsible for the initiation and propagation of intracellular Ca^{2+} signals (Coronado, 1994). In cardiac muscle, Ca^{2+} entry into cells directly triggers the opening of ryanodine receptors which results in release of Ca^{2+} from the main storage site, the sarcoplasmic reticulum (SR) (Stern, 1992). In skeletal muscle, membrane depolarization opens the ryanodine receptor without the participation of the Ca^{2+} current (McPherson, 1993). Another intracellular Ca^{2+} channel, the IP_3 receptor, has been implicated in stimulus-Ca^{2+} release coupling in many types of muscle and non-muscle cells (Mikoshiba, 1993). The presence of IP_3 receptors and ryanodine receptors in the same tissue has also been reported (Jorgensen, 1993; Moschella, 1993). In myocardium, the concerted action of IP_3 and ryanodine receptors may explain the α_1 receptor meadiated positive inotropic effect (Otani, 1988).

The availability of pharmacological agents to selectively block or enhance the activity of specific intracellular Ca^{2+} channels seems critical to evaluate their role in signal transduction. For the case of ryanodine receptors, the plant alkaloid ryanodine and it's related analogs (Humerickhouse, 1994), are the only agents that are sufficiently specific to permit inferences about the channel's contribution to Ca^{2+} signals *in situ* (Beuckelmann, 1988). The usefulness of ryanodine is rather limited since its slow association and dissociation kinetics make the onset of the pharmacological effect slow and essentially irreversible (Pessah, 1986). Furthermore, the effects of ryanodine are complex in that the receptor contains both high and low affinity binding sites for the alkaloid. Ryanodine binding to the high affinity site opens the channel to a low conductance state, while subsequent binding to the low affinity site results in channel closure (Wang, 1993). By depleting the SR of stored Ca^{2+}, ryanodine may also blunt Ca^{2+} signals initiated by non-ryanodine receptor pathways or channels.

Many studies have indicated that scorpion venoms are a rich source of peptides highly specific for a single type of channel (Strichartz, 1987; Miller, 1985). We turned to these venoms in search for a ligand specific for ryanodine receptors that could act quickly, reversibly and in a simple manner. We screened several scorpion venoms using a $[^3H]$ryanodine binding assay to rabbit skeletal muscle SR enriched in terminal cisternae. Ryanodine binds to the receptor irreversivebly and preferentially to the open state of the channel (Imagawa, 1987). Compounds such as caffeine and adenine nucleotides which activate the channel, increase the number of radiolabelled receptors per unit time whereas inhibitors such as Mg^{2+} and ruthenium red, decrease the number radiolabelled receptors per unit time (Pessah, 1987). Thus, by measuring the level of [^{3}H]ryanodine binding, at appropriate times before equilibrium is reached, we can correlate the changes in binding activity with the changes in the number of open channels produced by a given venom. We found that the venom of the North African scorpion *Buthotus hottentota* stimulated [^{3}H]ryanodine binding to rabbit skeletal muscle SR in a dose dependent manner. Furthermore, single channel recordings comfirmed that the venom activated ryanodine receptor channels incorporated into planar bilayers. In this chapter we describe the purification and characterization of ryanodine receptor-specific peptide components of *Buthotus hottentota* venom and discuss their potential uses to elucidate the structure and function of ryanodine receptors.

PURIFICATION OF TOXINS

Approximately 100 mg of lyophilized *Buthotus hottentota* venom were resuspended in 3 ml of deionized water and sonicated. The mucus was removed by aspiration yielding a water soluble fraction with a protein concentration of about 4 mg/ml. Active toxins were purified from the water soluble fraction by a total of three rounds of high performance liquid chromatography (HPLC) using preparative and analytical C_{18} reverse phase columns as described in the figure legends. Figure 15.1 corresponds to the purification scheme of two stimulatory toxins and Figure 15.2 corresponds to the purification scheme of two inhibitory toxins all from the same *Buthotus* venom. Bound peptides were eluted with acetonitrile gradients indicated by the dashed lines and monitored by light absorption at 254 nm. After each round of fractionation (HPLC I, II, IIIA, IIIB in Figure 15.1; HPLC I, A, B, C in Figure 15.2), the largest and best-resolved A_{254} peaks were collected, dried under vacuum centrifugation and resuspended in a minimal volume of distilled water. The resuspended fractions were tested for activity using the [^{3}H]ryanodine binding assay described below at approximately equivalent concentrations in A_{254} units/ml. Active fractions indicated by the arrows were subjected to subsequent rounds of fractionation until single homogeneous peaks were obtained. Table 15.1 summarizes the changes in binding activity relative to control produced by each of the collected HPLC fractions. Upon the fractionation of the whole venom we found peaks that, like the whole venom, stimulated [^{3}H]ryanodine binding. However, we also detected peaks which inhibited [^{3}H]ryanodine binding. For example, after the initial preparative HPLC (HPLC I), peak 5 stimulated specific [^{3}H]ryanodine binding approximately 600%, while peak 6 completely inhibited specific binding. As indicated in Table 15.1, other peaks from HPLC I had little or no effect on [^{3}H]ryanodine binding. In two subsequent rounds, we fractionated stimulatory peak 5 from HPLC I into two single active peaks labelled Buth$_A$-1 and Buth$_A$-2

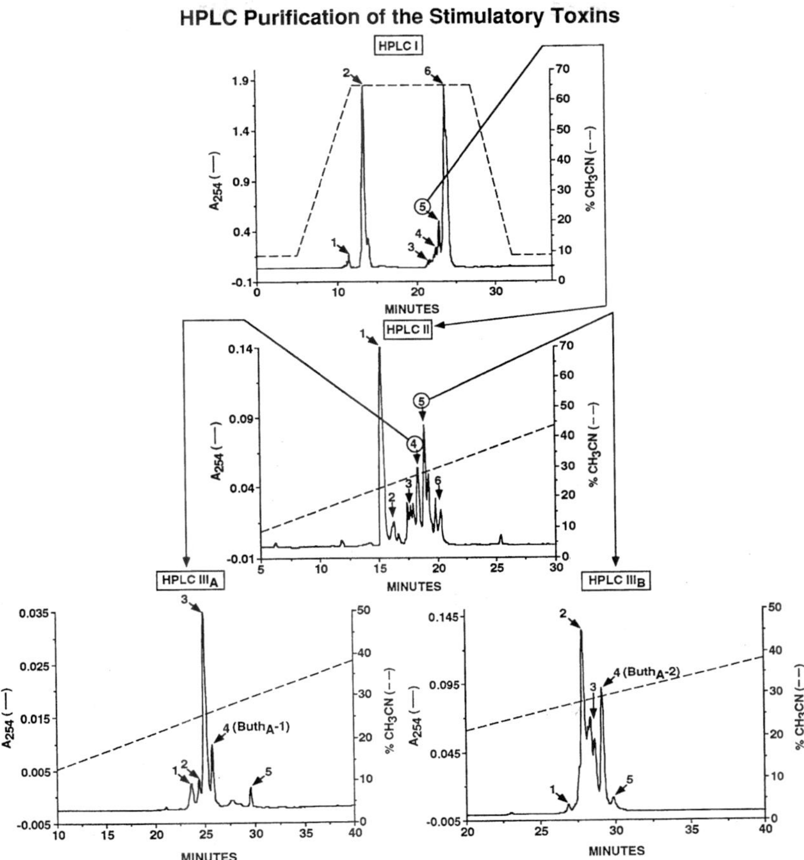

Figure 15.1 Purification scheme for the stimulatory toxins from *Buthotus hottentota* venom. *Top panel:* Chromatographic profile of HPLC I resulting from the injection of 500 μg of water soluble whole venom into a C_{18} reverse phase column (μBondapac-Waters, 0.78 × 30 cm). Fractions were eluted with a linear 8.75–65.0% acetonitrile gradient containing 0.1% trifluroacetic acid run at a flow rate of 1.0 ml/min for 60 min. *Middle Panel:* Chromatographic profile of HPLC II resulting from the injection of approximately 100 μg of stimulatory fraction 5 from HPLC I into a C_{18} reverse phase column (Vydac, 0.46 × 25 cm). Fractions were eluted with a linear 8.75–65.0% acetonitrile gradient containing 0.1% trifluroacetic acid run at a flow rate of 1.0 ml/min for 60 min. *Bottom Panel:* Chromatographic profiles resulting from the injection of approximately 5 mg of stimulatory fraction 4 from HPLC II (left) or 5 μg of stimulatory fraction 5 from HPLC II (right) into the Vyadac column. Fractions were eluted with a 8.75–42.5% acetonitrile gradient containing 0.1% trifluroacetic acid run at a flow rate of 1.0 ml/min.

Table 15.1 [³H]ryanodine binding activities of HPLC fractions resulting from the fractionation of the whole *Buthotus hottentota* venom. Control binding was labeled 100% and corresponds to 0.4 pmol/mg. Peaks within each HPLC run (HPLC I, II, IIIA, IIIB, A, B, C) were assayed at approximately the same concentration (0.05–0.1 A_{254} units/ml). Binding activities from peaks of different HPLC runs are not directly comparable as A_{254} units assayed were lower with increased toxin purity. Shaded peaks correspond to the pure toxin fractions.

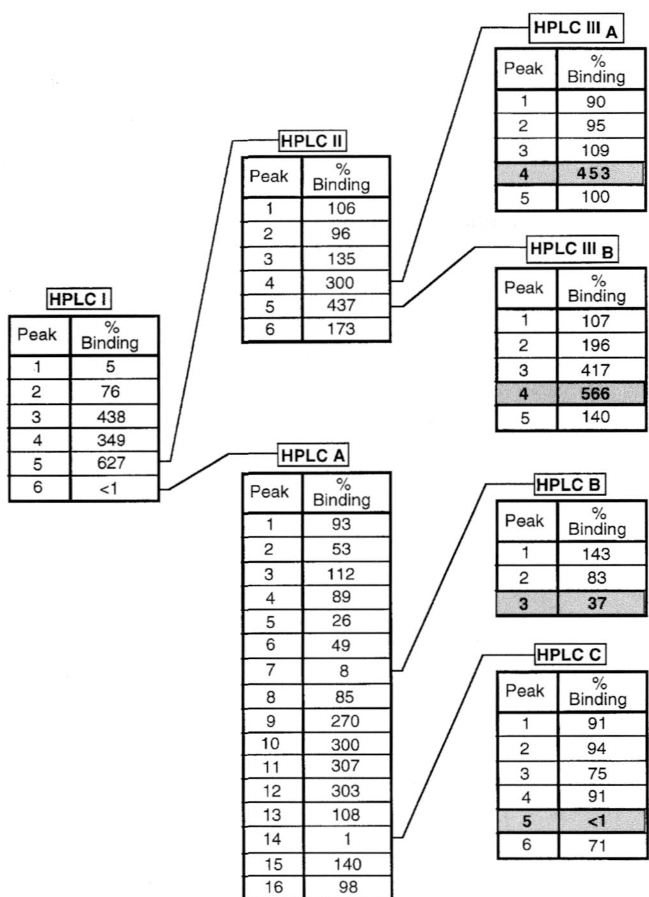

(Figure 15.1). Similarily, we fractionated the inhibitory peak 6 from HLPC I into two single active peaks labelled $Buth_r$-1 and $Buth_r$-2 (Figure 15.2). The peptide composition of the the four toxin fractions was determined by SDS-PAGE (sodium dodecyl sulfate-polyacrylamide gel electrophoresis) on 6-20% polyacrylamide gels.

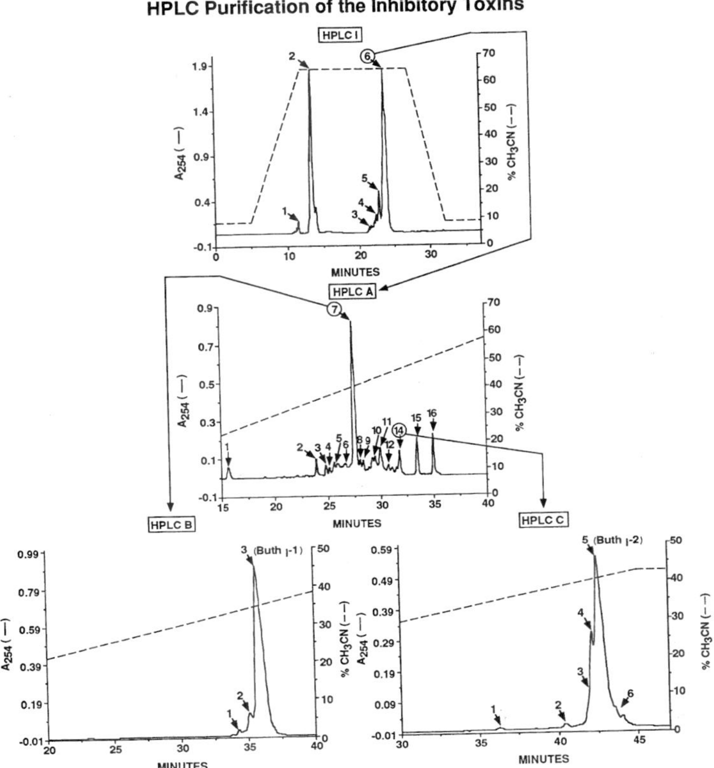

Figure 15.2 Purification scheme for the inhibitory toxins from *Buthotus hottentota* venom. *Top Panel*: Chromatographic profile of HPLC I resulting from the injection of approximately 500 μg of the water soluble whole venom into a C_{18} reverse phase column (μBondapac-Waters, 0.78 × 30 cm). Fractions were eluted with a linear 8.75–65.0% acetonitrile gradient containing 0.1% trifluroacetic acid run at a flow rate of 1.0 ml/min for 60 min. *Middle Panel*: Chromatographic profile of HPLC A resulting from the injection of approximately 500 mg of inhibitory fraction 6 from HPLC I into a C_{18} reverse phase column (Vydac, 0.46 × 25 cm). Fractions were eluted with a linear 8.75–65.0% acetonitrile gradient containing 0.1% trifluroacetic acid run at a flow rate of 1.0 ml/min for 60 min. *Bottom Panel*: Chromatographic profiles resulting from the injection of approximately 200 μg of the inhibitory fraction 7 from HPLC A (left) or 200 μg of inhibitory fraction 14 from HPLC A (right) into the Vyadac column. Fractions were eluted with a 8.75–42.5% acetonitrile gradient containing 0.1% trifluroacetic acid run at a flow rate of 1.0 ml/min.

As shown in Figure 15.3, all four toxin fractions eluted as single symmetrical peaks when further HPLC fractionation was attempted. Furthermore, the toxin fractions consisted of a single polypeptide with an estimated molecular weight of approximately 7, 12, 13 and 13 KDa for $Buth_A$-1, $Buth_A$-2, $Buth_I$-1 and $Buth_I$-2, respectively (Figure 15.4). The percentage of the toxins in the whole venom, estimated by the

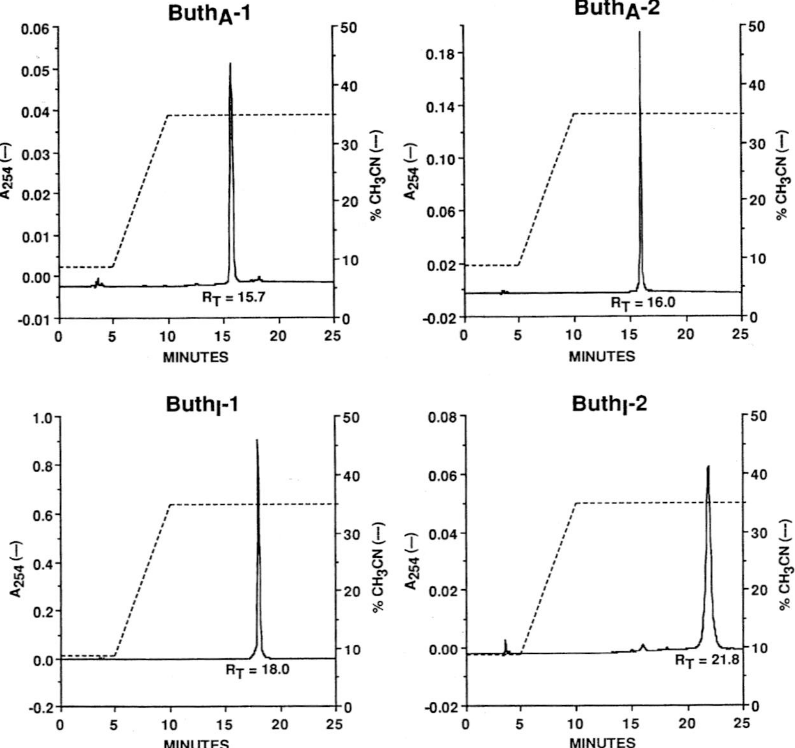

Figure 15.3 Chromatographic profile of the four pure *Buthotus hottentota* toxins. Injection of the peaks labelled Buth$_A$-1, Buth$_A$-2, Buth$_I$-1 and Buth$_I$-2 (Figure 15.1 and 15.2) into the Vydac column resulted in the elusion of single symetrical peaks. Pure toxins were eluted with an isocratic 35% acetonitrile gradient containing 0.1% trifluroacetic acid run at a flow rate of 1.0 ml/min. Retention times (R$_T$) are shown below each peak.

absorbance at 254 nm, was 0.04% for Buth$_A$-1, 0.35% for Buth$_A$-2, 17.7% for Buth$_I$-1 and 0.63% for Buth$_I$-2 (Table 15.2). The higher venom content of inhibitory toxins was surprising given that unfractionated venom stimulated [^{3}H]ryanodine binding.

CHARACTERIZATION OF TOXINS

Effects on [^{3}H]Ryanodine Binding

[^{3}H]ryanodine binding to rabbit skeletal muscle heavy SR has been described in detail elsewhere (Valdivia, 1991). Briefly, HPLC fractions (0.05–0.1 A$_{254}$ units/ml), rabbit skeletal muscle heavy SR (40 mg protein) and 7 nM [^{3}H]ryanodine were incubated for 90 minutes at 36°C in 0.1 ml of 0.2 M KCl, 1 mM Na$_2$EGTA (ethyl-

Table 15.2 Summary of the molecular weights, affinities and toxin content of the four ryanodine receptor toxins. ED_{50} values (concentration which produce 50% maximal stimulation or 50% maximal inhibition) are given in both mg/ml and molar units. Toxin contents were determined from the percent of the total absorbance at 254 nm contributed by each pure peak.

Peak	M.W (Da)	ED_{50}/IC_{50}	ED_{50}/IC_{50}	Content (% A.U. whole venom)
Buth$_A$-1	7,000	0.5 µg/ml	70 nM	0.04 %
Buth$_A$-2	11,750	10 µg/ml	0.9 µM	0.35 %
Buth$_I$-1	13,320	390 µg/ml	30 µM	17.7 %
Buth$_I$-2	13,320	90 µg/ml	6.9 µM	0.63 %

ene glycol-bis(aminoethyl ether) tetraacetic acid), 0.995 mM $CaCl_2$, 10 mM Na-PIPES (1,4-piperazine ethanesulfonic acid), pH 7.2. The calculated free Ca^{2+} concentration as determined with a computer program using the affinity constants of Fabiato (1988) was 10 µM. Samples were filtered on Whatman GF/B glass fiber filters, washed twice with 5 ml of distilled water and counted by scintillation. The nonspecific binding was measured in the presence of 10 µM unlabeled ryanodine and was subtracted from each sample. As shown in Figure 15.5A, whole *Buthotus* venom, as well as the purified stimulatory toxins, increased [³H]ryanodine binding in a dose dependent manner. On the other hand, the purified inhibitory toxins decreased [³H]ryanodine binding also in a dose dependent manner (Figure 15.5B). The concentrations which produced 50% maximal stimulation or 50% maximal inhibition (ED_{50}) were 70 nM and 0.9 µM for the stimulatory toxins Buth$_A$-1 and Buth$_A$-2, respectively and were 30 µM and 6.9 µM for the inhibitory toxins Buth$_I$-1 and Buth$_I$-2, respectively. The higher affinity of the stimulatory toxins may explain why the whole venom stimulated [3H]ryanodine binding even when inhibitory toxins were present in higher amounts (Table 15.2). It may also be possible that the inhibitory toxins were bound or inactivated in the whole venom by components later removed during purification.

In order to determined if the toxins interacted with ryanodine receptors in a specific manner, we performed receptor-radioligand binding assays in muscle and brain microsomes in the presence and absence of *Buthotus* toxins. We investigated the possible interactions of toxins with the dihydropyridine receptor and the voltage-dependent Na^+ channel in skeletal muscle transverse tubule membranes and with the IP_3 receptor in brain microsomes. Table 15.3 indicates that neither the unfractionated *Buthotus* venom nor the purified toxins interfered with the binding of [³H]PN200-110, [³H]saxitoxin or [³H]IP_3 to their respective receptors. This was significant since at the same concentration, the toxins produced a significant stimulation or inhibition of [³H]ryanodine binding to the skeletal ryanodine receptor. By this criterion, that is, the toxin's lack of interference with other ligand-receptor reactions, the *Buthotus* toxins appeared to be specifically targeted to the ryanodine receptor.

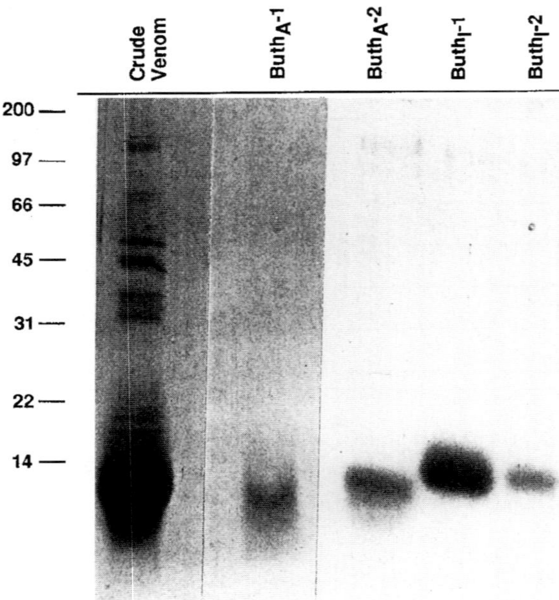

Figure 15.4   SDS polyacrylamide gel electrophoresis of the stimulatory and inhibitory *Buthotus* toxins. Samples were incubated for 10 min. at 80°C in 2.0% SDS (vol/vol), 2% β-mercaptoethanol (vol/vol), 10.0% glycerol (vol/vol) and 10 mM TRIS-Cl pH 6.8 and run on a 6–20% linear polyacrylamide gradient. Gels were stained using 0.05% coomassie blue in 10% (vol/vol) acetic acid. Lane 1: whole *Buthotus hottentota* venom, lane 2: 7,000 Dalton stimulatory toxin Buth$_A$-1, lane 3: 11,750 Dalton stimulatory toxin Buth$_A$-2, lane 4: 13,320 Dalton inhibitory toxin Buth$_I$-1 and lane 5: 13,320 Dalton inhibitory toxin Buth$_I$-2.

Effects on SR Ca^{2+} Release

The physiological consequences of the interaction between the toxins and the ryanodine receptor were investigated by determining the effect of the toxins on the Ca^{2+} permeability of SR vesicles. Changes in extravesicular free Ca^{2+} of a microsomal SR suspension were monitored with the Ca^{2+} indicator dye Fura-2 on a dual excitation wavelength Hitatchi F-2000 fluorescence spectrophotometer (Grynkiewicz, 1985). Approximately 150 mg of heavy SR vesicles were suspended in 400 ml of medium containing 0.1 M KCl, 5.0 mM phosphocreatine, 5.0 mg/ml creatine phosphokinase, 0.5 μM Fura-2 (free acid) and 20 mM HEPES (N-2-Hydroxyethyl piperazine-N′-2-ethanesulfonic acid)-TRIS (Tris[hydroxymethyl] aminomethane) pH 7.2. This solution was placed in a quartz cuvette in the spectrophotometer under constant stirring at 30°C. The free Ca^{2+} was in the range of 1–3 μM and was present as contaminant since no Ca^{2+} salt was added to this solution. As shown in Figure 15.6, SR vesicles were actively loaded with Ca^{2+} by stimu-

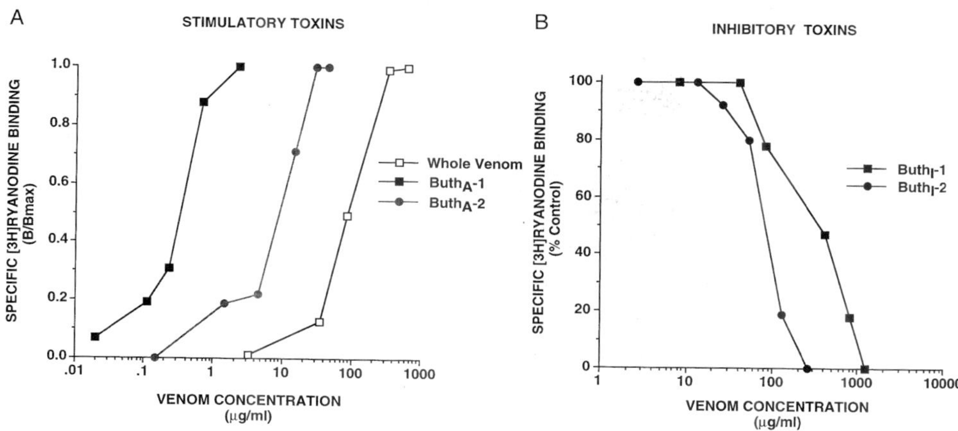

Figure 15.5 [^{3}H]ryanodine binding dose response curves for whole *Buthotus* venom and purified stimulatory and inhibitory toxins. A) Stimulation of 7 nM [^{3}H]ryanodine binding by whole venom (open squares), Buth$_A$-1 (filled squares) and Buth$_A$-2 (filled circles). In each case, the amount of binding over control (B) is expressed as a fraction of the maximal binding (B$_{max}$). Maximal binding was 1 pmol/mg for whole venom and 5 pmol/mg for Buth$_A$-1 and Buth$_A$-2. B) Inhibition of binding by Buth$_I$-1 (filled squares) and Buth$_I$-2 (filled circles). Binding is expressed as a percentage of control binding which was 0.4 pmol/mg.

Table 15.3 Receptor specificity of purified toxins. Binding of 7 nM [^{3}H]ryanodine to rabbit skeletal muscle SR was performed as described in methods. Binding of 5 nM [^{3}H]IP$_3$ to rat brain microsomes was performed according to *Supattapone et al.*, 1988. Binding of 0.4 nM [^{3}H]PN200–110 to rabbit skeletal muscle transverse tubule membranes was performed according to Fosset *et al.*, 1983. Binding of 0.7 nM [^{3}H]saxitoxin to rabbit skeletal muscle transverse tubule membranes was performed according to Catterall *et al.*, 1979.

			SPECIFIC BINDING (pmol/mg protein)					
LIGAND	CONC. ASSAYED	TISSUE	CONTROL	WHOLE VENOM	Buth$_A$-1	Buth$_A$-2	Buth$_I$-1	Buth$_I$-2
[^{3}H]RYANODINE	7 nM	Skeletal HSR	.50	1.6	2.5	2.0	.09	.03
[^{3}H]InsP$_3$	5 nM	Brain micosomes	.085	.054	.051	.082	.075	.082
[^{3}H]PN200-110	.4 nM	Skeletal t-tubule	1.0	.90	1.1	.96	.86	.97
[^{3}H]SAXITOXIN	.7 nM	Skeletal t-tubule	.70	.79	.67	.66	.66	.63

lating the Ca^{2+} pump with the addition of 1–2 mM MgATP to the cuvette. Ca^{2+} loading of SR vesicles was inferred from the decrease in extravesicular Ca^{2+} following the addition of ATP. Ca^{2+} release from the SR was inferred from the increase in extravesicular free Ca^{2+} stimulated by the addition of the ryanodine receptor activator caffeine, or by the addition of the stimulatory *Buthotus* toxins. In the control run shown in Figure 15.6A, approximately 500 nM Ca^{2+} was released upon the addition of 10 mM caffeine to Ca^{2+}-loaded SR. The caffeine-induced release could be blocked by the previous addition of the ryanodine receptor blocker ruthenium red (Figure 15.6B). In a similar manner, Figures 15.6C and D show that addition of either of the purified inhibitory toxins to the Ca^{2+}-loaded SR suspension prevented the release induced by caffeine. These results indicated that $Buth_I$-1 and $Buth_I$-2, like ruthenium red, were blocking the channel, thus preventing the activation of the channel by caffeine. This interpretation was consistent with that obtained from [³H]ryanodine binding, in that the toxins seemed to stabilize the channels in a closed conformation. On the other hand, Figures. 15.7A and B show that the stimulatory toxins, like caffeine, caused the release of 300–

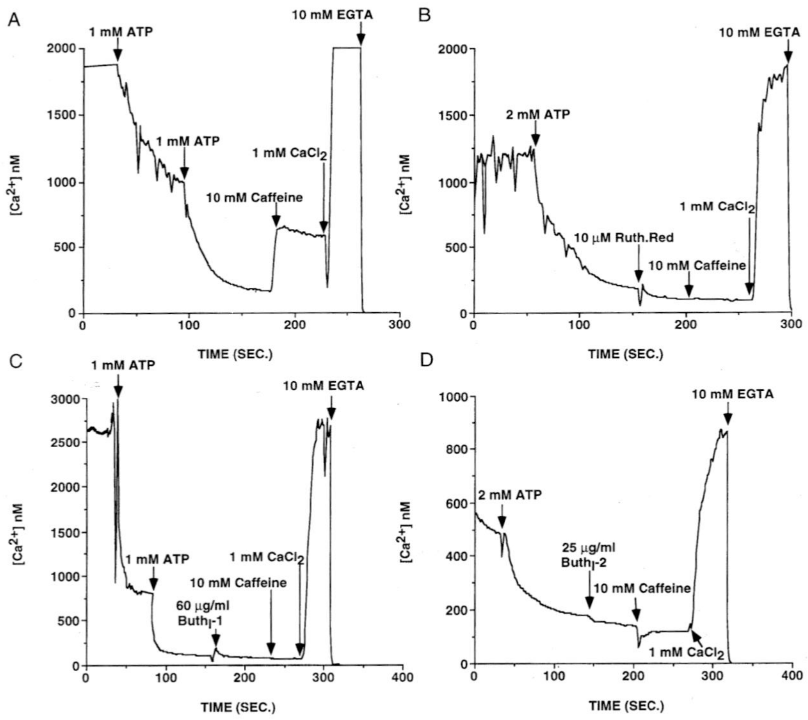

Figure 15.6 Block of caffeine-induced Ca^{2+} release from SR vesicles by the inhibitory toxins. A) Ca^{2+} release induced by 10 mM caffeine. Time of the addition is indicated by the arrow. B) The caffeine-induced Ca^{2+} release is blocked by the ryanodine receptor antagonist ruthenium red. C) and D) The caffeine-induced release is blocked by the inhibitory toxins, $Buth_I$-1 and $Buth_I$-2. $CaCl_2$ (1 mM) and Na_2EGTA (10 mM) were added at the end of each experiment in order to determine F_{max} and F_{min} which are necessary to calculate free Ca^{2+} using a ratiometric technique.

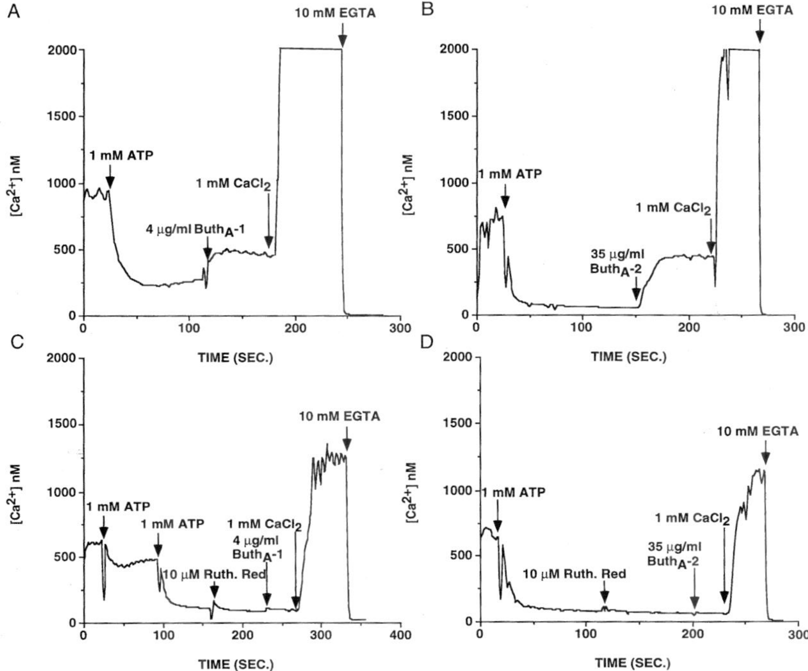

Figure 15.7 Stimulation of Ca^{2+} release from SR vesicles by the stimulatory toxins. A) and B) The stimulatory toxins, $Buth_A$-1 and $Buth_A$-2, induced a release of stored Ca^{2+}. C) and D) The toxin-induced release is blocked by ruthenium red.

600 nM Ca^{2+} following ATP-dependent uptake. The release was mediated by ryanodine receptors since the previous addition of ruthenium red blocked the toxin-induced release. These results indicated that $Buth_A$-1 and $Buth_A$-2 stabilized the channel in an open conformation thus stimulating the release of stored Ca^{2+}. A toxin-induced opening of the channel was also consistent with the stimulation of [^{3}H]ryanodine binding described above.

Effects of *Buthotus* Toxins On Single Channel Activity

The interactions of the *Buthotus* toxins with the ryanodine receptor were confirmed by planar bilayer recordings. Bilayer formation and recording was performed as previously described (Coronado, 1992). Briefly, lipid bilayers were cast using a 1:1 ratio of brain phosphatidylethanolamine and brain phosphatidylserine dissolved in decane at a concentration of 20 mg/ml. The lipid solution was spread over a 300 μM diameter aperature separating two aqueous chambers designated cis and trans. SR vesicles were added to the cis chamber containing 250 mM CsCl, 10 μM HEPES-TRIS pH 7.2. The free Ca^{2+} of the cis solution was between 1–3 μM as determined with a Ca^{2+} electrode. The trans chamber contained 50 mM CsCl, 10 mM HEPES-TRIS pH 7.2. The cis chamber was connected via an Ag/AgCl electrode to the headstage input of a List L/M EPC 7 amplifier while the trans chamber was held at ground. All recordings were taken at 10 mV, filtered at 1 kHz and digitalized at 3 kHz for analysis. Figure 15.8A shows records from an experiment in which two ryanodine receptor channels incorporated into the bilayer. With the addition of 31 μg/ml (2.3 μM) $Buth_I$-1 to the cis chamber, the probability of one channel open (P_{o1}) decreased from 0.42 in control to 0.04 in the pres-

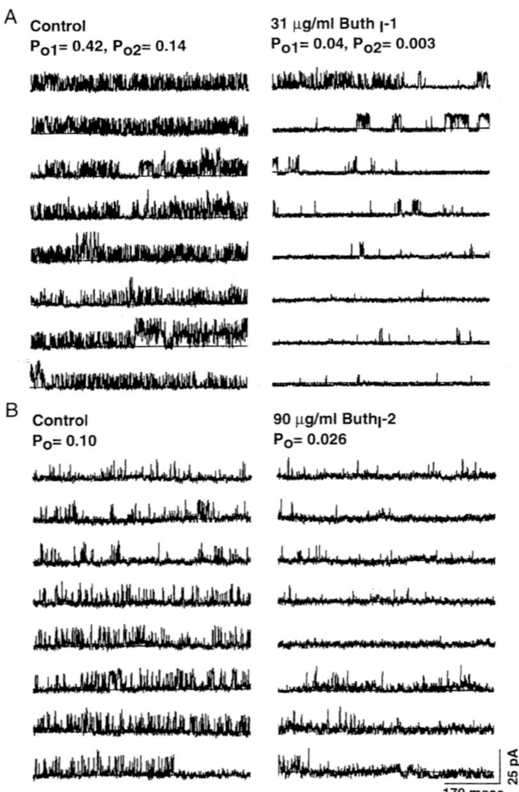

Figure 15.8   Inhibition of ryanodine receptor Ca^{2+} release channel of rabbit skeletal muscle by $Buth_I$-1 and $Buth_I$-2. Traces of ryanodine receptor channels incorporated into bilayers. Openings are indicated as upward deflections. Holding potential was +10 mV. Solutions were composed of cis 250 mM CsCl, trans 50 mM CsCl with approximately 1–3 μM free Ca^{2+} in both solutions. The solid line indicates baseline. A) Addition of $Buth_I$-1 to the cis solution caused significant inhibition of channel activity. P_{o1} is the probability of observing one open channel and P_{o2} is the probability of observing two open channels. B) Addition of $Buth_I$-2 to the cis solution caused significant inhibition of channel activity. The open probabilities are indicated next to each condition.

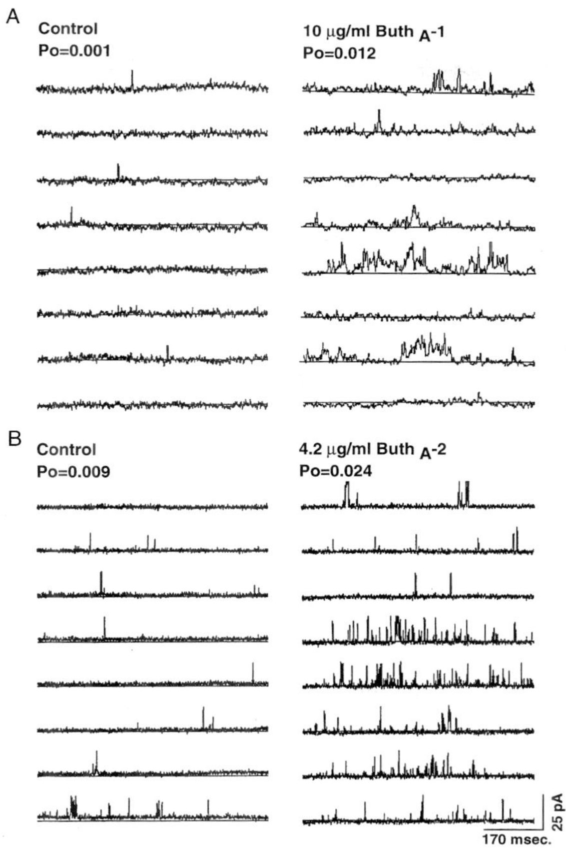

Figure 15.9 Stimulation of ryanodine receptor Ca^{2+} release channel of rabbit skeletal muscle by $Buth_A$-1 and $Buth_A$-2. Traces of ryanodine receptor channels incorporated into bilayers. Openings are indicated as upward deflections. Holding potential was +10 mV. Solutions were composed of cis 250 mM CsCl, trans 50 mM CsCl with 0.1 μM free Ca^{2+} on the cis side and approximately 1–3 μM on the trans side. The solid line indicates baseline. A) Addition of $Buth_A$-1 to the cis solution caused significant stimulation of channel activity. B) Additions of $Buth_A$-2 to the cis solution caused significant stimulation of channel activity. The open probabilities are indicated next to each condition.

ence of toxin. The probability of having both channels open (P_{o2}) decreased from 0.14 to 0.003 following the addition of toxin. The decrease in activity was quantified in Figure 15.10C in a plot of the cumulative P_o as a function of time. To construct this plot, the recording was divided into 683 msec intervals. The P_o of each interval was added to that of the previous interval and the cummulative sum was plotted as a function of time. Thus, the slope of this plot changes as the number of open events change. It was apparent that $Buth_I$-1 produced a constant inhibition of the channel that was not relieved with time. Similar results were obtained for a single channel recorded in the presence of Buth$_I$-2. Figure 15.8B shows records for a single channel before and immediately following the addition of 90 μg/ml (6.8 μM) Buth$_I$-2 to the cis chamber. The P$_o$ decreased from 0.1 to 0.026 following addition of toxin. As with Buth$_I$-1, the inhibition produced by Buth$_I$-2 was associated with a decrease in the number of open events per unit time and was persistent throughout the recording (Figure 15.10D). These results clearly indicate that the inhibitory toxins produce a closure of the channel which may therefore explain the inhibition of caffeine-induced release of SR-stored Ca^{2+} (Figure 15.6) and the inhibition of [^{3}H]ryanodine binding to SR (Figure 15.5).

The effects of the stimulatory toxins were also investigated in planar bilayer experiments. To better observe channel activation by the stimulatory toxins, Na$_2$EGTA was added to the cis chamber to decrease the free Ca^{2+} concentration to 0.1 μM and therfore, lower the channel's open probability. Figure 15.9A shows the records of a channel before (control) and after the addition of 10 μg/ml (1.4 μM) Buth$_A$-1 to the cis chamber. Following the addition of the toxin, the open probability increased from 0.001 to 0.012. The stimulation was associated with a increase in the number of open events per unit time and was persistent for the duration of the recording (Figure 15.10A). Similarily, the addition of 4.2 μg/ml (3.5 μM) Buth$_A$-2

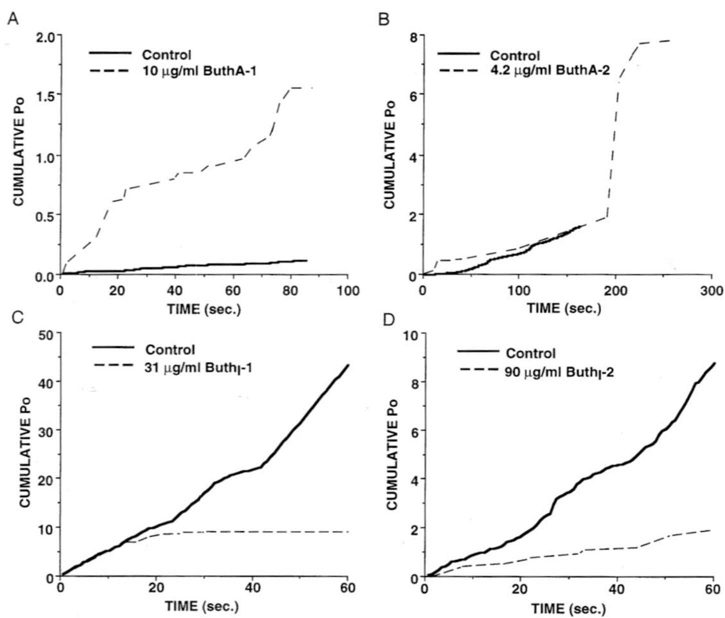

Figure 15.10 Plots of the cumulative open probability as a function of time. The complete channel recordings before (control) and after the addition of toxins were divided into 683 ms intervals. The P$_o$ of each interval was added to that of the previous one and the cumulative sum was plotted as a function of time. Thus the slope of the lines are proportional to the number of open events per unit time. After the addition of toxin, the time was reset to t = 0.

to the cis chamber increased the P_o from 0.009 in control to 0.024 in the presence of the toxin (Figure 15.9B). The stimulation was again associated with an increase in the number of open events per unit time (Figure 15.10B). However, the stimulation by $Buth_A$-2 was burst-like rather than persistent. Following the addition of toxin, the channel went through a burst of activity around 20 sec. and another more robust burst of activity around 200 sec. The bursts were followed by more quiescent control-like activity (Figure 15.10B). Such busting behavior was not seen during the control period where the channel was persistently in a state of low activity. This bursting followed by control-like activity may represent the binding and unbinding of $Buth_A$-2 to the channel.

POTENTIAL USES OF TOXINS

Given the intracellular location of ryanodine receptors, the elucidation of the mechanism of Ca^{2+} release from intracellular stores depends critically on the specificity of pharmacological agents to selectively alter a single intracellular Ca^{2+} channel type. We now know that a small fraction of venoms from scorpions and lizards have peptide toxins that appear to be specifically targeted against ryanodine receptors and are present in these venoms in biochemically-relevant quantities (Valdivia, 1991; 1992a; Morrissette, 1995). These toxins should be invaluable in relating the channel protein's structure to function and should be useful in understanding how intracellular Ca^{2+} signals are generated.

Pharmacological Dissection of Intracellular Ca^{2+} Release

Two Ca^{2+} pools, controlled by an IP_3-sensitive and a Ca^{2+}-sensitive channel, have been shown to operate during Ca^{2+} transients in secretory and non-excitable cells (Connor, 1993). Whether two Ca^{2+} pools are also activated in striated muscle during stimulus-contraction coupling is not known. However, this is an attractive model since the rate of Ca^{2+} induced Ca^{2+} release and the rate of IP_3-induced Ca^{2+} release in skeletal muscle membranes turned out to be equally fast (Valdivia, 1992). In the same cell, interactions between Ca^{2+} pools controlled by IP_3 and ryanodine receptors may serve several purposes. For example, it could serve to rapidly amplify a small release of Ca^{2+} produced by IP_3 into a much larger elevation of myoplasmic Ca^{2+} produced by Ca^{2+}-dependent activation of ryanodine receptors. The activation of phospholipase C (PLC) by Ca^{2+} (Baird, 1990), also makes possible an alternative cascade in which the transient increase in Ca^{2+} following voltage-activated Ca^{2+} release, would activate PLC and production of IP_3. There are numerous questions that may be addressed using ryanodine-receptor specific toxins. One of the most pressing ones seems to be which release pathway, IP_3 receptor, ryanodine receptor or both, is involved in Ca^{2+} waveforms in cells that have both types of channels? Furthermore, since both receptors are known to be Ca^{2+}-sensitive (Bezprozvanny, 1991; Finch, 1991; Smith, 1986) is the activation of one Ca^{2+} pool subsequently involved in the activation of the the other, resulting in synergistic Ca^{2+} release? With the availability of ryanodine receptor specific toxins, the ryanodine sensitive Ca^{2+} pool can be selectively depleted with stimulatory toxins or blocked with inibitory toxins such that it is should be possible to determine its relationship to the IP_3 sensitive Ca^{2+} pool.

Mapping of Regulatory Domains of Ryanodine Receptors

The large size of the ryanodine receptor (565 kDa, 5,037 amino acids in the skeletal receptor) makes the task of identifing domains of the receptor that are important for channel function quite difficult. Recently Chen *et al.* (1993), identified a functional Ca2+ activation site between amino acid residues 4489 and 4499 of the skeletal ryanodine receptor. A polyclonal antibody raised against this sequence increased the Ca2+ sensitivity of ryanodine receptors incorporated into lipid bilayers. Using a photoactivatible derivative of ryanodine, Witcher *et al.* (1994), have localized the high affinity ryanodine binding site to a 76 kDa C-terminal tryptic fragment of the skeletal receptor. Callaway *et al.* (1994), also localized the alkaloid binding site to the C-terminal portion of the receptor most likely between Arg-4475 and the end of the receptor. Scorpion toxins may prove useful in the further elucidation of domains critical to channel function. Since the *Buthotus* toxins effect channel gating, it seems likely that the binding sites for the toxins could be physically close to domains critically involved in the opening and closing of the channel. One possible approach is the use the toxins as potential substrates for the synthesis of photoaffinity labels. Once binding sites are identified, subsequent site directed mutagenesis of these domains should provide valuable information of the specific amino acids involved in channel gating.

Identification of Ryanodine Receptor Isoforms

Three different genes (RYR-1, RYR-2 and RYR-3) encoding for a skeletal, cardiac and brain isoform of the ryanodine receptor respectively, have been identified (Coronado, 1994; McPherson, 1993). However the expression of the three isoforms is not completely tissue specific. The receptor encoded by RYR-1 was found to be the predominate form expressed in skeletal muscle, but was also present in mouse cerebellum (Kuwajima, 1992). Likewise, the receptor encoded by RYR-2 was the predominate form expressed in the heart, but was also the predominate form expressed in the brain (Coronado, 1994, Otsu, 1990). Also found in the brain as well as in smooth muscle was the receptor encoded by RYR-3 (Hakamata, 1992). The cardiac receptor (RYR-2) and the brain receptor (RYR-3) were found to be 66% and 70% homologous to the skeletal receptor (RYR-1) (Otsu, 1990, Hakamata, 1992). Although there is a great deal of homology between the receptor isoforms, the skeletal and cardiac receptors are functionally different in terms of channel conductance, sensitivity to activation by Ca^{2+} and caffeine and inhibition by Mg^{2+} (Zimanyi, 1991). Furthermore, the cardiac and skeletal isoforms seem to be activated by different mechanisms during excitation-contraction coupling. In cardiac muscle, a small influx of Ca^{2+} across the sarcolemal membrane triggers a much more massive Ca^{2+} release via the ryanodine receptor (Stern, 1992). In skeletal muscle, activation of the sarcolemal dihydopyridine receptor by membrane depolarization leads to Ca^{2+} release via the ryanodine receptor, possibly through a mechanical coupling between the dihydropyridine and ryanodine receptors (McPherson, 1993).

Scorpion toxins that are isoform-specific should be quite useful for identification and characterization of ryanodine receptor isoforms. Valdivia *et al.* (1992a), have isolated a peptide toxin from the scorpion *Pandinus imperator*, named imperatoxin activator (IpTxa), that stimulates [³H]ryanodine binding and single channel activity of skeletal but not cardiac receptors. IpTxa was found to differentiate between the α (skeletal-like) and β (cardiac-like) ryanodine receptor isoforms in non-mammalian muscle (Valdivia, 1994). The stimulatory toxins from *Buthotus* described here, did not differentiate between receptor isoforms. Thus, a molecular comparison of the binding site for IpTxa and those of Buth$_A$-1 and Buth$_A$-2 may provide insight into the regions of the ryanodine receptor which confer the functional differences between receptor isoforms. Furthermore, isoform-specific toxins may prove useful in the selective purification of ryanodine receptor isoforms from tissues, such as the brain, that contain more than a single ryanodine receptor isoform.

Acknowledgments

Supported by National Institutes of Health grant GM 36852 and by grants from the American Heart Association and Muscular Dystrophy Association of America.

REFERENCES

Baird, J.G. and Nahorski S.R. (1990) Increased intracellular calcium stimulates [³H]-inositol polyphosphate accumulation in rat cerebral cortical slices, *J. Neurochem.*, **54**, 555–561.

Beuckelmann, D.J. and Wier W.G. (1988) Mechanism of release of calcium from sarcoplasmic reticulum of guinea-pig cardiac cells. *J. Physiol. (Lond.)*, **405**, 233–255.

Bezprozvanny, I., Watras, J. and Ehrlich, B.E. (1991) Bell-shaped calcium-response of Ins(1,4,5)P₃ and calcium-gated channels from endopolasmic reticulum of cerebellum. *Nature*, **351**, 751–754.

Callaway, C., Seryshev, jA., Wang, J-P., Slavik, K.J., Needleman, D.H., Cantu, C.III, Wu, Y., Jayaraman, T., Marks, A.R. and Hamilton, S.L. (1994) Localization of the high and low affinity [³H]ryanodine binding sites on the skeletal muscle Ca^{2+} release channel. *J. Biol. Chem.*, **269**, 15876–15884.

Catterall, W.A., Morrow, C.S. and Hartshorne, R.P. (1979) Neurotoxin binding to receptor sites associated with voltage-sensitive sodium channels in intact, lysed and detergent-solubilized brain membranes. *J. Biol. Chem.*, **254**, 11379–11387.

Chen, S.R.W., Zhang, L. and Maclennan, D.H. (1993) Antibodies as probes for Ca^{2+} activation site in the Ca^{2+} release channel (ryanodine receptor) of rabbit skeletal muscle sarcoplasmic reticulum. *J. Biol. Chem.*, **268**, 13414–13421.

Connor, J.A. (1993) Intracellular calcium mobilization by inositol 1,4,5-trisphosphate: Intracellular movements and compartmentalization. *Cell Calcium*, **14**, 185–200.

Coronado, R., Morrissette, J., Sukhareva, M. and Vaughan, D.M.V. (1994) Invited review: Structure and function of ryanodine receptors. *Am. J. Physiol.*, **266**, C1485–1504.

Coronado, R., Kawano, S., Lee, C.J., Valdivia, C. and Valdivia, H.H. (1992) Planar bilayer recording of ryanodine receptors of sarcoplasmic reticulum. *Meth. Enzymol.*, **207**, 699–707.

Fabiato, A., (1988) Computer programs for calculating total from specified free or free from specified total ionic concentrations in aqueous solutions containing multiple metals and ligands. *Meth. Enzymol.*, **157**, 387–417.

Finch, E.A., Turner, T.J. and Goldin, S.M. (1991) Calcium as a coagonist of inositol 1,4,5-triphosphate-induced calcium release. *Science*, **252**, 443–446.

Fosset, M., Jaimovich, E., Delpont, E. and Lazdunski, M. (1983) [^{3}H]nitrendipine receptors in skeletal muscle properties and preferential localization in transverse tubules. *J. Biol. Chem.*, **258**, 6086–6092.

Grynkiewicz, G., Poenie, G.M. and Tsien R.Y. (1985) A new generation of Ca^{2+} indicators with greatly improved fluorescence properties. *J. Biol. Chem.* **260**, 3440–3450.

Hakamata, Y., Nakai, J., Takeshima, H. and Iomoto, K. (1992) Primary structure and distribution of a novel ryanodine receptor/calcium release channel from rabbit brain. *FEBS Lett*, **312**, 229–235.

Humerickhouse, R.A., Bidasee, K.R., Gerzon, K., Emmick, J.T., Kwon, S., Sutko, J.L., Ruest, L. and Besch, H.R. Jr. (1994) High affinity C_{10}-O_{eq} ester derivatives of ryanodine. Activator-selective agonists of the sarcoplasmic reticulum calcium release channel. *J. Biol. Chem.*, **269**, 30243–30253.

Imagawa, T., Smith, J.S., Coronado, R. and Campbell, K.P. (1987) Purified ryanodine receptor from muscle sarcoplasmic reticulum is the Ca^{2+}-permeable pore of the calcium release channel. *J. Biol. Chem.*, **262**,16636–16643.

Jorgensen, A.O., Shen, A.C-Y., Arnold, W., McPherson, P.S. and Campbell, K.P. (1993) The Ca^{2+}-release channel/ryanodine receptor is localized in junctional and corbular sarcoplasmic reticulum in cardiac muscle. *J. Cell Biol.*, **120**, 969–980.

Kuwajima, G.A., Futatsugi, A., Niinobe, M., Nakanishi, S. and Mikoshiba, K. (1992) Two types of ryanodine receptors in mouse brain: skeletal muscle type exclusively in purkinje cells and cardiac muscle type in various neurons. *Neuron*, **9**, 1133–1142

McPherson, P.S. and Campbell, K.P. (1993) The ryanodine receptor/Ca^{2+} release channel. *J. Biol. Chem.*, **268**, 13765–13768.

Mikoshiba, K. (1993) Inositol 1,4,5-trisphosphate receptor. *Trends Pharmacol. Sci.*, **14**, 86–89.

Miller, C., Moczydlowski, E., Latorre, R. and Phillips, M. (1985) Charybdotoxin, a protein inhibitor of single Ca^{2+}-activated K+ channels from mammalian skeletal muscle. *Nature*, **313**, 316–318.

Morrissette, J.M., Kratzschmar, J., Haendler, B., El-Hayek, R., Mochca-Morales, J., Martin, B., Patel, J., Moss, R., Schleuning, W-D., Coronado, R. and Possani, L. (1995) Primary structure and properties of helothermine, a peptide toxin that blocks ryanodine receptors., *Biophys. J.*, **68**, 2280–2288.

Moschella, M.C. and Marks, A.R. (1993) Inositol 1,4,5-trisphosphate receptor expression in cardiac myocytes. *J. Cell Biol.*, **120**, 1137–1146.

Otani, H., Otani, H. and Das, D. (1988) α_1-Adrenoceptor-mediated phosphoinositide breakdown and inotropic response in rat left ventricular papillary muscles. *Circ. Res.*, **62**, 8–17.

Otsu, K., Willard, H.F., Khanna, V.K., Zorzato, F., Green, N.M. and Maclennan, D.H. (1990) Molecular cloning of cDNA encoding the Ca^{2+} release channel (ryanodine receptor) of rabbit cardiac muscle sarcoplasmic reticulum. *J. Biol. Chem.*, **265**, 13472–13483.

Pessah, I., Francini, A., Scales, D., Waterhouse, A. and Casida, J. (1986) Calcium-ryanodine receptor complex. Solubilization and partial characterization from skeletal muscle junctional sarcoplasmic reticulum vesicles. *J. Biol. Chem.*, **261**, 8643–8648.

Pessah, I.N., Stambuk, R.A. and Casida, J.E. (1987) Ca^{2+}-activated ryanodine binding: mechanism of sensitivity and intensity modulation by Mg^{2+}, caffeine and adenine nucleotides. *Mol. Pharmacol.*, **31**, 232–238.

Smith, J.S., Coronado, R. and Meissner, G. (1986) Single channel measurements of the calcium release channel from skeleat muscle sarcoplasmic reticulum. *J. Gen. Physiol.*, **88**, 573–588.

Stern, M.D. and Lakatta, E.G. (1992) Excitation-contraction coupling in the heart: the state of the question. *FASEB J.*, **6**, 3029–3100.

Strichartz, G., Rando, T. and Wang, G.K. (1987) An integrated view of the molecular toxicology of sodium channel gating in excitable cells. *Ann. Rev. Neurosci.*, **10**, 237–267.

Supattapone, S., Worley, P.F., Baraban, J.M. and Snyder, S.H. (1988) Solubilization, purification and characterization of an inositol triphosphate receptor. *J. Biol. Chem.*, **263**, 1530–1534.

Valdivia, H.H., Fuentes, O., El-Hayek, R., Morrissette, J. and Coronado, R. (1991) Activation of the ryanodine receptor Ca^{2+} release channel of sarcoplasmic reticulum by a novel scorpion venom. *J. Biol. Chem.*, **266**, 19135–19138.

Valdivia, H.H., Kirby, M.S., Lederer, W.J. and Coronado, R. (1992a) Scorpion toxins targeted against the sarcoplasmic reticulum Ca^{2+}-release channel of skeletal and cardiac muscle. *Proc. Natl. Acad. Sci. USA.*, **89**, 12185–12189.

Valdivia, C., Vaughan, D., Potter, B.V.L. and Coronado, R. (1992b) Fast release of $^{45}Ca^{2+}$ induced by inositol 1,4,5-trisphosphate and Ca^{2+} in the sarcoplasmic reticulum of rabbit skeletal muscle: Evidence for two types of Ca^{2+} release channels. *Biophys. J.*, **61**, 1184–1193.

Valdivia, H.H., Fuentes, O. and Block, B.A. (1994) Imperatoxin A, a selective activator of skeletal ryanodine receptors (RYR), distinguishes between α and β RYR isoforms in non-mammalian muscle. *Biophys. J.*, **66**, 418a (Abstr.)

Wang, J.P., Needleman, D.H. and Hamilton, S.L. (1993) Relationship of low affinity [3H]ryanodine binding sites to high affinity sites on the skeletal muscle Ca^{2+} release channel. *J. Biol. Chem.*, **268**, 20974–20982.

Witcher, D.R., McPherson, P.S., Kahl, S.D., Lewis, T., Bentley, P., Mullinnix, M.J., Windass, J.D. and Campbell, K.P. (1994) Photoaffinity labeling of the ryanodine receptor/Ca^{2+} release channel with an azido derivative of ryanodine. *J. Biol. Chem.*, **269**, 13076–13079.

Zimanyi, I. and Pessah, I.N. (1991) Comparison of [3H]ryanodine receptors and Ca++ release from rat cardiac and rabbit skeletal muscle sarcoplasmic reticulum. *J. Pharmacol. Exp. Ther.*, **256**, 938–946.

16. PROTEIN KINASE C — ACTIVATORS AND INHIBITORS

DAVID S. LESTER

*Food and Drug Administration, Center for Drug Evaluation and Research,
Office of Testing and Research, Laurel, MD 20708, USA*

INTRODUCTION

Protein phosphorylation is considered to be a vital mechanism of posttranslational modification. The catalyzing enzyme for this process is the protein kinase. There are well over 100 identified genes for different protein kinases and over 80 proteins with protein phosphotransferase activity have been identified (Hanks *et al.*, 1988). The second-messenger-dependent protein kinases have obtained the most attention due to their regulation by second-messengers, including Ca^{2+} and cyclic AMP. The Ca^{2+}/phospholipid-dependent protein kinase, protein kinase C (PKC) has the largest number of publications of all of the second-messenger-dependent protein kinases (Lester and Epand, 1992). This is due to the unique feature that it is considered to be active upon association with the intracellular surface of the plasma membrane (Lester, 1992). This particular enzyme, a long, single polypeptide of 78–82 kDa, is generally considered to be activated via transmembrane signal transduction pathways such as receptor activation (see Nishizuka, 1984). PKC can generally be divided into 2 major subunits; the regulatory subunit (the N-terminal) and the catalytic subunit (the C-terminal) (Figure 16.1). The enzyme can be proteolyzed into these two subunits by trypsinization (Huang and Huang, 1986), however, the physiological significance of this process has not been convincingly established. Interestingly, this proteolytic action has been described as a result of the action of tetanus toxin (see later section). The enzyme has multiple isoforms that show varying amounts of conserved sequences. The role of the isozymes is generally considered to be relative to different cell types or cellular locale (see Parker, 1992). It has been reported that the isozymes have different cofactor dependencies in some *in vitro* systems (Parker *et al.*, 1989), but this has not been verified in the lipid vesicle system nor has it been convincingly demonstrated *in vivo*.

The activators of protein kinase C are the two second messenger molecules, the calcium ion and the lipid product, diacylglycerol. It is generally considered that the diglyceride is the 1-stearyl-2-arachidonyl form released by the action of the receptor or calcium activated phosphatidylinositol specific phospholipase C (Meldrum *et al.*, 1991). However, it has recently been shown that there are different species of diglyceride which can be released by other phospholipase C forms (Kolesnick, 1990), that could be involved in the activation of protein kinase C. Another lipid product, the *cis*-unsaturated fatty acids, are also considered to play a significant role in certain PKC activation pathways, in particular, neuronal systems (Lester and Bramham, 1994). These lipid compounds are produced by the action of another phospholipase, phospholipase A_2. Besides these lipid products, a phospholipid support is required for production of a suitable lipid microenvironment. It is generally considered that the acidic phospholipid, phosphatidylserine (PS), is the required phospholipid (Bell and Burns, 1991). There have been reports that other acidic phospholipids, in particular, phosphatidic acid, may also be capable of acting as the phospholipid support (Epand *et al.*, 1992). Taking all of these factors into account, it can be concluded that PKC activation occurs at the membrane surface.

MECHANISM OF PROTEIN KINASE C ACTIVATION

In order to design and develop suitable PKC activators and inhibitors, it is important to understand the process of PKC activation. Unfortunately, this is still a subject of great contention and may explain why there has been such difficulty in finding specific pharmacological modulators. It was initially and still is considered that PKC activation occurs at the intracellular plasma membrane surface. There is now evidence that the enzyme may be active at other membrane surfaces. For example, there is a considerable concentration of the intrinsic PKC activator, diacylglycerol, in the membrane of the Golgi apparatus (Pagano and Longmuir, 1985). As γ-PKC antibodies have been found to colocalize with this intracellular organelle (Saito *et al.*, 1989), the Golgi is a potential site for PKC activity. Additionally, PKC may be involved with the membrane trafficking process (see Borner and Fabbro, 1992). More recently, PKC has been shown to be associated with the membrane cytoskeleton where there are a number of major PKC phosphoprotein substrates (see Jaken, 1992).

It is still considered that a decrease in the soluble/particulate cellular activity ratio (see Lester, 1989), that is, there is an increase in the particulate fraction, is indicative of PKC activation in the cell (Lester, 1992). It has been shown that the established PKC activation cofactors, Ca^{2+}, PS and diglyceride, are insufficient to promote long-term PKC membrane association (see Lester and Bramham, 1994). These three cofactors result in a transitory "translocation" that has been described in a number of systems (see Lester, 1992 for references and a discussion). It has been proposed that a fourth cofactor, the phospholipase A_2 product, the cis-unsaturated fatty acids, may interact with the other cofactors at a distinct binding site. This would induce tight, irreversible PKC membrane association and subsequent PKC constitutive activity (Bramham and Lester, 1994). This has been demonstrated in both *in vitro* (Lester *et al.*, 1991) and *in vivo* model systems (Lester *et al.*, 1991; Bramham *et al.*, 1994). All of these factors are critical in determining inhibitor and activator action.

BINDING SITES: TARGETS FOR ACTIVATORS AND INHIBITORS

Regulatory Site

It is well accepted that the two cysteine loop regions (see Figure 16.1) are the binding site for the PKC activators, the diglycerides and the class of tumor promoting agents known as the phorbol esters (Bell and Burns, 1991). One phorbol ester molecule may bind at each of the two loops, while the diglyceride binds only to one of the loops (Burns and Bell, 1991). The Ca^{2+}/PS binding site is considered to be in the conserved C2 region of the regulatory region. The pseudosubstrate site has been shown to interact with acidic phospholipids, such as PS (Mosior and Mclaughlin, 1992), which may play a role in the activation process (Orr *et al.*, 1992a). All of these sites have a number of charged amino acid residues that may account for the affinity to interact with lipids.

The best defined site of these three lipid binding sites is the diglyceride binding site. It has been shown that the interaction of the diglyceride with the protein depends on the chemistry of the diglyceride head group (Young and Rando, 1984). Extensive computer modeling has been done on this portion of the lipid activator. It has been correlated with other known PKC activators, such as the phorbol esters, the bryostatins, teleocidin and other compounds. A number of models have been proposed for the active pharmacophore structure and these have been well reviewed elsewhere (see Rando and Kishi, 1992). Considering this modeling, a number of different diglyceride analogs have been prepared with the hope of the development of some specific activator or inhibitor of protein kinase C (Molleyres and Rando, 1988; Kong *et al.*, 1990). However, these studies have just demonstrated the complexity of the diglyceride-PKC interaction and have not proven too successful. The second portion of the diglyceride molecule that has been largely overlooked in the preparation of PKC activators and inhibitors, is the

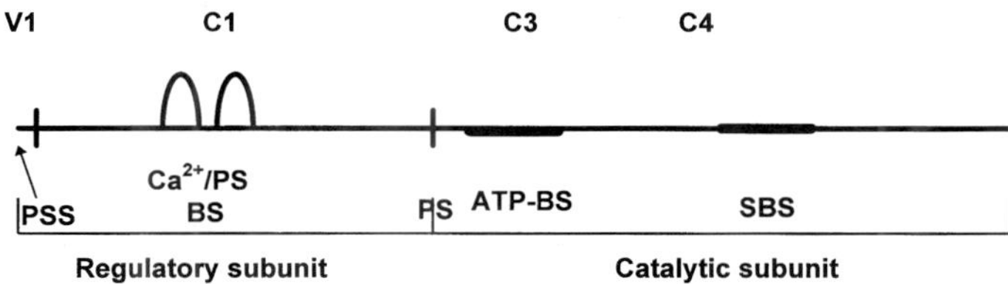

Figure 16.1   A schematic diagram of the major regulatory sites of the Ca^{2+}/phospholipid protein kinase C, PKC. The enzyme can be divided into two subunits, the regulatory and the catalytic subunits. The regulatory subunit contains two major regulatory regions, the pseudosubstrate binding site (PSS) and the Ca^{2+}/phospholipid binding site (Ca^{2+}/PS BS). The calcium/phospholipid binding site is represented by two zinc finger regions that are designated as the two loops. Separating the two regions is the trypsin proteolytic site (PS). The catalytic subunit, the longer of the two subunits, contains the substrate binding site (SBS) and the ATP binding site (ATP-BS) which are considered to be in close proximity.

fatty acid chain composition. In order to thoroughly examine these lipid molecules, it was necessary to use a well-defined lipid system, which was generally not done. Non-bilayer lipid mixes or detergent micelles were routinely used, but these systems are not comparable to the native membrane state (Boni, 1992). In this system, it was shown that the ability of the diglyceride to activate PKC depends on the head group and also the degree of unsaturation of the 2-acyl chain. The higher the degree of chain unsaturation the better the activating properties were demonstrated to be (Goldberg *et al.*, 1994). This correlated with the ability of the diglyceride to alter the biophysical properties of the membrane.

In fact, the action of a lipophilic, membrane-acting pharmacological reagent on PKC could be predicted based on its effects on the lipid bilayer, in particular, the ability to induce the hexagonal HII phase (Epand and Lester, 1990). The properties of the bilayer both at the membrane surface and within the acyl chain hydrocarbon chains determine the ability of the phospholipid support to activate or inhibit PKC. If the compound perturbs the bilayer such that it increases lipid spacing and subsequently promotes HII phase, it would be expected to have activating properties. Such reagents are the cis-unsaturated fatty acids (Lester, 1990; Lester *et al.*, 1991). However, these fatty acids are poor PKC activators on their own. When incorporated together with weak activating diglycerides (short chain or saturated acyl chains), they have a potent synergistic interaction (Lester *et al.*, 1991; D.S.Lester, unpublished data). Amphiphiles with positively charged headgroups are also capable of inducing HII phase when in high concentrations (Epand and Lester, 1990). As has been shown, these specific properties can be coordinated to develop a model that can predict whether an amphiphile is an activator or an inhibitor (Epand and Lester, 1990). Generally, the reagents that fit this model would be considered as nonspecific activators or inhibitors of PKC.

Ca^{2+}/Phospholipid Binding Site

It is not yet established whether the proposed conserved C3 region is the Ca^{2+}/phospholipid binding site. The evidence is based on studies and analyses made of the different isozymes and their lipid dependence (Huang *et al.*, 1988; Ono *et al.*, 1988). The chemistry and molecular modeling are unknown for this interaction. However, if this is a phospholipid binding site, pharmacological reagents that affect the lipid milieu would be expected to alter interactions between this peptide region and the lipid headgroups. In addition, the Ca^{2+} binding to the protein depends on the association of the protein with the lipid. This cooperative interaction has not been thoroughly examined nor has it been exploited in terms of activators or inhibitors of PKC.

Pseudosubstrate Site

This 35 peptide region is in the C1 domain of the regulatory region of PKC. It has been shown to play an integral role in the regulation of the activity state of the enzyme (Orr *et al.*, 1992a). In addition, the peptide has been shown to associate strongly with acidic phospholipids (Mosior and McLaughlin, 1991). Subsequently, this site has been proposed as a potential phospholipid binding site (Zidovetzki and Lester, 1992). While there have been no specific activators or inhibitors that have been directed to this site, it has been shown that the peptide is a potent inhibitor of the enzyme *in vitro* and *in vivo* (see House and Kemp, 1990). However, this could be another target site for amphiphiles.

Substrate Binding Site

Few studies have been made on the substrate binding site. The proposed site has a number of basic residues (O'Brian and Ward, 1992). This would explain why high salt has been shown to inhibit the PKC assay as it would interfere with electrostatic interactions that are proposed to be important in the association of the substrate and the enzyme (Brumfeld and Lester, 1990). A number of peptides representative of the phosphorylation sites of known PKC substrates have been shown to inhibit the enzyme (Ward and O'Brian, 1993). This is a competitive interaction between the substrate and the peptide for this enzyme substrate binding site.

ATP Binding Site

This site was originally proposed as "the PKC lipid binding site" due to the high concentration of basic amino acid residues in this region (Ono *et al.*, 1986). For a detailed review on the binding and hydrolysis of ATP, the reader is referred to an outstanding review by O'Brian and Ward (1992). It has been found that the majority of so-called PKC inhibitors bind at this site on the enzyme and interfere with the transferase reaction, that is, the hydrolysis of the γ-phosphate of ATP and its transfer to the protein substrate (Lester, 1995). Magnesium plays an important role at this site as it is the Mg^{2+}-ATP form of the substrate that binds to and is hydrolyzed by the enzyme (Ohno *et al.*, 1990). An understanding of the interactions that occur is limited as details of secondary structure are unknown. A comparison has been made between the PKC and the established cyclic AMP-dependent protein kinase crystal structure and is strongly recommended (see O'Brian and Ward, 1992).

PROTEIN KINASE C ACTIVATORS

Diglycerides

There have been a number of outstanding reviews that are available on the interactions between the diglyceride binding site and the lipid activator. Unfortunately, there is still considerable discussion as to the active pharmacophore site (see Rando and Kishi, 1992 for a review). The activation of PKC by the phosphatidylinositol product, diglyceride, was initially demonstrated by Nishizuka and colleagues (Takai *et al.*, 1979). The long chain, natural diglycerides are relatively insoluble and do not permeate the cell plasma membrane significantly. In the mid

1980s, two short chain diglyceride species were developed, sn-1,2-diocatonyl-(diC8)-glycerol and 1-oleoyl-2-acetylglycerol (Lapetina *et al.*, 1985). Both of these short chain diglyceride species have been extensively used in cellular studies of PKC activation (see Bramham *et al.*, 1994, for example). Unfortunately, both of these lipid activators do not activate PKC to an activation state observed for the native diglycerides (Goldberg *et al.*, 1994).

Phorbol Esters

The phorbol esters are a class of diterpenes that were originally purified from croton oil, extract from the seeds of the Euphorbiaceae and Thymelaeaceae plants. The oil and the purified phorbol ester are severe irritants and can prove toxic. In addition, they are potent tumor promoting agents (Hecker *et al.*, 1990). Nishizuka and colleagues were the first to demonstrate the activation of PKC by the phorbol esters (Castagna *et al.*, 1982). Since then, there have been an enormous number of studies examining the various phorbol ester species and their ability to activate PKC. There are at least 10 phorbol ester analogs capable of activating PKC *in vitro* and *in vivo*. They generally vary in the nature of the associated fatty acid chain. The most commonly used phorbol esters are: phorbol 12-myristate 13-acetate (PMA), phorbol 12, 13-dibutyrate (PdBu), phorbol 12, 13-didecanoate (PDD) and phorbol 12, 13-diacetate (PDA). These compounds, all derivatives of the parent diterpene phorbol, typically are active at nanomolar concentrations. When added to cellular or tissue preparations, they cause rapid translocation of cytosolic PKC to the plasma membrane resulting in enzyme activation and phosphorylation of phosphoprotein substrates (Nishizuka, 1984). Upon translocation, there is degradation of the enzyme via proteolysis, often called "down-regulation" (Nishizuka, 1986). As a negative control, the 4α form of the phorbol diacetate was designed by Hecker and co-workers (1990). The difference between the 4α PDD and PDD is the α-configuration of the hydroxyl group at position 4 (below the paper). This negative control has similar lipophilicity when compared to PDD, it has no PKC activity and is not a tumor promoter (Hecker *et al.*, 1990). A number of negative controls have been prepared so as to be used as controls with the different phorbol ester species.

In addition, there are numerous natural and synthetic deoxyphorbol diesters that have been identified as potent activators. Recent studies have shown that many of the PKC isozymes have different activation properties for these deoxyphorbol esters (Ryves *et al.*, 1991). While these studies were made in an *in vitro* system, they may explain some of the differential binding properties of the PKC isozymes in cellular systems (Dunn and Blumberg, 1983).

Teleocidin and Lyngbyatoxin A

The teleocidins are a class of mold-derived indole derivatives that have skin inflammatory and tumor promoting activities similar to the phorbol esters (Fujiki and Sugimura, 1987). They originally were purified and identified as compounds toxic to the teleost fish, *Oryzias latipes*, hence, their name. The toxins were found to come from *Streptomyces mediocidus* (for an excellent review see Fujiki and Sugimura, 1987). Structurally, they have little similarity to the phorbol esters, yet they bind to and activate PKC in a similar manner. In addition, their binding can be displaced by phorbol esters demonstrating that they bind at the same site (Fujiki and Sugimura, 1987). There are a number of teleocidin analogs, all potent tumor promoters and PKC activators. Lyngbyatoxin A, isolated from algae, was shown to be structurally identical to teleocidin (Sakai *et al.*, 1986). The basic structure of the teleocidins is the indolactam, an indole nucleus carrying a nine-membered lactam ring (Sugimura, 1982). The (−)-indolactam (stereoisomer causes negative rotation of polarized light) is nearly as potent as certain teleocidins but considerably less active than others. This may be due to differential sensitivity of different PKC isozymes (Ryves *et al.*, 1991). The (+)-indolactam stereoisomer can be used as a negative control as it is significantly less active than the (−) isomer.

Mezerein and Thymelatoxin

Mezerein is a very active antitumor compound isolated from the daphne species of the plant mezereon family, Thymelaeaceae (Barton *et al.*, 1989). It has structural similarity to phorbol myristate acetate and is a PKC activator with anti-leukemic properties (Adolf *et al.*, 1988). It is a second stage tumor promoter while the phorbol esters are first stage promoters. The difference in these tumor promoting properties when compared to phorbol ester can be accounted for by the differences in the relative binding capacities to PKC (Sharkey *et al.*, 1989). Thymelatoxin A is an analog of mezerein that was isolated from Thymelea hirsuta. It is relatively selective for the Ca^{2+}-dependent PKC species, α, β and γ, while having no activity for the Ca^{2+}-independent forms at concentrations of up to 1 mM (Ryves *et al.*, 1991).

Sapintoxins

The sapintoxins were isolated from unripe fruits, in particular, *Sapium indicum* and *S. sebiferum* (Brooks *et al.*, 1987). There are a number of sapintoxin analogs. Many of them are potent activators of PKC, active in the same range as many of the phorbol esters (Ryves *et al.*, 1991). However, distinct from the traditional phorbol esters, the sapintoxins are not tumor promoters at the concentrations where they activate PKC (Brooks *et al.*, 1989). They induce erythema and lymphocyte mitogenesis *in vitro* like many of the other phorbol esters. The sapintoxins do not activate the Ca^{2+}-independent PKC isozymes, similar to mezerein and thymelatoxin A (Ryves *et al.*, 1991). One of the sapintoxin analogs, sapintoxin A, is a fluorescent phorbol ester and has been considered to have promise as a possible imaging reagent for PKC activation (Brooks *et al.*, 1987). However, it has never been exploited for this purpose.

Aplysiatoxin and Debromoaplysiatoxin

Aplysiatoxin is a toxin that was isolated from the blue-green alga, *Lyngba majuscula*. This alga causes seaweed dermatitis (see Fujiki and Sugimura, 1987, for a review).

This class of compounds was identified in the early eighties as having potent PKC activity. It is a tumor promoter and like the phorbol esters, it causes expression of oncogenes. In terms of PKC activation, it has received less attention in recent years aside from the intense chemical and theoretical studies of Kishi and Rando who have concentrated on the active pharmacophore site of the debromoaplysiatoxins as a model for the pharmacophore of the PKC lipid activator (see Rando and Kishi, 1992, for a review).

Bryostatins

The bryostatins have been isolated from *Bugula meritana* and other marine bryozoans (Wender *et al.*, 1988). They are macrocyclic lactones distinct from other classes of PKC activators. These marine animal biosynthetic products only produce some of the biological responses induced by phorbol esters. They inhibit binding of phorbol esters to PKC at nanomolar concentrations, more potent than any other phorbol ester competitive agonists (Wender *et al.*, 1988). They cause rapid PKC translocation and subsequent down-regulation. Bryostatin induces differential translocation and subsequent down-regulation of different PKC isozymes, in contrast to the phorbol esters (Szallasi *et al.*, 1994). These differences were amplified further using a series of bryostatin analogs (Kazanietz *et al.*, 1994). However, unlike the phorbol esters, they do not induce cellular differentiation, nor are they tumor promoters in the mouse skin assay.

Uncharged Hydrophobic Activators

A number of uncharged hydrophobic activators have been shown to activate PKC. Organic solvents such as the alcohols and chloroform, which partition into the bilayer and perturb intrinsic structure can activate PKC at concentrations where they increase fluidity (Lester and Baumann, 1991). At higher concentrations where the bilayer structure is lost and hexagonal phase structures are formed (see Epand, 1992), the enzyme activity is lost (Lester and Baumann, 1991). As discussed previously, the cis-unsaturated fatty acids are poor activators of PKC. However, they act synergistically with the diglycerides to activate PKC. This is considered to be the mechanism of long-term activation of PKC (Lester and Bramham, 1994).

PKC INHIBITORS

Introduction

The PKC inhibitors can be classified in the general class as specific or nonspecific. Generally speaking, specific inhibitors acting at the regulatory site have not been identified, except, possibly for the sphingosines. Most of this class of compounds has been found to modulate hydrophobic interactions where they have effects on membrane surface charge or promote membrane bilayer disruption (see Epand and Lester, 1990 and Epand, 1992 for reviews). Thus, the regulatory site inhibitors can be classified as nonspecific. The catalytic site inhibitors are directed towards the substrate binding and phosphotransferase (ATP) sites. These sites have been shown to be relatively conserved for the various protein kinases (Hanks *et al.*, 1988). However, there are specific differences in the sequences of the various kinases that account for the "specificity" of the various kinase catalytic site inhibitors. Thus, while many of these classes of protein kinases inhibitors are often called nonspecific, this label is in relation to other kinases. The different PKC inhibitor classes will be discussed below.

Regulatory Site PKC Inhibitors

The majority of these reagents can be classified according to their charge or in the case of uncharged and zwitterionic compounds, on the basis of their effects on the phase behavior of model membranes (Epand and Lester, 1990). Reagents that affect the biophysical properties of the membrane bilayer have the potential of interfering with the interaction between the protein, PKC and the diglyceride or acidic phospholipid headgroups, or the acyl chain core. Membrane destabilizers that promote hexagonal phase formation can activate or inhibit depending on the membrane state. Reagents such as alcohols activate PKC at lower concentrations as they promote insertion of the enzyme into the bilayer and formation of an active protein state (Lester and Baumann, 1991). At higher concentrations, the membrane is destabilized, hexagonal phase formation occurs and the enzyme is inhibited as it cannot interact with its substrate nor do the activators have access to the activation site of the protein. Uncharged membrane destabilizers such as palmitoylcarnitine and quercetin inhibit PKC activity quite strongly (Epand and Lester, 1990). Alternately, positively charged compounds are also potent inhibitors of PKC. This includes such reagents as stearylamine, trifluoroperazine, polymixin B, polyamines such as spermine and spermidine, tamoxifen, amiloride, chlorpromazine and melittin. All of these reagents' actions can be described as altering the bilayer properties that may prevent binding of PKC to the suitable phospholipid milieu (Epand, 1992). These compounds are classified as nonspecific inhibitors of PKC and have not been used in *in vivo* and *in vitro* PKC assays for many years now. Occasionally, one sees these compounds appearing in more recent studies. One such compound is melittin. This will be dealt with in more detail in a later section of this chapter. Chlorpromazine and trifluoroperazine were originally classified as potent calmodulin antagonists (e.g. Hait *et al.*, 1985). A number of analogs of the phenothiazines have been examined for their PKC action and their effector site (Aftab *et al.*, 1991). While this was an interesting study in structure-function analysis, there was no indication that these compounds could be used as specific PKC inhibitors. There is a PKC membrane-associated form which is insensitive to the more established PKC inhibitors but is strongly inhibited by the phenothiazines (Orr *et al.*, 1992b). This lack of specificity by these charged compounds may explain some of the confusion that exists in the study of long-term potentiation in complex neuronal systems, where such compounds have been used. There is still much discussion as to whether this process involves PKC or the calmodulin-dependent protein kinase (Lester and Bramham, 1994).

Sphingosine

Sphingosine had received considerable attention over the previous years as a specific inhibitor of PKC (Bell and Burns, 1991). It has been considered to exert its action via the regulatory site of PKC (Hannun *et al.*, 1986). It was noticed some time ago that the inhibition of PKC *in vitro* by sphingosine, could be reversed by titrating this lipid with the acidic phospholipid, phosphatidylserine (Bazzi and Nelsestuen, 1987). Interestingly, sphingosine has properties similar to the positively charged inhibitors discussed above (Epand and Lester, 1990; Epand, 1992). Thus, the specificity of sphingosine for PKC is questionable. This could explain the recent gradual decline in the use of this compound in PKC-related studies.

Calphostin C

This compound was isolated from the fungus, *Cladosporium cladosporioides*. It is a relatively specific PKC inhibitor that supposedly interacts exclusively with the diglyceride/phorbol ester binding site of the enzyme. It has not been found to interfere with the Ca^{2+} or phospholipid interactions with the enzyme (Kobayashi and Nakano, 1989). In addition, its reactivity has been determined to be light dependent, that is, it needs to be exposed to light in order to promote its inhibitory properties (Tamaoki and Nakano, 1990). There has been some controversy as to the potency of this compound, however, it is the most specific of the PKC inhibitors acting at the regulatory site of the enzyme.

Binding Site PKC Inhibitors

The binding site is the region of the enzyme where the substrate binds to PKC such that the phosphate transfers from the ATP binding site. It is in the C4 region of the enzyme, a highly conserved region (Burns and Bell, 1992). This has also been proposed to be a potential lipid binding site (Zidovetzki and Lester, 1992). There are two classes of inhibitors that are considered to act at this site, peptides that compete with substrate binding and the inhibitor, chelerythrine chloride.

The inhibition of the enzyme by competing substrates was first reported by House and Kemp in their study of the pseudosubstrate inhibition of the enzyme (House and Kemp, 1987). More recently, it has been extensively analyzed by O'Brian and colleagues who have synthesized a number of peptides that inhibit PKC activity. These peptides are modified analogs of phosphorylation sites of known PKC phosphoproteins, such as the PKC phosphorylation site of the epidermal growth factor receptor (Ward and O'Brian, 1993). Interestingly, when the substrates are myristoylated at the N-terminal, they bind only to the free PKC molecule. If ATP is present, they will not bind to and subsequently inhibit the enzyme. This research group has made extensive modifications of the substrates and has developed an interesting list of peptide PKC inhibitors all acting at the PKC substrate binding site.

Chelerythrine chloride is a naturally occurring alkaloid that is at least 100 times more potent to PKC than other second-messenger-dependent protein kinases. It acts at the substrate binding site (Herbert *et al.*, 1990). It has fluorescent properties that may make it useful in cell imaging studies. While its specificity *in vitro* remains undeniable, its vast effects in physiological systems (Ko *et al.*, 1990; Walterova *et al.*, 1981) would suggest that it be used together with established PKC inhibitors. Interestingly, it has a net positive charge and it could also be classified in the class of PKC inhibitors that alter membrane biophysical properties thus hindering PKC-membrane association. It has been shown not to interfere with phorbol ester binding to PKC (Herbert *et al.*, 1990).

Catalytic Site PKC Inhibitors

There are numerous PKC specific inhibitors acting at the ATP catalytic site of the enzyme, however, most of them have some action on other related second-messenger protein kinases. This is not surprising considering the homology of consensus sequence of this portion of these protein kinases (Hanks *et al.*, 1988). Two specific inhibitor types have been found that interact on this site and natural and synthetic analogs of them have been the major direction for development of specific PKC inhibitors. These two groups are the H-7 isoquinolenes and the complex chemical toxin, staurosporine, isolated from *Streptomyces sp.* While this whole approach has been questioned particularly in relation to the "specificity" of these compounds (Ruegg and Burgess, 1989), they still remain the most specific of the commercially available PKC inhibitors. Some of the modified analogs of this compound have isozyme specific actions. However, due to the controversies in monitoring PKC activity in cellular preparations, care should be employed when using these compounds. One should not rely on just one class of inhibitor, but should use a number of inhibitors acting at different sites to conclusively demonstrate PKC action.

H-7

H-7 was the first identified PKC-specific inhibitor reported in 1984 by Hidaka and colleagues (see Hidaka and Hagiwara, 1987). This initial study demonstrated that it inhibited other second messenger-dependent protein kinase with similar potency, a feature often neglected by experimenters using this compound (Hidaka *et al.*, 1984). Care should be taken when using this compound particularly the source or supplier. Discrepancies in the PKC-dependent and -independent effects using this compound can be attributed in part to the chemical variants of this compound. The iso-form of this company found to be sold by certain suppliers is at least 4 fold less potent than H-7 (see LC Laboratories Catalog, Woburn, MA, 1993). In spite of these difficulties, H-7 is and will remain a standard PKC inhibitor in cell-free and cellular studies. However, as for other inhibitors and activators, it should not be used on its own. A modified analog of H-7, HA 100, has similar potency in relation to concentration and kinase specificity (Hagiwara *et al.*, 1987).

Staurosporine

Staurosporine was first identified as a potent PKC inhibitor by Tamaoki and colleagues in 1986 (Tamaoki *et al.*, 1986). It is a naturally occurring alkaloid isolated from Streptomyces that has been shown to act at the ATP binding site of a number of second messenger-dependent protein kinases. It was originally shown to have a K_i 10-fold lower for PKC than for other second-messenger-dependent protein kinases. In later years it was shown to interact and inhibit tyrosine kinases with lower potency (Fallon, 1990). This is significant as many researchers use this inhibitor in cellular studies at concentrations that would inhibit the tyrosine kinases also. For example, this could explain the PKC-independent induction of neurite extension in the PC12 cell line (Rasouly *et al.*, 1993). It crosses the membrane and inhibits intracellular PKC action (Blood and Zetter, 1989). Interestingly, it has been reported that staurosporine causes redistribution from the cytosol to the membrane of some of the Ca^{2+}-independent PKC forms (Kiley *et al.*, 1992). A fluorescein-labeled form of staurosporine, called FIM-1, has been shown to be a useful tool for monitoring PKC localization in living cells (Chen and Poenie, 1993).

At the same time that staurosporine was found to inhibit PKC, two compounds of similar structure, K-252a and K-252b, were isolated from the microbe, *Norcardiopsis sp.* (Kase *et al.*, 1986). These compounds are not as potent as staurosporine and they are considerably more expensive to purify, thus, they have not been as popular. However, they are particularly useful as the α-form penetrates cell membranes, while the β-form does not, providing an excellent control (Nagashima *et al.*, 1991).

In order to increase the specificity of staurosporine, a number of groups have prepared analogs of this moiety. The first group, the bisindoylmaleimides, are highly selective for PKC and membrane permeable (Toullec *et al.*, 1991). Bisindoylmeleimide 1, otherwise known as GF-109203X, is much more potent than staurosporine in cellular assays as it is a competitive inhibitor of the high-affinity ATP binding site (Toullec *et al.*, 1991). There are a number of analogs of this class of compounds, one of which is not active, thus, making it a useful negative control for PKC inhibition studies (Davis *et al.*, 1992). These compounds are not isozyme specific. Another staurosporine analog, Gö 6976, is a potent inhibitor of PKC that selectively inhibits the PKC α and β1 isozymes and has no effect on the Ca^{2+}-independent forms at 1000-fold greater levels (Martiny-Baron *et al.*, 1993). Ro-31-8220 is considered to be at least 20-fold more potent for PKC than Ca^{2+}/calmodulin-dependent protein kinases and has been shown to be very useful and specific in cellular studies (Coffey *et al.*, 1993). More recently, a hydroxylated staurosporine analog, UCN-01, has been found to be the most potent of the staurosporine-related compounds. In addition, it blocks tumor growth in cell culture and *in vivo* and was found to be 15–20 more potent in inhibiting the Ca^{2+}-dependent PKC isozymes (Seynaeve *et al.*, 1994; Mizuno *et al.*, 1995). Thus, these compounds appear to be promising directions for the development of specific PKC inhibitors.

However, our major limitation in this area is the lack of understanding of the mechanism of activation of the enzyme and its functioning in different cells under different physiological conditions. In addition, the situation is further complicated by the lack of uniform testing of these compounds in the *in vitro* and *in situ* conditions. Add to this the lack of experience in handling such compounds by the average biologist and the problem is further compounded. It is strongly suggested that the experimenter uses a variety of activators and inhibitors that act at different sites of the enzyme and under different cellular conditions. Also, negative controls and nonspecific agents should be employed to rule out their effects. All this suggests that in order to conclusively show PKC involvement or interaction with a pathway requires serious long-term and costly experimentation.

EXAMPLES OF PKC EFFECTS DUE TO TOXIC INSULT

When biological systems undergo stress, it is not unusual for PKC activities to be modified. Three systems are briefly presented here that relate to some toxic insult. While all show different effects on PKC, the three circumstances are a consequence of the toxic insult and in the mind of this author, cannot be considered as being direct effects due to toxicity. Further testing is required to validate for or against these claims, however, due to the problems involved with exogenous activators and inhibitors, this is not a simple task.

Melittin

Melittin has been demonstrated to inhibit PKC activity (Qi *et al.*, 1983). In the early studies, the melittin effect was considered to be similar to the polyamines. Its inhibitory action was considered to be due to the hydrophobic region of this polypeptide. This region has been found to bind to and purify calmodulin (Kincaid, 1987). In addition, it has a strong interaction with negatively charged phospholipids (Batenburg *et al.*, 1987). Kinetic studies have demonstrated that melittin binds to the substrate binding site in a manner similar to the PKC peptide inhibitors previously described (O'Brian and Ward, 1989). For this to occur, melittin must be directly exposed to the protein which would only occur in cell-free assays. Therefore, there are two potential membrane effects of melittin, a hydrophobic and a charge interaction. Thus, when melittin is used as a PKC inhibitor in cellular assays (see for example Robinson and Kendall, 1990), it cannot be assumed that the inhibitory effect is due to binding of the toxin with PKC. Rather, it may be due to some membrane perturbation that could prevent membrane-association and subsequent activation of PKC. Thus, the desired effect may occur, yet it is consequential. That is, it is not specific. This is further supported by evidence that melittin modulates other membrane-associated enzyme activities, such as calpain I (Brumley and Wallace, 1989), adenylate kinase (Wong and Chu, 1989) and phosphatidylinositol 4-kinase (Yang and Boss, 1994).

Tetanus Toxin

Tetanus toxin penetrates cells via translocation across the cell membrane. It has been shown in a number of systems that tetanus toxin causes translocation of PKC to the cell plasma membrane (Aguilera and Yavin, 1990; Aguilera *et al.*, 1990; Considine *et al.*, 1991; Marxen *et al.*, 1991; Presek *et al.*, 1992). Two studies have shown that there is a subsequent inactivation or down-regulation of PKC (Ho and

Klempner, 1988; Aguilera and Yavin, 1990). Interestingly, the ganglioside, GT1b, a putative receptor for tetanus toxin, prevents PKC activation and down-regulation (Aguilera *et al.*, 1993; Pitzurra *et al.*, 1993). Whether these effects can be considered to be a specific action of the tetanus toxin action is questionable. Other membrane-associated enzymes such as the phospholipases need to be examined and evaluated for effects. The translocation of the toxin across the membrane may cause significant membrane perturbation that could activate the phospholipases. Such an event would result in excessive diacylglycerol activation that leads to subsequent PKC translocation and eventual down regulation.

Cyanide

An established potent neurotoxin, cyanide, induces tonic seizures leading to violent convulsions and subsequent histotoxic hypoxia. It has been found that upon potassium cyanide treatment, there is activation of PKC as demonstrated by the translocation of PKC from the cytosol to the membrane (Rathinavelu *et al.*, 1994). This occurs specifically in hippocampal and cerebellar preparations, but not in cortical slices. When the NMDA glutamate receptor response is blocked by antagonists, the translocation of PKC is prevented, suggesting that this receptor mediates PKC activation (Rathinavelu *et al.*, 1994), a concept well established in NMDA-regulated processes (see Lester and Bramham, 1994). The molecular role of PKC has been suggested as regulating the phosphorylation state of the enzyme that regulates cyanide detoxification (Maduh and Baskin, 1994). This does not support an *in vivo* study indicating that PKC may be involved in the promotion of convulsions (Yamamoto, 1995).

These types of studies only further elucidate the problems in investigating the role of PKC in different pathways. As PKC is ubiquitous in every mammalian cell, it plays many different but specific cellular roles. Thus, while in one cell it may be activated leading to a particular response, it may already be active in another cell and inactive in a number of other cell types. Thus, besides the nonspecificity of many of the activators and the inhibitors, there is also considerable difficulty in evaluating results from *in vivo* studies due to the reasons previously described.

Acknowledgements

I would like to express my gratitude to Drs. Meir Shinitzky, Richard Epand, Harbay Bira and Raphael Zidovetzki for their collaborative efforts and important intellectual contributions.

Table 16.1 Some of the important properties of the most popular protein kinase C activators and inhbitors

Compound	Inhibitor (I) or Activator (A)	PKC specific	Isozyme specific	Cell permeable	Modulates PKC distribution	Reference
Diglycerides	A	Yes	No	+/−	++	Bell and Burns, 1991
Phorbol esters	A	Yes	Yes/No	++	+++	Hecker *et al.*, 1990
cis-unsaturated fatty acids	A	No	Yes/No	++	++	Lester and Bramham, 1994
Pseudosubstrate peptide	I	Yes	No	−	−	House and Kemp, 1990
Teleocidin (Lyngbyatoxin A)	A	Yes	Yes	++	++	Ryves *et al.*, 1991
Mezerein (Thymelatoxin)	A	Yes	Yes (Ca^{2+}-dependent)	++	++	Ryves *et al.*, 1991
Sapintoxin	A	Yes	Yes (Ca^{2+}-dependent)	++	++	Brooks *et al.*, 1989
Aplysiatoxin (debromo-aplysiatoxin)	A	Yes	No	++	++	Kishi and Rando, 1992
Bryostatin	A	Yes	Yes	++	+++	Wender *et al.*, 1988
Organic solvents	A/I	No	No	++	+	Lester and Baumann, 1991
Polyamines	I	No	No	−	+	Qi *et al.*, 1983
Melittin	I	No	No	−	−	O'Brian and Ward, 1989
Phenothiazine	I	No	No	−	−	Aftab *et al.*, 1991
Sphingosine	I	Yes/No	No	++	+	Hannun *et al.*, 1986

Table 16.1 *Continued*

Compound	Inhibitor (I) or Activator (A)	PKC specific	Isozyme specific	Cell permeable	Modulates PKC distribution	Reference
Calphostin C	I	Yes	No	++	+	Kobayashi and Nakano, 1989
Binding site peptides	I	Yes	No	−	−	Ward and O'Brian, 1993
Chelerythrine chloride	I	Yes	No	+	+	Herbert *et al.*, 1990
H-7	I	Yes/No	No	++	−	Hidaka *et al.*, 1984
Staurosporine	I	Yes/No	No	++	+	Tamaoki *et al.*, 1986
K-252a, K-252b	I	Yes/No	No	++	−	Nagashima *et al.*, 1991
GF-109203X	I	Yes	No	++	−	Toullec *et al.*, 1991
Gö 6976	I	Yes	Yes (Ca^{2+}-dependent)	++	−	Martiny-Baron *et al.*, 1993
Ro-31-8220	I	Yes/No	No	++	−	Coffey *et al.*, 1993
UCN-01	I	Yes	Yes (Ca^{2+}-dependent)	++	−	Seynaeve *et al.*, 1994

REFERENCES

Adolf, W., Seip, E.H., Hecker, E. and Dossaji, S.F. (1988) Irritant principles of the mezereon family (Thymelaeaceae), V. New skin irritants and tumor promoters of the daphnane and 1 alpha-alkyldaphnane type from Synaptolepsis kirkii and Synaptolepis retusa. *J.Nat.Prod.*, **51**, 662–674.

Aftab, D.T., Ballas, L.M., Loomis, C.R. and Hait, W.N. (1991) Structure-activity relationships of phenothiazines and related drugs for inhibition of protein kinase C. *Mol. Pharm.*, **40**, 789–805.

Aguilera, J. and Yavin, E. (1990) *In vivo* translocation and down-regulation of protein kinase C following intraventricular administration of tetanus toxin. *J. Neurochem.*, **54**, 339–342.

Aguilera, J., Lopez, L.A. and Yavin, E. (1990) Tetanus toxin-induced protein kinase C activation and elevated serotonin levels in the perinatal rat brain. *FEBS Lett.*, **263**, 61–65.

Aguilera, J., Padros,-Giralt, C., Habig, W.H. and Yavin, E. (1993) GT1b ganglioside prevents tetanus toxin-induced protein kinase C activation and down regulation in the neonatal brain *in vivo*. *J. Neurochem.*, **60**, 709–713.

Barton, K., Randall, G. and Sagone, Jr., A.L. (1989) The effects of the anti-tumor agent mezerein on the cytotoxic capacity and oxidative metabolism of human blood cells. *Invest. New Drugs*, **7**, 179–188.

Batenburg, A.M., van Esch, J.J., Leunissen-Bijvelt, J., Verkleij, A.J. and de Kruijff, B. (1987) Interaction of melittin with negatively charged phospholipids: consequences for lipid organization. *FEBS Lett.*, **223**, 148–154.

Bazzi, M.D. and Nelsestuen, G.L. (1987) Mechanism of protein kinase C inhibition by sphingosine. *Biochem. Biophys.Res.Comm.*, **146**, 203–207.

Bell, R.M. and Burns, D.J. (1991) Lipid activation of protein kinase C. *J. Biol.Chem.*, **266**, 4461–4464.

Blood, C.H. and Zetter, B.R. (1989) Membrane-bound protein kinase C modulates receptor affinity and chemotactic responsiveness of Lewis lung carcinoma sublines to an elastin-derived peptide. *J. Biol. Chem.*, **264**, 10614–10620.

Boni, L.T. (1992) PKC activity in micelles and vesicles. In*: Protein Kinase C: Current Concepts and Future Perspectives.* (Lester, D.S. and Epand, R.M., Eds.) Ellis Horwood, Chichester, 102–116.

Borner, C. and Fabbro, D. (1992) PKC and cellular transformation. In: *Protein Kinase C: Current Concepts and Future Perspectives.* (Lester, D.S. and Epand, R.M., Eds.) Ellis Horwood, Chichester, pp. 297–354.

Bramham, C.R., Alkon, D.L. and Lester, D.S. (1994) Arachidonic acid and diacylglycerol act synergistically through protein kinase C to persistently enhance synaptic transmission in the hippocampus. *Neuroscience*, **60**, 737–743.

Brooks, G., Evans, A.T., Aitken, A. and Evans, F.J. (1987) A fluorescent phorbol ester that is a potent activator of protein kinase C but is not a tumor promoter. *Cancer Lett.*, **38**, 165–170.

Brooks, G., Evans, A.T., Aitken, A. and Evans, F.J. (1989) Tumour-promoting and hyperplastic effects of phorbol and daphnane esters in CD-1 mouse skin and a synergistic effect of calcium ionophore with non-promoting activator of protein kinase C, sapintoxin A. *Carcinogenesis*, **10**, 283–288.

Brumfeld, V. and Lester, D.S. (1990) Penetration of protein kinase C into lipid bilayers. *Arch. Biochem. Biophys.*, **277**, 318–323.

Brumley, L.M. and Wallace, R.W. (1989) Calmodulin and protein kinase C antagonists also inhibit the Ca^{2+}-dependent protein protease, calpain I. *Biochem. Biophys. Res. Comm.*, **159**, 1297–1303.

Burns, D.J. and Bell, R.M. (1991) Protein kinase C contains two phorbol ester binding domains. *J. Biol. Chem.*, **266**, 18330–18338.

Castagna, M., Takai, Y., Kaibuchi, K., Sano, K., Kikkawa, U. and Nishizuka, Y. (1982) Direct activation of calcium-activated, phospholipid-dependent protein kinase by tumor promoting phorbol esters. *J. Biol. Chem.*, **257**, 7847–7851.

Chen, C.S. and Poenie, M.S. (1993) New fluorescent probes for protein kinase C. Synthesis, characterization and application. *J. Biol. Chem.*, **268**, 15812–15822.

Coffey, E.T., Sihra, T.S. and Nicholls, D.G. (1993) Protein kinase C and the regulation of glutamate exocytosis from cerebrocortical synaptosomes. *J. Biol. Chem.*, **268**, 21060–21065.

Considine, R.V., Handler, C.M., Simpson, L.L. and Sherwin, J.R. (1991) Tetanus toxin inhibits neurotensin-induced mobilization of cytosolic protein kinase C activity in NG-108 cells. *Toxicon*, **29**, 1351–1357.

David, P.D., Elliott, L.H., Harris, W., Hill, C.H., Hurst, S.A., Keech, E., Kumar, M.K., Lawton, G., Nixon, J.S. and Wilkinson, S.E. (1992) Inhibitors of protein kinase C. 2. Substituted bisindoylylmaleimides with improved potency and selectivity. *J. Med. Chem.*, **35**, 994–1001.

Dunn, J.A. and Blumberg, P.M. (1983) Specific binding of [20-3H] 12 deoxyphorbol 13-isobutyrate to phorbol ester receptro subclasses in mouse skin particulate preparations. *Cancer Res.*, **43**, 4632–4637.

Epand, R.M. and Lester, D.S. (1990) The role of membrane biophysical properties in the regulation of protein kinase C. *Trends Pharmacol. Sci.*, **11**, 317–320.

Epand, R.M. (1992) Effect of PKC modulators on membrane properties. In: *Protein Kinase C: Current Concepts and Future Perspectives*. (Lester, D.S. and Epand, R.M., Eds.) Ellis Horwood, Chichester, pp. 135–156.

Epand, R.M., Stafford, A.R. and Lester, D.S. (1992) Lipid vesicles which can bind to protein kinase C and activate the enzyme in the presence of EGTA. *Eur. J.Biochem.*, **208**, 327–332.

Fallon, R.J. (1990) Staurosporine inhibits a tyrosine kinase in human hepatoma cell membranes. *Biochem. Biophys. Res. Comm.*, **170**, 1191–1196.

Fujiki, H. and Sugimura, T. (1987) New classes of tumor promoters: Teleocidin, aplysiatoxin and palytoxin. *Adv.Cancer Res.*, **49**, 223–264.

Goldberg, E.M., Lester, D.S., Borchardt, D.B. and Zidovetzki, R. (1994) Effects of diacylglycerols and Ca^{2+} on structure of phosphatidylcholine/phosphatidylserine bilayers. *Biophys. J.*, **66**, 382–393.

Hagiwara, M., Inagaki, M., Watanabe, M., Ito, M., Onoda, K., Tanaka, T. and Hidaka, H. (1987) Selective modulation of calcium-dependent myosin phosphorylation by novel protein kinase inhibitors, isopquinolinesulfonamid derivatives. *Mol. Pharmacol.*, **32**, 7–12.

Hait, W.M., Grais, L., Benz, C. and Cadman, E.C. (1985) Inhibition of growth of leukemic cells by inhibitors of calmodulinL phenothiazines and melittin. *Cancer Chemother. Pharmacol.*, **14**, 202–205.

Hanks, S.K., Quinn, A.M. and Hunter, T. (1988) The protein kinase family: Conserved features and deduced phylogeny of the catalytic domains. *Science*, **241**, 45–52.

Hannun, Y.A., Loomis, D.R., Merrill, A.H., Jr. and Bell, R.M. (1986) Sphingosine inhibition of protein kinase C activity and of phorbol dibutyrate binding *in vitro* and in human platelets. *J. Biol. Chem.*, **261**, 12604–12609.

Hecker, E., Bhagavati, R., Sorg, B. and Lotter, H. (1990) Diterpene esters bioactive in cancerogenesis and their interactions with membrane receptor sites — new studies by computer-assisted molecular modeling (CAMM). *Stud. Biophys.*, **138**, 9–31.

Herbert, J.M., Augereau, J.M., Gleye, J. and Maffrand, J.P. (1990) Chelerythrine is a potent and specific inhibitor of protein kinase C. *Biochem. Biophys. Res. Comm.*, **172**, 993–999.

Hidaka, H., Inagaki, M., Kawamoto, S. and Sasaki, Y. (1984) Isoquinolinesulfonamides, novel and potent inhibitors of cyclic nucleotide-dependent protein kinase and protein kinase C. *Biochemistry*, **23**, 5036–5041.

Hidaka, H. and Hagiwara, M. (1987) Pharmacology of the isozuinoline sulfonamide protein kinase C inhibitors. *Trends Pharmacol. Sci.* **8**, 162–164.

Ho, J.L. and Klempner, M.S. (1988) Diminished activity of protein kinase C in tetanus toxin-treated macrophages and in the spinal cord of mice manifesting generalized tetanus intoxication. *J. Infect. Dis.*, **157**, 925–933.

House, C. and Kemp, B.E. (1987) Protein kinase C pseudosubstrate prototope in its regulatory domain. *Science*, **238**, 1726–1728.

House, C. and Kemp, B.E. (1990) Protein kinase C pseudosubstrate prototype: structure-function relationships. *Cell. Signal.*, **2**, 187–190.

Huang, K-P. and Huang, F.L. (1986) Immunochemical characterisation of rat brain protein kinase C. *J. Biol. Chem.*, **261**, 14781–14787.

Huang, K-P., Huang, F.L., Nakabayashi, H. and Yoshida, Y. (1988) Biochemical characterization of rat brain protein kinase C isozymes. *J. Biol. Chem.*, **263**, 14839–14845.

Jaken, S. (1992) PKC interactions with intracellular components. In: *Protein Kinase C: Current Concepts and Future Perspectives*. (Lester, D.S. and Epand, R.M., Eds.) Ellis Horwood, Chichester, pp. 237–254.

Kase, H., Iwahashi, K. and Matsuda, Y. (1986) K-252a, a potent inhibitor of protein kinase C from microbial origin. *J. Antibiot.*, *(Tokyo)* **30**, 1059–1065.

Kazanietz, M.G., Lewin, N.E., Gao, F., Pettitt, G.R. and Blumberg, P.M. (1994) Binding of [26-3H] bryostatin 1 and analogs to calcium-dependent and calcium-independent protein kinase C isozymes. *Mol. Pharm.*, **46**, 374–379.

Kiley, S.C., Parker, P.J., Fabbro, D. and Jaken, S. (1992) Selective redistribution of protein kinase C isozymes by thapsigargin and staurosporine. *Carcinogenesis*, **13**, 1997–2001.

Kincaid, R. (1987) The use of melittin-sepharose chromatography for gram-preparative purification of calmodulin. *Methods Enzymol.*, **139**, 3–19.

Ko, F.N., Chen, I.S., Wu, S.J., Lee, L.G., Huang, T.F. and Teng, C.M. (1990) Antiplatelet effects of chelerythrine chloride isolated from Zanthoxylum simulans. *Biochim. Biophys. Acta*, **1052**, 360–365.

Kobayashi, E. and Nakano, H. (1989) Calphostin C. A novel microbial compound is a highly potent and specific inhibitor of protein kinase C. *Biochem. Biophys. Res. Comm.*, **548**, 1989–1995.

Kolesnick, R.N. (1990) 1, 2-Diacylglycerols overcome cyclic AMP-mediated inhibition of phosphatidylcholine synthesis in GH3 pituitary cells. *Biochem. J.*, **267**, 17–22.

Kong, F., Kishi, Y., Perez-Sala, D. and Rando, R.R. (1990) The stereochemical requirement for protein kinase C activation by 3-methyldiglycerides matches that found in naturally occurring tumor promotor aplysiatoxins. *FEBS Lett.*, **274**, 203–206.

Lapetina, E.G., Reep, B., Ganong, B.E. and Bell, R.M. (1985) Exogenous sn-1,2-diacylglycerols containing saturated fatty acids function as bioregulators of protein kinase C in human platelets. *J. Biol. Chem.*, **260**, 1358–1361.

Lester, D.S. (1989) High-pressure extraction of membrane-associated protein kinase C from rat brain. *J. Neurochem.*, **52**, 1950–1953.

Lester, D.S. (1990) *in vitro* linoleic acid activation of protein kinase C. *Biochim. Biophys. Acta*, **1054**, 297–303.

Lester, D.S. and Baumann, D. (1991) Action of organic solvents on protein kinase C. *Eur. J. Pharmacol. Mol. Pharm.*, **206**, 301–308.

Lester, D.S., Collin, C., Etcheberrigaray, R. and Alkon, D.L. (1991) Arachidonic acid and diacylglycerol act synergistically to activate protein kinsae C *in vitro* and *in vivo*. *Biochem. Biophys. Res. Comm.*, **179**, 1522–1528.

Lester, D.S. (1992) Membrane-associated protein kinase C. In: *Protein Kinase C: Current Concepts and Future Perspectives.* (Lester, D.S. and Epand, R.M., Eds.) Ellis Horwood, Chichester, pp. 80–101.

Lester, D.S. and Epand, R.M. (1992) Editorial. In: *Protein Kinase C: Current Concepts and Future Perspectives.* (Lester, D.S. and Epand, R.M., Eds.) Ellis Horwood, Chichester, pp.ix–xii.

Lester, D.S. and Bramham, C.R. (1994) Persistent, membrane-associated protein kinase C: From model membranes to synaptic long-term potentiation. *Cell. Signal.*, **5**, 695–708.

Lester, D.S. (1995) Membrane protein kinase and phosphatases. In: *Handbook of Biomembranes.*, **Vol.3** (Shinitzky, M., ed.) VCH/Balaban Publishers, Heidelberg, pp. 253–281.

Maduh, E.U. and Baskin, S.I. (1994) Protein kinase C modulation of rhodanese-catalyzed conversion of cyanide to thiocyanate. *Res. Commun. Mol. Pathol. Pharmacol.*, **86**, 155–173.

Martiny-Baron, G., Kazanietz, M.G., Mischak, H., Blumberg, P.M., Kochs, G., Hug, H., Marme, D. and Schachtele, C. (1993) Selective inhibition of protein kinase C isozymes by the indolcarbazole Gö 6976. *J. Biol. Chem.*, **268**, 9194–9197.

Marxen, P., Bartels, F., Ahnert-Hilger, G. and Bigalke, H. (1991) Distinct targets for tetanus and botulinum A neurotoxins within signal transducing pathway in chromaffin cells. *Naunyn Schmiedebergs Arch. Pharmacol.*, **344**, 387–395.

Meldrum, E., Parker, P.J. and Carozzi, A. (1991) The PtdIns-PLC superfamily and signal transduction. *Biochim. Biophys. Acta*, **1092**, 49–71.

Mizuno, K., Noda, K., Ueda, Y., Hanaki, H., Saido, T.C., Ikuta, T., Kuroki, T., Tamaoki, T., Hirai, S. and Osada, S. (1995) UCN-01, an anti-tumor drug, is a selective inhibitor of the conventional PKC subfamily. *FEBS Lett.*, **359**, 259–261.

Molleyres, L.P. and Rando, R.R. (1988) Structural studies on the diglyceride-mediated activation of protein kinase C. *J. Biol. Chem.*, **263**, 14832–14838.

Mosior, M. and McLaughlin, S. (1991) Peptides that mimic the pseudosubstrate region of protein kinase C bind to acidic lipids in membranes. *Biophys. J.*, **60**, 149–159.

Mosior, M. and McLaughlin, S. (1992) Electrostatics and dimensionality can produce apparent cooperativity when protein kinase C and its substrates bind to acidic lipids in membranes. In: *Protein Kinase C: Current Concepts and Future Perspectives.* (Lester, D.S. and Epand, R.M., Eds.) Ellis Horwood, Chichester, 157–180.

Nagashima, K., Nakanishi, S. and Matsuda, Y. (1991) Inhibition of nerve growth factor-induced neurite outgrowth of PC12 cells by a protein kinase inhibitor which does not permeate the cell membrane. *FEBS Lett.,* **293**, 119–123.

Nishizuka, Y. (1984) The role of protein kinase C in cell surface signal transduction and tumor promotion. *Nature,* **308**, 693–695.

Nishizuka, Y. (1986) Studies and perspectives of protein kinase C. *Science,* **233**, 305-312.

O'Brian, C.A. and Ward, N.E. (1989) ATP-sensitive binding of melittin to the catalytic domain of protein kinase C. *Mol. Pharm.,* **36**, 355–359.

O'Brian, C.A. and Ward, N.E. (1992) The phosphotransferase reaction of protein kinase C: mechanistic studies. In: *Protein Kinase C: Current Concepts and Future Perspectives.* (Lester, D.S. and Epand, R.M., Eds.) Ellis Horwood, Chichester, pp. 117–134.

Ohno, S., Konno, Y., Akita, Y., Yano, A. and Suzuki, K. (1990) A point mutation at the putative ATP-binding site of protein kinase C-a abolishes the kinase activity and renders it down-regulation-insensitive. *J. Biol. Chem.,* **265**, 6296–6300.

Ono, Y., Kurokawa, T., Kawahara, K., Nishimura, O., Marumoto, R., Igarashi, K., Sugino, Y., Kikkawa, U., Ogita, K. and Nishizuka, Y. (1986) Coloning of rat brain protein kinase C complementary DNA. *FEBS Lett.,* **203**, 111–115.

Ono, Y., Fujii, T., Ogita, K., Kikkawa, U., Igarashi, K. and Nishizuka, Y. (1988) The structure, expression and properties of additional members of the protein kinase C family. *J. Biol. Chem.,* **263**, 6927–6932.

Orr, J.W., Keranen, L.M. and Newton, A.C. (1992a) Reversible exposure of the pseudosubstrate domain of protein kinase C by phosphatidylserine and diacylglycerol. *J. Biol. Chem.,* **267**, 15263–15266.

Orr, N., Yavin, E. and Lester, D.S. (1992b) Identification of two populations of protein kinase C activity in rat brain membranes. *J. Neurochem.,* **58**, 461–470.

Pagano, R.E. and and Longmuir, K.J. (1985) Phosphorylation, transbilayer movement and facilitated intracellular transport of diacylglycerol are involved in the uptake of a fluorescent analog of phosphatidic acid by cultured fibroblasts. *J. Biol. Chem.,* **260**, 1909–1916.

Parker, P.J. (1992) Protein kinase C: a structurally related family of enzymes. In: *Protein Kinase C: Current Concepts and Future Perspectives.* (Lester, D.S. and Epand, R.M., Eds.) Ellis Horwood, Chichester, pp. 3–24.

Parker, P.J., Kour, G., Marais, R.M., Mitchell, F., Pears, C.J., Schaap, D., Stabel, S. and Webster, C. (1989) Protein kinase C — a family affair. *Mol. Cell. Endocrinol.,* **65**, 1–11.

Pitzurra, L., Blasi, E., Puliti, M. and Bistona, F. (1993) Toxic effects of tetanus toxin on GG2EE macrophages: prevention of gamma interferon-mediated upregulation of lysozyme-specific mRNA levels. *Infect. Immun.,* **61**, 3605–3610.

Presek, P., Jessen, S., Dreyer, F., Jarvie, P.E., Findik, D. and Dunkley, P.R. (1992) Tetanus toxin inhibits depolarization-stimulated protein phosphorylation in rat cortical synaptosomes: effect on synapsin I phosphorylation and translocation. *J. Neurochem.,* **59**, 1336–1343.

Qi, D.F., Schatzman, R.C., Mazzei, G.J., Turner, R.S., Raynor, R.L., Liao, S. and Kuo, J.F. (1983) Polyamines inhibit phospholipid-sensitive and calmodulin-sensitive Ca^{2+}-dependent protein kinases. *Biochem. J.,* **213**, 281–288.

Rando, R.R. and Kishi, Y. (1992) The structural basis of protein kinase C activation by diacylglycerols and tumor promoters. In: *Protein Kinase C: Current Concepts and Future Perspectives.* (Lester, D.S. and Epand, R.M., Eds.) Ellis Horwood, Chichester, 41–61.

Rasouly, D., Rahamim, E., Lester, D., Matusda, Y. and Lazarovici, P. (1992) Staurosporine-induced neurite outgrowth in PC12 cells is independent of protein kinase C inhibition. *Mol. Pharm.,* **42**, 35–43.

Rathinavelu, A., Sun, P., Pavlakovic, G., Borowitz, J.L. and Isom, G.E. (1994) Cyanide induces protein kinase C translocation: blockade by NMDA antagonists. *J. Biochem.Toxicol.,* **9**, 235–240.

Robinson, J.P. and Kendall, D.A. (1990) No role for phospholipase A2 and protein kinase C in the potentiation by alpha-adrenoreceptors of beta-adrenoceptor-mediated cyclic AMP formation in rat brain. *J. Neurochem.,* **54**, 1082–1084.

Ruegg, U.T. and Burgess, G.M. (1989) Staurosporine, K-252 and UCN-01; potent but nonspecific inhibitors of protein kinases. *Trends Pharmacol. Sci.*, **10**, 218–220.

Ryves, W.J., Evans, A.T., Olivier, A.R., Parker, P.J. and Evans, F.J. (1991) Activation of the PKC-isotypes α, βI, γ, δ and ε by phorbol esters of different biological activities. *FEBS Lett,*. **288**, 5–9.

Sakai, S., Hitotsuyanagi, Y., Aimi, N., Fukiki, H., Suganuma, M., Sugimura, T., Endo, Y. and Shudo, K. (1986) Absolute configuration of lyngbyatoxin A (Teleocidin A-1) and teleocidin A-2. *Tetrahedron Lett.*, **27**, 5219–5220.

Saito, N., Kose, A., Ito, A., Hosoda, K., Mori, M., Hirata, M., Ogita, K., Kikkawa, U., Ono, Y., Igarashi, K., Nishizuka, Y. and Tanaka, C. (1989) Immunocytochemical localization of bII subspecies of protein kinase C in rat brain. *Proc. Natl. Acad. Sci. USA*, **86**, 3409–3413.

Seynaeve, C.M., Kazanietz, M.G., Blumberg, P.M., Sausville, E.A. and Worland, P.J. (1994) Differential inhibition of protein kinase C isozymes by UCN-01, a staurosporine analogue. *Mol. Pharm.*, **45**, 1207–1214.

Sharkey, N.A., Hennings, H., Yuspa, S.H. and Blumberg, P.M. (1989) Comparison of octahydromezerein and meserein as protein kinase C activators and as mouse skin tumor promoters. *Cancer Res.*, **10**, 1937–1941.

Sugimura, T. (1982) Potent tumor promoters other than phorbol ester and their significance. *Gann.*, **73**, 499–507.

Szallasi, Z., Smith, C.B., Pettitt, G.R. and Blumberg, P.M. (1994) Differential regulation of protein kinase C isozymes by bryostatin 1 and phorbol 12-myristate 13-acetate in NIH 3T3 fibroblasts. *J. Biol. Chem.*, **269**, 2118–2124.

Takai, Y., Kishimoto, A., Kikkawa, U., Mori, T. and Nishizuka, Y. (1979) Unsaturated diacylglycerol as a possible messenger for the activation of calcium-activated, phospholipid-dependent protein kinase system. *Biochim. Biophys. Res. Comm.*, **91**, 1218–1224.

Tamaoki, T., Nomoto, H., Takahashi, I., Kato, Y., Morimoto, M. and Tomita, R. (1986) Staurosporine, a potent inhibitor of phospholipid/Ca^{2+}-dependent protein kinase. Biochem. Biophys. Res. Comm., **135**, 397–404.

Tamaoki, T. and Nakano, H. (1990) Potent and specific inhibitors of protein kinase C of microbial origin. *Biotechnology*, **8**, 732–735.

Toullec, D., Pianetti, P., Coste, H., Bellevergue, P., Grand-Perret, T., Akajane, M., Baudet, V., Boissin, P., Boursier, E., Loriolle, F., Duhamel, L., Charon, D., Kirilowsky, J. (1991) The bisindolylmaleimide GF 1090203X is a potent and selective inhibitor of protein kinase C. *J. Biol. Chem.*, **264**, 15771–15781.

Walterova, D., Ulrichova, P., Preininger, V., Simanek, V., Lenfeld, J. and Lasovsky, J. (1981) Inhibition of liver alanine aminotransferase activity by some benzophenanthridine alkaloids. *J.Med.Chem.*, **24**, 1100–1103.

Ward, N.E. and O'Brian, C.A. (1993) Inhibition of protein kinase C by N-myristoylated peptide substrate analogs. *Biochem.*, **32**, 11903–11909.

Wender, P.A., Cribbs, C.M., Koehler, K.F., Sharkey, N.A., Herald, C.L., Kamano, Y., Pettit, G.R. and Blumberg, P.M. (1988) Modeling of the bryostatins to the phorbol ester pharmacophore on protein kinase C. *Proc. Natl. Acad. Sci. USA*, **85**, 7197–7201.

Wong, P.C. and Chu, D.Y. (1989) Evidence for a synaptic plasma membrane assoicated adenylate kinase in the rat brain. *Biochem. Int.*, **19**, 881–888.

Yamamoto, H. (1995) A hypothesis for cyanide-induced tonic seizures with supporting evidence. *Toxicology*, **95**, 19–26.

Yang, W. and Boss, W.F. (1994) Regulation of the plasma membrane type III phosphatidylinositol 4-kinase by positively charged compounds. *Arch. Biochem. Biophys.*, **313**, 112–119.

Young, N. and Rando, R.R. (1984) The stereospecific activation of protein kinase C. *Biochem. Biophys. Res. Comm.*, **122**, 818–823.

Zidovetzki, R. and Lester, D.S. (1992) The mechanism of activation of protein kinase C: a biophysical perspective. *Biochim. Biophys. Acta*, **1134**, 261–272.

17. THE USE OF RAPAMYCIN AND WORTMANNIN IN THE DISSECTION OF THE SIGNAL TRANSDUCTION PATHWAYS REGULATING THE PHOSPHORYLATION OF THE RIBOSOMAL PROTEIN S6

MICHAEL WHALIN and GORDON GUROFF

*Section on Growth Factors, National Institute of Child Health and
Human Development, National Institutes of Health,
Bethesda, MD 20892*

INTRODUCTION

Regulation of gene expression occurs at the transcriptional and translational levels. Most studies have focused on the mechanisms by which growth factors alter the transcription of genes involved in either cell division or differentiation. However, the transcription of genes is only meaningful if they are then translated into the proteins which mediate biochemical events in the cell. One of the early obligatory steps in the mitogenic response is a 2-3 fold increase in the rate of protein synthesis, which is regulated at the level of initiation. This increase in protein synthesis is accompanied by the phosphorylation of a number of specific proteins of the translational apparatus, including eukaryotic initiation factors 4B and 4E and the 40S ribosomal protein S6 (Hershey, 1989; 1991).

The majority of studies related to S6 have focused on the steps signaling mitogenesis which culminate in the phosphorylation of this protein. This event has become a diagnostic hallmark of mitogenic stimulation. To date, there has been little information regarding the role of S6 or of the S6 kinases in cell differentiation.

In this chapter we will review the putative function of S6 in regulating protein synthesis and examine growth factor-stimulated signal transduction pathways which are believed to be involved in the phosphorylation of S6. The use of modern molecular biological methodologies, such as the expression of mutated growth factor receptors and constitutively active or dominant negative inactive putative signaling molecules, has allowed substantial progress to be made in the dissection of signaling pathways leading to S6 phosphorylation. Furthermore, the elucidation of the signaling pathways regulating S6 phosphorylation has been greatly aided by the utilization of pharmacological tools such as the microbial toxins rapamycin and wortmannin (Figure 17.1).

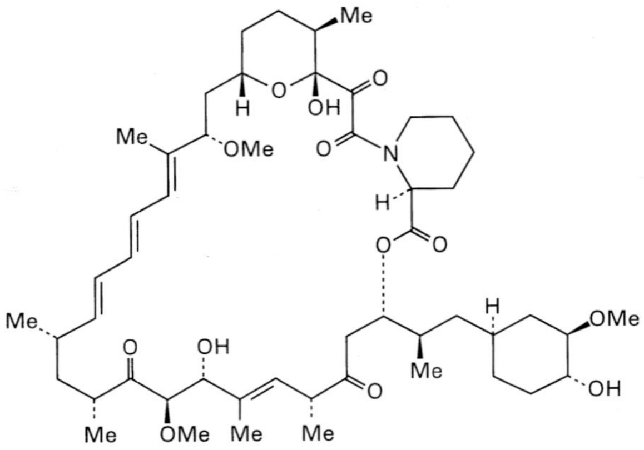

Rapamycin

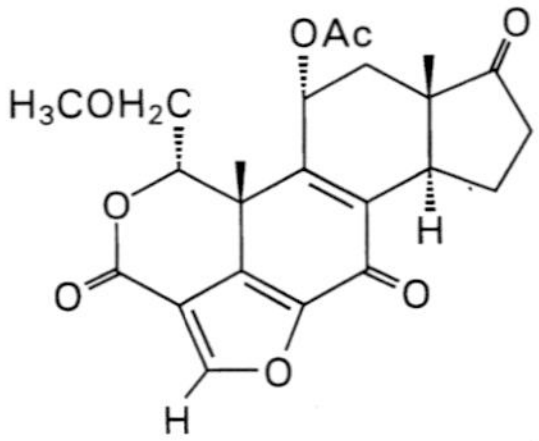

Wortmannin

Figure 17.1   Chemical structures of the fungal toxins rapamycin and wortmannin.

THE PHARMACOLOGICAL TOOLS USED IN THE DISSECTION

Cultures of a streptomycete isolated from an Easter Island soil sample were found to inhibit the yeast, *Candida albicans,* but had no activity against any of the gram-positive or gram-negative bacteria tested (Vezina *et al.,* 1975). The antibiotic-producing microorganism was characterized and identified as *Streptomyces hygroscopicus.* The antibiotic was named for the geographical location of its discovery, i.e, Easter Island is also called Rapa Nui, so the compound was named, rapa-mycin. Rapamycin was isolated from the mycelium of *Streptomyces hygroscopicus* NRRL 5491 by solvent extraction, followed by chromatography on silica gel and then crystallized as a colorless solid with a melting point of ~185°C (Sehgal *et al.,* 1975). As revealed by its characteristic ultraviolet absorption spectrum, rapamycin contains a triene. It is classified as a macrocyclic lactone. The compound is freely soluble in methanol, ethanol, acetone, chloroform, methylene dichloride, dimethylsulphoxide, but is practically insoluble in water. Acute toxicity of rapamycin in mice is low, i.e., LD_{50} (i.p.) is ~600 mg/kg and the LD_{50} (p.o.) is > 2,500 mg/kg.

The cytotoxicity of rapamycin on mammalian cells in culture has not been studied in depth, but no toxic effects have been noted in the human T cell line CTLL-2, with incubation times up to 20 hours and rapamycin concentrations up to 1 μM. On the other hand, rapamycin enhanced the cytotoxicity of and increased the sensitivity to the chemotherapeutic agent, cisplatin (Shi *et al.*, 1995). Inhibition of T cell proliferation by rapamycin, unlike inhibition by the steroid prednisolone, is not associated with cytotoxicity (Bansbach *et al.*, 1993). Most studies with mammalian cells in culture have utilized short treatment intervals of 10–30 minutes with concentrations of rapamycin of less than 1 μM. One pitfall when using rapamycin under tissue culture conditions is that the compound binds to serum proteins, thereby decreasing its effective dose.

Shortly after the discovery of rapamycin, it was demonstrated that the drug inhibited the immune response in rats (Martel *et al.*, 1978). Rapamycin prevented the development of two experimental imunopathies and it was suggested that the immunosuppressant activity was related to inhibition of the lymphatic system. It was subsequently demonstrated that rapamycin inhibited the proliferation of T cells by preventing their response to interleukin-2 (Dumont *et al.*, 1990; Bierer *et al.*, 1990). The low toxicity of rapamycin coupled with its immunosuppressive action has been exploited in transplantation biology; it is now the third and most recent addition to a group of antifungal antibiotics that have been utilized to prolong the survival of transplanted grafts (Morris, 1991). The exact molecular mechanisms involved in the actions of rapamycin are at the present time unknown, but may involve a biochemical system which includes the 40S ribosomal protein, S6.

Wortmannin (WT) was first isolated as an antibiotic produced by *Penicillium wortmanni* (Brian *et al.*, 1957). WT was purified by organic extraction followed by sequential silica gel and thin layer chromatography. The purified compound is soluble in dimethylsulfoxide, but has limited solubility in water. WT is very unstable in aqueous solutions at pH values between 3 and 8, a characteristic that makes it difficult to use. It was not until 1972 that the structure of WT was established (MacMillan *et al.*, 1972). In 1988, during a search for the agent causing mycotoxicosis in animals and humans, WT was isolated from *Fusarium oxysporum*, a predominant fungus found on many commodities and in soil in many food-producing areas (Abbas and Mirocha, 1988). Mycotoxicosis is described as tissue congestion and hemorrhage, food refusal, weight loss, vomiting, skin necrosis, and death and WT was identified as the hemorrhagic mycotoxin. It is a sterol-like compound that causes toxic effects in rats, including hemorrhages in the stomach, intestines, heart, and thymus. WT is also produced by *Penicillium wortmanni* (Brian *et al.*, 1957), *Penicillium funiculosum* (Haeflinger and Hauser, 1973), *Myrothecium roridum* (Petcher *et al.*, 1972) and *Talaromyces wortmanni* (Nakanishi *et al.*, 1992). It is unusual for more than one genus to produce the same mycotoxin. This finding may reflect an important evolutionary conservation in the production of this toxin.

As is the case with rapamycin, detailed cytotoxicity studies with WT have not been performed. Several reports have shown that human (HeLa and MCF-7) and murine (NIH-3T3) cells exposed to WT alone did not display significant cytotoxicity. However, if the cells were pretreated with radiation, a perturbation that causes DNA damage, WT potentiated the cytotoxicity of the ionizing radiation (Price and Youmell, 1996). This suggests that a WT-sensitive pathway may regulate the response of cells to DNA damage. Other studies showed that WT alone was cytotoxic to several human tumor cell lines *in vitro* (Schultz *et al.*,1995). The most sensitive cell lines included the GC3 colon carcinoma, the IGROV1 ovarian carcinoma, and the CCRF-CEM leukemia, all having IC_{50} values ranging from 0.7–2.1 μM. The cytotoxicity of WT was increased ~10-fold under serum-free conditions. This is directly due to the fact that WT binds to serum proteins which lowers its effective concentration. Therefore, caution should be used in interpreting the cellular effects of WT, because its effective concentration varies depending on the experimental conditions, such as time, temperature, pH, cell line used, concentrations of protein, and cell density.

THE CELLULAR SYSTEM TO BE DISSECTED

The eukaryotic ribosome is an 80S ribonucleoprotein complex comprised of two subunits of 60 and 40S. The 60S subunit contains ~45 different proteins and 5S, 5.5S, and 28S ribosomal RNAs (rRNA). The 40S subunit consists of one 18S rRNA molecule and ~30 different proteins (Wool and Stoffler, 1974), of which S6 is the only one known to be appreciably phosphorylated in cells treated with either mitogenic stimuli or differentiating agents.

Localization of S6

Chemical protection studies and crosslinking experiments have localized S6 to a region of the 40S ribosomal subunit which interfaces with the 60S subunit during 80S complex formation where it may be in direct contact with the 28S rRNA of the 60S subunit (Nygard and Nilsson, 1990). This is a region where many of the components of the translation apparatus, such as messenger RNA (mRNA) and transfer RNA (tRNA), as well as initiation, elongation, and termination factors interact (Morley and Thomas, 1991). Since ribosomes are highly flexible and dynamic collections of macromolecules, the localization of S6 to the interface between the 40 and 60S subunits may have functional significance and its phosphorylation state could be involved in the stabilization of conformational states of the ribosome thus influencing the transient association with translational proteins and mRNAs.

Phosphorylation of S6

The phosphorylation of S6 was first demonstrated during liver regeneration in rats (Gressner and Wool, 1974). The observation that there was a dramatic increase in protein synthesis during regeneration (Chaudhuri and Lieberman, 1968) prompted the speculation that the two events are correlated. The Gressner and Wool study (1974) was the first to implicate the phosphorylation of S6 as a regulatory mechanism and demonstrated the incorporation of 5 moles of phosphate into each mole of S6. The observation that during liver regeneration the remaining hepatocytes were induced to re-enter the cell cycle and to proliferate led to subsequent studies showing that the treatment of growth-arrested or quiescent cells in culture with serum or growth factors led to the multiple phosphorylation of S6.

More detailed biochemical studies showed that agents which elevate cyclic AMP (cAMP), such as the hormone glucagon, resulted in an increase in the phosphorylation of ribosomal proteins. Subsequent experiments indicated that both cAMP and glucagon increase the phosphorylation of S6 (Gressner and Wool, 1976). Differential phosphorylation of S6 was reported in 1982, when it was demonstrated, using isolated rat hepatocytes, that cAMP and glucagon increased the phosphorylation of S6 at sites which were distinct from those increased by insulin (Wettenhall et al., 1982). This finding suggested that differential phosphorylation of S6 may be involved in the regulation of glycogen metabolism, i.e., glucagon treatment, which activates catabolic pathways, results in the incorporation of two phosphates into S6, whereas insulin treatment, which activates anabolic pathways, results in more highly phosphorylated derivatives of S6.

The most remarkable aspect of the biochemistry of S6 is the extraordinary number and variety of agents and alterations in the cellular milieu that effect its phosphorylation. These include hormones (Gressner and Wool, 1976; Wettenhall et al., 1982; Smith et al., 1980), peptide growth factors (Thomas et al., 1982), toxic substances (Gressner and Greiling, 1977), viral infections (Kaerlein and Horak, 1976), fertilization (Nielsen et al., 1982), phorbol esters (Blenis and Erikson, 1985), heat shock (Scharf and Nover, 1982), inhibitors of protein synthesis (Blenis et al., 1991) and differentiating agents (Haleguoa and Patrick, 1980). However, a detailed analysis of the sites on which S6 is phosphorylated by each of these influences has not been performed.

FUNCTIONAL SIGNIFICANCE OF S6 PHOSPHORYLATION

Although the phosphorylation of S6 is a highly conserved response to signals that stimulate cell proliferation and differentiation and a wide variety of stimuli affect the phosphorylation state of the protein, the actual function of S6 has yet to be elucidated, let alone how that function may be altered by phosphorylation. Early studies provided only an indirect correlation between an increase in protein synthesis and an increase in S6 phosphorylation (Thomas et al., 1980; Nielsen et al., 1981; Duncan and McConkey, 1982; Burkhard and Traugh, 1983). However, more recently, the use of the microbial toxin, rapamycin, has yielded important information regarding the correlation between the phosphorylation of S6 and the repression of the translation of certain classes of mRNAs. Indeed, the translation of mRNAs containing a 5′ polypyrimidine motif and which encode proteins involved in the translation apparatus such as elongation factors (i.e., EF-2) and ribosomal proteins (i.e., S6) (Jeffries *et al.*, 1994; Terada *et al.*, 1994) is inhibited by rapamycin. It has also been shown that rapamycin inhibits the growth-dependent translation of the major 6.0-kb IGF-II mRNA (Nielsen *et al.*, 1995).

Biochemical Evidence

The reports showing a correlation between increased S6 phosphorylation and increased protein synthesis gave rise to the hypothesis that S6 phosphorylation is involved in the regulation of translation. The finding of an indirect correlation between the preferential utilization of phosphorylated 40S ribosomal subunits for the formation of the initiation complex suggested that an increased rate of initiation may be caused by covalent modification of the S6 protein (Duncan and McConkey, 1982). That is, following the addition of serum to quiescent cells, the phosphorylation of S6 is increased and ribosomes are recruited into polysomes, indicative of increased translational activity.

In studies using serum-deprived Hela and Swiss 3T3 cells, readdition of serum stimulated the G_o/G_1 cell cycle transition which correlated with inactive 80S ribosomes moving into actively translating polysomes and was coincident with an increase in S6 phosphorylation (Thomas *et al.*, 1980). Supporting studies investigating the mechanism by which S6 phosphorylation may alter protein synthesis showed that following serum addition to quiescent Hela cells, S6 became rapidly phosphorylated and there was a concomitant recruitment of mRNA and ribosomes into polysomes (Nielsen *et al.*, 1981). However, the level of S6 phosphorylation decreased to basal levels within 6 hours of serum addition, a time when protein synthesis was still at maximal rates. These findings led to the postulate that S6 phosphorylation increased the affinity of the 40S subunit for binding to mRNA and was involved in the initiation of translation, but not in its maintenance.

Studies in other biological systems also showed that high levels of S6 phosphorylation correlated with increased rates of protein synthesis, one of the earliest events required for cells in G_o to re-enter the cell cycle. For example, in immature *Xenopus laevis* oocytes, S6 is in a dephosphorylated state. However, when oocytes are treated with progesterone or insulin, agents which induce maturation, S6 becomes multiply phosphorylated (Maller *et al.*, 1977). This increase in S6 phosphorylation was subsequently correlated with a 2–4 fold increase in the rate of protein synthesis (Wasserman *et al.*, 1982).

The phosphorylation of S6 appears to be evolutionarily conserved. There is an increase in S6 phosphorylation that precedes the increase in the mitotic index of tomato cells during recovery from heat shock (Schraf and Nover, 1982). Similarly, S6 is completely dephosphorylated in the silk glands of *Bombyx mori* larvae during the molting period prior to the 5th larval instar. When the larvae resume feeding, protein synthesis in the silk glands increases in parallel with increased phosphorylation of S6 (Madjar and Fournier, 1987). Consistent with these findings, Thomas *et al.* (1982) showed that liver S6 was in a dephosphorylated state in rats starved for 24–36 hours. Refeeding the rats resulted in a burst of protein synthesis and an increase in S6 phosphorylation. It therefore appears that the increased phosphorylation of S6 is not strictly confined to the G_o/G_1 cell cycle transition, but is also a response of cells to a general up-regulation of their anabolic state.

In vitro studies using a reconstituted translational system showed that direct binding and translation of a synthetic message, poly AUG, was enhanced by the pre-phosphorylation of 40S ribosomal subunits by protease-activated kinase II (PAK II) (Burkhard and Traugh, 1983). In contrast, the pre-phosphorylation of 40S ribosomes by protein kinase A (PKA) resulted in a decrease in the binding and translation of the synthetic message. At the molecular level, the phosphorylation of S6 by PAK II resulted in the incorporation of up to 4 moles of phosphate per mole of S6, whereas phosphorylation of S6 by PKA resulted in the incorporation of only 2 moles of phosphate per mole of S6. This was the first study directly implicating the differential phosphorylation of S6 in the modulation of the binding and translation of mRNA to 40S ribosomal subunits.

A later series of experiments showed that the efficiency of translation of globin mRNA was altered depending on whether S6 was pre-phosphorylated with PKA or with PAK II (Palen and Traugh, 1987). That is, phosphorylation of 40S ribosomes by PKA, which resulted in the incorporatation of 2 moles of phosphate into each mole of S6, had no effect on the binding or translation of globin mRNA. In constrast, phosphorylation of 40S ribosomes by PAK II, which in this study led to the incorporation of 2.5 moles of phosphate per mole of S6, had no effect on globin mRNA binding, but did increase the translation of the mRNA by ~4 fold. This supported the concept that differential phosphorylation of S6 influenced the translation of mRNA and, furthermore, implied that the phosphorylation of S6 at multiple sites was associated with selective translation of specific mRNAs.

NUCLEAR FORMS OF S6 AND S6 KINASES

Most studies relating to S6 phosphorylation have focused on the cytoplasmic localization of the working ribosome. However, nuclear forms of phosphorylated S6 (Franco and Rosenfeld, 1990; Macfarlane and Gailani, 1990; Tang *et al.*, 1996) have been described, as have nuclear S6 kinases (Reinhard *et al.*, 1994; Coffer and Woodgett, 1994; Tang *et al.*, 1996). The function of nuclear forms of S6 and S6 kinases has yet to be elucidated. The absence of protein synthesis in the nucleus suggests a unique role of phosphorylated S6 in this compartment. One possibility is that phosphorylated S6 is involved in regulating ribosomal biosynthesis in the nucleolus. Another is that phosphorylated S6 is involved in the recognition and selective transport of essential mRNA transcripts. Alternatively, phospho-S6 may have no nuclear function and the nuclear S6 kinases may have substrates other than S6.

SITES OF PHOSPHORYLATION OF S6

The sites of S6 phosphorylation were first mapped to a 17-amino acid tryptic peptide thought to be the carboxy terminus of the protein (Wettenhall *et al.*, 1983;

Figure 17.2   Amino acid sequence of the carboxy terminal domain of mammalian S6 and the yeast homolog, S10. Arrows and bold letters in italics denote the serine residues that are known to be phosphorylated.

Wettenhall and Morgan, 1984). It was subsequently shown, using S6 isolated from livers of cycloheximide-treated rats, that all the sites of S6 phosphorylation did indeed reside in a carboxy terminal locale (Krieg *et al.*, 1988). A follow-up study demonstrated that serum stimulation of quiescent fibroblasts resulted in the phosphorylation of S6 at these same sites (Bandi *et al.*, 1993).

There are 15 seryl residues in S6, seven of which are found in a carboxy-terminal sequence of 15 amino acids (Figure 17.2) (Chan and Wool, 1988). Of the seven serine residues within this fragment, only serines 235, 236, 240, 244 and 247 were found to be phosphorylated in cells treated with agents that stimulate proliferation. This clustering of phosphorylation sites in protein kinase substrates is not an uncommon theme (Roach, 1991). For example, six of the seven phosphorylation sites in glycogen synthase are grouped together near the carboxy terminal of the protein. It has been shown that these sites are the substrates for five distinct protein kinases. Since S6 also has a clustering of phosphorylation sites in the carboxy terminal domain, it is tempting to speculate that this is of functional significance and, furthermore, that multiple kinases may act on the protein.

The cDNAs encoding S6 have been cloned from several mammalian species including rat (Chan and Wool, 1988), mouse (Lalanne *et al.*, 1987) and human (Heinze *et al.*, 1988; Pata *et al.*, 1992), as well as from *Drosophila melanogaster* (Spencer and Mackie, 1993) and *S. cerevisiae* (Leer *et al.*, 1982). Despite the high degree of conservation of S6 between species, there are important differences at the carboxy terminus of the protein where the phosphorylation sites have been mapped (Figure 17.2). These differences, particularly in yeast, have raised questions regarding the functional significance of the phosphorylation of S6. The yeast homolog of S6 termed S10, encodes a protein lacking the last ten carboxy terminal amino acids found in higher eukaryotes (Leer *et al.*, 1982). The only two amino acids in the yeast S6 that serve as kinase substrates are those corresponding to serines 235 and 236 in mammalian S6. Site-directed mutagenesis experiments in yeast, in which the serines were replaced with alanines, showed no altered phenotype suggesting that the phosphorylation state of S6 is not a major determinant of the growth of *S. cerevisiae* (Johnson and Warner, 1987). This finding, however, is consistent with the Palen and Traugh study (1987) which showed that phosphorylation of S6 at serines 235 and 236 by PKA did not enhance translation and also with studies which demonstrated that S6 containing phosphorylated serines at positions 235 and 236 did not have selective advantage in entering polysomes (Thomas *et al.*, 1982). These findings may reflect an evolutionary hierarchical control of gene expression at the translational level, with the carboxy terminal extension of S6 offering a selective advantage for higher metazoans in increasing the regulatory complexity of protein synthesis.

PROTEIN KINASES WHICH PHOSPHORYLATE S6

A number of protein kinases have been shown to phosphorylate S6 *in vitro* (Table 17.1). Phosphorylation of S6 by most of these kinases yields only mono- or di-phosphorylated derivatives. More highly phosphorylated derivatives of S6 have been found after treatment of various cell types with insulin or other growth-promoting compounds. The extent to which most of the enzymes in Table 17.1 phosphorylate S6 is limited and their modes of activation are not consistent with a mitogen-induced phosphorylation cascade. The most extensively characterized S6

Table 17.1

Kinase	Molecular Weight	Reference
Protein kinase A	41 kDa	Duvernay and Traugh (1978) Del Grande and Traugh (1982)
Protein kinase G	74 kDa	Del Grande and Traugh (1982)
Protein kinase C	82 kDa	Le Peuch *et al.* (1983) Parker *et al.* (1985)
Calmodulin-dependent protein kinase II	460–654 kDa	Gorelick *et al.* (1983)
Casein kinase I	49 kDa	Cobb and Rosen (1983)
Protease-activated kinase II	45–55 kDa	Lubben and Traugh (1983) Perisic and Traugh (1993)
H4/S6 Kinase	60 kDa	Donahue and Masaracchia (1984) Brandon and Masaracchia (1991) Dennis and Masaracchi (1993)
NGF-stimulated S6 kinase	45–50 kDa	Matsuda and Guroff (1987)
p90 S6 kinase	83–92 kDa	Erikson and Maller (1986) Erikson and Maller (1991) Lavoinne *et al.* (1991)
p70 S6 kinase	63–70 kDa αI) 85–90 kDa (αII)	Kozma et al. (1989) Price *et al.* (1989)
Insulin-stimulated S6 kinase	31 kDa	Hei *et al.* (1994)
Protein kinase N	45 kDa	Volonté and Greene (1992)

kinases have been the p70 S6 kinases (p70^{S6K}) and the p90 S6 kinases (p90^{rsk}) because both kinases are capable of multisite phosphorylation of S6 and both are activated in response to diverse mitogens *in vivo*.

To date, there has been little focus on the characterization of the S6 kinases involved in differentiation. A few reports have documented that nerve growth factor (NGF), a differentiation agent activates distinct kinases (Blenis and Erikson, 1986; Matsuda and Guroff, 1987). It has been demonstrated that mitogens such as epidermal growth factor (EGF) elicit differential phosphorylation of S6 compared to that observed with NGF (Mutoh *et al.*, 1992) and that separate kinases are responsible for the observed differential phosphorylation (Mutoh *et al.*, 1988). However, the functional role of this differential phosphorylation and its relationship to proliferation or differentiation has yet to be established.

The p90^{rsk} Family of S6 Kinases

The first S6 kinase purified to apparent homogeneity and partially characterized was from *Xenopus laevis* oocytes and termed S6 kinase II based on its chromatographic elution profile from DEAE-Sephacel (Erikson and Maller, 1986). The enzyme has an apparent molecular weight (M_r) of 92 kDa as estimated by SDS gel electrophoresis (SDS-PAGE). It was shown to be phosphorylated and activated *in vivo* during ooycte maturation (Erikson and Maller, 1989). The phosphorylation of the enzyme is a prerequisite for activation since it could be inactivated by phosphatase treatment (Erikson and Maller, 1989; Sturgill *et al.*, 1988). The inactivated enzyme was shown to be activated *in vitro* by mitogen-activated protein kinase (MAPK) (Sturgill *et al.*, 1988), suggesting that it may be a component of the MAPK cascade. A second *Xenopus* S6 kinase termed S6 kinase I has also been purified to apparent homogeniety and found to have an M_r of ~90 kDa (Erikson and Maller, 1991). This enzyme is also regulated by phosphorylation as evidenced by its inactivation by treatment with phosphatase. Immunological data and analysis of substrate specificity demonstrated that S6 kinase I is related to, but distinct from the S6 kinase II.

It was shown *in vitro* that purified S6 kinase II phosphorylated S6 on serine residues located in the carboxyl terminus of the protein (serines 235, 236, 240 and 244) which were also phosphorylated by S6 kinases from cycloheximide-treated rats (Wettenhall *et al.*, 1992). The same sites in *Xenopus* S6 were found to be phosphorylated *in vivo*, leading to the suggestion that S6 kinase II was a physiologically relevant enzyme mediating S6 phosphorylation in higher eukaryotes (Erikson and Maller, 1989). This postulate was strengthened by the fact that, at that time, no other S6 kinases had been detected in *Xenopus* oocytes. However, shortly after this report another laboratory described the presence of p70^{S6K} in *Xenopus* (Lane *et al.*, 1992).

The molecular nature of *Xenopus* S6 kinase II was revealed when two highly related cDNA clones which encode the *Xenopus* enzyme were obtained (Jones *et al.*, 1988). Shortly after this report, two homologous mammalian cDNAs were cloned from mouse and chicken (Alcorta *et al.*, 1989). Due to their sequence homology, the genes encoding these kinases were termed rsk for <u>r</u>ibosomal <u>S</u>6 kinase. Thus the mouse homolog is termed *rsk*^{mo}-1 and the chicken homolog termed *rsk*^{ch}. The partial cDNA encoding a second murine isoform (*rsk*^{mo}-2), which is 84% identical to *rsk*^{mo}-1 was also isolated. To date, no mammalian homolog of the *Xenopus* S6 kinase I has been identified.

A mitogen-responsive mammalian homolog of the *Xenopus* S6 kinase II was identified and characterized in chicken embryo fibroblasts and termed p90^{rsk} (Chen and Blenis, 1990). This mammalian enzyme has been the focus of the majority of recent investigations. Recently, a full length human cDNA encoding a novel p90^{rsk} isoform has been cloned and termed *RSK3* (Zhao *et al.*, 1995). This isoform is made up of 733 amino acids, has an apparent M_r of 83 kDa and contains a unique N-terminus which has a putative nuclear localization domain. RSK3 exhibits growth-stimulated autophosphorylation and kinase activation similar to other p90^{rsk} isoforms. However, its relative activity toward substrates differs. Moreover, the phosphatase-inactivated RSK3 is incapable of *in vitro* activation by MAPK. It therefore appears that the *rsk* kinases represent a family of isoenzymes which have different functions and may be regulated by distinct signaling cascades.

A unique characteristic of this enzyme family is that each member contains two unrelated protein serine/threonine kinase catalytic domains which are not a result of gene duplication (Alcorta *et al.*, 1989; Jones *et al.*, 1988). The C-terminal catalytic domain shares significant homology (30% identity) with the catalytic domain of phosphorylase *b* kinase. The N-terminal catalytic domain resembles the catalytic domains of protein kinase A, protein kinase C, and p70^{S6K}. Recently it has been demonstrated that the N-terminal catalytic domain is responsible for the ability of p90^{rsk} to phosphorylate substrates such as S6, c-Fos, SRF, and Nur77 (Fisher and Blenis, 1996). The C-terminal catalytic domain is thought to be involved in autophosphorylation, since substrates for this catalytic domain, other than p90^{rsk} itself, have not been identified. The results of this study suggest that the N-terminal catalytic domain activity is regulated by the activation of the C-terminal catalytic domain following its phosphorylation by MAPK. This speculation is supported by the observation that p90^{rsk} and MAPK interact physically (Hsiao *et al.*, 1994; Scimeca *et al.*, 1992) and by the mapping of this interaction to the C-terminal domain of the enzyme (Fisher and Blenis, 1996).

p90^{rsk} isoforms have been shown to be translocated to the nucleus (Chen *et al.*, 1992; Zhao *et al.*, 1995) and *in vitro* nuclear substates for p90^{rsk} include the orphan steroid receptor Nur77 or NGFI-B (Davis *et al.*, 1993), the serum response factor (SRF), a transcription factor which regulates the immediate early gene *c-fos* (Rivera *et al.*, 1993) and c-Fos (Chen *et al.*, 1993). These data imply that this form of S6 kinase has important functions in regulating nuclear phosphorylation events. Cytoplasmic substrates for p90^{rsk}, other than S6, include the glycogen-binding subunit of phosphatase 1 (Dent *et al.*, 1990; Lavoinne *et al.*, 1991) and glycogen synthase kinase 3 (GSK-3) (Sutherland *et al.*, 1993; Stambolic and Woodgett, 1994), an enzyme involved in glycogen metabolism, but which also phosphorylates another immediate early gene product, c-Jun (Nikolakaki *et al.*, 1993). These findings suggest that p90^{rsk} may be an important enzyme in regulating both nuclear events and other metabolic processes in cells.

The p70 S6 Kinases

S6 kinases with molecular weights ranging from 65–70 kDa, denoted p70^{S6K}, have also been purified to apparent homogeniety from different sources by several groups (Tabarini *et al.*, 1987; Blenis *et al.*, 1987; Jeno *et al.*, 1988; Jeno *et al.*, 1989; Price *et al.*, 1989; Gregory *et al.*, 1989; Lane and Thomas, 1991). The activity of p70^{S6K} is regulated by serine/threonine phosphorylation as evidenced by the demonstration that phosphorylated enzyme can be recovered from ^{32}P-labeled cells after stimulation with insulin (Price *et al.*, 1990), the absolute requirement for the addition of phosphatase inhibitors in extraction buffers to preserve its activity (Novak-Hofer and Thomas, 1984; Novak-Hofer and Thomas, 1985) and the finding that it is rapidly dephosphorylated and inactivated by *in vitro* treatment with protein phosphatases (Ballou *et al.*, 1988). Purified or recombinant p70^{S6K} has been shown to be phosphorylated by several insulin-stimulated, proline-directed protein kinases which were tentatively identified as cdc2 kinases, however activation of the enzyme by these phosphorylations was not observed (Mukhopadhyay *et al.*, 1992). It has also been shown that p70^{S6K} is not a substrate for purified PKA, casein kinase II, GSK-3, or insulin-stimulated MAPK (Price *et al.*, 1990). Therefore, the p70^{S6K} kinases remain unidentified.

It has been shown that purified p70^{S6K} incorporates only 4 moles of phosphate into S6 onto serines 235, 236, 240 and 244 (Ferrari *et al.*, 1991). It is interesting that in whole cells, S6 is phosphorylated on up to five sites (Figure 17.2) with the first four corresponding to the sites phosphorylated by p70^{S6K} *in vitro*. This finding could implicate yet other unidentified S6 kinases capable of phosphorylating all five sites in S6 or a subset of sites including serine 247.

Two p70^{S6K} cDNAs have been cloned from rat (Banerjee *et al.*, 1990; Kozma *et al.*, 1990) and human (Grove *et al.*, 1990) sources and termed p70 αI and p70 αII. The two cDNAs encode proteins with distinct molecular sizes; the p70 αI isoform has a predicted molecular weight of 59,200 daltons (based on the cDNA) and migrates as a molecule with an apparent molecular weight of ~85 kDa on SDS-PAGE and thus has been termed p85^{S6K}. The p70 αII isoform has a predicted molecular weight of 56,200 daltons and migrates as a ~70 kDa protein on SDS-PAGE and is termed p70^{S6K}. The p70^{S6K}/p85^{S6K}, collectively termed p70^{S6K}, appear to represent two isoforms of the same enzyme derived by differential splicing of a common gene (Reinhard *et al.*, 1992) and whose target (at least *in vitro*) is the 40S ribosomal protein S6.

The p85^{S6K} sequence is identical to that of p70^{S6K}, except for a 23-amino acid extension at its amino terminus. This extension contains a series of six basic arginine residues and is believed to be responsible for the anomalous migration behavior on SDS-PAGE. The amino terminal extension also harbors a putative nuclear signal sequence (Grove *et al.*,, 1991; Reinhard *et al.*, 1992) consistent with the finding of S6 phosphorylation in the nucleus of cells after mitogenic stimulation (Franco and Rosenfeld, 1990) or NGF treatment of PC12 cells (Tang *et al.*, 1996).

The p70^{S6K} appears to have a more selective substrate specificity than do the p90^{rsk} enzymes with S6 being the primary substrate. However, highly purified p70^{S6K} has been shown, *in vitro*, to phosphorylate other cytosolic proteins, including GSK-3 (Sutherland *et al.*, 1993) and several unidentified nuclear proteins (Price *et al.*, 1989). These results, coupled with the recent observation that p70^{S6K} can phosphorylate the cAMP response element modulator, CREMτ, a nuclear transcription factor (de Groot *et al.*, 1994) suggests that p70^{S6K} may have physiologic targets in the cell other than S6.

SIGNALING CASCADES IMPLICATED IN THE PHOSPHORYLATION OF S6

Early studies investigating the signaling elements responsible for the phosphorylation of S6 showed that treatment of Swiss mouse 3T3 cells with either EGF or sodium orthovanadate (an inhibitor of tyrosine phosphatases) resulted in the activation of S6 kinases (Novak-Hofer and Thomas, 1985). These findings implied that kinases responsible for the phosphorylation S6 were activated by tyrosine phosphorylation. This was further supported by the finding that in chicken embryo fibroblasts expressing an activated Rous sarcoma virus transforming gene product, the tyrosine kinase, p60$^{v\text{-}src}$, S6 kinase activity was elevated (Blenis and Erikson, 1985). The fact that no S6 kinase had been shown to be directly phosphorylated on tyrosine, prompted the search for coupling intermediates linking growth factor receptor tyrosine kinases to intracellular serine/threonine kinases responsible for the phosphorylation of S6.

Although mitogens that activate p70^{S6K} simultaneously activate the MAPK family of signal transduction kinases (Blenis, 1993), there is evidence that they are activated by distinct signaling pathways (Ballou *et al.*, 1991). The finding that p90^{rsk} and not p70^{S6K} can be phosphorylated and activated by MAPK supports this concept. Additional evidence of separate signaling pathways came from a report which showed that cycloheximide treatment of cells resulted in the activation of p70^{S6K}, but had no affect on either MAPK or p90^{rsk} activity (Blenis *et al.*, 1991).

Studies using the rat adrenal medullary tumor cell line, PC12, containing mutant signaling molecules, such as a dominant negative Ras have shown that p90^{rsk} lies in the Ras $\rightarrow$ Raf $\rightarrow$ MEK $\rightarrow$ MAPK pathway (Wood *et al.*, 1992). That is, expression of a defective Ras protein which cannot be activated by GTP, blocks the stimulation of MAPK and p90^{rsk} by growth factors, whereas p70^{S6K} activation is unaffected. It has also been shown that the expression of a constitutively active mutant of the MAPK activator, MEK, or of the MAPK kinase-kinase, Raf, does not result in the activation of p70^{S6K} (Mahalingam and Templeton, 1996). Other studies, using mutated growth factor receptors which cannot bind proteins which activate Ras, showed that p70^{S6K} activation by growth factors was not affected (Ming *et al.*, 1994). Therefore, the use of molecular biologically-engineered signal transduction molecules in certain cell lines has demonstrated that p90^{rsk} activation by growth factors is dependent on the Ras signaling cascade, whereas activation of the p70^{S6K} is not. These studies prompted an intensive effort to elucidate the molecular mechanisms by which p70^{S6K} is activated by growth factor receptor signaling cascades.

USE OF RAPAMYCIN AND WORTMANNIN IN DISSECTING THE SIGNAL TRANSDUCTION PATHWAYS REGULATING S6 PHOSPHORYLATION

Signaling pathways involved in G_o and G_1 phases of the cell cycle are dependent on the phosphorylation states of distinct proteins. In lymphocytes, these states can be modulated by complexes of three microbial natural products with their endogenous binding proteins. The natural products are cyclosporin A, FK506, and rapamycin which were discovered by their ability to suppress immune function. The binding proteins, termed immunophilins, are cyclophilin and FK506 binding protein (FKBP) (Schreiber, 1991). The discovery of these immunosuppressive drugs has revolutionized organ transplantation and the treatment of autoimmune diseases.

Rapamycin has been shown to inhibit the G_1/S cell cycle transition in T lymphocytes stimulated by interleukin 2 (IL-2) (Dumont *et al.*, 1990) and to irreversibly arrest *S. cerevisiae* in the G_1 phase of the cell cycle (Heitman *et al.*, 1991). These observations have made rapamycin a valuable tool for the study of the signaling events that convey mitogenic stimuli to the nucleus. The inhibition of cell cycle transition is mediated through the binding of rapamycin to the evolutionarily-conserved and ubiquitously-expressed immunophilin, FKBP. A specific immunophilin, FKBP12 is involved in rapamycin action as yeast strains lacking this particular protein are resistant to the action of rapamycin.

Interestingly, both rapamycin and the structurally related compound FK506 bind to FKBP12 and both agents have been shown to inhibit T cell proliferation. However, the inhibitory mechanisms involve separate signal transduction pathways with the FK506-FKBP complex inhibiting the calcium-dependent phosphatase, calcineurin (Liu *et al.*, 1991). Calcineurin has a role in regulating the nuclear transport of a transcription factor required for the expression of IL-2, a cytokine necessary for T cell proliferation (Flanagen *et al.*, 1991). The mechanism of action of the rapamycin-FKBP complex is just now being elucidated.

The products of two novel yeast genes termed TOR1 and TOR2 (target of rapamycin) have also been shown to be involved in rapamycin-mediated toxicity, with the mutation of either gene conferring rapamycin resistance (Heitman *et al.*, 1991). The cloning of TOR2 showed that it exhibited significant homology to two previously identified phosphatidylinositol 3-kinases (PI 3-kinase) (Kunz *et al.*, 1993).

An important finding in the quest to delineate the physiological signal transduction pathways leading to S6 phosphorylation came in 1992, when it was shown that rapamycin inhibited the phosphorylation and activation of p70^{S6K} stimulated by a host of agonists (Chung *et al.*, 1992). Half-maximal rapamycin-mediated inhibition of p70^{S6K} activation by serum growth factors in Swiss 3T3 cells was between 0.04–0.4 nM. Interestingly, pretreatment of cells with rapamycin was not necessary to block p70^{S6K} activation and addition of rapamycin to serum-stimulated cells containing maximally activated p70^{S6K} resulted in its rapid inactivation with a half-time of ~2 minutes. Enzymatic analysis showed that the phosphorylation and activation of p70^{S6K} was inhibited by rapamycin, while neither MAPK nor p90^{rsk} were affected. Furthermore, it was found using ^{32}P-labeled cells, that rapamycin inhibited S6 phosphorylation *in vivo*. This study suggested that all extracellular signaling culminating in S6 phosphorylation converged at a common point. It also strongly implied that p70^{S6K}, but not p90^{rsk} is a physiologically relevant S6 kinase. This original observation was followed by other reports showing rapamycin inhibition of p70^{S6K} stimulated by various growth factors and cytokines (Kuo *et al.*, 1992; Calvo *et al.*, 1992; Terada *et al.*, 1992; Price *et al.*, 1992). Although these studies show a clear correlation between the inhibition of p70^{S6K} as measured in *in vitro* immunocomplex kinase assays and the *in vivo* inhibition of S6 phosphorylation, they do not, however, rule out the presence of other yet unidentified rapamycin-sensitive S6 kinases.

The finding that rapamycin inhibited the activation of p70^{S6K} indirectly and had no direct effect on S6 phosphorylation by this enzyme *in vitro*, implied that other upstream components in the signaling cascade, presumably kinases, responsible for activation of p70^{S6K} are the target(s) of rapamycin (Chung *et al.*, 1992). This concept led to the use of rapamycin coupled to FKBP12 as an affinity ligand to purify proteins which are involved in its action. Using this approach, two independent reports (Brown *et al.*, 1994; Sabatini *et al.*, 1994) have described the isolation and sequencing of a protein termed either FRAP (FKBP-rapamycin-associated protein) or RAFT1 (rapamycin and FKBP12 target) which is a mammalian homolog of the *S. cerevisiae* TOR proteins. Several other groups subsequently purified and/or cloned the same protein (Chen *et al.*, 1994; Chiu *et al.*, 1994; Sabers *et al.*, 1995). The carboxy terminal 600 amino acid domain of FRAP/RAFT1 and the yeast TOR proteins is homologous to the p110 subunit of mammalian PI 3-kinase (Carpenter *et al.*, 1990), the yeast PI 3-kinase VPS34 (Schu *et al.*, 1993), two yeast PI 4-kinases (Flanagen *et al.*, 1993; Garcia-Bustos *et al.*, 1994) and a recently cloned mammalian PI 4-kinase (Wong and Cantley, 1994). This homology to known lipid kinases suggested that FRAP/RAFT1 may also be a PI kinase and that rapamycin exerts its effects by inhibiting this activity.

Indeed, the involvement of FRAP/RAFT1 in the activation of p70^{S6K} was recently demonstrated directly (Brown *et al.*, 1995). This study showed that FRAP/RAFT1 is capable of autophosphorylation *in vitro*, demonstrating that the protein has protein kinase activity. Mutation of residues in the putative catalytic domain of FRAP/RAFT1 inhibited its ability to autophosphorylate and abolished growth factor activation of p70^{S6K} (Brown *et al.*, 1995). Furthermore, a mutation within the FKBP-rapamycin binding domain of FRAP, rendering it unable to bind to FKBP-rapamycin, abolished rapamycin inhibition of growth factor stimulation of p70^{S6K}. These studies suggest that FRAP/RAFT1 is an upstream kinase in the cascade which activates p70^{S6K}. The scheme that emerges from these studies suggests that the rapamycin-FKBP12 complex binds to FRAP/RAFT1 and inhibits its ability to autophosphorylate, resulting in the inhibition of the activation of p70^{S6K}. The activating input stimulating the autophosphorylation and activation of FRAP/RAFT1 is unknown. Although none of these studies has shown that FRAP/RAFT1 directly phosphorylates p70^{S6K}, they do situate the protein in the S6 kinase signaling cascade.

It has recently been shown that FRAP/RAFT1 possess PI 4-kinase activity (Sabatini *et al.*, 1995). Therefore, this protein is able to phosphorylate both proteins, i.e., autophosphorylation, and lipids. A perplexing finding in the Sabatini study (1995) is that the lipid kinase activity of FRAP/RAFT1 is not inhibited by the rapamycin-FKBP complex. The inability of rapamycin-FKBP to inhibit the PI 4-kinase activity of FRAP/RAFT1 raises questions about the mechanism of rapamycin's pharmacological actions. This study, however, did not address the effect of rapamycin on the PI 4-kinase activity *in vivo*, so the lack of inhibition by rapamycin could reflect an *in vitro* artifact. Alternatively, rapamyin-FKBP12 may prevent FRAP/RAFT1 from interacting with substrates or perhaps alter the intracellular localization of the enzyme and prevent its access to lipid substrates associated with the membrane.

A sequence comparison of FRAP/RAFT1/TOR to related proteins show homology to a gene believed to be involved in ataxia telangiectasia (AT). AT is a human autosomal hereditary disease characterized by a wide spectrum of defects, including cerebellar degeneration, progressive retardation, uneven gait (ataxia), dilation of blood vessels (telangiectasia), immune deficiencies, premature aging and a ~100 fold increase in cancer susceptibility. Identification of the gene showed that most, if not all cases of this heterogeneous disease are caused by a mutation of a single gene called ATM (Savitsky *et al.*, 1995). Cloning and sequence analysis of ATM cDNA shows homology to the catalytic domains of several PI 3-kinases. This finding implies that the disruption of the S6 phosphorylation signaling cascade may have pathological sequelae.

Wortmannin (WT), originally discovered in 1957 as an antifungal antibiotic isolated from a culture of *Penicillium wortmanni* (Brian *et al.*, 1957), was "rediscovered" years later by a group of investigators searching for inhibitors of myosin light chain kinase (MLCK), an enzyme involved in smooth muscle contraction, in an attempt to obtain new pharmaceuticals which are vasodilators or bronchodilators (Nakanishi *et al.*, 1992). During the search for MLCK inhibitors, a compound termed MS-54 was isolated from the fungal strain *Talaromyces wortmanni KY12420*. It was subsequently found that MS-54 is identical to WT. These studies showed that WT inhibited the phosphorylation of myosin light chain and the contraction of isolated aortic rings at concentrations that did not inhibit other kinases. In these studies, micromolar concentrations of WT were required to inhibit MLCK, whereas nanomolar concentrations had biological actions in other cellular systems (Nakanishi *et al.*, 1995).

Subsequent work demonstrated that WT also has effects, in non-muscle cells, on MLCK-dependent exocytotic processes, such as catecholamine release from adrenal chromaffin cells (Ohara-Imaizumi *et al.*, 1992), release of serotonin during platelet activation (Yatomi *et al.*, 1992) and the budding of viral particles from host cells (Sasaki *et al.*, 1995). Other studies showed that WT inhibited histamine release from basophils (Kitani *et al.*, 1992) and the respiratory burst of neutrophils (Baggiolini *et al.*, 1987) at nanomolar concentrations. Because of these studies, WT received widespread attention and stimulated the search for physiologic cellular targets other than MLCK.

In the search for other cellular targets, it was discovered that PI 3-kinase was inhibited by WT at almost a 100-fold lower concentration than was necessary for

MLCK inhibition (Yano *et al.*, 1993; Okada *et al.*, 1994). This was the first indication that PI 3-kinase may be a physiological target of WT. At the molecular level it was shown that WT binds covalently to the 110 kDa subunit of PI 3-kinase and irreversibly inhibits the enzyme when added at nanomolar concentrations to cells (Yano *et al.*, 1993; Thelen *et al.*, 1994). At these concentrations the drug failed to affect other protein and lipid kinases that have been investigated to date. The finding that WT inhibits the classical PI 3-kinases at nanomolar concentrations has led to its use in dissecting signaling cascades which may involve this enzyme.

Although WT is an inhibitor of PI 3-kinase, it has also been shown to inhibit the activity of phospholipase A_2 in Swiss 3T3 cells at a half-maximal concentration of 2 nM, which makes its slightly more potent for this latter enzyme than for insulin-stimulated PI 3-kinase (~10 nM) (Cross *et al.*, 1995). Therefore, while WT is indeed a PI 3-kinase inhibitor, it is not completely specific and experimental conclusions based solely on its use should be treated with caution.

The molecular mechanism by which WT irreversibly inhibits PI 3-kinase involves the furan moiety of the molecule since derivatives of WT with an open or protected furan ring were ineffective in biological assays (Baggiolini *et al.*, 1987). Recently it has been demonstrated that the inhibition of PI 3-kinase occurs by the formation of a covalent enamine bond following the attack of lysine 802 of the enzyme on the furan ring of WT, indicating a crucial role for this residue in the phosphotransfer reaction of PI 3-kinase (Wymann *et al.*, 1996). These results provide the basis for the design of novel and specific inhibitors of the PI 3-kinase family of enzymes.

WT has been shown to inhibit both protein tyrosine kinase and G protein-$\beta\gamma$ subunit-regulated PI 3-kinases (Stephens *et al.*, 1994). Interestingly, WT does not inhibit a mammalian phosphatidylinositol-specific PI 3-kinase (Stephens *et al.*, 1994) and does not inhibit the lipid kinase activity of FRAP/RAFT1 (Sabatini *et al.*, 1995) at concentrations that are sufficient to inhibit known mammalian PI 3-kinases (Thelen *et al.*, 1994). It is only slightly inhibitory at 1 μM, a concentration reported to inhibit other PI 3-kinases (Stephens *et al.*, 1994), Type III PI 4-kinase (Stephens *et al.*, 1994) and a novel WT-sensitive PI 4-kinase (Nakanishi *et al.*, 1995). Therefore, FRAP/RAFT1 is not likely to be a direct target of WT *in vivo*. Although several PI 3-kinases have been shown to be inhibited by WT, one must keep in mind that all PI 3-kinases are not inhibited by the drug.

It is noteworthy to add that although many members of the PI 3-kinase family are bonafide lipid kinases, other members have been shown to possess protein kinase activity. For example, PI 3-kinase has been shown to autophosphorylate (Carpenter *et al.*, 1993) and also to phosphorylate the insulin receptor substrate, IRS-1 (Lam *et al.*, 1994). The most compelling evidence that members of this family of kinases have protein kinase activity is the recent report that the large catalytic subunit of the double-stranded DNA-dependent protein kinase (DNA-PK) lacks a conventional protein kinase catalytic domain, but has a PI 3-kinase domain (Hartley *et al.*, 1995). The kinase activity of DNA-PK is inhibited by WT, but at concentrations approaching those required for inhibition of MLCK.

Changes in PI 3-kinase activity have been correlated with cell growth (Cantley *et al.*, 1991; Varticovski *et al.*, 1994), but its downstream signal transducers are unknown. Since growth factor receptor tyrosine kinase activation has been shown to lead to the stimulation of p70[S6K] by a Ras-independent pathway, other pathways known to be activated by growth factors have been investigated. A report by Chung *et al.* (1994) implied that PI 3-kinase is involved in the activation of p70[S6K] in response to growth factors, such as platelet-derived growth factor (PDGF). In this study, a cell line lacking PDGF receptors was transfected with either mutated PDGF receptors which do not bind proteins known to associate with tyrosine kinase receptors and transduce signals, such as GAP-Ras, PLC-γ and PI 3-kinase, or with wild-type receptors. It was demonstrated that PDGF was unable to stimulate p70[S6K] in cells containing mutant PDGF receptors that cannot bind PI 3-kinase. This finding showed for the first time that PI 3-kinase is likely to be a link between growth factor receptor tyrosine kinases and the activation of p70[S6K]. This result was strengthened by the finding that WT inhibited PDGF-stimulated p70[S6K] with inhibition occurring at concentrations which inhibited PI 3-kinase. WT inhibition of growth factor activation of p70[S6K] occurred at a half-maximal concentration of ~10 nM. Furthermore, WT is able to inhibit maximally-activated p70[S6K] with a half-time of 15 minutes, as compared to 2–5 minutes with rapamycin. The differences in the rates of inactivation suggest that these reagents target different points along this signaling pathway.

Other studies have shown the phosphorylation and activation of p70[S6K] in cells expressing constitutively active PI 3-kinase (Weng *et al.*, 1995). These results markedly strengthen the observations with WT and clearly situate p70[S6K] downstream of a PI 3-kinase. Interestingly, PI 3-kinase does not directly phosphorylate p70[S6K] *in vitro* (Weng *et al.*, 1995). Therefore, at least one other intermediate kinase must be positioned between PI 3-kinase and p70[S6K].

A candidate for an upstream stimulatory input to p70[S6K] is the protein kinase c-Akt/PKB. It has recently been demonstrated that this kinase is activated by PI 3-kinase (Franke *et al.*, 1995). That is, it was shown that in cells expressing mutant PDGF receptors which cannot bind PI 3-kinase, PDGF was unable to stimulate c-Akt/PKB. However, in cells expressing wild-type PDGF receptors, PDGF was able to activate c-Akt/PKB. Furthermore, WT was shown to inhibit growth factor activation of c-Akt/PKB. These studies suggested that PI 3-kinase derived phosphorylated lipid products such as phosphatidylinositol 3,4,5-tris-phosphate were the activators of c-Akt/PKB. This contention was strengthened by the demonstration that c-Akt/PKB could be activated *in vitro* by phospholipids generated by PI 3-kinase action. This was the first report of a putative downstream target of PI 3-kinase. Another laboratory published similar results and also demonstrated that in cells expressing an activated form of c-Akt/PKB, p70[S6K] was activated, whereas MAPK was not (Burgering and Coffer, 1995). The activation of p70[S6K] in cells expressing activated c-Akt/PKB was not inhibited by rapamycin, indicating that c-Akt/PKB is situated upstream of, or parallel to, the site of rapamycin inhibition.

The combined use of rapamycin and WT, coupled with molecular biologically-engineered mutant p70[S6K], has shown that the activation of the enzyme involves multiple inputs from distinct pathways. This was demonstrated by generating deletion mutants of $p70^{S6K}$ which were still capable of being activated by growth factors, but which were differentially inhibited by either rapamycin or WT. That is, one mutant with an amino terminal deletion could be inhibited by WT, but was resistant to inhibition by rapamycin (Cheatham *et al.*, 1995; Weng *et al.*, 1995). These results suggest that at least two stimulatory pathways are required to activate $p70^{S6K}$.

INVOLVEMENT OF S6 KINASES IN CELL CYCLE REGULATION

In the last five years, intensive investigation into the molecular basis of eukaryotic cell cycle regulation has yielded important findings and has been the topic of several recent reviews (Heichman and Roberts, 1994; King *et al.*, 1994; Nurse, 1994; Sherr, 1994). Central cell-cycle regulatory proteins of this system include the cyclins, their associated kinases and several proteins which are cyclin-dependent kinase inhibitors (Sherr and Roberts, 1995). The phases of the cell cycle, i.e., mitosis (M), first gap (G_1), DNA synthesis (S), and second gap (G_2) have been defined by cellular DNA content and protein expression patterns which are characteristic of each phase. Progression through the cell cycle is precisely timed by the transient appearance of different cyclins which form multiprotein complexes with the cyclin-dependent kinases.

The original report of the apparent selective inhibition of $p70^{S6K}$ by rapamycin also showed that the compound delayed the entry of fibroblasts, and inhibited the entry of T cells, into S phase (Chung *et al.*, 1992). Although the growth of fibroblasts was slowed, the final saturation density was nearly the same as that of untreated cells. Thus, in fibroblasts, the $p70^{S6K}$ signaling system plays an ancillary and perhaps redundant role in the regulation of cell proliferation. In T cells, however, the $p70^{S6K}$ signaling system appears to play a more dominant role, as the rapamycin-mediated antagonism of $p70^{S6K}$ activation correlated with ~80% inhibition of cell proliferation.

The finding that rapamycin abrogates cytokine-dependent T cell proliferation by a mechanism involving inhibition of G_1/S cell cycle transition prompted several groups to examine the effect of rapamycin on the cell cycle machinery which governs this progression. A number of studies showed that treatment of various cell lines with rapamycin resulted in the inhibition of two cyclin-dependent kinase activities ($p34^{cdc2}$ and $p33^{cdk2}$) required for successful G_1/S transition (Morice *et al.*, 1993; Jayaraman and Marks, 1993). The molecular mechanism involved in the inhibition of kinase activity appeared not to be a direct block of the catalytic activity by rapamycin, but an inhibition of the expression of the cyclins necessary for the activation of the kinases. In particular, the expression of cyclin A, a cyclin involved in G_1/S and G_2/M transitions, was inhibited by rapamycin (Morice *et al.*, 1993).

It has been demonstrated that other cell cycle regulatory proteins are modulated by rapamycin. It was shown in T lymphocytes, that rapamycin selectively

blocked the expression of the proliferating cell nuclear antigen (PCNA), an oblig-
ate cofactor of DNA polymerase-δ and an important component of DNA replica-
tion (Feuerstein *et al.*, 1995). These studies showed that rapamycin potently
inhibited a cyclic AMP response element (CRE)-binding protein which is involved
in the transcription of the PCNA gene and suggested that this may serve as the
target in the rapamycin induction of immunosuppression and growth arrest at
G_1/S. Other cell cycle regulatory proteins modulated by rapamycin are the cyclin-
dependent kinase inhibitors, p21 and p27[Kip1] (Nourse *et al.*, 1994). It was shown
that during IL-2-stimulated T cell mitogenesis, p21 was induced and p27[Kip1] was
eliminated. Rapamycin blocked both the elimination of p27[Kip1] and the induction
of p21. It was postulated that the modulation of cyclin-dependent kinase inhibitors
is involved in rapamycin-mediated cell cycle arrest at G_1/S. Since a molecular
target of rapamycin is p70[S6K], these results suggested that p70[S6K] may be involved
in the expression of cycle cycle regulatory proteins and that its activity is required
for successful cell cycle progression.

Other recent studies add support for a regulatory role for p70[S6K] in cell cycle
transition. Microinjection of quiescent rat embyro fibroblasts with neutralizing
anti-p70[S6K] antibodies prevents serum-induced entry into S phase of the cell cycle
(Lane *et al.*, 1993). This effect was preceeded by almost a complete inhibition of
protein synthesis and of the expression of the immediate early gene product,
c-Fos. Another report demonstrated that inhibition of p85[S6K] by microinjection of
neutralizing antibodies into nuclei of cells also inhibited the G_1/S transition
(Reinhard *et al.*, 1994). These studies demonstrate that the p70[S6K] is important in
cell cycle regulation and suggest that the translation of proteins required for successful cell cycle transition is inhibited by the rapamycin-mediated inhibition of S6
phosphorylation. However, there is no direct evidence to sup- port a correlation
between the inhibition of G_o/G_1 transition and the state of S6 phosphorylation.

CONCLUSIONS AND PERSPECTIVES

In this chapter, we have tried to demonstrate the importance of the use of natural
products, in this case fungal toxins, in the dissection of signal transduction pathways which impinge on the 40S ribosomal protein, S6. The use of these pharmacological tools along with molecular biological methodologies has allowed us to
make great strides in elucidating the molecular components of this particular
signal transduction system. A proposed scheme of the growth factor signal transduction cascades involved in activating the known S6 kinases is presented in
Figure 17.3. The use of rapamycin and wortmannin have allowed the identification
of the true S6 kinases in the cell and the unmasking of enzymes that may act on S6
in cell-free preparations, but not in a physiological context. Continued use of
these agents gives promise of illuminating what appear to be several functions of
the phosphorylation of S6.

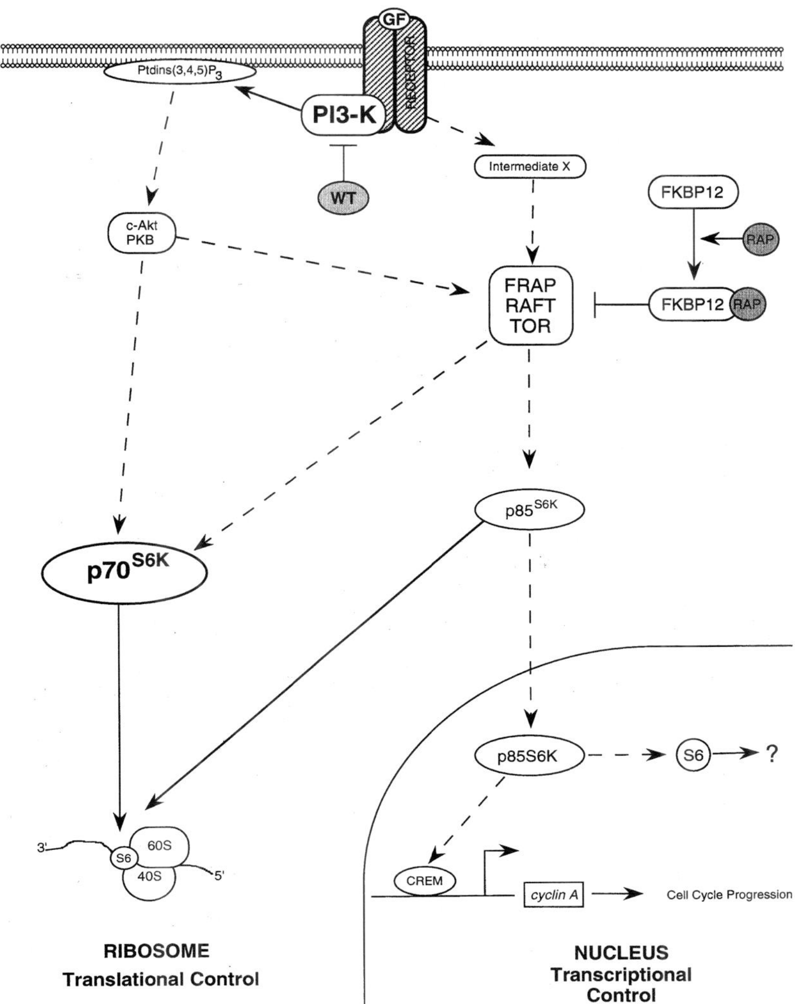

Figure 17.3 Proposed model of the signal transduction pathways by which growth factors regulate the activity of p70^{S6K}.

REFERENCES

Abbas, H.K. and Mirocha, C.J. (1988) Isolation and purification of a hemorrhagic factor (wortmannin) from *Fusarium oxysporum* (N17B). *Appl. Environ. Microbiol.*, **54**, 1268–1274.

Alcorta, D.A., Crews, C.M., Sweet, L.J., Bankston, L., Jones, S.W., and Erikson, R.L. (1989) Sequence and expression of chicken and mouse *rsk*: homologs of *Xenopus laevis* ribosomal S6 kinase. *Mol. Cell. Biol.*, **9**, 3850–3859.

Baggiolini, M., Dewald, B., Schnyder, J., Ruch, W., Cooper, P.H., and Payne, T.G. (1987) Inhibition of the phagocytosis-induced respiratory burst by the fungal metabolite wortmannin and some analogues. *Exp. Cell Res.*, **169**, 408–418.

Ballou, L.M., Jeno, P., and Thomas, G. (1988) Protein phosphatase 2A inactivates the mitogen-stimulated S6 kinase from Swiss mouse 3T3 cells. *J. Biol. Chem.*, **263**, 1188–1194.

Ballou, L.M., Luther, H., and Thomas, G. (1991) MAP2 kinase and 70K S6 kinase lie on distinct signalling pathways. *Nature*, **349**, 348–350.

Bandi, H.R., Ferrari, S., Krieg, J., Meyer, H.E., and Thomas, G. (1993) Identification of 40S ribosomal protein S6 phosphorylation sites in Swiss mouse 3T3 fibroblasts stimulated with serum. *J. Biol. Chem.*, **268**, 4530–4533.

Banerjee, P., Ahmad, M.F., Grove, J.R., Kozlosky, C., Price, D.J., and Avruch, J. (1990) Molecular structure of a major insulin/mitogen-activated 70 kDa S6 protein kinase. *Proc. Natl. Acad. Sci. USA*, **87**, 8550–8554.

Bansbach, C. C., Wancio, D., Caccese, R. G., Shen, C. F., and Sehgal, S. N. (1993) Rapamycin's inhibition of thymocyte proliferation, unlike that of cyclosporin A or prednisolone, is not associated with cytotoxicity. *Ann. N.Y. Acad. Sci.*, **685**, 114–116.

Bierer, B.E., Mattila, P.S., Standaert, R.F., Herzenberg, L.A., Burakoff, S.J., Crabtree, G., and Schreiber, S.L. (1990) Two distinct signal transduction transmission pathways in T lymphocytes are inhibited by complexes formed between an immunophilin and either FK506 or rapamycin. *Proc. Natl. Acad. Sci. USA*, **87**, 9231–9235.

Blenis, J. (1993) Signal transduction via the MAP kinases: Proceed at your own RSK. *Proc. Natl. Acad. Sci. USA*, **90**, 5889–5892.

Blenis, J. and Erikson, R.L. (1985) Regulation of a ribosomal protein S6 kinase activity by the Rous sarcoma virus transforming protein, serum or phorbol esters. *Proc. Natl. Acad. Sci. USA*, **82**, 7621–7625.

Blenis, J. and Erikson, R.L. (1986) Regulation of protein kinase activities in PC12 cells. *EMBO J.*, **5**, 3441–3447.

Blenis, J., Kuo, C.J., and Erikson, R.L. (1987) Identification of a ribosomal protein kinase regulated by transformation and growth-promoting stimuli. *J. Biol. Chem.*, **262**, 14373–14376.

Blenis, J., Chung, J., Erikson, E., Alcorta, D.A., and Erikson, R.L. (1991) Distinct mechanisms for the activation of the RSK kinases/MAP2 kinase/p90^{rsk} and pp70-S6 kinase signaling system are indicated by inhibition of protein synthesis. *Cell Growth Diff.*, **2**, 279–285.

Brandon, S.D. and Masaracchia, R.A. (1991) Multisite phosphorylation of a synthetic peptide derived from the carboxyl terminus of the ribosomal protein S6. *J. Biol. Chem.*, **266**, 380–385.

Brian, P.W., Curtis, P.J., Hemmings, H.G., and Norris, G.L.F. (1957) Wortmannin, an antibiotic produced by *Penicillium wortmanni*. *Trans. Br. Mycol. Soc.*, **40**, 365–368.

Brown, E.J., Albers, M.W., Shin, T.B., Ichikawa, K., Keith, C.T., Lane, W.S., and Schreiber, S.L. (1994) A mammalian protein targeted by G1-arresting rapamycin-receptor complex. *Nature*, **369**, 756–758.

Brown, E.J., Beal, P.A., Keith, C.T., Chen, J., Shin, T.B., and Schreiber, S.L. (1995) Control of p70 S6 kinase by kinase activity of FRAP *in vivo*. *Nature*, **377**, 441–446.

Burgering, B.M.Th. and Coffer, P.J. (1995) Protein kinase B (c-Akt) in phosphatidylinositol-3-OH kinase signal transduction. *Nature*, **376**, 599–602.

Burkhard, S.J. and Traugh, J.A. (1983) Changes in ribosome function by cAMP-dependent and cAMP-independent phosphorylation of ribosomal protein S6. *J. Biol. Chem.*, **258**, 14003–14008.

Calvo, V., Crews, C.M., Vik, T.A., and Bierer, B. (1992) Interleukin-2 stimulation of p70 S6 kinase activity is inhibited by the immunosupressant rapamycin. *Proc. Natl. Acad. Sci. USA*, **89**, 7571–7575.

Cantley, L.C., Auger, K.R., Carpenter, C.L., Duckworth, B., Graziani, A., Kapeller, R., and Soltoff, S. (1991) Oncogenes and signal transduction. *Cell*, **64**, 281–302.

Carpenter, C.L., Auger, K.R., Duckworth, B.C., Hou, W.M., Schaffhausen, B., and Cantley, L.C. (1993) A tightly associated serine/threonine protein kinase regulates phosphoinositide 3-kinase activity. *Mol. Cell. Biol.*, **13**, 1657–1665.

Carpenter, C.L., Duckworth, B.C., Auger, K.R., Cohen, B., Schaffhausen, B.S., and Cantley, L.C. (1990) Purification and characterization of phosphoinositide 3-kinase from rat liver. *J. Biol. Chem.*, **265**, 19704–19711.

Chan, Y-L. and Wool, I.G. (1988) The primary structure of rat ribosomal protein S6. *J. Biol. Chem.*, **263**, 2891–2896.

Chaudhuri, S. and Lieberman, I. (1968) Control of ribosome synthesis in normal and regenerating liver. *J. Biol. Chem.*, **243**, 29–33.

Cheatham, L., Monfar, M., Chou, M.M., and Blenis, J. (1995) Structural and functional analysis of pp70^{S6K}. *Proc. Natl. Acad. Sci. USA*, **92**, 11696–11700.

Chen, R-H. and Blenis, J. (1990) Identification of *Xenopus* S6 protein kinase homologs (pp90^{rsk}) in somatic cells: Phosphorylation and activation during initiation of cell proliferation. *Mol. Cell. Biol.* **10**, 3204–3215.

Chen, R-H., Sarnecki, C., and Blenis, J. (1992) Nuclear localization and regulation of *erk-* and *rsk*-encoded protein kinases. *Mol. Cell. Biol.*, **12**, 915–927.

Chen, R-H., Abate, C., and Blenis, J. (1993) Phosphorylation of the c-Fos transrepression domain by mitogen-activated protein kinase and 90 kDa ribosomal S6 kinase. *Proc. Natl. Acad. Sci. USA*, **90**, 10952–10956.

Chiu, M.I., Katz, H., and Berlin, V. (1994) RAFT1, a mammalian homolog of yeast TOR, interacts with the FKBP12/rapamycin complex. *Proc. Natl. Acad. Sci. USA*, **91**, 12574–12578.

Chung, J., Grammer, T.C., Lemon, K.P., Kazlauskas, A., and Blenis, J. (1994) PDGF- and insulin-dependent pp70^{S6k} activation mediated by phosphatidylinositol-3-OH kinase. *Nature*, **370**, 71–75.

Chung, J., Kuo, C.J., Crabtree, G.R., and Blenis, J. (1992) Rapamycin-FKBP specifically blocks growth factor- dependent activation of and signaling by the 70 kD S6 protein kinases. *Cell*, **69**, 1227–1236.

Cobb, M.H. and Rosen, O.M. (1983) Description of a protein kinase derived from insulin-treated 3T3-L1 cells that catalyzes the phosphorylation of ribosomal protein S6 and casein. *J. Biol. Chem.*, **258**, 12472–12481.

Coffer, P.J. and Woodgett, J.R. (1994) Differential subcellular localization of two isoforms of p70 S6 protein kinase. *Biochem. Biophys. Res. Commun.*, **198**, 780–786.

Cross, M. J., Stewart, A., Hodgkin, M. N., Kerr, D. J., and Wakelam, M. J. O. (1995) Wortmannin and its structural analog demethoxyviridin inhibit stimulated phospholipase A2 activity in Swiss 3T3 cells. Wortmannin is not a speciic inhibitor of phosphatidylinositol 3-kinase. *J. Biol. Chem.*, **270**, 25352–25355.

Davis, I.J., Hazel, T.G., Chen, R-H., Blenis, J., and Lau, L.F. (1993) Functional domains and phosphorylation of the orphan receptor Nur77. *Mol. Endocrinol.*, **7**, 953–964.

de Groot, R.P., Ballou, L.M., and Sassone-Corsi, P. (1994) Positive regulation of the cAMP-responsive activator CREM by the p70 S6 kinase: an alternative route to mitogen-induced gene expression. *Cell*, **79**, 81–91.

Del Grande, R.W. and Traugh, J.A. (1982) Phosphorylation of 40-S ribosomal subunits by cAMP-dependent, cGMP-dependent and protease-activated protein kinases. *Eur. J. Biochem.*, **123**, 421–428.

Dent, P., Lavoinne, A., Nakielny, S., Caudwell, F.B., Watt, P., and Cohen, P. (1990) The molecular mechanism by which insulin stimulates glycogen synthesis in mammalian skeletal muscle. *Nature*, **348**, 302–308.

Dennis, P.B. and Masaracchia, R.A. (1993) Activation of an S6 kinase from human placenta by autophosphorylation. *J. Biol. Chem.*, **268**, 19833–19841.

Donahue, M.J. and Masaracchia, R.A. (1984) Phosphorylation of ribosomal protein S6 at multiple sites by a cyclic AMP-independent protein kinase from lymphoid cells. *J. Biol. Chem.*, **259**, 435–440.

Dumont, F.J., Staruch, M.J., Koprak, S.L., Melino, M.R., and Sigal, N.H. (1990) Distinct mechanisms of suppression of murine T cell activation by the related macrolides FK-506 and rapamycin. *J. Immunol.*, **144**, 251–258.

Duncan, R. and McConkey, E.H. (1982) Preferential utilization of phosphorylated 40S ribosomal subunits during initiation complex formation. *Eur. J. Biochem.*, **123**, 535–538.

Duvernay, J.H. and Traugh, J.A. (1978) Two-step purification of the major phosphorylated protein in reticulocyte 40S ribosomal subunits. *Biochemistry*, **17**, 2045–2049.

Erikson, E. and Maller, J.L. (1989) *In vivo* phosphorylation and activation of ribosomal protein S6 kinases during *Xenopus* oocyte maturation. *J. Biol. Chem.* **264**, 13711–13717.

Erikson, E. and Maller, J.L. (1986) Purification and characterization of a protein kinase from *Xenopus* eggs highly specific for ribosomal protein S6. *J. Biol. Chem.*, **261**, 350–355.

Erikson, E. and Maller, J.L. (1991) Purification and characterization of ribosomal protein S6 kinase I from *Xenopus* eggs. *J. Biol. Chem.*, **266**, 5249–5255.

Ferrari, S., Bandi, H.R., Hofsteenge, J., Bussian, B.M., and Thomas, G. (1991) Mitogen- activated 70 kDa S6 kinase: Identification of *in vitro* 40S ribosomal S6 phosphorylation sites. *J. Biol. Chem.*, **266**, 22770–22775.

Feuerstein, N., Huang, D., and Prystowsky, M. B. (1995) Rapamycin selectively blocks interleukin-2-induced proliferating cell nuclear antigen gene expression in T lymphocytes. Evidence for inhibition of CREB/ATF binding activities. *J. Biol. Chem.*, **270**, 9454–9458.

Flanagan, C.A., Schnieders, E.S., Emerick, A.W., Kunisawa, R., Admon, A., and Thorner, J. (1993) Phosphatidylinositol 4-kinase: gene structure and requirement for yeast cell viability. *Science*, **262**, 1444–1448.

Flanagan, W.M., Corthesy, B., Bram, R.J., and Crabtree, G.R. (1991) Nuclear association of a T cell transcription factor blocked by FK-506 and cyclosporin A. *Nature*, **352**, 803–807.

Franco, R. and Rosenfeld, M.G. (1990) Hormonally inducible phosphorylation of a nuclear pool of ribosomal protein S6. *J. Biol. Chem.*, **265**, 4321–4325.

Franke, T.F., Yang, S., Chan, T.O., Datta, K., Kazlauskas, A., Morrison, D.K., Kaplan, D.R., and Tsichlis, P.N. (1995) The protein kinase encoded by the *Akt* proto-oncogene is a target of the PDGF-activated phosphatidylinositol 3-kinase. *Cell*, **81**, 727–736.

Garcia-Bustos, J.F., Marini, F., Stevenson, I., Frei, C., and Hall, M.N. (1994) PIK1, an essential phosphatidylinositol 4-kinase associated with yeast nucleus. *EMBO J.*, **13**, 2352–2361.

Gorelick, F., Cohn, J., Freedman, S., Delahunt, V., Gershoni, J., and Jamieson, J. (1983) Calmodulin-stimulated protein kinase activity from rat pancreas. *J. Cell Biol.*, **97**, 1294–1298.

Gregory, J.S., Boulton, T.G., Sang, B-C., and Cobb, M.H. (1989) An insulin-stimulated ribosomal S6 kinase from rabbit liver. *J. Biol. Chem.*, **264**, 18397–18401.

Gressner, A.M. and Wool, A.G. (1976) Influence of glucagon and cyclic adenosine 3′:5′-monophosphate on the phosphorylation of rat liver ribosomal protein S6. *J. Biol. Chem.*, **251**, 1500–1504.

Gressner, A.M. and Greiling, H. (1977) The phosphorylation of liver ribosomal protein S6 during the development of acute hepatic cell injury induced by D-galactosamine. *FEBS Lett.*, **74**, 77–81.

Gressner, A.M. and Wool, I.G. (1974) The phosphorylation of liver ribosomal proteins *in vivo*. *J. Biol. Chem.*, **249**, 6917–6925.

Grove, J.R., Banerjee, P., Balasubramanyam, A., Coffer, P.J., Price, D.J., Avruch, J., and Woodgett, J.R. (1991) Cloning and expression of two human p70 S6 kinase polypeptides differing only at their amino termini. *Mol. Cell. Biol.*, **11**, 5541–5550.

Haefliger, W. and Hauser, D. (1973) Isolierung und strukturaufklarung von II-desacetoxy-wortmannin. *Helv. Chim. Acta*, **56**, 2901–2904.

Halegoua, S. and Patrick, J. (1980) Nerve growth factor mediates phosphorylation of specific proteins. *Cell*, **22**, 571–581.

Hartley, K.O., Gell, D., Smith, G.C.M., Zhang, H., Divecha, N., Connelly, M.A., Admon, A., Lees-Miller, S.P., Anderson, C.W., and Jackson, S.P. (1995) DNA-dependent protein kinase catalytic subunit: A relative of phosphatidylinositol 3-kinase and the ataxia telangiectasia gene product. *Cell*, **82**, 849–856.

Hei, Y-J., Pelech, S.L., Chen, X., Diamond, J., and McNeill, J.H. (1994) Purification and characterization of a novel ribosomal S6 kinase from skeletal muscle of insulin-treated rats. *J. Biol. Chem.*, **269**, 7816–7823.

Heichman, K.A. and Roberts, J.M. (1994) Rules to replicate by. *Cell*, **79**, 557–562.

Heinze, H., Arnold, H.H., Fisher, D., and Kruppa, J. (1988) The primary sequence of the human ribosomal protein S6 derived from a cloned cDNA. *J. Biol. Chem.*, **263**, 4139–4144.

Heitman, J., Movva, N.R., and Hall, M.N. (1991) Targets for cell cycle arrest by the immunosuppressant rapamycin in yeast. *Science*, **253**, 905–909.

Hershey, J.W.B. (1989) Protein phosphorylation controls translation rates. *J. Biol. Chem.*, **264**, 20823–20826.

Hershey, J.W.B. (1991) Translational control in mammalian cells. *Annu. Rev. Biochem.*, **60**, 717–755.

Hsiao, K.-M., Chou, S-Y., Shih, S-J., and Ferrell, J.E., Jr. (1994) Evidence that inactive p42 mitogen-activated protein kinase and inactive Rsk exist as a heterodimer *in vivo. Proc. Natl. Acad. Sci. USA*, **91**, 5480–5484.

Jefferies, H.B.J., Reinhard, C., Kozma, S.C., and Thomas, G. (1994) Rapamycin selectively represses translation of the "polypyrimidine tract" mRNA family. *Proc. Natl. Acad. Sci. USA*, **91**, 4441–4445.

Jeno, P., Ballou, L.M., Novak-Hofer, I., and Thomas, G. (1988) Identification and characterization of a mitogen-activated S6 kinase. *Proc. Natl. Acad. Sci. USA*, **85**, 406–410

Jeno, P., Jaggi, N., Luther, H., Siegmann, M., and Thomas, G. (1989) Purification and characterization of a 40S ribosomal protein S6 kinase from vanadate-stimulated Swiss 3T3 cells. *J. Biol. Chem.*, **264**, 1293–1297.

Johnson, S.P. and Warner, J.B. (1987) Phosphorylation of the *Saccharomyces cerevisiae* equivalent of ribosomal protein has no detectable effect on growth. *Mol. Cell. Biol.*, **7**, 1338–1345.

Jones, S.W., Erikson, E., Blenis, J., Maller, J.L., and Erikson, R.L. (1988) A *Xenopus* ribosomal protein S6 kinase has two apparent kinase domains that are each similar to distinct protein kinases. *Proc. Natl. Acad. Sci. USA*, **85**, 3377–3381.

Kaerlein, M. and Horak, I. (1976) Phosphorylation of ribosomal proteins in Hela cells infected with *vaccinia* virus. *Nature*, **259**, 150–151.

King, R.W., Jackson, P.K., and Kirschner, M.W. (1994) Mitosis in transition. *Cell*, **79**, 563–571.

Kozma, S.C., Lane, H.A., Ferrari, S., Luther, H., Siegmann, D., and Thomas, G. (1989) A stimulated S6 kinase from rat liver: Identity with the mitogen-activated S6 kinase of 3T3 cells. *EMBO J.*, **13**, 4125–4132.

Kozma, S.C., Ferrari, S., Bassand, P., Siegmann, M., Totty, N., and Thomas, G. (1990) Cloning of the mitogen-activated S6 kinase from rat liver reveals an enzyme of the second messenger subfamily. *Proc. Natl. Acad. Sci. USA*, **87**, 7365–7369.

Krieg, J., Hofsteenge, J., and Thomas, G. (1988) Identification of the 40S ribosomal protein S6 phosphorylation sites induced by cycloheximide. *J. Biol. Chem.*, **263**, 11473–11477.

Kunz, J., Henriquez, R., Schneider, U., Deuter-Reinhard, M., Movva, N.R., and Hall, M.N. (1993) Target of rapamycin in yeast, TOR2, is an essential phosphatidylinositol kinase homolog required for G_1 progression. *Cell*, **73**, 585–590.

Kuo, C.L., Chung, J., Florentino, D.F., Flanagan, W.M., Blenis, J., and Crabtree, G.R. (1992) Rapamycin selectively inhibits interleukin-2 activation of p70 S6 kinase. *Nature*, **358**, 70–73.

Lalanne, J.L., Lucero, M., and Le Mouller, J.M. (1987) Complete sequence of mouse S6 ribosomal protein. *Nucl. Acids Res.*, **15**, 4990.

Lam, K., Carpenter, C.L., Ruderman, N.B., Friel, J.C., and Kelly, K.L. (1994) The phosphatidylinositol 3-kinase serine kinase phosphorylates IRS-1. Stimulation by insulin and inhibition by wortmannin. *J. Biol. Chem.*, **269**, 20648.

Lane, H.A., Morley, S.J., Doree, M., Kozma, S.C., and Thomas, G. (1992) Identification and early activation of a *Xenopus laevis* p70[S6k] following progesterone-induced meiotic maturation. *EMBO J.*, **11**, 1743–1749.

Lane, H.A. and Thomas, G. (1991) Purification and properties of mitogen-activated S6 kinase from rat liver and 3T3 cells. *Meth. Enzymol.*, **100**, 268–291.

Lane, H.A., Fernandez, A., Lamb, N.J.C., and Thomas, G. (1993) p70[S6k] function is essential for G1 progression. *Nature*, **363**, 170–172.

Lavoinne, A., Erikson, E., Maller, J.L., Price, D.J., Avruch, J., and Cohen, P. (1991) Purification and characterization of the insulin-stimulated protein kinase from rabbit skeletal muscle: Close similarity to S6 kinase II. *Eur. J. Biochem.*, **199**, 723–728.

Leer, R.J., van Raamsdonk-Duin, M.M.C., Molenaar, C.M.Th., Cohen, L.H., Mager, W.H., and Planta, R.J. (1982) The structure of the gene coding for the phosphorylated ribosomal protein S10 in yeast. *Nucl. Acids Res.*, **10**, 5869–5877.

Le Peuch, C., Ballester, R., and Rosen, O.M. (1983) Purified rat brain calcium- and phospholipid-dependent protein kinase phosphorylates ribosomal protein S6. *Proc. Natl. Acad. Sci. USA*, **80**, 6858–6862.

Liu, J., Farmer, J.D., Jr., Lane, W.S., Friedman, J., Weissman, I., and Schreiber, S.L. (1991) Calcineurin is a common target for cyclophilin-cyclosporin A and FKBP-FK506 complexes. *Cell,* **66**, 807–815.

Lubben, T.H. and Traugh, J.A. (1983) Cyclic nucleotide-independent protein kinases from rabbit reticulocytes. Purification and characterization of protease-activated kinase II. *J. Biol. Chem.,* **258**, 13992–13997.

Macfarlane, D.E. and Gailani, D. (1990) Identification of phosphoprotein NP33 as a nucleus-associated ribosomal S6 protein and its phosphorylation in hematopoietic cells. *Cancer Res.,* **50**, 2895–2900.

MacMillan, J., Vanstone, A.E., and Yeboab, S.K. (1972) Fungal products. Part III. Structure of wortmannin and some hydrolysis products. *J. Chem. Soc. Perkin Trans. I.,* **22**, 2898–2902.

Madjar, J-J. and Fournier, A. (1987) Starvation-induced alterations of ribosomal protein phosphorylation in *Bombyx mori. Eur. J. Biochem.,* **163**, 577–582.

Mahalingam, M. and Templeton, D.J. (1996) Constitutive activation of S6 kinase by deletion of aminoterminal autoinhibitory and rapamycin sensitivity domains. *Mol. Cell. Biol.,* **16**, 405–413.

Maller, J., Wu, M., and Gerhart, J.C. (1977) Changes in protein phosphorylation accompanying maturation of *Xenopus laevis* oocytes. *Dev. Biol.,* **58**, 295–312.

Matusuda, Y. and Guroff, G. (1987) Purification and mechanism of activation of a nerve growth factor-sensitive S6 kinase from PC12 cells. *J. Biol. Chem.,* **262**, 2832–2844.

Ming, X-F., Burgering, B.M.Th., Wennstrom, S., Claesson-Welsh, L., Heldin, C-H., Bos, J.L., Kozma, S.C., and Thomas, G. (1994) Activation of p70/p85 S6 kinase by a pathway independent of p21^{ras}. *Nature,* **371**, 326–329.

Morley, S.J. and Thomas, G. (1991) Intracellular messengers and the control of protein synthesis. *Pharmacol. Therap.,* **50**, 291–319.

Morris, R.E. (1991) Rapamycin: FK506's fraternal twin or distant cousin? *Immunol. Today,* **12**, 137–140.

Mukhopadhyay, N.K., Price, D.J., Kyriakis, J.M., Pelech, S., Sanghera, J., and Avruch, J. (1992) An array of insulin-activated, proline-directed serine/threonine protein kinases phosphorylate the p70 S6 kinase. *J. Biol. Chem.,* **267**, 3325–3335.

Mutoh, T., Rudkin, B.B., Koizumi, S., and Guroff, G. (1988) Nerve growth factor, a differentiating agent and epidermal growth factor, a mitogen, increase the activities of different S6 kinases in PC12 cells. *J. Biol. Chem.,* **263**, 15853–15856.

Mutoh, T., Rudkin, B.B., and Guroff, G. (1992) Differential responses of the phosphorylation of ribosomal protein S6 to nerve growth factor and epidermal growth factor in PC12 cells. *J. Neurochem.,* **58**, 175–185.

Nakanishi, S., Catt, K.J., and Balla, T. (1995) A wortmannin-sensitive phosphatidylinositol 4-kinase that regulates hormone-sensitive pools of inositol phospholipids. *Proc. Natl. Acad. Sci. USA,* 92, 5317–5321.

Nakanishi, S., Yano, H., and Matsuda, Y. (1995) Novel functions of phosphatidylinositol 3-kinase in terminally differentiated cells. *Cell. Signal.,* **7**, 545–557.

Nakanishi, S., Kakita, S., Takahashi, I., Kawahara, K., Tsukuda, E., Sano, T., Yamada, K., Yoshida, M., Kase, H., Matsuda, Y., Hashimoto, Y., and Nonomura, Y. (1992) Wortmannin, a microbial product inhibitor of myosin light chain kinase. *J. Biol. Chem.,* **267**, 2157–2163.

Nielsen, P.J., Thomas, G., and Maller, J.L. (1982) Increased phosphorylation of ribosomal protein S6 during meiotic maturation of *Xenopus* oocytes. *Proc. Natl. Acad. Sci. USA,* **79**, 2937–2941.

Nielsen, P.J., Duncan, R., and McConkey, E.H. (1981) Phosphorylation of ribosomal protein S6: relationship to protein synthesis in Hela cells. *Eur. J. Biochem.,* **120**, 523–527.

Nielsen, F.C., Ostergaard, L., Nielsen, J., and Christiansen, J. (1995) Growth-dependent translation of IGF-II mRNA by a rapamycin-sensitive pathway. *Nature,* **377**, 358–362.

Nikolakaki, E., Coffer, P.J., Hemelsoet, R., Woodgett, J.R., and Defize, L.H.K. (1993) Glycogen synthase kinase 3 phosphorylates Jun family members *in vitro* and negatively regulates their transactivating potential in intact cells. *Oncogene,* **8**, 833–840.

Nourse, J., Firpo, E., Flanagan, W. M., Coats, S., Polyak, K., Lee, M-H., Massague, J., Crabtree, G.R., and Roberts, J. M. (1994) Interleukin-2-mediated elimination of the p27^{Kip1} cyclin-dependent kinase inhibitor is prevented by rapamycin. *Nature,* **372**, 570–573.

Novak-Hofer, I. and Thomas, G. (1984) An activated S6 kinase in extracts from serum- and epidermal growth factor-stimulated Swiss 3T3 cells. *J. Biol. Chem.,* **259**, 5995–6000.

Novak-Hofer, I. and Thomas, G. (1985) Epidermal growth factor-mediated activation of an S6 kinase in Swiss mouse 3T3 cells. *J. Biol. Chem.*, **260**, 10314–10319.

Nurse, P. (1994) Ordering S phase and M phase in the cell cycle. *Cell*, **79**, 547–550.

Nygard, O. and Nilsson, L. (1990) Translational dynamics: Interactions between the translational factors, tRNA and ribosomes during eukaryotic protein synthesis. *Eur. J. Biochem.*, **191**, 1–17.

Ohara-Imaizumi, M., Sakurai, T., Nakamura, S., Nakanishi, S., Matsuda, Y., Muramatsu, S., Nonomura, Y., and Kumakura, K. (1992) Inhibition of Ca^{2+}-dependent catecholamine release by myosin light chain kinase inhibitor, wortmannin, in adrenal chromaffin cells. *Biochem. Biophys. Res. Commun.*, **185**, 1016–1021.

Okada, T., Sakuma, L., Fukui, T., Hazeki, O., and Ui, M. (1994) Blockage of chemotactic peptide-induced stimulation of neutrophils by wortmannin as a result of selective inhibition of phosphatidylinositol 3-kinase. *J. Biol. Chem.*, **269**, 3563–3567.

Palen, E. and Traugh, J.A. (1987) Phosphorylation of ribosomal protein S6 by cAMP-dependent protein kinase and mitogen-stimulated S6 kinase differentially alters translation of globin mRNA. *J. Biol. Chem.*, **262**, 3518–3523.

Parker, P.J., Katan, M., Waterfield, M.D., and Leader, D.P. (1985) Phosphorylation of eukaryotic ribosomal protein S6 by protein kinase C. *Eur. J. Biochem.*, **148**, 579–586.

Pata, I., Hoth, S., Kruppa, J., and Metspalu, A. (1992) The human ribosomal protein S6 gene: Isolation, primary structure and localization in chromosome 9. *Gene*, **121**, 387–392.

Perisic, O. and Traugh, J.A. (1993) Protease-activated kinase II mediates multiple phosphorylation of ribosomal protein S6 in reticulocytes. *J. Biol. Chem.*, **258**, 13998–14002.

Petcher, T.J., Weber, H.P., and Kis, Z. (1972) Crystal structure and absolute configuration of wortmannin and P-bromobenzoate. *J. Chem. Soc. Comm.*, **1332**, 1061–1062.

Price, D.J., Nemenoff, R.A., and Avruch, J. (1989) Purification of a hepatic S6 kinase from cycloheximide-treated rats. *J. Biol. Chem.*, **264**, 13825–13833.

Price, D.J., Gunsalls, R., and Avruch, J. (1990) Insulin activates a p70 kDa S6 kinase through serine/threonine specific phosphorylation of the enzyme polypeptide. *Proc. Natl. Acad. Sci. USA*, **87**, 7944–7948.

Price, D.J., Grove, J.R., Calvo, V., Avruch, J., and Bierer, B. (1992) Rapamycin-induced inhibition of the 70-kilodalton S6 protein kinase. *Science*, **257**, 973–977.

Price, B. D. and Youmell, M. B. (1996) The phosphatidylinositol 3-kinase inhibitor wortmannin sensitizes murine fibroblasts and human tumor cells to radiation and blocks induction of p53 following DNA damage. *Cancer Res.*, **56**, 246–250.

Reinhard, C., Thomas, G., and Kozma, S.C. (1992) A single gene encodes two isoforms of the p70 S6 kinase: Activation upon mitogenic stimulation. *Proc. Natl. Acad. Sci. USA*, **89**, 4052–4056.

Reinhard, C., Fernandez, A., Lamb, N.J.C., and Thomas, G. (1994) Nuclear localization of p85[S6k]: functional requirement for entry into S phase. *EMBO J.*, **13**, 1557–1565.

Rivera, V.M., Miranti, C.K., Misra, R.P., Ginty, D.D., Chen. R.,-H., Blenis, J., and Greenberg, M. (1993) A growth factor-induced kinase phosphorylates the serum response factor at a site that regulates its DNA-binding activity. *Mol. Cell. Biol.*, **13**, 6260–6273.

Roach, P.J. (1991) Multisite and hierarchal protein phosphorylation. *J. Biol. Chem.*, **266**, 14139–14142.

Sabatini, D.M., Pierchala, B.A., Barrow, R.K., Schell, M.J., and Synder, S.H. (1995) The rapamycin and FKBP12 target (RAFT1) displays phosphatidylinositol 4-kinase activity. *J. Biol. Chem.*, **270**, 20875–20878.

Sabatini, D.M., Erdjument-Bromage, H., Lui, M., Tempst, P., and Snyder, S.H. (1994) RAFT1: a mammalian protein that binds to FKBP12 in a rapamycin-dependent fashion and is homologous to yeast TORs. *Cell*, **78**, 35–43.

Sabers, C.J., Martin, M.M., Brown, G.J., Williams, J.M., Dumont, F.J., Wiederrecht, G., and Abraham, R.T. (1995) Isolation of a protein target of the FKBP-rapamycin complex in mammalian cells. *J. Biol. Chem.*, **270**, 815–822.

Sasaki, H., Nakamura, M., Ohno, T., Matsuda, Y., Yuda, Y., and Nonomura, Y. (1995) Myosin-actin interaction plays an important role in human immunodeficiency virus type 1 release from host cells. *Proc. Natl. Acad. Sci. USA*, **92**, 2026–2030.

Savitsky, K., Bar-Shira, A., Gilad, S., Rotman, G., Ziz, Y., Vanagaite, L., Tagle, D.A., Smith, S., Uziel, T., Sfez, S., Ashkenazi, M., Pecker, I., Frydman, M., Harnik, R., Patanjali, S.R., Simmons, A., Clines, G.A., Sartiel, A., Gatti, R.A., Chessa, L., Sanal, O., Lavin, M.F., Jaspers, N.G.J., Taylor, A.M.R., Arlett, C.F., Miki, T., Weissman, S.M., Lovett, M., Collins, F.S., and Shiloh, Y. (1995) A single ataxia telangiectasia gene with a product similar to PI 3-kinase. *Science*, **268**, 1749–1753.

Scimeca, J.C., Nguyen, T.T., Filloux, C., and Van Obbergen, E. (1992) Nerve growth factor-induced phosphorylation cascade in PC12 pheochromocytoma cells. Association of S6 kinase II with the microtubule-associated protein kinase, ERK1. *J. Biol. Chem.*, **267**, 17369–17374.

Sharf, S-D. and Nover, L. (1982) Heat-shock-induced alterations of ribosomal protein phosphorylation in plant cell cultures. *Cell*, **30**, 427–437.

Schreiber, S.L. (1991) Chemistry and biology of the immunophilins and their immunosuppressive ligands. *Science*, **251**, 283–287.

Schu, P.V., Takegawa, K., Fry, M.J., Stack, J.H., Waterfield, M.D., and Emr, S.D., (1993) Phosphatidylinositol 3-kinase encoded by yeast VPS34 gene essential for protein sorting. *Science*, **260**, 88–91.

Schultz, M.R., Merriman, R.L., Andis, S.L., and Bonjouklian, R. (1995) *In vitro* and *in vivo* antitumor activity of the phosphatidylinositol 3-kinase inhibitor, wortmannin. *Anticancer Res.* **15**, 1135–1139.

Sherr, C.J. (1994) G1 phase progression: Cycling on cue. *Cell*, **79**, 551–555.

Sherr, C.J. and Roberts, J.M. (1995) Inhibitors of mammalian G_1 cyclin-dependent kinases. *Genes & Dev.*, **9**, 1149–1163.

Shi, Y., Frankel, A., Radvanyi, L.G., Radvanyi, L.G., Penn, L.Z., Miller, R.G., and Mills, G.B. (1995) Rapamycin enhances apoptosis and increases sensitivity to cisplatin *in vitro*. *Cancer Res.*, **55**, 1982–1988.

Smith, C.J., Rubin, C.S., and Rosen, O.M. (1980) Insulin-treated 3T3-L1 adipocytes and cell-free extracts derived from them incorporate ^{32}P into ribosomal protein S6. *Proc. Natl. Acad. Sci. USA*, **77**, 2641–2645.

Spencer, T.A. and Mackie, G.A. (1993) The nucleotide sequence of a cloned cDNA encoding ribosomal protein S6 from *Drosophila melanogaster*. *Biochim. Biophys. Acta*, **1172**, 332–334.

Stambolic, V. and Woodgett, J.R. (1994) Mitogen inactivation of glycogen synthase kinase-3 beta in intact cells via serine 9 phosphorylation. *Biochem. J.*, **303**, 701–704.

Stephens, L., Cooke, F.T., Walters, R., Jackson, T., Volinia, S., Gout, I., Waterfield, M.D., and Hawkins, P.T. (1994) Characterization of a phosphatidylinositol-specific phosphoinositide 3-kinase from mammalian cells. *Curr. Biol.*, **4**, 203–214.

Stephens, L., Smrcka, A., Cooke, F.T., Jackson, T.R., Sternweis, P.C., and Hawkins, P.T. (1994) A novel phosphoinositide 3-kinase activity in myeloid-derived cells is activated by G protein $\beta\gamma$ subunits. *Cell*, **77**, 83–93.

Sturgill, T.W., Ray, L.B., Erikson, E., and Maller, J.L. (1988) Insulin-stimulated MAP-2 kinase phosphorylates and activates ribosomal protein kinase II. *Nature*, **334**, 715–718.

Sutherland, C., Leighton, I.A., and Cohen, P. (1993) Inactivation of glycogen synthase kinase-3 beta by phosphorylation: new kinase connections in insulin and growth factor signaling. *Biochem. J.*, **296**, 15–19.

Sweet, L.T., Alcorta, D.A., Jones, S.W., Erikson, E., and Erikson, R.L. (1990) Identification of mitogen-responsive ribosomal protein S6 kinase pp90[rsk], a homolog of *Xenopus* S6 kinase II in chicken embryo fibroblasts. *Mol. Cell. Biol.*, **10**, 2413–2417.

Tabarini, D., Garcia de Herreros, A., Heinrich, J., and Rosen, O.M. (1987) Purification of a bovine liver S6 kinase. *Biochem. Biophys. Res. Commun.*, **144**, 891–899.

Tang, T., Hirata, Y., Whalin, M., and Guroff, G. (1996) Nerve growth factor-stimulated nuclear S6 kinase in PC12 cells. *J. Neurochem.*, **66**, 1198–1206.

Terada, N., Lucas, J.J., Szepesi, A., Franklin, R.A., Takase, K., and Gelfand, E.W. (1992) Rapamycin inhibits the phosphorylation of p70 S6 kinase in IL-2 and mitogen-activated human T cell. *Biochem. Biophys. Res. Commun.*, **186**, 1315–1321.

Terada, N., Patel, H.R., Takase, K., Kohno, K., Nairn, A.C., and Gelfand, E.W. (1994) Rapamycin selectively inhibits translation of mRNAs encoding elongation factors and ribosomal proteins. *Proc. Natl. Acad. Sci. USA*, **91**, 11477–11481.

Thelen, M., Wymann, M.D., and Langen, H. (1994) Wortmannin binds specifically to 1-phosphatidylinositol 3-kinase while inhibiting guanine nucleotide-binding protein-coupled receptor signaling in neutrophil leukocytes. *Proc. Natl. Acad. Sci. USA*, **91**, 4960–4964.

Thomas, G., Siegmann, A., Kubler, A-M., Gordon, J., and Jimenez de Asua, L. (1980) Regulation of 40S ribosomal S6 phosphorylation in Swiss mouse 3T3 cells. *Cell*, **19**, 1015–1023.

Thomas, G., Martin-Perez, J., Siegmann, M., and Otto, A.M. (1982) The effect of serum, EGF, PGF2alpha and insulin on S6 phosphorylation and the initiation of protein and DNA synthesis. *Cell*, **30**, 235–242.

Varticovski, L., Harrison-Findik, D., Keeler, M.L., and Susa, M. (1994) Role of PI 3-kinase in mitogenesis. *Biochim. Biophys. Acta*, **1226**, 1–11.

Vézina, C., Kudelski, A., and Sehgal, S.N. (1975) Rapamycin (AY-22,989), a new antifungal antibiotic: I. Taxonomy of the producing streptomycete and isolation of the active principle. *J. Antibiotics*, 28, 721–726.

Volonté, C. and Greene, L.A. (1992) Nerve growth factor-activated protein kinase N: Characterization and rapid near homogeneity purification by nucleotide affinity-exchange chromatography. *J. Biol. Chem.*, **267**, 21663–21670.

Wasserman, W., Richter, J.P., and Smith, L.D. (1982) Protein synthesis during maturation promoting factor- and progesterone-induced maturation in *Xenopus* oocytes. *Develop. Biol.*, **89**, 152–158.

Weng, Q-P., Andrabi, K., Klippel, A., Kozlowski, M.T., Williams, L.T., and Avruch, J. (1995) Phosphatidylinositol 3-kinase signals activation of p70 S6 kinase *in situ* through site-specific p70 phosphorylation. *Proc. Natl. Acad. Sci. USA*, **92**, 5744–5748.

Weng, Q-P., Andrabi, K., Kozlowski, M.T., Grove, J.R., and Avruch, J. (1995) Multiple independent inputs are required for the activation of the p70 S6 kinase. *Mol. Cell. Biol.*, **15**, 2333–2340.

Wettenhall, R.E.H., Chesterman, C.N., Walker, T., and Morgan, F.J. (1983) Phosphorylation sites for ribosomal S6 protein kinases in mouse 3T3 fibroblasts stimulated with platelet-derived growth factor. *FEBS Lett.*, **162**, 171–176.

Wettenhall, R.E.H., Erikson, E., and Maller, J.L. (1992) Ordered multisite phosphorylation of *Xenopus* ribosomal protein S6 by S6 kinase II. *J. Biol. Chem.*, **267**, 9021–9027.

Wettenhall, R.E.H. and Morgan, F.J. (1984) Phosphorylation of hepatic ribosomal proteins on 80S and 40S ribosomes. *J. Biol. Chem.*, **259**, 2084–2091.

Wong, K. and Cantley, L.C. (1994) Cloning and characterization of a human phosphatidylinositol 4-kinase. *J. Biol. Chem.* **269**, 28878–28884.

Wood, K.W., Sarnecki, C., Roberts, T.M., and Blenis, J. (1992) Ras mediates nerve growth factor receptor-modulation of three signal-transducing protein kinases: MAP kinase, RAF-1 and RSK. *Cell*, **68**, 1041–1050.

Wool, I.G. and Stoffler, G. (1974) Structure and function of eukaryotic ribosomes. In M. Nomura, A. Tissieres, and P. Lengyel, (eds), *Ribosomes*, Cold Spring Harbor Laboratory, Cold Spring Harbor NY, pp. 417–460.

Wymann, M.P., Bulgarelli-Leva, G., Zvelebil, M.J., Pirola, L., Vanhaesebroeck, B., Waterfield, M.D., and Panayotou, G. (1996) Wortmannin inactivates phosphoinositide 3-kinase by covalent modification of Lys-802, a residue involved in the phosphate transfer reaction. *Mol. Cell. Biol.*, **16**, 1722–1733.

Yano, H., Nakanishi, S., Kimura, K., Hanai, N., Saitoh, Y., Fukui, Y., Nonomura, Y., and Matsuda, Y. (1993) Inhibition of histamine secretion by wortmannin through the blockage of phosphatidylinositol 3-kinase in RBL-2H3 cells. *J. Biol. Chem.*, **268**, 25846–25856.

Yatomi, Y., Hazeki, O., Kume, S., and Ui, M. (1992) Suppression by wortmannin of platelet responses to stimuli due to inhibition of pleckstrin phosphorylation. *Biochem. J.*, **285**, 745–751.

Zhao, Y., Bjorbaek, C., Weremowicz, S., Morton, C.C., and Moller, D.E. (1995) *RSK3* encodes a novel pp90[rsk] isoform with a unique N-terminal sequence: growth factor-stimulated kinase function and nuclear translocation. *Mol. Cell. Biol.*, **15**, 4353–4363.

18. TETANUS AND BOTULINUM TOXINS — ZINC PROTEASES — SYNAPTOBREVIN — SYNAPTOTAGMIN — EXOCYTOSIS

TORSTEN BINSCHECK and HANS H. WELLHÖNER

Institute of Toxicology, Medical School of Hannover, D-30623 Hannover

INTRODUCTION

During the last five years a far-reaching development has taken place and is still continuing in the fields of clostridial neurotoxins and molecular mechanisms of exocytosis, the most important event being the uncovering of the molecular mode of action of the toxins. These efforts have also provided a deeper insight into the molecular mechanisms of exocytosis itself, using the clostridial neurotoxins as specific tools for dissecting the components of the exocytotic machinery and assigning specific functions to them.

The group of clostridial neurotoxins (CNTs) consists of tetanus toxin (TeTx) and seven botulinum toxins, which are denominated BoNT/A, BoNT/B, BoNT/C₁, BoNT/D, BoNT/E, BoNT/F and BoNT/G. (For review see: Simpson, 1981; Habermann and Dreyer, 1986a; Wellhöner, 1992.) They share a high specificity for neuronal cells and disrupt neuronal transmission by blocking the exocytotic release of neurotransmitters stored in synaptic vesicles (Burgen *et al.*, 1949; Habermann *et al.*, 1980; Osborne and Bradford, 1973; Eisel *et al.*, 1986; Wellhöner, 1982; Wellhöner, 1989). Botulinum toxin C_2 is not dealt with in this article because it acts as ADP-ribosylating enzyme and, therefore, is not specifically neurotoxic. It disrupts the cytoskeleton by ADP-ribosylation of G-actin and interferes with regulatory G-proteins (Aktories *et al.*, 1990; Aktories *et al.*, 1988; Aktories and Wegner, 1989; Aktories and Just, 1990). The CNTs are the most potent toxins known and cause two distinct, life-threatening neurological disorders: TeTx induces tetanus which is characterized by severe muscle spasms (Habermann, 1978; Bleck, 1989). By contrast, the botulinum neurotoxins cause a flaccid paralysis of the skeletal muscles (Simpson, 1981). Without medical intervention both diseases are lethal. The dramatic clinical manifestations have sustained the interest of scientists in the CNTs for the last 120 years.

Effects of CNTs have been characterized in various parts of the nervous system (Wellhöner, 1982; Wellhöner, 1989; Bergey *et al.*, 1983; Bigalke *et al.*, 1983; Hagenah *et al.*, 1977), in various tissue preparations (Bigalke *et al.*, 1981a; Habermann *et al.*, 1988; Rabasseda *et al.*, 1988), on the cellular level (Bigalke *et al.*, 1978; Bergey *et al.*,

1987; Dolly *et al.*, 1987), in assays derived from subcellular preparations (e.g. synaptosomes) (Ashton and Dolly, 1988; McMahon *et al.*, 1992; Schweizer *et al.*, 1995) and also in neuroendocrine cells (Knight *et al.*, 1985; Knight, 1986; Penner *et al.*, 1986; Dayanithi *et al.*, 1990; Bittner and Holz, 1988; Habermann *et al.*, 1988). The quintessential result deduced from these experiments was that all CNTs specifically block the release of neurotransmitters from synaptic terminals, thereby disrupting synaptic transmission. (For reviews see Wellhöner, 1982; Habermann and Dreyer, 1986a; Wellhöner, 1989; Wellhöner, 1992). But it was not before this decade that the crucial investigations were made which allowed the complete understanding of the action of the CNTs at the molecular level. These toxins are Zn^{2+}-binding metalloendoproteases (Wright *et al.*, 1992; Schiavo *et al.*, 1992b; Schiavo *et al.*, 1992c; Montecucco and Schiavo, 1993; Villiers *et al.*, 1993) which specifically cleave distinct proteins involved in exocytosis in neuronal (Hohne Zell *et al.*, 1994; Schiavo *et al.*, 1994b) and also nonneuronal cells (McMahon *et al.*, 1993). As a windfall, a deeper insight into the molecular mechanisms of exocytosis was also gained.

Hence, the present article will focus on the molecular action of the CNTs on the exocytotic machinery which again allows conclusions with respect to the structural features of the toxins. It will summarize the present knowledge of exocytosis at the molecular level, because this is essential to the understanding of the toxin effects.

SOURCES AND STRUCTURES OF CLOSTRIDIAL NEUROTOXINS

Sources

Produced by *Clostridium tetani* under anaerobic conditions, TeTx, in the form of a single-chain protein (MW 150,000 Da) is released into the extracellular space (Habermann and Dreyer, 1986b; Weller *et al.*, 1988). The manufacture of the toxin depends on the presence of a plasmid containing the encoding gene (Laird *et al.*, 1980; Finn *et al.*, 1984). Loss of the plasmid results in non-toxigenic clostridial strains. For a detailed overview on this topic refer to Niemann (1991).

Seven distinct botulinum neurotoxins (A, B, C_1, D, E, F and G) have been classified on an immunological basis. They are produced by four different strains of *C. botulinum*. However, BoNT/E and BoNT/F are also secreted by certain strains of *C. butyricum* (Johnstone *et al.*, 1992) and *C. baratii* (Hall *et al.*, 1985; Gimenez *et al.*, 1992). Like TeTx the BoNTs are released as single-chain proteins (M_r 150,000 Da). However, some of them are associated with a haemagglutinin (M_r 450,000 Da) (Tsuzuki *et al.*, 1992). The role of this haemagglutinin is unclear. It may protect the toxin from digestive enzymes in the gastrointestinal tract (after ingestion).

Nucleotide Sequences

The nucleotide sequences for all CNTs have been published:

Toxin	Location	Reference
TeTx	Bacteriophage	(Eisel *et al.*, 1986) (Fairweather and Lyness, 1986)
BoNT/A	Chromosome	(Binz *et al.*, 1990b) (Willems *et al.*, 1993)
BoNT/B	Chromosome	(Whelan *et al.*,1992b)
BoNT/C$_1$	Bacteriophage	(Hauser *et al.*, 1990) (Kimura *et al.*, 1991)
BoNT/D	Bacteriophage	(Binz *et al.*, 1990a) (Sunagawa *et al.*, 1992)
BoNT/E	Chromosome	(Whelan *et al.*, 1992a) (Fujii *et al.*, 1992)
BoNT/F	Chromosome	(East *et al.*, 1992) (Thompson *et al.*, 1993)
BoNT/G	Plasmide	(Eklund *et al.*, 1988) (Campbell *et al.*, 1993)

All CNTs show a high degree (30–50%) of homology in their amino acid sequences, suggesting that they all may have evolved from a single ancestral gene (Eisel *et al.*, 1986; Niemann, 1991; Fairweather and Lyness, 1986; DasGupta and Datta, 1988). Interestingely the L-chains and the N-terminal parts of the H-chain, which consists of predominantly hydrophobic amino acid residues, are highly conserved among the different CNTs.

The knowledge of the nucleotide sequence not only allowed studies on homology but, by employing techniques of molecular biology, opened a new source of CNTs. Several groups of investigators succeeded in introducing and expressing genes encoding fragments of the CNTs (see below) in *E. coli* (Eisel *et al.*, 1986; Fairweather *et al.*, 1986; Fairweather *et al.*, 1987; Andersen Beckh *et al.*, 1989; Makoff *et al.*, 1989; Fairweather *et al.*, 1993).

Clinical Manifestations and Pathophysiology of Tetanus and Botulism

To understand the actions of CNTs at the cellular and molecular level it is useful to cast a look at the clinical characteristics of the diseases they cause. One important question arising from the pathophysiology of tetanus and botulism is: Why do TeTx and the BoNTs, despite the numerous molecular and functional features they have in common (Hagenah *et al.*, 1977; Habermann *et al.*, 1980; Bigalke *et al.*, 1981b; Simpson, 1988; Rabasseda *et al.*, 1988; Habermann, 1989), induce two clinically extremely different diseases? The underlying pathophysiology is well understood at the neurophysiological level (for TeTx: Wiegand and Wellhöner, 1979;

Bergey *et al.*, 1983; Dreyer *et al.*, 1983; Bigalke *et al.*, 1983; Habig *et al.*, 1986; Bergey *et al.*, 1987; Wellhöner, 1989; Wellhöner, 1992 for the BoNTs: Habermann and Heller, 1975; Hagenah *et al.*, 1977; Wiegand and Wellhöner, 1977; Simpson, 1980; Dolly *et al.*, 1987; Habermann, 1989)and partially reflects the distinct fates of TeTx and the BoNTs after gaining access to the nervous system. Some hypothetical reasons for the difference in intracellular routing between TeTx and BoNTs will be discussed below.

Spores of *C. tetani*, infecting lesions, germinate under anaerobic conditions and the emerging bacilli produce and release the toxin, which reaches neuromuscular junctions by diffusion or via lymphatic and blood vessels, binds to the presynaptic membrane (Wernig *et al.*, 1977) of motoneurons and is internalized. On entering the cytosol the toxin travels by retrograde intraaxonal transport to the nerve cell bodies located in the ventral grey matter of the spinal cord (Stoeckel *et al.*, 1975). Then it is released into the synaptic cleft and taken up, first by inhibitory interneurons and later probably also by other interneurons by transcytosis (Schwab and Thoenen, 1976; Wiegand *et al.*, 1976; Erdmann and Habermann, 1977; Büttner-Ennever *et al.*, 1981). In the terminals of inhibotory interneurons TeTx blocks the release of inhibitory neurotransmitters (GABA, glycine) (Osborne and Bradford, 1973; Bigalke *et al.*, 1981b; Bigalke *et al.*, 1983). As a consequence, the interneurons fail to restrain excitation of postsynaptic motoneurons. This disinhibition of α- and γ-motoneurons leads to extremely painful and — by action on the respiratory musculature- life-threatening muscle spasms (Sun *et al.*, 1994). (For review see: Habermann, 1978; Wellhöner, 1982). In some cases, when the amount of toxin is low and the port of entry is small, TeTx is not able to impair more than one or a few adjacent spinal segments and muscle spasms are confined to the corresponding segments (*local tetanus*). Since TeTx fails to cross the blood-brain-barrier, no impairment of higher brain functions (e.g. consciousness) is observed, even in severe tetanus. The spasms, involving simultaneously all agonistic and antagonistic skeletal muscles, are the visible consequence of disinhibition at the level of the spinal cord. For an artist's view of a tetanus patient refer to Figure 18.1.

The BoNTs follow another pathway to perform their detrimental action. In contrast to tetanus, which result from a (wound) infection with *C. tetani*, botulism, in most cases, is due to the ingestion of BoNTs. To reach their target structure — the presynaptic membranes of peripheral motoneurons — BoNTs have to be absorbed from the gastrointestinal tract. Only few data exist on the bioavailability of orally applied BoNTs and the mechanism of absorption is not clear. Some authors assume that these large protein molecules enter through micro-lesions of the gastrointestinal tract, whereas others favour unspecific pinocytosis by enterocytes. Rarely, botulism originates from wounds infected with toxigenic strains of *C. botulinum* (Burningham *et al.*, 1994). In infants, ingestion of spores (e.g. present in honey) may lead to relevant toxin production in the gastrointestinal tract after germination of the spores (Fenicia *et al.*, 1993). Independent of the initial conditions, BoNTs reach the blood and are distributed over the whole body. Like TeTx they are unable to cross the blood-brain-barrier; they bind to the presynaptic membranes of peripheral motoneurons (Dolly *et al.*, 1984) and retrograde intra-

Figure 18.1 Soldier of the Napoleonic wars exhibiting symptoms of severe tetanus [*Etching by Sir Charles Bell; published in: Essay on the anatomy and philosophy of expression, London (1824)*]

The typical posture of patients suffering from generalized tetanus is characterized by maximal contraction of all skeletal muscles. Due to the anatomical prevailance of muscles acting against gravitation, the trunk and legs are extended (*opisthotonus*), whereas arms, fingers and toes are bent. Patients are unable to open their mouths and have a typical "smiling" facial expression (*risus sardonicus*). Before the use of muscle relaxants and benzodiazepines, in combination with artificial respiration, became common practice in intensive care medicine, patients suffering from tetanus were doomed to die of respiratory failure.

axonal transport occurs as well (Wiegand *et al.*, 1976; Wellhöner, 1982). However, the latter phenomen does not contribute to the clinical picture of intoxication. BoNTs directly attack the exocytotic machinery of the cholinergic peripheral synapse (Burgen *et al.*, 1949; Simpson, 1981; Williams *et al.*, 1983; Eisel *et al.*, 1986). The most important clinical feature is the block of neuromuscular transmission causing a flaccid paralysis. Due to the haematogenic distribution of the BoNTs, all sceletal muscles are involved, but small muscles with dense innervation (e.g. ocular muscles) are more sensitive than large muscles with scattered innnervation (e.g. gluteal muscle) (Tacket and Rogawski, 1989).

Patients suffering from tetanus or botulism may experience full recovery after several weeks, indicating that neither TeTx or BoNTs lead to lasting structural damage to nervous tissues. The long duration of both diseases may indicate that (a) the toxins persist very long at the molecular locus of action (a pharmacokinetic reason) or (b) damaged cells need a long time to recover, independent of the presence of the CNTs (a pharmacodynamic reason).

Secondary Structure, Chains and Domains

During the last fifteen years a large amount of data has accumulated from biochemical experiments aimed at the elucidation of the molecular structure of the CNTs. Investigators succeeded in linking the structural data to functional features of the CNTs. All CNTs consist of a heavy chain (H-chain) containing the carboxy terminal and a light chain (L-chain) encompassing the amino terminal. The

chains are connected by a disulfide bond between two cystidine residues and a
peptide bond (Figure 18.2) Upon release by the bacterial cell, the single-chain
proteins (TeTx: 1315 amino acids) are nicked by various extracellular proteases
(TeTx: Habermann *et al.*, 1991, BoNTs: Simpson, 1981), resulting in the di-chain

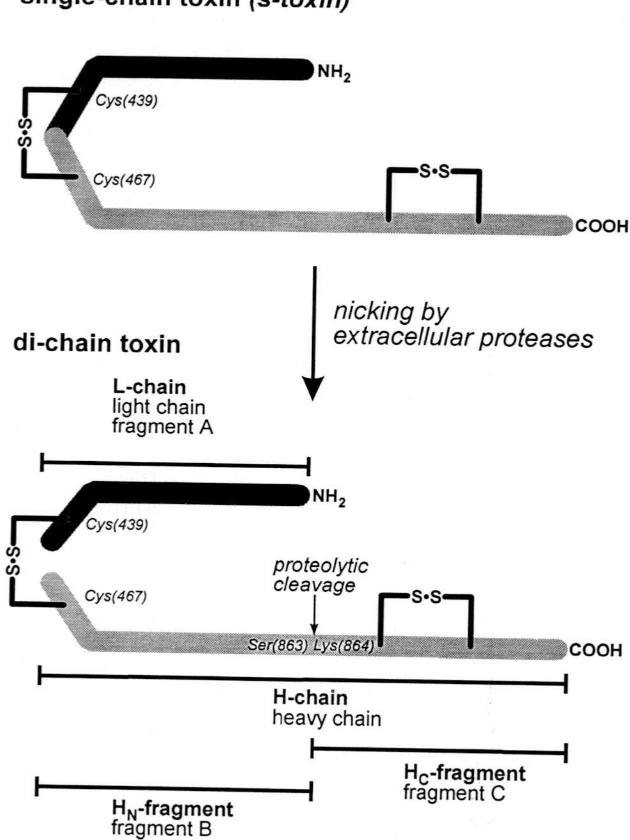

Figure 18.2   Molecular structure and nomenclature of CNTs

All CNTs possess similar molecular structures. They are secreted as single-chain toxins by
the bacteria and have to undergo a proteolytic cleavage separating the *light chain* (*dark
shaded*) from the *heavy chain* (*light shaded*), which increases toxicity in *in vitro* experiments.
The exact cleavage site depends on the attacking protease (e.g. trypsin, papain or clostri-
pain). Both chains remain interconnected by a intermolecular disulfide bridge, ionic and
hydrogen bonds. (Amino acid positions are given for TeTx). The H-chain can be further
degraded to two fragments by papain exposure (*proteolytic cleavage*), resulting in the amino
terminal H$_N$-*fragment* and the carboxy terminal H$_C$-*fragment*. The latter binds to gangliosides
and is involved in retrograde intraaxonal transport. The L-chain hosts the active site of
CNTs. All isolated chains and fragments are non-toxic *in vivo*, because the L-chain is not
internalized, whereas the H-chain has no enzymatic activity.

forms of the respective toxins. The N-terminal part represents the light chains (L-TeTx: M_r 52,000 kDa) and the C-terminal part represents the heavy chain (H-TeTx: M_r 98,300 kDa). Both chains remain linked together other by a disulfide bridge between Cys(439) on the L-chain and Cys(467) on the H-chain and by ionic interaction (Weller *et al.*, 1989; Schiavo *et al.*, 1990). The nicking site depends on the hydrolyzing protease but is always located between those cystidine residues. In consequence, the lengths of the L-chain and H-chain cannot be precisely defined. *In vivo* the single-chain TeTx is as potent as the nicked, di-chain form (Weller *et al.*, 1988; Kistner and Habermann, 1992). However, in permeabilized chromaffin cells only the di-chain form is active, whereas the single chain toxin is almost completely ineffective (Bergey *et al.*, 1986; Weller *et al.*, 1986; Ahnert Hilger *et al.*, 1989b; Bergey *et al.*, 1989). *In vivo* the nicking of TeTx is probably performed by unspecific proteases located in the tissue (Habermann *et al.*, 1991). From this it has been concluded that the single chain toxins have to be nicked to gain their biological activity.

In TeTx, a second disulfide bridge is located within the heavy chain between Cys(1076) and Cys(1092) (Krieglstein *et al.*, 1990). The H-chain of TeTx can be further cleaved by *papain*, hydrolyzing the peptide bond between Ser(864) and Lys(865) (Fairweather *et al.*, 1986; Eisel *et al.*, 1986; Krieglstein *et al.*, 1991). Two fragments are formed: the N-terminal part of the H-chain (H_N-fragment) and the C-terminal part of the H-chain (H_C-fragment). Similar to H-TeTx the heavy chains of the BoNTs could be further hydrolyzed by mild enzymatic digestion into the H_N-fragment and H_C-fragment (Sathyamoorthy *et al.*, 1988).

The di-chain form, characteristic of all CNTs, is also found in other protein toxins (e.g. *diphtheria* toxin, *ricin*, *cholera* toxin) (Ahnert-Hilger and Bigalke, 1995). All toxins possessing this structure are also designated as AB-toxins. One chain always carries the pharmacokinetic features (binding, internalization), whereas the other chain contains the pharmacodynamic part exerting the specific toxin action. A prerequisite for studies on the function of each chains was the biochemical isolation and purification of L- and H-chains and their fragments (Sathyamoorthy and DasGupta, 1985; Weller *et al.*, 1989). A different approach using molecular biological techniques, allowed the expression of isolated chains (Fairweather *et al.*, 1993) or toxin fragments (Halpern *et al.*, 1990) in *E. coli* or in nontoxigenic strains of *C. tetani* (Andersen Beckh *et al.*, 1989).

BINDING, INTERNALIZATION AND INTRACELLULAR PROCESSING

The H-chains of TeTx and the BoNts mediate binding to neuronal membranes, internalization and retrograd intraaxonal transport (Weller *et al.*, 1986). Isolated H-chains can compete with the holotoxins for the binding sites but they are completely untoxic. Further analysis of fragments of the H-chain has revealed that the binding moiety is located in the H_C-fragment, probably near the carboxy terminal of the protein chain (Bizzini *et al.*, 1981; Halpern and Loftus, 1993). The H_C-fragment is of special interest: It is responsible for the binding of toxin

molecules to neuronal cell membranes (Simpson, 1984; Simpson, 1985) and mediates the neuronal ascent into the spinal cord. Therefore, it could serve as a safe
anti-TeTx vaccine (Fairweather *et al.*, 1987) and it has also been used for carrying
other biological molecules into central neurons. The H_C-fragment (*fragment C*)
has been expressed in *E. coli*, for yielding highly purified preparations for that
purpose (Fairweather *et al.*, 1986; Jimenez *et al.*, 1993). Less is known about the N-
terminal part (H_N-fragment). Several publications report a pore-forming activity of
both H-TeTx and H-BoNT/A in black membranes (Donovan and Simpson, 1986),
lipid bilayers (Hoch *et al.*, 1985; Montecucco *et al.*, 1988; Gambale and Montal,
1988), liposomes (Högy *et al.*, 1992), spinal cord neurons (Beise *et al.*, 1994) and
chromaffin cells (own experiments). Pore formation probably results from the
action of the specific portion of the H-chains that contains an abundance of
hydrophobic amino acids.

The Three-Step Model

In 1981 Simpson proposed a model of BoNT/A action which was adopted for all
CNTs (Simpson, 1980; Simpson, 1981; Schmitt *et al.*, 1981). Acording to this
hypothesis the toxins do not act by means of a membrane-based receptor (in the
pharmacological sense of "receptor") but have a cytosolic target structure. The
model concludes three sequential steps: 1. The toxin molecule is bound to and
enriched on the plasma membrane. 2. The toxin molecule, or at least an active
component of it has to translocate through the membrane. 3. An intracellular
target structure is altered, resulting in the block of exocytosis. This model was
developed on the basis on results obtained with the *phrenic nerve-hemidiaphragm-
preparation*, which allowed the measurement of toxin action on the neuromuscular
junction. The following evidence supports the hypothesis: CNTs bind to neuronal
plasma membranes at low temperature, independent of calcium (Yavin *et al.*,
1983). Antibodies are able to neutralize bound toxin molecules. By contrast, internalization is temperature-dependent, indicating that translocation of the toxin is a
process consuming energy which has to be supplied by metabolism. The action of
translocated toxin cannot be prevented by external antibodies, because they are
unable to pass through the plasma membrane. Decreasing neuromuscular transmission, progressing to total blockade, is indicative of the intracellular action of
the toxin. Efforts have been made to modulate the action on neuromuscular transmission to gain insight into the intracellular mode of toxin action. They will be discussed below.

Binding of Clostridial Neurotoxins to Cytoplasmic Membranes

TeTx and the botulinum neurotoxins bind to neuronal membranes selectively
and with high affinity. Binding has been demonstrated in brain preparations
(Habermann and Heller, 1975; Pierce *et al.*, 1986), presynaptic membranes of
motoneurons (Kitamura, 1976; Dolly *et al.*, 1984), primary nerve cell cultures
(Critchley *et al.*, 1988) and neuroblastoma cell lines (DiMaggio *et al.*, 1994). In

contrast to other eukaryotic cells, membranes of neuronal cells are rich in poly-sialogangliosides which contain at least two neuraminic acid residues. Hence, these gangliosides appeared to be suitable candidates for putative toxin binding sites (Holmgren *et al.*, 1980; Kitamura *et al.*, 1980; Critchley *et al.*, 1986). CNTs bind to polysialogangliosides incorporated into lipid vesicles (Lazarovici *et al.*, 1987; Montecucco *et al.*, 1988), artifical membranes and plasma membranes of neuronal cells (Dimpfel *et al.*, 1977; Wernig *et al.*, 1977; Yavin *et al.*, 1987). Binding to poly-sialogangliosides can be enhanced by decreasing the ionic strength or pH of the incubation medium (Bakry *et al.*, 1991). Experiments with ganglioside-enriched chromaffin cells revealed different binding specifities for TeTx (GD_{1b} < GD_{1a} < GM_1) and BoNT/A (GD_{1a}) (Marxen *et al.*, 1989; Marxen and Bigalke, 1989).

In addition to gangliosides, membrane-bound (glyco)proteins might serve as toxin receptors. Mild trypsination of neuronal cells decreased the number of binding sites of TeTx (Lazarovici and Yavin, 1986; Yavin and Nathan, 1986) and BoNTs in different preparations. Binding of BoNT/B to the vesicular protein *synaptotagmin* and the gangliosides GT_{1b} and $GD_{1a,}$ forming a ternary complex, has been recently demonstrated (Nishiki *et al.*, 1994). The current hypothesis on binding of the CNTs to neuronal cells merges gangliosides and protein receptor to one scenario: Gangliosides may serve as low-affiniy binding sites accomplishing enrichment and fixation of toxin molecules. By lateral diffusion gangliosides could come into close contact with a protein binding CNTs, and they could pass the toxins to these receptors. Since *synaptotagmin*, which is an integral constituent of membranes of small synaptic vesicles (SSVs) and gangliosides coexist only in presynaptic membranes of nerve terminals, they may constitute the molecular cor-relate of CNT's binding specificity to this particular site.

Internalization Via the Endosomal Pathway and Differential Intracellular Sorting

To most proteins, the plasma membrane of eukaryotic cells represent an impene-trable barrier. The cell has specific mechanisms for the import and export of pro-teins. It is likely that the CNTs use one or more of these. There is strong evidence that the CNTs are internalized by adsorptive endocytosis. Internalization is acceler-ated by enhanced electrical activity of neurons (Critchley *et al.*, 1985). Depending on the presence or absence of polysialogangliosides, TeTx and BoNT/A have been detected in both clathrin-coated and uncoated smooth vesicles, the latter being indistinguishable from endosomes. When *mouse phrenic nerve-hemidiaphragm* prepa-rations, pretreated with BoNT/A and BoNT/B, are exposed to chloroquin, which decreases the endosomal pH-gradient, toxin action is inhibited. The same has been demonstrated for TeTx, although to a lesser extent. The importance of acidifica-tion for translocation is confirmed by experiments with *Bafilomycin A 1*, which inhibits endosomal acidification, elevates endosomal pH and thus prevents the action of TeTx on mouse spinal cord neurons (Williamson and Neale, 1994). In chromaffin cells, which naturally lack gangliosides, toxicity of TeTx is largely enhanced, when the cells are exposed to the toxin at pH 5 and it is further increased, if exposure to the toxin is preceded by enrichment of the plasma membrane with a ganglioside mixture. In these experiments at low pH the whole plasma membrane serves as an "endosomal" membrane, facilitating the translocation of toxin molecules. Remarkably, internalization of TeTx does not depend on the presence of gangliosides, suggesting that they are not a prerequisite for translocation. *Brefeldin A*, which inhibits the function of the *Golgi* apparatus, failed to block the action of TeTx in mouse spinal cord neurons (Williamson and Neale, 1994). It can be concluded that the CNTs are internalized from the extracellular space via endocytotic vesicles and the acidic endosomal compartment. This acidification appears to be an essential prerequisite for the toxin molecules to escape from the endosome into the cytoplasm. By contrast, the alternative internalization pathway, with endocytotic vesicles travelling to the endoplasmic reticulum and *Golgi* apparatus and subsequent generation of excretory vesicles, from which endocytosed substance could be released, can be ruled out.

Less is known about the physicochemical process allowing the toxin molecules to cross the vesicular membrane. The H-chains of TeTx and BoNTs are hydrophobic; in an acidic enviroment, as found in endosomes, hydrophobicity is further increased (Roa and Boquet, 1985). The H-chains interact with lipid bilayers (Hoch *et al.*, 1985; Gambale and Montal, 1988), liposomes (Montecucco *et al.*, 1988), synaptosomes (Högy *et al.*, 1992) and membranes of spinal cord neurons (Beise *et al.*, 1994). At least one α-helical region, consisting of hydrophobic amino acid residues in the H_N-fragment of TeTx, has been characterized (Montal *et al.*, 1992). This region is long enough to span the entire cytoplasmic membrane, thus being capable of forming a transmembraneous pore. Accordingly, pore formation by the H-chains and di-chain TeTx has been demonstrated in lipid vesicles (Boquet and Duflot, 1982), planar lipid bilayers (Donovan and Simpson, 1986), the membranes of mouse spinal cord neurons in culture as well as in bovine chromaffin cells and pores have been characterized by electrophysiological methods (Rauch *et al.*, 1990; Beise *et al.*, 1994). They are voltage-dependent and not selectively permeable for mono- and divalent ions. Electrophysiological experiments on chromaffin cells revealed that pore formation is independent of polysialogangliosides. However, voltage dependency is evident: Pores will form only if the toxin is applied into a compartment charged positively with respect to the plasma membrane. A negative potential or absence thereof inhibits pore formation in inside-out patches. It is tempting to assume that the pores introduced by the H-chains are the pathways by which the toxins or at least the L-chains enter the cytosol. The low unitary electrical conductance of 28–35 pS observed in spinal cord neurons and chromaffin cells indicates, however, that a translocation of the active L-chain through these pores is unlikely for steric reasons. There is no evidence from other sources whether the di-chain toxins or only the L-chains reach the cytosol.

Internalization is closely linked to the sorting process. Although there is retrograde intraaxonal transport not only of TeTx but also of several BoNTs, only TeTx acts primarily on the inhibitory interneurons. BoNTs, to exert an effect on the spinal cord, would have to be injected directly into the grey matter. Under physiological conditions, the arrival in the spinal cord of intraaxonally transported BoNTs does not appear to play a significant role. The BoNTs are about 1000 times as potent as TeTx in blocking exocytotic release of acetylcholine from peripheral nerve endings (e.g. at neuromuscular junctions). On the other hand, TeTx is 1000 times as potent as the BoNTs in disrupting the inhibitory synaptic transmission in central interneurons. One reason might be that the exocytotic machinery in inhibitory synapses is more sensitive to TeTx than to the BoNTs. So far, however, there is no evidence that excitatory synapses differ from inhibitory synapses at the molecular level. Another — pharmacokinetic — mechanism may offer an explanation: During uptake into the peripheral synapses TeTx and the BoNTs are handled in different ways. Whereas sufficient amounts of the BoNTs manage to rapidly escape into the synaptic cytosol, TeTx remains in the vesicular compartment and is quantitatively transported to the spinal cord. Without leaving its vesicular "shuttle" TeTx might undergo transcytosis from the motoneuron to the presynaptic terminal of the inhibitory interneuron. As a prerequisite to this sorting, the cell has to receive a signal which is decisive for the proper processing of the toxin-containing vesicles. Concerning the molecular structure of CNTs, the N-terminal part of the heavy chain (H_N-fragment) might contain a signalling sequence. Weller and coworkers (1991) found that chimeric toxins consisting of H-BoNT/A and L-TeTx resulted in a 6-fold increase in toxicity, as compared to native TeTx at the mammalian motor endplate. These results indicate that H-BoNT/A may have directed L-TeTx to the synaptic vesicles instead of sending it on the retrograde intraaxonal pathway.

It is still obscure, at what point during internalization the sorting decision is made. Moreover, no explanation of the molecular mechanism of sorting has been suggested. Experiments with fusion proteins from other bacterial toxins with known internalization pathways and the L-chains of CNTs could provide an answer.

While the process of internalization and intracellular sorting still awaits clarification, it is widely accepted that the H-chains have to be separated from the L-chains to activate the latter. Prior to dissociation of both components, reductive cleavage of the intermolecular disulfide bond has to occur. It is not yet clear, in which compartment reductive separation of both chains takes place. Purified thioredoxin reductase, which is abundant in the cytosol, has been demonstrated *in vitro* and in chromaffin cells to cleave TeTx and, at a lower rate, BoNT/A as well (Schiavo *et al.*, 1990; Kistner *et al.*, 1993). Western Blot analysis using a monoclonal antibody directed against L-TeTx provided evidence for the occurrence of L-TeTx molecules in the cytosol of intoxicated chromaffin cells (Erdal *et al.*, 1995). Capacitance measurements during calcium-stimulated exocytosis, starting two hours after internalizing equipotent amounts of TeTx or BoNT/A by electroporation, revealed a difference in the kinetics of the effect of both toxins. In chromaffin cells treated with BoNT/A exocytosis was completely inhibited within two hours after toxin entry. The blockade imposed by TeTx, however, increased gradually over 24 hrs, finally reaching the same intensity. This might reflect the L-chains of BoNT/A and TeTx being liberated from the di-chain forms at different rates. Given this, the dissociation of the di-chain toxins could be the rate-limiting step in intracellular toxin kinetics.

Retrograde Intraaxonal Transport

There is no doubt that TeTx and for that matter BoNT/A as well, does undergo retrograde intraaxonal transport to the somata of motoneurons (Bizzini, 1989). TeTx leaves the motoneurons to enter inhibitory interneurons where it blocks the release of inhibitory transmitters, producing the characteristic symptoms of tetanus. Histoautoradiographic studies demonstrated the accumulation in the ventral grey matter of the spinal cord of ^{125}I-labelled TeTx injected into the gastrocnemius muscle. Similar experiments were done using TeTx fragments and mildly toxoidated TeTx. The latter as well as isolated H_C-fragments representing the binding moiety of TeTx were transported to the spinal cord. H_C-TeTx was also used as a carrier to target conjugated enzymes and antibodies to neurons of the spinal cord (Fishman and Savitt, 1989).

**Degradation and Elimination of Clostridial Neurotoxins
From the Intracellular Compartment**

TeTx has a half-life of 6.5 days in the spinal cord and in cultured mouse spinal cord neurons. Its disappearance from the intracellular compartment may be due to degradation into small fragments undetectable by blotting techniques. However, there is also secretion of intact toxin molecules. Whereas some degradation of TeTx was detected in cultured mouse spinal cord neurons, NG108-15 cells were observed to release both TeTx and as fragments. In chromaffin cell cultures fragments of TeTx but also intact molecules, as well as L-chain were found in the incubation medium, as analysed by SDS-PAGE (Erdal *et al.*, 1995). Clinical experience and cell experiments show that the effects of TeTx and the BoNTs — once they have reached their intracellular target — can persist for several weeks, although the CNTs do not last nearly that long. One reason may be that the intracellular target damaged by the CNTs cannot be replaced so rapidly. Bartels and coworkers (Bartels and Bigalke, 1992) succeeded in stopping the intracellular action of TeTx by introducing neutralizing anti-TeTx antibodies into intoxicated chromaffin cells, employing the technique of electroporation. Even if one assumes that the antibodies had inactivated all toxin molecules, it still took 72 hours before the chromaffin cells had completely recovered. However, a few L-chain molecules, having escaped neutralization by antibodies, may suffice to sustain the action for so long.

MOLECULAR INTERACTION OF CLOSTRIDIAL NEUROTOXINS
WITH THE EXOCYTOSIC MACHINERY

The Active Moiety of the L-chains

In all CNTs, the L-chains contain the structures exerting the toxic effect. (L-TeTx: Stecher *et al.*, 1989; Ahnert Hilger *et al.*, 1989b; Poulain *et al.*, 1990; Stecher *et al.*, 1992; Dayanithi *et al.*, 1992), L-BoNTs: (Dayanithi *et al.*, 1990; de Paiva and Dolly,

1990; Rauch *et al.*, 1990; Lomneth *et al.*, 1991). Before this could become established, an important obstacle had to be surmounted. Isolated L-chains of CNTs, when applied to tissue preparations or intact single cells, are nontoxic, since the L-chains fail to enter the cytosol. Several techniques were used to introduce intact L-chain molecules into cells under experimental conditions. Direct injection of L-TeTx initiated inhibition of synaptic transmission in the buccal ganglion of *Aplysia californica* within 60 minutes (Mochida *et al.*, 1989). The blockage was complete after 120 minutes. This assay was also used to test the mRNA encoding L-TeTx by injecting it directly into the neurons. After successful translation of the mRNA into the gene product, synaptic transmission decreased until total blockage was achieved (Mochida *et al.*, 1990a; Mochida *et al.*, 1990b; Poulain *et al.*, 1990). In another experimental approach, chemically permeabilized chromaffin cells or PC12 cells in culture were used (Stecher *et al.*, 1989; Bittner *et al.*, 1989b; Bittner *et al.*, 1989a). The cell membranes were permeabilized with *digitonin* or *streptolysin O*, thereby creating membrane pores large enough to allow extracelullar L-TeTx or L-BoNT/A access to the cytosol (Ahnert Hilger *et al.*, 1989a). This technique demonstrated that the L-TeTx and L-BoNT/A were as potent as the di-chain forms in blocking the release of preloaded ^{3}H-norepinephrine triggered by stimulation with an cholinoceptor agonist (carbachol) in the presence of calcium. The main disadvantage of this method is the limited survival time of cells (<60 minutes). Another method allowing the introduction of the L-chains -as well as other molecules (e.g. antibodies)- into the cytosol is electroporation (Bartels and Bigalke, 1992). This procedure creates transient pores in plasma membranes when cells are exposed to a short-term high voltage electric field. After spontaneous pore closure the majority of cells recover and can be maintained in long-term cell culture, allowing experiments that elucidate the time course of intoxication and recovery. In this experimental set-up, the L-chains of TeTx and BoNT/A were almost as toxic as the corresponding di-chain forms.

Based on this evidence, scientific attention during the recent years was focused on the L-chains of CNTs. To analyze the specific functions of subdomains on the L-chains of TeTx and BoNT/A, site-directed mutagenesis of the encoding gene was achieved and the corresponding mRNA synthesized by *in vitro* transcription. For consecutive translation and testing of the toxicity of the gene products the transcription products were injected into presynaptic neurons of Aplysia *californica* (Mochida *et al.*, 1990a; Kurazono *et al.*, 1992). It turned out that the L-chains of all CNTs posses a highly conserved histidine-rich motif (HExxHxxH) and a glutamate residue which is spaced by 37 to 39 amino acid residues from the next histidine residue. The histidine-rich motif is located in an α-helical region of the molecule, with all histidines arranged on one side of the helix. Site-directed mutagenesis permitted the exchange of one histidine for other amino acids both in L-TeTx and BoNT/A. Except for the replacement of His(237) by proline, the latter probably disturbing the α-helical secondary structure, toxicity of all mutants was found to be conserved. However, replacing more than one histidine completely abolished toxicity of the gene products (Kurazono *et al.*, 1992; Niemann, 1991). Data base searches have shown that the histidine-rich motif of the L-chains is specific for metalloendoproteases like *thermolysin* (Jongeneel *et al.*, 1989). Wright and coworkers (1992) identified the HELIH-motif of L-TeTx as a low-affinity binding site for Zn^{2+} and Cu^{2+}. They found one zinc-binding site per molecule of L-chain with an association constant of 9–15 μM. Zinc and copper compete for the same binding site, the latter inducing metal-catalyzed oxidative cleavage of two peptide bonds within L-TeTx. The zinc content of L-TeTx as well as of BoNTs/A, B, C_1, E and F was confirmed by atomic absorption analysis (Schiavo *et al.*, 1992c). In this respect the CNTs resemble the active sites of *thermolysin*-like metalloendoproteases (Hooper, 1994). In these enzymes the zinc atom is bound in the centre of a tetrahedron, coordinated by two histidine residues and one water molecule, the latter being bound to a glutamate residue. Further experiments linked the zinc-binding HELIH motif to the toxicity of the L-chains. Chemical modification with diethylpyrocarbonate of histidine residues in L-BoNT/A and E (DasGupta and Rasmussen, 1984), or removal of zinc by incubating the L-chains with EDTA, ortho-phenantroline (Schiavo *et al.*, 1993c; Schiavo *et al.*, 1992c), or other zinc-chelating agents (Simpson *et al.*, 1993), leads to detoxification. Toxicity recovered after incubating the toxins in Zn^{2+} -containing buffer. Concerning the water molecule bound near the active site of thermolysin-like metalloproteases, computer-based modelling of the secondary and tertiary structure of L-TeTx confirmed the suspected importance of Glu(271). This residue is located close to the putative active site and therefore, is a candidate for binding a water molecule and completing the tetrahedron of the active site. Recently strucural studies on L-TeTx have provided evidence for this hypthesis using circular dichroism and fluorescence spectroscopy (De Filippis *et al.*, 1995).

Phosporamidon and captopril — both inhibitors of metalloendoproteases — were potent inhibitors of the action of L-TeTx *in vitro* (Schiavo *et al.*, 1993a; de Paiva *et al.*, 1993), in *Aplysia californica* (Schiavo *et al.*, 1992b; Schiavo *et al.*, 1994a) and neurohypophysial terminals, respectively (Dayanithi *et al.*, 1994). Moreover, the blockage of exocytosis, imposed by TeTx and BoNT/A on catecholamine-secreting chromaffin cells could be surmounted by electroporating intoxicated cells in the presence of an antibody raised against a HELIH-peptide (Bartels *et al.*, 1994).

Immediately after it became apparent that all L-chains of the CNTs are zinc-dependent metalloendoproteases, the search for their substrate(s) became the most important task. But before the toxins can reach their substrates, they have to overcome several obstacles.

To gain insight into the molecular mechanism of CNT action, many efforts have been made during the past decades, but so far, neither biochemical (Presek *et al.*, 1992; Facchiano *et al.*, 1993a; Facchiano *et al.*, 1993b) nor electrophysiological experiments on whole cells (Wiegand and Wellhöner, 1979; Dreyer *et al.*, 1987; Dreyer *et al.*, 1983; Evans *et al.*, 1986) or cell membranes have been able to reveal all details. On the level of cellular function all CNTs exert the same effect, which has been confirmed in various assays: they inhibit calcium-dependent exocytosis of neurotransmitters. Researchers were confronted with tremendous problems. The extremely high toxicity of all CNTs indicated that minute amounts of toxin were effective. This circumstance severely hampered approaches which tried to track the toxins in their intracellular paths. Moreover, the production and purification of toxins in amounts sufficiently large for biochemical modification was tedious and expensive. These efforts were revolutionized by applying the techniques of molecular biology to the CNTs. They allowed both sequencing of the CNTs and production of recombinant, highly purified fragments in appropriate amounts. But the main obstacle in discovering the molecular mode of action of the CNTs may have been that knowledge of the proteins and mechanisms involved in vesicle trafficking and the final steps of exocytosis was patchy. Hence a summary of our present understanding of the function and molecular composition of the exocytotic machinery, which now appears to be the site of action of all CNTs, will follow.

Different Types of Exocytosis

Cellular products designated for secretion are packaged into secretory vesicles. They are transported along the constitutive or regulated secretory pathways. Constitutive exocytosis is probably present in all eukaryotic cells. It links the endosomal compartment via the endoplasmic reticulum and *Golgi* apparatus to the plasma membrane, where fusion of vesicle and plasma membranes occurs during exocytosis. The release of substances by constitutive exocytosis appears to be unregulated; vesicles fuse with the plasma membrane immediately after arrival. Constitutive exocytosis may play an important role in the recycling of membrane-bound receptors or transporters (e.g. transferrin, glucose transporter). In contrast to regulated exocytosis, it is not controlled by the cytosolic calcium concentration.

Regarding the molecular mechanism of the CNTs, the second type of exocytosis has attracted more attention. According to the current concepts, exocytosis of small *synaptic vesicles* (SSVs) and their counterparts in neuroendocrine cells, the *large dense core vesicles* (LDCVs), is regulated. In contrast to constitutive exocytosis, vesicles undergoing regulated exocytosis are stored in specialized regions until the signal for exocytosis arrives. Most of the evidence has accumulated from investigations of the neuronal exocytosis of SSVs. They contain neurotransmitters like GABA, glycine, acetylcholine, noradrenaline or dopamine and undergo a continuous cycle of exo- and endocytosis. The trigger for the release of neurotransmitters is an elevation of the cytosolic Ca^{2+}-concentration. Calcium enters the cell through voltage-gated ω-conotoxin-sensitive channels which open during depolarisation in the course of an action potential. Thus, this system has to employ at least one calcium-sensitive component. Another important feature can be deduced from the extremely short latency between the elevation of the calcium concentration and the release of the vesicular, stored neurotransmitters, suggesting a location very close to the locus of calcium entry, because there may be little time for navigating SSVs to their docking and release position. It is probable that a special portion of SSVs, in close vincinity to the plasma membrane, is in a primed state in synaptic "active zones".

Components of the Exocytotic Machinery

A key role in probably all types of exocytosis is played by a N-ethylmaleimide-sensitive fusion protein (NSF), which is a soluble homotrimeric ATPase. NSF mediates the intracisternal Golgi transport and the vesicle transport from the rough endoplasmic reticulum via the Golgi complex to the plasma membrane. To become functional NSF has to bind soluble NSF attachment proteins (SNAPs). Three isoforms of SNAPs (termed α, β and γ-SNAP) have been identified. The NSF-SNAP complex binds to a specific receptor located on the vesicular membrane (v-SNAREs) and on the cytoplasmic side of the plasma membrane (t-SNAREs). Membrane fusion specificity appears to be defined by these SNARE molecules: A v-SNARE will bind only to its appropriate counterpart t-SNARE, which is presented in the correct target membrane. In SSVs and LDCVs, which participate in regulated exocytosis, synaptobrevin (also termed vesicle-associated membrane protein: *VAMP*), a 18 kD protein inserted into the vesicular membrane, has been identified as v-SNARE, whereas syntaxin and *a synaptosomal associated protein of 25 kD* (*SNAP-25*) represent t-SNAREs.

There are two isoforms of synaptobrevin expressed in neuronal tissue: *synaptobrevin I* and *II*. Both isoforms are found exclusively in the membranes of SSVs and LDCVs. A third isoform (*cellubrevin*) has been detected in membranes of transport vesicles of the constitutive pathway in non-neuronal cells (McMahon *et al.*, 1993; Chilcote *et al.*, 1995). Synaptobrevin II is a class-II integral membrane protein (116 amino acids) with only one membrane-spanning domain, a short intravesicular carboxyl terminal and a hydrophilic cytoplasmic amino terminal. Remarkably, there is a highly conserved region of 70 amino acid residues in the cytoplasmic domain.

SNAP-25 is a 25 kD protein of 206 amino acid residues located in synaptic terminals. By means of fatty acids associated with cystidine residues in the central region of the molecule it is loosely attached to the cytoplasmic side of the plasma membrane, thus representing a t-SNARE. It has recently been demonstrated that, apart from neurons, *SNAP-25* is also present in chromaffin cells (Roth and Burgoyne, 1994) and in pancreatic islets (Sadoul *et al.*, 1995), probably also serving as a t-SNARE.

Syntaxin has first been identified as the *retinal antigen HPC-1*. Two isoforms (A and B) are known. It is abundant in the CNS, where it is predominatly located in synaptic plasma membranes, but also associated to chromaffin granules (Tagaya *et al.*, 1995), serving as yet another t-SNARE. Like the synaptobrevins it is a class-II integral membrane protein. It consists of 288 amino acid residues and is fixed to the plasma membrane by a short transmembraneous domain closely located to the carboxy terminal. The amino terminal protrudes into the cytosol. It has recently been shown that both t-SNARES (*SNAP-25* and *syntaxin 1*) partially recycle with SSVs (Walch Solimena *et al.*, 1995).

In addition to molecules that represent v- and t-SNAREs other proteins are present in the membranes of synaptic terminals. The most important is the group of *synaptotagmins* (I–IV). They form a major protein constituent of the membranes of SSVs and LDCVs comprising 7–8% of total vesicle protein with synaptotagmin I predominanting. It contains two repeats homologous to the C-domain of protein kinase C and is able to bind to phospholipid membranes in a calcium-dependent manner (association constant 0.1–1 μM). The affinity to calcium makes *synaptotagmin* I the leading candidate for the calcium-sensitive structure during regulated exocytosis (Martin, 1994).

The Docking-Fusion Complex

The identification of several proteins as constituents of the exocytotic machinery has led to a unifying concept of the exoxytostic process itself (for reviews see Südhof *et al.*, 1993; Yamaguchi and Akagawa, 1994; Bajjalieh and Scheller, 1995; Südhof, 1995).

Syntaxin 1, synaptobrevin II and *SNAP-25* form a stable ternary complex in the absence of SNAP-NSF. The domains interacting in this complex are predominantly α-helical structures suggesting a coiled-coil structure and rendering the resulting complex resistant to SDS (Chapman *et al.*, 1994; Kee *et al.*, 1995; Oho *et al.*, 1995). This complex provides an additional binding site for synaptotagmin. The α-SNAP-NSF complex competes with *synaptotagmin* for the same binding moiety suggesting that, in the process of exocytosis, *synaptotagmin* is replaced by the α-SNAP-NSF which, ATP-dependently, disassembles the heterotrimeric SNARE-complex and initiates vesicle fusion (Hayashi *et al.*, 1994). This may be caused by changes in the conformation of *synaptotagmin* which reduce its affinity to the SNARE-complex, following binding of calcium ions. The function of synaptotagmin, then, would be that of a "clamp", inhibiting vesicle fusion in the absence of the calcium signal. Another hypothesis on regulation of exocytosis is based on the binding between *synaptobrevin* and *synaptophysins*, preventing the formation of the ternary docking-fusion complex (Edelmann *et al.*, 1995; Washbourne *et al.*, 1995). Furthermore the role of SNAP-25 in the formation of the complex is not clear (McMahon and Südhof, 1995).

Molecular Action of the CNTs

Rat synaptobrevin II (Schiavo *et al.*, 1992a; Link *et al.*, 1992) and cellubrevin (McMahon *et al.*, 1993; Link *et al.*, 1993) are hydrolysed by L-TeTx *in vitro*, whereas rat synaptobrevin I is resistant to the toxin. Cleavage was demonstrated by protein analysis of highly purified SSVs exposed to L-TeTx, followed by sequencing of synaptobrevin I and II. TeTx hydrolyses the peptide bond between Gln(76) and Phe(77) in synaptobrevin II. In cellubrevin the scissile bond is located between Gln(63) and Phe(64). In synaptobrevin I position 76 is occupied by valine, rendering this protein resistant to hydrolysis by Tetx. Synaptobrevin II is also a substrate for L-BoNT/B (East *et al.*, 1992), L-BoNT/D (Schiavo *et al.*, 1993a; Yamasaki *et al.*, 1994a), L-BoNT/ F (Schiavo *et al.*, 1993c; Yamasaki *et al.*, 1994a) and L-BoNT/G (Yamasaki *et al.*, 1994b). Interestingly, L-BoNT/B cleaves at the same site as does TeTx. BoNT/ D hydrolyses synaptobrevin II between positions Lys(59) and Leu(60), BoNT/F between Gln(58) and Lys(59) and BoNT/G between Ala(81) and Ala(82). All peptide bonds cleaved by the CNTs are located in the major cytoplasmic domain of the molecule. The toxins appeared to be proteolytic both in membrane-bound and in free recombinant synaptobrevin II. Studies on homologues and fragments of synaptobrevin II provided evidence that the CNTs mentioned above — unlike thermolysin — require at least 20–30 amino acid residues of the amino terminal part adjacent to the scissile bond to recognize their substrate. Similar to L-TeTx, L-BoNT/D and L-BoNT/F also hydrolyse cellubrevin in non-neuronal cells.

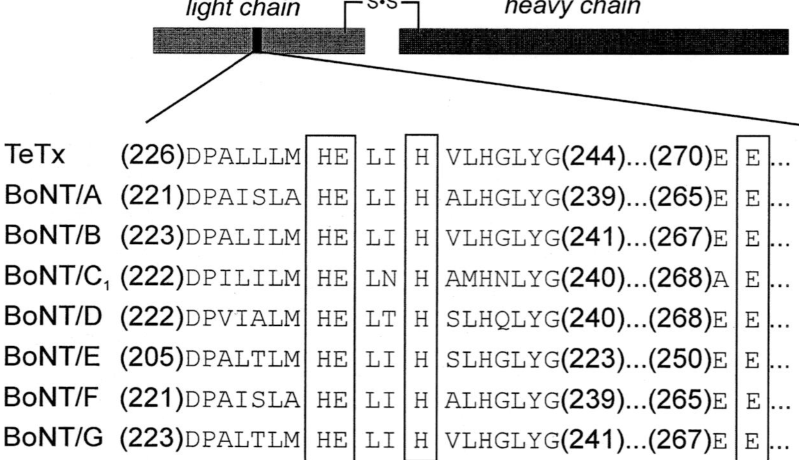

Figure 18.3 Comparison of the active sites of the CNTs
In addition to their structural similarities all eight CNTs share a histidine-rich motif in their L-chains, which is located within a highly conserved region of the molecule. The two histidine residues and the glutamate residue essential to toxicity are drawn in boxes. They bind one Zn^{2+} ion and an activated water molecule, thus resembling the active sites of other metalloendoproteases.

The t-SNARE *SNAP-25* is the target of L-BoNT/A (Blasi *et al.*, 1993; Binz *et al.*, 1994; Schiavo *et al.*, 1993b) and BoNT/E (Binz *et al.*, 1994; Schiavo *et al.*, 1993b), which cleave the molecule near its carboxy terminal between Gln(197) and Arg(198) and between Arg(180) and Ile(181), respectively, thus leaving a smaller molecule attached to the plasma membrane (Binz *et al.*, 1994). It is not clear whether this truncated molecule is completely nonfunctional. In some cellular assays involving neurosecretory cells, it has been demonstrated that BoNT/A can inhibit exocytosis only by approximately 50%. Furthermore, the toxin effect in chromaffin cells could partially be surmounted by excessive stimulation with carbachol or by elevated intracellular calcium concentrations (100 μM) during capacitance measurements. Perhaps the carboxy terminal fragment of SNAP-25, which remains attached to the membrane and encompasses 90% of the native molecule, is able to maintain a residuary function during exocytosis.

Both neuronal isoforms (A and B) of membrane-bound syntaxin are substrates for L-BoNT/C₁. The location of the scissile is between Lys(253) and Ala(254) for syntaxin A and between Lys(252) and Ala(253) for syntaxin B, respectively (Schiavo *et al.*, 1995). After cleavage a 4kD fragment remains inserted in the plasma membrane. Cleavage of syntaxin disrupts synaptic transmission in cultured neurons (Mochida *et al.*, 1995)

It is the current opinion that the cleavage of either synaptobrevin II, SNAP-25, or syntaxin is the molecular mechanism of the blockage of exocytosis induced by

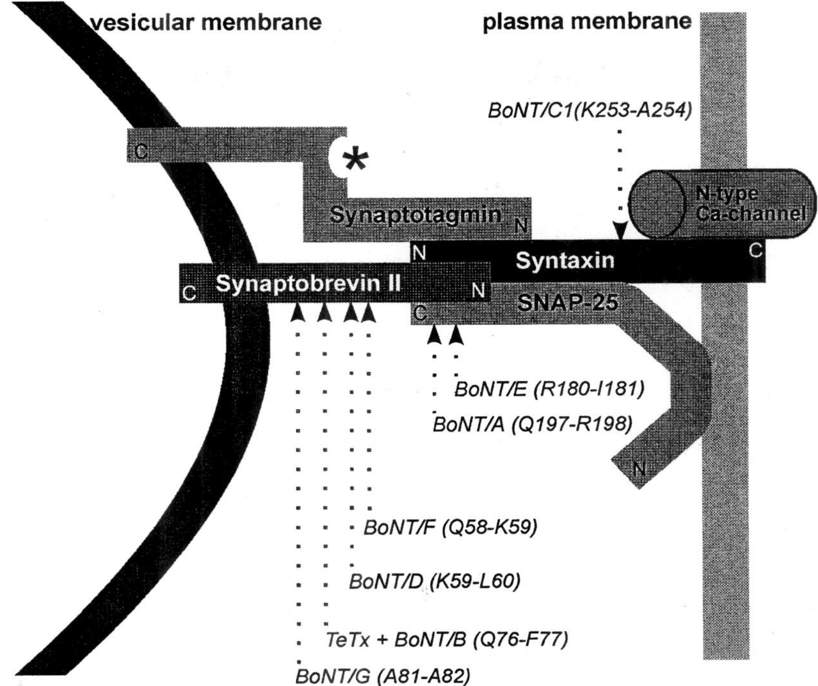

Figure 18.4   The docking-fusion complex and the cleavage sites of CNTs

According to the current hypothesis of exoctosis of SSVs and presumably LDCVs as well, the three proteins *synaptobrevin II, SNAP-25* and *syntaxin* form a specific ternary SDS-resistant core complex. Whereas *synaptobrevin II* serves as a *v-SNARE*, the latter are *t-SNAREs*. The complex formation is independent of the NSF-SNAP-complex and provides an additional binding site for *synaptotagmin*. This protein, which contains two homologous C2-domains, is anchored to the vesicular membranes of SSVs by means of its carboxy terminal. It has a binding site for calcium (*), which enters the cytosol through voltage-gated N-type calcium channels. These channel proteins closely interact with neurexins (not shown), *synaptotagmins* and *syntaxins*. It is assumed that SSVs in active zones are docked to the presynaptic membrane in this way and primed for the final step(s) of exocytotic release. These final steps could be initiated by the entry of calcium into the cell: Binding of calcium to *synaptotagmin* induces a conformational change reducing its affinity to the core complex. The *NSF-SNAP-complex* competes with *synaptotagmin* for the same binding side; therefore *synaptotagmin* is replaced by the latter. An ATP-dependent disassembly of the core complex by the *NSF-SNAP-complex* will finally lead to membrane fusion by an unknown mechanism.

The CNTs are potent inhibitors of neuronal exocytosis. They specifically hydrolyse distinct components of this docking-fusion complex. Whereas TeTx and BoNT/B, D, F and G cleave *synaptobrevin II* - both TeTx and BoNT/B sharing the same cleavage site — BoNT/A and BONT/F use *SNAP-25* as their substrate. *Syntaxin* is hydrolysed exclusively by BoNT/C₁. Scissile bonds in rat *synaptobrevin II* and *SNAP-25* are given in parenthesis. The molecular mechanism of the toxin action permits the conclusion that *synaptobrevin II* (cellubrevin in non-neuronal cells), *SNAP-25* and *syntaxin* play essential parts in the docking-fusion complex employed in neuronal exocytosis.

the various CNTs. The coincidence of cleavage and inhibited exocytosis has been reported from experiments involving *Aplysia californica*, squid neurons, neurosecretory and neuroendocrine cells and — with regard to cellubrevin — non-neuronal CHO cells. Although the latter fail to internalize TeTx, L-TeTx, delivered directly into the cytosol by overexpression of the encoding gene, cleaves cellubrevin and simultaneously reduces the density of transferrin receptors in the plasma membrane, hereby indicating an impairment of constitutive exocytosis. The direct link between the proteolysis of constituents of the exocytotic machinery and the blockade of exocytosis has been confirmed by the finding that IgA-protease from Neisseria gonorrhoe, introduced into chromaffin cells by electroporation, inhibits release of catecholamines (Binscheck *et al.*, 1995). IgA-protease cleaves synaptobrevin II, presumably at a proline-rich motif Pro(19)-Pro(20) ... Ala(21)-Pro(22) *in vitro.* Despite the fact that this cleavage site differs from those attacked by TeTx or the respective BoNTs, exocytosis is inhibited. So far, there is no evidence that exocytosis could be inhibited by the CNTs without cleavage of one of the proteins involved in the exocytotic process.

CLOSTRIDIAL NEUROTOXINS: PHARMACOLOGICAL AND CLINICAL TOOLS

The highly specific proteolysis of distinct proteins by the CNTs has allowed the identification of synaptobrevin, SNAP-25 and syntaxin as essential components of the docking-fusion complex. CNTs serve as tools to specifically dissect the molecular mechanism of exocytosis and have initiated successful efforts to assign functions to the distinct proteines involved. Since several proteins involved in regulated neuronal exocytosis also contribute to exocytosis in non-neuronal cells, the hypothesis has been advanced that regulated and constitutive exocytosis share many structural and functional features. From this, exocytosis may emerge as a universal cellular process independent of a particular cell type. The coding genes for L-TeTx were inserted in transgenic mice and Drosophila. Whereas in mice expression could only be detected in *Sertoli* cells and did not impair the nervous system, it caused behavioural defects in *Drosophila*, due to disrupted neuronal transmission (DiAntonio and Schwarz, 1994; Sweeney *et al.*, 1995). It has also been proposed to connect the gene to a promotor that is switched on only during certain periods of neuronal development. Thus, modulation of specific neuronal systems in the CNS might become possible.

Concerning the H-chains of the CNTs, they have already proven to be suitable carriers of substances which otherwise would not be able to gain access to the neuronal compartment and the synthesis of conjugates consisting of H-chain and other biologically active molecules is becoming widespread.

Since tetanus and botulism are long-lasting and life-threatening diseases, it is desirable to develop causative therapies based on the knowledge of the molecular action of CNTs. A hopeful approach would be to internalize specific inhibitors of metalloendoproteases (e.g. captopril) into neurons to stop the persistent attack of the L-chains on the exocytotic machinery. Captopril has prevented the action of L-TeTx in chromaffin cells (Hohne Zell *et al.*, 1994) and should be tested whether it is capable of curtailing the diseases. Since intracellularly applied antibodies, directed against the CNTs, are able to antagonize the toxic effect within a short period of time (Bartels *et al.*, 1994; Bartels and Bigalke, 1992), conjugates of H-chains and antibodies or their F(ab)$_2$-fragments might prove to be another therapeutic principle. The H-chains, which determine the specifity for neuronal cells, could, upon systemic application, direct the conjugate to the site of toxin action. Since they are non-toxic, the patients would not be exposed to additional risks.

The BoNTs, which block the release of acetylcholine particularly at the neuromuscular junction, are already used in the clinical treatment of several movement disorders (e.g. blepharospasm, hemifacial spasms, strabism, spasmodic torticollis and different focal dystonias) (Scott, 1989; Thomas *et al.*, 1993; Anonymous, 1991; Denislic *et al.*, 1994a). After injection close to the endplate region of the involved muscles during EMG-monitoring (Ostergaard *et al.*, 1994), a local flaccid, paralysis develops within several days. To prevent rapid BoNT degradation in the tissue, the haemagglutinin moiety must not be removed. After injection, the onset of the chemical denervation is delayed for some time until the toxin has diffused and reached the synaptic terminals (Borodic *et al.*, 1994). At least two problems arise from the therapeutic use of BoNTs. First, there is an unwanted spread of paralysis involving other muscle groups. If BoNTs are injected into the ocular muscles for treating blepharospasm and strabism, ptosis is a commom complication. After treatment of the sternocleidomastoid muscle a considerable number of patients experience dysphagia or even dysarthria, due to paralysis of the throat and larynx (Denislic *et al.*, 1994b). Unfortunately, there is no remedy for these unwanted effects. As a second complication, anti-BoNT antibodies are produced by some patients rendering the injected toxin ineffective. Since BoNT treatment is only symptomatic, patients have to undergo this therapy for years or even a life-time. The effect of BoNT-induced paralysis is limited to a few weeks, making regular injections inevitable. Various injection schemes have been tested in order to keep the number of patients with antibodies as low as possible. It turned out that the total number of injections received — but not the amount of toxin — is the crucial parameter for the immunologic response to BoNTs (Greene *et al.*, 1994). Fortunately, the antibodies raised against BoNTs, e.g. BoNT/A, which is the one used most often for clinical purposes, do not crossreact with other BoNTs. Therefore, if immunization has occurred, a different BoNT can be used to continue treatment (Greene and Fahn, 1993).

Acknowledgements

We would like to thank Dr. G. Erdmann and Dr. H. Bigalke for carefully reading the manuscript and helpful discussions.

REFERENCES

Anonymous. Clinical use of botulinum toxin. (1991) National Institutes of Health Consensus Development Conference Statement, November 12–14, 1990. *Arch. Neurol.*, **48**, pp. 1294–1298.

Ahnert Hilger, G., Bader, M.F., Bhakdi, S. and Gratzl, M. (1989a) Introduction of macromolecules into bovine adrenal medullary chromaffin cells and rat pheochromocytoma cells (PC12) by permeabilization with streptolysin O: inhibitory effect of tetanus toxin on catecholamine secretion. *J. Neurochem.*, **52**, pp. 1751–1758.

Ahnert Hilger, G., Weller, U., Dauzenroth, M.E., Habermann, E. and Gratzl, M. (1989b) The tetanus toxin light chain inhibits exocytosis. *FEBS Lett.*, **242**, pp. 245–248.

Ahnert-Hilger, G. and Bigalke, H. (1995) Molecular aspects of tetanus and botulinum neurotoxin poisoning. *Progress in Neurobiology* (in press).

Aktories, K., Just, I. and Rosenthal, W. (1988) Different types of ADP-ribose protein bonds formed by botulinum C2 toxin, botulinum ADP-ribosyltransferase C3 and pertussis toxin. *Biochem. Biophys. Res. Commun.*, **156**, pp. 361–367.

Aktories, K., Braun, U., Habermann, B. and Rösener, S. (1990) Botulinum ADP-Ribosyltransferase C3. In: *ADP-Ribosylating Toxins and G Proteins*, edited by Moss, J. and Vaughan, M. American Society for Microbiology, Washington, DC.

Aktories, K. and Just, I. (1990) Botulinum C2 Toxin. *In: ADP-Ribosylating Toxins and G Proteins*, edited by Moss, J. and Vaughan, M. American Society for Microbiology, Washington, DC.

Aktories, K. and Wegner, A. (1989) ADP-ribosylation of Actin by Clostridial Toxins. *J Cell Biol.*, **109**, pp. 1385–1387.

Andersen Beckh, B., Binz, T., Kurazono, H., Mayer, T., Eisel, U. and Niemann, H. (1989) Expression of tetanus toxin subfragments *in vitro* and characterization of epitopes. *Infect. Immun.*, **57**, pp. 3498–3505.

Ashton, A.C. and Dolly, J.O. (1988) Characterization of the inhibitory action of botulinum neurotoxin type A on the release of several transmitters from rat cerebrocortical synaptosomes. *J. Neurochem.*, **50**, pp. 1808–1816.

Bajjalieh, S.M. and Scheller, R.H. (1995) The biochemistry of neurotransmitter secretion. *J. Biol. Chem.*, **270**, pp. 1971–1974.

Bakry, N., Kamata, Y., Sorensen, R. and Simpson, L.L. (1991) Tetanus toxin and neuronal membranes: the relationship between binding and toxicity. *J. Pharmacol. Exp. Ther.*, **258**, pp. 613–619.

Bartels, F., Bergel, H., Bigalke, H., Frevert, J., Halpern, J. and Middlebrook, J. (1994) Specific antibodies against the Zn(2+)-binding domain of clostridial neurotoxins restore exocytosis in chromaffin cells treated with tetanus or botulinum A neurotoxin. *J. Biol. Chem.*, **269**, pp. 8122–8127.

Bartels, F. and Bigalke, H. (1992) Restoration of exocytosis occurs after inactivation of intracellular tetanus toxin. *Infect. Immun.*, **60**, pp. 302–307.

Beise, J., Hahnen, J. Andersen Beckh, B. and Dreyer, F. (1994) Pore formation by tetanus toxin, its chain and fragments in neuronal membranes and evaluation of the underlying motifs in the structure of the toxin molecule. *Naunyn Schmiedebergs Arch. Pharmacol.*, **349**, pp. 66–73.

Bergey, G.K., Habig, H.W. and Lin, C. (1986) Nicking of tetanus toxin increases activity. *Ann. Neurol.*, **20**, pp. 137–138.

Bergey, G.K., Bigalke, H. and Nelson, P.G. (1987) Differential effects of tetanus toxin on inhibitory and excitatory synaptic transmission in mammalian spinal cord neurons in culture: A presynaptic locus of action for tetanus toxin. *J. Physiol*, **57**, pp. 121–131.

Bergey, G.K., Habig, W.H., Bennett, J.I. and Lin, C.S. (1989) Proteolytic cleavage of tetanus toxin increases activity. *J. Neurochem.*, **53**, pp. 155–161.

Bergey, K.G., MacDonald, R.L., Habig, W.H., Hardegree, M.C. and Nelson, P.G. (1983) Tetanus toxin: Convulsant action on mouse spinal cord neurons in culture. *J. Neurosci.*, **3**, pp. 2310–2323.

Bigalke, H., Dimpfel, W. and Habermann, E. (1978) Suppression of ^{3}H-acetylcholine release from primary nerve cell cultures by tetanus and botulinum-A toxin. *Naunyn Schmiedeberg's Arch. Pharmacol.*, **303**, pp. 133–138.

Bigalke, H., Ahnert-Hilger, G. and Habermann, E. (1981a) Tetanus toxin and botulinum A toxin inhibit acetylcholine release from but not calcium uptake into brain tissue. *Naunyn Schmiedeberg's Archives of Pharmacology*, **316**, pp. 143–148.

Bigalke, H., Heller, I., Bizzini, B. and Habermann, E. (1981b) Tetanus toxin and botulinum A toxin inhibit release and uptake of various transmitters, as studied with particulate preparations from rat brain and spinal cord. *Naunyn Schmiedeberg's Archives of Pharmacology*, **316**, pp. 244–251.

Bigalke, H., Bergey, K.G. and Nelson, P.G. (1983) Inhibition of synaptic transmission by tetanus toxin in cultured mammalian spinal cord neurons. *Naunyn Schmiedeberg's Arch. Pharmacol.*, **S322**, pp. R 61.

Binscheck, T., Bartels, F., Bergel, H., Bigalke, H., Yamasaki, S., Hayashi, T., Niemann, H. and Pohlner, J. (1995) IgA protease from Neisseria gonorrhoeae inhibits exocytosis in bovine chromaffin cells like tetanus toxin. *J. Biol. Chem.*, **270**, pp. 1770–1774.

Binz, T., Kurazono, H., Popoff, M.R., Eklund, M.W., Sakaguchi, G., Kozaki, S., Krieglstein, K., Henschen, A., Gill, D.M. and Niemann, H. (1990a) Nucleotide sequence of the gene encoding Clostridium botulinum neurotoxin type D. *Nucleic. Acids. Res.*, **18**, p. 5556.

Binz, T., Kurazono, H., Wille, M., Frevert, J., Wernars, K. and Niemann, H. (1990b) The complete sequence of botulinum neurotoxin type A and comparison with other clostridial neurotoxins. *J. Biol. Chem.*, **265**, pp. 9153–9158.

Binz, T., Blasi, J., Yamasaki, S., Baumeister, A., Link, E., Südhof, T.C., Jahn, R. and Niemann, H. (1994) Proteolysis of SNAP-25 by types E and A botulinal neurotoxins. *J. Biol. Chem.*, **269**, pp. 1617–1620

Bittner, M.A., DasGupta, B.R. and Holz, R.W. (1989a) Isolated light chains of botulinum neurotoxins inhibit exocytosis. Studies in digitonin-permeabilized chromaffin cells. *J. Biol. Chem.*, **264**, pp. 10354–10360.

Bittner, M.A., Habig, W.H. and Holz, R.W. (1989b) Isolated light chain of tetanus toxin inhibits exocytosis: studies in digitonin-permeabilized cells. *J. Neurochem.*, **53**, pp. 966–968.

Bittner, M.A. and Holz, R.W. (1988) Effects of tetanus toxin on catecholamine release from intact and digitonin-permeabilized chromaffin cells. *J. Neurochem.*, **51**, pp. 451–456.

Bizzini, B., Grob, P. and Akert, K. (1981) Papain derived fragment II$_c$ of tetanus toxin: Its binding to isolated and retrograd axonal transport. *Brain Res.*, **210**, pp. 291–299.

Bizzini, B. (1989) Axonal transport and transsynaptic movement of tetanus toxin. In: *Botulinum Neurotoxin and Tetanus Toxin*, edited by Simpson, L.L. Academic Press, New York.

Blasi, J., Chapman, E.R., Link, E., Binz, T., Yamasaki, S., De Camilli, P., Südhof, T.C., Niemann, H. and Jahn, R. (1993) Botulinum neurotoxin A selectively cleaves the synaptic protein SNAP-25 [see comments]. *Nature*, **365**, pp. 160–163.

Bleck, T.P. (1989) Clinical Aspects of tetanus. In: *Botulinum Neurotoxin and Tetanus Toxin*, edited by Simpson, L.L. New York, Academic Press.

Boquet, P. and Duflot, E. (1982) Tetanus toxin forms channels in lipid vesicles at low pH. *Proc. Natl. Acad. Sci. USA*, **79**, pp. 7614–7618.

Borodic, G.E., Ferrante, R., Pearce, L.B. and Smith, K. (1994) Histologic assessment of dose-related diffusion and muscle fiber response after therapeutic botulinum A toxin injections. *Mov. Disord.*, **9**, pp. 31–39.

Burgen, A.S.V., Dichens, F. and Zatman, L.J. (1949) The action of botulinum toxin on the neuromuscular junctions. *Journal of Physiology*, **109**, pp. 10–24.

Burningham, M.D., Walter, F.G., Mechem, C., Haber, J. and Ekins, B.R. (1994) Wound botulism. *Ann. Emerg. Med.*, **24**, pp. 1184–1187.

Büttner-Ennever, J.A., Grob, P., Akert, K. and Bizzini, B. (1981) Transsynaptic retrograde labelling in the oculomotor system of the monkey with (^{125}I) tetanus toxin B-II$_b$ fragment. *Neurosci. Lett.*, **26**, pp. 233–238.

Campbell, K., Collins, M.D. and East, A.K. (1993) Nucleotide sequence of the gene coding for Clostridium botulinum (Clostridium argentinense) type G neurotoxin: genealogical comparison with other clostridial neurotoxins. *Biochim. Biophys. Acta*, **1216**, pp. 487–491.

Chapman, E.R., An, S., Barton, N. and Jahn, R. (1994) SNAP-25, a t-SNARE which binds to both syntaxin and synaptobrevin via domains that may form coiled coils. *J. Biol. Chem.*, **269**, pp. 27427–27432.

Chilcote, T.J., Galli, T., Mundigl, O., Edelmann, L., McPherson, P.S., Takei, K. and De Camilli, P. (1995) Cellubrevin and synaptobrevins: similar subcellular localization and biochemical properties in PC12 cells. *J. Cell Biol.*, **129**, pp. 219–231.

Critchley, D.R., Nelson, P.G., Habig, H.W. and Fishman, P.H. (1985) Fate of tetanus toxin bound to the surface of primary neurons in culture: evidence for rapid internalization. *J. Cell Biol.*, **100**, pp. 1499–1507.

Critchley, D.R., Habig, H.W. and Fishman, P.H. (1986) Reevaluation of the role of gangliosides as receptors for tetanus toxin. *J. Neurochem*, **47**, pp. 213–222.

Critchley, D.R., Parton, R.G., Davison, M.D. and Pierce, E.J. (1988) Characterization of tetanus toxin binding by neuronal tissue. In: *Neurotoxins in neurochemistry*, edited by Dolly, J.O. Wiley and Sons, London.

DasGupta, B.R. and Datta, A. (1988) Botulinum neurotoxin type B (strain 657): partial sequence and similarity with tetanus toxin. *Biochimie*, **70**, pp. 811–817

DasGupta, B.R. and Rasmussen, S. (1984) Effects of diethylpyricarbonate on the biological activities of botulinum neurotoxins types A and E. *Arch. Biochem. Biophys.*, **232**, pp. 172–178.

Dayanithi, G., Ahnert Hilger, G., Weller, U., Nordmann, J.J. and Gratzl, M. (1990) Release of vasopressin from isolated permeabilized neurosecretory nerve terminals is blocked by the light chain of botulinum A toxin. *Neuroscience*, **39**, pp. 711–715.

Dayanithi, G., Weller, U., Ahnert Hilger, G., Link, H., Nordmann, J.J. and Gratzl, M. (1992) The light chain of tetanus toxin inhibits calcium-dependent vasopressin release from permeabilized nerve endings. *Neuroscience*, **46**, pp. 489–493.

Dayanithi, G., Stecher, B., Hohne Zell, B., Yamasaki, S., Binz, T., Weller, U., Niemann, H. and Gratzl, M. (1994) Exploring the functional domain and the target of the tetanus toxin light chain in neurohypophysial terminals. *Neuroscience*, **58**, pp. 423–431.

De Filippis, V., Vangelista, L., Schiavo, G., Tonello, F. and Montecucco, C. (1995) Structural studies on the zinc-endopeptidase light chain of tetanus neurotoxin. *Eur. J. Biochem.*, **229**, pp. 61–69.

de Paiva, A., Ashton, A.C., Foran, P., Schiavo, G., Montecucco, C. and Dolly, J.O. (1993) Botulinum A like type B and tetanus toxins fulfils criteria for being a zinc-dependent protease. *J. Neurochem.*, **61**, pp. 2338–2341.

de Paiva, A. and Dolly, J.O. (1990) Light chain of botulinum neurotoxin is active in mammalian motor nerve terminals when delivered via liposomes. *FEBS Lett.*, **277**, pp. 171–174.

Denislic, M., Pirtosek, Z., Vodusek, D.B., Zidar, J. and Meh, D. (1994a) Botulinum toxin in the treatment of neurological disorders. *Ann. NY Acad. Sci.*, **710**, pp. 76–87.

Denislic, M., Pirtosek, Z., Vodusek, D.B., Zidar, J. and Meh, D. (1994b) Botulinum toxin in the treatment of neurological disorders. *Ann. NY Acad. Sci.*, **710**, pp. 76–87.

DiAntonio, A. and Schwarz, T.L. (1994) The effect on synaptic physiology of synaptotagmin mutations in Drosophila. *Neuron*, **12**, pp. 909–920.

DiMaggio, D.A., Farah, J.M., Jr. and Westfall, T.C. (1994) Effects of differentiation on neuropeptide-Y receptors and responses in rat pheochromocytoma cells. *Endocrinology*, **134**, pp. 719–727.

Dimpfel, W., Huang, R.T.C. and Habermann, E. (1977) Gangliosides in nerve tissue cultures and binding of ^{125}I-labelled tetanus toxin, a neuronal marker. *J. Neurochem.*, **29**, pp. 329–334.

Dolly, J.O., Black, J.D., Williams, R.S. and Melling, J. (1984) Acceptors for botulinum neurotoxin reside on motor nerve terminals and mediate its internalization. *Nature*, **307**, pp. 457–460.

Dolly, J.O., Lande, S. and Wray, D.W. (1987) The effects of *in vitro* application of purified botulinum neurotoxin at mouse motor nerve terminals. *J. Physiol. (London)*, **386**, pp. 475–484.

Donovan, J.J. and Simpson, L.L. (1986) Ion-conducting channels produced by botulinum toxin in planar lipid membranes. *Biochemistry*, **25**, pp. 2872–2876.

Dreyer, F., Mallart, A. and Brigant, J.L. (1983) Botulinum A toxin and tetanus toxin do not effect presynaptic membrane current in mammalian nerve endings. *Brain Res.*, **270**, pp. 373–375.

Dreyer, F., Rosenberg, F., Becker, C., Bigalke, H. and Penner, R. (1987) Differential effects of various secretagogues on quantal transmitter release from mouse motor nerve terminals treated with botulinum A and tetanus toxin. *Naunyn Schmiedeberg's Archives of Pharmacology*, **335**, pp. 1–7.

East, A.K., Richardson, P.T., Allaway, D., Collins, M.D., Roberts, T.A. and Thompson, D.E. (1992) Sequence of the gene encoding type F neurotoxin of Clostridium botulinum. *FEMS Microbiol. Lett.*, **75**, pp. 225–230.

East, A.K., Thompson, D.E. and Collins, M.D. (1992) Analysis of operons encoding 23S rRNA of Clostridium botulinum type A. *J. Bacteriol.*, **174**, pp. 8158–8162.

Edelmann, L., Hanson, P.I., Chapman, E.R. and Jahn, R. (1995) Synaptobrevin binding to synaptophysin: a potential mechanism for controlling the exocytotic fusion machine. *EMBO J.*, **14**, pp. 224–231.

Eisel, U., Jarausch, W., Goretzki, K., Henschen, A., Engels, J., Weller, U., Hudel, M., Habermann, E. and Niemann, H. (1986) Tetanus toxin: primary structure, expression in E. coli and homology with botulinum toxins. *EMBO J.*, **5**, pp. 2495–2502.

Eklund, M.W., Poysky, F.T., Mseitif, L.M. and Strom, M.S. (1988) Evidence for plasmid-mediated toxin and bacteriocin production in Clostridium botulinum type G. *Appl. Environ. Microbiol.*, **54**, pp. 1405–1408.

Erdal, E., Bartels, F., Binscheck, T., Erdmann, G., Frevert, J., Kistner, A., Weller, U., Wever, J. and Bigalke, H. (1995) Processing of tetanus and botulinum A neurotoxins in isolated chromaffin cells. *Naunyn Schmiedeberg's Archives of Pharmacology*, **in press**.

Erdmann, G. and Habermann, E. (1977) Histoautoradiography of the central nervous system in rats with generalized tetanus due to ^{125}I-toxin. *Naunyn Schmiedeberg's Arch. Pharmacol.*, **301**, pp. 135–138.

Evans, D.M., Williams, R.S., Shone, C.S., Hambleton, P., Melling, J. and Dolly, J.O. (1986) Botulinum neurotoxin type B. Its purification, radioiodination and interaction with rat-brain synaptosomal membranes. *Eur. J. Biochem.*, **154**, pp. 409–416.

Facchiano, F., Benfenati, F., Valtorta, F. and Luini, A. (1993a) Covalent modification of synapsin I by a tetanus toxin-activated transglutaminase. *J. Biol. Chem.*, **268**, pp. 4588–4591.

Facchiano, F., Valtorta, F., Benfenati, F. and Luini, A. (1993b) The transglutaminase hypothesis for the action of tetanus toxin. *Trends. Biochem. Sci.*, **18**, pp. 327–329.

Fairweather, N.F., Lyness, V.A., Pickard, D.J., Allen, G. and Thomson, R.O. (1986) Cloning, nucleotide sequencing and expression of tetanus toxin fragment C in *Escherichia coli*. *J. Bacteriol.*, **165**, pp. 21–25.

Fairweather, N.F., Lyness, V.A., Pickard, D.J., Allen, G. and Thomson, R.O. (1987) Immunization of mice against tetanus with fragments of tetanus toxin fragment C synthesized in Escherichia coli. *Infect. Immun.*, **55**, pp. 2541–2545.

Fairweather, N.F., Sanders, D., Slater, D., Hudel, M., Habermann, E. and Weller, U. (1993) Production of biologically active light chain of tetanus toxin in *Escherichia coli*. Evidence for the importance of the C-terminal 16 amino acids for full biological activity. *FEBS Lett.*, **323**, pp. 218–222.

Fairweather, N.F. and Lyness, V.A. (1986) The complete nucleotide sequence of tetanus toxin. *Nucleic Acids Res.*, **14**, pp. 7809–7812.

Fairweather, N.F. and Lyness, V.A. (1986) The complete nucleotide sequence of tetanus toxin. *Nucleic Acids Res.*, **14**, pp. 7809–7812.

Fenicia, L., Ferrini, A.M., Aureli, P. and Pocecco, M. (1993) A case of infant botulism associated with honey feeding in Italy. *Eur. J. Epidemiol.*, **9**, pp. 671–673.

Finn, C.W.J., Silver, R.P., Habig, W.H., Hardegree, M.C., Zon, G. and Garon, C.F. (1984) The structural gene for tetanus neurotoxin is on a plasmid. *Science*, **224**, pp. 881–884.

Fishman, P.S. and Savitt, J.M. (1989) Transsynaptic transfer of retrogradely transported tetanus protein-peroxidase conjugates. *Exp. Neurol.*, **106**, pp. 197–203.

Fujii, N., Kimura, K., Yashiki, T., Tsuzuki, K., Moriishi, K., Yokosawa, N., Syuto, B. and Oguma, K. (1992) Cloning and whole nucleotide sequence of the gene for the light chain component of botulinum type E toxin from Clostridium butyricum strain BL6340 and Clostridium botulinum type E strain Mashike. *Microbiol. Immunol.*, **36**, pp. 213–220.

Gambale, F. and Montal, M. (1988) Characterization of the channel properties of tetanus toxin in planar lipid bilayers. *Biophys. J.*, **53**, pp. 771–783.

Gimenez, J.A., Gimenez, M.A. and DasGupta, B.R. (1992) Characterization of the neurotoxin isolated from a Clostridium baratii strain implicated in infant botulism. *Infect. Immun.*, **60**, pp. 518–522.

Greene, P., Fahn, S. and Diamond, B. (1994) Development of resistance to botulinum toxin type-A in patients with torticollis. *Mov. Disord*, **9(N2)**, 217.

Greene, P.E. and Fahn, S. (1993) Use of botulinum toxin type F injections to treat torticollis in patients with immunity to botulinum toxin type A. *Mov. Disord.*, **8**, pp. 479–483.

Habermann, E. (1978) Tetanus. In: *Handbook of clinical neurology: Infections of the nervous system*, edited by Vinken, P.J. and Bruyn, G.W. North Holland Publishing Company, Amsterdam/NewYork/Oxford.

Habermann, E., Dreyer, F. and Bigalke, H. (1980) Tetanus toxin blocks the neuromuscular transmission *in vitro* like botulinum A toxin. *Naunyn Schmiedeberg's Archives of Pharmacology*, **311**, pp. 33–40.

Habermann, E., Muller, H. and Hudel, M. (1988) Tetanus toxin and botulinum A and C neurotoxins inhibit noradrenaline release from cultured mouse brain. *J. Neurochem.*, **51**, pp. 522–527.

Habermann, E. (1989) Clostridial neurotoxins and the central nervous system: Functional studies on isolated preparations. In: *Botulinum Neurotoxin and Tetanus Toxin*, edited by Simpson, L.L. Academic Press, New York.

Habermann, E., Weller, U. and Hudel, M. (1991) Limited proteolysis of single-chain tetanus toxin by tissue enzymes, in cultured brain tissue and during retrograde axonal to the spinal cord. *Naunyn Schmiedebergs Arch. Pharmacol.*, **343**, pp. 323–329.

Habermann, E. and Dreyer, F. (1986a) Clostridial neurotoxins: Handling and action at the cellular and molecular level. *Current Topics in Microbiology and Immunology*, **129**, pp. 93–179.

Habermann, E. and Dreyer, F. (1986b) Clostridial Neurotoxins: Handling and action at the cellular and molecular level. *Curr. Top. Microbiol. Immunol.*, **129**, pp. 93–179.

Habermann, E. and Heller, I. (1975) Direct evidence for the specific fixation of Cl. botulinum A neurotoxin to brain matter. *Naunyn Schmiedeberg's Arch. Pharmacol.*, **287**, p. 9.

Habig, W.H., Bigalke, H., Bergey, G.K., Neale, E.A., Hardegree, M.C. and Nelson, P.G. (1986) Tetanus toxin in dissociated spinal cord cultures: Long-term characterization of form and action. *Journal of Neurochemistry*, **47**, pp. 930–937.

Hagenah, R., Benecke, R. and Wiegand, H. (1977) Effects of Type Botulinum Toxin on the Cholinergic Transmission at Spinal Renshaw Cells and on the Inhibitory Action at Ia Inhibitory Interneurons. *Naunyn Schmiedebergs Arch. Pharmacol*, **299**, pp. 367–272.

Hall, J.D., McCroskey, L.M., Pincomb, B.J. and Hatheway, C.L. (1985) Isolation of an organism resembling *Clostridium baratii* which produced type F botulinal toxin from an infant with botulism. *Clin. Microbiol.*, **21**, pp. 654–655.

Halpern, J.L., Habig, W.H., Neale, E.A. and Stibitz, S. (1990) Cloning and expression of functional fragment C of tetanus toxin. *Infect. Immun.*, **58**, pp. 1004–1009.

Halpern, J.L. and Loftus, A. (1993) Characterization of the receptor-binding domain of tetanus toxin. *J. Biol. Chem.*, **268**, pp. 11188–11192.

Hauser, D., Eklund, M.W., Kurazono, H., Binz, T., Niemann, H., Gill, D.M., Boquet, P. and Popoff, M.R. (1990) Nucleotide sequence of Clostridium botulinum C1 neurotoxin. *Nucleic. Acids. Res.*, **18**, p. 4924.

Hayashi, T., McMahon, H., Yamasaki, S., Binz, T., Hata, Y., Südhof, T.C. and Niemann, H. (1994) Synaptic vesicle membrane fusion complex: action of clostridial neurotoxins on assembly. *EMBO J.*, **13**, pp. 5051–5061.

Hoch, D.H., Romera-Mira, M., Ehrlich, B.E., Finkelstein, A., DasGupta, B.R. and Simpson, L.L. (1985) Channels formed by botulinum, tetanus and diphteria toxins in planar lipid bilayers: relevance to translocation of proteins across membranes. *Proc. Natl. Acad. Sci.*, **82**, pp. 1692–1696.

Hohne Zell, B., Ecker, A., Weller, U. and Gratzl, M. (1994) Synaptobrevin cleavage by the tetanus toxin light chain is linked to the inhibition of exocytosis in chromaffin cells. *FEBS Lett.*, **355**, pp. 131–134.

Holmgren, J., Elwing, H., Fredman, P. and Svennerholm, L. (1980) Polystyrene-adsorbed gangliosides for investigations of the tetanus and cholera toxin receptor. *Eur. J. Biochem.*, **106**, pp. 371–379.

Hooper, N.M. (1994) Families of zinc metalloproteases. *FEBS Lett.*, **354**, pp. 1–6.

Högy, B., Dauzenroth, M.E., Hudel, M., Weller, U. and Habermann, E. (1992) Increase of permeability of synaptosomes and liposomes by the heavy chain of tetanus toxin. *Toxicon*, **30**, pp. 63–76.

Jimenez, R.R., Lopez, M.G., Sancho, C., Maroto, R. and Garcia, A.G. (1993) A component of the catecholamine secretory response in the bovine adrenal gland is resistant to dihydropyridines and omega-conotoxin. *Biochem. Biophys. Res. Commun.*, **191**, pp. 1278–1283.

Johnstone, S.A., Hubaishy, I. and Waisman, D.M. (1992) Phosphorylation of annexin II tetramer by protein kinase C inhibits aggregation of lipid vesicles by the protein. *J. Biol. Chem.*, **267**, pp. 25976–25981.

Jongeneel, C.V., Bouvier, J. and Bairoch, A. (1989) A unique signature identifies a family of zinc-dependent metallopeptidases. *FEBS Lett.*, **242**, pp. 211–214.

Kee, Y., Lin, R.C., Hsu, S.C. and Scheller, R.H. (1995) Distinct domains of syntaxin are required for synaptic vesicle fusion complex formation and dissociation. *Neuron*, **14**, pp. 991–998.

Kimura, K., Fujii, N., Tsuzuki, K., Murakami, T., Indoh, T., Yokosawa, N. and Oguma, K. (1991) Cloning of the structural gene for Clostridium botulinum type C1 toxin and whole nucleotide sequence of its light chain component. *Appl. Environ. Microbiol.*, **57**, pp. 1168–1172.

Kistner, A., Sanders, D. and Habermann, E. (1993) Disulfide formation in reduced tetanus toxin by thioredoxin: the pharmacological role of interchain covalent and noncovalent bonds. *Toxicon*, **31**, pp. 1423–1434.

Kistner, A. and Habermann, E. (1992) Reductive cleavage of tetanus and botulinum neurotoxin by the thioredoxin system from brain. *Naunyn Schmiedeberg's Archives of Pharmacology*, **345**, pp. 227–234.

Kitamura, M. (1976) Binding of botulinum neurotoxin to the synaptosome fraction of rat brain. *Naunyn Schmiedeberg's Arch. Pharmacol.*, **295**, pp. 171–175.

Kitamura, M., Iwamori, M. and Nagai, Y. (1980) Interaction between Clostridium botulinum neurotoxin and gangliosides. *Biochim. Biophys. Acta*, **628**, pp. 328–335.

Knight, D.E., Tonge, D.A. and Baker, P.F. (1985) Inhibition of exocytosis in bovine adrenal medullary cells by botulinum toxin type D. *Nature*, **317**, pp. 719–721.

Knight, D.E. (1986) Botulinum toxins types A, B and D inhibit catecholamine secretion from bovine adrenal medullary cells. *FEBS Lett.*, **207**, pp. 222–226.

Krieglstein, K., Henschen, A., Weller, U. and Habermann, E. (1990) Arrangement of disulfide bridges and positions of sulfhydryl groups in tetanus toxin. *Eur. J. Biochem.*, **188**, pp. 39–45.

Krieglstein, K.G., Henschen, A.H., Weller, U. and Habermann, E. (1991) Limited proteolysis of tetanus toxin. *Eur. J. Biochem.*, **343**, pp. 41–51.

Kurazono, H., Mochida, S., Binz, T., Eisel, U., Quanz, M., Grebenstein, O., Wernars, K., Poulain, B., Tauc, L. and Niemann, H. (1992) Minimal essential domains specifying toxicity of the light chains of tetanus toxin and botulinum neurotoxin type A. *J. Biol. Chem.*, **267**, pp. 14721–14729.

Laird, J.W., Aaronson, W., Silver, R.P., Habig, W.H. and Hardegree, M.C. (1980) Plasmid-associated toxigenicity in Clostridium tetani. *J. Infect. Dis.*, **142**, pp. 623.

Lazarovici, P., Yanai, P. and Yavin, E. (1987) Molecular interactions between micelar polysialoganglio-sides and affinity-purified tetanotoxin in aqueous solution. *J. Biol. Chem.*, **262**, pp. 2645–2651.

Lazarovici, P. and Yavin, E. (1986) Affinity purified tetanus neurotoxin interaction with synaptic membranes: Properties of a protease-sensitive receptor component. *Biochemistry*, **25**, pp. 7047–7054.

Link, E., Edelmann, L., Chou, J.H., Binz, T., Yamasaki, S., Eisel, U., Baumert, M., Südhof, T.C., Niemann, H. and Jahn, R. (1992) Tetanus toxin action: inhibition of neurotransmitter release linked to synaptobrevin proteolysis. *Biochem. Biophys. Res. Commun.*, **189**, pp. 1017–1023.

Link, E., McMahon, H., Fischer von Mollard, G., Yamasaki, S., Niemann, H., Südhof, T.C. and Jahn, R. (1993) Cleavage of cellubrevin by tetanus toxin does not affect fusion of early endosomes. *J. Biol. Chem.*, **268**, pp. 18423–18426.

Lomneth, R., Martin, T.F. and DasGupta, B.R. (1991) Botulinum neurotoxin light chain inhibits nor-epinephrine secretion in PC12 cells at an intracellular membranous or cytoskeletal site. *J. Neurochem.*, **57**, pp. 1413–1421.

Makoff, A.J., Oxer, M.D., Romanos, M.A., Fairweather, N.F. and Ballantine, S. (1989) Expression of tetanus toxin fragment C in *E. coli*: high level expression by removing rare codons. *Nucleic. Acids. Res.*, **17**, pp. 10191–10202.

Martin, T.F. (1994) The molecular machinery for fast and slow neurosecretion. *Curr. Opin. Neurobiol.*, **4**, pp. 626–632.

Marxen, P., Fuhrmann, U. and Bigalke, H. (1989) Gangliosides mediate inhibitory effects of tetanus and botulinum A neurotoxins on exocytosis in chromaffin cells. *Toxicon*, **27**, pp. 849–859.

Marxen, P. and Bigalke, H. (1989) Tetanus toxin: Inhibitory action in chromaffin cells is initiated by specific types of gangliosides and promoted in low ionic strenth solution. *Neuroscience Letters*, **107**, pp. 261–266.

McMahon, H.T., Foran, P., Dolly, J.O., Verhage, M., Wiegant, V.M. and Nicholls, D.G. (1992) Tetanus toxin and botulinum toxins type A and B inhibit glutamate, gamma-aminobutyric acid, aspartate and met-enkephalin release from synaptosomes. Clues to the locus of action. *J. Biol. Chem.*, **267**, pp. 21338–21343.

McMahon, H.T., Ushkaryov, Y.A., Edelmann, L., Link, E., Binz, T., Niemann, H., Jahn, R. and Südhof, T.C. (1993) Cellubrevin is a ubiquitous tetanus-toxin substrate homologous to a putative synaptic vesicle fusion protein [see comments]. *Nature*, **364**, pp. 346–349.

McMahon, H.T. and Südhof, T.C. (1995) Synaptic core complex of synaptobrevin, syntaxin and SNAP25 forms high affinity alpha-SNAP binding site. *J. Biol. Chem.*, **270**, pp. 2213–2217.

Mochida, S., Poulain, B., Weller, U., Habermann, E. and Tauc, L. (1989) Light chain of tetanus toxin intracellularly inhibits acetylcholine release at neuro-neuronal synapses and its internalization is mediated by heavy chain. *FEBS Lett.*, **253**, pp. 47–51.

Mochida, S., Poulain, B., Eisel, U., Binz, T., Kurazono, H., Niemann, H. and Tauc, L. (1990a) Molecular biology of clostridial toxins: Expression of mRNA encoding tetanus and botulinum neurotoxins in Aplysia neurons. *J. Physiol. (Paris)*, **84**, pp. 278–284.

Mochida, S., Poulain, B., Eisel, U., Binz, T., Kurazono, H., Niemann, H. and Tauc, L. (1990b) Exogenous mRNA encoding tetanus or botulinum neurotoxins expressed in Aplysia neurons. *Proc. Natl. Acad. Sci. USA*, **87**, pp. 278–284, pp. 7844–7848.

Mochida, S., Saisu, H., Kobayashi, H. and Abe, T. (1995) Impairment of syntaxin by botulinum neurotoxin C1 or antibodies inhibits acetylcholine release but not Ca^{2+} channel activity. *Neuroscience*, **65**, pp. 905–915.

Montal, M.S., Blewitt, R., Tomich, J.M. and Montal, M. (1992) Identification of an ion channel-forming motif in the primary structure of tetanus and botulinum neurotoxins. *FEBS Lett.*, **313**, pp. 12–18.

Montecucco, C., Schiavo, G., Gao, Z., Bauerlein, E., Boquet, P. and DasGupta, B.R. (1988) Interaction of botulinum and tetanus toxins with the lipid bilayer surface. *Biochem. J.*, **251**, pp. 379–383.

Montecucco, C. and Schiavo, G. (1993) Tetanus and botulism neurotoxins: a new group of zinc proteases. *Trends. Biochem. Sci.*, **18**, pp. 324–327.

Niemann, H. (1991) Molecular Biology of Clostridial Neurotoxins. In: *Sourcebook of Bacterial Protein Toxins*, edited by Alouf, J.E. and Freer, J.H. Academic Press, London.

Nishiki, T., Kamata, Y., Nemoto, Y., Omori, A., Ito, T., Takahashi, M. and Kozaki, S. (1994) Identification of protein receptor for Clostridium botulinum type B neurotoxin in rat brain synaptosomes. *J. Biol. Chem.*, **269**, pp. 10498–10503.

Oho, C., Seino, S. and Takahashi, M. (1995) Expression and complex formation of soluble N-ethylmaleimide-sensitive factor attachment protein (SNAP) receptors in clonal rat endocrine cells. *Neurosci. Lett.*, **186**, pp. 208–210.

Osborne, R. and Bradford, H. (1973) Tetanus toxin inhibits amino acid release from nerve endings *in vitro. Nature, New Biology*, **244**, pp. 157–158

Ostergaard, L., Fuglsang Frederiksen, A., Werdelin, L., Sjo, O. and Winkel, H. (1994) Quantitative EMG in botulinum toxin treatment of cervical dystonia. A double-blind, placebo-controlled study. *Electroencephalogr. Clin. Neurophysiol.*, **93**, pp. 434–439.

Penner, R., Dreyer, F. and Neher, E. (1986) Intracellularly injected tetanus toxin inhibits exocytosis in bovine adrenal chromaffin cells. *Nature*, **324**, pp. 76–77.

Pierce, E.J., Davison, M.D., Parton, R.G., Habig, H.W. and Critchley, D.R. (1986) Characterization of tetanus toxin binding to rat brain membranes. *Biochem. J.*, **236**, pp. 845–852.

Poulain, B., Mochida, S., Wadsworth, J.D.F., Weller, U., Habermann, E., Dolly, J.O. and Tauc, L. (1990) Inhibition of transmitter release by botulinum neurotoxins and tetanus toxin at Aplysia synapses: Role of the constituent chains. *J. Physiol. (Paris)*, **84**, pp. 247–261.

Presek, P., Jessen, S., Dreyer, F., Jarvie, P.E., Findik, D. and Dunkley, P.R. (1992) Tetanus toxin inhibits depolarization-stimulated protein phosphorylation in rat cortical synaptosomes: effect on synapsin I phosphorylation and translocation. *J. Neurochem.*, **59**, pp. 1336–1343.

Rabasseda, X., Blasi, J., Marsal, J., Dunant, Y., Casanova, A. and Bizzini, B. (1988) Tetanus and botulinum toxins block the release of acetylcholine from slices of rat striatum and from the isolated electric organ of Torpedo at different concentrations. *Toxicon*, **26**, pp. 329–336.

Rauch, G., Gambale, F. and Montal, M. (1990) Tetanus toxin channel in phosphatidylserine planar bilayers: conductance states and pH dependence. *Eur. Biophys. J.*, **18**, pp. 79–83.

Roa, M. and Boquet, P. (1985) Interaction of Tetanus Toxin with Lipid Vesicles at Low pH. *J. Biol. Chem.*, **260**, **(No. 11)**, pp. 6827–6835.

Roth, D. and Burgoyne, R.D. (1994) SNAP-25 is present in a SNARE complex in adrenal chromaffin cells. *FEBS Lett.*, **351**, pp. 207–210.

Sadoul, K., Lang, J., Montecucco, C., Weller, U., Regazzi, R., Catsicas, S., Wollheim, C.B. and Halban, P.A. (1995) SNAP-25 is expressed in islets of Langerhans and is involved in insulin release. *J. Cell Biol.*, **128**, pp. 1019–1028.

Sathyamoorthy, V., DasGupta, B.R., Foley, J. and Niece, R.L. (1988) Botulinum neurotoxin type A: cleavage of the heavy chain into two halves and their partial sequences. *Arch. Biochem. Biophys.*, **266**, pp. 142–151.

Sathyamoorthy, V. and DasGupta, B.R. (1985) Separation, purification, partial characterization and comparison of the heavy and light chains od botulinum neurotoxins types A, B and E. *Biol. Chem.*, **260**, pp. 10461–10466.

Schiavo, G., Papini, E., Genna, G. and Montecucco, C. (1990) An intact interchain disulfide bond is required for the neurotoxicity of tetanus toxin. *Infect. Immun.*, **58**, pp. 4136–4141.

Schiavo, G., Benfenati, F., Poulain, B., Rossetto, O., Polverino de Laureto, P., DasGupta, B.R. and Montecucco, C. (1992a) Tetanus and botulinum-B neurotoxins block neurotransmitter release by proteolytic cleavage of synaptobrevin [see comments]. *Nature*, **359**, pp. 832–835.

Schiavo, G., Poulain, B., Rossetto, O., Benfenati, F., Tauc, L. and Montecucco, C. (1992b) Tetanus toxin is a zinc protein and its inhibition of neurotransmitter release and protease activity depend on zinc. *EMBO J.*, **11**, pp. 3577–3583.

Schiavo, G., Rossetto, O., Santucci, A., DasGupta, B.R. and Montecucco, C. (1992c) Botulinum neurotoxins are zinc proteins. *J. Biol. Chem.*, **267**, pp. 23479–23483.

Schiavo, G., Rossetto, O., Catsicas, S., Polverino de Laureto, P., DasGupta, B.R., Benfenati, F. and Montecucco, C. (1993a) Identification of the nerve terminal targets of botulinum neurotoxin serotypes A, D and E. *J. Biol. Chem.*, **268**, pp. 23784–23787.

Schiavo, G., Santucci, A., DasGupta, B.R., Mehta, P.P., Jontes, J., Benfenati, F., Wilson, M.C. and Montecucco, C. (1993b) Botulinum neurotoxins serotypes A and E cleave SNAP-25 at distinct COOH-terminal peptide bonds. *FEBS Lett.*, **335**, pp. 99–103.

Schiavo, G., Shone, C.C., Rossetto, O., Alexander, F.C. and Montecucco, C. (1993c) Botulinum neurotoxin serotype F is a zinc endopeptidase specific for VAMP/synaptobrevin. *J. Biol. Chem.*, **268**, pp. 11516–11519.

Schiavo, G., Rossetto, O., Benfenati, F., Poulain, B. and Montecucco, C. (1994a) Tetanus and botulinum neurotoxins are zinc proteases specific for components of the neuroexocytosis apparatus. *Ann. NY Acad. Sci.*, **710**, pp. 65–75.

Schiavo, G., Rossetto, O., Benfenati, F., Poulain, B. and Montecucco, C. (1994b) Tetanus and botulinum neurotoxins are zinc proteases specific for components of the neuroexocytosis apparatus. *Ann. NY Acad. Sci.*, **710**, pp. 65–75.

Schiavo, G., Shone, C.C., Bennett, M.K., Scheller, R.H. and Montecucco, C. (1995) Botulinum neurotoxin type C cleaves a single Lys-Ala bond within the carboxyl-terminal region of syntaxins. *J. Biol. Chem.*, **270**, pp. 10566–10570.

Schmitt, A., Dreyer, F. and John, C. (1981) At least three sequential steps are involved in the tetanus toxin-induced block of neuromuscular transmission. *Naunyn Schmiedeberg's Arch. Pharmacol.*, **317**, pp. 326–330.

Schwab, M.E. and Thoenen, H. (1976) Electron microscopic evidence for transsynaptic migration of tetanus toxin in soinal cord motoneurons: an autoradiographic and morphometric study. *Brain Res.*, **105**, pp. 213–227.

Schweizer, F.E., Betz, H. and Augustine, G.J. (1995) From vesicle docking to endocytosis: intermediate reactions of exocytosis. *Neuron*, **14**, pp. 689–696.

Scott, A.B. (1989) Clostridial toxins as therapeutic agents. In: *Botulinum Neurotoxin and Tetanus Toxin*, edited by Simpson, L.L. Academic Press, New York.

Simpson, L.L. (1980) Kinetic studies on the interaction between botulinum toxin A and the cholinergic neuromuscular junction. *J. Pharmacol. Exp. Ther.*, **212**, pp. 16–21.

Simpson, L.L. (1981) The origin, structure and pharmacological activity of botulinum toxin. *Pharmacol. Rev.*, **33**, pp. 155–188.

Simpson, L.L. (1984) Fragment C of tetanus toxin antagonizes the neuromuscular blocking properties of native tetanus toxin. *J. Pharmacol. Exp. Ther.*, **228**, pp. 600–604.

Simpson, L.L. (1985) Pharmacological experiments on the binding and internalization of the 50,000 Dalton carboxyterminus of tetanus toxin at the cholinergic neuromuscular junction. *J. Pharmacol. Exp. Ther.*, **234**, pp. 100–105.

Simpson, L.L. (1988) Use of pharmacologic antagonists to deduce commonalities of biologic activity among clostridial neurotoxins. *J. Pharmacol. Exp. Ther.*, **245**, pp. 867–872.

Simpson, L.L., Coffield, J.A. and Bakry, N. (1993) Chelation of zinc antagonizes the neuromuscular blocking properties of the seven serotypes of botulinum neurotoxin as well as tetanus toxin. *J. Pharmacol. Exp. Ther.*, **267**, pp. 720–727.

Stecher, B., Weller, U., Habermann, E., Gratzl, M. and Ahnert Hilger, G. (1989) The light chain but not the heavy chain of botulinum A toxin inhibits exocytosis from permeabilized adrenal chromaffin cells. *FEBS Lett.*, **255**, pp. 391–394.

Stecher, B., Hens, J.J., Weller, U., Gratzl, M., Gispen, W.H. and De Graan, P.N. (1992) Noradrenaline release from permeabilized synaptosomes is inhibited by the light chain of tetanus toxin. *FEBS Lett.*, **312**, pp. 192–194.

Stoeckel, K., Schwab, M.E. and Thoenen, H. (1975) Comparison between the retrograde axonal transport of nerve growth factor and tetanus toxin in motor, sensory and adrenergic neurons. *Brain Res.*, **99**, pp. 1–16.

Südhof, T.C., DeCamilli, P., Niemann, H. and Jahn, R. (1993) Membrane fusion machinery: Insights from synaptic proteins. *Cell*, **75**, pp. 1–4.

Südhof, T.C. (1995) The synaptic vesicle cycle: a cascade of protein-protein interactions. *Nature*, **375**, pp. 645–653.

Sun, K.O., Chan, Y.W., Cheung, R.T.F., So, P.C., Lu, Y.L. and Li, P.C.K. (1994) Management of tetanus — a review of 18 cases. *J. Roy. Soc. Med.*, **87(N3)**, pp. 135–137.

Sunagawa, H., Ohyama, T., Watanabe, T. and Inoue, K. (1992) The complete amino acid sequence of the Clostridium botulinum type D neurotoxin, deduced by nucleotide sequence analysis of the encoding phage d-16 phi genome. *J. Vet. Med. Sci.*, **54**, pp. 905–913.

Sweeney, S.T., Broadie, K., Keane, J., Niemann, H. and O'Kane, C.J. (1995) Targeted expression of tetanus toxin light chain in Drosophila specifically eliminates synaptic transmission and causes behavioral defects. *Neuron*, **14**, pp. 341–351.

Tacket, C.O. and Rogawski, M.A. (1989) Botulism. In: *Botulinum Neurotoxin and Tetanus Toxin*, edited by Simpson, L.L. New York, Academic Press.

Tagaya, M., Toyonaga, S., Takahashi, M., Yamamoto, A., Fujiwara, T., Akagawa, K., Moriyama, Y. and Mizushima, S. (1995) Syntaxin 1 (HPC-1) is associated with chromaffin granules. *J. Biol. Chem.*, **270**, pp. 15930–15933.

Thomas, R., Mathai, A., Rajeev, B., Sen, S. and Jacob, P. (1993) Botulinum toxin in the treatment of paralytic strabismus and essential blepharospasm. *Indian J. Ophthalmol*, **41**, pp. 121–124.

Thompson, D.E., Hutson, R.A., East, A.K., Allaway, D., Collins, M.D. and Richardson, P.T. (1993) Nucleotide sequence of the gene coding for Clostridium barati type F neurotoxin: comparison with other clostridial neurotoxins. *FEMS Microbiol. Lett.*, **108**, pp. 175–182.

Tsuzuki, K., Kimura, K., Fujii, N., Yokosawa, N. and Oguma, K. (1992) The complete nucleotide sequence of the gene coding for the nontoxic-nonhemagglutinin component of Clostridium botulinum type C progenitor toxin. *Biochem. Biophys. Res. Commun.*, **183**, pp. 1273–1279.

Villiers, M.B., Gabert, F.M., Jacquier, M.R., Villiers, C.L. and Colomb, M.G. (1993) Involvement of the Zn-binding region of tetanus toxin in B and T recognition. Influence of Zn fixation. *Mol. Immunol.*, **30**, pp. 129–136.

Walch Solimena, C., Blasi, J., Edelmann, L., Chapman, E.R., von Mollard, G.F. and Jahn, R. (1995) The t-SNAREs syntaxin 1 and SNAP-25 are present on organelles that participate in synaptic vesicle recycling. *J. Cell Biol.*, **128**, pp. 637–645.

Washbourne, P., Schiavo, G. and Montecucco, C. (1995) Vesicle-associated membrane protein-2 (synaptobrevin-2) forms a complex with synaptophysin. *Biochem. J.*, **305**, pp. 721–724.

Weller, U., Taylor, C.F. and Habermann, E. (1986) Quantitative comparison between tetanus toxin, some fragments and toxoid for binding and axonal transport in the rat. *Toxicon*, **24**, pp. 1054–1063.

Weller, U., Mauler, F. and Habermann, E. (1988) Tetanus toxin: biochemical and pharmacological comparison between its protoxin and some isotoxins obtained by limited proteolysis. *Naunyn Schmiedebergs Arch.Pharmacol.*, **338**, pp. 99–106.

Weller, U., Dauzenroth, M.E., Meyer zu Heringdorf, D. and Habermann, E. (1989) Chains and fragments of tetanus toxin. Separation, reassociation and pharmacological properties. *Eur. J. Biochem.*, **182**, pp. 649–656.

Wellhöner, H.H. (1982) Tetanus Neurotoxin. *Rev. Physiol. Biochem. Pharmacol.*, **93**.

Wellhöner, H.H. (1989) Clostridial toxins and the central nervous system: Studies on *in situ* tissues. In: *Botulinum Neurotoxin and Tetanus Toxin*, edited by Simpson, L.L. Academic Press, New York.

Wellhöner, H.H. (1992) Tetanus and Botulinum Neurotoxins. In: *Handbook of Experiemental Pharmacology*, *Vol. 102; Selective Neurotoxicity*, edited by Herken, H. and Hucho, F. Springer-Verlag, Berlin.

Wernig, A., Stover, H. and Tonge, D. (1977) The labelling of motor endplates in skeletal muscle of mice with ^{125}I-tetanus toxin. *Naunyn Schmiedeberg's Arsch. Pharmacol.*, **298**, pp. 37–42.

Whelan, S.M., Elmore, M.J., Bodsworth, N.J., Atkinson, T. and Minton, N.P. (1992a) The complete amino acid sequence of the Clostridium botulinum type-E neurotoxin, derived by nucleotide-sequence analysis of the encoding gene. *Eur. J. Biochem.*, **204**, pp. 657–667.

Whelan, S.M., Elmore, M.J., Bodsworth, N.J., Brehm, J.K., Atkinson, T. and Minton, N.P. (1992b) Molecular cloning of the Clostridium botulinum structural gene encoding the type B neurotoxin and determination of its entire nucleotide sequence. *Appl. Environ. Microbiol.*, **58**, pp. 2345–2354.

Wiegand, H., Erdmann, G. and Wellhöner, H.H. (1976) ^{125}I-labeled botulinum A neurotoxin: Pharmacokinetics in cats after intramuscular injection. *Naunyn Schmiedeberg's Arch. Pharmacol.*, **292**, pp. 161–165.

Wiegand, H. and Wellhöner, H.H. (1977) The action of botulinum A neurotoxin on the inhibition by antidromic stimulation of the lumbar monosynaptic reflex. *Naunyn Schmiedberg's Arch. Pharmacol.*, **298**, pp. 235–238.

Wiegand, H. and Wellhöner, H.H. (1979) Electrical excitability of motoneurons in early local tetanus. *Naunyn Schmiedeberg's Arch. Pharmacol.*, **308**, pp. 71–76.

Willems, A., East, A.K., Lawson, P.A. and Collins, M.D. (1993) Sequence of the gene coding for the neurotoxin of Clostridium botulinum type A associated with infant botulism: comparison with other clostridial neurotoxins. *Res. Microbiol.*, **144**, pp. 547–556.

Williams, R.S., Tse, C.K., Dolly, J.O., Hambleton, P. and Melling, J. (1983) Radioiodination of botulinum neurotoxin type A with retention of biological activity and its binding to brain synaptosomes. *Eur. J. Biochem.*, **131**, pp. 437–445.

Williamson, L.C. and Neale, E.A. (1994) Bafilomycin A1 inhibits the action of tetanus toxin in spinal cord neurons in cell culture. *J. Neurochem.*, **63**, pp. 2342–2345.

Wright, J.F., Pernollet, M., Reboul, A., Aude, C. and Colomb, M.G. (1992) Identification and partial characterization of a low affinity metal-binding site in the light chain of tetanus toxin. *J. Biol. Chem.*, **267**, pp. 9053–9058.

Yamaguchi, K. and Akagawa, K. (1994) Exocytosis relating proteins in the nervous system. *Neurosci. Res.*, **20**, pp. 289–292.

Yamasaki, S., Baumeister, A., Blasi, J., Link, E., Cornille, F., Roques, B., Fykse, E.M., Südhof, T.C., Jahn, R., Niemann, H. and Binz, T. (1994a) Cleavage of members of the synaptobrevin/VAMP family by type-D and type-F botulinal neurotoxins and tetanus toxin. *J. Biol. Chem.*, **269(N17)**, pp. 12764–12772.

Yamasaki, S., Binz, T., Hayashi, T., Szabo, E., Yamasaki, N., Eklund, M., Jahn, R. and Niemann, H. (1994b) Botulinum neurotoxin type-G proteolyses the ALA(81)-ALA(82) bond of rat synaptobrevin-2. *Biochem. Biophys. Res. Commun.*, **200(N2)**, pp. 829–835.

Yavin, E., Yavin, L. and Kohn, L. (1983) Temperature-mediated interaction of tetanus toxin with cerebral neuron cultures: characterization of a neuraminidase-insensitive toxin-receptor complex. *J. Neurochem.*, **40**, pp. 1212–1219.

Yavin, E., Lazarovici, P. and Nathan, A. (1987) Molecular interactions of ganglioside receptors with tetanotoxin on solid supporters, aqueous solutions and natural membranes. In: *Membrane Receptors, Dynamics and Energetics*, edited by Wirtz, K.A.W. Plenum Press, New York.

Yavin, E. and Nathan, A. (1986) Tetanus toxin receptor on nerve cells contain a trypsin-sensitive component. *Eur. J. Biochem.*, **154**, p. 403–407.

GLOSSARY

Acellular vaccine — vaccine composed of specific antigens isolated from a pathogenic organism.

adenylate cyclase — an enzyme that hydrolyzes ATP to form the second messenger cAMP.

ADP-ribosylation — post-translantional modification of proteins through attachment of an ADP-ribose moiety. ADP-ribose derived from NAD is covalently linked to proteins by either bacterial or endogenous ADP-ribosyltransferases to an electrophilic acceptor, such as the guanidino group of arginine.

ADP-ribosylation factors (ARFs) — a family of ~20 kDa monomeric GTP-binding proteins, identified initially as stimulators of cholera toxin and subsequently of PLD; also shown to be important for vesicular trafficking in mammalian systems.

ω-Agatoxin IVA — is isolated from the venom of the spider *Agelenopsis aperta*. At nanomolar concentrations it is a selective blocker of P-type Ca^{2+} channels and at micromolar concentrations it also inhibits Q-type Ca^{2+} channels.

Amphiphilic — describes molecules containing groups with both polar and nonpolar domains.

Anaphylatoxin — a chemical that produces symptoms of an anaphylactic reaction (allergic response).

ANP — atrial natriuretic peptide.

ApUp — adenylyl-(3′,5′,)-uridine 3′-monophosphate, an endogenous dinucleotide that binds in the catalytic site of diphtheria toxin.

ARD — ARF domain protein.

ARF-like Protein (ARL) — ADP-ribosylation factor-like proteins; ~40–50% identical to ARF proteins, but do not display the activities of ARF (i.e., CT or PLD activation).

Autophosphorylation — the capacity of an enzyme molecule to catalyze self-incorporation of inorganic phosphate at specific amino acid sites or for two like enzyme molecules to induce the phosphorylation of one another upon activation.

βARK — β-adrenergic receptor kinase, a serine/threonine kinase phosphorylating only agonist activated receptors.

Batrachotoxin — a steroidal alkaloid (lipid soluble neurotoxin) isolated from the skin of the Colombian frog *Phyllobates asrataenia* and related species. Batrachotoxin shifts the voltage-dependent activation of sodium channels to more negative membrane potentials and inhibits their inactivation, causing persistent activation of sodium channels. Batrachotoxin shares receptor site 2 on sodium channels with other alkaloid neurotoxins.

Bay K 8644 — is a 1,4-dihydropyridine derivative that selectively activates the L-type Ca^{2+} channels.

Botulinum toxins — a family of neurotoxins, which block transmitter release as a result of the proteolytic cleavage of synaptobrevin (B,P,F,G), SNAP-25 (A,E) or syntaxin (C1).

Brevetoxin (PbT) — a series of structurally-related lipid-soluble compounds, produced by the marine dinoflagellate *Plychodiscus brevis* (previously *Gymnodinium breve*, that produce "red tides"). Brevetoxins are complex multiring methylated polyether neurotoxins, highly toxic to fish and human. Brevetoxins, similarly to ciguatoxins, depolarize nerve membranes by shifting inactivation of sodium channels to a more negative membrane potential and slowing the kinetics of both activation and inactivation. Brevetoxins (and ciguatoxins) share receptor site 5 on sodium channels.

α-bungarotoxin (α-BTX) — a snake venom neurotoxin with curare-like activity which binds with very high affinity to the nicotinic AChR.

Cdc42 — small GTPases that regulate the *Saccharomyces cerevisiae* cell division cycle. The mammalian counterpart is involved in lamellipodia formation.

Chemotaxis — the movement of cells towards or away from chemical substances.

Cholera toxin (CT) — proteinaceous toxin secreted by *Vibrio cholerae*. It catalyzes the mono-ADP-ribosylation of arginine or guanidino containing molecules such as α-subunits of Gs and Gt.

Clostridium botulinum ADP-ribosyltransferase C3 — this toxin transfers a single ADP-ribose moiety from NAD to Rho proteins. C3 ADP-ribosyltransferases are not related to clostridial neurotoxins which inhibit exocytosis by proteolytic cleavage of vesicular proteins.

Clostridium botulinum toxin C2 — this toxin exerts its effect through ADP-ribosylation of actin.

Clostridium difficile toxin A and B — these toxins modify Rho proteins and Cdc42 by attachment of a glucose moiety.

Clostridial neurotoxins (CNT) — a group of clostridial neurotoxins consisting of tetanus and several botulinum toxins. High specificity for neuronal cells. They block neuronal transmission by proteolytic cleavage of synaptic proteins.

δ-conotoxin — a family of conotoxins, peptides of 20–30 amino acid residues cross-linked by 3 disulfide bridges, derived from *Conus* snail venoms. Their main effect is inhibition of sodium current inactivation. δ-conotoxin competes in binding assays with δ-conotoxin TxVIA for binding to receptor site 6 on sodium channels.

μ-conotoxin — a family of conotoxins derived from the marine snail *Conus geographus* that inhibit sodium conductance, similarly to TTX and STX, and compete in binding studies with TTX and STX for binding to receptor site 1. Unlike TTX and STX, which bind with high affinity to both muscle and neuronal sodium channels, μ-conotoxins bind selectively only to vertebrate skeletal muscle.

ω-conotoxin GVIA — isolated from the venom of the piscivorous marine snail *Conus geographus*. It is a highly selective irreversible blocker of N-type Ca^{2+} channels.

ω-conotoxin MVIIA — is a toxin isolated from the piscivorous marine snail *Conus magus* and blocks selectively but reversibly N-type Ca^{2+} channels.

ω-conotoxin MVIIC — is a toxin isolated from the piscivorous marine snail *Conus magus* and blocks N-,P- and Q-type Ca^{2+} channels.

CREM (cyclic AMP response element modulator) — a transcription factor which binds to the cyclic AMP response element (CRE) found in the promoter of many genes.

CRM — cross-reacting material, mutants of diphtheria toxin that are recognized by antibodies raised against the native toxin.

Cyclin A — a regulatory subunit of several kinases whose activity is required for successful cell cycle transition.

Dbl — the Dbl (<u>d</u>iffuse <u>B</u>-cell <u>l</u>ymphoma) gene product stimulates guanine nucleotide exchange at RhoA, Racl and Cdc42Hs.

1,4-Dihydropyridines — selective blockers of L-type Ca^{2+} channels; the prototype is nifedipine.

DRAP 27/CD9 — diphtheria toxin receptor-associated protein. DRAP 27 is a 27 kDa protein that closely associates with proHB-EGF to make up the diphtheria toxin receptor complex. DRAP 27 is the monkey equivalent of human CD9, and although it does not actually bind the toxin, it serves to upregulate the number of functional receptors and thus increase toxin binding.

Diphtheria toxin — catalyzes the transfer of ADP-ribose from NAD to EF-2 thus inactivating EF-2 and inhibiting protein synthesis.

Endothelins (ETs) — a family of potent vasoactive peptides composed of 21 amino-acids and two disulfide bridges. They act through different 7 TM receptors to activate various G-proteins.

Eukaryotic — pertaining to eukaryote; an organism whose cells have a true nucleus.

Excitation Contraction Coupling (EC coupling) — the process by which membrane depolarization leads to opening of ryanodine receptors and Ca^{2+} release from the SR.

Exotoxin — a toxic substance formed by species of certain bacteria that is found outside the bacterial cell, or free in bacterial cell culture medium. Exotoxins are heat-labile and protein in nature.

FRAP (FKBP-rapamycin-associated protein) — a 220 kDa protein which has been shown to be involved in the rapamycin inhibition of $p70^{S6K}$.

G — guanine-nucleotide-binding heterotrimeric protein consisting of an α subunit (guanine nucleotide binding) and a $\beta\gamma$ subunit. G proteins couple receptor binding to the production of second messengers in a process that involves binding and hydrolysis of guanine nucleotides such as GTP.

G_o — a stage of the cell cycle when cells are in a growth-arrested or quiescent state.

Ganglioside — a lipid compound composed of ceramide, sugars and sialic acids.

Monosialogangliosides: GM1-galactosyl-N-acetylgalactosaminyl (N-acetylneuraminyl) galactosylglucosyl-ceramide.

Trisialogangliosides: GT1b-N-acetylneuraminylgalactosyl-N-acetylgalactosaminyl (N-acetylneuraminyl-N-acetylneuraminyl) galactosylglucosylceramide.

GAP — GTPase activating proteins. They stimulate GTP hydrolysis by small guanine nucleotide binding proteins. E.g. RhoGAP activates specifically Rho proteins.

GDI — guanine nucleotide dissociation inhibitors for small GTPases. e.g. RhoGDI prevents dissociation of GDP from Rho, Rac and Cdc42.

GDS — guanine nucleotide dissociation stimulators enhance dissociation of the nucleotide from small GTP-binding proteins.

GEF — guanine nucleotide exchange factors are identical to GDS proteins (cf. GDS).

glycerophospholipid — an amphipathic lipid with a glycerol backbone: fatty acids are ester-linked to C-1 and C-2 of glycerol, and a polar alcohol is attached through phosphate to C-3.

GM3 — N-acetylneuraminylgalactosylglucosylceramide.

Grb2 — growth factor receptor binding protein, possessing SH2 and SH3 domains.

Growth factor receptors — membrane receptors endowed with tyrosine protein kinase, activated by dimerization following agonist binding.

Guanylate cyclase — an enzyme that may be part of plasma membrane receptors or located within the cytosol. Activation of this enzyme catalyzes the formation of the second messenger cGMP from GTP.

Heat-labile enterotoxin (LT) — heat-labile enterotoxin is produced by *Escherichia coli*; highly similar in amino-acid sequence, enzymatic activity and structure to cholera toxin.

hexagonal phase — refers to the inverted hexagonal or HII phase in which phospholipid molecules are arranged as water-filled cylinders with a hydrophobic exterior.

K252 compounds — indole [2,3-alpha] carbazole alkaloids isolated from filamentous bacteria which inhibit protein kinases by competition with ATP *in vitro* and are selective *in vivo* inhibitors of neurotrophins-trk receptor tyrosine kinases.

Major histocompatibility complex (MHC) — cell surface markers that when coupled with antigen is recognized by T cells of the immune system to provoke an immune response.

MAPK — mitogen activated serine/threonine protein kinase.

MEK — MAPK kinase; a dual function, serine/threonine and tyrosine kinase, which activates MAPK by dual phosphorylation.

Mono-glucosylation — post-translational modification of proteins through attachment of a glucose moiety. Cosubstrate for either bacterial or endogenous mono-glucosyltransferases in UDP-glucose.

MTx1, MTx2 — 67 amino acid polypeptide toxins, isolated from *Dendroaspis angusticeps* snake venom, which act as selective, irreversible agonists of the m1 muscarinic receptors.

MTx3 — a 65 amino acid polypeptide toxin, homologous to MTx1 and MTx2, isolated from the *Dendroaspis angusticeps* snake venom, which is highly potent and very selective agonist for m4 muscarinic receptors.

N-ethylmaleimide sensitive factor (NSF) — trimeric ATPase required for *in vitro* membrane fusion during vesicular transport.

Neurexins — nerve terminal cell surface proteins with more than 1000 isoforms generated by alternative splicing for three genes. Neurexins include the receptor for α-latrotoxin and may function in cell-cell recognition in the nervous system.

Neurotrophin — a low molecular weight, basic polypeptide growth factor responsible for the survival, maintenance and differentiation of neurons.

NO — nitric oxide, a free radical; also stimulates soluble guanylyl cyclase.

NOS — NO synthase — two types are known; constitutive (cNOS), always present in the cells concerned, but activated by a rise in Ca_i; inducible (iNOS), absent from the cells concerned, synthesized following exposure to specific cytokines, like interferon γ. In iNOS the CaM is bound to the enzyme at $[Ca_i]$ of cells at rest.

Pardaxin — a potent ionophore toxic polypeptide isolated from the skin secretion of the fish *Pardachirus marmoratus.*

Pertussis toxin — a bacterial exotoxin produced by *Bordetella pertussis* that can ADP-ribosylate the α-subunits of G_t, G_i, and G_o.

PH — domain of structural (rather than amino acid sequence) homology to pleckstrin (a cytoskeletal protein) adapted for protein-protein and acid phospholipid interaction.

Phorbol esters — a class of diterpene ester tumor promoters, originally purified from croton oil, that specifically bind to and activate protein kinase C.

Phosphatidylinositol-glycan — a membrane-associated molecule comprised of phospholipid, phosphatidylinositol and a polysaccharide. This lipid attaches covalently to certain membrane proteins, anchoring them to the lipid bilayer. Anchored proteins are released from the membrane by the hydrolytic action of enzymes such as phospholipase C.

Phosphatidylinositol 3-kinase (PI 3-kinase) — a kinase which phosphorylates phosphatidylinositol at the D-3 position of the inositol ring.

phosphoinositides — phospholipids with inositol as the polar headgroup.

Phospholipase A2 (PLA2) — a phosphatide 2-acylhydrolase enzyme that catalyzes the hydrolysis of phospholipids, to generate lysophospholipid and polyunsaturated (arachidonic) fatty acid.

Phospholipase C (PLC) — a phosphodiesterase that catalyzes the hydrolysis of ester bonds located in phospholipids to produce diacylglycerol and the polar headgroup of the phospholipid.

Phospholipase D (PLD) — an enzyme that catalyzes the hydrolysis of phospholipids resulting in the formation of phosphatidic acid and a product dependent on the specific lipid head group (e.g., phosphatidylcholine $\rightarrow$ phosphatidic acid + choline); specific isoforms are stimulated by ARF *in vitro.*

PKA — cAMP activated protein kinase.

PKC — a family of serine/threonine protein kinases activated by phospholipids, diacylglycerol (or cis- unsaturated fatty acids), calcium dependent or independent.

PKG — cGMP activated protein kinase.

pro HB-EGF — a membrane-bound form of the heparin-binding epidermal growth factor-like precursor that serves as the diphtheria toxin receptor. Pro HB-EGF interacts with DRAP 27/CD9 as well as with other molecules to form the diphtheria toxin receptor complex.

Protein kinase — an enzyme that transfers a phosphate group from a phosphate donor (usually adenoside trisphosphate, ATP) to an acceptor amino acid in a substrate protein.

protein phosphatase — an enzyme that catalyzes the hydrolytic release of inorganic phosphate from phosphorylated proteins.

P-site — polyphosphate-binding site composed of a cluster of basic residues in the R domain of diphtheria toxin. The toxin is folded so that the P-site is located in close proximity to the nucleotide binding site of the A fragment and the receptor-binding domain of the B fragment.

pseudosubstrate — a peptide portion of protein kinase that binds to the substrate binding site of the same enzyme.

Rac — Rac proteins (Ras-related C3 botulinum toxin substrate) are small GTPases and form a functional part of the oxygen radical producing NADPH oxidase complex in neutrophils. Despite their name they are poor or no substrates for C3 ADP-ribosyltransferase.

raf — serine/threonine kinase activated by ras which phosphorylates MEK.

RAFT1 (rapamycin and FKBP12 target) — a protein which is identical to FRAP.

Rapamycin — a fungal toxin which can be classified as a macrocyclic lactone. This compound was instrumental in the elucidation of the signal transduction pathways leading to growth factor-stimulated S6 kinase activation.

Retrograde transport — processes of intra-axonal transport(s) in neurons, carrying molecules from the synapse to the soma of the neurons at different speeds, utilized by clostridial neurotoxins to access intracellular compartments of different neurons and delivery from the periphery to the central nervous system.

Rho proteins — small guanine nucleotide binding and hydrolyzing proteins (Ras homolog) which play a role in organization of the cytoskeleton and cell divisions and are substrates for C3-like ADP-ribosyltransferases.

Ricin — a poisonous substance which is derived from the seed of the castor oil plant.

RTX toxins — a group of pore-forming, cytolytic toxins that have disseminated widely among Gram-negative bacteria. They are characterized by a repeat region which consists of several highly conserved units of nine amino acids (repeats in toxins).

Ryanodine — a plant alkaloid that binds to the ryanodine receptor Ca^{2+} release channel in the sarcoplasmatic membrane (SR).

Ryanodine Receptors — high molecular weight proteins in the SR which form a Ca^{2+} channel activated during EC coupling.

S6 — the major phosphoprotein associated with the 40S ribosome. Its differential phosphorylation state is believed to be involved in the translational regulation of critical proteins required for successful cell cycle progression and proteins necessary for the differentiation of cells.

S6 kinase — a protein kinase which phosphorylates the 40S ribosomal protein, S6.

Sarafotoxins (SRTx) — potent vasopressors, 21 amino acids, polypeptide toxins which are isolated from the venom of the barrowing asp *Atractaspis engadensis*.

Sarcoplasmic Reticulum (SR) — intracellular Ca^{2+} storing organelle.

Saxitoxin (STX) — a heterocyclic guanidine (like TTX), produced by dinoflagellates of the genus *Gonyaulax* and it is found in clams, mussels and other shellfish that feed on them. For action and binding see *Tetrodotoxin*.

α-scorpion toxin — a family of polypeptide neurotoxins derived from scorpion venoms that are selective to sodium channels and bind to a single receptor site, designated as receptor site 3 on sodium channels. Like other scorpion toxins active on sodium channels, the α-scorpion toxins are composed of a single chain of 60–70 amino acid residues cross-linked by 4 disulfide bridges. Their binding result in prolonged action potentials due to the inhibition or slowing of the inactivation of the sodium current.

β-scorpion toxin — scorpion toxins selective to sodium channels (see *α-scorpion toxin*) which bind the receptor site 4 on sodium channels. The main result of their binding is induction of repetitive firing activity ('trains' of action potentials) and depolarization of the nerve membranes due to a shift in the activation of sodium channels to more negative membrane potentials and a slowing of the kinetics of activation and inactivation.

secretagogue — an agent that stimulates secretion/exocytosis.

Serine/threonine kinases — protein kinases which selectively transfer a phosphate group to an alcohol group in the amino acids serine and threonine in proteins (E.C. 2.7.10).

SH domains — referred to as SH2 and SH3, or src homology domains. These highly conserved sequences of amino acids were initially described based on their resemblance to regions in the non-catalytic, regulatory domains of non-receptor tyrosine kinases including products of the src gene.

SH2 — domain of src homology 2 endowing the protein with the capacity to bind to tyrosine phosphorylated proteins.

SH3 — domain of src homology 3 constituting a steric structure with high affinity for proline-rich sequences.

Signal transduction — amplifying cascades of enzymatic interactive pathways which transfer information from the plasma membrane to certain biological processes using receptors, second messengers, coupling proteins and effector enzymes.

SNAP-25 — abundant synaptic terminal membrane protein that is palmitoylated. It is cleaved by botulinum toxins A and E and binds to syntaxins.

α/β/γ SNAPs — soluble NSF-attachment proteins required to recruit NSF to membranes in an ATP-dependent manner.

SNARE — synaptosomal associated proteins participating in exocytosis.

Synaptobrevins (VAMPs) — vesicle associated membrane proteins that are cleaved by tetanus toxin and botulinum toxins B,D,F and G.

Synaptotagmins — synaptic plasma membrane glycoproteins, with at least eight isoforms, that bind calcium and phosholipids and interact with syntaxin, neurexins, AP2, etc.

Syntaxin (originally named retinal antigen HPC-1) — ubiquitous synaptic plasma membrane proteins that are cleaved by botulinum toxin C1 and bind to synaptotagmins, SNAP-25, synaptobrevins, SNAPs, calcium channels etc.

Taicatoxin — an olygmeric toxin isolated from the Australian-Taipan venom of the snake *Oxyuranus s. scutellatus* which reversibly, selectively and voltage-dependently blocks the L-type, voltage gated calcium channels in excitable membranes.

TCR — T-cell membrane receptor activated by antigen binding.

Tetanus toxin — a 150 kD protein neurotoxin from *Clostridium tetani* bacteria, potent inhibitor of neurotransmitter release by proteolytic cleavage of synaptobrevins.

Tetrodotoxin (TTX) — a heterocyclic guanidine (like saxitoxin, water-soluble neuro-toxins) found in puffer fish organs of the suborder *Gymnodontes* and in some other animals. Its pharmacological and pathological effects derive from the inhibition of action potential generation in nerve and muscle cells. TTX binds to the extracellular vestibule of the sodium channel pore at receptor site 1, also shared by saxitoxin. Their binding blocks the sodium conductance through the pore.

7TM — receptors with seven transmembrane segments coupled to G proteins in the transduction sequence.

transducin — Gt, the G protein involved in visual transduction that couples the photoreceptor to cGMP phosphodiesterase.

trk — a family of 145 kD proteins representing the high affinity receptors for neur-otrophins. Possesses intrinsic tyrosine kinase activity.

Tyrosine kinases — protein kinases which selectively transfer a phosphate group to the phenolic group of the aromatic amino acid tyrosine in proteins (E.C.2.7.11).

Veratridine — a steroidal alkaloid (lipid-soluble neurotoxin) produced by some plants of the family *Lilacea*. Veratridine has a broad spectrum of pharmacologi-cal activities including repetitive firing and muscle contraction, attributed to its alteration of sodium channel properties leading to hyperexcitability and depol-arization of the excitable cell membrane. Veratridine, similarly to other alkaloid neurotoxins, binds to receptor site 2 on sodium channels and shifts the activa-tion to a more negative membrane potential and inhibits the inactivation process, causing persistent activation of the sodium current.

Wortmannin — a sterol-like fungal toxin which has been shown to be a potent inhibitor of phosphatidylinositol 3-kinase. This compound has also been used in the dissection of signalling cascades involved in the phosphorylation of the 40S ribosomal protein, S6.

Xenopus — genus name for toads, whose oocytes are used in genetic research.

ZAP-70 — Zeta associated protein tyrosine kinase.

INDEX